RACE AND ETHNIC RELATIONS
American and Global Perspectives

MARTIN N. MARGER
Michigan State University

Sixth Edition

Australia • Canada • Mexico • Singapore • Spain • United Kingdom • United States

THOMSON
™
WADSWORTH

Senior Executive Editor: *Sabra Horne*
Sociology Editor: *Bob Jucha*
Assistant Editor: *Stephanie Monzon*
Editorial Assistant: *Matthew Goldsmith*
Technology Project Manager: *Dee Dee Zobian*
Marketing Manager: *Matthew Wright*
Marketing Assistant: *Michael Silverstein*
Project Manager, Editorial Production: *Paula Berman*
Print/Media Buyer: *Tandra Jorgensen*

Permissions Editor: *Stephanie Keough-Hedges*
Production Service and Compositor: *Hespenheide Design*
Copy Editor: *Christianne Thillen*
Cover Designer: *Patrick Devine*
Cover Image: *PhotoDisc*
Text and Cover Printer: *Phoenix Color Corp/Book Technology Park*

For more information about our products, contact us at:
Thomson Learning Academic Resource Center
1-800-423-0563
For permission to use material from this text, contact us by:
Phone: 1-800-730-2214 **Fax:** 1-800-730-2215
Web: http://www.thomsonrights.com

Wadsworth/Thomson Learning
10 Davis Drive
Belmont, CA 94002-3098
USA

Asia
Thomson Learning
60 Albert Street #15-01
Albert Complex
Singapore 189969

Australia
Nelson Thomson Learning
102 Dodds Street
South Melbourne, Victoria 3205
Australia

Canada
Nelson Thomson Learning
1120 Birchmount Road
Toronto, Ontario M1K 5G4
Canada

Europe/Middle East/ Africa
Thomson Learning
Berkshire House
168–173 High Holborn
London WC1V 7AA
United Kingdom

Library of Congress Control Number: 2002103125
ISBN 0-534-53686-7

 This book is printed on acid-free recycled paper.

BRIEF CONTENTS

CONTENTS

Chapter 8: Jewish Americans   227

Chapter 9: African Americans   258

Chapter 15: Canada: Ethnic Model of the Future?   493

Chapter 16: Northern Ireland: Enduring Ethnic Conflict?   534

PREFACE

RACE AND ETHNIC RELATIONS: American and Global Perspectives is intended to explore race and ethnic relations in a global context, while covering extensively groups and issues in the United States. The need for such a comparative approach seems more critical than ever in light of the increasing ethnic diversity of most contemporary societies as well as the prominence of ethnic conflicts in virtually all world regions. Indeed, with the end of the Cold War, the new world order seems one in which ethnic nationalism has become the major source of inter- and intrasocietal conflict. Though they are usually uninformed about complexities, most Americans have become at least vaguely aware of ethnic conflict in societies as distant and exotic as Rwanda and Kosovo, as well as those closer geographically and culturally, such as Canada and Northern Ireland. Moreover, Americans are forced to acknowledge that in their own society unresolved issues of race and ethnicity remain the most critical domestic concern. Although ethnic harmony may typify few multiethnic societies, there are encouraging instances of the diminishment of racial and ethnic conflict, as in South Africa, and the lessening of ethnic discrimination in the United States and other advanced industrial societies. Nonetheless, it seems clear that racial and ethnic issues and the conflicts they give rise to will continue to engage our attention well into the twenty-first century.

Curiously, American social scientists have not always kept pace in adapting to the global context of race and ethnic relations. Some continue to focus almost exclusively on the United States, paying only incidental attention to ethnic patterns and events in other societies. In line with this view, texts in the field of race and ethnic relations have ordinarily provided no more than cursory coverage to ethnic issues outside the American sphere—if at all. Students, therefore, often continue to think of racial and ethnic, or minority, issues as uniquely American phenomena.

Many other social scientists, however, have come to see the utility and relevance of a more cross-national approach to the study of race and ethnicity. Such an approach distinguished *Race and Ethnic Relations* from other texts in the field when it was first published in 1985. Its objective was to provide readers with a comparative perspective without sacrificing a strong American component. That objective was retained in subsequent editions and remains unchanged in this, the sixth edition. The book's overriding theme is the global nature of ethnicity and the prevalence of ethnic conflict in the modern world.

At the same time that an international perspective on race and ethnicity seems more compelling than ever, a close and careful analysis of American ethnic relations is surely imperative. For better or worse, the United States, the most diverse of multiethnic societies, more often than not is a global pacesetter in ethnic relations. More important, most readers of *Race and Ethnic Relations* continue to be American students, who require a solid understanding of their own society, which subsequently can be used as a comparative frame of reference. *Race and Ethnic Relations*, therefore, provides thorough coverage of America's major ethnic groups and issues. My own teaching experience has confirmed, however, that American students commonly acquire a broader and richer comprehension of ethnic relations and issues in the United States when these are presented in a global context and can be viewed from a comparative perspective.

The number of American college and university courses with ethnic content has grown enormously in recent years. This, I believe, is a reflection of the pressing problems and commanding issues of race and ethnicity in America and the growing awareness of ethnic divisions and inequalities in an increasingly diverse society. The content of *Race and Ethnic Relations* is comprehensive and thus appropriate for a variety of courses that may be differently titled and structured (for example, "race and ethnicity," "minority relations," "ethnic stratification," "multiculturalism") but that all deal in some fashion with ethnic issues.

The theoretical and conceptual thrust of this edition will be familiar to past readers: a power-conflict perspective, emphasizing the power dynamics among ethnic groups. Race and ethnic relations are seen as manifestations of stratification and of the competition and conflict that develop over societal rewards—power, wealth, and prestige. In accord with this perspective, I have emphasized the structural, or macrolevel, patterns of race and ethnic relations rather than the social-psychological patterns, though the latter are interspersed throughout.

OVERVIEW OF THE BOOK

The intent of the chapters that make up Part I, THE NATURE OF ETHNIC RELATIONS, is to introduce the principal terms, concepts, and theories of the field of race and ethnic relations. These chapters are designed to serve as an analytic framework within which U.S. racial and ethnic groups, as well as those of other multiethnic societies, can be systematically examined. Although illustrations of ideas and concepts and applications of theory are related primarily to the United States, whenever appropriate they are presented in an international context.

Part II, ETHNICITY IN THE UNITED STATES, focuses on American society, examining the formation of its ethnic system and its major racial and ethnic populations. Chapter 5 traces the development of the American ethnic configuration and hierarchy, and is followed in Chapters 6 through 11 by descriptions and analyses of Native Americans, Euro-Americans—specifically Italian Americans and Jewish Americans—African Americans, Hispanic Americans, and Asian Americans. These are presented within the framework of theories and concepts introduced in Part I. Chapter 12 is intended to present what I believe are the major ongoing issues of race and ethnic relations in the United States: the persistent gap between Euro-Americans and racial-ethnic groups; policies designed to address that gap (in particular, affirmative action); and large-scale immigration.

Part III, ETHNIC RELATIONS IN COMPARATIVE PERSPECTIVE, examines several societies that stand as intriguing comparisons vis-à-vis the United States. Chapter 13 deals with South Africa, not long ago the most rigidly racist society on the globe. Today South Africa is in the midst of a radical transformation and serves as an important case for students of contemporary ethnic relations, illustrating how oppressive systems of ethnic inequality are capable of fundamental change in a surprisingly brief time with a minimum of violence. Brazil, examined in Chapter 14, is ideally suited to a comparison with the United States, given its past history of slavery and its multiracial composition. Canada, the focus of Chapter 15, is often seen as a northern replica of the United States, but in its ethnic ideology and policies, students will find that it stands in sharp contrast to American society. In a number of ways, Canada may represent a future model for multiethnic societies. Chapter 16 examines Northern Ireland, a society that has been plagued by a deadly ethnic conflict for generations, but particularly during the past forty-five years. Students will find it a riveting case, given that the groups in conflict are racially indistinguishable and culturally closer than is usually assumed. Chapter 17 explores the global nature of contemporary ethnic conflict, focusing heavily on the breakup of the former Yugoslavia and the resultant ethnic wars that have been waged throughout the past decade. If South Africa serves as an example of ethnic conflict reduction, the former Yugoslavia informs students of ethnic relations of how in the modern world multiethnic societies can disintegrate while creating, in the process, a monstrous scene of torture chambers and killing fields.

Instructors may want to cover all of these cases or may choose to focus on only one or two. All lend themselves to easy comparison with the United States and comparative points are drawn throughout each chapter. No prior familiarity with any of these societies is assumed, either on the part of instructors or students.

NEW TO THE SIXTH EDITION

In the relatively brief span of time since the publication of the fifth edition of *Race and Ethnic Relations*, much has changed in the United States and in other multiethnic societies that bears directly on the study of race and ethnicity. In response to those changes, the sixth edition has been revised with the latest data and the most relevant examples from a variety of societies, and includes references to important new studies. Several chapters have been thoroughly rewritten and all others have been substantively revised. In addition, nine tables and three figures are entirely new. The total of these additions and updates represents the most substantial revision since the book's initial publication in 1985.

The structure and coverage of Part I remain basically unchanged, but the most current research findings have been cited in each chapter. In Chapter 3, for example, more recent work on ethnic stereotyping is discussed, updating the pioneering Princeton studies of 1933, 1951, and 1967.

Several important changes have been made in Part II. Chapter 5 has been substantially rewritten. It describes the formation of the United States as a multiethnic society, mostly through immigration, and the development of its ethnic hierarchy. The establishment of Anglo dominance and subsequent patterns and processes of immigration are traced historically, culminating in the most recent immigrant wave, which continues to reconfigure the society's ethnic order. Middle Eastern Americans are discussed in this chapter as one of the society's growing and increasingly prominent ethnic populations. This chapter also contains a discussion of shifting racial and ethnic classifications, as well as the emergence of a so-called "mixed race" category.

Chapter 6 is now devoted exclusively to Native Americans who, in previous editions, had been covered as part of Chapter 5. This chapter constitutes a virtually new and significantly expanded presentation of materials.

Several topics are noticeable additions to other chapters in Part II. Chapter 7 contains updated research findings on the persistence of ethnic stereotypes of Italian Americans, as well as additional demographic data on this group. An expanded section on the tenuous nature of Jewish identity in America is included in Chapter 8. Chapter 9 contains a discussion of emergent demographic trends among African Americans, including increasing suburbanization and remigration to the South, as well as sustained residential segregation. Also new to this chapter is a discussion of expanding black ethnic diversity in America. Chapter 10 includes coverage of population developments among Hispanic Americans—namely, rapid growth and increasing geographic dispersion—as well as emergent Hispanic political power. An enhanced discussion of the internal stratification of the Asian-American population is contained in Chapter 11, as well as new findings regarding Americans' attitudes toward Asian Americans. Also new material in this chapter expands the discussion of the competing demands of ethnic retention and assimilation. New to Chapter 12 is a discussion of the different modes of adaptation of America's latest immigrants, compared to those of immigrants of earlier eras. Also, the coverage in this chapter of the affirmative action controversy reflects the most recent issues and legal cases.

Part II, then, is now made up of eight chapters, and comprises descriptions of the major U.S. ethnic groups, bracketed by a chapter that examines the shaping of the American ethnic hierarchy and a chapter that explores the society's current major racial and ethnic issues.

Each of the societies examined in Part III continues to be a case of major significance in the study of race and ethnic relations in the contemporary world. Important events and processes during the past few years, however, have produced changes in all, especially South Africa, Northern Ireland, and the former Yugoslavia. In the case of South Africa, massive obstacles remain in the path toward a just and equitable multiethnic society, but structural changes that seemed unimaginable only a few years ago continue to evolve. Though plagued with severe internal problems, South Africa today remains, as the journalist Anthony Lewis has called it, "a land of possibility." In Northern Ireland, too, although a long-awaited peace accord hangs fragilely in the balance, events of recent years have contributed to a reduction of violence and have given hope to an eventual cessation of more than three decades of ethnically inspired hostility. The area comprising the former Yugoslavia, by contrast, tragically endured an ever more frightful expansion of ethnic conflict during the 1990s and early 2000s, resulting in more tightly drawn ethnic divisions throughout the region.

ANCILLARIES

Instructor's Manual with Test Bank

Written by Martin Marger, this instructor supplement contains chapter outlines, key terms and concepts, multiple choice questions, essay and discussion questions, and transparency masters.

 ### Web Site

Virtual Society: The Wadsworth Sociology Resource Center

http://sociology.wadsworth.com

At Wadsworth's Sociology Resource Center, you will find a career center, Sociology in the News, Research Online, InfoTrac® College Edition, a Race and Ethnicity Virtual Exploration, and many other selections. From this site you can also access the newly enhanced book-specific Web site which contains chapter-related Internet links, MicroCase Online exercises, InfoTrac College Edition search terms, online practice quizzes, and more.

CNN Today Video Series: Race and Ethnic Relations, Volumes I-III

This video series is an exclusive series jointly created by Wadsworth and CNN. Each video in the series consists of approximately 45-minutes of footage originally broadcast on CNN within the last several years and selected specifically to accompany your race and ethnic relations course. The videos are broken into short segments, perfect for classroom use as lecture launchers or to illustrate key sociological concepts. An annotated table of contents accompanies each video with descriptions of the segments. Special adoption conditions apply.

ACKNOWLEDGMENTS

As with past editions of *Race and Ethnic Relations*, I wish to acknowledge those who helped prepare the sixth: At Wadsworth, Eve Howard, Stephanie Monzon, and Paula Berman smoothly guided the manuscript through its various stages, minimizing along the way the complications that inevitably arise in a project of this scope. Chris Thillen provided thorough and precise copyediting. Also to be acknowledged are the reviewers for this edition who contributed helpful insights and suggestions: Nicole Isaacson, Rutgers University; Frank Weed, University of Texas-Arlington; Deidre Tyler, Salt Lake Community College; Roger Guy, Texas Lutheran University; Rosemary D'Apolito, Youngstown State University; and Satoshi Ito, College of William and Mary. My wife Connie, as always, responded unselfishly to my pleas for help—especially on matters involving the mysteries (to me) of computer technology—despite the burdens of her own work and studies.

ABOUT THE AUTHOR

Martin N. Marger received his bachelor's degree from the University of Miami, his master's from Florida State University, and his doctorate from Michigan State University. In addition to his research and writing in the field of race and ethnic relations, his work includes studies in social inequality and political sociology. He is the author of *Social Inequality: Patterns and Processes*, 2d ed. (McGraw-Hill, 2002), *Elites and Masses: An Introduction to Political Sociology*, 2d ed. (Wadsworth, 1987), and coeditor, with the late Marvin Olsen, of *Power in Modern Societies* (Westview, 1993). His articles have appeared in many sociological journals, including *Social Problems, Ethnic and Racial Studies,* and the *International Journal of Comparative Sociology*. He is currently affiliated with the Center for Integrative Studies in Social Science at Michigan State, where he is also associate director of the Canadian Studies Center.

THE NATURE OF
ETHNIC RELATIONS

The following four chapters erect a framework for analyzing ethnic relations in the United States (in Part II) and in five other societies (in Part III). Each chapter in Part I deals with a particular dimension of interethnic relations. Chapter 1 introduces some basic concepts and terminology of the field; in particular, ethnicity and race, the latter an especially misunderstood term in everyday usage. Chapter 2 concerns the structure of inequality, which seems to be an inevitable outcome when diverse ethnic groups live side by side under a common political and economic system. Chapter 3 explains the techniques by which this system of inequality is maintained. Finally, Chapter 4 introduces the notion of conflict and explains it as an inherent characteristic of interethnic relations. However, there are differing degrees of group inequality and differing levels of group conflict, and these variations create different types of multiethnic societies; several of these types are described.

One objective of Part I is to explain the sociological approach to race and ethnic relations. This approach is fundamentally different from the manner in which relations among racial and ethnic groups are commonly viewed and interpreted. Sociologists see everyday social occurrences differently from laypersons, and they describe them differently as well. They go beneath the superficial to uncover the unseen and often unwitting workings of society, frequently exposing the erroneousness of much of what is considered well-established knowledge. Sociologist Peter Berger has put it well: "It can be said that the first wisdom of sociology is this—things are not what they seem" (1963:23). This is particularly so in the study of race and ethnic relations.

As an example, most people, if asked, could attempt an explanation of why black-white relations in the United States have been customarily discordant, and they might even venture to explain why conflict is also so commonplace among ethnic groups in other parts of the world. They would probably explain that humans are belligerent "by nature" or that there are "inherent" differences among groups, creating unavoidable fear and distrust. Although these explanations are direct and apparently simple to comprehend, they do not necessarily stand up when subjected to sociological analysis. Groups with different cultural origins and physical traits may indeed clash quite commonly, but as we will see, there are social factors more significant than any "innate" tendencies that account for that discord.

The subject matter of sociology—or any of the other social sciences—is not the abstruse world of physics or chemistry but the everyday life of people. Because the objects of their study are so much a part of common human experience, sociologists often seem to make unnecessarily complex what appears to be quite simple. But the application of rigorous theory and methods and the use of precise terminology are the chief distinguishing features of the sociological approach, in contrast to the more unencumbered ways of problem solving that most people employ. In short, sociologists apply a scientific approach to analyzing human relations. And in doing so, they find that much of what is taken for granted as commonsensical is not so simple or common and perhaps not at all "sensical." In studying race and ethnic relations, therefore, it is necessary to establish more precise terms for various racial and ethnic phenomena and to become aware of the major theories and research findings that underlie the sociological approach to this field.

Race and ethnic relations cannot be explained with a single analytic tool or with one general theory. Thus, we will find that sociologists are not in agreement on all issues in the field. At times, in fact, they may be diametrically opposed. But it is necessary to understand that the very nature of scientific inquiry makes such a lack of consensus an almost foregone conclusion. Scientists, whether physical or social, recognize no absolute explanations or unchanging theories. Like any science, sociology poses more questions than it answers. In the following chapters, therefore, we will find no explanations that have not been questioned, tested, and retested. Though we can obtain no final answers to many puzzling questions using the sociological approach, we can sharpen immeasurably our insight into the whys and wherefores of relations among different racial and ethnic groups. But we must be prepared to accept the frustration that often accompanies examining new ideas about what is old and familiar and employing new methods of observing what has customarily been seen uncritically.

1

INTRODUCTION:
SOME BASIC CONCEPTS

ON THE AFTERNOON of October 3, 1995, most Americans paused momentarily to listen to the verdict in the trial of O. J. Simpson, in what had become, arguably, the most celebrated judicial case in U.S. history. The defendant had been charged with the murder of his former wife and her acquaintance. The trial and the jury's verdict, however, went well beyond the arguments presented by the prosecution and the defense, and captivated the nation for much of an entire year. Most important, they touched upon the society's most basic ethnic cleavage, with public opinion polarized along ethnic lines. The trial and the verdict were interpreted differently by blacks and whites. Most blacks saw the judicial process in the case as tainted and suspect; most whites saw it as fair. Much of blacks' dubiousness stemmed from their perception of the criminal justice system, particularly the police, as incapable of treating blacks with the same probity as whites. As to the "not guilty" verdict, public opinion polls indicated that most blacks saw it as just, while most whites saw it as a perversion of justice (Newport and Saad, 1997a). Nothing could better demonstrate the divergent perspectives than televised pictures of blacks spiritedly celebrating the verdict's announcement while whites stared glumly in disbelief. The trial and its aftermath dramatized the depth of division between black and white segments of multiethnic America and, in the view of many, reinforced and even sharpened that division.

At the dawn of the twenty-first century, the relations between blacks and whites remain the most obdurate and vexing issue in American society. Although there can be no argument that the degree of inequality and the levels of prejudice and discrimination between these two components of the U.S. population have changed dramatically in the past forty years, it is equally clear that deep divisions and inequalities remain. Moreover, public opinion polls indicate consistently that members of each group do not see the content and direction of intergroup relations similarly. Most whites see a society in which race and ethnicity no longer carry significant social and economic consequences for people; blacks, by contrast, see race and ethnicity as still very much a major determinant of their life chances.

The division between blacks and whites, however, no longer dominates so thoroughly the content of race and ethnic relations in the United States, as it had for most of the twentieth century. Almost from its very beginnings, American society was ethnically diverse. But for the past several decades, large-scale immigration has transformed it into a profoundly more heterogeneous nation. Literally dozens of new groups, markedly varied in racial and ethnic characteristics, have reconfigured the American population. Virtually no region of the country and no community has been unaffected by this newest influx of immigrants. With increasing ethnic diversity has come public debate—often passionate and shrill—over issues of affirmative action, multiculturalism, and immigration itself. With the burgeoning of American ethnic diversity have also come new problems of ethnic conflict.

These were no better illustrated than in the public response to the terrorist attack on the United States on September 11, 2001. When it was realized that the perpetrators of the attack were associated with Arab Islamic terror networks, Arab Americans, one of America's newest ethnic populations, became widespread targets of verbal and, in some cases, physical, abuse. In the aftermath of the destruction of the World Trade Center in New York—in which almost 3,000 people were killed—threats were made against stores owned by Arabs and against Muslim schools, several mosques were firebombed, and people who appeared to be Arab were set upon. In several cases airline passengers and flight crews refused to fly with Arab passengers, who were removed from their flights (Harden and Sengupta, 2001). People were verbally or physically attacked who were not even Arab or Muslim, but were mistaken for them. Sikhs, for example, followers of an Indian religion that has no relation to Islam, had to struggle to explain to an uncomprehending public that despite their turbans and beards, they were not followers of Islam and were in no way associated with the views of Muslim extremists (Goodstein and Lewin, 2001). A national public opinion poll found that over one-third of Americans felt they now had less trust in Arabs living in the United States, and a majority supported requiring people of Arab descent, even if they were U.S. citizens, to be subjected to special, more intensive security checks at airports (Gallup Organization, 2001c).

Most Americans today acknowledge the seriousness of division and conflict between people of various racial and ethnic groups in the United States and the strains they create in the social fabric. Indeed, many would readily assert that problems stemming from race and ethnicity are the most severe, persistent, and irresolvable facing the society. In recognizing the critical nature of these problems, however, many forget or fail to observe that they are not unique to the United States. Most of the world's most deadly conflicts involve clashes of one ethnic group with another. Ongoing conflicts in the former Yugoslavia and in Northern Ireland represent two of the more prominent and well-reported of these. But serious conflicts with racial or ethnic dimensions, marked by periodic outbursts of violence, also continue to occur in Sri Lanka, India, Burundi, Rwanda, Angola, Sudan, Nigeria, Lebanon, Israel, Spain, Indonesia, Russia, and several of the former republics of the Soviet Union. Other ethnic-related conflicts, less volatile or sizable, occur in many other countries as well. Moreover, much of western Europe, like the United States and Canada, has been affected by a greatly increased flow of immigration from poorer and politically unstable societies, made up of people sharply different racially and ethnically from the native population. This has produced serious incidents of ethnic violence and anti-immigrant political movements. As one looks at the contemporary world, it becomes evident that ethnic conflict in some degree is a basic feature of almost all modern societies with diverse populations.[1]

Societies comprising numerous racial, religious, and cultural groups can be described as *multiethnic*. It is mainly these societies with which we will be concerned. Although racial and ethnic diversity in the modern world is most apparent in those societies in which ethnic conflict has been continual for many generations or in which it has recently emerged with great intensity, ethnic heterogeneity in contemporary societies is commonplace, not exceptional. In fact, few nations today are not multiethnic. Only a handful of the more than 180 member countries of the United Nations are ethnically homogeneous. "Multiethnicity," notes sociologist Robin Williams, "is the rule" (1994:50). Moreover, the extent of diversity within these societies is very great. In most, the largest group is not even half the society's population (Connor, 1972).

Table 1.1 indicates the relative ethnic diversity of some contemporary societies. Ethnic diversity includes differences among groups in language, religion, race, and culture. As can be seen, ethnic heterogeneity characterizes societies on every continent and in various stages of socioeconomic development.

Social scientists had maintained for many years that industrialization and the forces of modernization would diminish the significance of race and ethnicity in heterogeneous societies (Deutsch, 1966). They felt that with the breakdown of small, particularistic social units and the emergence of large, impersonal bu-

---

[1]The seriousness of ethnic conflict in the modern world is highlighted by the fact that, since 1945, perhaps as many as twenty million people have died as a result of ethnic violence (Williams, 1994).

Table 1.1 Ethnic Diversity of Selected Nation-States

HIGH	MEDIUM	LOW
United States	United Kingdom	Sweden
Canada	Australia	Irish Republic
Russia	Nigeria	Egypt
India	Argentina	Colombia
Israel	Malaysia	Japan

reaucratic institutions, people's loyalty and identity would be directed primarily to the national state rather than to internal racial and ethnic communities. The opposite trend, however, seems to characterize the contemporary world.[2]

During the past several decades in the industrialized nations, ethnic groups thought to be well absorbed into the national society have reemphasized their cultural identity, and new groups have demanded political recognition. In the United States, the emergence of black ethnic consciousness in the 1960s stimulated ongoing social movements among other ethnic groups, including Hispanic Americans, Asian Americans, American Indians, and even those of European origin. A new wave of immigration begun in the 1970s further heightened American ethnic issues. In Western Europe, ethnically based political movements have been evident in Britain, France, Spain, and Belgium. Throughout Eastern Europe, the late 1980s brought massive economic and political change, rekindling ethnic loyalties that had been suppressed for several decades. And in Canada, the traditional schism dividing English- and French-speaking groups has in recent years widened, threatening to break up the Canadian nation.

In the developing, or Third World, nations, too, the ethnic factor has emerged with great strength. World War II marked the end of several centuries of imperialist domination of non-Western peoples by European powers, and many new nations were created, the political boundaries of which were often carved out of the administrative districts of the old colonial states. Often these artificial boundaries were drawn up with little consideration of the areas' ethnic composition. As a result, the new nation-states found themselves faced with the problem of integrating diverse cultural groups, speaking different languages and even maintaining different belief systems, into a single national society. The upshot has been numerous and constant ethnic conflicts in which one group refuses to

---

[2]Connor (1972), in fact, contends there is much evidence to support the thesis that modernization results in increasing demands for ethnic separatism. Blumer (1965) has also shown that industrialization does not necessarily lead to more benign ethnic relations or to displacement of the established ethnic order.

recognize the political authority of the central government, which is usually dominated by members of a rival group. In 1947, for example, following independence from Britain, the Indian subcontinent was thrown into violent discord between Hindus and Muslims, the area's two major religious groups. This conflict was resolved only by the eventual partitioning of the region into two autonomous nation-states, India and Pakistan. Perhaps the most frightful of ethnic conflicts among developing societies occurred in Rwanda, a tiny central African nation, in 1994. Ongoing hostilities between the society's two major ethnic groups, Tutsis and Hutus, culminated in the slaughter of perhaps half a million— no one can be certain of the exact number—within just a few weeks.

It must be remembered, however, that these conflicts represent only the most severe and devastating. Others, more moderate and less violent, typify ethnic relations in multiethnic societies. In short, racial and ethnic forces, though varying in scope and intensity, are important bases of both group solidarity and cleavage in the modern world. Moreover, their impact is not likely to diminish in the foreseeable future.

RACE AND ETHNIC RELATIONS

The study of race and ethnic relations is concerned generally with the ways in which the various groups of a multiethnic society come together and interact over extended periods. As we proceed in our investigation, we will be looking specifically for answers to four key questions.

Basic Questions

1. *What is the nature of intergroup relations in multiethnic societies?* As we will see, ethnic relations commonly take the form of conflict and competition. Indeed, we can easily observe this by following the popular media accounts of ethnic relations in the United States and other nations. These are usually descriptions of hostility and violence. However, intergroup relations are never totally conflictual. If this were not so, ethnic groups would exist in a perpetual whirlwind of discord and strife. Cooperation and accommodation also characterize ethnic relations. Just as we will be concerned with understanding why conflict and competition are so common among diverse groups, we will also investigate harmonious conditions and the social factors that contribute to them.

2. *How are the various ethnic groups ranked, and what are the consequences of that ranking system?* In all multiethnic societies, members of various groups are

treated differently and receive unequal amounts of the society's valued resources—wealth, prestige, and power. In short, some get more than others and are treated more favorably. Moreover, this inequality is not random but is well established and persistent for many generations. A structure of inequality emerges in which one or a few ethnic groups, called the *dominant group* or groups, are automatically favored by the society's institutions, particularly the state and the economy, whereas other ethnic groups remain in lower positions. These subordinate groups are called ethnic *minorities*. We will be concerned with describing this hierarchy and determining how such systems of ethnic inequality come about.

3. *How does the dominant ethnic group in a multiethnic society maintain its place at the top of the ethnic hierarchy, and what attempts are made by subordinate groups to change their positions?* The dominant group employs a number of direct and indirect methods (various forms of prejudice and discrimination) in protecting its power and privilege. This does not mean, however, that subordinate groups do not attempt to change this arrangement from time to time. In fact, organized movements may arise, such as the black civil rights movement of the 1960s in the United States, in which concerted efforts are made by subordinate groups to change their position in the ethnic hierarchy. One of our chief concerns, then, will be the ways in which systems of ethnic inequality are maintained and how they change.

4. *What are the long-range outcomes of ethnic interrelations?* When ethnic groups exist side by side in the same society for long periods, either they move toward some form of unification or they maintain or even intensify their differences. These various forms of integration and separation are called *assimilation* and *pluralism*. Numerous outcomes are possible, extending from complete assimilation, involving the cultural and physical integration of the various groups, to extreme pluralism, including even expulsion or annihilation of groups. Usually, less extreme patterns are evident, and groups may display both integration and separation in different spheres of social life. Again, our concern is not only with discerning these outcomes but also with explaining the social forces that favor one or the other.

A Comparative Approach

The study of race and ethnic relations has a long tradition in American sociology, beginning in the 1920s with the research of Robert Park, Everett Hughes, and Louis Wirth. These scholars were among the first to focus attention on the relations among ethnic groups, particularly within the ethnic mélange of large American cities like Chicago and New York. The sociology of race and ethnic relations has progressed enormously since that time so that it now constitutes one of the chief subareas of the sociological discipline.

With few exceptions, however, American sociologists have continued to concentrate mainly on American groups and relations, often neglecting the analysis of similarities and differences between the United States and other heterogeneous societies. But if we are to try to understand the general nature of race and ethnic relations, it is necessary to go beyond the United States—or any particular society—and place our analyses into a comparative, or cross-societal, framework. It is now obvious that ethnic diversity, conflict, and accommodation are worldwide phenomena, not unique to American society. However, because most research in race and ethnic relations has been the product of American sociologists dealing with the American experience, we are often led to assume that patterns evident in the United States are the same in other societies.

This book adopts a comparative perspective in which the United States is seen as one among many contemporary multiethnic societies. This requires paying close attention to patterns of race and ethnicity in non-American societies. Because readers are likely to be most familiar with the United States, however, focus will fall most intently on American groups and relations. Even in those chapters in which cases outside the United States are the major topic (Part III), some room will be allotted for comparisons with the United States.

A comparative approach will not only enable us to learn about race and ethnicity in other societies but will also provide us with a sharper insight into race and ethnicity in the United States. It has often been observed that we cannot begin to truly understand our own society without some knowledge of others. Moreover, in addition to the differences revealed among societies, similarities may also become apparent. As sociologists Tamotsu Shibutani and Kian Kwan point out, comparing American ethnic relations with those of other societies "reveals that patterns of human experience, though infinitely varied, repeat themselves over and over in diverse cultural contexts" (1965:21). Discovering such generalizable patterns of the human experience is the ultimate aim of all sociological efforts.

ETHNIC GROUPS

Although they are now familiar, the terms *ethnic group* and *ethnicity* are relatively new, not even appearing in standard English dictionaries until the 1960s (Glazer and Moynihan, 1975). Groups generally referred to today as "ethnic" were previously thought of as races or nations, but these are terms clearly different in meaning.

Characteristics of Ethnic Groups

Unique cultural traits Basically, ethnic groups are groups within a larger society that display a unique set of cultural traits. As sociologist Melvin Tumin defined it, an ethnic group is "a social group which, within a larger cultural and social system, claims or is accorded special status in terms of a complex of traits (ethnic traits) which it exhibits or is believed to exhibit" (1964:243). Ethnic groups, then, are subcultures, maintaining certain behavioral characteristics that, in some degree, set them off from the society's mainstream, or modal, culture. Such unique cultural traits are not trivial but are fundamental features of social life such as language and religion.

Unique cultural traits, however, are not sufficient alone to define ethnic groups. Can we speak of physicians as an ethnic group? or truck drivers? or college students? Obviously, we would consider none of these categories "ethnic" even though they are groups that exhibit some common behavioral traits setting them off from the larger society. Clearly, we need further qualifications for distinguishing ethnic groups.

Sense of community In addition to a common set of cultural traits, ethnic groups display a sense of community among members; that is, a consciousness of kind or an awareness of close association. In simple terms, there exists a "we" feeling among members. Sociologist Milton Gordon (1964) suggests that the ethnic group serves above all as a social-psychological referent in creating a "sense of peoplehood." This sense of community, or oneness, derives from an understanding of a shared ancestry, or heritage. Ethnic group members view themselves as having common roots, as it were. When people share what they believe to be common origins and experiences, "they feel an affinity for one another," writes sociologist Bob Blauner, "a 'comfort zone' that leads to congregating together, even when this is not forced by exclusionary barriers" (1992:61).

Such common ancestry, however, need not be real. As long as people regard themselves as alike by virtue of their perceived heritage, and as long as others in the society so regard them, they constitute an ethnic group, whether such a common background is genuine or fictive (Shibutani and Kwan, 1965). Sociologists Everett and Helen Hughes have perceptively observed that "an ethnic group is not one because of the degree of measurable or observable difference from other groups; it is an ethnic group, on the contrary, because the people in it and the people out of it know that it is one; because both the ins and outs talk, feel, and act as if it were a separate group" (1952:156). Ethnic groups, then, are social creations wherein ethnic differences are basically a matter of group perception. Groups may be objectively quite similar but perceive themselves as very different, and the converse is equally true.

Sociologists have debated the relative significance of the cultural element and the sense of community as most critical to the formation of an ethnic group (Dorman, 1980). The argument boils down to a question of whether ethnic groups are objective social units that can be identified by their unique culture or merely collectivities that people themselves define as ethnic groups. Whereas some view the cultural features of the group as its key distinctive element, others argue that stressing its unique culture minimizes the importance of the subjective boundaries of the group that people themselves draw (Barth, 1969). Most simply, the latter maintain that if people define themselves and are defined by others as an ethnic group, they *are* an ethnic group, whether or not they display unique cultural patterns. If this is the case, the cultural stuff of which the ethnic group is composed is unimportant.

Although this may seem like a relatively minor theoretical point, it is of importance when ethnic groups in a society begin to blend into the dominant cultural system. Sociologists have traditionally assumed that as groups integrate into the mainstream society, the basis of retention of ethnicity diminishes. But whether people continue to practice ethnic ways may matter little as long as they continue to define themselves and are defined by others in ethnic terms. Many Americans continue to think of themselves as ethnics even though they exhibit little or no understanding of or interest in their ethnic culture. Do third- or fourth-generation Irish Americans, for example, really share a common culture with their first-generation ancestors? Wearing a button on St. Patrick's Day proclaiming "I'm proud to be Irish" is hardly a display of the traits of one's Irish-American forebears. Yet an ethnic identity may remain intact for such persons, and they may continue to recognize their uniqueness within the larger society. Thus, despite the lack of a strong cultural factor, the sense of Irish-American identity may be sufficient to sustain an Irish-American ethnic group.

We thus have two views of the ethnic group: (1) it is an objective unit that can be identified by a people's distinct cultural traits, or (2) it is merely the product of people's thinking of and proclaiming it as an ethnic group. To avoid the extreme of either of these views, sociologist Pierre van den Berghe defines the ethnic group as both an objective and a subjective unit: "An ethnic group is one that shares a cultural tradition *and* has some degree of consciousness of being different from other such groups" (1976:242). As he points out, it is foolish to think that ethnic groups simply arise when people so will it. Fans of a particular football team may feel a sense of commonality and even community, but they surely do not compose an ethnic group. In short, there must be some common cultural basis and sense of ancestry to which ethnic group members can relate. As van den Berghe notes, "There can be no ethnicity (or race) without some conception and consciousness of a distinction between 'them' and 'us.' But these subjective perceptions do not develop at random; they crystallize around clusters of objective characteristics that become badges of inclusion or exclusion"

Part I: The Nature of Ethnic Relations

(1978:xvii). Although ethnic boundaries are very flexible, they are always founded on a cultural basis. At the same time, however, an ethnic group cannot exist in an objective sense independent of what its members think and believe. There must be a sense of commonality, and such a feeling of oneness arises generally through the perception of a unique cultural heritage.

Ethnocentrism The "we" feeling of ethnic groups ordinarily leads naturally to *ethnocentrism*, the tendency to judge other groups by the standards and values of one's own. Inevitably, this produces a view of one's own group as superior to others. The ways of one's own group (*in-group*) become "correct" and "natural," and the ways of other groups (*out-groups*) are seen as "odd," "immoral," or "unnatural." Sociologists and anthropologists have found the inclination to judge other groups by the standards of one's own and to view out-groups as inferior or deficient to be a universal practice.

In multiethnic societies, such feelings of group superiority become a basis for group solidarity. In addition to fostering cohesiveness within one group, however, ethnocentrism also serves as the basis of conflict between different groups. As Bonacich and Modell have explained, "Ethnicity is a communalistic form of social affiliation, depending, first, upon an assumption of a special bond among people of like origins, and, second, upon the obverse, a disdain for people of dissimilar origins" (1980:1). Here we can begin to understand why conflict among ethnic groups is so pervasive and intractable.

Ascribed Membership Ethnic group membership is ordinarily ascribed. This means that one's ethnicity is a characteristic acquired at birth and not subject to basic change. Being born a member of an ethnic group, one does not leave it except in unusual circumstances. One might change ethnic affiliation by "passing"—that is, by changing one's name or other outward signs of ethnicity—or by denying group membership. But it is extremely difficult to divest oneself completely of one's ethnic heritage. Through the socialization process, individuals come to learn their group membership early and effectively and to understand the differences between themselves and members of other groups. So well internalized is this group identification that one comes to accept it almost as naturally as accepting one's gender. As Hughes and Hughes have suggested, "If it is easy to resign from the group, it is not truly an ethnic group" (1952:156). This understanding of ethnic descent is what creates such sharp and, at times, deadly divisions among ethnic groups (Williams, 1994).

Those who attempt to shed their ethnic identity find that the society will rarely permit this fully. Gordon notes in this regard that in American society a person who attempts to relinquish his or her ethnic identity finds "that the institutional structure of the society and the set of built-in social and psychological categories with which most Americans are equipped to place him—to give him a 'name'—are loaded against him" (1964:29). Once the ethnic categories in a society are set, explains Gordon, placing people into them is almost automatic and is by no means subject entirely to people's volition.

In multiethnic societies where ethnic boundaries are not rigid and where there is much marriage across ethnic lines, like the United States and Canada, the voluntary nature of ethnicity becomes more salient. The ethnic origins of third- or fourth-generation Euro-Americans, for example, may be quite varied. Individuals therefore make decisions about "who they are" ethnically, some in a contrived fashion and others almost unconsciously. A person whose family origins may contain Italian, Polish, and Irish elements might emphasize the Italian part and identify himself as "Italian American," disregarding his other ancestral links. In such cases, the volitional component of ethnic identity is strong. For those whose ethnic identity is based also on physical, or racial, characteristics, however, the capacity to choose becomes more limited. For such people, ascription is paramount.

Territoriality Ethnic groups often occupy a distinct territory within the larger society. Most of the multiethnic societies of Europe consist of groups that are regionally concentrated. Basques and Catalans in Spain, Welsh and Scots in Britain, and Flemings and Walloons in Belgium are groups that maintain a definable territory within the greater society. Such multiethnic societies are quite different from the United States, Canada, or Australia, where ethnic groups have for the most part immigrated voluntarily and, though sometimes concentrated in particular areas, are not regionally confined.

When ethnic groups occupy a definable territory, they also maintain or aspire to some degree of political autonomy. They are, in a sense, "nations within nations." In some societies, the political status of ethnic groups is formally recognized. Each group's cultural integrity is acknowledged, and provision is made for its political representation in central governmental bodies. Such societies are best referred to not only as multiethnic but as multinational (van den Berghe, 1981).

In other societies, where such multinationality is not formally recognized, certain ethnic groups may seek greater political autonomy or perhaps even full independence from the national state, usually dominated by other ethnic groups. The Spanish Basques represent such a case. With a culture and language distinct from other groups in Spain, the Basques have traditionally seen themselves as a separate nation and have negotiated with the Spanish government at various times to promote their sovereignty. A Basque nationalist movement has been evident since the late nineteenth century, but in the past thirty years it has taken on a particularly virulent and often violent form (Anderson, 2001; Kurlansky, 1999; Ramirez and Sullivan, 1987). Similar nationalist movements, perhaps not so hostile, have, during most of the post-

World War II era, typified many multiethnic societies. We will look at one such movement, among French-speaking Canadians in Quebec, in Chapter 15.

Where ethnic groups do not continue to maintain significant aspects of culture (like language) for many generations and where they are geographically dispersed rather than concentrated, such nationalist movements do not ordinarily arise. This is the case in the United States, where ethnic groups are scattered throughout the society, have generally taken on the major cultural ways of the dominant white Anglo-Saxon Protestant, or "WASP," group after a generation or two, and seek greater power within the prevailing political system. Nonetheless, even where ethnic groups are no longer territorially based, they usually retain sentimental ties to their society of origin.

Ethnicity as a Variable

Each of these characteristics—unique culture, sense of community, ethnocentrism, ascribed membership, and territoriality—will be displayed to a different degree by different ethnic groups. These traits are variables and will not only differ from group to group but also change at various historical times within any single ethnic group. Thus, we should not expect to find all ethnic groups in a society equally unique in cultural ways, strongly self-conscious and recognized by out-groups, or even ethnocentric. The extent to which ethnic groups are noticeable and maintain a strong consciousness among members depends on both in-group and out-group responses. Some ethnic groups seek rapid assimilation and are accepted into the society's mainstream relatively quickly. Others, however, may retain their group identity for many generations because of rejection by the dominant group, their own desire to maintain the ethnic community, or a combination of these two. Jews, for example, have maintained a strong group consciousness in most societies, largely as a result of the historically consistent hostility to which they have been subjected but also because they have consciously sought to preserve their group identity.

Here we might consider sociologist Robin Williams's distinction between what he calls ethnic "collectivities" and ethnic "categories" (1964, 1979, 1994). The former are made up of people who have a recognizable culture, are aware of themselves as a unit, are recognized as a unit by outsiders, interact with other group members, and feel a sense of obligation to support and defend the group. Ethnic categories, in contrast, are merely groupings of people with one or a number of similar characteristics but with little sense of membership and little interaction among them. At different times and in different social contexts, ethnic groups may be both collectivities and categories. No single Italian American, for example, can interact with every other Italian American. At the level of interaction, then, there are really many smaller Italian-American ethnic groups or communities. All Italian Americans, however, can recognize similarities of culture and behavior among themselves, can identify themselves as Italian Americans, and can maintain a sense of group solidarity. Members of out-groups may also recognize these similarities and identify an Italian-American ethnic group. There is, then, in addition to numerous ethnic communities, one larger, more abstract collectivity called Italian Americans. Going one step further, journalists, pollsters, or even social scientists may enumerate a group referred to as Italian Americans, regardless of the degree of ethnic identification of the millions of individuals who make up this category.

For persons in multiethnic societies, the ethnic group becomes a key source of social-psychological attachment and serves as an important referent of self-identification. Put simply, people feel naturally allied with those who share their ethnicity and identify themselves with their ethnic group. Their behavior is thus influenced by ethnicity in various areas of social life.

However, just as ethnicity will differ in scope and consequence at the group level, so too for individuals it will play a varying role. For some, it is a major determinant of behavior, and most social relations will occur among those who are ethnically similar. For others, ethnicity may be insignificant, and they may remain essentially devoid of ethnic consciousness. For most people in multiethnic societies, however, the ethnic tie is important in shaping primary relations—those that occur within small, intimate social settings such as the family and the peer group. These relations include one's choice of close friends, marital partner, residence, and so on.

There are certainly other groups in modern societies to which people feel a sense of attachment and that provide a source of identification. These include one's social class, gender, age, and occupation. Like ethnic groups, they also become bases of solidarity and societal cleavage. But in multiethnic societies, ethnicity is a primary base of loyalty and consciousness for most people and thus serves as a strong catalyst for competition and conflict. Moreover, ethnicity is usually interrelated and overlaps with these other sources of group identification and attachment.

Most important, ethnicity is a basis of ranking, in which one is treated according to the status of his or her ethnic group. In no society do people receive an equal share of the society's rewards, and in multiethnic societies, ethnicity serves as an extremely critical determinant of who gets "what there is to get" and in what amounts. In this sense, ethnicity is a dominant force in people's lives whether or not they are strongly conscious of their ethnic identity and regardless of the degree to which ethnicity shapes their interrelations with others.

RACES

Without question, *race* is one of the most misunderstood, misused, and often dangerous concepts of the modern world. It is not applied dispassionately by laypeople or even, to a great extent, by social scientists. Rather, it arouses emotions such as hate, fear, anger, loyalty, pride, and prejudice. It has also been used to justify some of the most appalling injustices and mistreatments of humans by other humans.

The idea of race has a long history, extending as far back as ancient civilizations. It is in the modern world, however—specifically, the last two centuries—that the notion has taken on real significance and fundamentally affected human relations. Unfortunately, the term has never been applied consistently and has meant different things to different people. In popular usage, it has been used to describe a wide variety of human categories, including people of a particular skin color (the Caucasian "race"), religion (the Jewish "race"), nationality (the British "race"), and even the entire human species (the human "race"). As we will see, none of these applications is accurate and meaningful from a social scientific standpoint. Much of the confusion surrounding the idea of race stems from the fact that it has both biological and social meanings. Let us look briefly at the difference between the two.

Race as a Biological Notion

The essential biological meaning of race is a population of humans classified on the basis of certain hereditary characteristics that differentiate them from other human groups. Races are, in a sense, pigeonholes for categorizing human physical types. Efforts at classification, however, have created a virtually hopeless disagreement among social and biological scientists. The biological understanding of race has led to an enormous variation in thought and almost no accord among biologists, geneticists, physical anthropologists, and physiologists concerning either the term's meaning or its significance. Although it is impossible to do justice to the controversies surrounding the notion of race in a few pages, several of the more apparent problems attached to this most elusive of ideas can be briefly explored.

Genetic Interchangeability To begin with, the difficulty in trying to place people into racial categories on the basis of physical or genetic qualities stems from the fact that all members of the human species, Homo sapiens, operate within a genetically open system. This means that humans, regardless of physical type, can interbreed. If genes of different human groups were not interchangeable, the idea of race as a biological concept might have some useful meaning. But because this is not the case, we see an unbounded variety of physical types among the peoples of the world.

The Bases of Racial Classification That a person from one genetic population can interbreed with a person from any other population creates a second difficulty in dealing with the notion of race: answering the question, "What are the characteristics that differentiate racial types?" Again, there has been little agreement among social scientists. Physical anthropologists distinguish major categories of human traits as either *phenotypes*—visible anatomical features such as skin color, hair texture, and body and facial shape—or *genotypes*—genetic specifications inherited from one's parents. Races have traditionally been classified chiefly on the basis of the most easily observable anatomical traits, like skin color; internal and blood traits have been deemphasized or disregarded. Today, the study of human biological diversity rejects the approach that stresses the classification of individuals into specific categories based on assumed common ancestry and emphasizes, instead, efforts to explain human differences (Kottak and Kozaitis, 1999). This has occurred mainly because there is simply no agreement on which traits should be used in defining races. Moreover, attempts to clearly categorize humans have proved futile because differences among individuals of the same group (or "racial type") are greater than those found between groups (Marks, 1995).

Anthropologist Ruth Benedict fittingly remarked that "in all modern science there is no field where authorities differ more than in the classifications of human races" (1959:22). If researchers are in agreement about anything concerning race, it is that racial classification systems are by and large arbitrary and depend on the specific objectives of the classifier. All agree that "pure" races do not exist today, and some even question whether they have ever existed (Dunn, 1956; Fried, 1965; Pettigrew, 1964; Tobias, 1995).

Physical differences among people obviously exist, and these differences are statistically clear among groups. It is true that, through a high degree of inbreeding over many generations and as adaptations to different physical environments, groups with distinctive gene frequencies and phenotypic traits are produced. There are evident differences, for example, between a "typical" black person and a "typical" white person in the United States. People may be said, therefore, to fall into statistical categories by physical type.

But these statistical categories should not be mistaken for actual human groupings founded on unmistakable hereditary traits. Racial categories form a continuum of gradual change, not a set of sharply demarcated types. Physical differences between groups are not clear-cut but instead tend to overlap and blend into one another at various points. Petersen aptly notes that humans are not unique in this regard: "It follows from the theory of evolution itself that all biological divisions, from phylum through subspecies, are always in the process of change, so there is almost never a sharp and permanent boundary setting one off from the next" (1980:236).

The popular division of the human population into three major racial groupings—Caucasoid, Mongoloid, and Negroid—is thus imprecise and largely arbitrary. Moreover, this scheme excludes large populations that do not easily fit into a simple tripartite arrangement. Where, for example, should East Indians, a people with Caucasian features but with dark skin, be placed? Or where do groups thoroughly mixed in ancestry, like most Indonesians, fit? Because all human types are capable of interbreeding, there are simply too many marginal cases like these that do not easily conform to any particular racial scheme, regardless of its complexity. In studying human differences, most modern investigators have divided the human species into populations, or subpopulations, as the unit of analysis, not racial groupings (Gould, 1984; Marks, 1995; Tobias, 1995).

Race and Social Traits Perhaps the most troublesome aspect of the idea of race concerns the relation of racial inheritance to social and personality traits. As with other features of the race controversy, there is no unanimity of thought on the issue of whether and to what degree one's racial inheritance affects intelligence, temperament, and other individual characteristics.

Particular controversy has arisen over the issue of race and intelligence. This argument reflects the more general debate concerning the effects of heredity and environment on human behavior, or what has been referred to as the "nature versus nurture" question. Although it is generally understood that heredity and environment are interrelated and affect individual behavior and aptitude in complementary ways, the dispute revolves around the degree to which each is consequential. As to intelligence, whereas hereditarians argue that differences among human groups are attributable most basically to genetic factors, environmentalists contend that social variables such as class, family, language, and the development of cognitive skills are most significant. The debate is made relatively meaningless, however, because there is little consensus regarding the very meaning of intelligence and how it can be adequately measured. Do we mean by intelligence innate potential, educability, or the capacity to think abstractly? And how can any of these be measured correctly by "intelligence" tests? None of these questions has been answered satisfactorily.

In any case, the issue of the effects of heredity and environment on intelligence must remain unresolvable if for no other reason than that environmental factors are never exactly the same for any two individuals. People of any society are products of different social classes and subcultures and thus are afforded different life chances, education in particular. Some are born into well-to-do families and thereby receive the maximum opportunities to develop their talents and skills, and others are born into less fortunate circumstances, resulting in fewer chances to fully realize their capacities.

If all people were to start at an equal point in the social hierarchy and were afforded truly equal opportunities for social advancement, the differences they subsequently displayed might be attributed to nonenvironmental causes. However, the opportunity structure of any society is hardly equitable, and people do not begin their quest for social success with the same resources. If runners begin a race at different points on the track, the winner is not necessarily the fastest athlete. The same is true in trying to determine intelligence. That some prove themselves more intelligent than others may indicate only that they have begun with more favorable opportunities. Thus, until the opportunity structure can be made equitable—a remote possibility in any society—the relative effects of heredity and environment on intelligence will not be scientifically measurable with much accuracy.

The issue of race and intelligence was given new impetus when educational psychologist Arthur Jensen (1969) published a report suggesting that heredity was the major determinant in accounting for the collective differences in IQ between blacks and whites. Jensen's theory, methods, and interpretation of findings were all quickly and thoroughly challenged (Deutsch, 1969; Montagu, 1975; Rose and Rose, 1978).

More recently, the publication in 1994 of Richard J. Herrnstein and Charles Murray's book, *The Bell Curve*, refueled the debate regarding the impact of race on mental ability and set off much controversy. Herrnstein and Murray present a kind of social Darwinistic thesis: Intelligence, as measured by IQ, is in large part genetic. They see strong relationships between IQ and various social pathologies. Thus, those with lower IQs have a greater proclivity toward poverty, crime, illegitimacy, poor educational performance, and other social ills. Because IQ is mostly genetic, they argue, there is no way to change the condition of those with low intelligence through educational reforms or welfare programs. Because lower-intelligence people are reproducing much faster than higher-intelligence people, the society is faced with the possibility of a growing underclass, increasingly dependent on the more intelligent and productive classes.

Perhaps the most controversial aspects of Herrnstein and Murray's book concern the linkage of IQ and race. The racial dimension is introduced when the authors point out that the average IQ of blacks is fifteen points lower than the average IQ of whites—a difference, they claim, that holds regardless of social class. Moreover, even when change in average IQ for groups is acknowledged, the differential remains intact. The implication, then, is clear: Blacks are inferior to whites and are thus apt to remain in a state of dependency on the nonpoor and continue to engage in antisocial activities. Herrnstein and Murray therefore question the value of welfare payments, remedial educational programs, affirmative action, and other efforts designed to raise the social level of the poor, who, in the United States, are disproportionately black.

As with Jensen earlier, Herrnstein and Murray's thesis, findings, and conclusions have been overwhelmingly rejected by mainstream social scientists, who claim that their methods are flawed and their reasoning specious (Fischer et al., 1996; Fraser, 1995; Jacoby and Glauberman, 1995). Moreover, many have viewed the book as much a statement of the political leanings of the authors as a work of social science. Specifically, Herrnstein and Murray are challenged on a number of points. For one, IQ has been shown to measure only certain kinds of intelligence. Furthermore, IQ is not fixed but is subject to variation within one's lifetime and, for groups, subject to change over generations. Perhaps most important, the authors do not place sufficient weight on the environmental factors that enable people to express their intelligence, regardless of IQ. Also, the authors reify "race," referring to whites and blacks as if these were clear-cut, distinct genetic groups, ignoring the countless variations within designated racial categories.

Theoretical views like Jensen's and Herrnstein and Murray's have appeared periodically, but the understanding that environment, not racial inheritance, is the key influence in shaping social behavior clearly predominates in social science thinking today (Marks, 1995; Montagu, 1972). As we will see, however, this has not always been the case.[3]

It is important to bear in mind that the issue of race and social behavior is significant only to the extent that a society makes it so. As Berry and Tischler note, "It only assumes importance if people have racist views and want to use the arguments as ammunition for their cause" (1978:84). It is of significance, in other words, only where people are dealt with not as individuals but as members of racial categories. This distinction suggests the social meaning of race.

The Social Meaning of Race

Many popular ideas are of very dubious scientific validity; race is certainly among these. But as André Béteille has pointed out, "Sociological analysis is concerned not so much with the scientific accuracy of ideas as with their social and political consequences" (1969:54). Whether the idea of race is meaningful in a biological sense remains a controversial and seemingly unresolvable issue. But whatever its biological validity, the importance of race for the study of intergroup relations clearly lies in its social meaning.

Most simply, people attach significance to the concept of race and consider it a real and important division of humanity. And, as long as people *believe* that differences in selected physical traits are meaningful, they will act on those beliefs, thereby affecting their interrelations with others. Sociologist W. I. Thomas observantly asserted that "if men define situations as real they are real in their

---

[3]For a succinct discussion of the controversy surrounding race and intelligence, see Berry and Tischler (1978:63-86). The issue of intelligence tests and differences among racial groups is well covered in Samuda (1975) and Gould (1983). A simplified explanation of Jensen's controversial position and a rejoinder by a prominent geneticist are found in Jensen (1973) and Dobzhansky (1973). The arguments surrounding *The Bell Curve* are detailed in Fraser (1995) and Jacoby and Glauberman (1995).

consequences" (Thomas and Znaniecki, 1918:79). If, for example, those classified as black are deemed inherently less intelligent than those classified as white, people making this assumption will treat blacks accordingly. Employers thinking so will hesitate to place blacks in important occupational positions; school administrators thinking so will discourage blacks from pursuing difficult courses of study; white parents thinking so will hesitate to send their children to schools attended by blacks; and so on.

The creation of such categories and the beliefs attached to them generate what sociologists have called the "self-fulfilling prophecy" (Merton, 1968). This refers to a process in which the false definition of a situation produces behavior that, in turn, makes real the originally falsely defined situation. Consider the aforementioned case. If blacks are considered inherently less intelligent, fewer community resources will be used to support schools attended primarily by blacks on the assumption that such support would only be wasted. Poorer-quality schools, then, will inevitably turn out less capable students, who will score lower on intelligence tests. The poorer performance on these tests will "confirm" the original belief about black inferiority. Hence the self-fulfilling prophecy. The notion of black inferiority is reinforced, and continued discriminatory treatment of this group rationalized.

Anthropologist Robert Redfield has noted that "it is on the level of habit, custom, sentiment, and attitude that race, as a matter of practical significance, is to be understood. Race is, so to speak, a human invention" (1958:67). The scientific validity of race, then, is of little consequence; rather, it is the belief system of a society that provides its significance. Each heterogeneous society takes whatever are perceived as important physical differences among people and builds a set of racial categories into which those people are placed. But these categories are fully arbitrary. Different societies will use different criteria with which to assign people racially, thereby creating classification systems that may have little or no correspondence from one society to the next. Omi and Winant use the term *racial formation* to describe "the process by which social, economic, and political forces determine the content and importance of racial categories, and by which they are in turn shaped by racial meanings" (1986:61). The social meaning of race, they explain, is constantly subject to change through political struggle.

The arbitrariness of racial categorizing can be seen easily when we compare different societies, each with numerous physical types. The same individual categorized as "black" in the United States, for example, might be categorized as "white" in Brazil. The racial classification systems in these two societies do not coincide. As we will see in Chapter 14, Brazilians do not see or define races in the same way that Americans do, nor do they necessarily use the same physical characteristics as standards with which to categorize people (Harris, 1964; Pitt-Rivers, 1987; van den Berghe, 1978). Obviously, different criteria and different categories of race are operative in each society. Indeed, so subject to cultural definition is the idea of race that the selected physical attributes used to classify people need not even be obvious, only the *belief* that they are evident. In Northern Ireland, for example, both Protestants and Catholics sometimes say they are able to identify members of the other group on the basis of physical differences, despite their objective similarity. In the United States, too, European-origin ethnic groups that today are seen as racially the same (Italians, Poles, Jews, and so on) were defined in the past as racially distinct. As historian Matthew Frye Jacobson has written, "…entire races have disappeared from view, from public discussion and from modern memory, though their flesh-and-blood members still walk the earth" (1998:2).

What is perhaps most important regarding the social classification of races is that the perceived physical differences among groups are assumed to correspond to social or behavioral differences. Thus, blacks are assumed to behave in certain ways and to achieve at certain levels because they are black; whites are assumed to behave and achieve in other ways because they are white; and so on. As van den Berghe notes, "What makes a society multiracial is not the presence of physical differences between groups, but the attribution of social significance to such physical differences as may exist" (1970:10). Redfield has drawn an apt analogy: "If people took special notice of red automobiles, and believed that the redness of automobiles was connected inseparably with their mechanical effectiveness, then red automobiles would constitute a real and important category" (1958:67).

It is most critical, then, to look not simply at the racial categories that different societies employ but also at the social beliefs attached to those categories. Such beliefs are the product of racist thinking, which we will consider shortly.

Race and Ethnicity: A Synthesis

As should now be obvious, the term "race" is so charged and misconceived that it is very difficult to employ in a useful analytic manner. To add to the confusion, many groups today defined as ethnic groups were in previous historical periods defined as races. As noted above, European immigrant groups in the United States, for example, representing different nationalities or religions, were classified as races during the early part of the twentieth century by laypersons and many social scientists as well.

Another difficulty arising from the use of the term race is that it is often applied too sweepingly, encompassing many specific ethnic groups. For example, to speak of "blacks" in the United States in the aggregate is to assume a homogeneity that does not exist. Overlooked is the ethnic variety among U.S. blacks. Although African Americans, those whose ancestry is traceable to American slavery, predominate, today Haitians, Jamaicans, other Caribbeans, many Latinos, and even a growing population of African immigrants are also part of the American "black" population. Culturally these groups may have little in common. Moreover, to speak of a global racial commonality is even more misleading. As Glazer and Moynihan have noted,

> it is hardly likely that Moslem, Swahili-speaking blacks of Zanzibar would find much in common with the black institutions and culture that are now being built up in this country. They would not have any predilection for soul music or soul food, would find the styles of dress, hair, walk, and talk that are now popular as defining blackness distinctly foreign (1970:xxxix).[4]

Likewise, the term "white" has no significance beyond a reference to white ethnic groups in the aggregate. To speak of Polish Americans, Jewish Americans, and Irish Americans as part of a common group is to falsely meld groups whose cultural traditions are quite distinct.

Because of its confusing usage and its questionable scientific validity, many sociologists and anthropologists have dispensed entirely with the term race and instead use "ethnic group" to describe those groups commonly defined as racial (Berreman, 1972; Gordon, 1964; Patterson, 1997; Schermerhorn, 1970; Shibutani and Kwan, 1965; Williams, 1979). In the United States, African Americans, American Indians, Chinese Americans, Japanese Americans, and Mexican Americans have all the earmarks of ethnic groups—unique culture, consciousness of kind, ascriptive membership, and in some cases even territoriality—at the same time that most members of these groups are physically distinct from Americans of European origin. Classifying all these groups as ethnic seems most reasonable because, in addition to their physical traits, there are always consistent and significant cultural traits that set them off from other groups. Sociologist Richard Alba has offered a definition of race as a variant of ethnicity: "A racial group is...an ethnic group whose members are believed, by others if not also by themselves, to be physiologically distinctive" (1992:576).

For the sake of simplicity and clarity, then, the term *ethnic group* will be used in this book in a broad manner so as to include groups and organizations identified by national origin, cultural distinctiveness, racial characteristics, or religious affiliation. As we will see, ethnic groups in most modern societies comprise combinations of these national, cultural, physical, and religious traits. For those groups that are particularly divergent physically from the dominant group, such as African Americans, Richard Burkey (1978) has suggested the term *racial-ethnic group*, and that term will be used accordingly.

---

[4]From Nathan Glazer and Daniel P. Moynihan, *Beyond the Melting Pot*, 2d ed. Copyright (c) 1970 by MIT Press, Cambridge, Mass. Reprinted by permission.

Though the impact of hereditary features on social behavior has been shown by social science to be at the most, questionable, and at the least, minimal, many people continue to be guided by racist thinking. As Benedict put it, "Any scientist can disprove all its facts and still leave the *belief* untouched" (1959:99).

Racist thinking involves principles that lead naturally and inevitably to the differential treatment of members of various ethnic groups. As we will see in the next chapter, in no society are valued resources distributed equally; in all cases, some get more than others. But in multiethnic societies, ethnicity is used as an important basis for determining the nature of that distribution. Ethnic groups are ranked in a hierarchy, and their members are rewarded accordingly, creating a system of ethnic inequality. Groups at the top compound their power and maintain dominance over those lower in the hierarchy. Such systems of ethnic inequality require a belief system, or ideology, to rationalize and legitimate these patterns of dominance and subordination, and racism has usually served this function.

The Ideology of Racism

As a belief system, or ideology,[5] racism is structured around three basic ideas:

1. Humans are divided naturally into different physical types.

2. Such physical traits as they display are intrinsically related to their culture, personality, and intelligence.

3. On the basis of their genetic inheritance, some groups are innately superior to others (Banton, 1970; Benedict, 1959; Montagu, 1972; Shibutani and Kwan, 1965).

In sum, *racism* is the belief that humans are subdivided into distinct hereditary groups that are innately different in their social behavior and mental capacities and that can therefore be ranked as superior or inferior. The presumed superiority of some groups and inferiority of others is subsequently used to legitimate the unequal distribution of the society's resources, specifically, various forms of wealth, prestige, and power.

Racist thinking presumes that differences among groups are innate and not subject to change. Intelligence, temperament, and other primary attitudes,

---

[5]On varying definitions of ideology, see Lane (1962), Mannheim (1936), Plamenatz (1970), and Shils (1968).

beliefs, and behavioral traits are thus viewed as not significantly affected by the social environment. The failures of groups at the bottom of the social hierarchy are interpreted as a natural outcome of an inferior genetic inheritance rather than of social disadvantages that have accumulated for the group over many generations. In the same manner, the achievements of groups at the top of the social hierarchy are seen as a product of innate superiority, not of favorable social opportunities.

Racist thought is inherently ethnocentric. Those espousing racist ideas invariably view ethnic out-groups as inferior. Moreover, such thought naturally leads to the idea that ethnic groups must be kept socially and, especially, physically apart. To encourage social integration is to encourage physical integration, which, it follows, contributes to the degeneration of the superior group.

Ideologies do not necessarily reflect reality; indeed, they are largely mythical. They comprise beliefs that, through constant articulation, become accepted as descriptions of the true state of affairs. We have already discussed some of the scientifically erroneous or dubious principles concerning race. As anthropologist Manning Nash (1962) has explained, racist ideologies depend on three logical confusions: (1) the identification of racial differences with cultural and social differences; (2) the assumption that cultural achievement is directly, and chiefly, determined by the racial characteristics of a population; and (3) the belief that physical characteristics of a population limit and define the sorts of culture and society they are able to create or participate in.

Racist thought is most prevalent in societies in which physical differences among groups are pronounced, such as differences between blacks and whites in the United States or South Africa. But racism describes any situation in which people's social behavior is imputed to innate, or hereditary, sources. Racist beliefs are therefore not limited to ideas about groups commonly referred to as "races" but can apply to any ethnic group, whether distinguishable by race or culture. Racism can pertain to Jews, Italian Americans, Northern Irish Catholics, or French Canadians as much as to African Americans, North American Indians, or other physically salient groups. Members of ethnic groups identifiable primarily by culture rather than physical traits are often described as displaying behavioral characteristics "natural" to their group. Jews in Germany during the 1930s, for example, were physically indistinct from other Germans, but this did not prevent the creation by the Nazis of an elaborate racial ideology pertaining to the Jews. Claims that Italians are by nature flamboyant or that Poles are innately slow-witted are similar manifestations of racist thought.

A better term might be *ethnicism,* because racism applies to all types of ethnic groups. However, racism has become so well established in the sociological lexicon and in everyday language that its continued usage seems inevitable.[6]

Certain beliefs regarding behavioral and personality differences between men and women are based on the same mode of thinking as racism. Men are assumed to be innately better qualified for certain social roles on the basis of their masculinity, and women are seen as quite naturally occupying other roles suited to their femininity. Although most of the behavioral differences between men and women are attributable more to social learning than to biology, the beliefs regarding the congenital nature of these differences are accepted uncritically by many. The belief in the innate behavioral differences between men and women has been referred to as sexism, but it is obvious that the foundation of this ideology is similar to that of racism.

The Functions of Racism

The belief in innate differences among groups is used to justify the unequal distribution of a society's rewards. The place of groups at the top of the social hierarchy and of those at the bottom is explained quite simply as "natural." Racist ideology, then, promotes an ethnic status quo in which one group predominates in the society's economy, polity, and other key institutions and thus receives the greatest share of the society's wealth and power.

Why, for example, do African Americans occupy a disproportionately low number of top positions in all important institutions and own a disproportionately low share of wealth by comparison with Euro-Americans? Racist ideology explains such inequalities as resulting from the inherent inferiority of blacks and superiority of whites. It is asserted that something in the character of black people themselves is at the root of their socially subordinate position, just as something in the character of white people leads very naturally to their social dominance. The same explanation might be used by Protestants in Northern Ireland to account for their dominance over Catholics in various areas of social life or by South African whites to rationalize their past dominance of nonwhites.

In the racist mode of thought, the different social and cultural environments of groups are not of major importance in accounting for their differences in social achievement. And, as we noted earlier, the perpetuation of beliefs in group superiority and inferiority gives rise to the substantiation of those beliefs through the self-fulfilling prophecy.

---

[6]In recent years, the term *racism* has often been used in a sweeping and imprecise fashion, describing almost any negative thought or action toward members of a racial minority or any manifestation of racial inequality, not simply as an ideology. On the different conceptualizations and applications of the notion of racism, see Blauner (1992).

If groups are effectively portrayed as inferior, they can be not only denied equal access to various life chances but in some cases enslaved, expelled, or even annihilated with justification. Slave systems of the eighteenth and nineteenth centuries in the Americas were rationalized by racial belief systems in which blacks were seen as incapable of ever attaining the level of civilization of whites. Similarly, the enactment in the 1920s of strict U.S. immigration quotas, favoring northwestern European groups and discriminating against those from southern and eastern Europe and Asia, was impelled by an intricate set of racist assumptions. Northwestern Europeans were seen as innately more adaptable to the American social system, and other groups were viewed as naturally deficient in favorable social and moral qualities. The latter were believed to represent inferior human breeds, whereas northwestern Europeans embodied the most desirable traits of the species.

The Development of Racism

Although beliefs in the superiority and inferiority of different groups have been historically persistent in human societies (recall the universality of ethnocentrism), the belief that such differences are linked to racial types is a relatively new idea, one that did not arise forcefully until the eighteenth century in Europe. At that time a number of political and scientific factors came together that seemed to inspire the ideology of racism.

First, European peoples began to have contacts with peoples of the Americas and Africa, who were not only culturally alien but physically distinct as well. During the Age of Discovery, starting in the fifteenth century, lands were conquered by Spain, Portugal, England, France, and Holland, and white Europeans encountered large numbers of nonwhite peoples for the first time. At first, the justification for subjecting these groups to enslavement or to colonial repression lay not so much in their evident physical differences as in what was seen as their cultural primitiveness, specifically, their non-Christian religions (Benedict, 1959; Gossett, 1963).

It was the later development of "scientific" racism, however, that gave impetus to the view that European peoples were superior to nonwhites because of their racial inheritance. Eighteenth-century scholars of various disciplines, including medicine, archaeology, and anthropology, had begun to debate the origin of the human species, specifically, the question of whether the species was one or many. Although inferiority of all non-European cultures was assumed, up to this time, most had viewed all human types as subdivisions of a single genus (Rose, 1968; Smedley, 1999). Thought now turned to the much earlier but generally discounted theory of *polygenesis*, the notion that human groups might be derived from multiple evolutionary origins. No resolution of this debate came until the publication in 1859 of Darwin's theory of evolution, *The Origin of Species* (Benedict, 1959; Gossett, 1963). Darwin was clear in his explanation that differences among humans were superficial and that their more general similarities nullified any idea of originally distinct species or races. Although he recognized racial differences, it was left to others to vigorously pursue the measurement of these differences and to attach social meaning to them (Gossett, 1963).

Scientists of the nineteenth century investigating the idea of race were heavily influenced by Darwin even though he had said little about race per se. Many who studied his ideas, however, drew inferences to human societies from what he had postulated about lower animal species. Darwin's idea of natural selection was now seen as a mechanism for producing superior human societies, classes, and races. Expounded by sociologists such as Herbert Spencer and William Graham Sumner, these ideas became the basis of Social Darwinism, popularly interpreted as "survival of the fittest." Early racist thought, then, was supported by an element of what was considered scientific validity.

With the scientifically endorsed belief that social achievement was mostly a matter of heredity, the colonial policies of the European powers were now justified. "Race," writes anthropologist Audrey Smedley, "became a worldview that was extraordinarily comprehensive and compelling in its explanatory powers and its rationalization of social inequality" (1999:253). Native people of color were seen as innately primitive and incapable of reaching the level of civilization attained by Europeans. Economic exploitation was thus neatly rationalized. And, because nonwhites represented a supposedly less developed human evolutionary phase, the notion of a "white man's burden" arose as justification for imposing European cultural ways on these people.

In short, the idea of race appropriately complemented the political and economic designs of the European colonial powers. Sociologist Michael Banton notes that "the idea that the Saxon peoples might be biologically superior to Celts and Slavs, and white races to black, was seized upon, magnified, and publicized, because it was convenient to those who held power in the Europe of that day" (1970:20). Banton adds that the coincidental appearance of these theories with the demise of slavery in the early and mid-nineteenth century provided a new justification to some for subordinating former slaves.

Racism should not be thought of as an opportune invention of European colonialism. As van den Berghe explains, racist thinking "has been independently discovered and rediscovered by various peoples at various times in history" (1978:12).[7] But it was in the colonial era that the idea was received most enthusiastically and firmly established as a social doctrine.

---

[7]On the intellectual perils of interpreting past racist thought and action in terms of current understandings of the idea, see Banton (1983).

The idea that race was immutably linked to social and psychological traits continued to be a generally accepted theory in Europe and America during the early years of the twentieth century, aided by the unabashedly racist ideas of writers such as Houston Stewart Chamberlain and Count Arthur de Gobineau in Europe and Lothrop Stoddard and Madison Grant in the United States. Grant's book, *The Passing of the Great Race*, published in 1916, was essentially a diatribe aimed at restricting immigration of southern and eastern European groups, mainly Catholics and Jews, who were at that time most numerous among American immigrants. Grant divided the European population into three racial types—Alpines, Mediterraneans, and Nordics—and he attributed to the latter (from northwestern Europe) the most desirable physical and mental qualities. Consequently, he warned of the degeneration of the American population through the influx of southern and eastern Europeans, whom he classified as inferior Mediterranean and Alpine types.

The impact of Grant's and Stoddard's books was extended by the development of intelligence testing during World War I by American psychologists, whose findings seemed to confirm the notion of inherent racial differences. On these tests, Americans of northwestern European origin outscored, on the average, all other ethnic groups. This result was taken by many as empirical evidence of the superiority of the Nordic "race." Intelligence was now seen as primarily hereditary even though little attention was paid to environmental factors in interpreting test results. That northern blacks outscored southern whites, for example, was attributed not to social factors such as better educational opportunities in the North but to the selective migration of more intelligent blacks out of the South. This argument proved false, however, because the same differences were shown between northern and southern whites (Montagu, 1963; North, 1965; Rose, 1968).

During the 1920s, cultural anthropologists began to question the biological theories of race and to maintain that social and cultural factors were far more critical in accounting for differences in mental ability. Chief among these theorists was Franz Boas, who pointed out the lack of evidence for any of the common racist assertions of the day. Gossett explains Boas's critical role in reversing social scientific thinking on the notion of race: "The racists among the historians and social scientists had always prided themselves on their willingness to accept the 'facts' and had dismissed their opponents as shallow humanitarians who glossed over unpleasant truths. Now there arose a man who asked them to produce their proof. Their answer was a flood of indignant rhetoric, but the turning point had been reached and from now on it would be the racists who were increasingly on the defensive" (1963:430).

Racism in Modern Thought With few exceptions (such as Jensen's controversial findings in 1969 and Herrnstein and Murray's in 1994), the social scientific emphasis on environmental factors as determinants of most social and mental characteristics has remained firm. The preeminent view is that genetic differences among human groups are of minimal significance as far as behavior and intelligence are concerned. Those scholars not subscribing to this view have, in van den Berghe's words, "been voices in the wilderness (except in societies like Nazi Germany, South Africa, or the American South) and their views have not been taken seriously" (1978:xxii). Yet as William Newman (1973) points out, ironically, science itself created the myth of race that it is today still attempting to dispel. Moreover, for many decades few sociologists questioned the prevailing ideas about racial inequality. "Both popular and educated beliefs," notes James McKee, "provided an unqualified confidence in the biological and cultural superiority of white people over all others not white" (1993:27).

Just as scientific thinking on the issue of race has changed, so, too, has lay thinking. Most people now view the discrepancies in social achievement among ethnic groups more as cultural differences, not as biogenetic ones (Banton, 1970; Schuman, 1982). Because of an inability to adapt to the dominant culture, it is argued, economic and social handicaps persist among certain ethnic groups. These inabilities are traced to a people's way of life—its culture—which hinders conformity to the norms and values of the dominant group.[8]

Even though popular thinking has drifted away from the old biological racist ideas, much inconsistency and confusion remain. As a result, the prevailing social science explanation stressing environment is not necessarily accepted. Instead, many attribute low achievement among certain ethnic groups to lack of individual motivation while still denying both biological *and* environmental arguments (Schuman, 1982). In sum, racism is a belief system that has proved tenacious, though modifiable in form and content, in multiethnic societies. It is a social phenomenon whose consequences continue to be felt by both dominant and subordinate ethnic groups.

SUMMARY

The major aspects of the study of race and ethnic relations are (1) the nature of relations among ethnic groups of multiethnic societies; (2) the structure of inequality among ethnic groups; (3) the manner in which dominance and

---

[8]Although the ideology of cultural deprivation is seemingly more benign than "old-fashioned" racism, focusing on biological superiority and inferiority, critics have pointed out that the difference between the two is not so sharp (Ryan, 1975). They argue that cultural deprivation, like racist notions, emphasizes individual and group shortcomings rather than a social system that, through subtle discrimination, prevents the bulk of minority group members from attaining economic and social parity with the dominant group.

subordination among ethnic groups are maintained; and (4) the long-range outcomes of interethnic relations; that is, either greater integration or greater separation.

An *ethnic group* is a group within a larger society that displays a common set of cultural traits, a sense of community among its members based on a presumed common heritage, a feeling of ethnocentrism among group members, ascribed group membership, and, in some cases, a distinct territory. Each of these characteristics is a variable, differing from group to group and among members of the same group.

Race is an often misused notion having biological and social meanings. Biologically, a race is a human population displaying certain hereditary features distinguishing it from other populations. The idea is essentially devoid of significance, however, because there are no clear boundaries setting off one so-called race from another. Hence, there can be no agreement on what the distinguishing features of races are, much less how many races exist. Further, the weight of scientific evidence refutes any meaningful relationship of intelligence and race. The sociological importance of race lies in the fact that people have imputed significance to the idea despite its questionable validity. Hence, races are always socially defined groupings and are meaningful only to the extent that people make them so.

Racism is an ideology, or belief system, designed to justify and rationalize racial and ethnic inequality. The members of socially defined racial categories are believed to differ innately not only in physical traits but also in social behavior, personality, and intelligence. Some "races," therefore, are viewed as superior to others.

The ideology of racism emerged most forcefully in the age of colonialism, when white Europeans confronted nonwhites in situations of conquest and exploitation. The idea that race was linked to social and psychological traits was given pseudoscientific validity in the late nineteenth and early twentieth centuries by American and European observers, a view that was not basically reversed until the 1930s. Today, scientific thought has given little credence to racial explanations of human behavior, stressing instead environmental causes.

Suggested Readings

Benedict, Ruth. 1959 (1940). *Race: Science and Politics*. New York: Viking. A brief work, by one of the most eminent American anthropologists, that has had a profound impact on social scientific thinking about race.

Davis, James. 1991. *Who Is Black? One Nation's Definition*. University Park, Pa.: Pennsylvania State University Press. Deals with the issue of racial identification and the variable, often arbitrary, nature of racial classification in the United States as well as other societies.

Glazer, Nathan, and Daniel P. Moynihan (eds.). 1975. *Ethnicity: Theory and Experience*. Cambridge, Mass.: Harvard University Press. A collection of essays by prominent social scientists that explores the concept of ethnicity.

Isaacs, Harold R. 1989. *Idols of the Tribe: Group Identity and Political Change*. Cambridge, Mass.: Harvard University Press. A finely written account of the strength of ethnic identity and its political consequences in contemporary societies.

Jacobson, Matthew Frye. 1998. *Whiteness of a Different Color: European Immigrants and the Alchemy of Race*. Cambridge, Mass.: Harvard University Press. Through a historical analysis, explains how the idea of "whiteness" as a racial identity in American society has been subject to numerous changes.

Marks, Jonathan. 1995. *Human Biodiversity: Genes, Race, and History*. New York: Aldine de Gruyter. Examines the biological basis of human differences, demonstrating the speciousness of the division of the world's people into racial categories.

McKee, James B. 1993. *Sociology and the Race Problem: The Failure of a Perspective*. Urbana: University of Illinois Press. Traces the development of the sociology of race relations, particularly in the earlier part of the twentieth century.

Montagu, Ashley. 1974. *Man's Most Dangerous Myth: The Fallacy of Race*. 5th ed. New York: Oxford University Press. A good explanation of the mythical nature of race. Demonstrates that many alleged differences among so-called racial groups do not, in fact, exist or are socially insignificant.

Omi, Michael, and Howard Winant. 1986. *Racial Formation in the United States: From the 1960s to the 1980s*. New York: Routledge & Kegan Paul. Illustrates the idea of race as a social construct, emphasizing how race has been differently conceptualized in the United States in response to changing political circumstances.

Smedley, Audrey. 1999. *Race in North America: Origin and Evolution of a Worldview*. 2d ed. Boulder, Colo.: Westview. A thorough exposition of the origins and social applications of racial ideas and racist ideologies. Shows, through historical analysis, how race, until the last few decades, had been an effective social construction that rationalized the power of dominant groups over others.

van den Berghe, Pierre. 1981. *The Ethnic Phenomenon*. New York: Elsevier. One of many significant works on race and ethnic relations by van den Berghe. Includes his controversial sociobiological interpretation of ethnicity.

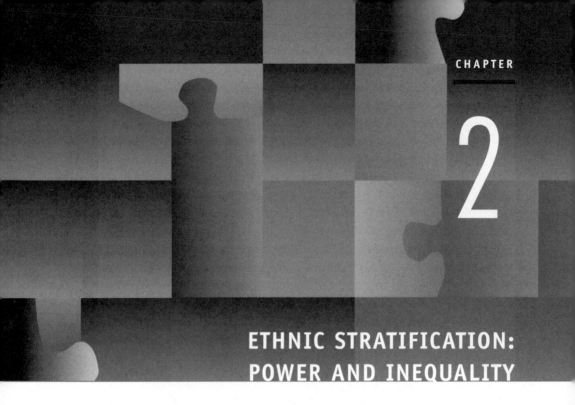

2

ETHNIC STRATIFICATION: POWER AND INEQUALITY

A PROMINENT SOCIOLOGIST has suggested that the first questions asked by sociology were these: "Why is there inequality among men? Where do its causes lie? Can it be reduced, or even abolished altogether? Or do we have to accept it as a necessary element in the structure of human society?" (Dahrendorf, 1968:152). Such questions remain fundamental to sociological inquiry and are particularly critical in the study of race and ethnic relations.

Humans are unequal, of course, in many ways. They differ in physical features and in mental capacities, talent, strength, musical aptitude, and so on. All these inequalities are a product of both social learning and genetic inheritance though the significance of each of these factors is, as we have now seen, not always clear. Perhaps more important, however, people are also unequal in their access to social rewards, that is, various forms of wealth, power, and prestige. These differences, all primarily of social origin, are of greatest consequence in accounting for who we are and who we ultimately may be as members of our society. It is these inequalities with which sociologists are most concerned.

STRATIFICATION SYSTEMS

In all societies, people receive different shares of what is valued and scarce. This unequal distribution of the society's resources creates a system of stratification.

People are grouped on the basis of how much of the society's rewards they receive, and these groups, or strata, are arranged in a rank order, or hierarchy. Those at the top receive the most of what there is to get, and those at the bottom the least. Societies may comprise any number of strata, but in all cases, this system of inequality is structured. This means that stratification is not random, with groups and individuals occupying different positions by chance; rather, social institutions such as government, the economy, education, and religion operate to assure the position of various groups in the hierarchy. Moreover, the system of stratification in all societies is legitimized by an ideology that justifies the resultant inequality. The pattern of stratification in a society therefore remains stable for many generations.

Modern societies are stratified along several dimensions, the most prominent of which is *class* stratification, in which groups are ranked on the basis of income, wealth, and occupation. Multiethnic societies are also stratified on the basis of ethnicity, and it is on this dimension that we will primarily focus. So that we can better understand the specific nature of ethnic stratification, however, we look first at some of the more general characteristics of stratification systems.

Dimensions of Stratification

Beginning with Karl Marx, many have stressed the economic dimension of social stratification. A social class, according to Marx, comprises those who stand in a common position with regard to the society's productive system. Although he recognized many subclasses, Marx saw the eventual emergence of two encompassing classes: those who own and control the means of production and those who do not. Those who have access to the society's productive resources—in industrial societies, factories, banks, transportation, communications, and so on—constitute a ruling class. Those who own no productive property, the working class, can offer only their human labor in exchange for material needs. The different economic interests of these two classes become the basis of political struggle and societal change.

Weber's Multidimensional Model Marx's view of stratification is essentially unidimensional: Classes are founded on economic criteria. One of the first scholars to suggest a more elaborate, multidimensional model of stratification was Max Weber. In addition to a hierarchy based on economic factors, Weber denoted hierarchies based on status and on political power, the latter referred to as "party" (Gerth and Mills, 1946).

Weber's notion of class is similar to Marx's: A class comprises those who stand in a similar position with regard to their opportunities to acquire the society's economic rewards. It is in the status and power dimensions, however, that Weber's model is innovative.

The status dimension, according to Weber, consists of groups that display a particular lifestyle and that are aware of differences between themselves and other status groups. Such status communities manifest themselves in common consumption patterns, club memberships, residential areas, schools, and ultimately intermarriage. Weber was careful to explain that economic class and status are generally overlapping dimensions because a particular lifestyle necessitates commensurate wealth. Most sociologists today use the term *socioeconomic status*, a combination of Weber's class and status dimensions, in place of social class.

Weber's third dimension, party, denotes political rank—one's standing in a collectivity or organization whose "action is oriented toward the acquisition of social 'power,' that is to say, toward influencing a communal action" (Gerth and Mills, 1946:194). Such organizations may represent either class or status group interests or perhaps both. In addition to formal political organizations like parties, examples in modern industrial societies are labor unions, consumer groups, business and professional associations, and ethnic organizations; in short, any group designed for or capable of involvement in political action. This element of Weber's scheme is often misinterpreted as describing power as a distinct dimension of stratification. Weber notes, however, that all three dimensions—class, status, and party—are "phenomena of the distribution of power within a community" (Gerth and Mills, 1946:191).

Both the Marxian and the Weberian models of stratification are valuable, and they are necessary to fully understand structured inequality in societies. The basic difference between them, however, is that with Weber's multidimensional model, stratification can be seen as emanating from several sources, not simply one's economic position. Though Weber, like Marx, recognized the fundamental nature of people's place in the productive system in determining their overall position in the stratification system, Weber's model provides for cases in which other factors enter into the picture. This is particularly important in multiethnic societies, where people are ranked on the basis of their ethnic group membership in addition to their economic position. Although, as we will see, there is generally a close correlation between ethnicity and economic class, the two do not always coincide.

Lenski's Multidimensional Model   Gerhard Lenski (1966) has offered another multidimensional model of stratification, which disentangles much of the confusion of stratification analysis and is particularly suited to our purposes. Lenski recognizes several class hierarchies, which he calls "class systems," each based on some key social criterion—wealth, occupation, education, political authority, ethnicity, and so on. Each class system comprises a hierarchy of classes that are made up of individuals who enjoy an approximately equal measure of authority, wealth, ethnicity, or whatever criterion of rank is applied. These various hierarchies, or class systems, together make up what Lenski calls the "distributive system," which can be interpreted essentially as the total socioeconomic structure of the society. If we visualize a social system as comprising individuals, classes, class systems, and the distributive system, each represents a different level of organization. They can be seen as telescoping in such a way that individuals are at the basic level and constitute the units within classes, which in turn are the units within class systems. Together, these class systems make up the distributive system. This model is shown in Figure 2.1.

Lenski's model of class systems is advantageous to our purposes for several reasons. First, it shows us the relationship between ethnicity and other

Figure 2.1 The Distributive System of American Society. *A, B,* and *C* Represent Three Individuals

THE DISTRIBUTIVE SYSTEM			
The political class system	The property class system	The occupational class system	The ethnic class system
The elite	The upper class	Capitalists	
The bureaucracy ⒜	The upper-middle class ⒜	Professionals, managers, entrepreneurs ⒜	Anglo-Americans ⒜
The electorate ⒝	The middle class ⒝	Skilled workers, technicians ⒝	Other Euro-Americans ⒝
The apolitical populace ⒞	The working class		Asian Americans
	The poor ⒞	Unskilled workers ⒞	Hispanic Americans
	The underclass	The unemployed	African Americans ⒞
			American Indians

SOURCE: Adapted from Lenski, 1966.

dimensions of stratification. How much consistency across the various class systems do individuals and groups display? Do those who rank high in the ethnic dimension rank equally high in others, and if there are inconsistencies, what accounts for them? In most multiethnic societies, there is a clear-cut relationship between one's place in the ethnic hierarchy and in other hierarchies such as occupation and political authority. Those ranking high in the ethnic hierarchy ordinarily rank high in the others. Exceptions do arise, however, among both ethnic groups and their individual members.

A second advantage to Lenski's model is that it may help us determine in different societies the relative importance of each class system or hierarchy, particularly the ethnic one. In some societies, like Brazil and Mexico, ethnicity plays a less consequential role in determining people's life chances than in societies like the United States or South Africa, where it is enormously significant. In the latter, people's ethnic rank may loom large in determining which occupations they will perform, how much income and wealth they can be expected to acquire, and how much political power they will be able to wield.

Power and Stratification In a basic sense, power underlies all forms of stratification. Whether social classes are seen as outgrowths of a society's productive system (à la Marx) or as derivative of other criteria as well (à la Weber or Lenski), they are essentially founded on and maintained by differentials of power (Parkin, 1971). Just as differences in wealth, education, occupation, and prestige are mirrors of a society's power arrangement, so, too, are differences in rank among a society's ethnic groups. Social stratification, then, is a system of unequal distribution of a society's rewards, determined above all by power differentials. In simple terms, those at the top get more of what is valued because they are more powerful; they possess greater power resources in the form of wealth, property, political office, arms, control of communications, and knowledge. The position of others in the stratification system is determined accordingly on the basis of their ability to amass and apply power resources.

Stratification and Ideology

The power of a dominant class is not simply the power of force but also the power to propound and sustain an ideology that legitimizes the system of inequality. Although coercion is always at the root of obedience to authority, and all dominant groups use force when the need arises, coercive techniques are commonly used only in societies where the prevailing system is not accepted by a significant part of the populace. In South Africa under apartheid or in the antebellum U.S. South, for example, the powerful traditionally enforced their will through blatant forms of repression.

The use of raw force alone, however, cannot be effective in prompting compliance with a system of inequality over long periods. The stability of systems that rely primarily on coercion is always precarious. For government and other supportive institutions of the dominant group to establish and sustain a ruling system that is popularly supported over many generations requires that power be legitimized in less repressive and less direct ways. In protecting their privileges, dominant groups try to engender loyalty and respect in subordinates, not fear (Jackman, 1994). People must come to see the inequalities in power and wealth as just and even socially beneficial. Only then do systems of social inequality attain stability. When this is accomplished, ruling groups need no longer resort to force as the principal means of assuring their power and privilege. Such long-range stability and legitimacy require the development of an effective ideology and its communication through socialization.

Despite their acceptance—usually reflexively—by both ruling groups and masses, the fundamental ideological values tend to accommodate mostly the interests of the society's ruling groups. In the United States, for example, the dominant explanation for social inequality centers on the belief that the society's opportunity structure is open, providing equal chances for all to achieve material success or political power, regardless of their social origins. This presumably being the case, each person controls his or her placement in the social hierarchy. Social success, then, is explained as the result of one's willingness to work hard; failure is the product of lack of ambition or desire to improve oneself. Differences in wealth and power are not denied, but they are seen as a result of individual factors rather than the workings of a class system that ordinarily engenders success for the wellborn and failure for the poor. In reality, however, the opportunity structure is hardly equal, and the dominant values of individualism, competition, and achievement favor those who are wealthy and can easily avail themselves of the opportunities for success. The social system, then, tends to favor those who are already wealthy and powerful.

Just as there are ideologies that explain and rationalize inequalities in societal wealth, occupation, prestige, and political authority, so, too, there are ideologies that explain the differential treatment of groups on the basis of ethnicity. These ideologies generally comprise some form of racism (or ethnicism), the basic components of which were discussed in Chapter 1. And just as most of those who are not wealthy and successful still accept much of the dominant class ideology, so, too, members of minority ethnic groups may sometimes accept the dominant ethnic group's explanations of superiority and inferiority. They may internalize attitudes of inferiority and even manifest feelings of group self-hatred. Anti-Semitism has at times been observed among Jews, for example, in which individuals attempt to disguise their Jewish identity and identify instead with the dominant ethnic group by changing their name, altering their behavior, and even changing their physical appearance. The acceptance of dominant ethnic group values is also illustrated in the once-common adoption by American blacks of white standards of beauty. For many generations, blacks used various devices such as hair straighteners and skin bleaches in conforming to white ideals. Women, too, in the past generally accepted prevailing explanations of their inferiority vis-à-vis men, and only in recent decades have they questioned the inevitability of male dominance in most areas of social life. Such acceptance by minority groups of the prevailing standards, values, and beliefs further solidifies the power of the dominant group and contributes to the maintenance of the status quo.

To summarize, social stratification is a system of structured inequality in which people receive different amounts of the society's valued resources. This inequality is relatively stable and gives rise to social classes—groups of people of approximately equal income and wealth. In multiethnic societies, ethnicity becomes an additional basis of stratification. Differential power underlies all forms of inequality, and the system is underwritten by an ideology—propounded by the dominant group but generally accepted by other classes—that justifies differences in social rewards.

ETHNIC STRATIFICATION SYSTEMS

Ethnic stratification, like other forms of stratification, is a system of structured social inequality. In almost all multiethnic societies, a hierarchical arrangement of ethnic groups emerges in which one establishes itself as the dominant group, with maximum power to shape the nature of ethnic relations. Other, subordinate ethnic groups exert less power, corresponding to their place in the hierarchy, extending down to the lowest-ranking groups, which may wield little or no power.

Group rank is determined mainly on the basis of distance from the dominant group in culture and physical appearance. Those most like the dominant group are more highly ranked, and those exceedingly different are ranked correspondingly low. A system of *ethnic stratification*, then, is a rank order of groups, each made up of people with presumed common cultural or physical characteristics interacting in patterns of dominance and subordination. Sociologists ordinarily refer to ethnic stratification systems as *majority-minority*, or *dominant-subordinate*, systems. Let's look more closely at the nature of both minority and majority groups.

Minority Groups

Differential Treatment Minority groups are those groups in a multiethnic society that, on the basis of their physical or cultural traits, receive fewer of the society's rewards. In a classic definition, Louis Wirth defined a minority group as "a group of people who, because of their physical or cultural characteristics, are singled out from the others in the society in which they live for differential and unequal treatment, and who therefore regard themselves as objects of collective discrimination" (1945:347). Members of minority groups disproportionately occupy poorer jobs, earn less income, live in less desirable areas, receive an inferior education, exercise less political power, and are subjected to various social indignities. These inequalities are the result of their social mark—the physical or cultural features that distinguish them. Moreover, as Wirth pointed out, minority group members are conscious of the fact that they are differentially treated.

In addition to the fact that they commonly receive fewer of the society's rewards, several additional features are fundamental to minority groups.

Social Definition First, the physical or cultural traits on which minority status is based are socially defined. Thus, any characteristic may serve as the basis of minority status as long as it is perceived as significant. Let us suppose that hair color was considered a meaningful distinguishing feature. If blond hair was deemed more desirable than black or brown, those with black or brown hair might be singled out accordingly and treated differentially for no other reason than that their hair was not blond.

Differential Power Second, minority groups are afforded unequal treatment because they lack the power to negate or counteract that treatment. Blond-haired persons could continue to treat nonblonds as less than desirable types and withhold various forms of social rewards only if they maintained sufficient power to do so. As Yetman notes, "Majority-minority relations do not appear until one group is able to impose its will on another" (1991:1). Minority status, then, is above all a reflection of differential power. James Geschwender's definition of minority includes this important power factor. He defines a minority as "any group that is socially defined as different from the dominant group in society, is at a power disadvantage, receives less than its proportionate share of scarce resources due to its power disadvantage, and finds its differential treatment justified in terms of socially defined differences" (1978:17).

Categorical Nature Third, minority denotes a group, not an individual, status. All those who are classified as part of the group will experience differential treatment, regardless of their personal characteristics or achievements. Thus, people cannot voluntarily remove themselves from their minority position.

Sociological and Numerical Meaning Finally, it is important to understand that the sociological meaning of minority is not the same as the mathematical definition. Numbers have no necessary relation to a group's minority status. For example, nonwhites in South Africa, as we will see in Chapter 13, make up more than 85 percent of the population, yet constituted a sociological minority until only the past few years. They had almost no access to political and economic power, were assigned the lower occupational positions, and were afforded grossly inferior opportunities in all areas of social life by comparison with South African whites, with less than 15 percent of the population. Rather than relative size, it is a group's marginal location in the social order that defines it as a minority.

Types of Minorities

In its sociological meaning, the term *minority* can be applied to a variety of social groups. In addition to ethnic traits, many other physical or behavioral characteristics are sufficient to set off groups of people from the society's mainstream, resulting in differential treatment. To equate minority groups only with highly visible ethnic groups fails to account for the many other types of minorities found in complex societies.

Sex, for example, is a clearly distinguishable physical characteristic that in most societies serves to single out one group—ordinarily, women—for differential treatment. It is in this sense that women may be said to constitute a minority group. Traditionally, women have rarely occupied positions of great political or economic power, have been barred from entrance into many occupations, and have been excluded from numerous areas of social life. Only in recent years have these deeply rooted patterns of sex discrimination begun to change.

People with physical disabilities—the blind, the deaf, those confined to wheelchairs, and so on—are another evident minority. On the basis of their physical distinctions, they are singled out and given differential treatment in many social contexts. Not until recent years had efforts been made to accommodate such people in public buildings, for example, and they had rarely been afforded equal educational and occupational opportunities.

Age constitutes another physical feature that serves to set groups apart for differential and unequal treatment. In all societies, of course, certain limitations are placed on people according to their age. The very young are not expected to fulfill adult roles, and at the opposite end of the life cycle, the aged are not expected to perform as younger persons do. But in some cases, this necessary age differentiation exceeds any rational explanation, and people of a particular age are singled out for differential treatment merely on the basis of their age. Often workers are forced into retirement at 65, even if they remain capable of carrying out their occupational duties and wish to do so. Or certain occupations are denied those who exceed an arbitrarily selected age requirement. Cases such as these constitute age discrimination.

Certain groups are also singled out for differential treatment on the basis of their past or current behavior. Homosexuals are a very obvious example, as are those labeled mentally ill and ex-convicts. Those acknowledging homosexuality, past treatment for mental illness, or a prison record risk social exclusion and discrimination by employers, landlords, and law enforcement agencies.[1]

For each of these groups, a belief system, or ideology, explains and justifies their differential treatment. Women are alleged to be physically or mentally inferior to men or to be unable to perform at the level of men; the elderly are perceived as slow, incompetent, or senile; homosexuals are "sick" or "depraved"; and so on. Although there may be little or no truth in these assertions, they serve as devices to sustain unequal treatment.

Ethnic Minorities

Our chief concern is with ethnic minorities, those groups singled out and treated unequally on the basis of their cultural or physical differences from the dominant group. Ordinarily, attention is directed to the most conspicuous ethnic groups, those with especially marked differences in skin color or those that maintain unusually divergent cultural beliefs and behavior. In the United States today, these are blacks, American Indians, Asians, and Hispanics. These are the groups that experience the most blatant and consequential forms of discriminatory treatment and about whom mythical belief systems have been created to justify such treatment.

But we should not think of ethnic minorities as always indelibly marked or as only those who experience overt and severe prejudice and discrimination. There are also ethnic minorities whose identification is not so clear because they do not experience such flagrant, institutionalized forms of discrimination and because many individuals in these groups have gained access to important occupational and political positions. Ruth Glass refers to such groups as "hidden minorities" who "do not wear a striking label of inferiority." But as she explains, "If one looks closely, one can see that there is still a blank patch on the collar where such a label can be pinned on; and at any time of general social stress, it may indeed be stuck on again" (1964:150). The case of Japanese Americans illustrates this idea. Although they had long been targets of particularly vicious prejudice and

---

[1]On these nonethnic minorities, see Sagarin (1971).

discrimination in West Coast communities, as a collectivity they had made a strikingly successful adjustment to American society. But in 1942, thousands of Japanese Americans were forcibly removed from their residences in California, Oregon, and Washington and interned in prison camps, where they were kept almost until the end of World War II. Similarly, Jews often experience the manifestations of such underlying hostility. German Jews in the 1930s, for example, were considered the most thoroughly assimilated among all European Jews; but this did not prevent the Nazi Holocaust, which resulted in the almost total destruction of the German Jewish population.

In any case, the treatment of ethnic minorities will vary from group to group and from time to time. Some groups may be consistently singled out and deprived of social rewards, and others may experience only minimal discriminatory treatment. Even for any single group, minority status is not a simple matter to explain. Prejudice and discrimination may be quite strong at certain times and diminished at others depending on various economic, political, and social conditions. Moreover, members of a particular group may experience rejection and denial in some areas of social life but not in others.

One feature of ethnic minorities that distinguishes them from other types of minorities is *endogamy*, which refers to marriage within one's social group and is regulated on the basis of social class, educational level, religion, or other social characteristics. Middle-class people are encouraged to marry other middle-class people, Catholics are encouraged to marry other Catholics, and so on. But especially strong enforcement of endogamy in multiethnic societies concerns ethnicity, particularly among physically distinct groups.

Such endogamous patterns may be enforced by the dominant group or by the minority group itself. Jews, on the one hand, have traditionally been more endogamous than other ethnic groups, and much of the compulsion to marry within the group has been dictated by the group itself. Prohibition of interracial marriage in the United States, on the other hand, has been rigidly enforced primarily by the dominant white group. In fact, not until 1967 did the Supreme Court strike down statutes proscribing marriage between blacks and whites in many southern states.

Dominant Groups

In the study of ethnic relations, it has been most common for sociologists to focus primarily on minority groups. Ethnic relations, however, involve not only the problems of those at the bottom of the ethnic hierarchy but also the manner in which those at the top maintain their dominance. Obviously, the existence of minority groups implies a majority group. In this regard,

Hughes and Hughes have noted: "It takes more than one ethnic group to make ethnic relations. The relations can be no more understood by studying one or the other of the groups than can a chemical combination by study of one element only, or a boxing bout by observation of only one of the fighters. Yet it is common to study ethnic relations as if one had to know only one party to them" (1952:158).

In the United States, for example, for many years (and even today) black-white relations were portrayed as "the black problem in America," and most studies dealt with the social and psychological problems faced by blacks because of their minority status. Only in the 1960s did sociologists begin to show that intergroup relations between blacks and whites were mostly a *white* problem because it was the dominant white group that controlled the character and course of those relations more than did blacks themselves. Our major concern is not the unique character of particular ethnic groups per se as much as it is the social situation within which different groups interact. It is thus necessary to look carefully at the nature and functions of dominant as well as minority ethnic groups.

Most sociologists refer to a society's majority group as the *dominant group*, and that usage is adopted here. This avoids the tendency to think in numerical terms in cases where the dominant group may not be a numerical majority. For consistency, we might also refer to minority groups as *subordinate groups*, implying the existence of a power relationship rather than a numerical one. However, because the term *minority* is so firmly fixed in the sociological literature as well as in popular usage, we will continue to use it.

Political, Economic, and Cultural Dominance Schermerhorn defines a dominant group as "that collectivity within a society which has preeminent authority to function both as guardians and sustainers of the controlling value system, and as prime allocators of rewards in the society" (1970:12-13). The term *dominant*, or majority, then, has both cultural and political connotations, in each case denoting the power of one group over others. Let's look more closely at each element of the dominant group's power.

Most simply, the dominant ethnic group is the group at the top of the ethnic hierarchy, with maximal access to the society's power resources, particularly political authority and control of the means of economic production. This power advantage in the political and economic realms enables it to acquire a disproportionate share of the society's valued resources and thereby to further sustain its dominance. This does not mean, of course, that all those classified as part of the dominant ethnic group enjoy equally great power advantages. Rather, it means only that members of the dominant ethnic group disproportionately occupy such positions. In the United States, for example, white Anglo-Saxon

Protestants (WASPs), or Anglo-Americans, historically have been, and to a great extent remain, the dominant ethnic group.[2] But not all members of this category are part of the decision-making orders of the polity or the economy. They are, however, disproportionately holders of the most important positions in these institutions and are able to render decisions that favor the interests and values of Anglo-Americans generally. Thus, compared with other ethnic groups, Anglo-Americans own a greater share of the society's wealth, earn more income, acquire more and better education, work at higher-ranking and more prestigious occupations, and generally attain more of the society's valued resources.

In addition to the dominant ethnic group's greater economic and political power, its norms and values prevail in the society as a whole. The cultural characteristics of the dominant group become the society's standards. As Schermerhorn explains, "When we speak of a 'dominant group' we mean that group whose historical language, traditions, customs, and ideology are normative for the society; their preeminence is enforced by the folkways or by law, and in time these elements attain the position of cultural presuppositions" (1949:6).

The cultural supremacy of the dominant ethnic group in a multiethnic society applies to *major* norms and values rather than to every element of the society's culture. In any heterogeneous society, certain cultural traits of minority ethnic groups are bound to seep into the mainstream. In the United States, for example, egg rolls, bagels, tacos, and pizza are standard "American" fare. But items such as foods represent only minor cultural components. There is no similar acceptance of variety regarding the more basic aspects of culture such as language, religious values, political practices, and economic ideology. On these counts, ethnic groups are expected to acculturate to the dominant group's customs and ideals. In the United States, for example, it is assumed that all groups will speak English, will maintain Judeo-Christian ethics, will abide by democratic principles, and will accept capitalist values. Those who do not remain outside the societal mainstream and may be subjected to opprobrium. Although changes even in these areas of culture are usually evident over the long run, such changes are slow, deliberate, and accompanied by controversy and open conflict. Bilingual education or bilingual government services in the United States, for example, have often met with strong resistance.

Control of Immigration The dominant ethnic group, given its political, economic, and social power, is able to regulate the flow and composition of further immigration to the society and to determine the social treatment of other

---

[2]In Chapter 5, the specific makeup of the dominant ethnic group in the United States will be discussed in greater detail.

groups after they have entered the society. Once its preeminence has been established, the dominant group becomes synonymous with the "host" or "receiving" society, or what Porter calls the "charter group":

> In any society which has to seek members from outside there will be varying judgements about the extensive reservoirs of recruits that exist in the world. In this process of evaluation the first ethnic group to come into previously unpopulated territory, as the effective possessor, has the most say. This group becomes the charter group of the society, and among the many privileges and prerogatives which it retains are decisions about what other groups are to be let in and what they will be permitted to do. (1965:60; see also Gordon, 1964)

No matter how liberal the immigration policies of a multiethnic society may be, certain restrictions or quotas are always imposed on groups that are highly unlike the dominant group in culture or physical appearance. And in the same manner, those immigrants whose origins are culturally and physically close to the dominant group's enjoy not only easier entrance but also more rapid and less impeded upward social mobility.[3]

It must be emphasized that dominant group power is never absolute. Although the dominant group is at the top of the ethnic hierarchy, its power and influence are relative, not complete. Members of that group are *disproportionately* in positions of power and influence. Likewise, cultural dominance should not be interpreted as complete. The dominant group has a disproportionate influence on shaping the society's cultural mold; but as already noted, minority groups obviously make contributions and compel changes. The United States today is hardly a WASP model in literature, music, the arts, politics, education, and most other aspects of culture. On the really fundamental aspects of culture, however—such as language, law, and religion—the dominant group's influence remains much stronger and less vulnerable to minority influences.

It is equally important to understand that the ethnic hierarchy is not fixed. Rather it is always under challenge, to some degree; and therefore groups' positions may change and their power either enhanced or diminished. To speak, for example, of the Anglo group as dominant in American society today is not to imply that its position is comparable to what it was in an earlier time when its

---

[3]Petersen (1980) notes that the dominant group may not always be large and powerful enough to act as a "host" to immigrants. He points to modern Israel as a case in which immigrants have been acculturated not to a host population but to an ideology of Zionism. Similarly, Argentina's development in the past 100 years has been marked by the economic and political ascendance of immigrants.

power was far more thorough. As minority groups over the years have acquired more economic, political, and social resources, WASP power has been diluted.

Middleman Minorities

Certain ethnic groups in multiethnic societies sometimes occupy a middle status between the dominant group at the top of the ethnic hierarchy and subordinate groups in lower positions. These have been referred to as *middleman minorities* (Blalock, 1967; Bonacich, 1973; Bonacich and Modell, 1980; Turner and Bonacich, 1980; Zenner, 1991).

Middleman minorities often act as mediators between dominant and subordinate ethnic groups. They ordinarily occupy an intermediate niche in the economic system, being neither capitalists (mainly members of the dominant group) at the top nor working masses (mainly those of the subordinate groups) at the bottom. They play such occupational roles as traders, shopkeepers, moneylenders, and independent professionals. Middleman minorities therefore serve a function for both dominant and subordinate groups. They perform economic duties that those at the top find distasteful or lacking in prestige, and they frequently supply business and professional services to members of ethnic minorities who lack such skills and resources.

Given their intermediate economic position, such groups find themselves particularly vulnerable to out-group hostility, emanating from both dominant and subordinate groups. In times of stress they are, as Blalock (1967) describes them, natural scapegoats. They are numerically and politically lacking in power and therefore must appeal to the dominant group for protection, which will be provided as long as it is felt that their economic role is necessary. But the role may still be seen as tainted, thus prompting feelings of revulsion and discriminatory actions. Subordinate groups will also view middleman minorities with disdain because they often encounter them as providers of necessary business and professional services. Such entrepreneurs therefore come to be seen as exploiters. In the United States, for example, Jews often operated businesses in black ghetto areas of large cities—a role increasingly assumed today by new immigrants, especially Koreans. Conflict between these businesspeople and their neighborhood customers has been common and, at times, severe.

Because they stand in a kind of social no-man's-land, middleman minorities tend to develop an unusually strong in-group solidarity and are often seen by other groups as clannish (Bonacich and Modell, 1980). Such in-group solidarity, as well as their business success, creates resentment and antipathy, which in turn prompt continued ethnic solidarity.

A number of groups that seem to fit the characteristics of middleman minorities have been evident in almost all parts of the world. Jews in Europe have historically been a classic illustration. As moneylenders in medieval times, they were a generally despised group. They assumed this economic role, however, because for Christians, moneylending was regarded as sinful. Such moneylending activity was nonetheless necessary, and as non-Christians, Jews naturally came to fill this occupational place.

The ethnic Chinese in various Southeast Asian societies, sometimes called the "overseas Chinese," have played a similar role (Dobbin, 1996; Freedman, 1955; Zenner, 1991). In the Philippines, for example, the Chinese have traditionally been highly successful in business but have also been the target of much prejudice and discrimination by Filipinos (Hunt and Walker, 1974). Originally, the Chinese acted as commercial intermediaries between the native Filipinos and, first, the Spanish and, then, the American colonialists. Their business activities were protected by the dominant group because they were valuable to the development of the colonial economy. Following Philippine independence, the Chinese held a significant place in the society's commerce; and as a culturally despised and politically powerless minority, they became a readily available target of both official and informal discrimination (Dobbin, 1996).

In Indonesia, the overseas Chinese have been especially prominent in the middleman role. Although they make up less than 4 percent of the population of 200 million, they play a dominant role in the economy, accounting for perhaps 70 percent of all private economic activity (Schwarz, 1994). This has bred much resentment on the part of indigenous Indonesians, who view the ethnic Chinese as a cohesive and clannish group—despite the fact that the Chinese have been in Indonesia since the nineteenth century and in recent years have absorbed much of Indonesian culture. Episodes of violence against the ethnic Chinese have been regular occurrences in the past few decades. In 1998, the economic crisis in Indonesia provoked a wave of rioting that resulted in the looting and burning of Chinese businesses and homes (Kristof, 1998; Mydans, 1998).

Obviously, many individuals within groups that have been labeled middleman minorities do not conform to the preeminent characteristics of this type; and even the group as a whole may, at different times, display few of them. But the idea of middleman minorities forces us to consider the many variations among minority groups of a multiethnic society. Moreover, as Bonacich and Modell suggest, it "makes us look at the relationship between an ethnic group and its societal context" (1980:24).

The Relativity of Dominant and Minority Status

Whether a person is part of the dominant group or a minority group depends, in all cases, on the social situation. In some instances, people will be part of one; and in other instances, they may find themselves part of another. For example, Jews in American society are ordinarily designated a minority group. Yet from the standpoint of blacks, who generally rank lower in economic class and prestige, Jews are part of the dominant white group. Similarly, Appalachian whites who have migrated to northern cities are often discriminated against and derisively referred to as "hillbillies"; they are, then, a minority in that context even though they are usually white, Protestant, and Anglo-Saxon in origin, that is, indisputably part of the society's dominant ethnic group (Killian, 1985; Philliber and McCoy, 1981).

Group membership in a modern, complex society is rarely a simple matter of "either-or" but instead takes the form of combinations that often yield confusing and ambiguous statuses. Different ethnic designations alone can produce equivocal statuses. For example, are Irish-American Catholics responded to more generally as whites (that is, part of the dominant group) or as Catholics (that is, part of a minority group)? In different social contexts, either response may be expected. Among blacks, Irish Catholics will be seen and interacted with primarily as white, and their religion or national origin will be of no significance. Among white Protestants, however, their religious identity will be stressed, and their racial identity unimportant.

THE ORIGINS OF ETHNIC STRATIFICATION

How does ethnic stratification arise? And why does it seem so inescapable in multiethnic societies? In other words, what are the forces that seem to lead almost always to patterns of dominance and subordination when ethnic groups come into contact with one another? Sociologists have suggested several factors critical to the emergence of ethnic inequality.

Forms of Contact

To begin with, all systems of ethnic stratification are products of the contact of previously separated groups (Shibutani and Kwan, 1965). Put simply, the composition of multiethnic societies depends on diverse groups coming together in some manner. This entails the movement of people from one area to another but, more fundamentally, from one political unit to another. People may move great distances and remain within the confines of the same political unit, as when a family in the United States moves from New York to California. Ethnic stratification systems, however, are created by the movement of people across national boundaries, usually bringing with them different languages and cultural systems, or by the establishment of new political boundaries. Multiethnic societies are formed through one or a combination of several contact patterns.

Conquest Conquest is a form of contact in which people of one society subdue all or part of another society and take on the role of the dominant group.

European colonialism of the eighteenth and nineteenth centuries best exemplifies this pattern. Through greater military technology, the British, French, Spanish, Dutch, Belgians, and Portuguese conquered peoples in a variety of world areas, including Asia, Africa, Australia, and North and South America. Indigenous groups were brought under colonial rule and became economic appendages of the mother country.

Annexation A political occurrence in which a part or possibly all of one society is incorporated into another is annexation. As Burkey explains, "If the incorporating society has a dominant group, then the ethnic groups within the incorporated, or annexed, society become subordinate at the point that sovereignty is transferred" (1978:72). Such annexation may occur in a peaceful or a violent manner. The United States' acquisition of the Louisiana Purchase from France in 1803 illustrates a peaceful annexation. More commonly, however, such acquisitions of territory are made through successful military ventures. For example, the United States acquired most of what is today its Southwest through an expansionist war with Mexico in 1846. After the transfer of territory at the conclusion of the war, Mexicans living in these annexed areas became a minority, subject to the dominance of American cultural and political institutions.

Voluntary Immigration The most common patterns by which ethnic groups come into contact involve immigration. The migration of peoples from one society to another may be either voluntary or involuntary. Voluntary immigration has been the chief source of ethnic heterogeneity in the United States, Canada, Australia, and New Zealand. All white ethnic groups in these societies have been the product of voluntary immigration from European societies during the nineteenth and twentieth centuries. In the United States and Canada, immigrants from Asia, Latin America, and the Caribbean have, in recent decades, contributed even greater diversity to these societies' ethnically varied populations. Other societies that have relied heavily on immigration include Argentina, Brazil, and Israel. In addition, many previously ethnically homogeneous societies, especially those of Western Europe, have been infused with new racial and ethnic groups as a result of voluntary immigration since the end of World War II. Patterns of immigration to the United States and other societies will be discussed in later chapters.

Involuntary Immigration Involuntary immigration involves the forced transfer of peoples from one society to another. Such forced movements are best exemplified by the slave trade of the seventeenth, eighteenth, and nineteenth centuries, which brought millions of blacks from Africa to work the cotton and sugar plantations of the United States, Brazil, and the West Indies. Slavery in the Western Hemisphere will be discussed further in Chapters 9 and 14.

Consequences of Contact

Lieberson's Model The nature by which diverse ethnic groups initially meet has been shown to be a critical factor in explaining the emergence of ethnic inequality and the specific patterns it subsequently takes. Sociologist Stanley Lieberson (1961) distinguishes two major types of contact situations: those involving subordination of an indigenous population by a migrant group and those involving subordination of a migrant population by an indigenous racial or ethnic group. The first type, *migrant superordination,* is illustrated by various colonial conquests in which a technologically and organizationally more powerful migrant group subdues the native population. The second, *indigenous superordination*, is characteristic of most voluntary and involuntary immigrations such as those to North America; in these cases, the arriving groups are initially made subordinate to a resident dominant group.

Lieberson maintains that long-term conflict is more likely in societies where the indigenous population at initial contact is subordinate. Native groups less powerful than the arriving colonials are left with few options other than resistance to the new social order imposed on them. This hostility is further strengthened when the conquering group, over time, becomes itself an indigenous group.

In each of these situations, it is the relative power of the migrant and indigenous groups that determines the eventual nature of ethnic stratification. Where an invading group is successful in subduing the native population, the political and economic systems of the new group are imposed, and warfare and general conflict are likely to result quickly. Van den Berghe (1976) also points out that in contact situations between a conquering group and a weaker native group, there is usually a territorial factor, with the defeated group retaining an indigenous area. Indian reservations in North America are an example. Such a territorial base may provide the foundation for a separatist movement, an option not available to immigrant groups who enter as subordinates and are typically dispersed.

Situations in which the native group wields greater power and immigrant groups enter as subordinates produce less overt conflict initially. The indigenous group retains control over the size and character of immigration and may encourage quick assimilation, as in the case of most European immigrants to the United States. Moreover, conflict is diminished by the fact that if the immigration is voluntary, dissatisfied immigrants may return to their society of origin.

Noel's model Although the nature of initial group contact may be important in giving rise to and shaping the eventual system of ethnic stratification, Donald Noel (1968) has pointed out three additional factors: ethnocentrism, competition for scarce societal resources, and an unequal distribution of power. Let's look more closely at Noel's theory.

Part I: The Nature of Ethnic Relations

On initial contact, divergent groups will judge each other in terms of their own culture, that is, ethnocentrically. Given the nature of ethnocentrism, such evaluations will usually be negative. The extent of these negative judgments will, however, depend on the degree of difference between the groups: The more dissimilar they are, the more negative the judgment. Studies measuring the degree of acceptance of members of different ethnic groups in the United States, for example, have consistently shown those groups closest in culture and physical appearance to the dominant Anglo group (such as northwestern Europeans) to be ranked more favorably than southern and eastern Europeans such as Italians and Poles and considerably more favorably than African Americans, Mexican Americans, and American Indians (Bogardus, 1959). When culturally or physically dissimilar groups meet, then, *ethnocentrism* can be expected to typify intergroup attitudes.

However, ethnocentrism alone, explains Noel, is not sufficient to produce ethnic stratification. Groups may view one another negatively without the necessary emergence of dominant-subordinate relations among them. An additional prerequisite is *competition*, structured along ethnic lines. When groups strive for the same scarce resources, their interrelations take on the characteristics of competition and conflict. Noel posits that the more intense such competition, the greater the likelihood of the emergence of ethnic stratification. Within the competitive arena, those groups with the greatest capacity to adapt to the social and physical environment will end up higher in the ethnic hierarchy. Thus, in the American case, groups emigrating from Europe arrived with different skills, which initially determined their occupational place and subsequently enabled them to climb upward at different rates. In no case, however, did any European group place below blacks, because the latter were kept in a basically noncompetitive situation.

Differential power among the various groups is, according to Noel, the final prerequisite for the development of ethnic stratification. Unless one can overpower another, there is no basis for a stable rank order of ethnic groups, even if there is competition and ethnocentrism among them. When there is a particularly wide power gap between competing and ethnocentric groups, the emergent stratification system is likely to be quite durable. Power breeds more power, and once established, the dominant group uses its power to obstruct the competition of other groups and to solidify its dominance. In the end, then, differential power among the various groups is the most critical of the requirements for the emergence of ethnic stratification (see also Wilson, 1973).

In sum, Noel's theory postulates that competition for scarce resources provides the motivation for stratification, ethnocentrism channels this competition along ethnic lines, and differential power determines whether one group will be able to subordinate others (Barth and Noel, 1972).

Ethnic Strata: Clarity and Mobility

Ethnic stratification, like socioeconomic forms of stratification, is founded on the power of one group over others. The ensuing relations among groups are, therefore, ordinarily relations of conflict (Olsen, 1970). But if the theme of differential power is common to ethnic and socioeconomic forms of stratification, there are also several important differences between the two; namely, the clarity of and mobility between strata.

Mobility Between Strata In modern industrial societies, socioeconomic classes are relatively open, with boundaries between them porous. Possibilities thus exist for individuals to move upward from a lower to a higher class or vice versa. Sociologists refer to such movement as *social mobility*. In reality, social mobility from one generation to the next or within one's own lifetime is ordinarily limited. People inherit their class position, and most do not move extensively upward or downward. Nonetheless, equalitarian ideologies proclaim and encourage class mobility. Those born into lower economic classes are, at least theoretically, able to advance.

Ethnic stratification, however, is a system in which the boundaries between strata (ethnic groups) are far more distinct. For most people ethnicity is, as we have seen, an ascribed status; ethnic group membership is assigned at birth and is ordinarily not subject to fundamental change. As a result, ethnic consciousness is more strongly developed among members, and the competition and conflict among ethnic groups are more sharply focused.

In most multiethnic societies, the degree of mobility between ethnic strata is minimal. Particularly where physical differences, like skin color, are pronounced, extremely wide schisms develop between groups, and the lines of ethnic division remain rigid and relatively impermeable for many generations. This is so even when cultural differences are slight. African Americans are well assimilated into the dominant culture, yet they remain more rigidly segregated than others. In general, the more visible the differences between ethnic groups, the more clear-cut and inflexible will be the divisions between them. Moreover, physical differences ordinarily relate closely to the degree of inequality between the groups. Wilson notes that "there are no known cases of racial groups in advanced nation states having established equalitarian relationships" (1973:18).

Though it is nowhere common, ethnic mobility differs in degree in various multiethnic societies. In a few, the movement from one ethnic group to another is not unusual. In Mexico and Peru, for example, people who may be Indians can move into the mestizo group, the culturally and politically dominant group, merely by dropping the use of their Indian language and adopting Spanish in its place, by wearing shoes rather than the peasant *huaraches*, or sandals, and generally by practicing non-Indian ways (van den Berghe, 1978, 1979b). Physical visibility is not a critical factor that prevents such movement. To a lesser degree, the same is true of Brazil. Even in those societies with more rigid boundaries between ethnic groups, mobility may be evident among those close to the dominant group in culture and physical appearance, and such groups may be absorbed into the dominant group over a few generations. But in general, where groups remain culturally or physically distinct, mobility between ethnic strata is limited.

Caste In some cases, ethnic stratification takes on the characteristics of a *caste system,* the most rigidly static type of stratification, in which movement from one group to another is highly restricted by custom and law. The caste system of India is perhaps the most noted, but it is marked not so much by physical distinctions between groups as by people's social descent (Béteille, 1969). As it is used in non-Indian settings, the idea of caste more generally refers to "a major dichotomous division in a society between pariahs and the rest of the members of the society" (Berreman, 1966:292; see also van den Berghe, 1981). Pariahs are those who are stigmatized on the basis of some physical or cultural feature and, in most multiethnic societies, that stigma is skin color.

Where ethnic groups are racially defined, relations among them tend toward caste. Endogamy within castes is strongly enforced, and interaction between them in intimate social settings such as peer groups, clubs, and neighborhoods is minimized. Subordinate castes are usually exploited occupationally by the dominant group and experience little or no change in their collective social position.

The United States and South Africa provide the best examples of castelike systems. In both societies, the lines of division between black and white groups are sharply drawn on the basis of perceived physical differences, and endogamy is the norm. Relations between blacks and whites remain chiefly of a secondary nature, that is, in settings that do not call for close personal and intimate contacts.

One interesting case of caste exists in Japan, where a pariah group, the Burakumin, are indistinguishable physically from the rest of the population. Nonetheless, the extreme discrimination imposed on this group has been justified in terms of race. As De Vos and Wagatsuma explain, there is a commonly shared social myth that the Burakumin "are descendants of a less human 'race' than the stock that fathered the Japanese nation as a whole" (1966:xx). In the past, Burakumin were required to wear unique identifying garb, were strictly segregated residentially, and were limited to low-status occupations (Aoyagi and Dore, 1964). Although they are today not as stigmatized as in the past, their social and economic standing remains low (Kristof, 1995). Here, then, is an example of how presumed group differences—though objectively slight or even nonexistent—can form the basis of strong ethnic divisions.

Despite cases such as the Burakumin, placement and subsequent mobility within an ethnic stratification system are ordinarily affected greatly by visibility. Generally, those who are very distinct culturally and, most important, physically from the dominant group tend to remain at the bottom of the ethnic hierarchy. Those who can be stigmatized on the basis of easily perceived differences are more easily discriminated against than those whose differences are relatively indistinct. It is for this reason that the Nazis forced German Jews, physically indistinguishable from the rest of the German population, to wear yellow Stars of David on their clothing.

MINORITY RESPONSES TO SUBORDINATION

Not all minority groups in a multiethnic society react similarly to their subordinate status. Some may begrudgingly accept their place and wait for a more just world in the future, and others may struggle relentlessly to reverse their position. Wirth (1945) suggested four types of minority response: pluralistic, assimilationist, secessionist, and militant.

Pluralistic Minorities

Pluralistic minorities seek to maintain their cultural ways at the same time as they participate in the society's major political and economic institutions. Groups that enter multiethnic societies as voluntary immigrants usually adopt this position, at least for some time after their arrival, and appeal to the dominant group to tolerate their differences.

Some groups may carry the pluralistic idea further, opting out almost completely from the larger cultural, economic, and political systems. Certain religious groups in the United States and Canada such as Hutterites, Amish, and Hasidic Jews have chosen to segregate themselves even though they have not necessarily been rejected by the dominant group. These groups see contact with the larger society as a kind of contamination and a threat to their cultural integrity. Thus, they may not use the public schools or participate in mainstream political processes. Even economic matters may be largely self-contained within these groups (Hostetler, 1993; Kephart and Zellner, 1994; Poll, 1969). Such groups have achieved an accommodation with the dominant group in their society, permitting them to practice cultural patterns that are clearly aberrant in terms of the mainstream culture. Consider, for example, this description of the Amish:

> At best, the Amish seek accommodation, or a state of equilibrium in which working arrangements can be developed whereby they may maintain their

unique group life without conflict; in short, they seek a kind of antagonistic cooperation. The Amish group's aim, then, is merely tolerance for their differences, and they are not at all interested in assimilation, "Americanization," or anything that would tend to merge them with the American culture and society (Smith, 1958:226).

Assimilationist Minorities

By contrast with pluralistic minorities, *assimilationist minorities* seek integration into the dominant society. Wirth explained that such groups crave "the fullest opportunity for participation in the life of the larger society with a view to uncoerced incorporation in that society" (1945:357-58). Thus, whereas the pluralistic minority will usually insist on endogamy and will enforce adherence to the norms and values of the in-group, assimilationist minorities aim for eventual absorption into the larger society.

Most European ethnic groups in American society have maintained assimilation as their long-range objective, though some have been more desirous than others of retaining their ethnic heritage for several generations. Those groups closely similar to the dominant group in culture and physical appearance will ordinarily assimilate more easily and thoroughly than others. Some may even reach the point of complete absorption into the dominant group. In the United States, for example, northwestern European ethnic groups (Dutch, Scandinavian, and German) have been so completely assimilated over several generations that today they are virtually indistinguishable as ethnic units. However, such cases are exceptional. Rarely are minority ethnic groups completely absorbed into the dominant group, totally relinquishing their culture or physical identity. Moreover, even where assimilation is the group's objective, the prejudice and discrimination the group encounters will often retard such efforts. In Chapter 4, we will discuss at greater length the process of assimilation and the factors that promote it.

Secessionist Minorities

Secessionist minorities, according to Wirth, desire neither assimilation nor cultural autonomy. Their aim is a more complete political independence from the dominant society. Such groups are usually what were earlier referred to as nations, not only aspiring to some degree of political autonomy but also maintaining territorial integrity. The separatist movement in the Canadian province of Quebec represents a contemporary secessionist minority. In this case, a substantial element of the Quebec populace seeks not only to retain the French language and French-Canadian culture but also to establish an independent Quebec state.

Most nationalist movements are the outgrowths of dissatisfaction on the part of ethnic groups stemming from their minority status and treatment.

In addition to seeking separation from the larger society, Wirth pointed out, a secessionist minority may desire integration with another group or society to which it feels a closer cultural and political similarity. For example, as we will see in Chapter 16, many among Northern Ireland's minority Catholics seek to unite the six counties of Ulster (Northern Ireland) with the Catholic-dominated Irish Republic.

Militant Minorities

Militant minorities seek as their ultimate goal not withdrawal, as do secessionist minorities, but rather establishing themselves as the society's dominant group. The case of Latvia, a small Baltic country, illustrates. During the period when Latvia was ruled by the Soviet Union, natives were in some ways reduced to minority status, particularly regarding language and culture, which were Russian-dominated. With independence from the USSR, a shift in political power to the native Latvian group occurred. As a result, Latvians not only replaced the Russian political elite, but imposed language and citizenship measures that made non-Latvians, including ethnic Russians, minorities.

Wirth maintained that these four types of response represent general stages in the life cycle of a minority group. At first, a group will seek toleration for its cultural differences; if successful it may then, over several generations, seek incorporation into the dominant group. Blockage of such assimilation, however, will produce secessionist tendencies, which in turn may lead to the objective of domination, abetted by militant tactics.

It is important to consider that these variable responses of minority groups are not totally or, in some cases, even largely voluntary. Rather, they very much depend on the power of the dominant group to accept or reject minority group aims (Schermerhorn, 1970). It is not only a question of what a minority group initially desires, but what it desires in conjunction with what the dominant group desires for it. A minority group may seek assimilation, for example, but be repelled by the dominant group's aim to keep it isolated. Minority objectives may also change depending on the dominant group's responses. Continual frustration of assimilationist goals, for example, is likely to eventually create a pluralistic or secessionist response.

One shortcoming of Wirth's typology is its concentration on the cultural elements of ethnicity and its neglect of other critical factors, particularly race, as they shape dominant-minority relations. The nature of ethnic relations is not dependent simply on whether and in what degree the minority ethnic group becomes culturally "like" the dominant group. Chapter 4 will show that assimilation and pluralism are multifaceted processes. They involve not only the group's adoption or rejection of mainstream cultural traits but also its power relations and other interactions with members of the dominant group. As Schermerhorn (1970) points out, in multiethnic societies where groups are distinguished on the basis of physical as well as cultural traits, the dominant group is often concerned not with whether the minority group will adopt its cultural ways but only with maintaining sufficient social distance between itself and the out-group. Ethnic groups may assimilate culturally yet remain excluded from significant participation in the dominant institutional structure. This is the case, for example, for African Americans.

THE RELATIONSHIP BETWEEN CLASS AND ETHNICITY

Although ethnicity and class are distinct dimensions of stratification, they are closely interrelated. In almost all multiethnic societies, people's ethnic classification becomes an important factor in the distribution of social rewards and, hence, in their economic and political class positions. Put simply, people receive different amounts of what is valued—jobs, education, wealth, and so on—in some part as a result of their ethnicity.

That ethnic groups are hierarchically ordered would mean nothing if that rank order were not tied to the distribution of the society's wealth, power, and prestige. The fact that members of one or a few ethnic groups maintain most of the important positions of political and economic power, own an inordinate share of the society's wealth, and enjoy the most social prestige is due neither to chance nor to their greater motivation or innate capabilities. Rather, it is a consequence of the integral link between ethnic stratification and other forms of stratification in the general distributive system. In short, the ethnic and class systems are in large measure parallel and interwoven. Where people begin their quest for the society's rewards and what they ultimately achieve very much depend on their ethnicity.

Success is by no means assured by a favored ethnic background. People's class position at birth, even for those of the dominant ethnic group, is an overarching factor in determining their eventual wealth, power, and prestige. Thus, not all Anglo-Americans, for example, are doctors, lawyers, corporation executives, and high-ranking politicians. But for dominant group members, the ethnic factor is removed as an impediment to upward mobility; other factors, both individual and structural, will affect their fortunes, but ethnicity will not.

In the same way, of course, we should not think minority status means that a person is automatically relegated to the bottom rungs of the wealth, occupational, educational, political, and other class hierarchies. But for minorities, the chances of winding up at the bottom are much greater. As we proceed down the ethnic hierarchy, we find increasing political powerlessness, lack of economic opportunity, and social discrimination and exclusion. The closer to the bottom of the hierarchy, the more difficult the path to social success in whatever form, regardless of other nonethnic social traits.

In sum, the effect of ethnicity is that minority ethnic group members encounter barriers to the attainment of the various rewards of their society that dominant ethnic group members do not face.

This does not mean, however, that all minorities are affected to the same extent. For those who are members of low-ranking groups with high ethnic visibility—African Americans, for example—minority status may be the overriding determinant of one's economic and political class position. For other minorities more highly ranked and less visible—for example, Polish Americans or Irish Americans—ethnicity will be much less consequential. It is important to remember that there are *degrees* of minority status. The ethnic hierarchy in any society is rarely a simple two-part structure with a dominant group at the top and subordinate groups at the bottom. Rather, minority groups occupy places on a continuum, and the impact of ethnicity varies among them. The question for minority group members, then, is: To what extent does ethnicity become a factor in the allocation of jobs, education, wealth, political power, and other life chances?

Minority status does not necessarily indicate where in the class hierarchy a group's members enter the society but rather where they remain long after they have entered. Indeed, most groups, even those who immigrate voluntarily, enter collectively at or near the bottom. Some may rise very quickly, whereas others remain trapped at the bottom, generation after generation.

Let's look more closely at how ethnic groups as collectivities relate to other forms of stratification.

Ethnicity and Economic Class

Members of various ethnic minorities are found at different points in a society's occupational hierarchy. In the United States, for example, we find African-American doctors, Jewish-American factory workers, Mexican-American engineers, and Anglo-American restaurant workers. More commonly, however, we find Jewish-American doctors, Mexican-American restaurant workers, Anglo-American engineers, and African-American factory workers. The relationship between economic class and ethnicity is usually very evident in the society's occupational patterns. Ethnic groups tend to concentrate in particular areas of work, especially as they enter the society. In the United States in the 1830s and 1840s, Irish workers built the canal system. In the second half of the nineteenth century and the early twentieth century, Italians were heavily represented in building construction, Jews in the garment trades, and Slavs in the steel industry. Racial-ethnic groups like blacks remained overwhelmingly in the least advanced sectors of the economy as agricultural and service workers. Though ethnic groups today are more occupationally dispersed, the remnants of these earlier patterns are still evident. Newer ethnic groups, moreover, display much occupational concentration.

In addition to occupation, other aspects of economic class—education, income, and wealth—are closely linked to ethnicity. Those most highly educated, those earning the highest incomes, and those possessing the greatest wealth are statistically more likely to be members of the dominant ethnic group or members of ethnic groups closest to the dominant group in culture and physical appearance.

Members of minority ethnic groups are never randomly scattered throughout the economic class structure but are disproportionately clustered at certain points. If ethnicity had no effect on the allocation of wealth and income, we would find a broadly proportionate distribution of members of various ethnic groups in all economic classes. If, for instance, a group constitutes 10 percent of the general population, we would ideally expect to find 10 percent of each economic class (lower, working, upper-middle, upper, and so on) made up of members of that group. Of course, other social variables such as age, sex, and region also affect people's class position, but ethnicity is a particularly critical factor. In Part II, we will see this class-ethnicity relationship more clearly as it pertains to specific American ethnic groups.

A similar class-ethnicity relationship characterizes most other multiethnic societies. The differentials among groups may be more or less extreme than what is seen in the United States; but in most cases, the pattern is readily apparent.

Ethnicity and Societal Power

Because all aspects of stratification are founded on power, whatever changes occur in the distribution of wealth, income, education, and other life chances necessarily depend on the makeup of the society's power elites—those who formulate policies, guide the activities, and decide the significant issues of the government, the corporation, education, and other major societal institutions. In analyzing the relationship between decision-making power and ethnicity, the key question to be asked is: How open are the power elites to members of different ethnic groups? To the extent that they are closed, the status of these groups in other dimensions of stratification will remain consistently low.

In most multiethnic societies, there is an apparent relationship between ethnicity and access to important power positions. Some groups are favored over others, and much ethnic conflict centers on the process of filling these elite posts. In the United States, the ideology of equal opportunity does not always conform to the reality of ethnic discrimination in elite recruitment. With slight changes from time to time and from one elite sector to another, historically, the dominance of white Protestants has been the general rule at the highest levels of government and the economy.

Similar patterns can be discerned in other multiethnic societies. Israel, for example, is a multiethnic society in which Jews of European and American origin are generally higher ranked in the society's income, occupational, educational, and political hierarchies than Jews from North African and Middle Eastern societies (Cohen, 1968; Goldscheider, 1996; Simon, 1978; Smooha, 1978). Israeli Arabs, about 18 percent of the Israeli population, are also severely limited in access to positions of influence and prestige.

Other cases demonstrating the relationship between societal power and ethnicity will be examined in Part III.

Class and Ethnicity Reconsidered

At this point, observations of the social standing of some members of dominant and minority groups might give rise to wonderment at what ethnic dominance and subordination really mean. Do we not see in the United States, for example, Jews in important power positions or blacks in high-ranking occupations? And, conversely, are not some Anglo-Americans continually at the bottom of the economic class hierarchy? If members of certain minority ethnic groups achieve upper-class standing or significant power, and similarly, if members of the dominant ethnic group remain at the bottom of the economic and political hierarchies, does this not negate the idea of ethnic dominance and subordination? How, in short, can we account for these inconsistent cases?

First, it must be emphasized that the link between class and ethnicity in any society is not perfect; obviously, some do achieve significant upward mobility despite the handicap of low ethnic rank. But we cannot generalize about groups as collectivities on the basis of the achievement of an outstanding few. Popular interpretations of the relative success or failure of ethnic groups, however, often rely on such faulty generalizations.

For example, a widespread explanation for the inability of certain ethnic groups to rise collectively in economic and political standing is "self-motivation" (Schuman, 1982). It is assumed that individuals themselves, not social forces, are mainly responsible for their social placement. The poor are poor, in this view, because they lack the motivation to improve themselves, and the wealthy are successful because they have greater incentive. That a few from low-ranking minority groups do actually achieve great social success, and that many from the dominant group do not, serve to "confirm" this oversimplistic explanation.

Individual talent, motivation, and ability are, of course, important factors in social success; but only after the competitive field has been severely thinned out by ascribed characteristics such as class origin, gender, and ethnicity. Those with advantages by birth are assured relatively high achievement regardless of their individual capacities. Many in low-ranking groups may have abilities that are necessary to high social achievement, but unless they can gain access to the proper staging of those abilities through quality education, a good first job, and the nurturing of key social connections, they will go unnoticed. One of the basic features of minority groups, it must be remembered, is the social discrimination their members face in important life chances such as these.

Second, no matter how many exceptional cases sneak through the ethnic barriers, minority groups remain underrepresented in the society's top wealth, power, and prestige classes. This underrepresentation will vary in degree from group to group, but the lack of equal access to the society's rewards is precisely what defines all such groups as minorities. Glass (1964) suggests that *minority* designates a group not totally part of the society's mainstream. The extent to which a group is "outside" will vary, of course. Italian Americans, for example, are obviously less removed from the mainstream of American society than are African Americans. Thus, we would expect to (and in fact, do) find more Italian Americans penetrating the higher ranks of wealth, prestige, and power than African Americans.

Moreover, even when members of a minority group are increasingly scattered throughout the socioeconomic and political class systems, the status of individual members at the higher levels is associated with their ethnic group membership and is therefore tenuous. As Glass explains, the economic and political status of a minority ethnic group may be neither uniformly high nor uniformly low; but in any case, it is "invariably a vulnerable one, associated with notions of inferiority" (1964:142). In the United States, a city mayor who happens to be black, for example, is not seen or responded to simply as a mayor but always as a *black* mayor.

Third, in accessing the society's rewards, members of the dominant ethnic group—even those at the bottom of the economic and political hierarchies—often retain privileges not enjoyed by members of minority groups, even those of upper-class standing. For example, before the 1960s in most cities of the American South, black entertainers receiving huge salaries still could not stay at the very hotels in which they were performing. A white person, regardless of income or occupation, would not have experienced such exclusion on the basis of skin color. Today, African Americans with high-status occupations—doctors, lawyers, managers—commonly describe the suspicion their presence arouses in suburban, primarily white, neighborhoods or prestigious shopping areas.

Finally, the empirical link between class and ethnicity is complex and subject to frequent change. The class position of particular groups or large segments of those groups may be altered significantly from one generation to another. For example, most members of white ethnic groups in the United States today are part of the middle class or working class, though most of their forebears entered the society at an appreciably lower level. A large proportion of

African Americans have also displayed a substantial change in class standing in the past three decades.

Given such changes and the intricacies of the relationship between class and ethnicity, we should not expect to find all members of a particular ethnic group with the same economic and political rank or the same economic and political concerns and attitudes. There is a range of classes *within* each ethnic group even though, for the most part, group members remain concentrated at particular levels of the society's general class system.

SUMMARY

Social stratification is a system of structured social inequality in which groups receive different amounts of the society's wealth, power, and prestige and are hierarchically arranged accordingly. In multiethnic societies, ethnicity is an additional basis of stratification. There is, therefore, a hierarchy of ethnic groups in which one, the dominant group, maintains maximum power to determine the nature of interethnic relations. Minority, or subordinate, groups rank in different places below the dominant group, depending on their cultural and physical distance from it.

Dominant-minority relations are relations of power. Although there are other bases for minority status such as gender and age, our focus is on ethnicity. On the basis of their physical and cultural characteristics, ethnic minorities are distinguished from others in the society and given differential treatment. The power of the dominant group is based on its disproportionate control of the society's political and economic resources and its ability to shape the society's major norms and values.

Ethnic stratification is the product of contact between previously separate groups. Initial contact may be in the form of conquest, annexation, voluntary immigration, or involuntary immigration. The manner in which ethnic groups meet is a critical factor in explaining the shape of the system of ethnic inequality that ordinarily ensues. Following contact, groups engage in competition; view one another ethnocentrically; and ultimately, one imposes its superior power over the others, emerging as the dominant group.

There are four types of ethnic minorities: pluralistic minorities, seeking to maintain some degree of separation from the larger society; assimilationist minorities, aiming for full integration into the dominant society; secessionist minorities, seeking political autonomy from the dominant society; and militant minorities, trying to establish dominance themselves.

Ethnic rank commonly parallels closely rank in the society's class system. Members of minority ethnic groups are never randomly scattered throughout the economic and political class hierarchies but tend to cluster at specific points.

Suggested Readings

Béteille, André. 1977. *Inequality Among Men*. Oxford: Basil Blackwell. Examines various forms of stratification, including racial and ethnic, using examples from different societies.

Kessner, Thomas. 1977. *The Golden Door: Italian and Jewish Immigrant Mobility in New York City 1880-1915*. New York: Oxford University Press. A sociohistorical comparison of the upward mobility of two ethnic groups, Italians and Jews, following their entrance into American society.

Jackman, Mary R. 1994. *The Velvet Glove: Paternalism and Conflict in Gender, Class, and Race Relations*. Berkeley: University of California Press. Analyzes how dominant groups use persuasion, in the form of ideology, rather than force, to create legitimacy and to sustain systems of social inequality.

Lenski, Gerhard. 1966. *Power and Privilege: A Theory of Social Stratification*. New York: McGraw-Hill. A classic work that examines the basis of structural inequality in its various forms.

Shibutani, Tamotsu, and Kian M. Kwan. 1965. *Ethnic Stratification: A Comparative Approach*. New York: Macmillan. Develops a theory of ethnic stratification, drawing on case studies from around the world.

Steinberg, Stephen. 1989. *The Ethnic Myth: Race, Ethnicity and Class in America*. Updated and expanded ed. New York: Atheneum. Deals with the issue of ethnic versus class factors in group mobility, leaning strongly toward the latter.

Wilson, William J. 1973. *Power, Racism, and Privilege: Race Relations in Theoretical and Sociohistorical Perspectives*. New York: Free Press. Using a power-conflict approach, explains the development of racial stratification with a comparison of the United States and South Africa to illustrate.

Zenner, Walter P. 1991. *Minorities in the Middle: A Cross-Cultural Analysis*. Albany, N.Y.: State University of New York Press. Examines middleman minorities in a number of historical and cultural settings.

TECHNIQUES OF DOMINANCE:
PREJUDICE AND DISCRIMINATION

WE HAVE NOW seen that in most multiethnic societies, ethnic groups are arranged in a hierarchy in which the dominant group receives a disproportionate share of the society's rewards because of its greater power; specifically, political authority and control of the means of economic production. It is not enough, however, to simply proclaim that the dominant group is more powerful though this is, of course, the crux of the matter. We must also look at the specific techniques by which dominance and subordination are maintained and stabilized in systems of ethnic stratification.

To enforce its power and sustain its privileges, the dominant ethnic group employs certain tools, which can be subsumed under the categories of prejudice and discrimination. Widely held beliefs and values regarding the character and capacities of particular groups are necessary to assure the long-range durability of ethnic inequality. These beliefs and values take the form of *prejudices*, that is, negative ideas regarding subordinate ethnic groups and ideas expressing the superiority of the dominant group. These beliefs often come together in a cohesive ideology of racism or another deterministic notion; but at other times, they are applied to groups in a somewhat disparate, unsystematic fashion. In addition to prejudices, the dominant group applies various actions against minority ethnic groups, including avoidance, denial, threat, or physical attack. At different times, all these forms of coercion may be used, depending on how threatening the minority group is perceived to be. These actions are collectively termed *discrimination*.

In this chapter, we look at some theories and research findings regarding particular forms of prejudice and discrimination in multiethnic societies. This is an area of ethnic relations that has been the focus of much research, yielding a great amount of empirical data.

In looking at prejudice and discrimination, we will necessarily investigate some of the social psychology of ethnic relations, that is, ethnic relations at the individual, or interpersonal, level. To this point we have been concerned primarily with ethnic relations at the group, or structural, level; and our approach to prejudice and discrimination will continue to emphasize the group dynamics and consequences of these social phenomena. But here, more than in other aspects of ethnic studies, we must pay closer attention to individual attitudes and motives, recognizing that, as C. Wright Mills so aptly put it, "Neither the life of an individual nor the history of a society can be understood without understanding both" (1956:3).

PREJUDICE

Most simply, *prejudice* can be defined as an arbitrary belief or feeling toward an ethnic group or its individual members. Specifically, prejudice involves a judgment, "based on a fixed mental image of some group or class of people and applied to all individuals of that class without being tested against reality" (Mason, 1970:52). It is in other words a generalized belief, usually unfavorable and rigid, applied to all members of a particular group. Although often defined as a prejudgment or preconcept founded on inadequate evidence (Klineberg, 1968), prejudice is, as Berry and Tischler (1978:235) have pointed out, "more emotion, feeling, and bias than it is judgment."

Ethnic prejudices are characterized by several specific features. First, they are *categorical*, or generalized, thoughts. Individuals are judged on the basis of their group membership, not their personal attributes. A prejudicial attitude may be directed at a particular person, but it is the person's group and its alleged traits that evoke this attitude rather than his or her individual actions and qualities. Once the person's group is known, his or her behavioral traits are inferred. Thus, prejudice violates "the norms of rationality" (Pettigrew, 1980:821).

Second, ethnic prejudices are *inflexible*. As Gordon Allport explains, "Prejudgments become prejudices only if they are not reversible when exposed to new knowledge" (1958:9). A prejudice is not simply an error in thought, but one not subject to correction. Individuals develop emotional attachments to certain beliefs and will not discard them in the light of contrary evidence. People may proclaim that "some of their best friends" are members of a particular ethnic group that they generally view adversely. The implication is that such persons do not exhibit the negative qualities ordinarily attributed to members of their ethnic group. But instead of refuting the belief, which logic would dictate, they serve only as "exceptions that prove the rule." Such contrary evidence is recognized but excluded from the generalization; it thus has no correcting effect (Allport, 1958).

Third, ethnic prejudices are usually *negative* in content. That is, the specific traits ascribed to targeted groups are inferior and socially undesirable. Of course, prejudices may be positive as well as negative. Ethnic group members maintain an overly favorable image of their own group, just as they maintain overly unfavorable images of certain out-groups. Indeed, all ethnic groups express ethnocentric notions regarding their unique character. Moreover, what is often interpreted as prejudice against out-groups may in fact be more a matter of in-group bias; that is, favoritism toward members of one's own group (Brewer, 1979; Jones, 1996). In studying ethnic relations, however, sociologists and psychologists have been concerned almost exclusively with negative prejudices.

Finally, prejudices are based on erroneous or inadequate group images called *stereotypes*. Because they compose the chief content of ethnic prejudices, let us look more closely at these generalized group images.

Stereotypes

Stereotypes were first suggested in 1922 by Walter Lippmann, who described them as "pictures in our heads" that we do not acquire through personal experience. In the case of ethnic stereotypes, distinctive behavioral traits of an ethnic group are selected by out-group members who exaggerate them to construct what Shibutani and Kwan call "a shorthand depiction" of the group (1965:86). Ehrlich refers to stereotypes as "a special language," which functions to "reinforce the beliefs and disbeliefs of its users, and to furnish the basis for the development and maintenance of solidarity for the prejudiced" (1973:21). Mackie defines stereotypes as "those folk beliefs about the attributes characterizing a social category on which there is substantial agreement" (1973:435).

Rarely will people claim to dislike someone of another ethnic group merely because he or she is a member of that group. Instead, the adverse view will be couched in "rational" terms. One dislikes Jews because they are "shrewd," blacks because they are "lazy," or Italians because they are "loud and uncouth." These mental images of groups thus serve as supports for the negative beliefs that constitute prejudice. Once we learn the stereotypes attached to particular groups, we tend to subsequently perceive individual members according to those generalized images. To stereotype someone, then, "is to attribute to that person some characteristics which are seen to be shared by all or most of his or her fellow group members" (Brown, 1995:82).

Clearly, however, generalizing about groups or objects is a most common pattern of thought. Indeed, it is the very mental technique that facilitates social interaction, particularly in modern, complex societies where we cannot possibly know all the personal characteristics of those whom we encounter daily. Thus, on the basis of some identifying marks such as ethnicity, sex, age, or occupation, we generalize and make judgments about people. In a sense, generalizing on the basis of group membership is a kind of predictive mechanism we use in various social situations.

Consider an example familiar to college students. On the first day of classes, students meet Professor X. Professor X is known only as part of the group called professors, about which there are certain general ideas. Professors ordinarily lecture, give examinations, grade students, and so on. Armed with this understanding of the category *professors*, students naturally expect Professor X to behave accordingly. The chances are very great that Professor X will conform to their expectations. In the same way, of course, Professor X may meet his or her students for the first time knowing nothing more about them but that they are students. With this bit of knowledge, however, Professor X can "predict" the actions of the students, using the same technique of categorizing as they have used. In this situation, people are expected to perform the roles of professor and student according to the social "script" attached to each.

In most cases our expectations or predictions about how people will act in different social situations will be correct. As a result, we are enabled to interact with people about whom we know nothing more than their social identifications, that is, their group memberships. Individuals occasionally fail to perform in the manner expected, and our predictions are then incorrect. But these are not the usual cases. For the most part, we understand correctly what is expected of people in various social roles—professor, student, doctor, father, Catholic, and so on—and pattern our behavior toward them accordingly. This is the nature of rational thought and behavior. How, then, do ethnic stereotypes differ from these common forms of generalized or predictive thought?

Ethnic stereotypes differ from rational generalizations in that they are oversimplistic and overexaggerated beliefs about a group, generally acquired secondhand and resistant to change (Harding, 1968; Pettigrew, 1980). Thus, ethnic stereotypes are sustained despite numerous individual cases that clearly refute their validity.

The characteristics attributed to various ethnic groups are established in popular beliefs and are reinforced through selective perception. This means that people take note of those cases that confirm their stereotypical pictures and overlook or ignore those that refute them. Those who believe blacks to be lazy or Jews to be deceitful will take special notice of those blacks or Jews who do in fact exhibit such traits but will fail to notice the many more blacks and Jews who do not. Moreover, these stereotypical traits will be inferred, even if they are not evident, so that the observers of an ethnic group will interpret the actions of group members based on their preconceived image.

In addition, stereotypes fail to show how members of the dominant group may share the same negative traits imputed to minority groups; or how the dominant group, through the self-fulfilling prophecy, may contribute to the very creation of these unfavorable traits (Simpson and Yinger, 1972).

Studies of Stereotyping The pioneer study of ethnic stereotyping in the United States was conducted in 1933 by Katz and Braly, who questioned 100 Princeton undergraduates on the prevailing stereotypes of various ethnic groups. Their findings indicated a very high degree of uniformity, in some cases as high as 75 percent. Jews, for example, were consistently described as "shrewd," "mercenary," and "ambitious"; blacks as "superstitious," "happy-go-lucky," and "lazy"; and Germans as "scientifically minded," "industrious," and "stolid." Katz and Braly noted that the students had had little or no contact with members of most of the groups they described, an indication that they had acquired the stereotypes by absorbing the dominant culture.

Gilbert replicated this study in 1951 and found a prevalence of many of the same stereotypes. However, he also detected a marked change in attitude. Many students now expressed reluctance to categorize people whom they did not know. A third generation of Princeton students was questioned in 1967, and the researchers in that case also encountered irritation and resentment among those asked to generalize about ethnic groups. Further, they found that the actual content of the stereotypes had now changed considerably, with some groups such as blacks and Jews being assigned more favorable traits than had been the case in 1933 (Karlins et al., 1969). But the uniformity of the application of stereotypes had not changed. In other words, some traditional stereotypes had declined in frequency, but they had been replaced by others, some more favorable than previously. The researchers concluded that although the students in 1967 appeared to be more tolerant and less receptive to stereotyping, the application of ethnic stereotypes was still evident.

In a more recent study, Devine and Elliot (1995) suggested that the three Princeton studies all suffered from several methodological shortcomings. First, they did not distinguish between stereotypes and personal beliefs. Knowing an ethnic stereotype does not necessarily mean that one *believes* that these traits actually characterize members of the group. Also, the Princeton studies made no assessment of respondents' level of prejudice, nor did they employ stereotypical terms that are more prevalent today than they were in 1933, when the study was first conducted. Devine and Elliot sought to correct these shortcomings, although they measured changes only in the stereotypes of blacks, not the full range of groups dealt with in the Princeton studies. They found that there is today a clear and consistent negative stereotype of blacks, but that personal beliefs about blacks are equally consistently positive. They concluded that the earlier Princeton studies were actually measuring personal beliefs, not stereotypes, which may or may not be congruent. While the stereotype is well understood by all, only highly prejudiced people actually endorse it.

An important question posed by Devine and Elliot's study is why stereotypes persist if they don't represent what a great many people believe about a group. Perhaps most important is their durability as part of the society's cultural fabric. Social norms regarding overt expressions of prejudice have changed considerably, yet black stereotypes persist in more subtle and covert, yet effectual, ways.

One noteworthy study of an international nature was conducted by Buchanan and Cantril in the early 1950s. Respondents from eight countries were asked to describe people of other countries by choosing from a list of descriptive adjectives. (This was similar to the technique used by Katz and Braly and others.) The general findings of this study were that people in all eight countries displayed a tendency to use stereotypes in describing other national groups, that their own compatriots were always described in flattering terms, and that the choice of either complimentary or derogatory adjectives depended largely on the current state of relations between the nations.

Based on their evidence, Buchanan and Cantril also concluded that stereotypes, rather than preceding people's reaction to a certain group, ordinarily do not exist until objective events demand their creation. Thus they stressed that stereotypes should not be thought of as causative but as symptomatic. As they put it, "Perhaps their important function is the wartime one of providing a rationale within which men are able to kill, deceive, and perform other acts not sanctioned by the usual moral code" (1953:57).

A clear illustration of the creation of stereotypes as a means of rationalizing events is the dramatic change after World War II in the stereotypes of Japanese and Germans held by Americans. During the war, negative images of these two groups—as evil, hostile, and cruel—prevailed; but by the 1960s the groups were seen as clean, efficient, and industrious. The image of Russians was also altered, but in the opposite direction. As World War II allies, the Russians were portrayed in generally positive terms, but this image was changed in the late 1940s with the onset of the Cold War and the emergence of the Soviet Union as the United States' chief ideological foe. Note in Table 3.1 the striking increases in the adjectives *conceited* and *cruel* and the decrease in *brave*.

Table 3.1 Stereotypes of Russians Held by Americans

TRAIT	1942	1948
Hardworking	61%	49%
Intelligent	16	12
Practical	18	13
Conceited	3	28
Cruel	9	50
Brave	48	28
Progressive	24	15

SOURCE: Buchanan and Cantril, 1953. Reprinted by permission.

In the same way that wartime stereotypes are used to rationalize hostility toward enemy nations, stereotypes are used by dominant groups in multiethnic societies to sustain their competitive advantage over challenging or threatening ethnic groups. The negative images of blacks held by whites in the United States or South Africa have their counterparts in Northern Ireland, where Protestants hold adverse images of Catholics, or in Israel, where Jews of European origin often express negative beliefs about North African or Middle Eastern Jews. In each of these cases, the persistence of negative group images can be explained as a rationale on the part of the dominant group for keeping the minority ethnic group or groups in a subordinate position. As long as groups are perceived as undeserving, their social disadvantages can be justified (Devine and Sherman, 1992).

The competitive use of stereotypes may often explain their very content. Simpson and Yinger note, for example, that those groups that have successfully competed with the dominant group cannot be labeled lazy or unintelligent, "so they are pictured as too ambitious, and with a crafty kind of self-interested intelligence" (1972:155). This can be seen clearly in the case of Jews in the United States and other societies where they have exhibited an uncommon ability to achieve economic success. Allport (1958) compares the admirable traits of Abraham Lincoln with the disliked traits of Jews and finds them quite the same. Both are generally described as thrifty, hardworking, eager for knowledge, ambitious, devoted to the rights of the average man, and eminently successful in climbing the ladder of opportunity. The key difference, explains Allport, is that the terms used to describe the Jews are often disparaging. Thus, thrifty becomes "tight-fisted," hardworking becomes "overambitious," ambitious becomes "pushy," and concerned about human rights becomes "radical" (1958:184). Much the same semantic reversals of positive traits have been used in describing Chinese and East Indians in various Asian and African societies where they constitute economically successful minority groups (Hunt and Walker, 1974; Kristof, 1998). A related notion is "ambivalent stereotypes" (Fiske et al., 1999). The idea is that one can envy and respect high-status groups for their competence, but dislike them. By the same token, one can disrespect low-status groups for their incompetence, but like them.

The content of stereotypes may change periodically and sometimes radically, depending on different economic, political, and social circumstances. Whatever their content, however, stereotypes are conveyed in subtle but effective ways through various socialization agencies, including the family, the school, and the mass media. Thus, stereotypes cannot be seen as irrational pictures of ethnic groups held by a numerically insignificant part of the general population. Rather, they are part of the society's heritage, and as Ehrlich (1973) explains, no person can grow up in a society without having learned them.

The Mass Media and Stereotypes In modern societies, the mass media—television, radio, newspapers, magazines, and motion pictures—are extremely critical agents of socialization. They are one of the chief means by which people interpret the structures and events of their society, and they provide role models to which individuals might not otherwise be exposed. Because their techniques are so pervasive, the media as socializers may supersede in some ways even the family and the school. Although there is still debate over the nature of the media's exact effect on socialization, no one denies that they play an important role for most people (Graves, 1999; Marger, 1993; Wright, 1986).

With their enormous impact, the media—television and motion pictures, in particular—are key conveyors of ethnic stereotypes. Until recently, blacks in American films, for example, traditionally played subservient characters and rarely were given starring roles except in all-black productions (Brown, 1981; Sterngold, 1998a). Other racial and ethnic minorities have fared little better. American Indians have been portrayed customarily as savages and Hispanics as untrustworthy villains (Coward, 1999; Engelhardt, 1975; Moore and Pachon, 1985). In American television, the 1950s were marked by an almost total exclusion of minorities from the screen except for stereotyped roles in programs like "Amos 'n' Andy." The absence of blacks was due mainly to the fear held by networks and program sponsors of offending white viewers, especially in the South (U.S. Commission on Civil Rights, 1977; Sterngold, 1998a). This situation was modified somewhat in the 1960s with the inclusion of a few blacks and other minorities as actors and as news reporters.

The 1970s marked a significant change in television's presentation of blacks, who were increasingly incorporated into situation comedies and even became the chief protagonists in several. Many of these roles presented a new racial consciousness to television, priding themselves "not on their similarity to the mainstream, but on their differences from it" (Lichter et al., 1987). Moreover, these programs attracted white as well as black audiences (Sterngold, 1998b). Critics pointed out, however, that blacks in these programs remained stereotyped and were almost always portrayed as comical figures, not featured in serious parts (MacDonald, 1983; U.S. Commission on Civil Rights, 1979). Moreover, minorities other than blacks—particularly Hispanic Americans, American Indians, and Asian Americans—were rarely seen in any role on American television.

The heightened racial and ethnic diversity of the 1970s gave way in the 1980s to more integrated casts and characters. Communications researcher Robert Lichter and his associates (1987) reported that from 1975 to 1986, nearly one in ten characters on television were black. Moreover, blacks played in more starring roles and portrayed more positive characters than in the past. Bill Cosby, for example, was among the most prominent television personalities of the decade. Programming with blacks in major roles, especially sitcoms, continued to proliferate in the 1990s, though marketing considerations seemed to dictate a shift away from black shows in prime time (Farhi, 1999; Sterngold, 1998b). Hispanics were given far less exposure in prime-time roles, but the rapidly growing Hispanic-American population augured a similar expanded presence for them in television programming (Wilson and Gutiérrez, 1995). In news reporting as well, by the late 1990s, the representation of blacks and Hispanics was proportionate to or greater than their presence in the population (Greenberg and Brand, 1998).

Most attention has been focused on how the mass media sustain negative ethnic stereotypes. Media researcher Robert Entman has demonstrated how television news upholds the common stereotype of African Americans, for example, as "poor." The linkage of "black" and "poor" is so strong in the presentation of televised news, explains Entman, that "the white public's perceptions of poverty appear difficult to disentangle from their thinking about African Americans" (1997:33). The stereotypical linkage of crime and violence with African Americans is similarly bolstered by television news, especially at the local level.

The mass media, however, may also convey exaggerated positive images. De Roche and de Roche (1991) studied episodes of five police series popular in the 1980s, comparing characterizations of black and white officers. They found that black men in these dramas had been "counterstereotyped." That is, not only were these men not described with negative ghetto stereotypes, but they were portrayed in overly favorable terms in comparison with white officers. "In the collection of characters on which we focused, black men are modeled for us as more bourgeois than their white counterparts, more self-directed and affectively well-managed, and more tasteful. . . . They clearly confirm the traditional ideal of a stable, financially responsible husband/father, and this in an era when female-headed, single-parent families are becoming increasingly common in all social sectors" (De Roche and de Roche, 1991:86).[1]

Although most observers agree that the electronic media are important transmitters of ethnic stereotypes and attitudes, both positive and negative, studies have indicated that their effect may be only to reinforce ideas already acquired before exposure. Vidmar and Rokeach's study (1974) of viewers of a popular television sitcom of the 1970s, "All in the Family," demonstrates this point. This program set a precedent in its candid use of ethnic humor and its general treatment of ethnic-related issues. The chief protagonist, Archie Bunker, is a humorous bigot who voices all the well-established ethnic slurs and stereotypes and expresses blatant racist ideas. The producers of the program argued that by openly dealing with bigotry and making the major character a ridiculous figure, it would serve to reduce societal prejudice. Poking fun at bigotry, it was felt, would force viewers to recognize their own prejudices and thus reduce them. Others, however, argued that the program encouraged prejudice by introducing people, particularly children, to ideas they might not otherwise be exposed to.

Vidmar and Rokeach hypothesized that a process of selective perception would cause viewers to react differently to Archie Bunker's bigotry, depending on their prior attitudes. Viewers who were not prejudiced to begin with, they felt, would see Bunker as a bigot and would understand the show's satirical messages, whereas prejudiced viewers would see Bunker as "telling it like it is." Two groups of respondents—one made up of American adolescents, the other of Canadian adults—were asked their reactions to the program; they were also given a set of attitudinal questions designed to measure their level of prejudice. The study's findings confirmed the selective-perception hypothesis: Reactions to the program varied in relation to prior attitudes. Those less prejudiced recognized Archie Bunker as bigoted, rigid, and domineering; and those more highly prejudiced recognized him as down-to-earth, honest, and hardworking.

Vidmar and Rokeach also hypothesized that those high in prejudice would not watch the program as frequently as those less prejudiced. This hypothesis was based on the well-researched notion that people expose themselves to social stimuli that are compatible with their established views and attitudes. For example, political scientists have demonstrated that campaign speeches by political candidates are listened to or watched mainly by those who are already committed to those candidates. In much the same way, it was thought that because the show depicted satirical bigotry, the more frequent viewers would be those low in prejudice. However, Vidmar and Rokeach suggested

---

[1] The researchers found, however, that blacks in these police dramas tended to remain support characters, not as qualified as whites to head up their organizations.

that if many viewers did not see the program as satire, it would be just as reasonable to predict the opposite; that is, that the more frequent viewers would be the most highly prejudiced. Their data supported the latter view: The program seemed more appealing to those more highly prejudiced.

This and other studies (see, for example, Sarlin and Tate, 1976) suggest that television may only bolster ethnic stereotypes and attitudes already conveyed in other socialization settings. Though the electronic media are unarguably pervasive, their effectiveness in conveying the same messages to all viewers is by no means certain. Moreover, the extent and nature of television's effect on how children of different racial groups interact with each other, as well as how it affects racial attitudes and knowledge, are still debatable (Graves, 1999).

Social Distance

Prejudice involves not simply mental perceptions of ethnic groups (sociologists and psychologists refer to this as the cognitive dimension of prejudice) but also emotions and a preparedness to act in a certain way toward members of those groups (the affective and conative dimensions). If people believe blacks to be lazy and shiftless, for example, they will also probably feel resentment over welfare payments to blacks, who it is felt obviously do not deserve such assistance. Similar payments to poor whites, however, may go unnoticed or may be seen as merited.

The affective and conative dimensions of prejudice are reflections of what sociologists call *social distance*. Park (1924) first defined social distance as a degree of intimacy that people are prepared to establish in their relations with others. Feelings of social distance are, according to Williams, "feelings of unwillingness among members of a group to accept or approve a given degree of intimacy in interaction with a member of an outgroup" (1964:29). It is, in a sense, an indication of how acceptable or objectionable are various ethnic groups.

In 1925, sociologist Emory Bogardus constructed a technique to measure social distance between specific American ethnic groups. Subsequently, this social distance scale has commonly been used by sociologists as a general measure of ethnic prejudice. Respondents are asked to indicate whether they would accept a member of an ethnic out-group in varying social contexts, extending from very close encounters to very remote ones. Bogardus asked his respondents to indicate their willingness to interact with members of particular groups in the following situations:

- as close kin by marriage

- as fellow club members

- as neighbors

- as workers in my occupation

- as citizens of my country

- as visitors to my country

- as persons to be excluded from my country

Each situation represents a lower degree of social intimacy, and assigning an increasing numerical value to each allows a score to be computed for each ethnic group. American social distance scales compiled by Bogardus and others over several generations reveal the relative consistency of the group ranking from year to year. WASPs (white Anglo-Saxon Protestants) are at the top of the scale, followed by other northwestern European groups and, in descending order, southern and eastern Europeans, Jews, and various racial-ethnic groups (Bogardus, 1925a, 1925b, 1959; Owen et al., 1981).

Studies of social distance in non-American societies indicate the similar construction of a hierarchy (Bogardus, 1968; Lever, 1968; Pettigrew, 1960), but the basis of social distance may vary from one society to another. Whereas ethnic differences, particularly those with a physical basis, are the most significant criteria of social distance for Americans, in other societies, religion, class, or political ideology may be more important factors in separating people (Banton, 1967).

Feelings of social distance and ethnic prejudicial attitudes in general are difficult to accurately determine, for in real situations individuals do not always think or act as their verbal expressions seem to indicate. In short, attitudes are elusive and not always subject to clear-cut measurement, no matter how sophisticated the technique. What people tell pollsters, for example, may reflect only what they think are socially acceptable responses, not necessarily their real beliefs and feelings. That is, the pressure to conform to community and other group standards may force individuals to express attitudes that are not genuine. Moreover, the way in which poll questions are worded on racial and ethnic issues may strongly influence the results (Langer, 1989).

A study by Richard LaPiere (1934) demonstrated the unreliability of people's statements regarding ethnic prejudice. Between 1930 and 1932, LaPiere traveled throughout the United States with a Chinese couple and was refused hotel and dining service only one time. Shortly after his travels, he sent a questionnaire to hotels and restaurants asking whether they would accept Chinese as guests. To his surprise, LaPiere received mainly negative replies from those who responded. A similar finding was obtained in the early 1950s using a black couple as the test case (Kutner, Wilkens, and Yarrow, 1952).

In sum, prejudice is multidimensional, and the negative attitudes people maintain toward various ethnic groups vary not only in intensity but also in consistency. As we will see, changing situational contexts may force people to change their attitudes accordingly.

DISCRIMINATION

Whereas prejudice is the attitudinal element in enforcing ethnic stratification, discrimination is the active, or behavioral, element. *Discrimination* involves behavior aimed at denying members of particular ethnic groups equal access to societal rewards. Thus, discrimination goes well beyond merely thinking unfavorably about members of certain groups. Feagin and Feagin define discrimination as "actions or practices carried out by members of dominant groups, or their representatives, which have a differential and negative impact on members of subordinate groups" (1978:20–21). They further note that discrimination entails both effects and mechanisms. That is, there are impacts of discriminatory behavior, and there are modes of operation or techniques by which discrimination is carried out.

Although there are links between the two, prejudice and discrimination must be dealt with as distinct phenomena. There is ordinarily a tendency for prejudicial attitudes to accompany discriminatory behavior; but as we will see, one may be evident without the other. Moreover, although they may be causally related in some instances, in others there need be no cause-effect relationship between the two. In any case, as Simpson and Yinger (1972) note, prejudice and discrimination are most frequently mutually reinforcing.

It is important to stress that, like prejudice, discrimination is applied on the basis of group membership, not individual attributes. Antonovsky explains that discrimination is a situation in which "individuals are denied desired and expected rewards or opportunities for reasons related not to their capacities, merits, or behavior, but solely because of membership in an identifiable out-group" (1960:81). Or, as Pettigrew defines it, discrimination is "an institutional process of exclusion against an out-group, racial or cultural, based simply on who they are rather than on their knowledge or abilities" (1980:821).

The discriminatory actions that create disadvantages for minority group members may vary widely both in form and degree. These variations may be seen on a scale, or continuum (Figure 3.1). The use of derogatory labels in referring to members of ethnic groups (kike, wop, nigger) or phrases with pejorative ethnic references (to Jew down, to gyp, Indian giver) is a relatively minor form of ethnic discrimination; in many cases the language may not even be understood by the user as disparaging. Nonetheless, such terms and phrases contribute to the perpetuation of ethnic stereotypes and render psychological damage of some nature to those who are their subjects.

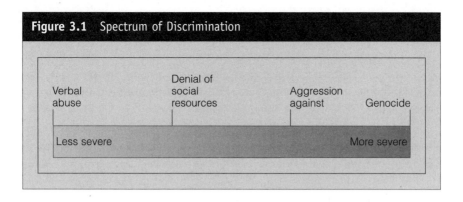

Figure 3.1 Spectrum of Discrimination

Verbal abuse

Denial of social resources

Aggression against

Genocide

Less severe

More severe

More serious forms of discrimination with much greater injury to minorities involve the denial of access to various life chances such as jobs, housing, health, education, justice, and political participation. Minority ethnic groups are placed in disadvantageous positions with regard to these societal rewards and end up receiving less than they would if no ethnic barriers were established. Indeed, the very notion of minority implies differential treatment in vital areas of social life.

The most severe forms of discrimination involve acts of aggression against minority ethnic groups, ranging from isolated incidents of violence to the deliberate destruction of an entire group. Examples of the full range abound in the modern world. In the United States, attacks on ethnic minorities have a long tradition, occurring as both selective actions undertaken by individuals or communities (such as lynchings, beatings, and bombings) and calculated public policies (such as Indian removal in the nineteenth century and Japanese-American internment in the 1940s). But the United States holds no monopoly on ethnic aggression. Comparable actions typify the history, current and past, of most multiethnic societies. Indeed, the nineteenth and twentieth centuries have witnessed extremes of ethnic violence and destruction on a massive scale in a variety of world areas including, among many others, the annihilation of native peoples in Australia, South Africa, and North America; the slaughter of over a million Armenians by Turks; the systematic murder by the Nazis of six million Jews[2]; and more recently, massacres of entire populations in Rwanda and Bosnia.

Discrimination is not always overt, nor does it always entail intentional actions of denial or aggression. Furthermore, there is a vast difference between isolated actions of individuals and the rational policies of institutions in creating

[2]The literature on the Nazi holocaust is enormous, but two of the more comprehensive works are Dawidowicz (1975) and Hilberg (1979). On theories and policies of genocide, see Kuper (1981) and Fein (1993).

and sustaining patterns of discrimination. The behavior of one café owner in refusing to serve members of a particular ethnic group is hardly the equivalent of a state policy requiring separate eating facilities for members of that group. The different social contexts in which discrimination may occur and the often concealed and unintentional forms it may take require a more precise outline of discrimination.

To simplify matters, we can classify two general types of discrimination: individual and institutional. Actions carried out by individuals or small groups constitute one type, and actions carried out as a result of the norms and structures of organizations and institutions constitute the other. The latter may be further broken down into direct and indirect types.

Individual Discrimination

Actions taken by individuals or groups of limited size to injure or deny members of minority ethnic groups are perhaps the most easily understood form of discrimination. The employment manager who refuses to hire Asians, the judge who metes out unusually harsh sentences to blacks, and the homeowners' group that agrees not to sell houses in the neighborhood to Jews are examples of discriminators at this level. In these cases, actions are taken by one or a few with the intent to harm or restrict in some way members of minority groups. Notice that the actors are not part of a large-scale organization or institution but operate within a relatively isolated context.

In cases of individual discrimination, the actions taken against minority group members are intentional. Moreover, they appear to be the implementation of prejudicial attitudes. At first glance we might assume that the employment manager thinks unfavorably of Asians, the judge of blacks, and the homeowners of Jews. This may in fact be the motivating force behind the discrimination in all these cases, but we cannot be certain until we understand more fully the context in which these actions occur. The employment manager, for example, may have no ill feeling toward Asians but may feel compelled to carry out what he perceives to be the unwritten yet generally understood company policy of not hiring Asians. The judge may feel that sentencing blacks more harshly will gain her votes among her predominantly white constituency in the next election. And the members of the homeowners' group may simply be responding to what they feel are neighborhood pressures to conform. Thus, prejudice need not be at the root of even such blatant instances of intentional discrimination. In any of these instances, of course, whether or not the actors' beliefs and attitudes are consistent with their actual behavior does not negate the detrimental effect on those who are the victims of the discriminatory actions: The Asian is still not hired; the black still serves a longer sentence; and the Jew is still denied a home.

Institutional Discrimination

The preceding cases pertain to the actions of individuals and small groups usually acting in violation of the society's norms. Discrimination, however, may be legal or customary, in which case it is not socially unexpected or disapproved but is legitimized. Such direct institutional discrimination is not limited to specific cases of negative actions taken against members of particular groups but is firmly incorporated in the society's normative system. Social conventions exist in which members of particular groups are legally or customarily denied equal access to various life chances.

In the United States before the 1960s, a well-institutionalized system of discrimination served to effectively block the access of blacks to the same economic, political, and social opportunities afforded whites. In the South, an elaborate system of custom and law legitimized segregated and unequal schools, housing, transportation, and public facilities; kept blacks in low-paying and less desirable jobs; and essentially prevented them from participating in the political process either as voters or officials. Most of this system was formally established and maintained by specific discriminatory laws, but much of it was also based on the development of customary practices.

South Africa is another case in which a formal and entrenched system of racial segregation and discrimination traditionally served to assure the power, wealth, and prestige of one group—whites—at the expense of another—nonwhites. As we will see in Chapter 13, this system, called apartheid, has been officially renounced but many of its essential features are still supported by customs that validate and enforce discriminatory policies and practices. Clearly, cases such as South Africa under apartheid are exceptional in the modern world. In most multiethnic societies, discrimination is more subtle, less obvious, and more indirect in application.

The form of discrimination most difficult to detect in modern multiethnic societies is indirect institutional discrimination, or structural discrimination. That it is so covert also makes it particularly difficult to eradicate. Unlike other forms of discrimination, it is unintended. It cannot be attributed to the prejudicial beliefs or conforming pressures of individuals, nor to the deliberate establishment of a set of rules seeking to withhold privileges or injure members of particular ethnic groups. Rather, it exists as a product of the normal functioning of the society's institutions. Because of past discrimination of an overt, intentional nature, or because of the spillover effect of direct discrimination in one institutional area into another, certain groups find themselves perpetually at a disadvantage in the society's opportunity and reward structures.

The structural form of discrimination is difficult to perceive because it does not use ethnicity as the subordinating mechanism, but uses other devices only indirectly related to ethnicity. This can be illustrated with a few current examples from the United States.

In recent decades, most new jobs have been created not in central cities, where they had been concentrated in the past, but in outlying and suburban areas, where shopping centers and factories are built on large expanses of land, where transportation lines, particularly highways, are more accessible, and where taxes are lower. The outlying location of these jobs, however, handicaps African Americans who might qualify for them but who reside mainly in central cities. Qualified African Americans are therefore less likely to secure these jobs, but for reasons that do not necessarily involve direct discriminatory practices by employers. Merchants may choose to locate their stores in the suburbs not because they wish to avoid hiring African Americans but simply because it is economically more advantageous. When combined with the more overt discriminatory practices in housing that create the concentration of African Americans in central cities, the outlying location of commerce and industry has the effect of discriminating against African Americans. In other words, the more overt discrimination in one institutional area—housing—has created an indirect discriminatory effect in another—employment.

Feagin and Feagin refer to this kind of situation as "side effect discrimination." As they explain, "Intentional discrimination by persons in one sphere can result in unintentional discrimination by those in another sphere, because most societal spheres (the economy, the polity, etc.) are intimately linked to one another. Discrimination has an inter-institutional character" (1978:32).

This interinstitutional nature of discrimination can also be seen clearly in the area of education. Given the poorer quality of inner-city schools, commonly attended primarily by African Americans, those who attend them are automatically placed at a disadvantage in qualifying for well-paying and promising jobs. Similarly, entrance into top-ranking colleges and universities will be more difficult for them because they will not be adequately prepared to meet the rigid academic requirements. In both cases, there is no necessary intention to discriminate. Decisions by employment managers or college admissions officers may be quite rational, made not on the basis of the applicants' skin color but on objective employment and academic standards. What is overlooked here is that African Americans are, as a group, less qualified than whites because inner-city public schools have inadequately prepared them for better jobs and colleges.

Such unintentional, yet effective, ethnic discrimination is repeated in a variety of areas. Bankers who hesitate to lend to African-American entrepreneurs because they fear they are poor credit risks may be simply engaging in practical business tactics. The grocery chain that charges higher prices in its inner-city stores than in its suburban stores will explain such practices as necessary to offset the higher operating costs in its central-city stores, due to higher rents or insurance costs. Similar patterns emerge in health services, administration of justice, and other key areas of social life.

Perhaps most significant is that such discriminatory practices are not only unintentional but also largely unconscious. As Baron notes, "The individual only has to conform to the operating norms of the organization and the institution will do the discriminating for him" (1969:143). Discrimination, in other words, does not depend in these cases on the actions of specific individuals or even organizations. Instead, it is simply a function of the normal operating procedures of societal institutions.

Because it is unintentional and largely unwitting, structural discrimination remains difficult to detect; and even when it does become apparent, it is not easy to determine who is ultimately responsible. Paradoxically, it is ordinarily carried out by individuals and groups who do not consider themselves discriminators. Thus, many who have purged their own behavior of discriminatory actions may come to feel that it is minority ethnic groups themselves who are mostly responsible for their subordinate status. The burden of responsibility for social problems, then, is placed on the individual or the group, not on patterns of discrimination that may be built into the institutional structure. Some have referred to this as "blaming the victim" (Ryan, 1975).

THEORIES OF PREJUDICE AND DISCRIMINATION

How do prejudice and discrimination arise, and why are they seemingly so inevitable in societies where diverse ethnic groups live side by side? Common-sense explanations, and even some scientific thought of earlier times, have accounted for out-group antipathy as a natural, or innate, pattern of human thought and action. The pervasiveness of conflict between different cultural and racial groups and the universality of ethnocentrism among peoples of the world seem to validate the notion that antagonism toward strangers or members of out-groups is simply part of the human psyche. To dislike or fear those who are different from us is, according to this view, both natural and unavoidable.

Today, however, there is general agreement among sociologists and psychologists that prejudice and discrimination are not innate human characteristics. A closer look at historical evidence indicates that people of different cultures and physical attributes do not always relate to each other antagonistically and that, when they do, the nature and intensity of their animus are highly variable. Given the variety of form, content, and extent of prejudice and discrimination, efforts to explain them as universal phenomena fall short. Further evidence that ethnic antagonism is not innate lies in social-psychological studies that have traced the various phases of socialization from early childhood through later stages of the life cycle.[3] All indicate that prejudice and discrimination are learned patterns of thought and action.

But if they agree that prejudice and discrimination are not inborn human traits, sociologists and psychologists differ in their explanations of how these thoughts and behaviors arise and are sustained in multiethnic societies. The psychological tradition emphasizes prejudice as the key component in ethnic interrelations; prejudicial thought, therefore, is the focus of most psychological theories. Sociologists have stressed situational factors and power structures as the bases of both negative thought and action toward ethnic groups; prejudice is thus seen not necessarily as the source of hostility but as its outgrowth. Whereas psychologists are likely to see prejudice and, by implication, discrimination as characteristic of certain personalities, sociologists tend to see them as part of the society's normative order, to which individuals are socialized. In the remainder of this chapter we look at several theories, both psychological and sociological. Because prejudice and discrimination are multifaceted, each theory may contribute to our understanding of these phenomena.

Psychological Theories

Psychological theories of prejudice have focused on the manner in which antipathy toward out-groups either satisfies certain psychic needs or complements the general personality structure of certain people. In each case, the source of prejudice is traced primarily to individuals rather than to the social forces weighing on them or the groups within which they interact.

Frustration-Aggression One of the earliest psychological theories explains prejudice as a means by which people express hostility arising from frustration. This has also been referred to as scapegoating. The essential idea is that a person who is *frustrated* in his or her efforts to achieve a highly desired goal tends to respond with a pattern of *aggression*. Because the real source of frustration is either unknown or too powerful to confront directly, a substitute is found on whom the aggression can be released. The substitute target is a scapegoat, a person or group close at hand and incapable of offering resistance. The aggressive behavior is thus displaced (Dollard et al., 1939). Minority groups in multiethnic societies have served as convenient and safe targets of such displaced aggression. Allport (1958) explains that racial, religious, or ethnic groups can be blamed unfairly for a variety of evils because they are permanent and stable and can be easily stereotyped. Therefore, they serve as more of an "all-purpose" scapegoat than those groups or individuals who are blamed for specific frustrations.

---

[3]For an explanation and discussion of these processes, see Allport (1958:271–321).

The frustration-aggression, or scapegoating, theory of prejudice seems at first glance convincing. It is, as Allport notes, easily understood because of "the commonness of the experience" (1958:330). All of us at times suffer frustrations of our needs and desires, whether they are simple ones like seeing our favorite football team lose a key game or significant ones like failing to gain a job promotion. When these events occur, we are sometimes inclined to strike out at substitute targets such as our spouses or children. If the frustration is continuous, we may begin to blame more remote groups or institutions like "the government," "bureaucrats," "blacks," "Jews," or "gays."

Despite the common experience of frustration-aggression, however, this theory leaves many questions unanswered. First, under what conditions will frustration *not* lead to aggression? Psychologists have shown that the frustration-aggression sequence is not necessarily inevitable (Allport, 1958; Ehrlich, 1973). Nor is aggression always displaced; sometimes it is directed inward, and at other times it may be thrust on the real source of frustration. This theory does not tell us how and why these responses may be produced.

Another obvious shortcoming of the frustration-aggression theory is its failure to explain the choice of scapegoats or targets of displaced aggression. Why are some groups chosen rather than others? The notion that scapegoats are always "safe goats"—defenseless and easily used—does not hold up when we consider that minority group members may themselves harbor great antipathy for the dominant group or for other minorities. A national survey in 1995, for example, documented the strong prejudices of American minority groups toward each other. It indicated, in fact, that ethnic minorities are more likely than whites to agree to negative stereotypes about other minority groups (National Conference, 1995).

Finally, we must consider the fact that the displacement of aggression on substitute targets can bring at best a very short-lived relief of an individual's anxiety. Unless the actual source is attacked, the feeling of frustration will quickly recur or even intensify. As Allport has put it, concerning frustration, "Nature never created a less adaptive mechanism than displacement" (1958:332).

The Authoritarian Personality The question of why prejudice exerts a strong influence on certain individuals but plays a relatively minor role for others is addressed by the theory of the authoritarian personality. The essence of this theory is that there is a personality type prone to prejudicial thought. In the same way, according to this view, there are basically tolerant personalities.

The authoritarian personality theory was developed after World War II by a group of social scientists who were determined to trace the psychological foundations of such destructive and regressive movements as Nazism. It subsequently became one of the most widely tested and debated ideas in the social sciences (see Kirscht and Dillehay, 1967). In their studies, T. W. Adorno and his associates (1950) found evidence to support the notion that prejudice and political extremism are more generally characteristic of a definite personality type. Such people, they maintained, are highly conformist, disciplinarian, cynical, intolerant, and preoccupied with power. They are particularly authority-oriented and are thus attracted to sociopolitical movements that require submission to a powerful leader. Such personality traits extend well beyond people's political beliefs and are reflected in all aspects of their social life. In the family, for instance, authoritarians will subject their children to strong disciplinary action; and in their religious beliefs, they will emphasize submission and obedience. In sum, such people strongly support conservative values and resist social change. They are thus more likely to display prejudicial thought and to discriminate when given the opportunity.

Other scholars, although not necessarily subscribing to the notion of an authoritarian personality, suggest that prejudice is a general way of thinking for some people. Hartley (1946), for example, found that when purely fictitious groups were presented in a social distance test, people prejudiced toward other groups tended to express prejudice toward the fictitious groups as well. Allport also maintains that "the cognitive processes of prejudiced people are in general different from the cognitive processes of tolerant people" (1958:170).

In any case, theories suggesting that certain personality types are generally prone to prejudice and discrimination suffer several critical shortcomings. Like the frustration-aggression theory, they fail to tell us how ethnic prejudice and discrimination arise in the first place. The authoritarian personality theory has fallen out of use and is regarded as flawed because it reduces authoritarianism to a personality trait. We see the prejudiced person in action, but not the social conditions that create *ethnic* hostility rather than other forms (Duckitt, 1989).

Moreover, the emphasis of this genre of studies has been on patently and intensely prejudiced people, such as members of the Ku Klux Klan. Most prejudice and discrimination, however, are more subtle and less intense and are characteristic of people who cannot be categorized as extremists. Indeed, in some societies prejudice can be consensual, evident among an entire cross section of individuals with different personality traits (Brown, 1995). For example, studies have shown that the Nazi movement had appeal across a wide social spectrum of German society (Gerth, 1940; Peukert, 1987). Undoubtedly, prejudice and discrimination play a vital part in the thought and actions of many people; but it is necessary to carefully delineate the varying degrees and forms of ethnic antipathy displayed by different individuals.

The chief criticism of psychological theories in general is that the situations within which people think and act are not given sufficient attention as variables that fundamentally affect the nature of that thought and action. Schermerhorn cautions us to consider that "if research has confirmed anything in this area, it is that prejudice is a product of *situations*, historical situations, economic situations, political situations; it is not a little demon that emerges in people simply because they are depraved" (1970:6).

However, psychological theories such as frustration-aggression and the authoritarian personality have been popular not only because they are more easily understood than sociological theories emphasizing structural conditions, but precisely because they focus the blame for ethnic antagonisms on disturbed individuals—those who are pathological or overtly irrational in behavior and thought. This focus deflects attention from the society's normally functioning institutions, which may compel people to think and act negatively toward members of particular ethnic groups. Indeed, as will be seen shortly, prejudice and discrimination are in most cases *conforming* thoughts and actions, not those of a few maladjusted persons. As long as ethnic antipathy is thought to be characteristic only of the sick few rather than a proper response to the expectations of the community or society as a whole, it can be seen as eradicable simply by treating those few, not by painfully reexamining established societal institutions.

Whatever their effect on prejudice and discrimination, psychological components must be seen in conjunction with political and economic structures out of which intergroup relations develop and are sustained. William Wilson asserts that psychological explanations of prejudice and discrimination "prove to have little predictive value when social factors are taken into account" (1973:39). These social factors are the basis of normative theories, to which we now turn.

Normative Theories

Why do we often compliment a friend on his or her new hairdo or clothes when in fact we think they are quite unattractive? Similarly, why do we many times feel obliged to contribute to a class discussion or a business meeting when we really have nothing meaningful to say? Sociologists explain such apparently inconsistent thought and action as a product of situational norms by which we feel compelled to abide. We understand that such actions are expected of us, and in most cases we conform to those expectations even when we have a real desire to ignore or disobey them. *Norms* are group standards that define how people are expected to act in particular social situations. There are positive sanctions for conforming to and negative sanctions for deviating from them. Because there are norms pertaining to all social situations in which we find ourselves, these social "rules" enable us to predict others' behavior, and in doing so they facilitate interaction. In a real sense, they provide the society with order.

Prejudice and discrimination can be explained within the framework of social norms. Rather than the thoughts and actions of a deviant few, they are conforming responses to social situations in which people find themselves. When negative thoughts about particular ethnic groups and discriminatory behavior toward them are expected, individuals will feel compelled to think and act accordingly. Thus it is to individuals' social environment—the groups they belong to, the cultural and political norms operative in their society and community, and the processes of socialization—that prejudice and discrimination can be traced. Obviously, such explanations are considerably different from psychological theories, which focus not on the group contexts of individual thought and action but on the individuals themselves.[4]

In this view, bigots emerge out of the social experiences to which they are exposed. Frank Westie has succinctly explained the essence of normative theory: "Individuals are prejudiced because they are raised in societies which have prejudice as a facet of the normative system of their culture. Prejudice is built into the culture in the form of normative precepts—that is, notions of 'ought to be'—which define the ways in which members of the group ought to behave in relation to the members of selected outgroups" (1964:583–84). Normative theories thus concentrate primarily on the transmission of ethnic prejudices through the socialization process and on the social situations that compel discriminatory behavior.

Socialization Prejudice and discrimination can be seen as part of a society's social code, which is passed down from generation to generation. Fear of, dislike for, and antipathy toward one group or another are learned in much the same way that people learn to eat with a knife and fork rather than with their bare hands or to respect others' privacy in personal matters. These standards of behavior are the product of learning processes of which we ordinarily have little cognizance. Socialization is subtle and works in a largely unconscious manner. Thus, prejudice and discrimination need not be taught directly and intentionally. If these are the norms and values of the society or community within which the individual interacts, the chances are very great that they will be adopted with little specific instruction. As Westie puts it, "The child soon learns which people are 'good' and which are 'not so good' in the eyes of his society" (1964:583).

Parents are sometimes puzzled by certain expressions and attitudes of their children because they are sure they did not impart these. In their puzzlement, they discount the informal and often undetected ways in which the society's culture is transmitted by the various agents of socialization outside the family. Most learning is accomplished not through direct teaching methods but through observation and imitation. Children—and adults—take cues from their peers and other important reference groups as well as the mass media. Social psychologists have shown that among American children 4 years of age, ethnic values and attitudes are already beginning to crystallize (Aboud, 1987; Goodman, 1964;

---

[4]Dean and Rosen (1955) and Williams (1964) cite many studies showing that people's behavior toward different ethnic groups is dictated by social circumstances rather than their personal or psychodynamic makeup.

Porter, 1971; Ramsey, 1987). By this age, children have been exposed to the society's ethnic hierarchy, particularly the white-over-black element.

Prejudice and discrimination, then, are no more indicators of a defective personality than one's taste in food or fashion. They are simply products of socialization. If prejudice and discrimination are pervasive in the society or community, the more logical question may not be, "Why do some people display prejudice and discrimination?" but rather, "Why do some people *fail* to display prejudice and discrimination?"

Consider South Africa under apartheid or the pre-1960s American South. Using the normative approach, white prejudice and discrimination against blacks in those settings can be explained not as the product of deviant individuals but as the natural outgrowth of a whole system of racist norms, learned early and thoroughly, that guided people's actions and attitudes. In that system, blacks were not to be thought of or treated in the same way as whites. For whites to avoid almost any contact with blacks beyond the most purely functional (supervising them in a work situation, for example) was correct, expected, and positively sanctioned by societal norms. For whites to deviate from such behavior would have been unusual and responded to with negative sanctions. John Stone (1973), in a study of British immigrants to South Africa in the 1970s, discovered that they frequently changed their attitude toward the segregationist policy of apartheid. Before leaving Britain, the majority was either opposed to or had no opinion about it; after living in South Africa for a time, however, an even larger majority stated that it favored the policy. Stone concluded that the change reflected not the manifestation of latent racist personalities but the need of immigrants to adapt to the ways of their new society: "We are not witnessing the mass attraction of bigoted racialists to a segregationalist's dream, rather we are observing how ordinary people, confronted by a particular social structure, will tend to conform to the attitudes, values, and norms implicit in it" (1973:253).

Reference Groups Even where societal norms dictate fairness toward different groups, prejudice and discrimination may typify the behavior of some people who have been exposed to reference groups that strongly prescribe such behavior. Reference groups are those that provide individuals with standards by which they shape their own patterns of action and from which they adopt important beliefs and values. In a sense, they serve as models of thought and action. We ordinarily think of the family as a reference group; but there are many others to which we may look for behavioral guidance, even some of which we are not members. In the latter case, we may aspire to membership and thus take on the ways and attitudes of the group. This is most apparent in the early stages of socialization, when children begin to identify with particular occupational groups—firefighters, doctors, nurses, and the like—or with sports and entertainment groups they admire. As might be expected, studies have shown that individuals tend to adopt beliefs and values congruent with those of the groups with which they identify.

Applying the concept of the reference group to prejudice and discrimination, we see such thoughts and actions as normal responses of individuals when called for by their reference groups. No one is immune to the pressures to conform—applied by family, friendship cliques, or other significant groups. The fear of group rejection is constantly present and serves as an effective disciplinary mechanism. Again, this process is subtle, and the individual may not see such conformity as a response to external coercion. As Vander Zanden explains, "Through conformity individuals achieve pride, self-identity, a sense of security—products of belonging to a group. It is *their* group, *their* norms" (1983:104).

If the person's reference groups change, attitudes and actions can be expected to change accordingly. College students, for example, often face a challenge to their well-formed values when they encounter new ideas from their instructors and classmates. These new ideas often involve social issues like ethnic prejudice. Pearlin's (1954) study at a southern women's college demonstrated the effect of a change in reference groups on students' racial attitudes. Pearlin started with the assumption that white students would find in the college environment sentiments more favorable to blacks than they had found in their precollege experiences. But merely being exposed to new and positive ideas about blacks, he believed, would not in itself reduce prejudicial attitudes. Rather, modification of attitudes was believed to depend more on changes in social relationships. Thus, attitudinal changes were likely to come about only if the students began to identify with new groups holding those favorable attitudes toward blacks. Pearlin's hypotheses were confirmed: Those most prejudiced toward blacks remained most strongly affiliated with their precollege membership groups, and those least prejudiced experienced a weakening of such ties and an increasing identification with their new college groups. In short, Pearlin's findings showed that people tend to take the attitudes of those groups with which they most strongly identify.

Merton's Paradigm Given the social, rather than the personality, origins of ethnic prejudice and discrimination, changing social situations can produce fluctuations in individual thought and behavior. Ethnic prejudice and discrimination are thus not constant and unchanging but variable, depending on a number of situational factors: the person's definition of the situation, the compulsion to conform to societal and reference group norms, and the rewards—economic, prestige, political—to be gained by acting and believing in such a manner. That attitudes and actions toward members of particular ethnic groups may fluctuate within different social contexts is demonstrated by sociologist Robert Merton (1949) in a well-known paradigm (Figure 3.2).

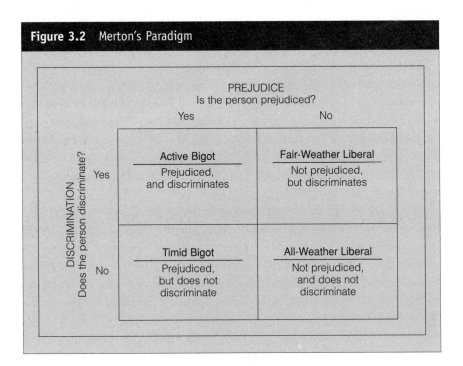

Figure 3.2 Merton's Paradigm

PREJUDICE
Is the person prejudiced?

		Yes	No
DISCRIMINATION Does the person discriminate?	Yes	**Active Bigot** Prejudiced, and discriminates	**Fair-Weather Liberal** Not prejudiced, but discriminates
	No	**Timid Bigot** Prejudiced, but does not discriminate	**All-Weather Liberal** Not prejudiced, and does not discriminate

By combining the prejudicial attitudes or lack of such attitudes with the propensity either to engage in discriminatory actions or to refrain from them, Merton suggested four ideal types. First, he denoted unprejudiced nondiscriminators, whom he called "all-weather liberals." These are people who accept the idea of social equality and refrain from discriminating against ethnic minorities. Their behavior and attitudes are thus consistent. A second type, also consistent in behavior and attitude, is prejudiced discriminators, whom Merton labeled "active bigots." Such people do not hesitate to turn their prejudicial beliefs into discriminatory behavior when the opportunity arises. Members of organizations such as the Ku Klux Klan or neo-Nazi parties in the United States or the National Party in Britain exemplify such people. Both of these types, consistent as they are in belief and behavior, might indicate by themselves support for the psychological perspective; that is, there are prejudiced people or tolerant people, who may be expected to act accordingly.

Merton's third and fourth types, however, demonstrate the situational context and the effect it may have on people's behavior. In these cases, behavior and attitude are not consistent. Prejudiced nondiscriminators, or "timid bigots," as Merton called them, maintain negative beliefs and stereotypes toward ethnic minorities but are precluded from acting out those beliefs by situational norms. If a situation requires fair treatment toward ethnic groups who are viewed negatively by such people, fair treatment will mark their behavior. For example, whites traveling from the American South to northern states before the 1960s would find that the laws and customs of the North required them to interact with blacks in a manner unheard of in their home states. Lewis Killian described how white working-class southerners who had migrated to Chicago responded to blacks in their new environment: "The 'hillbillies' constantly praised the southern pattern of racial segregation and deplored the fact that Negroes were 'taking over Chicago.' In most of their behavior, however, they made a peaceful, if reluctant, accommodation to northern urban patterns" (1953:68). The hotel and restaurant keepers encountered by LaPiere in his previously cited study would also fall into this category.

Unprejudiced discriminators, whom Merton called "fair-weather liberals," also adjust their behavior to meet the demands of particular circumstances. When discrimination is normative in the group or community, such people abide by those patterns of behavior even though they may harbor no prejudicial feelings toward members of the targeted group. To do otherwise would jeopardize their social standing and might even constitute violations of the law. Such a situation was faced by many whites in the pre-1960s American South who did not share the racial animosity of their neighbors. In a 1987 case, the owner of a pharmacy in Tifton, Georgia, dismissed a black student pharmacist who had been placed in the store as part of her training at the University of Georgia's College of Pharmacy. She was dismissed, the owner explained, because he feared negative customer reaction (*New York Times*, 1987).

It must be remembered that each of Merton's four cases is an ideal type and thus does not reflect perfectly any individual's behavior and attitudes. More realistically, people can be expected to display higher or lower degrees of each. We should also remember that prejudice and discrimination directed at one ethnic group do not necessarily imply the same attitudes and behavior toward others. Those who are antiblack are not necessarily anti-Catholic, and so on. One may be a fair-weather liberal in one instance and a timid bigot in another.

Situational explanations of prejudice and discrimination like Merton's demonstrate that there is no necessary causal relationship between the two. Common thought has generally assumed that prejudice leads to or causes discrimination. Abundant sociological evidence, however, has shown not only that this sequence need not occur but also that the very opposite is more usual (Pettigrew, 1979; Raab and Lipset, 1971). "What we call prejudices," writes anthropologist Marvin Harris, "are merely the rationalizations which we acquire in order to prove to ourselves that the human beings whom we harm are not worthy of better treatment" (1964:68). Prejudice, then, is used to rationalize discriminatory behavior *after* the fact.

This is an important observation, for it seriously challenges the idea that eliminating discrimination requires a change in attitude, that is, the elimination of prejudice. This was the generally shared opinion of both scholars and policymakers in the United States before the 1960s. Reeducating people was therefore the most frequently proposed remedy for alleviating ethnic hostility. It was thought that if people's faulty ideas about race and ethnicity could be corrected, they would, as a result, be induced to change their behavior. In the past few decades, however, it has become obvious that prejudicial attitudes may have little or no bearing on whether people discriminate against particular groups. Instead, people appear to be motivated to change their behavior toward ethnic groups by laws and other social mechanisms that seriously alter their social situation vis-à-vis those groups. After such situational changes occur, individuals seem to adjust their ideas to fit these new modes of behavior (Ehrlich, 1973; Pettigrew, 1980). Hence, changes in ethnic relations must be impelled not by trying to change attitudes but by changing the structure of those relations.

Van den Berghe (1997) has suggested a model of discrimination akin to Merton's, that links such behavior to stereotyping. He posits that stereotypes are most often used in a rational way, enabling people to make quick decisions about others in interactive situations where they have minimum information. "Statistical discrimination," in which people are treated negatively on the basis of beliefs about the category of which they are members, is distinct from "categorical discrimination," in which people are treated negatively simply on the basis of being part of a socially assigned category. The former is rational in that it is responsive to counterevidence, while the latter form of discrimination is not. Van den Berghe contends that statistical discrimination is more common than categorical discrimination because it is based on self-interest, whereas the latter is nonrational. Hence, stereotypes are not to be seen as evidence necessarily of prejudice, but are simply "guidelines in making statistical discriminations in situations of imperfect information" (5). Where information is in short supply and costly, people will rely on stereotypes.

For example, because blacks appear to commit a disproportionate percentage of violent crimes, whether due to a racist criminal justice system or to class bias, whites stereotype blacks as "criminally inclined" and thus may discriminate against blacks with whom they interact. In reality, the relationship of crime to such factors as social class and age is stronger than the relationship of crime to race. But most whites, living as they do apart from blacks, are not as attuned to class differences among blacks. It is, then, less costly to discriminate on the basis of race—which is more readily apparent—than class. Banks may make decisions about loans using much the same calculus based on racial criteria, as a result granting fewer loans to blacks or charging them higher interest rates. Again, the decision is a rational one: "Money-lenders are in the business of evaluating risk as cheaply as possible and of adjusting interest rates in direct proportion to assessed risk. . . . [F]rom the perspective of the bank ignoring race would be a costly mistake, unless the bank can develop a more valid and equally cheap discriminator" (12). Racial profiling, whereby police select someone for investigation or stronger action on the basis of race or ethnicity, may be seen in the same way.

Like Merton, van den Berghe demonstrates how discrimination is not necessarily motivated by prejudice. "Modern industrial societies," he notes, "are hot-beds of stereotypy, not because they are populated by bigots, but because people have little else to structure their relationships" (13). Where race and ethnicity create barriers to communication and where members of different groups live in segregated areas, the chances of changing the criteria of statistical discrimination are reduced.

Power-Conflict Theories

Though the normative theories of prejudice and discrimination appear to go well beyond the earlier psychological theories, both approaches are deficient in one important regard: They basically explain only how these social phenomena are transmitted and sustained, not how or why they arise in the first place. To begin to understand the origins of ethnic antagonism, we must turn to power-conflict theories.

Most simply, these theories view prejudice and discrimination as emerging from historical instances of intergroup conflict (Bernard, 1951; Newman, 1973). In this view, discrimination serves as a means of injuring or neutralizing out-groups that the dominant group perceives as threatening to its position of power and privilege. Negative beliefs and stereotypes, in turn, become basic components of the dominant group's ideology, which justifies differential treatment of minority ethnic groups. When prejudice and discrimination are combined, they function to protect and enhance dominant group interests. And once established, prejudice and discrimination are used as power resources that can be tapped as new conflict situations demand.

In short, prejudice and discrimination, in this view, are products of group interests and are used to protect and enhance those interests. To understand negative ethnic beliefs and behavior thus requires a focus not on individual personalities or even on the constraints and demands of different social situations but on the economic, political, and social competition among groups in a multiethnic society.

Economic Gain Chief among power-conflict theories are those emphasizing the economic benefits that derive from prejudice and discrimination. Simply put, prejudice and discrimination, in this view, yield profits for those who engage in them. Different groups may be targeted because they present—or are perceived as presenting—a threat to the economic position of the dominant group. In addition to the economic advantages emanating from such thought and behavior, benefits may also take the form of enhanced prestige or political gain.

Colonial and slave systems, buttressed by elaborate racist ideologies, are obvious cases in which economic benefits accrued to a dominant group from the exploitation, both physical and mental, of minority groups. We need not look at such historically distant examples, however, to understand the relation between ethnic antagonism and economic gain. In his study of black-white relations in the American South of the 1930s, John Dollard showed that in every sphere of social life—work, health, justice, education—the white middle class realized substantial gains from the subordination of blacks. In exploiting blacks, explained Dollard, southern whites were simply "acting as they have to act in the position within the social labor structure which they hold, that is, competing as hard as they can for maximum returns" (1937:115). Later studies (Glenn, 1963, 1966; Thurow, 1969) concluded that prejudice and discrimination against blacks in the United States continued to benefit at least some segments of the white population.

The strong resistance to school busing, residential desegregation, and affirmative action during the past three decades can be interpreted as the negative reaction of whites who see their economic gains threatened by blacks and other minorities seeking upward social mobility. Similar reactions can be expected whenever economic advantages appear to be challenged by lower-ranking or more recently arrived ethnic groups. In 1981, for example, white fishermen in the shrimping grounds in Galveston Bay, Texas, encountered competition from immigrant Vietnamese fishermen. About 100 Vietnamese shrimpers had come to the area during the previous two years, challenging the economic dominance of the whites. Although the situation was eventually resolved, for several months white fishermen, with support from the Ku Klux Klan, engaged in acts of intimidation against the Vietnamese, including physical attacks and arson.

Marxian Theory Class theorists, in the tradition of Marx, have conventionally held that in capitalist societies, ethnic antagonism serves the interests of the capitalist class—those who own and control the means of economic production—by keeping the working class sufficiently fragmented and thus easier to control. The basic idea is "divide and rule." One ethnic element of the working class is pitted against another, and as long as this internal discord can be maintained, the chances that the working class will unite in opposition to the interests of the capitalists are reduced. Capitalists are able to foster ethnic division and ethnic consciousness among the workers, thereby curtailing the development of worker solidarity and class consciousness.

In the United States, for example, the conflict between black and white workers has been construed by Marxists as deflecting attention from the common anticapitalist interests of both groups (Allen, 1970; Cox, 1948; Reich, 1978; Szymanski, 1976). Ethnic prejudice, therefore, is viewed as a means of sustaining a system of economic exploitation, the benefits of which accrue to the capitalist class. Oliver C. Cox described prejudice as "the social-attitudinal concomitant of the racial-exploitative practice of a ruling class in a capitalistic society" (1948:475). Though capitalists may not consciously conspire to create and maintain racist institutions, they nonetheless reap the benefits of racist practices and therefore do not seek to completely dismantle them.

The Split Labor Market Theory Whereas conventional Marxist thought holds that the profits of ethnic hostilities redound primarily to the owners of capital, others maintain that it is workers of the dominant ethnic group who are the chief beneficiaries of prejudice and discrimination. If ethnic minorities are kept out of desired occupations, the favored workers, rather than the capitalists, are viewed as gaining the most from discriminatory institutions. This is the crux of sociologist Edna Bonacich's split labor market theory (1972, 1976).

According to Bonacich, there are three key groups in a capitalist market: businesspeople (employers), higher-paid labor, and cheap labor. One group of workers controls certain jobs exclusively and gets paid at one scale, and the other group is confined to jobs paid at a lower rate. Given the imperatives of a capitalist system, employers seek to hire workers at the lowest possible wage and therefore turn to the lower-paid sector when possible as a means of maximizing profits. Recent immigrants or ethnic groups migrating from rural areas in search of industrial jobs ordinarily make up this source of cheap labor. These groups can be used by employers as strikebreakers and as an abundant labor supply to keep wages artificially low. Because these groups represent a collective threat to their jobs and wages, workers of the dominant ethnic group become the force behind hostile and exclusionary movements aimed at curtailing the source of cheap labor. Thus wage differentials do not arise through the efforts of capitalists to prevent working-class unity by favoring one group over another, but through the efforts of higher-paid laborers to prevent lower-paid workers—mainly lower-ranking ethnic groups—from undercutting their wages and jobs. This goal is effected by various forms of prejudice and discrimination.

The split labor market theory is supported by historical evidence in American society. Successive waves of European immigrants during the nineteenth and early twentieth centuries traditionally served as a source of cheap labor and became the targets of nativist movements, usually backed strongly by labor unions. Following the cessation of European immigration, northward-migrating blacks from the rural South assumed a similar role, touching off periodic racial violence in many cities. Depending on how threatening they were perceived by native workers, various groups at different times were the objects of worker-inspired hostility. For example, efforts in the nineteenth century to restrict the Chinese to particular occupations and to limit their immigration was spurred largely by white workers fearing a deluge of cheaper labor. Lyman notes that after 1850, "It was the leadership of the labor movement that provided the most outrageous rhetoric, vicious accusations, and pejorative demagoguery for the American Sinophobic movement" (1974:70).

Makabe (1981) studied Japanese immigrants in Brazil and Canada and found support for the split labor market theory. Japanese immigrants entering Canada, specifically British Columbia, in the pre–World War II years experienced an extremely harsh reception from native Canadians. Makabe explains that among white workers the rejection of the Japanese was unusually cruel. This is accounted for by the fact that the Japanese entered the Canadian economy at the bottom, enabling employers to pay them lower wages and thereby undercutting the more highly paid native workers. Striving for upward mobility, Japanese found themselves in direct competition with those immediately above them in economic position—white workers. The result was discrimination against the Japanese in the workforce and pressure to halt all Japanese immigration. In Brazil, however, the Japanese experienced a significantly different situation. Rather than entering the labor force in competition with higher-paid workers, they found themselves with skills and financial resources superior to those of most native workers, who themselves were mostly severely disadvantaged former slaves. Little competition and, hence, little conflict arose between them because they did not seek similar occupational positions.

Illustrations of the notion that prejudice and discrimination arise out of economic competition can also be seen at the higher levels of the class hierarchy. Williams (1964) found the greatest frequency of prejudice against Jews among people of higher occupational rank. He found this compatible with the supposition that prejudice reflects a sense of threat. Upper-class Gentiles are likely to see Jews as more threatening to their economic position than other minorities less affluent and less educated. McWilliams (1948), too, concluded that the development of strong anti-Semitism in the late nineteenth century in the United States was largely a creation of the wealthy—specifically, the industrial magnates of the day, who used prejudice and discrimination against Jews as a "mask for privilege." Anti-Semitism will be discussed in some detail in Chapter 8.

Status Gains In addition to economic benefits, status privileges may derive from ethnic antagonism. People may enjoy more prestige simply from being a member of the dominant group, regardless of their social class. In the American South, working- and lower-class whites could traditionally take comfort in knowing that they were part of the dominant ethnic group even though economically they were in much the same position as blacks. As Dollard described it, white subordination of blacks consisted of "the fact that a member of the white caste has an automatic right to demand forms of behavior from Negroes which serve to increase his own self esteem" (1937:174). Van den Berghe described the same well-understood racial etiquette in South Africa: "Non-Europeans are expected to show subservience and self-deprecation, and to extend to the whites the titles of 'Sir,'

'Madam,' or 'baas.' The Europeans, as a rule, refuse to extend the use of titles and other forms of elementary courtesy to nonwhites, and call the latter by first names (real or fictitious), or by the terms 'boy' and 'girl'" (1967:142).

Political Gain Prejudice and discrimination directed at certain ethnic groups may also be the result of political circumstances, regardless of the economic or status relationship between dominant and minority groups. In the United States, ethnic antagonism has often been tied closely to domestic and foreign political issues. Anti-German feelings were extremely high before and during World War I, for example, with charges of disloyalty leveled at German-American immigrants and even those of the second generation. This case demonstrates how prejudicial beliefs and discriminatory behavior can change both rapidly and radically. Before 1915, when the anti-German campaign began, Germans had been considered the most assimilable and reputable of all immigrant groups and were repeatedly praised for their law-abiding and patriotic behavior (Higham, 1963).

The political basis of negative feelings and actions toward members of particular ethnic groups is well illustrated by the anti-Arab sentiments expressed by Americans in recent years. Starting in the 1970s, a series of key political events—namely, the rise of the Organization of Petroleum Exporting Countries (OPEC) cartel, the activities of Palestine Liberation Organization (PLO) terrorists, and the holding of American hostages in Iran—created a generally anti-Arab public mood and, in the case of Iranians, hastily conceived government policies to expel those living or studying in the United States.[5] The development of negative Arab stereotypes was now evident, with the popular media portraying Arabs as people given to ostentatious displays of wealth or to political fanaticism (see Said, 1981; Shaheen, 1984). Arab Americans, though directly involved in none of these events, were nonetheless affected by the adverse feelings they aroused. The Persian Gulf War and the bombing of the World Trade Center in New York in 1993 reinforced anti-Arab sentiment and negative stereotypes. When the federal building in Oklahoma City was bombed in 1995, the perpetrators were assumed almost immediately by the media and public officials to be Arab terrorists. Although this assumption quickly proved false, it exposed how firmly this stereotype is applied to Arabs. The terrorist attack on the World Trade Center in New York in 2001 bolstered the negative stereotypes of Arab Americans more firmly than ever.

Political leaders have long recognized the value of exploiting ethnic divisions for attaining and enhancing their power. For example, until blacks became an electoral factor of some significance in the 1970s, racist politicians in the

---

[5]Though technically not part of the Arab world, Iran was generally linked with the Arab states because of its Muslim heritage, its location in the Middle East, and its political involvement with the Arab nations.

American South effectively manipulated white fears of blacks to their own ends. After losing the Alabama gubernatorial election in 1958 to a candidate even more avowedly racist than himself, George Wallace declared that he would not be "out-nigguhed again" (Frady, 1968). In the 1988 presidential campaign, George Bush (Sr.) used the case of Willie Horton, a convicted murderer who raped a woman while on furlough from a Massachusetts prison, to portray his Democratic opponent, Michael Dukakis, as soft on crime. As a black man, Horton's image was intended to elicit white fears of black crime. Similarly, in recent years, rightist politicians in France, Germany, and other western European countries have stirred anti-immigrant feelings, particularly against Muslims from North Africa and the Middle East, in appealing to voters.

Power-Conflict Theories in Sum In all these cases, prejudicial beliefs and discriminatory actions are used by elements of the dominant ethnic group, sometimes directly and other times indirectly, to secure its power and privileges. Herbert Blumer (1958) has pointed out that prejudice is always a protective device used by the dominant group in a multiethnic society in assuring its majority position. When that group position is challenged, prejudice is aroused and directed at the group perceived to be threatening.

Wilson (1973) also notes that when the system of ethnic stratification is challenged, that is, when minority groups no longer accept their group position, strong prejudices founded on a racist ideology emerge. Through this ideology, members of the dominant group can, as Wilson explains, "claim that they are in a superior position because they are naturally superior, that subordinate members do not possess qualities enabling them to compete on equal terms" (1973:43). The dominant ideology, incorporating key negative stereotypes of minority ethnic groups, thus reinforces the sense of group position, aids in maintaining patterns of subordination, and serves as a philosophical justification for exploitation.

This does not mean that in all multiethnic societies the nature and intensity of prejudice will be the same. It may differ, depending on such factors as the size of the targeted group, its cultural and physical characteristics, how much of a threat it is perceived to be, the ideology that has grown up around it, and the historical circumstances out of which its position in the ethnic hierarchy has been determined. Discrimination, too, as we have seen, can vary in degree and form. In some cases it may consist of little more than name-calling or the telling of ethnic jokes; but in others, the negative actions may be more serious, including the denial of various life chances to members of certain groups and even the physical destruction of human life and property.

Paradoxically, prejudice and discrimination may serve certain functions for minority groups themselves. Sociologists have recognized that conflict between groups has a unifying effect on the members of each. External threats tend to strengthen group ties and create a sense of solidarity that might not otherwise exist (Coser, 1956). The ability of Jews to survive in various societies in which they were persecuted, for example, has often been attributed to the continued anti-Semitic hostility itself. As constant targets of antagonism, Jews have strengthened their resolve to maintain a group identity and cohesiveness.

Continued prejudice and discrimination directed at a minority group may also contribute to a sense of psychological security for its members. Even though their place is at the bottom, they may take consolation in the certainty and predictability of their social relationships with the dominant group (Levin and Levin, 1982). Moreover, the minority individual's self-esteem may be protected by attributing personal failures to abstract notions like "the system" or "racism," rather than to individual shortcomings.

Prejudice may also serve as a release of frustration for minorities, just as it may for those of the dominant group. Indeed, prejudice should not be seen as characteristic only of dominant groups. Although sociologists have been reluctant to deal with it, prejudice is commonly displayed by minority groups as well, not only toward the dominant group—which seems entirely logical—but also toward other minority groups. Recall the strong antipathy toward each other expressed by American minority ethnic groups, mentioned earlier.[6] If prejudice is normative in the society, minority group members socialized to those norms will be affected in much the same way as members of the dominant group.

The benefits to minority ethnic groups that derive from prejudice and discrimination, however, should not be overdrawn. Clearly, the primary beneficiaries of ethnic antagonism are members of the dominant group.[7]

Theories of Prejudice and Discrimination: An Assessment

As we have now seen, the explanation for prejudice and discrimination in multiethnic societies is complex and by no means agreed on by theorists and researchers. It may very well be that a full investigation of these phenomena requires a multidimensional approach using different aspects of psychological, normative, and power-conflict theories. All may have some validity, depending on which aspect or level of ethnic antagonism is focused on.

---

[6]Westie (1964) asserts that the prejudice of minority group members has been largely ignored by social scientists. He suggests that this may be a result of the sympathy social scientists usually display for social underdogs. Moreover, interethnic conflict is usually perpetrated by members of the dominant group, and the prejudices of minority group members are seen mainly as responses to these actions. Westie maintains that, however well intended this view may be, it has produced social science literature "which gives the impression that the minority person can 'do no wrong'" (1964:605).

[7]There are also certain negative effects of prejudice and discrimination on the dominant group. See, for example, Bowser and Hunt (1981).

In this book, power-conflict theories of prejudice and discrimination are favored because the structural rather than the psychological or small-group dynamics of race and ethnic relations are emphasized. At this level of intergroup analysis, psychological and normative theories are less appropriate. In Parts II and III, therefore, the analysis of prejudice and discrimination in the United States and other multiethnic societies will view these ethnic attitudes and actions as tools of dominance, developed and used by one group over others in competition for the society's resources.

Although power-conflict theories are stressed, it is important to bear in mind that prejudice and discrimination are multifaceted, and therefore other theories cannot be disregarded. Power-conflict theories will not entirely explain, on the one hand, why some people will not discriminate even when it is profitable to do so or, on the other hand, why some will continue to discriminate when it is no longer beneficial. For such cases, psychological or normative theories may offer additional insight. As Simpson and Yinger point out, not all prejudice and discrimination can be explained by structural variables alone; individuals' responses to group influences are conditioned by their personality, and vice versa. Therefore, "The task is to discover how much of the variance in prejudice and discrimination can be explained by attention to personality variables, how much by social structural variables, and how much by their interaction" (1972:29).

SUMMARY

Prejudice and discrimination are techniques of ethnic dominance. Prejudice is the attitudinal dimension of ethnic antagonism. Prejudices are categorical, inflexible, and negative attitudes toward ethnic groups, based on simplistic and exaggerated group images called stereotypes. Discrimination is the behavioral dimension and involves actions designed to sustain ethnic inequality. Discrimination takes various forms, ranging from derogation to physical attack and even extermination. Two types of discrimination are individual and institutional: The former is carried out by single persons or small groups, usually in a deliberate manner; the latter is rendered as a result of the norms and structures of organizations and institutions, often in an unwitting and unintentional manner.

There are three major theoretical traditions in explaining the origins and patterns of prejudice and discrimination. Psychological theories focus on the ways in which group hostility satisfies certain personality needs; prejudice and discrimination, in this view, are traced to individual factors. Normative theories explain that ethnic hostilities are conforming responses to social situations in which people find themselves. Power-conflict theories see prejudice and discrimination as products of group interests and as tools used to protect and enhance those interests. Focus is placed not on the individual or even the immediate group, but on the dynamics of political, economic, and social competition among a society's ethnic groups. The latter view has been adopted in this book.

Suggested Readings

Allport, Gordon W. 1958. *The Nature of Prejudice*. Garden City, N.Y.: Doubleday. Presents perhaps the most complete examination of the origins and forms of prejudice, stressing psychological factors. A classic work that has stood the test of time.

Duckitt, John. 1992. *The Social Psychology of Prejudice*. New York: Praeger. A thorough conceptual and theoretical analysis of the causes and intergroup dynamics of prejudice.

Gioseffi, Daniela (ed.). 1993. *On Prejudice: A Global Perspective*. New York: Anchor. A collection of essays and literary pieces by social scientists, writers, and poets focusing on ethnic and racial prejudice and its consequences.

Knowles, Louis L., and Kenneth Prewitt (eds.). 1969. *Institutional Racism in America*. Englewood Cliffs, N.J.: Prentice-Hall. Describes historical and modern-day patterns of institutional discrimination in the United States.

Levin, Jack, and William Levin. 1982. *The Functions of Discrimination and Prejudice*. 2d ed. New York: Harper & Row. Examines prejudice and discrimination by focusing on the consequences of these thoughts and actions for both dominant and minority groups.

Schuman, Howard, Charlotte Steeh, Lawrence Bobo, and Maria Krysan. 1997. *Racial Attitudes in America: Trends and Interpretations*. Rev. ed. Cambridge, Mass.: Harvard University Press. Through a review of relevant survey data, traces the changing racial attitudes of Americans during the past five decades.

Simpson, George Eaton, and J. Milton Yinger. 1972. *Racial and Cultural Minorities: An Analysis of Prejudice and Discrimination*. 4th ed. New York: Harper & Row. Explains the psychological and sociological origins and effects of prejudice and discrimination, with a comprehensive review of significant studies regarding these phenomena.

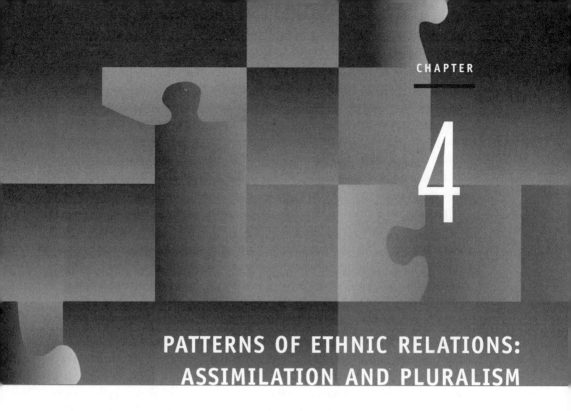

PATTERNS OF ETHNIC RELATIONS:
ASSIMILATION AND PLURALISM

THE CHIEF QUESTION we pursue in this chapter is, "What is the nature of relations among diverse groups in a multiethnic society?" Specifically, when various ethnic groups meet, what is the outcome of that contact? We have already seen that the emergence of a system of ethnic stratification, with its attendant prejudice and discrimination, is most common. But there are other dimensions and processes of intergroup relations that create a variety of patterns between dominant and minority ethnic groups.

CONFLICT AND ORDER

Judging from news accounts of violence and general hostilities in multiethnic societies, it would seem that ethnic relations are as a rule conflictual. Indeed, almost all theorists have viewed conflict as a fundamental, if not permanent, feature of relations between ethnic groups. Donald Young's observation of over seven decades ago seems no less accurate today: "Group antagonisms seem to be inevitable when two peoples in contact with each other may be distinguished by differentiating characteristics, either inborn or cultural, and are actual or potential competitors" (1932:586).

Though it may not always be overt and extreme, conflict is the primary underlying facet of systems of ethnic inequality. In this regard, Berry and Tischler note that "if we recognize the fact that conflict involves also those subtle, restrained forms of interaction wherein one seeks to reduce the status of one's opponents, and not to eliminate them entirely from the conflict, then perhaps it is true that conflict invariably occurs when unlike peoples meet" (1978:123). No matter how apparently stable and peaceful dominant-minority relations may seem, potential conflict always remains beneath the surface and reveals itself in periodic and unanticipated eruptions. Ethnic relations, then, are by definition relations of conflict. The dominant group may mobilize its power through force, ideology, or both to assure its dominance, and minority groups will respond with counterforce, accommodation, or submission.

Although conflict does characterize interethnic relations as a rule, certain qualifications to this maxim must be kept in mind. First, ethnic conflict is not maintained at a constant rate, does not take the same form, and is not based on the same factors from one society to another or even within the same society. It is obviously more intense and sustained in some cases than in others. Ethnic conflict can be reduced, and some societies, though they are rare, may even achieve a system whereby groups live side by side for long periods in a generally harmonious state (see, for example, Gillen, 1948; Redfield, 1939). Switzerland provides such an example. Here four languages are spoken, and the society is further subdivided along two main religious lines. Yet linguistic and religious tolerance have characterized Swiss ethnic relations, and serious conflict has been avoided for many generations. Later chapters, in Part III, will investigate several societies that represent a broad range of ethnic relations, extending from relative tranquillity to deep-seated hostility among groups.

It should also be understood that conflict is not characteristic only of multiethnic societies, nor is *ethnic* conflict the only or even chief form in all societies, even those that are ethnically heterogeneous. Societal conflict is based on power relations that stem from differences in class, age, gender, and numerous other social factors in addition to ethnicity.

Finally, no matter how discordant or peaceful they may be, intergroup relations in multiethnic societies do not remain fixed. Internal migrations, immigration from other societies, political events, and varying economic conditions continually force revisions in public policies and create the social conditions that lead to change. At particular times, societies may move in the direction of either more harmonious relations among groups or greater separation and more intense conflict.

Sociologists have often proposed theoretical models that describe patterns of intergroup relations in multiethnic societies (Barth and Noel, 1972; Burkey, 1978; Marden and Meyer, 1978; Schermerhorn, 1970) or cycles of relations through which such societies presumably pass (Bogardus, 1930; Gordon, 1964; Park, 1924). These theories have usually suggested that ethnic groups follow one of two paths: They either increasingly blend together or remain segregated.

Most simply, groups may become more alike culturally and interact with one another more freely—this is *assimilation*—or they may remain culturally distinct and socially segregated—this is *pluralism*. The latter is of two types, depending on the distribution of political and economic power among groups. In the first, culturally and structurally distinct groups are relatively balanced and proportionate in political and economic resources; in the second, they are unequal politically and economically.

As models of interethnic relations, assimilation and pluralism are related to the two broad theoretical paradigms in sociology, *order* and *conflict*. These paradigms pertain not simply to specific parts or aspects of societies (like ethnic relations) but to *all* social structures and relations. They are intended to explain generally how societies are sustained and how they change.

Order theorists, whose tradition is most heavily influenced by the late-nineteenth-century French sociologist Émile Durkheim, see society as a relatively balanced system made up of differently functioning but interrelated parts. In this view, society is held together and social order maintained through a consensus of values among its groups and through the imperatives of functional interdependence. In contrast, conflict theorists, beginning with Karl Marx, have seen societies as held together not by broad agreements among groups but by the power of dominant classes and ruling elites to impose their will on others. Stability and order are maintained through coercion, not consensus. Whereas order theorists stress the manner in which societies maintain cohesion and balance, conflict theorists emphasize the disintegrative aspects of societies and the manner in which they change. As many have pointed out, societies are of course neither wholly ordered nor wholly in conflict, and we must therefore account for both (Dahrendorf, 1959). Nonetheless, sociologists have favored one or the other of these broad theoretical perspectives in their analyses.

Assimilation and pluralism, as models of interethnic relations, correspond quite closely with these two general sociological paradigms. Order theorists have stressed the assimilation side, emphasizing the ways in which different groups progressively become more unified and indistinct. Conflict theorists, in contrast, emphasize the inequality among ethnic groups and the patterns of dominance and subordination that develop among them. They have thus preferred a pluralistic model of ethnic relations that underscores the persistence of group differences in culture and structure.

For many years, American sociologists traditionally favored the assimilationist model (Metzger, 1971). The prevailing assumption was that multiethnic societies like the United States tend to gradually but inevitably move toward a fusion of diverse groups. American and world patterns of ethnic conflict in recent decades, however, seriously undermined this assimilationist bias. Sociologists today increasingly recognize the complexity of interethnic relations and seem to have accepted the inevitability of some degree of pluralism and conflict.

The investigations of American ethnic relations in Part II and of several other multiethnic societies in Part III generally employ a conflict approach; thus, the pluralistic aspects of these societies are highlighted. Theoretical frameworks in ethnic relations vary in utility, however, depending on the specific problems to be explained. An assimilationist model may therefore be applicable to some societies and groups and a pluralistic model to others. In general, assimilation, stressing progressive cohesion, applies to groups in multiethnic societies that have entered as voluntary immigrants, whereas some form of pluralism is relevant to those groups that have entered through involuntary immigration, conquest, or expansion. In the remainder of this chapter, we look more closely at these two general processes and the more specific levels and forms of each.

ASSIMILATION

Most simply, assimilation means increasing similarity or likeness (Brubaker, 2001). As sociologist J. Milton Yinger defines it, assimilation is "a process of boundary reduction that can occur when members of two or more societies or of smaller cultural groups meet" (1981:249). Similarly, Harold Abramson defines it as "the processes that lead to greater homogeneity in society" (1980:150). Each of these definitions stresses *process*. Rather than a fixed condition or state of relations, assimilation is best seen as a path, or trajectory, on which ethnic groups may move.

The end point of this homogenizing process is, for an ethnic group, "the disappearance of an ethnic and racial distinction and the cultural and social differences that express it" (Alba and Nee, 1999:159). Or, for the society as a whole, it is "the biological, cultural, social, and psychological fusion of distinct groups to create a new ethnically undifferentiated society" (Barth and Noel, 1972:336). Ideally, then, at the point of complete assimilation, there are no longer distinct ethnic groups. Rather, there is a homogeneous society in which ethnicity is not a basis of social differentiation and plays no role in the distribution of wealth, power, and prestige. This does not mean, of course, that other forms of social differentiation and stratification such as age, gender, and class do not exist; it means only that the ethnic forms are no longer operative. In essence, a society in which all groups have perfectly assimilated is no longer a multiethnic society.

This complete form of assimilation, however, is rarely achieved either for a society as a whole or for specific groups and individuals. Instead, assimilation takes different forms and is evident in different degrees. In other words, it is a variable that "can range from the smallest beginnings of interaction and cultural exchange to the thorough fusion of the groups" (Yinger, 1981:249). Therefore, in examining the assimilation of ethnic groups, the question that must concern us most is not simply, "Are groups becoming more alike?" but, "To what extent and in what ways are they becoming more alike?"

Dimensions of Assimilation

Assimilation can be seen at four distinct, though related, levels or dimensions: the cultural, the structural, the biological, and the psychological. Our concern is mostly with the first two levels, but let's briefly examine each of the four.

Cultural Assimilation The cultural dimension of assimilation involves the adoption by one ethnic group of another's cultural traits—language, religion, diet, and so on. Some have referred to this process as *acculturation* (Gordon, 1964; Yinger, 1981). Almost always, weaker (that is, minority) groups take on the cultural traits of the dominant group, though there is ordinarily at least some exchange in the opposite direction as well. This results not only because of the dominant group's superior power but also because of the social advantages for the subordinates in adapting to the dominant group's ways. As van den Berghe explains, "It often pays to learn the ways of the rich, the powerful and the numerous; in the process one becomes more like them and, by that token, often becomes more acceptable to them" (1981:215). The end point of the process of cultural assimilation implies a situation in which the previously distinct cultural groups are no longer distinguishable on the basis of their behavior and values.

Yinger (1981) has noted that when groups are not highly antagonistic or culturally very disparate, acculturation can be additive rather than substitutive. That is, one group may augment its native culture with selective elements of the other's rather than substituting entirely. If the most basic cultural components (like language and religion) are not exchanged, however, the assimilation process can proceed only to a minimal point.

Structural Assimilation Whereas cultural assimilation refers to a blending of behaviors, values, and beliefs, structural assimilation refers to an increasing degree of social interaction among different ethnic groups. Specifically, with structural assimilation, members of minority ethnic groups are dispersed throughout the society's various institutions and increasingly enter into social contacts with members of the dominant group.

Structural assimilation may occur at two distinct levels of social interaction: the *primary* (or informal) and the *secondary* (or formal). Primary relations are those occurring within relatively small and intimate groups, in particular, the family and friendship cliques. Relations among members of these groups are affective, and the group's purposes extend well beyond merely instrumental goals. There is, most simply, an emotional bond among group members, and relations are therefore close and long lasting. Secondary relations, by contrast, are chiefly within large, impersonal groups like the school, the workplace, or the polity. These groups are purposeful, designed to fulfill some practical and specific social need; relations among members are thus formal and nonaffective.

Structural assimilation at the primary level implies interaction among members of different ethnic groups within personal networks—entrance into clubs, neighborhoods, friendship circles, and ultimately marriage. In short, people interact in close, personal relations without regard for one another's ethnic identity. Other social traits, such as class, and individual characteristics become more critical than ethnicity. To measure the degree of informal, or primary, structural assimilation of a particular group, such indicators as rate of intermarriage, club memberships, and residential patterns would be used. The higher the level of interaction with members of the dominant group in these areas of social life, the greater the extent of structural assimilation.

At the secondary level, structural assimilation entails equality of access to power and privilege within the society's major institutions—the economy, the polity, education, and so on. That is, jobs, housing, schooling, and other key life chances are not distributed with regard to people's ethnic affiliation. We might say, then, that structural assimilation at the secondary level involves, in its ultimate stage, the elimination of minority status. To measure the degree of secondary structural assimilation, we would look at the extent to which a minority ethnic group is approaching parity with the dominant group in the allocation of income and wealth, political power, and education (Hirschman, 1975).

Secondary structural assimilation has often been referred to as *integration* (Burkey, 1978; Davis, 1978; Hunt and Walker, 1974; Simpson, 1968; Vander Zanden, 1983), and we can consider these terms to be synonymous. The essential idea is that people of diverse ethnic groups come to interact in the institutions of the larger society free of the constraints of ethnicity. Hunt and Walker refer to integration as a situation in which "all citizens of the nation, or possibly even all members of the society regardless of citizenship, participate freely in all forms of social interaction without concern for ethnic affiliation" (1974:8).

This level of assimilation involves essentially a legal termination of group discrimination based on ethnicity. Inequality exists, of course, but is founded on bases other than ethnic group membership. With integration, ethnic groups may remain distinct, and members may continue to identify with them. Similarly, out-groups may continue to recognize and respond to them as ethnic groups. What has been achieved, however, is a measure of political and economic equality.[1]

The distinction between primary and secondary levels of structural assimilation is important because it is clear that the entrance of ethnic minorities into formal relations with the dominant group must precede relations within intimate social settings. Groups may achieve a significant degree of secondary structural assimilation without moving beyond this level into the primary type. Most Euro-American ethnic groups in American society, for example, are approaching a point at which they enjoy relatively equal access to jobs, political authority, and other important life chances. They have, in other words, begun to enter into full participation in all institutional areas, even though for some, primary relations may remain limited largely to the ethnic group. For African Americans and other racial-ethnic groups, the distinction between secondary and primary structural assimilation is even more obvious. Although in recent years they have begun to achieve substantial integration in the economy, polity, and education, that has not been the case in clubs, cliques, neighborhoods, and intermarriage.

Biological Assimilation Biological assimilation, or *amalgamation*, represents the ultimate stage in the assimilation process. At this point, intermarriage has occurred to such an extent that there is a biological merging of formerly distinct groups. They are indistinguishable not only culturally and structurally but physically as well. Some degree of amalgamation is a common by-product of group contact and interaction. But the complete biological fusion of diverse groups is an unusual outcome of interethnic relations. Mexico, for one, is a society that seems to have moved far along toward this long-range objective (van den Berghe, 1978); and Brazil, as will be seen in Chapter 14, has also progressed in this direction.

Psychological Assimilation Our focus is primarily on how various groups interrelate; our concern, therefore, is chiefly with the cultural and structural

---

[1] Schermerhorn (1970) has used the term *integration* in a somewhat different but related way. As he defines it, integration is "a process whereby units or elements of a society are brought into an active and coordinated compliance with the ongoing activities and objectives of the dominant group" (1970:66). The key to integration in this sense is a mutual acceptance of the scope and nature of activities and group objectives by both the dominant group and minority groups. Thus societies may be integrated around a system in which extreme segregation and inequality are accepted by all parties. Schermerhorn postulates that "when the ethos of the subordinates has values common to those in the ethos of the superordinates, integration (coordination of objectives) will be facilitated; when the values are contrasting or contradictory, integration will be obstructed" (1970:172). See also Kuper (1968).

dimensions of assimilation. But there is an individual dimension of assimilation, in which attention falls on particular members of an ethnic group rather than on the group as a whole. This is part of the social psychology of ethnic relations and concerns the extent to which individuals have been absorbed into the larger society and identify with it.

With psychological assimilation, members of an ethnic group undergo a change in self-identity. To the extent that individuals feel themselves part of the larger society rather than an ethnic group, they are psychologically assimilated. As psychological assimilation proceeds, people tend to identify themselves less in ethnic terms. Whereas first-generation immigrants and their children will define themselves as Italian American or Irish American, by the third generation, ethnicity for most is no longer a major component in the response to the question, "Who am I?" This level of assimilation consists not simply of becoming culturally like members of the mainstream society but also of "accepting that society as the home base, the prime focus of allegiance and the place where personal ambitions are formed, achieved, and enjoyed" (Wilkie, 1977:88).

Psychological assimilation is a process of change not only in self-identification but also in identification by others. Although individual members of an ethnic group may see themselves as simply part of the larger society rather than as ethnics, outsiders may continue to identify them as members of their group, thereby impeding psychological—and structural—assimilation. As Yinger notes, "Prejudice on the part of a dominant group may prevent the granting of full membership in a society to members of minority groups, even though the latter think of themselves only in terms of the larger society" (1981:253). Visibility is, of course, critical here. Those with salient marks of ethnic identity—skin color, in particular—are unable to fully achieve out-group recognition as "nonethnics."

Some individuals may find themselves unable to feel fully part of the larger society *or* the ethnic group. Such people have been referred to as "marginal men" (Park, 1928; Stonequist, 1937). First-generation immigrants often find themselves in a situation where they are pulled in the direction of the culture of the new society but remain culturally and psychologically tied to the old. Those who are not clearly part of one racial category or another also may be marginal people. In American society in recent years, such individuals have begun to be recognized by themselves and others as constituting a unique "mixed-race" or "multiracial" category (Wright, 1994).

Although sociologist Robert E. Park, in a classic work (1928), described marginal people as experiencing psychological turmoil, not all of them find difficulty in social adaptation. Some successfully adapt to two cultures, shuttling between them, and others may use their marginality to advantage by serving as middlemen between dominant and minority groups. Moreover, where they constitute a large population, marginal people may form communities of their own, occupying an in-between status in the society's ethnic hierarchy (Shibutani and Kwan, 1965). Recall the discussion of middleman minorities in Chapter 2.

The social psychology of ethnic identity is a complex matter, the details of which go well beyond our present purposes.[2] Suffice it to say that within any multiethnic society, individuals may vary widely in the extent to which they identify with an ethnic group or choose to disregard ethnicity as part of their self-identity. Much depends not only on visibility but also on the political and social conditions that may affect the individual costs and values of an ethnic identity. Most people are permitted much latitude in their control of personal identity. Thus, for any single person, ethnic identity "expands and contracts, it changes over time, it looms large or small depending on the particular context and possibly on the advantage to be gained from manipulating the ethnic label" (Elkin and Handel, 1978:101–2).

Two Theories of Assimilation

Most analysts of assimilation have dealt with societies like the United States or Australia, in which sizable groups of voluntary immigrants, mainly from Europe, have made up the largest element of the ethnic mosaic (Taft, 1963). As a result, their findings and the hypotheses they suggest are applicable only to a narrow range of societies and specific ethnic groups. Two assimilation theories that, although American in origin, have had somewhat broader applications and have been especially influential are Park's and Gordon's.

Park's Race Relations Cycle American sociologists in the 1920s, led by Robert Park, were concerned with the ethnic polyglot that made up cities of the Northeast and Midwest, and the focus of their studies became the forces that brought these groups together or sustained their differences. Park was one of the first to suggest a cycle of race or ethnic relations through which groups would pass in a sequence of stages, leading ultimately to full assimilation.

Park explained that groups first come into *contact* through migration and subsequently engage in *competition*, often characterized by conflict. Out of such competition eventually emerges some form of *accommodation* among the groups, leading finally to *assimilation*. Park maintained that this four-stage cycle pertained to race relations everywhere, not simply the United States. Moreover, he saw the sequence as "apparently progressive and irreversible" (1950:150).

A view of ethnic assimilation not unlike Park's race relations cycle has been generally popular in American society. The prevailing thought is that over several generations, group boundaries break down, and the society becomes more

---

[2]The issues of ethnic identity at the individual level are detailed in De Vos and Romanucci-Ross (1975), Horowitz (1975), Romanucci-Ross and De Vos (1995), and Shibutani and Kwan (1965).

homogeneous. However, Park's model has been subject to much sociological criticism over the years. Some have noted the cycle's lack of applicability to many groups (Lyman, 1968a). Although it has seemed to describe fairly accurately the experiences of most European immigrant groups in the United States (as well as Australia, Canada, Argentina, and other immigrant societies), it does not conform to the patterns displayed by more salient—that is, physically distinct—groups and those that have entered involuntarily. Critics of Park's and other cyclical theories of ethnic relations have also pointed out that such cycles are rarely complete; that is, there are too many truncated instances (Shibutani and Kwan, 1965). As Barth and Noel (1972) point out, interethnic contact can lead to stable outcomes—such as exclusion, pluralism, or continued ethnic stratification—that do not lead inevitably to assimilation. Finally, some have criticized the model's claim of irreversibility (Barth and Noel, 1972; Berry and Tischler, 1978). They note that the cycle may be terminated at any point, and groups may even revert to earlier stages.[3]

Gordon's Stages of Assimilation Park's model, despite its shortcomings, stands as a precursor to subsequent, more sophisticated theories of the assimilation process. Perhaps the most precise and compelling of these is Milton Gordon's (1964).

Like Park, Gordon explains assimilation as a series of stages, or steps, through which various groups pass. But rather than following a straight line leading from contact to eventual absorption, groups may remain indefinitely at one or another of these stages. The seven stages that Gordon outlines extend from cultural assimilation, the least intense, to civic assimilation, the most complete (Table 4.1). The first two, cultural and structural assimilation, are the most important.

As noted earlier, cultural assimilation, or acculturation, denotes the adoption by a minority ethnic group of the dominant group's (or, as Gordon prefers, the host society's) cultural patterns—language, political beliefs, and so on. But acculturation, though a prerequisite, does not assure movement to the next phase. Gordon explains that groups may become very much like the dominant group in behavior and values but still remain structurally segregated.

Structural assimilation, the second stage, is essentially what was previously referred to as primary, or informal, structural assimilation. This stage is the most critical, explains Gordon, for it is the key to all subsequent stages: "Once structural assimilation has occurred, either simultaneously with or subsequent to acculturation, all of the other types of assimilation will naturally follow"

---

[3]Geschwender (1978) disputes the interpretation of Park's race relations cycle as absolute, maintaining instead that it is "situationally specific." He sees Park's cycle as an ideal type, not to be taken as inevitable.

Table 4.1	Gordon's Stages of Assimilation
STAGE	CHARACTERISTICS
Cultural or behavioral assimilation (acculturation)	Change of cultural patterns to those of host society
Structural assimilation	Large-scale entrance into cliques, clubs, and institutions of host society on primary group level
Marital assimilation (amalgamation)	Large-scale intermarriage
Identificational assimilation	Development of sense of peoplehood based exclusively on host society
Attitude receptional assimilation	Absence of prejudice
Behavior receptional assimilation	Absence of discrimination
Civic assimilation	Absence of value and power conflict

SOURCE: Milton M. Gordon, *Assimilation in American Life: The Role of Race, Religion, and National Origins,* New York: Oxford University Press, 1964, p. 71. Copyright © 1964 by Oxford University Press, Inc. Reprinted by permission.

(1964:81). Presumably, as people of minority and dominant ethnic groups interact within close, intimate social settings, the other stages of assimilation necessarily occur, much like falling dominoes: Minority group members increasingly intermarry with those of the dominant group, relinquish their ethnic identity, no longer encounter prejudice and discrimination, and fully agree with the dominant group on issues involving values and power conflicts. Indeed, the remaining stages (three through seven) in Gordon's scheme can be subsumed under structural assimilation. The dominant group's acceptance of members of a minority ethnic group into primary relations implies, for example, the absence of prejudice and discrimination and the likelihood of increased intermarriage.

It is with structural assimilation, then, that full assimilation involving all other stages becomes inevitable (though Gordon suggests no timetable or further conditions affecting the rate at which they are achieved). Cultural assimilation, the first stage, does not necessarily lead to structural assimilation; it may take place independently of any of the others and may continue to be the extent of assimilation for many generations (Gordon, 1964:77; see also van den Berghe, 1981:216; Wagley and Harris, 1958:288). Minorities may take on all or most of the cultural ways of the dominant group but still be refused entry into primary relations with its members. African

Americans are an evident illustration of this. Though they have adopted in large measure the major elements of the dominant culture, they remain unassimilated at the structural level (specifically the *primary* structural level). In short, groups may become culturally alike yet remain in relatively segregated subsocieties.

Several criticisms have been made of Gordon's assimilation model. One serious shortcoming lies in its understanding of structural assimilation, stage two, as entailing interaction with the dominant group only at the primary level. Intergroup relations, however, occur at the secondary level as well and indeed are antecedent to primary relations in any significant degree. To what extent do members of a minority ethnic group enter into positions of power in the society's economic, political, and other key institutions? To what extent are they afforded equal opportunities in employment and education? These are significant measures of structural assimilation—at the secondary level—that Gordon does not consider (Marger, 1979). African Americans, for example, appear to have realized substantial integration in recent years in the areas of work and government. They have, in other words, experienced increasing secondary structural assimilation. That they have not accomplished an equivalent level of interaction with whites at the primary level would, in Gordon's view, imply that little structural assimilation had occurred.

Moreover, Gordon seems to suggest that if minorities do not enter into primary relations with the dominant group, it is because the dominant group has held them out. But such social segregation may be largely voluntary.

Finally, Gordon's model has, like Park's, been criticized by some as implying a unidirectional movement whereby groups, though perhaps delayed indefinitely at one stage, do not return to earlier stages. The assimilation process is, in this view, invariably progressive in direction.

Despite these omissions, Gordon's model, with modification, is valuable to the analysis of interethnic relations primarily because it spells out the intricacies of the assimilation process and the various forms and degrees it manifests. Assimilation, as Gordon demonstrates, is clearly not a simple, straightforward movement, as earlier theories had seemed to imply.

Factors Affecting Assimilation

Why do some groups in multiethnic societies display a rapid and almost complete assimilation into the larger society, whereas others remain segregated and are the constant targets of prejudice and discrimination? Where assimilation is the prevailing model of intergroup relations, several factors are important in shaping the experience of different groups. These are how and when a group enters a society, its size and dispersion, its cultural similarity to the dominant group, and its visibility.

Manner of Entrance As we saw in Chapter 2, the way in which a group enters the society is critical in determining its place in the ethnic hierarchy; it is also important in accounting for the nature of the group's long-range societal adjustment. Except for those groups that maintain unfaltering pluralistic aims, those that enter voluntarily always make a less conflict-ridden adjustment than those that enter involuntarily or those that are conquered by more powerful invaders. Involuntary immigrants or conquered groups remain in a condition of segregation to one degree or another.

The United States presents a clear illustration of this tendency. European ethnic groups, having entered by choice, were able to assimilate culturally at a pace determined in large part by themselves, no matter how strong the pressures to conform. Furthermore, the option of returning to their society of origin was always present and was, in fact, taken by many. African Americans and American Indians were absorbed into the society involuntarily; as a result they could not follow similar paths no matter what their intentions or long-range goals (Blauner, 1972). As noted in Chapter 2, indigenous groups who are made subordinate at the entrance of an invading group (for example, North American Indians) or who enter involuntarily (for example, African Americans) are left with few options other than resistance to the new social order that is imposed on them. Long-term conflict is thus the usual outcome. Subordinate voluntary immigrants (like Europeans to America), in contrast, are more rapidly assimilated (Lieberson, 1961; Schermerhorn, 1970; van den Berghe, 1976; Wilkie, 1977).

Time of Entrance In general, the more recent a group's entry into the society, the more resistance there is to its assimilation (Mack, 1963). Other things being equal, the simple factor of time will tend to ease the fear and suspicion that accompany the entrance of strangers. Groups with alien ways are seen differently after they have lived in the society for several generations. Examples abound in the United States, beginning with the influx of large numbers of Irish immigrants in the early nineteenth century. Given the very substantial assimilation of Irish Americans today (such that they are approaching ethnic extinction, save for their Catholicism), the virulent and rabidly hostile reception encountered by the first generation is easily forgotten.

Demographic Factors The degree and rate of assimilation for minority ethnic groups is also affected by their size and the concentration of their population (Blau, 1977; Frisbie and Neidert, 1977). The entrance and assimilation of groups relatively small in number will be resisted less forcefully than that of groups representing a competitive threat. Van den Berghe (1981) suggests that smaller groups are assimilated more easily because they have fewer resources and therefore depend on the larger society and because they necessarily interact more frequently with out-group members.

The concentration or dispersion of ethnic groups also bears on their assimilation. Concentration in particular neighborhoods and geographical areas or in certain occupations tends to retard assimilation because the group is better able to retain its cultural ways and resist intrusions of the dominant group (van den Berghe, 1981). Dispersal, on the other hand, leads to unavoidable contact and interaction with the dominant group and thus speeds up the assimilation process.

Cultural Similarity No matter how or when the group enters the host society, or what its demographic patterns, assimilation is favored if the group is culturally similar to the dominant group (Berry and Tischler, 1978; van den Berghe, 1981). Those groups in the United States that have followed the assimilation route furthest have, predictably, been culturally closest to Anglo Protestants. In general, the more compatible the culture of the minority group with the dominant group's, the greater will be the force and speed of assimilation.

Visibility In almost all multiethnic societies, the most critical factor in determining the degree and rate of assimilation of ethnic groups is visibility. Where physical differences are obvious, manner of entrance, temporal factors, demographic patterns, and even cultural similarity are of much less consequence. For racial-ethnic groups, structural separation remains far more persistent than for groups that are only culturally distinct (Wirth, 1945; Yinger, 1981). Observers of American ethnic relations have long interpreted the retarded structural assimilation of blacks, for example, as a product chiefly of visibility (Park, 1950; Warner and Srole, 1945). It is, of course, not visibility per se that deters assimilation of racial-ethnic groups but rather the beliefs connected with their physical distinctness.

In short, physical differences delay the process of assimilation more than other factors. Harold Isaacs has poignantly described this dilemma: "An individual can change his name, acquire a new language, ignore or conceal his origins, disregard or rewrite his history, abandon his ancestral religion or convert to another one, adopt a different nationality, embrace new mores, ethics, philosophies, styles of life. But there is not much he can do to change his body" (1989:46). Thus, the more visible the group or individual, the longer and more difficult is the process of structural assimilation.

Assimilation as Public Policy

Assimilation can be viewed as a goal, or ideal, for which multiethnic societies aim. As such, it is sometimes the basis of public policies designed to reduce the cultural and structural divisions between groups. In the United States, for example, measures intended to lessen segregation in various public spheres (housing, schools, work) and to equalize access to power and privilege (affirmative action programs, voting rights) can be understood as outgrowths of a societal commitment to the eventual achievement of complete assimilation for all ethnic groups. However, the form and ultimate objective of assimilation may vary in the minds not only of policymakers but also of both dominant and minority group members.

Abramson (1980) points out that there are three possible forms of complete assimilation for which societies may strive, each involving a different path and a somewhat different objective. First, minority ethnic groups may assimilate into the dominant ethnic group. This has been the chief form and process of assimilation in the United States. Second, groups may assimilate into an entirely new ethnicity. This is the popular notion of the "melting pot," in which all groups surrender their ethnic heritage but in the process create a hybrid society with no dominant group as such (see Gordon, 1964). Israel is a modern multiethnic society on such a course. Third, minority ethnic groups may assimilate into another nondominant ethnic background or group. For example, some have suggested a "triple melting pot" in the United States, in which numerous ethnic groups representing diverse national origins have seemed to converge into three encompassing religious bodies: Protestantism, Catholicism, and Judaism (Herberg, 1960; Kennedy, 1944, 1952).

It is the first form of complete assimilation, however, that remains the major objective in most multiethnic societies where groups are encouraged to pursue an assimilationist course. The answer to the question, "Assimilation to whose values and traits and to whose institutions?" is almost always, "The dominant group's culture and social structure." This is nothing more than a manifestation of power. As long as one group disproportionately controls power resources in the economic and political realms, that group is enabled to dictate the shape and direction of minority group adjustment. It is the dominant group that, in the main, mandates or approves policies of acculturation and integration. In the United States, for example, there has never been any question of whose language, whose religious and political principles, and whose institutions will prevail. All have been those of the dominant Anglo core group and continue to be so. As William Newman has succinctly put it, "Regardless of the number of groups involved, the theory of assimilation is a theory of majority conformity" (1973:53). Ideologically, societies may advocate some kind of ethnic melting pot wherein all groups contribute in proportionate amounts to form a new social system; but such a cultural and, particularly, structural fusion is a chimera.

As previously noted, however, we should not think that assimilation is a one-way process whereby minority ethnic groups seek out and become like the dominant group, with no change occurring in the opposite direction. Obviously, the assimilation process will be to some extent mutual, with many aspects of minority cultures becoming part of the dominant culture (Greeley, 1974; Wagley and Harris, 1958); even social interaction is at times initiated by members of the dominant group. But the exchange is far short of equal. On the major elements of culture—language, religion, political beliefs, economic practices, and so on—there is little evidence of mutuality, and at the structural level the exchange is even more lopsided. These aspects of culture and social structure remain firmly controlled by the dominant group, and minority ethnic groups must adjust to them. Indeed, so powerful is the dominant group's model that it is often internalized by minority group members who, in the process of assimilation, "tend to conceive of themselves as *inferior* objects and look with envy upon members of the dominant group" (Shibutani and Kwan, 1965:507).

Segmented Assimilation Some contemporary immigration theorists have explained that assimilation is more varied than the classical models have presumed. They point out that assimilation is not a single process and may occur in different domains entailing different reference populations (Brubaker, 2001; Fernández Kelly and Schauffler, 1996; Waters, 1999). There are, as a result, diverse outcomes of the adaptation process of immigrants and, especially, their children; hence, the term *segmented assimilation* (Portes and Zhou, 1993; 1994). Ordinarily it has been assumed that movement toward assimilation implies adoption of the dominant group's culture and incorporation into the mainstream (i.e., "white, middle-class") social structure. However, for immigrant groups today, these assumptions do not necessarily pertain. Most are not European in origin and thus are perceived as racially different; for many, this blocks the path of rapid assimilation into the middle class. Also, economic conditions no longer provide robust opportunities for upward mobility through jobs in the industrial sector, as they did for immigrants of earlier periods.

Given these conditions, the children of immigrants may still move in the traditional path of assimilation, adopting the dominant culture and integrating into the dominant society. But it also possible for them to assimilate to an impoverished minority culture, thus effectively eliminating any chance of upward social or economic mobility. And in a third variant of assimilation, under certain conditions immigrant families may advance economically, yet deliberately seek to preserve their ethnic culture and solidarity. In such cases, by retaining the ethnic culture, parents try to protect their children from assimilation into an inner-city culture that may lead to downward mobility (Hirschman et al., 1999).

Like assimilation, pluralism entails several dimensions and forms.[4] In all cases, however, the retention or even strengthening of differences among ethnic groups is presumed. Thus, in a general sense, pluralism is the opposite of assimilation. Abramson defines pluralism as "conditions that produce sustained ethnic differentiation and continued heterogeneity" (1980:150). In short, pluralism is a set of social processes and conditions that encourages group diversity and the maintenance of group boundaries.

Just as assimilation occurs in different degrees and at different stages, so too pluralism must be understood as a variable for groups and societies. Ethnic pluralism never entails an absolute separation of groups. Recall the definition of *ethnic group* as a distinguishable group *within a larger society*. Thus, in a pluralistic society there is always some common political or economic system that binds various ethnic groups together. If this were not so, there would be not a multiethnic society but several distinct societies in themselves. Beyond the recognition of a common political or economic system, however, groups may differ widely.

As with assimilation, several dimensions of pluralism can be delineated, the most important being the cultural and the structural. *Cultural* pluralism implies the maintenance of many varied cultural systems within the framework of the larger sociocultural system (Gordon, 1964). This dimension of pluralism has been referred to in the United States in recent years as "multiculturalism." *Structural* pluralism connotes not simply differences in culture but also the existence in some degree of segregated ethnic communities within which much of social life occurs for group members. These ethnic subsocieties, or communities, comprise institutions—schools, businesses, churches, and the like—that duplicate to some extent those of the dominant group. Notice how cultural and structural pluralism are basically opposing counterparts of cultural and structural assimilation.

Two forms of ethnic pluralism are evident in modern multiethnic societies. In the first, groups maintain cultural and structural autonomy but remain relatively equal in political and economic power; moreover, their separation is mainly voluntary. This form is *equalitarian pluralism*. In the second, groups maintain structural segregation and perhaps cultural distinctness as well but are unequal in political and economic power; further, group separation in these cases is ordinarily involuntary. This form is *inequalitarian pluralism*.

---

[4]Pluralism as applied to political systems refers to the relative dispersion of power among various interest groups in a society. This is different from its usage in describing ethnic relations. See Marger (1987).

Equalitarian Pluralism

Where equalitarian pluralism characterizes interethnic relations, groups retain their cultural and, for the most part, structural integrity while participating freely and equally within common political and economic institutions (Barth and Noel, 1972; Shibutani and Kwan, 1965). Some have referred to this condition as *accommodation*, wherein the minority group "desires equality with, but separation from, the dominant group and the dominant group agrees to this arrangement" (Kurokawa, 1970:131). Technically, of course, if equality with the dominant group is reached, there are no longer dominant-minority relations; there are relations among ethnic groups, but they are not hierarchical and invidious.

Equalitarian pluralism corresponds, like assimilation, to the order model of society, in which balance and cohesion are emphasized. Differences among groups are recognized and even encouraged, but within the framework of a larger set of agreed-on principles. All groups give allegiance to a common political system, participate in a common economic system, and understand a common set of broad ethical values (Williams, 1977). In a sense, ethnic groups become political interest groups that compete for the society's rewards (Glazer and Moynihan, 1970). But these competitive differences do not lead necessarily to serious cleavages and conflict; rather, they are dealt with by a reasonable give-and-take within the context of the consensual rules of the society.

There may be vast cultural differences among ethnic groups, but their essence is not threatened because tolerance of such differences is integral to the social order. Ronald Taft explains that "pluralism implies that beyond the acceptance of supraordinate national values essential to the nation's existence there need be no agreement between immigrants and native citizens excepting that their cultural differences be mutually tolerated and preserved" (1963:279). Relations between ethnic groups are thus confined mainly to functional areas such as government and the marketplace, not affective ones like the family or friendship circles.

The equalitarian form of pluralism has been manifested somewhat differently in the United States than in other multiethnic societies.

Cultural Pluralism: The American Case Order theorists of pluralism have concentrated chiefly on the United States and its system of ethnic heterogeneity, in which groups continue to express elements of their ethnic cultures despite the forces of assimilation. Those who have analyzed American society with such a model have noted the persistence of ethnic groups over many generations after their immigration and the tendency for individuals to continue to conduct their primary group relations within an ethnic context. The choice of close friends or marital partners, it is pointed out, is still strongly dictated by ethnicity.

Ethnicity in such a setting is more a matter of individual choice than a collective imperative. Gordon (1975, 1981) has referred to this condition as "liberal pluralism," in which there is no formal recognition of ethnicity in the allocation of government offices or economic rewards and in which individuals are free to express their ethnic identity to whatever extent they choose.

Many have also advocated cultural pluralism, rather than assimilation, as a desirable end product of American interethnic relations. Gordon explains:

> The presumed goal of the cultural pluralists is to maintain enough subso-cietal separation to guarantee the continuance of the ethnic cultural tradi-tion and the existence of the group, without at the same time interfering with the carrying out of standard responsibilities to the general American civic life. In effect, this demands keeping primary group relations across ethnic lines sufficiently minimal to prevent a significant amount of inter-marriage, while cooperating with other groups and individuals in the secondary relations areas of political action, economic life, and civic responsibility. Within this context the sense of ethnic peoplehood will re-main as one important layer of group identity while, hopefully, prejudice and discrimination will disappear or become so slight in scope as to be barely noticeable (1964:158).[5]

Although this ideal has fairly accurately described the process by which some degree of ethnic pluralism has persisted in American society for Euro-American groups, it does not account for the maintenance of gross inequalities in the plu-ralistic system, particularly as it works for racial-ethnic groups. Moreover, though this form of American pluralism is most commonly referred to as cul-tural pluralism, it is more accurately *structural* pluralism because most groups have adopted the key elements of the mainstream culture after the second im-migrant generation.

In the United States, structural pluralism exists to a greater or lesser extent depending on either the visibility of the ethnic group or the desire of the group to remain apart from the larger society. Particularly at the level of primary re-lations, ethnic groups continue to display at least minimal internal cohesion by encouraging endogamy, a sense of community, and at least in symbolic form, ethnic identification. For the more salient groups, particularly African Americans, separation extends well into the level of secondary relations, where segregated neighborhoods and schools persist and job opportunities are often circumscribed. There is, then, a severely unequal outcome among ethnic groups despite the society's adherence to an ideology of political equality and equality of economic opportunity. As we will see in Part II, the United States has retained many characteristics of an ethnically pluralistic society, but in ways closer to the inequalitarian form than to the equalitarian form stressed by order theorists.

---

[5]From Milton M. Gordon, *Assimilation in American Life: The Role of Race, Religion, and National Origins.* Copyright © 1964 by Oxford University Press, Inc.; renewed 1992 by Milton M. Gordon. Reprinted by permission.

Corporate Pluralism: Non-American Cases American sociologists have paid far less attention to other pluralistic societies, in which a much greater degree of structural and cultural separation is sustained by political authorization. In societies such as Switzerland, Belgium, Malaysia, and to some extent, Canada, the structural and cultural differences among ethnic groups are protected by the state, and institutional provisions are made to encourage an ethnically proportionate distribution of societal rewards. Gordon (1975, 1981) refers to such cases as *corporate* pluralism.

In corporate pluralistic systems, ethnic units are formally recognized by the government, and political and economic power is allocated on the basis of an ethnic formula. Thus, in the political arena legislative seats and other government offices may be apportioned on the basis of ethnicity. Not only is proportionality in the distribution of political benefits assumed, but most important, there is cooperation among leaders of all the significant segments of the plural society (Lijphart, 1977). Moreover, on local matters each group maintains a great deal of political autonomy. In the economic realm as well, the objective is an equal distribution of income and jobs among the various groups, proportional to their makeup in the national population. Such systems are sometimes referred to as *consociational* (Lijphart, 1977; McRae, 1974; van den Berghe, 1981).

In this type of pluralism, cultural and structural separation are emphasized, not discouraged, and multilingualism is officially sanctioned. As Gordon explains, the position of corporate pluralism is "that the various racial and ethnic groups have the right and, indeed, should be encouraged to retain their ancestral languages, that there is no reason why there must be only one official language, and that all members of the national polity should be encouraged, perhaps even compelled, to become bilingual or multilingual" (1981:187). Switzerland, with its four official languages—German, French, Italian, and Romansh (the latter spoken only by a very small number)—is perhaps the most obvious and successful multilingual system (McRae, 1983). Canada, with its French- and English-speaking groups, is another familiar multilingual (in this case, bilingual) society, though, as we will see in Chapter 15, multilingualism has not been accepted enthusiastically and without serious problems.

In societies characterized by corporate pluralism, ethnic groups consist mainly of homogeneous, territorially concentrated peoples who have long historic roots in their native area. They have become part of a larger national society either through conquest or by voluntarily relinquishing sovereignty to a central state in order to secure economic and political benefits. Such societies are not at all similar to multiethnic societies like the United States, in which ethnic groups have been formed primarily by voluntary immigrants who came from distant societies, severed most of their native roots, and scattered geographically throughout the society.

The cultural divisions among ethnic groups in corporate pluralistic systems are much sharper than those among groups in the United States, where the English language tends to become a great commonizing factor. Structural separation is also stronger, given the territorial concentration of groups. In Switzerland, for example, each ethnic group remains geographically distinct; and in Canada over 80 percent of French-speaking people reside in one province, Quebec. By contrast, ethnic groups in the United States are more dispersed, though regional concentrations of particular groups are noticeable (such as Mexican Americans in the Southwest or Swedish Americans in the upper Midwest). Ethnic concentrations in the United States more often take the form of urban pockets or ghettos.

With their retention of language and their territorial base, ethnic groups in such corporate pluralistic systems are essentially "subnations" (Petersen, 1980). They are integrated only in their mutual allegiance to a larger national government and the need to participate in a national economic system. Ideally, no single group is dominant, and each is afforded an approximately proportionate share of the society's rewards. As Wagley and Harris note, "A satisfactory pluralistic adjustment implies not only that the dominant group will accept a minority's distinctive traits without prejudice, but also that the minority will have a more equal share in the material and nonmaterial wealth of the nation" (1958:288). No corporate pluralistic society, of course, has met this ideal, though some have achieved a greater degree of group equality than others.

Belgium is a society that is organized along such corporate pluralistic, or consociational, lines. It is divided mainly on the basis of language: Walloons speak French, Flemings speak Dutch. Each group is also regionally concentrated. Traditionally, French was the dominant language, and Flemings, if they expected to rise in social standing, had to adopt that language as well as other aspects of Francophone culture. This changed in the post–World War II era as the economic position of the Flemish region markedly improved. The language division became more acute, however, and ethnic separation more pronounced. In response, Belgian social and political institutions, already with consociational features, were further adjusted to accommodate this division (McRae, 1986). Language and educational rights were afforded each group and, most important, a political power-sharing formula was created. The national cabinet, for example, must comprise equal numbers of Dutch- and French-speaking members (Heisler, 1991).

Switzerland, with its four linguistic groups, is another striking example of ethnic autonomy and balance. Some have explained the Swiss success in maintaining concordant interethnic relations as due to the lack of contacts between groups and the decentralization of decision making in matters of ethnic relevance, such as the use of language (Connor, 1976).

Equalitarian pluralism is more a product of group choice than the largely involuntary structural pluralism of racial-ethnic minorities in the United States or of the inequalitarian form of pluralism, which is discussed later. But if we can judge from contemporary societies that have attempted this path to ethnic harmony, the results are not always more benign. The breakup of Yugoslavia is a tragic illustration. Here, several ethnic nationalities, living for the most part in distinct territories, made up a state that provided roughly proportional political power for each. But the system collapsed in 1991, giving rise to one of the most brutal episodes of ethnic warfare that has been witnessed in modern Europe. This case will be discussed in more detail in Chapter 17. Another such case is Canada, where, as we will see in Chapter 15, French and English groups have sustained conflict (though nonviolent) for two centuries. Cases like Belgium and Switzerland, where interethnic relations are less discordant, appear to be exceptional, not common; and even in those societies periodic group conflict is evident. The assertion of Wagley and Harris (1958) that pluralistic aims perpetuate some degree of conflict and the subordination of one group by another is well taken. Indeed, pluralism, no matter how seemingly benign, seems almost always productive of conflict.

Inequalitarian Pluralism

Both assimilation and equalitarian pluralism are most characteristic of societies that have to some extent blunted the acuteness of ethnic stratification and have made the reduction of ethnic inequality a well-established commitment of the state, legitimated by an equalitarian ideology. As public policies, assimilation and equalitarian pluralism are aimed at reducing the level of ethnic conflict. Inequalitarian pluralism, however, describes outcomes and processes that are clearly inequitable for the society's various ethnic groups. Although most characteristic of classic colonial societies and modern racist regimes like South Africa during the period of apartheid, inequalitarian pluralism can also be seen to some degree in other multiethnic societies where assimilation or equalitarian pluralism is the prevailing type of interethnic relations. This model of ethnic relations is favored by power-conflict theorists, who stress the divisiveness of multiethnic societies and the competition among groups.

Under inequalitarian pluralism, ethnic groups are not only separated structurally and perhaps culturally but also exist in a state of highly unequal access to power and privilege. Indeed, the authority and power of the dominant group are the key coordinating mechanisms of such systems (Kuper and Smith, 1969; van den Berghe, 1978). In an equalitarian pluralistic system, the various ethnic groups are held together through the consensual allegiance to a common state. In inequalitarian pluralistic societies, however, the state holds the different groups together not through a mutually recognized legitimacy but through coercion. Basically, the state acts to protect the interests of the dominant group. Anthropologist M. G. Smith (1969:33) explains that in a plural society of this type, the state is the agent of the ruling group only. Others have no rights or protection. The majority of people are "subjects, not citizens." Whereas equalitarian pluralism assumes a progressive equalization of political and economic power among groups, the assumption here is quite the opposite. Sustained or increased inequality among groups is a built-in feature of the system, with the dominant group retaining all political authority and the bulk of material wealth.

Social relations between dominant and minority groups are typified by extreme polarization, supported by high levels of prejudice and discrimination. Basically, it is only within the impersonal confines of the economic and political systems that dominant and minority group members interrelate, and these relations are limited to purely functional contacts such as work and government administration. As Kuper and Smith (1969:11) describe it, "Economic symbiosis and mutual avoidance, cultural diversity and social cleavage" characterize the social basis of this type of plural society.

Although the dominant group holds sway, ethnic groups are in a state of economic interdependence. The dominant group needs subordinate ethnic groups to perform physical and menial laboring tasks necessary to economic production; and given their relative powerlessness, subordinate groups must meet these demands. Occupational roles are assigned strictly on the basis of ethnicity, with the subordinate group or groups delegated those most onerous and least prestigious (Rex, 1970). The dominant group, as van den Berghe explains, "rationalizes its role in an ideology of benevolent despotism and regards members of the subordinate group as childish, immature, irresponsible, exuberant, improvident, fun-loving, good humored, and happy-go-lucky; in short, as inferior but lovable as long as they stay in their place" (1978:27).

Institutional separatism and duplication characterize inequalitarian pluralistic systems (Smith, 1965; van den Berghe, 1978). This means that each group maintains its own schools, churches, businesses, and so on. In each case, however, there is a great disparity in the quality of dominant and minority group institutions.

In its most extreme form, inequalitarian pluralism resembles a caste system in which strict segregation is enforced in all areas of social life. J. S. Furnivall describes it as follows:

Each group holds by its own religion, its own culture and language, its own ideas and ways. As individuals they meet, but only in the marketplace, in buying and selling. There is a plural society, with different sections of the community living side by side but separately, within the same political unit. Even in the economic sphere there is a division of labour along racial lines. . . . There is, as it were, a caste system, but without the religious basis that incorporates caste in social life in India. (1948:304–5)

Where the extreme of caste is reached, the social distance between groups is maximized in all social situations. What emerges is a system of what van den Berghe (1978) calls "paternalistic" race relations. All people understand their social place and, as long as the subordinates do not deviate from their ascribed role, stability is assured. In such systems there is at least some acquiescence of subordinate groups to their inferior status. Actual physical distance need not be enforced in all situations so long as the *social* distance between groups is understood and adhered to. Slavery in the American South, for example, was characterized by a good deal of physical proximity, particularly among slaves who performed household duties (Stampp, 1956).

Competitive Race Relations   Inequalitarian pluralistic relations are realized in full only in slave or classic colonial systems (Rex, 1970), neither of which is evident in the contemporary world (though South Africa, until recently, closely approximated them). In those industrialized societies where ethnic divisions are based primarily on race rather than culture, ethnic stratification bears some resemblance to a caste or colonial system but is less extreme in the segregation of social institutions and relations. Van den Berghe (1978) has referred to such cases as "competitive" race relations. With the society's economic base changed from agrarianism to industrialism, such competitive relations replace paternalistic relations. Industrialization requires that social roles be assigned more through competition than ascription. As a result, there are no longer "master-servant" relations as in a caste or paternalistic system; however, competition between the subordinate group or groups and the working-class element of the dominant group emerges (van den Berghe, 1978).

The gap between ethnic groups in such a competitive system is narrowed, and each group itself becomes more class diversified. In a slave or colonial situation, ethnicity and class are essentially synonymous: One group has almost all the society's power, wealth, and prestige, and the others have relatively none. In a competitive system, however, class hierarchies emerge within each ethnic group. Of course, as noted in Chapter 2, members of particular groups continue to cluster at well-defined points in the overall class system of the society.

Because the norms of social distance, so well institutionalized in the paternalistic condition, now break down, physical segregation becomes more rigid. Ethnic groups tend to reside and carry out much of their daily activities in homogeneous ghettos. The dominant group's image of the subordinate group or groups also changes, explains van den Berghe, "from one of backward but ingratiating grownup children to one of aggressive, insolent, 'uppity,' clannish, dishonest, underhanded competitors for scarce resources and challengers of the status quo" (1978:30). Accordingly, conflict and bigotry become more virulent.

As we will see in Chapter 9, the plight of African Americans following slavery generally conformed to this pattern.

Internal Colonialism Internal colonialism is a type of inequalitarian pluralism characteristic of societies whose ethnic relations otherwise follow the assimilation model, such as the United States. Here, racial-ethnic groups are treated in a colonial fashion. In the traditional form of colonialism practiced by the European powers in the seventeenth, eighteenth, and nineteenth centuries, a conquering group—a numerical minority—established political and economic domination over a native people. Internal colonialism, by contrast, is a condition in which both the dominant group and subordinate groups are indigenous; moreover, the dominant group is a numerical majority. Classic colonialism, explains Robert Blauner, "involved the control and exploitation of the majority of a nation by a minority of outsiders. Whereas in America the people who are oppressed were themselves originally outsiders and are a numerical minority" (1969:395).

Although the situations from which they emerge are different, classic and internal colonialism nonetheless have common features. Blauner (1969:396) states four components of the "colonization complex": (1) colonization begins with a forced, involuntary entry; (2) the colonizing power alters basically or destroys the indigenous culture; (3) members of the colonized group tend to be governed by representatives of the dominant group; and (4) the system of dominant-subordinate relations is buttressed by a racist ideology.

Using this model, Blauner and others (Blackwell, 1976; Carmichael and Hamilton, 1967; Clark, 1965) have interpreted the African-American experience in the United States as basically a colonial one, and ghetto uprisings as anticolonial movements. This internal colonialism model has also been applied to other racial-ethnic groups in the United States—namely, Mexican Americans and American Indians (Blauner, 1972; Moore, 1970; Snipp, 1986; Thomas, 1966)—and, with some variation, to groups in other societies as well (Zureik, 1979).

Particularly in societies like the United States, where in addition to voluntary immigration, conquest or involuntary immigration has contributed to the society's multiethnic character, the internal colonial model represents an alternative for explaining the inability of racially distinct groups—those that have usually entered involuntarily or who have been conquered—to follow a course of gradual assimilation like that followed by European immigrant groups.[6]

---

[6]A somewhat different application of the idea of internal colonialism is made by Michael Hechter (1975), who describes it as a condition in which a dominant ethnic group in the industrialized core of a nation subjugates an industrially backward ethnic group in the nation's periphery.

Annihilation or Expulsion Inequalitarian pluralism may reach an extreme form in the expulsion or even the annihilation of minority ethnic groups. Neither of these outcomes is without precedent in recent Western history. The deportation of Chinese immigrants from the United States in the nineteenth century and of Germans from Canada during World War I, and the internment of Japanese in both the United States and Canada during World War II, are notable examples of expulsion. As for annihilation, destruction of native groups by white settlers in North America, Australia, and South Africa in the nineteenth century as well as in Latin America earlier are all relevant examples. A more deliberate and methodical case is the genocidal policy of the Nazis, which resulted in the destruction of the European Jews. In the 1930s, German Jews were systematically subjected to an almost complete expulsion from every phase of the society's life. Later, Jews were impelled to leave Germany through a terroristic campaign that included physical attacks and the appropriation of their homes, businesses, and wealth. Finally, Nazi policies culminated in the establishment of death camps to which the German and subsequently other European Jewish populations were sent to be slaughtered. The campaign of "ethnic cleansing" by Serbs against Muslims in Bosnia and ethnic Albanians in Kosovo are more recent cases of a combination of expulsion and annihilation.

THE VARIABILITY OF ETHNIC RELATIONS

The three major patterns of ethnic relations we have examined are outlined in Figure 4.1. Several points should be kept in mind in looking at each.

First, in no case does any of these models by itself perfectly characterize what occurs in any multiethnic society. Realistically, most societies will exhibit features of two or possibly even all three simultaneously. The United States illustrates this quite well. Cultural assimilation has progressed steadily for most Euro-American groups, so that only symbolic traces of the ethnic culture remain in evidence past the third immigrant generation. Structural assimilation has also proceeded apace for most of these groups, though more slowly and far less thoroughly. For African Americans and other racial-ethnic groups, however, inequalitarian pluralism, specifically the internal colonial model, seems appropriate in some degree. African Americans in particular remain highly segregated socially despite substantial cultural assimilation. Thus in societies like the United States, Canada, and South Africa, where physically distinct minorities exist alongside those differing only in culture, there may be two or more sets of ethnic relations in place, all requiring different theoretical models for analysis.

Second, these different processes and outcomes of interethnic relations should not be thought of as mutually exclusive, even for a single group in the so-

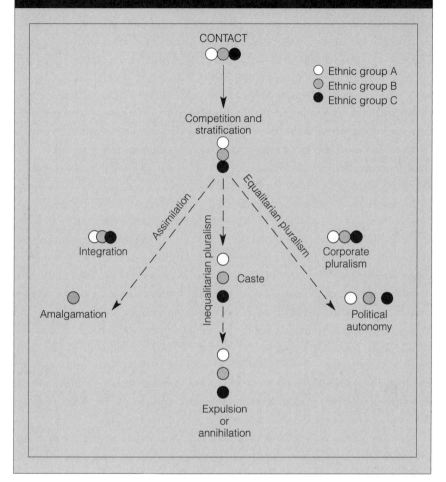

Figure 4.1 Relations among Three Ethnic Groups, Beginning with (1) Initial Contact, Followed by (2) Emergence of Competition and Stratification and (3) Movement in One of Three General Directions

CONTACT

○ Ethnic group A
◐ Ethnic group B
● Ethnic group C

Competition and stratification

Assimilation

Equalitarian pluralism

Inequalitarian pluralism

Integration

Corporate pluralism

Caste

Amalgamation

Political autonomy

Expulsion or annihilation

ciety. Groups can move along two or more paths concurrently; as noted earlier, African Americans have accomplished a high level of acculturation yet continue to display features of caste. Or groups may vacillate, moving in one direction and then another at different times, depending on various factors. As Newman (1973) points out, assimilation and pluralism are not linear.

Third, none of these three processes or conditions are irreversible. Recall Wirth's four types of minorities (Chapter 2), each with different goals; all "contain within themselves tendencies and movements in which we can discern the characteristic features of one or more of these types" (1945:364). Group objectives may thus change at different historical moments. Traditional cyclical theories like Park's or even Gordon's assimilation variables do not deal with groups that initially resist assimilation but then accept it, nor with groups that reassert their distance from the dominant group following substantial assimilation. Such theories cannot easily explain, for example, the shifting goals of acculturation and cultural nationalism among African Americans (Wilson, 1973). Nor can they explain the apparent revitalization of ethnicity in the 1960s and 1970s among Euro-American groups, which for several generations had seemingly traveled well down the path of both cultural and structural assimilation.

Fourth, some degree of conflict is characteristic of all three general patterns. Although it is obvious in the case of inequalitarian pluralism, conflict is a feature in more subdued fashion of assimilation and equalitarian pluralism as well. It must be remembered that these types of interethnic relations are movements in the direction of either homogeneity or ethnic separation with equality, not realized conditions. In either case, some degree of ethnic stratification is evident, and conflict, whether potential or active, is therefore present as well. In societies where assimilation or equalitarian pluralism is more prevalent, the ethnic hierarchy will, of course, be less inflexible and the divisions between groups less acute than in societies where inequalitarian pluralism is the predominant form.

Fifth, any of these processes and outcomes, whether a form of assimilation or pluralism, ultimately depends on the objectives of both the dominant group and minority groups. As noted in Chapter 2, minority groups have different goals regarding their place in the society and the nature of their relations with the dominant group, but this is only one side of the issue. Perhaps more important are the goals the dominant group wishes for the minorities (Schermerhorn, 1970). When the objectives of dominant and minority groups are congruent, conflict is reduced; when they are in opposition, conflict is unavoidable.

Schermerhorn (1970) has suggested that multiethnic societies may experience centripetal and centrifugal forces in regard to dominant-minority group relations. Centrifugal forces are those that drive groups apart, whereas centripetal forces thrust them toward closer relations. In both cases, there may or may not be agreement among dominant and minority groups. An assimilationist goal of a minority group, for example, may be at odds with the segregationist goal of the dominant group. Such opposing goals have been at the crux of black-white relations for most of the past century in the United States, resulting in continual conflict of one form or another. The reverse situation may also be evident. That is, the dominant group may insist on assimilation, and the minority group may resist relinquishing its ethnic culture and social structure. The relationship between American Indians and whites has been typified by this set of incongruent goals. Indians did not seek assimilation, desiring to remain on their traditional lands as independent nations. But whites responded to Indian objectives either by destroying the social and economic base on which their communities were built or by demanding assimilation into Euro-American values and lifestyles to which Indians did not aspire.

Moreover, the dominant group may use manipulation of ethnic relations for political advantage and thus dictate sharp reversals of policy. As Leo Kuper notes, "Politicians readily find in racial and ethnic cleavages a resource for political exploitation" (1969:485). Societies that advocate and display harmony and progressive assimilation among ethnic groups may revert to pluralistic policies when political conditions demand. The Nazis, for instance, determinedly reversed the quite thorough assimilation of Jews into the mainstream German society. Such policy reversals may occur in the opposite direction as well. For example, following the civil rights movement in the American South, many politicians, ordinarily staunchly segregationist, advocated moderately integrationist positions when it became obvious that the black vote might be a decisive electoral factor.

It must be understood, of course, that ethnic groups do not speak and act as one; thus the implementation of goals and policies may not always reflect the wishes of all group members. In all cases, then, we can expect some individuals of both the dominant group and minority groups to deviate from general group processes and objectives.

In sum, all these processes and outcomes of ethnic contact are relations of power—power of minority groups and, of course, power of the dominant group. In any situation of ethnic contact, one must first ask, "Who has power and to what ends can they apply it?" On that question hinges the eventual nature of relations among groups.

A TYPOLOGY OF MULTIETHNIC SOCIETIES

In Chapter 2, we explored the manner in which multiethnic societies are stratified, noting that the degree of ethnic stratification can vary from society to society. These different levels of ethnic inequality can now be combined with the processes of assimilation and pluralism to shape a typology of multiethnic societies (Table 4.2). These descriptions are not meant to serve as pigeonholes into which particular societies can be conveniently placed; any society will probably exhibit some features of all. In broad terms, however, a society will be more characteristic of one or the other. The typology will, then, point out major features of multiethnic societies and will serve as a reference that we can consult in our analyses of specific groups and societies in Parts II and III.

Table 4.2 Three Types of Multiethnic Society

	TYPE OF SOCIETY		
FEATURE	*Colonial*	*Corporate Pluralistic*	*Assimilationist*
Initial contact between dominant and minority groups	Conquest of indigenous groups by dominant group or involuntary migration of minorities	Annexation or voluntary immigration	Mainly voluntary immigration but involuntary immigration and conquest for salient minorities
Relations between dominant and minority groups	Paternalistic or competitive	Ideally equalitarian, but often competitive	Competitive
Nature of stratification	Caste or castelike; caste and ethnicity overlap closely	Class hierarchy within each ethnic group	Class; class and ethnicity generally overlap, but minority group members are dispersed throughout general class system
Segregation between groups	Very rigid; explicitly defined and enforced by tradition and law	Voluntarily rigid; groups ordinarily concentrated in distinct territories	Mild and largely voluntary for groups culturally and physically similar to dominant group; rigid and involuntary for salient minorities
Institutional separation among ethnic groups	High except in economy	High except in economy and central government	Low in polity and economy; variable in other areas
Physical and cultural differences between dominant and minority groups	Sharp physical differences; sharp cultural differences, at least initially	Usually slight or no physical differences; key cultural difference usually language or religion	Broad range of physical types; sharp cultural differences initially
Main objectives of ethnic policy	Inequalitarian pluralism	Equalitarian pluralism	Assimilation; some degree of unofficial structural pluralism for salient minorities
Degree of conflict among ethnic groups	High eventually though usually subdued for long periods	Relatively low except on matters pertaining to cultural and political rights	Variable but generally high between racially distinct groups
Examples	Antebellum U.S. South, colonial India, South Africa under apartheid	Switzerland, Malaysia, Canada (partially)	United States, Brazil, Israel

Colonial Societies

Inequalitarian pluralism is the chief feature of societies that can be called colonial. A dominant group exerts maximum political and economic power and is thereby enabled to shape the nature of interethnic relations in such a way as to sustain its interests.

In these societies, the dominant group has ordinarily entered as a conqueror of physically distinct indigenous groups or has brought the minority groups to the society as slaves. Such societies are ordinarily agrarian or preindustrial, with labor-intensive economies calling for a large supply of unskilled workers; minority ethnic groups assume that role. Undergirding this exploitative system is a racist ideology in which the subordinate place of minorities is deemed natural.

Social segregation between dominant and minority groups is maximized, and the rules of interaction are explicitly defined and enforced by both tradition and law. The dominant group and minority groups develop parallel and duplicative institutions (education, religion, recreation), and only within the polity and economy do they interact with frequency. In these contacts, however, the dominant group is clearly in command.

The socioeconomic and ethnic hierarchies closely overlap in these societies so that, for the most part, defining one's ethnicity is tantamount to defining one's class. Indeed, stratification is more castelike, with members of groups automatically afforded privileges (in the case of the dominant group) or inferior status and low access to the society's rewards (in the case of minorities). There is little or no mobility for individuals or groups within this system.

Given the extremely wide power differentials and the well-entrenched racist ideology, conflict usually remains submerged for long periods, and attempts to upset the system are rare and short-lived. Eventually, however, the forces of both coercion and ideology break down, and the degree of internal conflict becomes great.

The colonial societies and slave societies of the seventeenth, eighteenth, and nineteenth centuries are, of course, the prime examples of this type. In recent times, among the world's nations only South Africa has reflected some of the characteristics of a colonial society.

Corporate Pluralistic Societies

Corporate pluralistic societies are also characterized by cultural and physical separation of ethnic groups, but the key difference between these and colonial societies is the extent to which one group is dominant. In corporate pluralistic societies, the groups are relatively balanced in political and economic power, and no single one is able to exert its will on all vital societal issues. Moreover, group segregation is mainly voluntary.

A basic assumption on which such societies are founded is that ethnicity is not to be discouraged; on the contrary, ethnic groups are expected to retain their identity, uniting only within a common political and economic system. Indeed, the pluralistic basis of the society is usually built into the political framework. Territorial concentration of groups contributes to the retention of ethnic differences and interests.

Given the territorial clustering of groups and the relatively balanced power situation among them, conflict in these societies is ordinarily held to a minimum. Where issues of cultural or territorial integrity arise, however, competition and conflict may become quite intense. In Canada, for example, the dominant English-speaking group is not entirely comfortable with the linguistic rights of French-speaking Quebecers. This has created constant tension between the two groups, at times subdued and at other times heightened.

Assimilationist Societies

The assimilationist society differs from either of the first two types in that there is no recognized obligation or objective in protecting the retention of ethnicity. If ethnic groups survive, they do so because of voluntary actions by the groups themselves or because of informal patterns of prejudice and discrimination—not through the designs of political institutions, as in a corporate pluralistic society, or the segregationist dictates of a dominant group, as in a colonial society.

In these societies, there is a dominant group and a large number of minority ethnic groups, the latter in various stages of cultural and structural assimilation. The dominant group encourages cultural assimilation—to its ways—and most groups retain only expressive aspects of their ethnic cultures beyond two or three generations. Structural assimilation, however, is a different matter. Much social interaction, particularly at the primary level, continues to take place within ethnic subsocieties. Highly visible ethnic groups are most segregated in social relations and thus least structurally assimilated. Those closer to the dominant group in culture and physical appearance are in more advanced stages of structural assimilation, including intermarriage.

Ethnic relations are competitive. Groups compete in the political and economic arenas for power and jobs, as well as in other areas of social life. The nature of socioeconomic stratification is class, not caste; members of various ethnic groups can thus rise (or fall) on their individual merits, at least theoretically. In reality, ethnic groups tend to cluster at certain points in the socioeconomic and political hierarchies so that class and ethnicity overlap to a greater extent than the society's ideology may proclaim. Indeed, for some groups—namely, racial-ethnic groups—castelike features of stratification are often apparent.

In assimilationist societies, a racist ideology explaining the superiority of the dominant group is more restrained in expression and is not accepted by minority group members. Any claims to its legitimacy therefore create conflict.

Because of the more open and competitive stratification system and the lack of an officially sanctioned racist ideology, conflict among ethnic groups may be more prevalent in assimilationist societies than in either of the other types. In the colonial society, people "know their place" and stick to it; in the corporate pluralistic society, conflict is held in check by a relative balance of political and economic power. But in the assimilationist society, minority ethnic groups compete with the dominant group for social positions and power. In this competition, of course, the dominant group retains substantial power resources and thus great advantages.

SUMMARY

What is the nature of relations among different groups in a multiethnic society? Although intergroup conflict seems intrinsic, groups follow one of two general paths: increasing integration or increasing separation. The former is called *assimilation* and the latter, *pluralism*. Each is a process through which groups pass as well as an outcome of group relations.

Assimilation can be viewed at four distinct but related levels: cultural, structural, biological, and psychological. For our purposes, emphasis is placed on the first two. *Cultural* assimilation is the adoption by one ethnic group of another's (usually the dominant group's) cultural traits. *Structural* assimilation is the increasing social interaction between different ethnic groups at both the primary and secondary levels.

Pluralism occurs in two different forms in modern multiethnic societies, *equalitarian* and *inequalitarian*. In the former, groups retain their cultural and much of their structural distinctness but participate on an equal basis in a common political and economic system. In the latter, ethnic groups are also structurally separated; but they do not participate in the society's political and economic institutions in an equitable manner.

Multiethnic societies can be placed in a typology on the basis of several characteristics: the form of contact that brought the different groups together, the character of relations among them, the main objectives of the society's ethnic policies, and the degree of conflict among groups. *Colonial* societies are those in which groups meet through conquest or involuntary migration of minorities. Paternalistic relations within a castelike stratification system are in effect, and segregation between groups is rigid. In *corporate pluralistic* societies, annexation of territory or voluntary migration brings previously separate groups together. Segregation between them is high, usually in distinct areas, though it is mainly voluntary. *Assimilationist* societies emerge mainly from voluntary immigration of groups, though other contact situations may be apparent as well. The degree of segregation between groups varies on the basis of cultural and physical visibility, as does the level of institutional separation and intergroup conflict. Assimilation is the long-range societal objective.

Suggested Readings

Banton, Michael. 1983. *Racial and Ethnic Competition*. Cambridge: Cambridge University Press. Explores the competitive relations among racial and ethnic groups using different societies as illustrative cases. Proposes an application of rational choice theory, commonly used by economists, to race and ethnic relations.

Blalock, Hubert M., Jr. 1967. *Toward a Theory of Minority-Group Relations*. New York: Wiley. Constructs theoretical propositions that explain competition and conflict between dominant and minority groups.

Gordon, Milton M. 1964. *Assimilation in American Life: The Role of Race, Religion, and National Origins*. New York: Oxford University Press. The most complete explanation of ethnic assimilation in its various forms. Presents a theoretical model that remains influential in the study of race and ethnic relations.

Hirschman, Charles, Philip Kasinitz, and Josh DeWind (eds.). 1999. *The Handbook of International Migration: The American Experience*. New York: Russell Sage Foundation. Contains a number of essays by immigration researchers offering new perspectives on assimilation and pluralism.

Horowitz, Donald. 1985. *Ethnic Groups in Conflict*. Berkeley: University of California Press. Examines the sources of ethnic conflict in multiethnic societies, and the possibilities for reducing it, with special emphasis on the developing nation-states of Asia and Africa.

Schermerhorn, R. A. 1970. *Comparative Ethnic Relations: A Framework for Theory and Research*. New York: Random House. Not an easy book to read, but a valuable theoretical work that tries to analyze race and ethnic relations with a set of universally relevant principles.

van den Berghe, Pierre. 1978. *Race and Racism: A Comparative Perspective*. 2d ed. New York: Wiley. A brief, but significant, work that presents a theoretical framework and a useful typology for analyzing race relations. Four multiracial societies are compared: the United States, Mexico, Brazil, and South Africa.

II

ETHNICITY IN
THE UNITED STATES

NOW THAT the basic concepts and theories of ethnic relations have been outlined, we investigate several contemporary multiethnic societies. In Part II, our focus is the United States. So ethnically diverse is American society that many volumes would be required even to begin describing its various groups and the relations among them. For this reason, following a description of the shaping of the American ethnic hierarchy in Chapter 5, several groups, each of them broadly representative of a particular American ethnic experience, are dealt with in depth. Chapter 6 deals with the society's indigenous peoples, American Indians, followed by chapters devoted to Italians, Jews, African Americans, Latinos, and Asians. These five (the latter two are panethnic categories that each include a number of distinct ethnic groups) represent a range of the American ethnic mosaic in their immigration and settlement patterns, their past and present place in the society's stratification system, the extent to which they have been targets of prejudice and discrimination, and the degree to which they have assimilated into the mainstream society.

As discussed in Chapter 2, how an ethnic group enters the society has enormous implications for its subsequent status and the character of relations between it and the dominant group. Those who enter involuntarily, who are conquered, or whose territory is annexed remain in a subordinate position for many generations and maintain a high level of conflict with the dominant group. Relations between the dominant group and subordinate groups who enter voluntarily, however, are less conflictual and result in more rapid and complete assimilation. The American cases examined in the following chapters demonstrate this principle quite clearly. Except for the Asians, whose physical mark proved a more enduring handicap, the voluntary immigrants progressed steadily toward assimilation, though at noticeably different rates. Conflict with the dominant group for them was often profound, but their place in the society was not questioned after a generation or two. Dominant-minority relations have been far more tortuous and conflictual for the groups entering involuntarily (African Americans) or through conquest (American Indians).

A second key factor accounting for differences in group placement in the stratification system and the severity of conflict with the dominant group is physical and cultural visibility. The more distant the group is in appearance and behavior from the dominant group, the lower its social standing will be and the more intense the hostilities it must confront. American ethnic groups display a vast range of cultural and physical traits, and the historical patterns of conflict and accommodation among them are very clear. The highest levels of prejudice and discrimination and the greatest resistance to the attainment of political, social, and economic equality with the dominant group have been experienced by racial-ethnic groups—blacks, Indians, Asians, and to some extent, Latinos.

Even some European groups were racially perceived when they entered the society—namely, those from southern and eastern Europe—and their treatment for a time, especially for the immigrant generation, was similar in form to that afforded nonwhites. But how they were perceived changed as their social position improved. As sociologist Herbert Gans has observed, "Although Italians, Jews, and other participants in the 'new immigration' were treated as dark races when they first came here, their swarthy complexions turned, or were perceived to turn, whiter as their incomes rose" (1982:291).

The analysis of each group in the following chapters is organized around four key issues: the group's sociohistorical development as a minority, its status based on several measures of stratification, current and past patterns of prejudice and discrimination directed against it, and the extent to which pluralist and assimilationist models are applicable to it.

Part II concludes with Chapter 12, which discusses the currently most salient ethnic issues of American society.

5

FOUNDATIONS OF THE AMERICAN ETHNIC HIERARCHY

IN ITS ETHNIC DIVERSITY, the United States is unmatched in the contemporary world. It is a society whose population derives from virtually every region of the world, encompassing people of every imaginable culture, displaying equally varied physical characteristics. Ethnic diversity, however, is not a new American phenomenon. Although its heterogeneity has expanded in recent decades, from the outset of European settlement in the 1600s, a continuous flow of newcomers added fresh ingredients to the American potpourri. It is not an exaggeration to describe the United States as the world's greatest experiment in social relations. No society before it and few since have attempted to blend such a varied and steadily changing mix of human elements.

That America is a nation of immigrants is now a well-worn cliché. The historian Oscar Handlin professed that when he began to write a history of immigrant groups in the United States, he discovered that the immigrants *were* American history. With the exception of American Indians and some Mexican Americans, all ethnic groups in the United States trace their origins to other societies. Even those exceptional groups, of course, were at one time migrants to North America. But how the various groups entered American society, the ways they adapted and were responded to, and their initial placement in and subsequent movement along the ethnic hierarchy all differed enormously.

In this chapter, we trace the formation of the United States as a multiethnic society and the development of its ethnic hierarchy. As explained in Chapter 2, groups may make initial contact in several ways, including conquest, annexation, and voluntary and involuntary immigration. And, that initial form of contact is critical in determining each group's subsequent rate and manner of absorption into the mainstream society. In the case of American Indians, conquest was the nature of contact; and for a few Hispanics, annexation. These cases are discussed in greater detail in Chapters 6 and 10. It has been immigration, however—both voluntary and involuntary—that has been the way most groups entered American society and subsequently took their place in the ethnic hierarchy.

THE ESTABLISHMENT OF ANGLO DOMINANCE

Despite the human mix that characterized the United States almost from its very founding, before the 1830s, the scope of ethnic diversity was not yet great. When the first European settlers arrived, they encountered an indigenous population that, as we will see in Chapter 6, was both culturally diverse and geographically dispersed. By the third decade of the nineteenth century, the indigenous peoples had been reduced to subordination and no longer posed a serious threat to European dominance.

The only other significant non-European group that was part of the American ethnic picture was composed of black slaves who had been brought from Africa beginning in the late seventeenth century. When blacks first entered the society, the slave system was not yet institutionalized. Most blacks were indentured servants, contracted to serve their masters for a certain period of time, usually several years (Franklin, 1980). This was not essentially different from the status of many white indentured servants who had come from England, Scotland, and Ireland. By the late 1600s, however, black slavery was established in several colonies and the slave trade was firmly in place. Slavery in the American colonies at first grew slowly but increased rapidly during the eighteenth century. On the eve of the American Revolution, slaves accounted for over 20 percent of the colonial population (Dinnerstein et al., 1996). Following independence, however, that percentage would begin to decline as Europeans began to enter the society in larger numbers.

Because of their cultural—but especially their physical—distinctness from the European settlers, blacks and American Indians (as well as Mexicans in the Southwest) were, from the first, relegated to the bottom of the emerging ethnic hierarchy; their position would not basically change afterward.

Although their numbers were small, immigrants were not unwelcome. In fact, following the establishment of the American state, a general tolerance for European immigrants was quite evident, and their national origins and even religion were not of great concern. George Washington, speaking to newly arrived Irish immigrants in 1783, expressed this attitude clearly in proclaiming that "The bosom of America is open to receive not only the opulent and respectable stranger, but the oppressed and persecuted of all nations and religions" (quoted in Muller, 1993:19).

It was the British-origin, or Anglo, group—whose defining ethnic features were its northwestern European and Protestant origins—that became the host, or dominant, group. They subsequently set the cultural tone of the society and established its major economic, political, and social institutions. All subsequent groups would be required to adapt to this Anglo-Protestant social and cultural framework. It became the standard, notes historian Arthur Schlesinger Jr., "to which other immigrant nationalities were expected to conform, the matrix into which they would be assimilated" (1992:28). Glazer and Moynihan vividly describe the establishment of a dominant ethnic group in American society with the power to select those who came after it:

> The original Americans became "old" Americans or "old stock," or "white Anglo Saxon Protestants," or some other identification which indicated they were not immigrants or descendants of recent immigrants. These original Americans already had a frame in their minds, which became a frame in reality, that placed and ordered those who came after them. It was important to be white, of British origin, and Protestant. If one was all three, then even if one was an immigrant, one was really not an immigrant, or not for long (1970:15).[1]

Creation of the Dominant Ethnic Culture

Although the Anglo core group has been diluted by successive immigrant waves over many generations and its cultural imprint likewise colored by other groups, its social dominance has remained unwavering throughout American history. "The white Anglo-Saxon Protestant," notes sociologist Lewis Killian, "remains the typical American, the model to which other Americans are expected and encouraged to conform" (1975:16). The societal power of the Anglo core group also remains evident, as indicated by its continued overrepresentation among political and economic elites. It is true nonetheless that its power has been diminished in the past few decades by the increasing penetration of non-Anglo groups into these positions (Alba and Moore, 1982; Davidson et al., 1995).

The makeup of the dominant group itself, however, has changed somewhat as other white Protestant groups have melded with it. Today the dominant ethnic group in American society may be said to broadly comprise white Protestants,

---

[1]Nathan Glazer and Daniel P. Moynihan, *Beyond the Melting Pot*, 2d ed. Copyright © 1970 by MIT Press. Reprinted by permission.

because those Protestants from other northwestern European societies—namely, the Scandinavian nations, Germany, and the Netherlands—have blended almost imperceptibly with those of British origins. White Protestants, then, have varied national roots, but their common Protestantism and racial character have neutralized any meaningful national differences among them. References to the WASP, Anglo, or Anglo-American group, then, should be understood to mean "white Protestants of various national origins." These terms are used interchangeably to denote the dominant American ethnic group and its core culture.

Some might question whether white Protestants constitute an ethnic group in the sense that other more culturally or physically distinct groups do. On this point, Charles Anderson has noted that "as Catholics, Jews, and blacks press in around them, and as internal differences decline, white Protestants increasingly perceive that they, too, constitute a definable group with distinctive social and familial networks, and psychological and cultural moorings" (1970:xiii–xiv).

Dominant Cultural Values Ideologies and public policies concerning American ethnic relations have historically reflected the cultural preferences and the economic and political interests of the WASP, or Anglo, core group in several ways.

To begin with, the preeminence of Anglo cultural values has consistently underlain public policies in education, language, law, welfare, and religion. Thus the ascendancy of the English language, the English legal system, and with few exceptions, the Christian faith, was never seriously challenged. From the beginning, the expectation held sway that entering groups—immigrant, conquered, or enslaved—would conform to this core culture.

At certain times—especially in the early years of the twentieth century, when ethnic diversity seemed to reach a high point—the idea of the *melting pot* called into question the inevitability as well as the desirability of Anglo cultural dominance. Some believed that the cultural differences among the many immigrant groups, so conspicuous at that time, might somehow disappear through a gradual blending process. Different ethnic cultures, it was felt, would eventually fuse into a single "American" culture as a kind of hybrid creation. Although the notion of a melting pot remained symbolically popular for many decades, it never found manifestation in public policy or gained widespread allegiance. In any case, the idea was incomplete because the place of nonwhites in this ideal social brew was never fully dealt with. The melting pot was belied as well by the fact that the boundaries among the European groups did not disappear, nor were their cultures ever entirely diluted by WASP norms and values.

Another competing ethnic ideology that has gained prominence at times is *cultural pluralism*, the notion that various ethnic cultures and communities should be tolerated and protected within a system of political equality. As explained in

Chapter 4, the objective of cultural pluralism, rather than a fusion of diverse ethnic groups, is the preservation of each. In the past, cultural pluralism in the United States never found more than token acceptance at the level of public policy. In recent times, however, schools, universities, and government agencies have acknowledged the pluralistic nature of American society and have made efforts to affirm *multiculturalism* as the society's dominant ethnic ideology (Glazer, 1997).

A more prevalent ethnic ideology, expressed most forcefully in governmental policies for most of U.S. history, is assimilation into the dominant group, or what Gordon (1964) has called *Anglo-conformity*. Groups are expected to shed their ethnic uniqueness as quickly and as completely as possible and take on the ways of the core culture. This expectation always guided the prevailing social thought and policy regarding new ethnic groups; it continues to do so even today, despite the pluralistic rhetoric of recent years.

Some have argued that today there is, in reality, no core culture per se. That is, given the rapidly increasing ethnic heterogeneity of American society in recent decades, it has become increasingly difficult to identify a universally acknowledged core culture (Brubaker, 2001). Here it should be remembered that when ethnic groups meet, there is always an exchange, to some degree, between them. While there is little debate about the increasing diversity of American society, it is equally apparent that, although its effect is waning and it is becoming more indistinct, a core cultural base consisting of a few major elements—language, religion, political norms and values—remains in place.

Dominant Economic and Political Interests The prevalent social thought and policy regarding ethnic groups has most commonly been an expression of the dominant group's economic and political interests. Thus, pressures on ethnic groups to assimilate culturally have not been complemented by corresponding pressures to assimilate structurally, that is, to enter into full and equal relations with the dominant group in all institutional areas. Although Anglo-conformity has generally been expected of ethnic groups in their behavior and appearance, a pluralistic outcome has more often than not typified the dominant group's expectations of interethnic relations. The more culturally and/or physically distinct a group, the more its members have been encouraged to remain "among their own" in the intimate areas of social life—marriage, residence, social clubs, and the like (Table 5.1).

Most important, the efforts of minority ethnic groups to attain economic and political equality have traditionally met with resistance, often of an official nature. Through immigration quotas and exclusionary measures, Indian-removal acts, slave laws, institutionalized segregation, antilabor regulations, voting restrictions, and an array of other measures, dominant interests were protected. Of course, these policies all generated great controversy and conflict, and mi-

Table 5.1 Degree of Assimilation of American Ethnic Groups

GROUP	CULTURAL ASSIMILATION	SECONDARY STRUCTURAL ASSIMILATION[a]	PRIMARY STRUCTURAL ASSIMILATION[b]
Anglo-Protestants (core group)	High	High	High
Northwestern Europeans	High	High	High
Germans	High	High	High
Irish Catholics	High	High	Moderate
Southern and Eastern European Catholics	Moderate	Moderate	Moderate
Jews	Moderate	Moderate	Moderate
Asians	Moderate	Moderate	Moderate
Hispanics	Moderate	Low	Moderate
American Indians	Moderate	Low	Moderate
Blacks	High	Low	Low

[a]As indicated by entrance into economic, political, and other social institutions at various levels.
[b]As indicated by entrance into primary relations with the dominant group in areas such as residence, club memberships, and intermarriage.

nority challenges often met with success. But concessions wrested from the dominant group have always been slow, costly, and incremental.

In sum, for most of American history, public policies and publicly proclaimed ideologies have generally corresponded to the dual aims of, on the one hand, cultural assimilation (specifically, Anglo-conformity); and on the other, structural pluralism, the latter in a form assuring Anglo dominance. These objectives have reflected the American system of ethnic stratification. Those groups least visibly different from the Anglo core group were received with less hostility and presented with an opportunity structure more open and less limited than were those distinctly nonwhite or non-Protestant.

EUROPEAN IMMIGRATION

Except for the Indian and slave populations, the vast majority of Americans in the early nineteenth century were of English and Protestant background. Other, smaller, groups were mainly northwestern Europeans—Scots, Welsh, Scotch-Irish, Dutch, Scandinavians, and German Protestants—all of them culturally and racially close to this core group and, as a result, relatively quickly absorbed into it. Thus the core group represented a blend of relatively similar cultural and racial elements, with the English unmistakably its majority component.

In the third and fourth decades of the nineteenth century, the relative homogeneity that characterized the United States in its formative years began to change with the onset of large-scale European immigration. The variety of groups that would subsequently enter was quite astonishing and would fundamentally reshape the society from that point forward. Indeed, the start of immigration to America marked the beginning of the creation of a multiethnic society whose heterogeneity was unprecedented. In the 100 years to follow, more than 30 million immigrants would come to the United States, making it, as the historian Maldwyn Jones called it, "the classic country of immigration" (1960:2).

After the colonial period and before the modern period of immigration starting in the late 1960s, the United States experienced two major waves of European immigration. Both contributed great portions of the white population, but the principal groups composing each were markedly different in their national origins and their sociological characteristics. As a result, not all immigrant groups were received by the dominant, or core, group in the same way, and their ensuing absorption into the society also varied.

The "Old Immigration": The First Great Wave

With immigration before 1820 at a relatively low level, we can think of the six decades between 1820 and 1880 as the first great wave, or as historians often refer to it, the "old immigration." This period marks the beginning of large-scale immigration to America, bringing millions who would begin to populate every region of the country. Equally important is that for the first time, non-British groups of significant size began to enter the society, although immigration from Britain continued unabated.

Push and Pull Factors One school of sociological analysis of immigration has theorized that immigration is the result of a combination of "push" and "pull" factors. Push factors are those economic, social, and political conditions in the origin society that exert a stimulus for people to leave. Pull factors are conditions that exert a magnetic attraction to a target society. Much of the immigration of this period seems to conform closely to this model.

Immigration during these years was driven primarily by a combination of labor needs in the expanding American economy and the dislocations occurring in Europe due to the breakdown of the agrarian order and the rise of industrialism. In addition to the declining agricultural sector and emergent industrialization, the Napoleonic Wars created much political and economic instability in Europe, providing an additional catalyst for immigration. The United States held out promise of a new beginning for those who were displaced from their land, who were unable to find a place in the newly emergent economic and social orders, or who were fleeing political revolt. These strong push conditions were accompanied by a growing demand for labor, both agricultural and industrial, in the New World. The demand for workers was made more pressing by the abolition of the African slave trade. Without new settlers, the vast resources of the United States could not be exploited and, as a result, efforts to recruit immigrants were made by state and territorial governments, railroads, and emigrant aid societies.

Scope and Variety The initial flow of immigration increased considerably during the two decades preceding the Civil War. Between 1840 and 1860, more than 4 million people entered the United States—equal to about 30 percent of the total free population of the nation in 1840 (Levine, 1992). In proportion to the total population, it was the greatest influx of immigration in American history. By 1860, the population of most of America's largest cities was nearly half foreign-born. Much of this immigration was driven by the establishment of regular steamship voyages between Europe and America. What had been a dangerous and painful—for many, fatal—voyage of weeks by sailing ship became a relatively uneventful crossing of days by steamer. Moreover, this new form of trans-Atlantic travel provided mental comfort to immigrants who were no longer faced with what had seemed an irreversible decision. Now immigrants knew they could more easily return if America failed to meet their expectations or if they simply could not adapt to life removed from the old country. "Steamships," writes historian William Van Vugt, "made the world smaller so that emigration was no longer as great a step as it had been" (1999:13).

The immigrants' origins were varied; they included not only those from Britain and other northwest European societies but also, in the West, Chinese; in New England, French Canadians; and in the Southwest, Mexicans. The two most sizable groups arriving during this era, however, were the Germans and the Irish. After the Civil War, most immigrants continued to come from Germany, Ireland, and Britain, but a substantial number came from Scandinavia as well. Altogether, more than 2 million Swedes, Norwegians, and Danes came to the United States after 1860 (Dinnerstein et al., 1996).

This period marked the entrance not only of non-British people in large numbers but also, for the first time, significant numbers of non-Protestants. Most of the Irish were Catholics, as were about half the Germans. As a result, nativist activities, specifically anti-Catholic actions and rhetoric, became widespread. Catholics were seen by many as a seditious element and Catholicism as basically incompatible with American ways (Higham, 1963). Growing fears of immigration generally and the fear of Catholicism specifically gave rise to a nativist political movement, led especially by the American, or Know-Nothing

Party (after members' refusal to answer questions about their political activities). Its single issue was opposition to immigration and, as the largest non-Protestant group, its clearest target was the Irish. Catholicism was seen as a creeping menace, transmitted by the Irish, that would eventually destroy the society's political and religious liberties. Many in the Know-Nothing Party were respectable and sincere reformers who believed that the Irish Catholic presence would have a corrosive effect on American society (McCaffrey, 1997).

By mid-century, however, anti-Catholic nativism had begun to wane as national attention turned to the issue of slavery and as immigration was temporarily halted by the Civil War. Moreover, as the American industrial economy surged, immigration generally was welcomed by industrialists and others of the business community who saw immigrants as a source of cheap labor and a growing consumer market.

Despite early prejudice and discrimination toward the Catholics among them, particularly the Irish, these northwestern European groups were, by comparison with groups that would follow them, advantaged in two ways: they were close enough culturally to the Anglo core group to assimilate within a relatively short historical time, and they were physically indistinct from the Europeans who had preceded them to America, thereby avoiding long-lasting imputations of racial identity. Thus, their integration into the mainstream society was not severely constrained, and the resistance of the dominant group to them was, in historical perspective, relatively mild. Let's look briefly at the two major groups of this period of immigration.

The Irish Irish immigration to America was marked by almost classic push-pull factors. A disastrous potato blight swept Ireland in the late 1840s, driving the rural peasantry off the land. Famine and disease were rampant, resulting in perhaps a million deaths (Miller, 1985). For most, the choice was simple: emigrate or perish. As historian Lawrence McCaffery has written, most Irish Catholics arriving in the United States during this period "were running away from destitution and oppression rather than rushing toward freedom and opportunity" (McCaffery, 1997:7). Though not as compelling as the push of a deteriorating rural economy, the pull of the United States, and to a lesser degree Canada, nonetheless seemed irresistible. North America, with seemingly unlimited labor opportunities, was for these emigrants a logical destination. The response was overwhelming. Between 1845 and 1855, about one-fourth of Ireland's entire population emigrated overseas, almost 1.5 million sailing to the United States and another 340,000 to Canada (Miller, 1985).

Most Irish immigrants were illiterate peasants who found work in railroad and canal construction, as stevedores on docks at major ports, and in the various building projects of the expanding cities. For Irish immigrant women, domestic service became the most common occupation. Wherever they settled, the Irish were disproportionately concentrated in the lowest-paid, least-skilled, and most dangerous and insecure jobs (Glazer and Moynihan, 1970).

In the decades following the Great Famine, push and pull factors stimulating Irish immigration—enduring rural poverty in Ireland and the promise of jobs in North America—did not essentially change. As a result, Irish immigration to North America continued with the characteristics established earlier. Most immigrants, women as well as men, were poor, from rural areas, young, unskilled, and unmarried. With the continued flow of Irish immigration, large Irish-American communities were established, which in turn financed the passage of future immigration from the island.

By the end of the American Civil War, the Irish constituted 7 percent of the white U.S. population. The large and relatively rapid entrance of Irish Catholics between 1830 and 1850 was the most significant factor that fueled the nativism of this period. As noted earlier, anti-Catholicism was the core of nativism, and the Irish were portrayed as the carriers to America of a potentially polluting influence. Although no group escaped the anti-immigrant sentiment, as Catholics and as mostly poor peasants, the Irish were viewed with particular disdain. They were subjected to virulent and often militant prejudice and discrimination, expressed by the familiar "No Irish Need Apply" notices displayed by landlords and employers. Much of the anti-Irish sentiment in America was simply a carryover of the intensely scornful view of English Protestants toward Irish Catholics. "Nativists in Britain and the United States," writes McCaffrey, "despised Irish Catholics as cultural and political subversives, treacherous agents of authoritarianism, ignorance, and superstition, the key ingredients of popery." (McCaffrey, 1997:2). The fact that the Irish in America were overwhelmingly poor, uneducated, and living in sordid urban conditions reinforced these prejudices. Additional resentment of the Irish arose from native workers, who accused them of depressing their wages and threatening their jobs.

It is interesting to consider how the Irish, the first among many immigrant groups to come to the United States who did not closely conform to the dominant group, were viewed in terms of race and ethnicity. Their case demonstrates the fluid nature of "race" and how it can be altered. The racial status of the Irish was indeterminate for many decades as they shared a place at the bottom of the labor hierarchy with African Americans. Historian Noel Ignatiev writes that "Strong tendencies existed in antebellum America to consign the Irish, if not to the black race, then to an intermediate race located socially between black and white" (1995:76). Only later were the Irish assigned to the "white race."

The Germans Germans had settled in America as early as the seventeenth century; they were, in fact, the largest non-British group during the colonial period. But German immigration developed on a massive scale beginning in the 1830s, with the largest numbers coming in two great waves. In just seven years, from 1850 to 1857, almost 1 million Germans arrived; and between 1865 and 1885, 2 million more. By the late nineteenth century, Germans were the second largest ethnic group in the society, exceeded only by the British. Whereas most of the Irish remained in eastern cities, particularly New York and Boston, German immigrants dispersed westward throughout the Northern tier of states from New York to the Mississippi River. Many settled in several large Midwestern cities, including Cincinnati, Milwaukee, St. Louis, Detroit, and Chicago. There they established ethnically homogeneous neighborhoods—so-called *kleindeutschlands,* or Little Germanys—that would grow with the continued arrival of new immigrants (Dinnerstein et al., 1996; Rippley, 1984).

The class origins of the Germans were more varied than those of the Irish, who tended to be poor and unskilled. Most Germans were tradesmen, industrial workers, small shopkeepers, farmers, and professionals; only a small percentage were unskilled laborers (Keil, 1991). With the growth of manufacturing, German craftsworkers became the core of the burgeoning U.S. industrial working class (Levine, 1992). Many of the craftsmen and merchants took up residence in small towns while farmers, attracted by rich lands, settled in large numbers in Midwestern states like Wisconsin, Illinois, and Ohio. In religion, too, the Germans were a more diverse immigrant population. Roughly half were Protestant and half Catholic, but there was a small German-Jewish element as well.

In addition to economic circumstances, some German emigrants were driven by political crisis. One small but very influential segment of the German immigration of this period was a group known as "Forty-eighters." Forced out of Germany because of their participation in an unsuccessful revolution in 1848, they took political refuge in the United States. They subsequently exerted an inordinate influence not only on German-American affairs, but on national issues as well. They were vehemently antislavery, as well as anticlerical, and they sowed the seeds of the American labor movement that emerged in the late nineteenth century (Levine, 1992).

The "New Immigration": The Second Great Wave

Although substantial German and Irish immigration had presented the first real challenge to WASP dominance, the polyglot nature of American society took shape most forcefully through what has been called the "new immigration." During a period extending roughly from the early 1880s until the outbreak of World War I, almost 25 million European immigrants came to the United States. This influx represents the archetypal American immigration, and it is this period that we most often think of when we visualize traditional immigrant life. It is from this wave of immigration that most of today's non-Protestant Euro-Americans are descended. Indeed, these are the groups ordinarily understood to be, and often referred to as, "white ethnics."

These new immigrants not only represented the numerical peak of European immigration to the United States but also radically changed the ethnic character of the society. To begin with, most were from southern and eastern European societies, whose cultures were alien to those of Britain and northwestern Europe. Most important, they were non-Protestant—primarily Catholic and secondarily Jewish. Although many of the earlier Irish were Catholic, in other respects they were not so culturally distant from the Anglo-Americans. Likewise, the Germans may have spoken a different language, but over half were Protestant, and they were not looked on by the dominant Anglo group as cultural inferiors. Finally, the new immigrants were mainly people whose class origins were lower than those of previous groups, except perhaps the Irish. They were chiefly peasants from the poorer states of Europe—southern Italy, Russia, Poland, Greece, and the many parts of the Austro-Hungarian Empire.

The shift in immigrant origins was brought about by a number of factors. First, the growth of industry in Britain and Germany furnished more labor opportunities for their populations. Hence, the pressures to migrate were reduced. German immigration to the United States, for example, dropped from 1.5 million in the 1880s to 0.5 million in the 1890s (Dinnerstein and Reimers, 1988). At the same time, problems experienced in northern European countries earlier now spread to southern and eastern Europe, creating food shortages, high unemployment, and an incentive to leave.

Pull factors also changed. With the closing of the Western frontier, land was no longer the chief lure of immigrants to the United States. The expanding American industrial economy, however, now exerted an even stronger pull. Factories, mines, and mills demanded ever greater numbers of workers, and immigrants supplied the major part of this need.

One important characteristic of the new immigration was that many who went to America expected to return to their origin societies. Consequently, a major portion of the immigration comprised men unaccompanied by their families. Their objective was to earn enough money to enable them to return to their native towns or villages and purchase land or establish themselves in an enterprise. Many did just that. Although records are nonexistent and only rough estimates can be made, perhaps one-third of all immigrants during this period remigrated back to their homelands or to other societies (Thistlethwaite, 1991).

The largest of the new immigrant groups were Italians, mostly from the south of Italy, and eastern European Jews, mostly from Poland and Russia. Both tended to concentrate in cities of the Northeast, especially New York. Another major component of the new immigration was Slavic groups, including Russians, Ukrainians, Slovaks, Slovenes, Croatians, Serbs, Bulgarians, and especially Poles.

Together, these groups made up about 4 million of the new immigrants. Other central European groups, including Hungarians, Czechs, and Armenians, also were part of the immigration of this era. Although like Italians and Jews, many stayed close to where they disembarked, most went further inland to the industrial cities of the Great Lakes and the coal-mining regions of western Pennsylvania and Ohio.

Settlement Patterns The settlement patterns of the new immigrants differed in some respects from the old. Whereas many immigrants of the earlier period were attracted by the opportunity to own land and thus settled in rural areas, the new immigrants were mainly industrial workers who settled in cities and towns that supported those industries. As a result, rather than dispersing geographically, the new immigrants tended to concentrate in urban areas, particularly of the Northeast and Midwest. It was here that industrial expansion was occurring, and the most alluring work opportunities were being offered. The growth of these industrial centers was rapid and profound. From 1900 to 1920, Milwaukee's population, for example, increased by more than 60 percent; Cleveland's by over 120 percent; and Detroit's by almost 250 percent (U.S. Census Bureau, 1920). Although it was aided by a steady rural-urban movement of farmers and lumbermen, the primary elements of this population increase were the foreign-born. As seen in Table 5.2, the foreign-born comprised over 30 percent of the population of seven of the fourteen largest cities in 1900.

Once they entered the United States, the different immigrant groups migrated to various cities and regions, determined in large measure by the groups' occupational proclivities. Particular groups seemed to be channeled to specific industries. The Polish immigration illustrates this tendency. Census labor reports show that the overwhelming majority of Polish immigrants were classified as laborers or workers in manufacturing; few were found in agriculture or the professions, in clerical or sales positions, or in building trades (Hutchinson, 1956). In a sense, the Polish immigrant to the United States was a migrant laborer, not so much an immigrant as a "temporary worker," intent on returning home. Historian Caroline Golab (1977) explains that prior to their emigration, Poles already had much experience in the industrial sphere because they had been migrating to various European societies during the late nineteenth century. "It was rare," she writes, "to find a Polish immigrant who had not migrated to some other part of Europe or within his own country before coming to America" (1977:74). Hence, Poles were familiar with town and city life and with machines and factories. Quite logically, then, upon arriving in the United States they migrated to the same industries that had used them in Europe, such as heavy metals and mining. This accounts for the readiness of Poles to settle in cities of the Great Lakes, like Chicago, Buffalo, Detroit, and Milwaukee, as well

Table 5.2 Foreign-Born in the Fourteen Largest U.S. Cities, 1900

CITY	TOTAL POPULATION	TOTAL FOREIGN-BORN	PERCENT FOREIGN-BORN
New York	3,437,202	1,270,080	37.0
Chicago	1,698,575	587,112	34.6
Philadelphia	1,293,697	295,340	22.8
St. Louis	575,238	111,356	19.4
Boston	560,892	197,129	35.2
Baltimore	508,957	68,600	13.5
Cleveland	381,768	124,631	32.7
Buffalo	352,387	104,252	29.6
San Francisco	342,782	116,885	34.1
Cincinnati	325,902	57,961	17.8
Pittsburgh	321,616	84,878	26.4
New Orleans	287,104	30,325	10.6
Detroit	285,704	96,503	33.8
Milwaukee	285,315	88,991	31.2

SOURCE: Twelfth U.S. Census, 1900.

as in the anthracite region of western Pennsylvania, where these industries were most prevalent at the turn of the century and in the several decades afterward. As we will see in Chapters 7 and 8, similar factors operated for immigrant Italians and Jews in accounting for their occupational inclinations.

The new immigrants in the cities tended to settle in depressed areas, inhabiting run-down tenements that had been vacated by earlier ethnic groups such as the Irish. The concentration of immigrants in these areas was partially the result of simple economic necessity; such housing was available and affordable. The creation of ethnic enclaves in large cities, however, was also the product of the natural attraction of immigrants to areas inhabited by those who had preceded them to America and came from the same towns and regions. New immigrants sought the security and support of coethnics and thus gravitated to common areas. "Life in the Italian quarter," writes Virginia Yans-McLaughlin, "provided a coherence and familiarity which drew immigrants irresistibly toward it" (1982:118). It was much the same for other groups. As the skeleton of the ethnic community was shaped, others were attracted, creating in turn the need for larger and more complex ethnic institutions. Thus, in large measure, immigration was a chain phenomenon.

More than ethnic needs, however, residential clustering and the building of ethnic communities was mostly a product of the attraction of work opportunities in specific locations. In their classic study of Polish immigrants, Thomas and Znaniecki (1918) explained that in large cities, the desire to live near work was a stronger consideration than the desire to live near one's fellow ethnics. Thus, institutions catering to immigrants—churches, groceries, banks—followed the residential concentration created by employment opportunities. Moreover, one should not exaggerate the ethnic homogeneity of these urban areas, as movies and novels have so often portrayed them. Although there certainly were neighborhood concentrations of immigrants and their children from particular countries, these pockets were never as ethnically pure as has been commonly assumed, nor did they endure much beyond the first two immigrant generations (Hershberg et al., 1979).

Reemergent Nativism and the End of the Second Wave The sharp cultural and class differences of the new immigrants provoked a strongly antagonistic reaction from the dominant group. Americans questioned how the immigrants would ultimately fit into the society and, abetted by popular notions of social Darwinism, widely assumed their cultural inferiority (Jacobson, 1998).

Prompted by the continued influx, nativist feelings and racist theories reached an apex by the early 1920s, resulting in restrictive quotas that curtailed large-scale European immigration once and for all. The quotas were deliberately designed to favor the sending countries of northwestern Europe and to limit immigration from southern and eastern European countries, the source of most immigrants of this era. Previously, with no quotas and few stipulations other than health requirements, an open door, in effect, had existed for Europeans. The restrictive legislation represented a symbolic closing of the door. Further curtailing immigration from Europe was the outbreak of World War I in 1914. Immigration resumed at high levels after the war, but the quotas that were established in 1924 changed the makeup of immigration back to mainly northwestern Europeans. Moreover, immigration to the United States of all groups dropped precipitously from 700,000 in 1924 to fewer than 300,000 the following year. Not until the early 1970s would the number again approach the level of immigration reached prior to the mid-1920s.

In addition to restrictive quotas, immigration was discouraged by the Great Depression of the 1930s. Unemployment reached record levels, and the United States was therefore no longer seen as an attractive destination. Fewer immigrants arrived during the 1930s than during any previous decade in a hundred years. In fact, more people left the United States during this time than entered as immigrants. As the world again plunged into war in the late 1930s, a further obstacle to immigration was imposed. The currents of American immigration can be seen in Figure 5.1.

Table 5.3 shows the major ancestry groups that currently make up the Euro-American population, which derives primarily from the two main periods, old and new, of European immigration.[2]

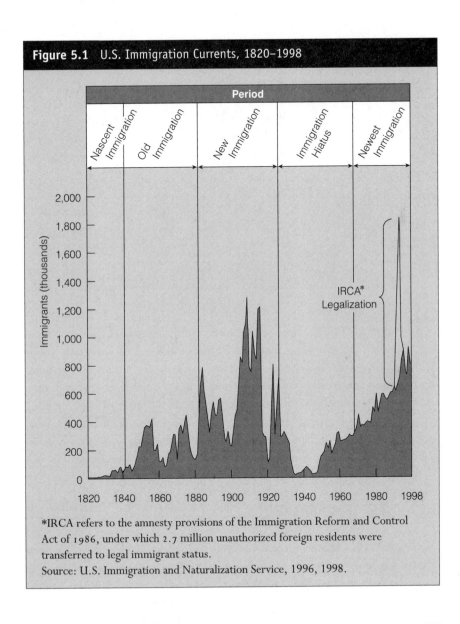

Figure 5.1 U.S. Immigration Currents, 1820–1998

*IRCA refers to the amnesty provisions of the Immigration Reform and Control Act of 1986, under which 2.7 million unauthorized foreign residents were transferred to legal immigrant status.
Source: U.S. Immigration and Naturalization Service, 1996, 1998.

---

[2]The diversity of the European component of the American ethnic configuration is actually greater than shown here, because only the largest ancestry groups are included.

Table 5.3 Largest Euro-American Ancestry Groups

ANCESTRY GROUP	NUMBER	PERCENT U.S. POPULATION
German	46,452,000	17.0
Irish	33,027,000	12.1
English	28,255,000	10.3
Italian	15,904,000	5.8
French	9,768,000	3.6
Polish	9,050,000	3.3
Scottish	5,419,000	2.0
Scotch-Irish	5,223,000	1.9
Dutch	5,219,000	1.9
Norwegian	4,547,000	1.7
Swedish	4,322,000	1.6
Russian	2,987,000	1.1

SOURCE: U.S. Census Bureau, 2001i.

Immigration to Other Societies The immigration beginning in the early nineteenth century and extending into the first few decades of the twentieth was by no means limited to the United States. Other societies that attracted large numbers of immigrants from Europe during this great age of international migration included Brazil, Australia, Canada, and Argentina. In the latter two countries, in fact, immigration rendered an even more profound demographic effect, proportionately, than in the United States (Thistlethwaite, 1991). However, in absolute numbers and ethnic variety, American immigration was far greater.

THE NEWEST IMMIGRATION

As we will see in Chapter 9, the civil rights movement of the 1960s proved to have far-reaching consequences for American society, resulting in fundamental changes in the society's ethnic order. No groups were unaffected. But something else occurred in the 1960s that, arguably, had an equally significant impact on ethnic relations in the United States. This was the revision of immigration laws, encapsulated in the Hart-Cellar Act of 1965, abolishing the restrictive quota system that had effectively limited immigration to those from Europe, especially northwestern Europe. The new legislation not only erased the country-of-origin quota system that had virtually prohibited certain groups from immigrating to the United States, but it also created a system of criteria for admission. Priority was given to those with family ties to U.S. citizens or permanent residents and to those with needed occupational skills.

Several driving forces provoked this truly radical change in immigration laws. First, the civil rights movement itself provided an initial impetus. As an adjunct to the mounting pressure for an end to discriminatory measures against African Americans, public sentiment now seemed prepared to do the same with immigrants, bringing the reality of immigration laws into line with the rhetoric of equal opportunity. Other factors contributing to the acceptability of change were the sentiments and efforts of the new presidential administration that entered Washington in 1961. John F. Kennedy, in his 1960 campaign for the presidency, had explicitly pledged to work toward the removal of discriminatory immigration regulations. As with much of the civil rights legislation of the mid-1960s, the revision of immigration laws was enacted in large measure as a tribute to Kennedy after his assassination in 1963. In 1965, when the revision was passed, few predicted the massive numbers that would enter the United States as a result, nor did they foresee the profound impact this legislation would subsequently have on the ethnic makeup of the society.

The ensuing massive and immensely diverse immigration has led to an unprecedented level of heterogeneity of the U.S. population. The two previous periods of substantial immigration, occurring in the nineteenth and early twentieth centuries, had been primarily European in origin. Though resisted by nativist forces, these immigration periods did not essentially threaten the basic cultural and racial makeup of the society. Until the last two decades, the connotation of *immigrant* in the United States was "of European descent." What has occurred in the modern period of immigration, however, is the onset of a movement which assures that the society's ethnic configuration, and thus its culture, will not be in the future what it has been for most of its history. To distinguish it from previous large-scale immigrations to the United States, the most recent wave may be referred to as the *newest immigration*.

Characteristics of the Newest Immigration

Scope   During the three decades between 1960 and 1990, 15 million people either entered the United States as legal immigrants or were granted permanent residence. Nine million came during the 1980s alone, exceeding the total of the first decade of the 1900s, when immigration to the United States was at its peak (Figure 5.1). Another 8 million were admitted between 1990 and 1997 (Martin and Midgley, 1999). Today legal immigration to the United States fluctuates between 700,000 and 900,000 per year, a number larger than the total immigration to all other developed countries together.

The figures for legal immigrants and refugees, of course, do not take into account those who enter illegally. No one can be certain of their numbers, but government estimates are in the range of 300,000 each year. As of 1996, an estimated 5,275,000 immigrants were residing in the United States illegally (U.S. Immigration and Naturalization Service, 2001). More than half were from Mexico alone, but undocumented immigrants come from the same broad range of countries as legal immigrants.

The foreign-born population of the United States has today reached an all-time high. Even during the peak immigration years of the early twentieth century, immigrants numbered only about half of what they do today (Martin and Midgley, 1999). However, when looked at in proportionate, not absolute, terms, immigrants today do not have as great a demographic impact as they did in the early 1900s. At that time, immigrants constituted about 15 percent of the total population, compared to 10 percent today. Even adding illegal immigrants would not increase that figure significantly. Nonetheless, it is very obvious that immigration in all forms continues to play a large role in U.S. population growth and will play an even greater role in the future (Heer, 1996; Martin and Midgley, 1999; Smith and Edmonston, 1997).

Ethnic Makeup   Numbers alone do not tell the whole story of the current period of American immigration. Perhaps more important, the character of immigration has been radically changed from that of past eras. Whereas the vast majority of immigrants throughout the nineteenth and early twentieth centuries were from European societies, today most are from Latin America and Asia (Figure 5.2). The ramifications of that change have had a ripple effect throughout virtually all economic, political, and social institutions.

Since the 1970s, the leading countries of origin of legal immigrants to the United States have been Mexico, the Philippines, Korea, and four small Caribbean nations—the Dominican Republic, Jamaica, Haiti, and Guyana (a South American country that considers itself Caribbean in its economic and social structure). But the sheer diversity of non-European immigrants is quite astonishing. Among the Asians, for example, are Filipinos, Koreans, Indians, Pakistanis, Chinese, Vietnamese, Cambodians, Laotians, and Hmong. From the Middle East have come Palestinians, Iraqis, Iranians, Lebanese, Syrians, Egyptians, and Israelis. Those from Africa include Somalis, Ethiopians, and Nigerians. And from Latin America, the national origins of the immigrants cover the entire Central and South American regions, in addition to the Caribbean. Figure 5.3 shows the leading countries of origin of immigrants in 1999.

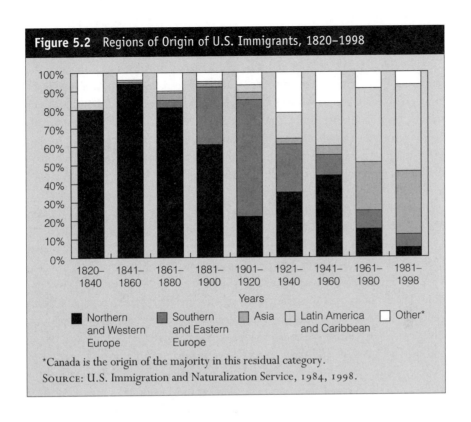

Figure 5.2 Regions of Origin of U.S. Immigrants, 1820–1998

Years

- ■ Northern and Western Europe
- ▨ Southern and Eastern Europe
- ▦ Asia
- ☐ Latin America and Caribbean
- ☐ Other*

*Canada is the origin of the majority in this residual category.

SOURCE: U.S. Immigration and Naturalization Service, 1984, 1998.

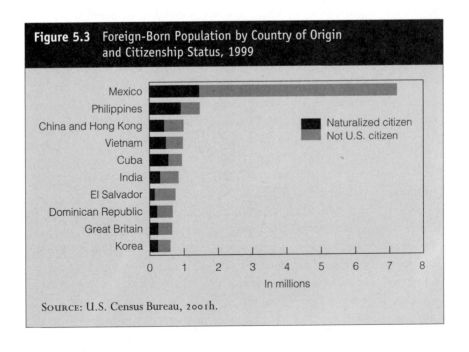

Figure 5.3 Foreign-Born Population by Country of Origin and Citizenship Status, 1999

Mexico
Philippines
China and Hong Kong
Vietnam
Cuba
India
El Salvador
Dominican Republic
Great Britain
Korea

- ■ Naturalized citizen
- ▨ Not U.S. citizen

In millions

SOURCE: U.S. Census Bureau, 2001h.

Motivating Forces What has prompted this newest large-scale immigration to the United States? Obviously the changes in immigration law in 1965 provided the favorable legal conditions and thus the necessary vehicle. But essentially the same push-pull factors that motivated past immigrations to America have been functioning for the newest wave. Although many have come as political refugees (mostly from Southeast Asia, Cuba, and Central America), most immigrants have responded to the population and economic pressures of their societies of origin. Most, in other words, seek economic betterment and improved social conditions. A national survey asked foreign-born adults living in the United States what they preferred about America in comparison with their homeland. Some answered political freedom, fair treatment under laws, safety from crime, or moral values. But the most common answer was job opportunities (Saad, 1995).

Contemporary immigration to the United States, as well as to other industrialized societies of North America, Europe, and Australia, accelerated in the 1960s in large measure because of the widening gap between rich and poor nations. The economies of the latter, many of them newly independent states, were generally incapable of supporting rapidly growing populations, thereby creating a migration push. Most of the world's population growth has occurred since the end of World War II, and the overwhelming majority of that growth has taken place in the less developed countries. Moreover, the gains in economic development made by many of these countries during the 1960s and 1970s slowed or reversed in the 1980s due to economic recession, growing debt, and internal political conflicts. These conditions induced further pressures to migrate to the wealthier countries, like the United States, whose industrialized economies promised better employment opportunities and a superior quality of life.

Poverty and overpopulation, however, cannot alone explain the emergence and steady flow of migration streams to the United States and other industrialized countries. Additional factors help induce people to migrate. In the U.S. case, large inflows among the newest immigrants in recent years have been sustained by the changing structure of the U.S. economy, especially the expansion of low-wage jobs (Sassen, 1998). The newest immigration, then, must be seen as part of the internationalization of capitalist economies and the resultant globalization of the workforce (Castles and Miller, 1998; Chaney, 1979).

Ironically, much of the migration to the United States from underdeveloped societies has been induced by American political and economic influences. As American-based multinational corporations bring capital and technology to these developing economies, new wealth is created, but it remains highly maldistributed. Thus, the expectations of work and consumption opportunities are frustrated, leading to a natural gravitation to the United States, which promises to deliver these advantages. "It is only natural," notes Alejandro Portes, "that many trek North in search of the means to acquire what transnational firms and the mass media have so insistently advertised for years" (1979:434).

Geography Although the newest immigration has affected virtually all regions and states, it still is most evident in a few states, namely California, New York, Illinois, Texas, and Florida. Similarly, though immigrant groups have dispersed widely, specific groups remain concentrated in particular regions. Asians, for example, have settled mostly in California; Mexicans, in the Southwest; Cubans and other Caribbeans in Miami; and almost all new groups have large communities in New York City. So great has the change been in California that the state has been transformed into one where non-Hispanic whites are no longer the numerical majority.

Like so many American immigrants of past generations, many today come with few expectations of remaining permanently in the United States. A great number do leave, and many (particularly from Mexico) engage in a continual back-and-forth movement. But also like their predecessors, many who at first see their immigration as temporary become firmly ensconced in their new society. The result has been, as in earlier eras of immigration, the flourishing of ethnic enclaves in cities that have attracted large numbers of new immigrants. Much as New York, Chicago, and other cities of the East and Midwest served in the nineteenth and early twentieth centuries as meccas of European immigration, cities of the Sun Belt—Los Angeles, Miami, Houston—are the focal points of the newest immigration. The immigrant population in Los Angeles, for example, is almost a third of the metropolitan area's total, equivalent numerically to the foreign-born population of New York City during the 1920s (Muller, 1993; U.S. Census Bureau, 2001m). The ethnic diversity of the metropolitan area is reflected in the fact that students in the Los Angeles School District speak more than eighty languages. In Miami the trend is much the same; over 40 percent of the metropolitan area's population is foreign-born. In 1969, 19 percent of Miami-Dade County's students were Hispanic; in 2000 they were more than half, with non-Hispanic whites only 12 percent (Nazareno, 2000).

Sun Belt cities, however, have by no means monopolized the newest immigration. New York remains a traditional port of entry for new ethnic groups. The city has absorbed especially large numbers of newcomers from the Caribbean basin, including English-speaking West Indians from Jamaica, Trinidad and Tobago, Barbados, and Guyana; French and Creole-speaking Haitians; and Spanish-speaking groups from the Dominican Republic, Cuba, Colombia, and Mexico. Hispanics from virtually every Caribbean, Central American, and South American country comprise the largest component of New York's newest immigrants, but large numbers of Asians have also settled in the city, including Chinese, Koreans, Filipinos, and South Asians from India, Pakistan, and Bangladesh. Even Europeans entered in significant numbers in the late 1970s and early 1980s, most notably thousands of Soviet Jews (Foner, 1987; Muller, 1993; Winnick, 1990). So great has been the immigrant influx that Euro-Americans now constitute less than half of the city's total population (Mumford Center, 2001c). Other eastern and

Midwestern cities have also attracted large new immigrant populations. Table 5.4 shows the growing ethnic diversity of the largest U.S. metropolitan areas, most of which has been the result of increasing levels of immigration.

Although most of the newest immigrants have focused on living in the largest cities, virtually no U.S. city or town has been unaffected. Many immigrants have leapfrogged over the central cities and settled in suburban communities, a pattern not common in the past. As an example, consider Bridgewater, New Jersey, an affluent township forty miles from New York City. Seemingly immune to ethnic changes occurring in the City and its surrounding communities, Bridgewater had been a solidly white, middle-class enclave. Its Asian population had been almost invisible in 1980, but twenty

Table 5.4 Ethnic Composition of the Largest U.S. Metropolitan Areas

METROPOLITAN AREA	WHITE	BLACK	HISPANIC	ASIAN	AMERICAN INDIAN
New York	57.4%	19.3%	16.8%	6.2%	0.3%
Los Angeles	42.0	8.4	37.8	11.1	0.7
Chicago	63.5	19.3	13.1	3.9	0.2
Washington-Baltimore	64.3	25.7	4.9	4.8	0.3
San Francisco	53.9	8.7	18.6	18.1	0.7
Philadelphia	73.0	19.4	4.7	2.7	0.2
Boston	85.6	5.7	5.1	3.4	0.2
Detroit	74.4	21.0	2.3	1.9	0.4
Dallas–Fort Worth	66.7	14.2	15.1	3.4	0.6
Houston	52.9	18.3	23.6	4.8	0.4
Atlanta	68.4	25.8	3.0	2.6	0.2
Miami–Ft. Lauderdale	41.8	19.6	36.6	1.8	0.2
Seattle	82.0	5.0	4.0	7.7	1.3
Cleveland	79.8	16.5	2.2	1.3	0.2
Minneapolis–St. Paul	89.2	4.4	2.0	3.4	1.0
Phoenix	71.9	4.0	19.5	2.2	2.4
San Diego	57.4	6.4	25.1	10.2	0.9
St. Louis	79.7	17.6	1.3	1.2	0.2
Pittsburgh	90.0	8.3	0.7	0.9	0.1
Denver	77.2	5.2	14.0	2.8	0.8

SOURCE: U.S. Census Bureau, 1998a.

years later it was 10 percent Asian, and Bridgewater schools were teaching students with forty-one native languages (Chen, 1999).

Class Range Another evident feature of the newest immigration is its class range. Unlike past great waves of immigration, in which most immigrants were either poor or of modest means, the newest influx is made up of individuals and families who span the entire class spectrum, from low-level, unskilled workers to highly educated professionals. The result of this socioeconomic variety has been equally diverse patterns of settlement and adaptation among immigrants. These issues are discussed in Chapters 10, 11, and 12.

Gender During previous large-scale immigration periods, male immigrants outnumbered females among most groups; the newest immigration is more gender balanced. Females actually comprise a majority among immigrants from some countries. Over half of all immigrants from the Philippines, Korea, Jamaica, and the Dominican Republic—each a source country sending large numbers to the United States—are women (Rumbaut, 1996). Immigrant women have disproportionately filled jobs in the low-wage service sector—like hotel and restaurant workers—and in certain labor intensive industries. In San Francisco, for example, the garment industry is the largest manufacturing sector, employing over 25,000 workers; 90 percent are women, most of them Asian immigrants (Louie, 2000).

Middle Eastern Americans

In Chapters 9, 10, and 11, we discuss many elements of the newest immigration, particularly those who are part of the Hispanic-American and Asian-American panethnic categories. A group that is beginning to emerge with greater clarity as part of the American ethnic mix, but one that has not been clearly classified as part of any of the society's broad ethnic categories, includes peoples from the Middle East. They serve as an excellent case to demonstrate how the United States has accommodated new and, from the standpoint of native-born Americans, culturally distant groups during the past four decades.

Small communities of people from Middle Eastern societies have actually been in the United States since the late nineteenth century, and most Middle Eastern Americans today trace their origins to immigration waves occurring prior to World War II. But, like other small groups with a relatively long U.S. history, their numbers swelled starting in the 1970s as part of the newest immigration. Although the 1990 U.S. Census described the Arab-American population as 1 million, estimates range as high as 3 million Americans with Arab origins (Telhami, 2002). Because not all Middle Eastern Americans are Arab in origin, the total Middle Eastern population is also larger than the official count.

Before World War II, most immigrants from the Middle East were poor, uneducated, and took jobs as unskilled workers. Most also were Christian, not Muslim. Many came from Lebanon and blended almost imperceptibly into the U.S. mainstream, moving rapidly into the middle class as entrepreneurs and professionals (Kayal and Kayal, 1975; Naff, 1983; Walbridge, 1999). The newest Middle Eastern immigrants, however, are far more diverse in geographical origin, ethnicity, religion, and social class. And, their absorption into American society has been far more torturous.

The most recent immigrants from the Middle East have come from over twenty countries in Southwest Asia and North Africa (Suleiman, 1999). The largest numbers have come from Iran, Egypt, Iraq, and the Palestinian territories. Some have been motivated to immigrate to the United States for reasons of economic betterment; but many have been impelled to leave their origin countries because of political unrest. The Middle East has been plagued by regional conflicts and civil wars over the past four decades, and many immigrants have sought refuge in America.

Most Americans usually perceive all Middle Easterners simply as "Arabs" and assume they are all Muslim. Neither view is correct. In fact, there are significant ethnic and religious differences among the numerous groups now comprising the Middle Eastern population in the United States. Although they are predominantly Arab, Middle Easterners comprise a diverse ethnic population. Iranians, for example, the largest component of the newest immigration from the Middle East, are not ethnically Arab, nor do they speak Arabic. As to religion, while most of the newest immigrants are Muslim, some are Christian, and a few are even Jewish (Bozorgmehr et al., 1996). Even among the Muslim immigrants, however, there are significant differences in belief and practice. For example, those from Yemen are devoutly religious, while those from Iran are predominantly secular (Walbridge, 1999). When considered in total—not simply the newest immigrants—Middle Eastern Americans are overwhelmingly Christian, not Muslim (Abraham, 1983).

Like other ethnic populations of the newest immigration, Middle Easterners are represented in a broad range of social classes. Some are poor and unskilled, but many are highly educated professionals—doctors, lawyers, engineers—and entrepreneurs who are relatively well-off. Only their geographical origin gives them commonality.

Middle Eastern communities can be found in all regions of the United States, but their major concentrations are in a few large urban areas—namely Detroit, Los Angeles, and New York. Detroit, with the largest settlement of Middle Easterners, is home to an estimated 300,000, mostly Arabs; in suburban Dearborn, Arabs now make up about a quarter of the city's population, and almost 60 percent of the students in its schools are from Arab families (Hamada, 1990; Singer, 2001; Warikoo, 2001).

Because Muslims predominate among the latest cohort of Middle Easterners, they have, more than other immigrant groups, run headlong into the dominant American culture. Islam is a religion that few Americans are familiar with, and they commonly harbor misconceptions about its tenets and rituals. Many Muslim women, for example, feel an obligation to wear a head scarf in public. This practice has made them an easy target for harassment and discrimination. Young Muslim girls may face ostracism at school, and dozens of companies have refused to employ women who insist on wearing the scarf (Goodstein, 1997). Perhaps more serious is the persistence of negative stereotyping that has plagued Middle Easterners in the United States. The activities of Arab terrorists in the Middle East and elsewhere have created a sinister image of Arab and other Middle Eastern groups—an image that was greatly exacerbated by the attack on the World Trade Center in 2001. This image reaches back much further than simply the last few years, however. Political events in the Middle East during the past four decades—as well as the oil crisis of the 1970s, induced by the mainly Arab-led OPEC oil cartel—caused many Americans to view Middle Easterners in the United States with suspicion and disdain. One researcher in the 1980s documented more than 100 different pop-entertainment programs, cartoons, and documentaries relative to Arabs and concluded that negative stereotypes were virtually the only TV images of the group. The TV portrayal of Arabs, he found, perpetuated four basic myths: "they are all fabulously wealthy; they are barbaric and uncultured; they are sex maniacs with a penchant for white slavery; and they revel in acts of terrorism" (Shaheen, 1984:4). A later study (Lind and Danowski, 1998) found little change in these images, except that no longer did the media portray Arabs as fabulously wealthy. The researchers concluded that the U.S. electronic media continued to give little coverage to Arabs in general. This, of course, changed radically after the events of September 11, 2001.

The Middle Eastern population is one of the most diverse and least understood among the newest ethnic elements of U.S. society. Because the majority are Arabs, their racial/ethnic classification remains uncertain and debated both within and outside the community (Naber, 2000). Moreover, the cultural confrontation of devout Muslims with dominant American norms and values has led to a stronger sense of isolation among them than has been the case with most immigrants to the United States. The determination of many Arab Muslim immigrants to uphold their religious convictions in the face of an overwhelming Judeo-Christian tradition and population has intensified that separateness.

Contemporary Immigration as a Global Phenomenon

As vast and diverse as immigration to the United States has been during the past three decades, it has hardly been a uniquely American phenomenon. Indeed, looking at the world's industrial societies reveals that almost all have been affected in a similar way. The numbers may be smaller than in the United States, but the relative impact has been no less deeply felt and, in some cases, has produced even greater social strain. In addition to the traditional immigrant-receiving societies like Canada, Australia, and New Zealand, many western European nations have experienced large immigration movements and, in the process, emerged as multiethnic societies. For countries such as Britain, France, and Germany, this has represented a profound transformation and a wrenching societal experience, the consequences of which are still unfolding. Even relatively homogeneous societies like Italy, Sweden, and Japan have begun to evince greater ethnic diversity due to immigration. And, in all of these societies, the changing ethnic composition has brought forth heated public debate, just as in the United States. Are immigrants becoming too numerous? Is the society's cultural integrity threatened? Are immigrants taking jobs from native workers? These and questions like them are heard today in many nations of the developed world.

To recognize that we are in an age of immigration should not lead us to think that immigration is a uniquely modern phenomenon. In fact, immigration has been ongoing since the very emergence of human societies. When we extend our historical perspectives far enough, we discover that the "original" inhabitants of any land were preceded by others. In a basic way, then, all societies have taken shape by migration from other societies. Nonetheless, it would hardly be exaggerating to declare that today we are in a period in which the dimensions of immigration are perhaps greater, and its consequences loom larger, than in any previous era in human history. And there appears no likelihood that the tide of international migration will recede in the foreseeable future. As long as there is a global stratification system, well-off nations with strong economies will attract people from nations with weak economies and severe population pressures. It seems apparent, then, that the issues and social consequences of large-scale immigration and increasing global diversity will follow us for many decades. We will revisit these issues in Chapter 17.

THE AMERICAN ETHNIC CONFIGURATION

It is now a commonplace that the newest immigration has thrust the United States into a period in which ethnic heterogeneity is historically unprecedented. Never before has America been more diverse, and that diversity will in all likelihood continue to expand. The traditional ethnic dominance of Euro-Americans, therefore, is no longer a given. "The United States," demographer Leon Bouvier has suggested, "is inexorably on its way to becoming a society with no one predominant group" (1992:148). As shown in Figure 5.4, if current demographic trends continue, non-European ethnic groups will make up barely half the American population by the middle of the twenty-first century.

"Official" Racial and Ethnic Classification

The demographic projections in Figure 5.4 must be studied with some caution. To begin with, these are merely estimates demographers have constructed on the basis of current and, presumably, future demographic trends. But demographers have no way of predicting what kind of fertility and mortality rates might be unleashed by economic and social forces. Few were able to foresee the post–World

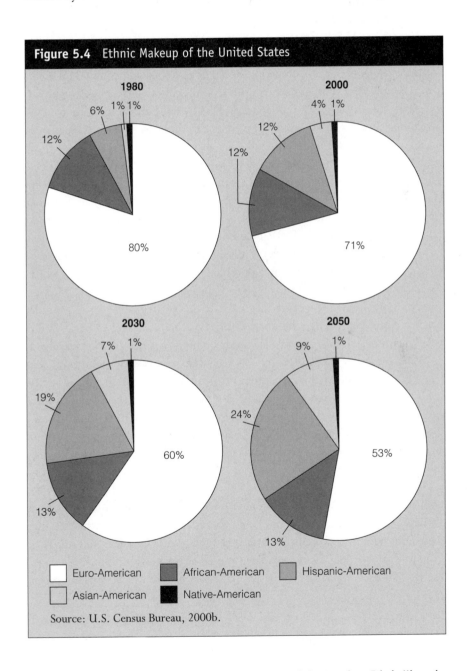

Figure 5.4 Ethnic Makeup of the United States

1980
6% 1% 1%
12%
80%

2000
4% 1%
12%
12%
71%

2030
7% 1%
19%
13%
60%

2050
9% 1%
24%
13%
53%

Euro-American African-American Hispanic-American
Asian-American Native-American

Source: U.S. Census Bureau, 2000b.

War II baby boom, for example. Nor can they be sure of future immigration rates, since these are determined mostly by federal legislation, making them a key political issue subject to all kinds of political influences.

But another, perhaps more significant, factor makes these projections less compelling than they initially seem. Even if demographers were capable of making accurate predictions, these estimates do not take into consideration that the categories currently used to classify the population by race and ethnicity are likely to change in the future.

In 1977 the U.S. federal government, as a means of monitoring and enforcing civil rights legislation and affirmative action programs, adopted a set of ethnic categories by which people could be classified and counted (Wright, 1994). These were subsequently adopted by virtually all government agencies, including the U.S. Census Bureau, as well as by educational institutions, corporations, and the mass media. Five racial/ethnic categories were put in place, and they have been changed only in minor ways since 1977: white, black, Hispanic, Asian/Pacific Islander, and American Indian/Alaskan Native.[3] These have become familiar categories, appearing on various educational and employment applications, census forms, and surveys. Unfortunately, rather than clarifying the conceptual morass of race and ethnicity in American society, these categories have further confounded the issue and contributed to even greater confusion. Consider, for example, that Hispanic, an ethnic designation, is a separate category, alongside "racial" categories—black, white, Asian, American Indian; or that extremely broad ranges of cultures and physical types are lumped together into each of these categories; or that Middle Eastern Americans, reflecting a wide variety of cultural and physical differences, are a population whose place in this scheme has not yet been resolved.

What the current U.S. classification system represents, then, is an extremely crude and bewildering attempt to count the population by ethnicity. It also demonstrates the arbitrariness of racial or ethnic classification. Race and ethnicity, it should be remembered, are social constructs; they are not biologically fixed divisions of the human population or of the population of any particular society. The ways that societies choose to subdivide and classify their populations can and do change.

An even more important caveat to be considered in looking at these projections is that—as we will see in the next few chapters—much intermarriage is occurring across racial and ethnic lines, the effect of which is to further blur these already fuzzy and politically influenced categories. Consider golf star Tiger Woods as an example. Woods has been touted as the newest black sports hero. But is he "black?" His father is African American, his mother is Asian American,

---

[3]In the 2000 U.S. Census, Pacific Islanders were separated from Asian Americans and constituted a category of their own.

and his ancestry also includes Euro-American and Native American. As Woods himself explains, "Growing up, I came up with this name: I'm a 'Cablinasian,'" a blend of Caucasian, black, Indian, and Asian (Fletcher, 1997).

As the population becomes more diverse, cases such as Tiger Woods's are becoming more unexceptional. Indeed, Woods may be an exemplar of America's ethnic future. The result is that the familiar racial/ethnic categories—white, black, Hispanic, Asian—no longer make much sense (if they ever did) in classifying people and simply no longer work in accurately describing American society.

Although the current classification system may be far less than satisfactory, we remain thoroughly reliant on data gathered by the U.S. Census Bureau, as well as by other public and private agencies and organizations that employ the official racial/ethnic scheme, for an understanding of the ethnic makeup of the United States. Until sociologists and other social scientists can collaborate with policymakers to construct a more rational and meaningful way of conceptualizing ethnic boundaries, we are constrained to use the categories currently in place.

"Mixed Race" People The failure of the commonly used racial/ethnic classification scheme to accurately reflect the increasingly mixed blend of the U.S. population has led in recent years to calls for the creation of a "mixed race" category. Presumably this would account for those individuals who fall through the cracks of the current system, that is, those whose parentage reflects more than one racial/ethnic category. The 2000 U.S. Census stopped short of creating such a category; though it did, for the first time, give people the choice of identifying themselves with more than one racial category—an option chosen by 2.4 percent of the population (Lee, 2001). The creation of a mixed-race category would more comfortably fit a growing segment of the population, but it would do nothing to make the racial/ethnic classification scheme any less arbitrary; it would only create another category that carries no more real meaning than any of the others. Moreover, creating a mixed-race category raises the implication that there are "pure" races to begin with. As explained in Chapter 1, nothing could be further from the truth. As one researcher has put it, "insofar as there is no biological foundation for the three or four acknowledged races, neither is there a biological foundation for mixed-race groups" (Zack, 2001).

In reality, of course, there has been a mixed-race population in American society from the very beginning. In earlier times, when the essential racial division was black/white, the so-called one-drop rule arbitrarily crystallized this racial dichotomy. In effect, one had to be either black or white; there were no in-between categories.[4] Blacks were defined as those having black ancestors, that

---

[4]In fact, the U.S. Census during the nineteenth century did provide an "intermediate" racial category, "mulatto," which was dropped in 1920 when *black* was defined as "anyone with any black ancestry" (Davis, 1991).

is, "one drop of black blood." But the percentage of black ancestry (what fraction constituted one drop?) varied from state to state. In some it was one-fourth; in others, one-eighth, one-sixteenth, or even one-thirty-second (Davis, 1978). At the turn of the last century, the one-drop rule was essentially the law of the land; even today it affects racial classification. Most of the U.S. black population, of course, has some white ancestry, and a small portion of the white population has some black ancestry (Blackburn, 2000; Smedley, 1999).

One unmistakable fact does emerge from the confusion accompanying attempts to classify and quantify the American population by race and ethnicity: The United States continues to evolve into an ever-more heterogeneous society. In the twenty-first century, America's ethnic character will be more culturally and physically diverse than it was in the twentieth century. And despite the efforts of some to preserve groups' social boundaries, the trend toward a fading of those racial and ethnic lines seems irreversible.[5]

The American Ethnic Hierarchy Today

The analysis that follows in the next six chapters suggests that the American ethnic hierarchy today can be viewed as divided into three comprehensive tiers:

1. The top tier comprises white Protestants of various national origins, for whom ethnicity has no real significance except to distinguish them from the remainder of the ethnic hierarchy.

2. The intermediate tier comprises white Catholics of various national origins, Jews, and many Asians, for whom ethnicity continues to play a role in the distribution of the society's rewards and continues to influence social life—but in both instances, decreasingly so.

3. The bottom tier comprises racial-ethnic groups—blacks, Hispanics, American Indians, and some Asians—for whom ethnicity today has great consequences and for whom it continues to shape many of the basic aspects of social life.

---

[5]Despite the virtually unparalleled heterogeneity of the United States today, it is interesting—if bemusing—to consider how misinformed most Americans are about the ethnic makeup of their society. Most cannot describe the actual size of various ethnic groups with anything even approaching accuracy. This has been demonstrated by national surveys over the past decade, in which Americans grossly exaggerate the size of racial and ethnic minorities. The typical American estimates the size of African-American, Hispanic-American, and Asian-American populations to be more than twice as large than they actually are (Carroll, 2001). This seriously distorted picture of the American ethnic configuration is bound to influence people's attitudes and actions regarding ethnic-related issues like immigration, affirmative action, and multiculturalism.

This ethnic hierarchy has remained relatively stable for the past century and a half. Although the distance between many of the groups has been reduced, their rank order has with few exceptions not been basically altered.

Arranging ethnic groups in such a rank order, of course, masks the more specific class and cultural differences among them, as well as the internal differences within each. But in a very general sense, each of the country's ethnic groups can be placed in one of these tiers, based on a combination of the factors we will look at in upcoming chapters: the collective place of the group in the society's economic and political hierarchies; the extent to which prejudice and discrimination remain significant for the group; and the extent to which group members have entered into full social participation with members of the dominant group.[6]

Perhaps the most important aspect of this three-part American ethnic hierarchy is that the gap between the bottom tier of groups and the other two is much greater than the gap between the top and intermediate tiers. Thus, except for Asians and some Hispanics, the American ethnic system today seems to be increasingly dichotomized—white and nonwhite. The overriding issues of American ethnic relations, therefore, remain focused on the economic, political, and social disparities between whites and nonwhites and the policies intended to reduce them. The crux of the matter is relatively simple: Will the gap between whites and nonwhites continue in its present form, will it be reduced substantially, or will it perhaps widen? The issues of American race and ethnic relations are largely subsumed by this pivotal question.

SUMMARY

Following its conquest of the indigenous peoples and enslavement of blacks, the Anglo core group established its social and political dominance and exercised the power of selection over those who came afterward. American ideologies and public policies regarding ethnic groups and issues have consistently reflected Anglo cultural preferences and Anglo economic and political interests. Based on these thoughts and actions, an ethnic hierarchy has been shaped, and the rank order of groups has not essentially changed.

The ethnic diversity of the American population blossomed forcefully during the approximately 100 years between the 1820s and 1920s. Two massive waves

---

[6]Using a combination of these factors means that some will offset others. For example, although Jews and Asians rank higher in their aggregate economic class position than northwestern Europeans, they still do not fully interact at the primary level with those groups, and the extent of prejudice and discrimination directed against them is much greater. Hence, when all these factors are considered, Jews and Asians are part of the intermediate tier of groups rather than the top.

of European immigration took place. The "old immigration," for the first fifty years of that period, brought mainly groups from northern and western Europe, the largest among them the Irish and Germans. During the second fifty years, the "new immigration" consisted primarily of groups from southern and eastern Europe. The cultural distinctiveness of the latter provoked a nativist movement that ultimately resulted in the passage of restrictive immigration laws, ending, in effect, further large-scale European immigration.

The contemporary period of immigration, referred to as the "newest immigration," was launched by the passage of immigration reforms in 1965, which ended country-of-origin restrictions. Whereas previous periods of heavy immigration had consisted mostly of European groups, the newest immigration is made up in largest part of groups from Asia and Latin America. The scope of the newest immigration has exceeded even that of the early 1900s, previously the heaviest period of immigration to the United States. As a result, an ethnic reconfiguration of the U.S. population is occurring. Asian and Latin American groups are growing, while Euro-American groups are declining in size.

The American ethnic hierarchy today is made up essentially of three comprehensive tiers: white Protestants of various national origins at the top; white Catholics and Jews of various national origins, along with most Asian Americans, in the middle; and Hispanic Americans, African Americans, and American Indians at the bottom.

Suggested Readings

Anderson, Charles H. 1970. *White Protestant Americans: From National Origins to Religious Group.* Englewood Cliffs, N.J.: Prentice-Hall. Describes the changing social composition and status of what has been the dominant ethnic group in American society.

Baltzell, E. Digby. 1964. *The Protestant Establishment: Aristocracy and Caste in America.* New York: Vintage. A fascinating social history of the WASP upper class, especially its efforts to maintain its ethnic purity. Argues that the WASP upper class has become a castelike group, whereas societal power has passed to a more ethnically diverse elite.

Dinnerstein, Leonard, and David M. Reimers. 1988. *Ethnic Americans: A History of Immigration.* 3d ed. New York: Harper & Row. A succinct history of the several waves of voluntary immigration to the United States, describing social, economic, and political conditions experienced by immigrants, as well as the society's response to them.

Glazer, Nathan, and Daniel P. Moynihan. 1970. *Beyond the Melting Pot.* 2d ed. Cambridge, Mass.: MIT Press. An influential book at the time of its writing; its major theme is that American ethnic groups do not completely assimilate but evolve into political units that compete for resources in the urban environment. Contains vivid comparisons of the different groups making up the American ethnic hierarchy, using New York as a backdrop.

Higham, John. 1963. *Strangers in the Land: Patterns of American Nativism 1860–1925*. 2d ed. New York: Athenaeum. Examines the different expressions of nativism, anti-Semitism, and anti-Catholicism that characterized American thought and action during the period of massive European immigration to the United States.

Lieberson, Stanley. 1980. *A Piece of the Pie: Blacks and White Immigrants Since 1880*. Berkeley: University of California Press. Addresses the issue of why European ethnic groups and African Americans have experienced different rates of economic and political mobility.

Portes, Alejandro, and Rubén Rumbaut. 1996. *Immigrant America: A Portrait*. 2d ed. Berkeley: University of California Press. Examines the sociological issues of America's newest immigrants, such as occupational adaptation, urban settlement patterns, political participation, and language.

Ungar, Sanford J. 1995. *Fresh Blood: The New American Immigrants*. New York: Simon and Schuster. Drawing on extended interviews, describes many of the new ethnic communities arising in different parts of America. The author sees the newest immigration as having generally positive societal effects.

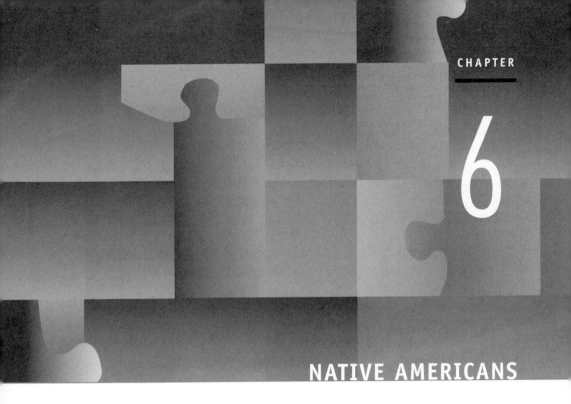

NATIVE AMERICANS

IF THE UNITED STATES is a country of immigrants, what can be said of American Indians (or Native Americans)? As the original inhabitants of the continent, surely they cannot be considered immigrants in the conventional sense, like Europeans of the nineteenth and early twentieth centuries, or like Asians and Hispanics of today. But in a sense, even these native peoples can be seen as an immigrant group when considered in a long-range historical context. From 12,000 to 50,000 years ago, humans reached the Western Hemisphere from Northeast Asia with the help of glaciers that, from time to time, covered most of what is today Canada and Alaska, forming a land bridge between the two hemispheres. Some may have come by water as well as land. These migrations probably occurred over thousands of years, and it is unlikely that the migrants came from similar racial origins (Goodman and Heffington, 2000). How humans first came to inhabit what is today North and South America remains an unresolved question among archaeologists. But what is no longer much debated is that American Indians originated in another geographic locale.

Realistically, of course, it would seem absurd to refer to American Indians as an "immigrant" group, despite their ancient migration from another part of the planet. Their cultures and societies developed here, not on another continent. Clearly, American Indians are North America's indigenous peoples, and in this they maintain a unique status among the ethnic populations of the United States and Canada. As Ronald Wright has put it, "Native Americans . . . are American in a way that no others can be. Even if we suppose that their ancestors arrived 'only' 15,000 years ago . . . they have been here thirty times longer than anyone else. If we call that time a month, Columbus came yesterday" (1992:x).

CONQUEST AND ADAPTATION

Because American Indians inhabited the continent prior to the explorations of Europeans, they occupy a singular place in the shaping and development of the American ethnic hierarchy. Unlike any other group, their placement into the society's ethnic scheme and their subsequent adaptation were shaped by conquest.

European Contact

When Europeans crossed the Atlantic beginning in the late fifteenth century, they found two continents inhabited by perhaps 75 million people.[1] Estimates of the indigenous population north of the Rio Grande at the time of Columbus's arrival range from 1 million to 18 million (Thornton, 1995; Wax, 1971). These "Indians," as the Europeans called them, were by no means homogeneous but were organized into hundreds of tribes or nations, some quite advanced in social organization and technology. Different they were from the Europeans, however, and the white invaders ethnocentrically interpreted these physical and cultural differences as proof of inferiority and lack of civilization.

Like the more numerous natives of what is today Latin America, these peoples spoke many different languages, conformed to disparate cultural systems, and were organized in a variety of economic and social forms. There were, then, no "American Indians" per se but rather a great number of diverse ethnic groups, viewing themselves as different from one another as from the European invaders (Lurie, 1991).

Emergence of an Ethnic Hierarchy The establishment of European supremacy over the Indians conforms closely to the theories of the emergence of dominance and subordination explained in Chapter 2. As a conquered population, the subordination of American Indians occurred through a highly conflictual process. As noted earlier, an indigenous group invaded by a more powerful settler group can only capitulate or, more commonly, resist. This ordinarily makes for a high level of conflict and rejection of assimilation into the dominant group.

---

[1] The number of indigenous people inhabiting North and South America at the time of European discovery is disputed among anthropologists and historians. Estimates range from less than 10 million to more than 100 million (Utter, 1993). As one researcher has declared, "Prior to the late nineteenth century, all numbers are uncertain, and a uniquely correct number of natives never will be had" (Jaffe, 1992:2).

The four major conditions for the emergence of ethnic stratification—contact, ethnocentrism, competition, and differential power—are clearly seen in the case of Indian-white relations at the outset of European settlement and afterward.

The Native Americans' physical and cultural distance from European standards inevitably produced an ethnocentric white view. Of particular importance to the North American settlers was that the Indians were "heathen." Moreover, they were preliterate and at a technological level far below that of the Europeans. A combination of these traits contributed to the image of Indians as savage and inferior and, consequently, to the rationale for their exploitation and annihilation. As Alvin Josephy has written, "In the Western Hemisphere, Indians, to the whites, were all the same, and the newcomers disagreed among themselves only over the extent to which the native populations differed from, or seemed to be inferior to, Europeans" (Josephy, 1969:279).

Contact and even ethnocentrism among alien ethnic groups will not in themselves create dominant-subordinate relations, however. Competition over valued resources is a third factor, and in the case of Indian-white relations, land provided the necessary ingredient. The settlers' objective was to secure land for agricultural production. From the outset, divergent cultural perceptions of "property" led to misunderstandings and hostilities between Indian and settler. Indian lands were held communally, not as private property. Whites consequently perceived them as basically unoccupied. Furthermore, most Indians did not engage in European-style farming, which led to the view that these lands were underutilized and thus justifiably occupied by those who would cultivate them in a "proper" manner.

In the end, however, it was the differential power between Indians and settlers that proved the crucial element in establishing dominance and subordination. In particular, the deciding factor in the European conquest of indigenous peoples in all parts of the Western Hemisphere was the Europeans' technological skills, which gave them a tremendous advantage in war. When the vastly superior arms of the colonialists were combined with the inability of Indians to resist the diseases introduced by the Europeans, it was inevitable that whites would prevail in the struggle for resources and native populations would decline.

Displacement and Depopulation

From the beginning of European settlement to the end of the eighteenth century, Indian-white relations in North America centered primarily on the fur trade. During that time, three major colonial powers vied for the economic resources of the Western Hemisphere: Britain, France, and Spain. Each had different colonial objectives. Spain, chiefly concerned with exploiting the mineral wealth of its possessions, concentrated its activities primarily in the southern part of the hemisphere. In North America, the Spanish influence was eventually confined to the present-day American Southwest. Indigenous peoples in what are today Mexico, Central and South America, and parts of the Caribbean were brutally exploited as a labor force, first through enslavement and later through debt peonage (Cumberland, 1968).

France's economic activities, by contrast, focused mostly on supplying the European market with furs. Its activities extended throughout most of the northeastern and central parts of the North American continent, including the St. Lawrence waterway, eastern Canada, the Great Lakes, the Ohio Valley, and the Mississippi River Valley. As a result, the French necessarily nurtured close relations with North American Indian nations who aided them in their economic goals. Through the fur trade, Indians were integrated into the emergent North American economy.

The British Triumph The British, too, were engaged in the fur trade and, like the French, entered into alliances with Indian nations. The British-French rivalry in the fur trade led to deadly conflict among the Indian nations attached to each country. Moreover, as a result of their economic and political relations with the Europeans, native Indians' cultural institutions and forms of social organization were radically transformed in numerous ways. More than the Spanish and French, the British valued land as a resource, and their colonial activities eventually became focused on that goal. Britain was interested in having colonists permanently settle the land and farm it rather than simply exploit it for minerals and furs.

With the eventual triumph of the British in North America in the late eighteenth century, the crux of Indian-white relations shifted from the fur trade to issues of land. The British colonial government as well as its successor, the newly independent United States, sought to control the acquisition of lands by American settlers and to protect the rights of Indians, but both governments failed. The quest for land by British colonists proved insatiable, leading ultimately to a displacement of the indigenous population.

At the start of the nineteenth century, Indian-white relations evolved into a confrontation in which the rapacious quest of white settlers for land led to evergreater efforts to dispossess Indians of lands they were occupying. Two paths to this objective became apparent: Indians would be separated from their lands either through assimilation—by encouraging them to abandon their communal patterns of landholding in favor of private property—or through removal, that is, relocating Indians through negotiated purchase of land or through conquest (Cornell, 1988). These two paths were not mutually exclusive; rather, they were complementary, not only in a practical sense but also as a way of satisfying the objectives of settlers—who wanted land—and reformers—who sought to "civilize" the Indians with American culture. This dual objective, notes historian Robert Berkhofer, was a constant pattern throughout most of American history: "From the founding of the nation until recent times, and some would include today as well, United States policy makers placed two considerations above all others in the nation's relation with Native Americans as Indians: the extinction of native title in favor of White exploitation of native lands and resources and the transformation of native lifestyles into copies of approved White models" (1979:135).

By the late nineteenth century, most of the remaining Native population had been forcibly resettled in reservations west of the Mississippi. It was not coincidental that reservations were located on lands that whites deemed virtually worthless for farming or grazing.

Depopulation Indian societies were physically reduced through a combination of exposure to European diseases, armed conflict, starvation, and the breakup of cultural systems that had traditionally provided for social and material needs. Though Indians did make many cultural adaptations to the white presence, doing so did not prevent the eventual destruction of their social forms and the succession of white dominance (Wax, 1971). By the time Americans had gone beyond the Appalachian Mountains, Indians had already adopted several new technologies from the Europeans. Firearms now aided in hunting and, on the plains, the horse, introduced by the Spanish in the Southwest, had thoroughly transformed the Indians' way of life. Despite these adaptations, however, Indians remained politically fractured into many scattered tribes. This made them vulnerable to American settlers' intrusions and, ultimately, divestiture.

As for diseases, perhaps nothing more severely affected the Indians' physical condition than the epidemics brought by Europeans, which decimated many tribes from the very outset of their contact with settlers and explorers. These epidemics continued, repeatedly, throughout the nineteenth century (Thornton, 1995; Washburn, 1975). Native populations lacked immunity to diseases like smallpox, measles, cholera, typhoid, malaria, and others; their effects were devastating.

Government Policies and American Indians

Stephen Cornell (1988) has suggested that the "Indian problem" for Euro-Americans from the time of settlement to the present has had three aspects: economic—how best to secure Indian resources, especially land[2]; cultural—how best to assimilate Indians into the dominant, non-Indian, culture; and political—

---

[2] Cornell (1990) explains that at the outset of contact between Indians and Europeans, it was not land but other resources—namely, furs—that Europeans sought. From the late eighteenth century on, however, land was the major objective.

how best to control Indians in order to bring about solutions to the first two problems. The counterpart to the Indian problem was a "Euro-American problem" for Indians. "In its essence," notes Cornell, "this problem seems to have been tribal survival: the maintenance of particular sets of social relations, more or less distinct cultural orders, and some measure of political autonomy in the face of invasion, conquest, and loss of power" (1988:7). It is around these two conflicting problems that Indian-white relations have revolved for the past three centuries.

The solution to the Euro-Americans' economic, cultural, and political "Indian problem" was displacement and depopulation of Indian tribes, as well as cultural decimation through forced assimilation. All of these strategies were underwritten by government edict. Several phases of the history of relations between American Indians and the federal government may be delineated, each corresponding to shifting objectives and specific measures, always undertaken with the implicit goals of eliminating the Native-American population as an impediment to western settlement and the needs of an expanding American economy, and eradicating Native American cultures and political forms.

Separation The earliest U.S. policy toward Indians grew out of the struggle for land that had characterized relations between Indians and the English settlers. Various agreements between Indians and colonists essentially held that the Indian nations were independent political entities and were to be dealt with on that basis. This policy continued after U.S. independence. Specific bounded territories were promulgated, separating Indian nations from Americans. Thus, the paradoxical status of Indians was, as Edward Spicer describes it, "a people within the territory of the new nation and yet outside the processes of its political life" (1982:179).

The legal status of Indian lands, however, quickly became hazy as settlers coveted these lands and began to infringe on them. Indian lands were ceded irreversibly by treaties, usually entered into by Indians under duress or in ignorance of their meaning, through fraudulent schemes perpetrated by whites, or by sheer force. Moreover, treaties themselves provided no protection of those lands left to the Native Americans, because renegotiations and violations became standard practice whenever disputes arose. Where there was desirable land, whites eventually took it. Clashes occurred repeatedly; the solution, from the American perspective, was to engage the Indians in negotiation and land exchanges for their removal to lands further west, where it was assumed these border issues would no longer arise.

Removal The Indian Removal Act of 1830 called for the relocation of all tribes living in the eastern United States to lands set aside for them west of the Mississippi River. This policy was pursued vigorously by President Andrew Jackson, who represented the sentiments of settlers who saw Indian occupation of desired lands as an obstacle to the development of white dominance. Removal was rationalized by the view that Indians were not using land as "God had intended"; that is, to farm in the way of the white man. As Governor William Henry Harrison asked, "Is one of the fairest portions of the globe to remain in a state of nature, the haunt of a few wretched savages, when it seems destined by the Creator to give support to a large population and to be the seat of civilization?" (quoted in Hagan, 1961:69).

This ten-year movement marked a time of great hardship and suffering for Indians and resulted in the virtual devastation of many tribes, particularly those that resisted removal. Perhaps the most infamous episode involved the forced migration of the Cherokees in 1838, an affair referred to as the "Trail of Tears." In the summer of that year, the U.S. Army rounded up all 16,000 Cherokees and held them for months in disease-infested camps. The march west to Oklahoma extended over the autumn and winter, and by its end, 4,000 had died (Wright, 1992). As one historian has described it, the removal of Indians, at best, "resembled closely the pioneering experience of thousands of their white contemporaries" and, at worst, "approached the horrors created by the Nazi handling of subject peoples" (Hagan, 1961:77). The Indian Removal Act was perhaps the most devastating single action taken by the federal government in destroying Indian cultures and societies.

The Western Indians After the American Civil War, the U.S. government took steps to put an end to Indian resistance that stood in the way of western development. Thus, the locus of Indian affairs shifted to the tribes of the Plains, the Southwest, and the Far West. The result was the further dispossession of Indian lands through military campaigns as well as the signing of treaties that were quickly broken by the government when their terms impeded settlement.

The culmination of this effort was the suppression of the Ghost Dance religious movement. The Ghost Dance religion had been founded by a prophet, Wovoka, of the Paiute tribe. Believers were convinced that they could drive whites from Indian lands and restore their traditional way of life by following certain rituals, particularly a ceremonial dance. In the late 1880s, the movement spread across the plains. The U.S. government viewed the Ghost Dance as a serious threat; in 1890, following the killing of Sitting Bull, one of the movement's leaders, the government ordered troops to move on the Ghost Dancers. This led to the slaughter of hundreds of Sioux Indians at Wounded Knee near the Pine Ridge Reservation in South Dakota (Brown, 1972). This incident represented the culmination of military force against Indians, and from that point forward, Indian resistance ceased. Wounded Knee became a symbol of the federal government's destructive anti-Indian policies.

In addition to military force, the Plains Indians were decimated by disease and by the extermination of the buffalo (or bison), which had been the very heart of their way of life. They depended on it not only for food, but for clothing and shelter as well. The herds were drastically reduced when white hunters shot them, harvesting meat and hides. In a matter of only a few years, as many as 15 million buffalo were virtually extinguished. The passing of the buffalo further diminished the power of Western Indians to resist white settlement. The choice they now faced was between depending on government handouts and adopting the Euro-American way of life (Wexler, 1995).

Allotment and Coercive Assimilation By the latter part of the nineteenth century, armed resistance to federal policy had ended. Government officials and reformers now advocated a policy that they felt would once and for all solve the Indian problem: Reservation lands would be broken up and allotted to individual tribal members. Unallotted lands would then be sold to non-Indians. This policy was formalized in 1887 with congressional passage of the Dawes Act (also called the General Allotment Act). The policy of allotment further diminished those lands remaining in Indian possession, leaving them with even fewer productive resources. Furthermore, the administrative effects of the Dawes Act led to an even greater intrusion of the BIA (the Bureau of Indian Affairs, the government's major agency for dealing with Indians) into Indian affairs, further diminishing native cultures and weakening tribal political institutions.

As a means of coercing Indians into an acceptance of their now virtual powerlessness in the face of the federal government, a policy of forced assimilation was put into place. The policy of breaking up Indian lands and assigning title to individuals was, by itself, intended to make over Indian families, using white settlers as the model: "hard working and economically motivated" (Spicer, 1982:183). But part of the policy also consisted of specific attempts to assimilate Indians into the mainstream society. In cooperation with religious groups, efforts were made to eradicate native cultures and impose the American way of life. This meant destroying native religions and putting Christianity in their place, denying Indians native languages and substituting English, and generally reducing education to an effort to "civilize" Indians. Native-American children were placed into government-run boarding schools, which indoctrinated them with the belief that Euro-American culture was vastly superior to "primitive" tribal cultures. Separated from their families for years at a time, the children were taught to speak English, wear Western clothing, and pray as Christians; failure to behave accordingly merited stern punishment from school authorities (Adams, 1995). The BIA commissioner in 1889 bluntly stated his bureau's intent: "The Indians must conform to 'the white man's ways,' peaceably if they will, forcibly if they must" (quoted in Cornell, 1988:56).

At the start of the twentieth century, then, Indians were impoverished and virtually at the mercy of the federal government, whose paternalistic policies continued to reflect white ethnocentrism. By the end of the century, the Indian population had been reduced to fewer than 250,000. Indians were now officially wards of the federal government and were thus no longer seen or dealt with as separate nations.

A New Approach to Indian Affairs In the 1930s, an effort was made to reverse the decades of neglect and exploitation that had characterized government policies vis-à-vis American Indians. It was also an attempt to create a more pluralistic system of Indian-white relations. John Collier, enlightened and sympathetic toward Native causes, was appointed to head the BIA during President Franklin D. Roosevelt's first term. Collier proposed major changes in Indian policy that included tribal self-rule and efforts to encourage Native cultural preservation. Most important was his proposal to end the allotment system created by the Dawes Act, which had resulted in the loss of much Indian land. Congress passed the Indian Reorganization Act (IRA) in 1934, which included a number of Collier's recommendations. It was not met with enthusiasm by all Indians, however. Some were committed to assimilation, and others were not prepared to accept at face value any measure offered by the federal government. Still others saw the act as simply one more form of paternalism. Most tribes, however, approved the act. The IRA, though failing to fundamentally alter the status and condition of Native Americans, did result in halting some of the more egregious aspects of past Indian policy. It prohibited further allotments of tribal lands and encouraged Indians to establish self-governing systems within their tribes. Moreover, Collier reduced the heavy-handed control of Indian life by the BIA and reduced efforts aimed at forced cultural assimilation.

Termination and Its Reversal Once again, in the 1950s, a turnabout in the government's approach to Indian affairs emerged with a new policy known as "termination." In essence the objective was, once again, assimilation. Termination essentially meant the end of the federal government's responsibility to provide various social, educational, and economic services to Indians; it also ended government protection of Indian lands and property held in trust for the tribes (Senese, 1991). In a sense, Indians would now be entirely on their own. Moreover, the tribal organization of Indian life was seen as a relic of the past, and it was felt that Indians should be dealt with on an individualistic, not a communal, basis. The idea was that they should be treated no differently from other citizens. To accomplish this, it was proposed that the reservation system be dismantled and the federal government's unique role toward Indians be severed. No longer would treaty obligations be recognized, and the sovereign status of tribes would be eliminated.

Most Indians strongly resisted termination, realizing that the ultimate result would be the loss of tribal lands as well as Native cultures. The failure of termination led to a return to a policy more in line with the IRA during the next three decades. Reversing its position again, federal policy toward

American Indians in the 1970s moved in the direction of greater autonomy for tribal governments. During this decade, the Red Power movement put pressure on the federal government to address Indian issues and needs and to reaffirm Indian rights. In 1975 Congress passed the Indian Self-Determination and Education Assistance Act, permitting tribes to take control of numerous federal programs on reservations. This marked a revolutionary break with the past, releasing Indian tribes from the strict control and supervision of the BIA (Josephy et al., 1999). The effect was a renewal of tribal sovereignty and, in effect, a reversal of the policy of termination. Many of the reservations' social and economic functions were now assumed by tribal governments.

In the 1980s, the Reagan administration made severe cuts in federal assistance to American Indians. As with all other federal agencies, the Reagan approach was to remove the federal government from activities and turn these over to the private sector. In 1983 alone, aid was cut by more than one-third, affecting programs on all reservations (Worsnop, 1992). These policies reflected a philosophical rejection of self-determination that had seemed to hold sway in the 1970s. But they were short-lived. By the late 1980s and throughout the 1990s, self-determination once again dominated government policy as Congress and the courts made decisions that strengthened the sovereignty of Indian nations (Josephy et al., 1999). In 1990, for example, Congress enacted the Native American Languages Act, intended to promote the use and development of Indian languages.

Self-Determination   Recent politics of American Indians have focused primarily on political self-determination and on the protection of remaining Indian lands. Indians have, in the past two decades, generally experienced greater empowerment than at any time since the nineteenth century. Much of this power has been the result of the leverage that tribes have been able to exert in their control of valuable minerals and other resources that are now part of Indian-controlled lands. Increasingly they have made effective legal challenges to what they have interpreted as incursions into their lands. Moreover, many tribes have become adept at business affairs and have developed numerous enterprises on reservations.

A certain theme seems to run through the history of government policy toward Native Americans. At all times, whatever the issues or whichever administration formulated policy, Indians themselves lacked the political power to defend their interests. They were always reactors to, never initiators of, policies affecting their welfare. At certain times the object of government measures was to assimilate Indians to the cultural ways of whites, whereas at other times, Indians were seen as unassimilable and thus to be left alone. In either case— with few exceptions—gross neglect typified the attitudes of policymakers, and each succeeding generation of Native Americans has felt the consequences of that neglect.

Red Power   During the late 1960s and early 1970s, Native Americans put pressure on the federal government to address Indian issues and needs and to reassert Indian rights. Moreover, as with other ethnic groups, American Indians during that time began to reaffirm their ethnic identity. These ideas and actions, in total, came to be known as the Red Power movement. Specifically, the aim of Red Power was to end the federal policy of termination, return Indians to traditional cultural ways, and review and revitalize Indian communities. As it took shape, Red Power in many ways emulated similar movements occurring at that time among African Americans and Latinos. It brought the grievances and general plight of Native Americans to public awareness through numerous incidents—marches, protests, sit-ins, and demonstrations—including the occupation of Alcatraz Island in 1969, the takeover of the BIA headquarters in 1972, and the FBI confrontation with Indian activists in 1973 at Wounded Knee, South Dakota.

Perhaps the most publicized and attention-grabbing event of the Red Power movement was the 19-month occupation of Alcatraz Island, an abandoned federal prison in San Francisco Bay, beginning in 1969. Many Indian activists saw this as the spark that ignited the Red Power movement. Following Alcatraz, several other occupations took place in various locations around the country. The siege at Wounded Knee in the Pine Ridge Reservation in South Dakota in 1973, however, overshadowed the others. The conflict actually began as an internal power struggle among the Oglala Sioux tribe. Eventually it involved not only the two sides of the tribe, but federal law enforcement officials, the BIA, local citizens, nationally prominent entertainers who lent their celebrity endorsement, national religious and legal organizations, and the national news media (Josephy et al., 1999). One of the major groups involved was AIM, the American Indian Movement. AIM had been founded in 1968 by a group of young Indian men, including Dennis Banks, and had been one of the major organizing units of the earlier Alcatraz occupation.

Wounded Knee culminated in a standoff between AIM and federal officials that continued for over two months, with sporadic violence resulting in several deaths and many injuries. Banks and Russell Means, another activist leader, were subsequently arrested and charged with inciting to riot and conspiracy. The choice of Wounded Knee was hardly random. This location held great symbolic significance for American Indians because it had been the site where the U.S. cavalry had massacred hundreds of Sioux men, women, and children in 1890. After Wounded Knee, the militance of Red Power subsided relatively quickly, however, and was replaced with a more legalistic approach to preserving and enhancing Indian cultural and economic interests.

American-Indian Demographics Today

It is difficult today to accurately measure the Indian population, because definitions of who is "Indian" are not clear and unequivocal. As with other ethnic groups, the U.S. Census counts anyone as Indian who declares himself or herself so. Within the Native-American population itself, Indian identity is left to each tribe to define. Thus a variety of criteria may be used in establishing a person's tribal membership (Snipp, 1997). Often blood quantum is a qualification; but in some cases, one need only trace his or her lineage to earlier tribal members (Thornton, 1987).

Population Growth Despite the problems in accurately determining Indian identity, there is little doubt that Indians are a relatively young and growing ethnic population. The "official" Indian population today (including Alaska natives) is 2.5 million, which is less than 1 percent of the total U.S. population. Nonetheless, this represents a 38 percent increase over the 1980 recorded Indian population of 1.4 million and a quadrupling from 550,000 in 1960. As noted in Chapter 5, the 2000 U.S. Census enabled people to declare more than one racial identity. If those who declared some Indian ancestry are included with those who identified themselves as American Indian only, the population of this category jumps to 4.1 million, or 1.5 percent of the U.S. population (U.S. Census Bureau, 2002).

Much of the increase in the Native-American population is the product of a high birthrate, but a larger part is accounted for by the growing tendency for people to declare their Indian affiliation (Sandefur et al., 2001). By the late 1960s, "Indianness" had risen in popularity. As a result of the Red Power movement, Indian activism inspired new pride among Native Americans, as did the more favorable portrayal of Indians by the mass media (Nagel, 1995). In addition to the new positive public perception of this group, many now claimed Indian ancestry in order to be included in lucrative judgments being awarded to tribes in land-claims cases. "Every announcement of a large judgment," notes historian William Hagan, "seemed to trigger the memories of some Americans that their family trees included an Indian, usually a chief's daughter, a princess" (1992:284). Declaring oneself "Indian," many discovered, also resulted in eligibility for various federal assistance programs, including health and education. Indian identity, then, remains problematic, with no systematic way to verify the validity of genuine Indianness. This is hardly a recent problem, however; it extends well back in American history (Hagan, 1992).

Today the federal government officially recognizes more than 300 Indian tribes in the United States; over 100 other tribes are seeking official recognition. Not all American Indians, however, are actually enrolled members of these tribes. Figure 6.1 lists the ten largest American Indian tribes.

Reservation Indians The American Indian population is split between those living on and those living off the 278 federal Indian reservations (U.S. Department of the Interior, 1991). The largest of these reservations is the Navajo, covering 16 million acres of land in Arizona, New Mexico, and Utah, with a population of 143,000. Most other reservations, however, are much smaller. The second largest is the Pine Ridge Reservation in South Dakota and Nebraska, with a population of only 11,000.

Urban Indians Most non-reservation Indians live in urban areas. Indeed, the movement of American Indians to cities has been a continuous process. Federally sponsored urban relocation programs begun during the 1950s impelled much of this movement. In 1960 only about one-quarter of American Indians lived in urban areas; by 1990, over half did so.

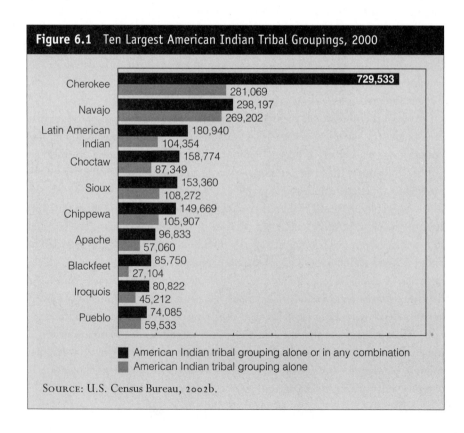

Figure 6.1 Ten Largest American Indian Tribal Groupings, 2000

Cherokee — 729,533 / 281,069
Navajo — 298,197 / 269,202
Latin American Indian — 180,940 / 104,354
Choctaw — 158,774 / 87,349
Sioux — 153,360 / 108,272
Chippewa — 149,669 / 105,907
Apache — 96,833 / 57,060
Blackfeet — 85,750 / 27,104
Iroquois — 80,822 / 45,212
Pueblo — 74,085 / 59,533

■ American Indian tribal grouping alone or in any combination
▨ American Indian tribal grouping alone

SOURCE: U.S. Census Bureau, 2002b.

While urban Indians are less impoverished than reservation Indians, they remain concentrated near the bottom of the class system, experiencing high rates of unemployment and difficulty in adjusting to urban economies that no longer can absorb workers with low skills and education. The urban migration of American Indians is not likely to decline and thus will continue to render profound effects on social and cultural aspects of Indian life.

Tables 6.1 and 6.2 show those states and metropolitan areas where Indians are most populous. As is evident, Indians are concentrated in a few states, mostly in the West. The largest populations are in California, Oklahoma, Arizona, and New Mexico.

An interesting recent demographic development is the increasing repopulation of the Great Plains by American Indians. This is an area of the United States, wedded largely to agriculture, from which whites are out-migrating as farming becomes less profitable. The only significant population gains are occurring among American Indians, who continue to return. States like Nebraska, Montana, and the Dakotas continue to lose white population, while the overall Indian population has grown since the mid-1980s. Although they still make up

Table 6.1 States with the Largest American-Indian and Alaskan-Native Populations, 2000*

STATE	NUMBER	PERCENT OF TOTAL POPULATION
California	333,346	1.0
Oklahoma	273,230	7.9
Arizona	255,879	5.0
New Mexico	173,483	9.5
Texas	118,362	0.6
North Carolina	99,551	1.2
Alaska	98,043	15.6
Washington	93,301	1.6
New York	82,461	0.4
South Dakota	62,283	8.3
Michigan	58,479	0.6
Montana	56,068	6.2
Minnesota	54,967	1.1

SOURCE: U.S. Census Bureau, 2002b.
*American Indian or Alaskan Native alone, not in combination with other racial categories.

Table 6.2 Metropolitan Areas with the Largest American-Indian Populations

METROPOLITAN AREA	AMERICAN-INDIAN POPULATION	PERCENT OF TOTAL METRO POPULATION
Los Angeles–Riverside–Orange County, CA	113,000	0.7
Phoenix–Mesa, AZ	65,000	2.4
New York–Northern NJ–Long Island	61,000	0.3
Tulsa, OK	50,000	6.6
San Francisco–Oakland–San Jose, CA	49,000	0.7
Oklahoma City, OK	48,000	4.7
Seattle–Tacoma–Bremerton, WA	45,000	1.3
Albuquerque, NM	38,000	5.7
Flagstaff, AZ	34,000	28.4
Minneapolis–St. Paul, MN	28,000	1.0

SOURCE: U.S. Census Bureau, 1998a.

only a small fraction of the total population of these states, Indians are being lured back to the reservations because of the promise of jobs, mainly in the casinos that are now commonly part of many reservations' landscape. Moreover, the bison is making a comeback in the northern Plains, with more than 30 tribes controlling large herds (Egan, 2001).

AMERICAN INDIANS IN THE CLASS SYSTEM

Socioeconomic Status

Wherever American Indians reside, on reservations or in urban areas, they are below the national average on most socioeconomic measures—income, education, occupation, employment, health care, mortality, and housing.

Economic Measures Table 6.3 indicates some of the economic discrepancies between Indians and the general population. As can be seen, median household income is well below that of the total population, and the percent of families in poverty is almost three times greater. The social and economic conditions on reservations are especially dismal and remain resistant to significant change. Three of the ten poorest counties in the United States are Sioux reservations in South Dakota.

Table 6.3 The Socioeconomic Status of American Indians, 1990

	AMERICAN INDIANS	TOTAL POPULATION
Education		
Percent high school graduates	65.6	77.6
Percent college graduates	9.4	21.3
Median family income	$21,619	$39,066
Percent families in poverty	27.2	10.1

SOURCE: U.S. Census Bureau, 1998a.

Of particular consequence is the serious Native-American unemployment rate. A combination of poor education and geographical isolation has created extremely severe employment problems both on reservations and in urban areas. The rate of unemployment on Indian reservations nationally is almost 50 percent (Kilborn, 1997). And in urban areas, too, Native Americans commonly find themselves without the skills and qualifications to secure steady jobs. Thus the impoverished status of many Native Americans continues from one generation to the next.

Education Regarding education, Native Americans again find themselves at or close to the bottom of the social hierarchy. For example, Native students have the highest high school dropout rate among all ethnic groups. Achievement levels are among the lowest as well (U.S. Department of Education, 1991). More than any other factor, the low level of human capital among Native Americans, primarily in the form of education, accounts for the socioeconomic discrepancy between Indian and white populations (Waters and Eschbach, 1995).

From very early on, the federal government had had some role in Indian education. Missionary schools had been established by the early nineteenth century, and the government assisted these schools in their efforts. The BIA operated its own school system with schools on or near reservations. After the Civil War, many boarding schools were established for Indians. As noted earlier, mental and physical abuse were common in these schools, and suppression of Indian cultural ways was a fundamental objective. The failure of the boarding school system was eventually recognized, and Indian students were increasingly channeled into Indian day schools and the public school system (Thornton, 2001).

By the mid-twentieth century most observers agreed that the system of educating Indian children was pathetically ineffective and in many ways perpetuated the low social and economic status of Native Americans (Josephy et al., 1999). A U.S. Senate report in 1969, summarizing a lengthy investigation of the Indian education system, concluded that "our national policies for educating American

Indians are a failure of major proportions. They have not offered Indian children—either in years past or today—an educational opportunity anywhere near equal to that offered the great bulk of American children" (cited in Josephy et al., 1999:186). Federal control has undergone slow change over the past several decades. By the late 1960s, several states had assumed responsibility for educating Indian children. Today American Indians have gained a greater measure of control over the education of their children, and this trend is likely to continue (Thornton, 2001). Nonetheless the educational gap between Indians and the larger society remains a problem of serious proportions.

On a positive note, the numbers of Native-American students attending college and pursuing graduate and other professional degrees have steadily risen. Between 1984 and 1993, for example, American Indians attending college increased 45 percent from 84,000 to 122,000 (U.S. Census Bureau, 1995). Still, a Department of Education task force studying Indian education concluded that "It is evident that the existing educational systems, whether they be public or federal, have not effectively met the educational, cultural, economic, and social needs of Native communities" (U.S. Department of Education, 1991:12).

Health Economic insecurity and low educational achievement translate into other measures of deprivation. Although improved, the health of American Indians is still worse than that of the general population by almost every indicator. About 33 percent of Indians die before age 45, compared to 11 percent for the U.S. population as a whole (Garrett, 1994). Much of the lower health standards among Indians is the result of seriously insufficient health-care delivery. On reservations, there are 96 doctors and 251 nurses per 100,000 people, compared with 208 doctors and 672 nurses per 100,000 of the general population (Worsnop, 1992). Poor diet also contributes mightily to Indian health problems, particularly on reservations. The rate of diabetes, for example, has reached epidemic proportions. Further, suicide and alcoholism rates are many times higher among Native Americans than among the general population. Alcoholism is particularly problematic (McCoy, 1994). From 1991 through 1993, the mortality rate from alcoholism among American Indians was over six times the rate among whites (U.S. Department of Health and Human Services, 1996).

Low economic and educational standing, when combined with alcohol abuse, relate to high crime rates among American Indians. They make up about 2.5 percent of federal prison inmates, a number far out of proportion to their population size (Grobsmith, 1994).

Although the differences between Indians and the remainder of the population are wide, progress has been made in recent years in improving health and educational standards among the Native American population. Life expectancy of Indians has risen significantly and is now about the same as the national average (Kington and Nickers, 2001). Nonetheless, in specific cases, conditions remain especially dire for Native Americans. Among the Oglala Sioux on the Pine Ridge Reservation in South Dakota, for example, the infant mortality rate is three times the national average (Kilborn, 1997).

One of the reasons that changes in the status of Indian tribes over the past two decades have not had a profound impact on long-term problems of American Indians is that more than half of all Indians today do not live on reservations, but are urban dwellers. New policies and favorable court decisions have clearly improved economic and social conditions for American Indians vis-à-vis the larger American society. However, because most government policies pertain to tribes on reservations, the majority of American Indians may not be affected (Nichols, 1992). Moreover, in no major urban area do Indians represent a population base large enough to translate into effective political power. Hence, other urban minority populations generally receive proportionately greater benefits from government programs (Green, 1999).

Recent Economic Projects

Indian activism of the past three decades has been met with measures designed mostly to quell militance rather than to address the unique socioeconomic problems of Native Americans. Moreover, as vital resources such as oil, natural gas, coal, and uranium have been discovered in recent years on federal Indian reservations, Native Americans have experienced new pressures threatening their control over their remaining lands. Indian lands contain about one-twentieth of U.S. petroleum reserves and one-third of low sulfur strippable coal deposits (Goodman and Heffington, 2000). These resources, however, have also served as valuable political tools for particular Indian groups in pursuing political and economic goals.

In recent years, many Indian tribes have become more aggressive in pursuing projects designed to foster greater economic development and independence. Because American Indian nations are considered sovereign, the U.S. government does not tax tribes or their wealth. This has provided opportunities to establish businesses of various kinds, some of which have been extremely successful and have resulted in significant economic benefits to Indian reservations. The Lac du Flambeau Chippewa tribe in Wisconsin, for example, distributes Ojibwe brand pizza in seven states, and the St. Regis Mohawk tribe of New York produces widely sold fishing products. In Arizona and New Mexico, the Navajo reservation now grows produce that it sells in twenty-one states as well as Mexico and Canada, with annual sales of more than $32 million (Trueheart and McAuliffe, 1995).

Indian Casinos   The largest and most significant of these enterprises, however, is legalized gambling ("gaming"). The economic revenues accumulated and the jobs created by gaming have far exceeded those of other projects. In many cases, gaming has turned what were economically dormant Indian nations into communities with a solid economic base from which other aspects of improvement have been launched. In 1988, the federal government enacted the Indian Gaming Regulatory Act (IGRA), which provided for the operation of gaming by Indian tribes, with the objective of promoting tribal economic development and self-sufficiency. Since then, about 200 tribes in 24 states have set up casinos and bingo parlors, which in 1994 earned an estimated $4.2 billion in annual profits (Johnson, 1995; McAuliffe, 1996; Trueheart and McAuliffe, 1995). In New Mexico, for example, eleven of the state's twenty-two tribes operate casinos, taking in over $350 million each year in revenue (Pulley, 1999). Ironically, the success of some of these operations has redounded to non-Indians by providing jobs as well as tax revenues to states. The Mashantucket Pequots of Connecticut are a community of only a few hundred, but they operate a casino that employs 10,000 people and is exceeded only by the federal government as a contributor to the state's treasury (Johnson, 1995). Many Indian nations have used their profits from gambling enterprises as a springboard to other forms of tribal economic development. The Pequots, for example, are buying back their land and are investing in nongaming businesses. Moreover, the emergence of a thriving gaming industry on many reservations has led to the return migration of many Native Americans who had been living in urban areas (Johnson, 1999).

The boom in casino gambling has resulted in an unanticipated twist of good fortune for some Indian tribes, which are shifting suddenly from near destitution to substantial wealth. Funds have been used to construct schools, health clinics, and housing, and to capitalize other businesses. Moreover, some of the monies newly acquired have been used in efforts to recover tribal relics and other elements of Indian culture that had been neglected or appropriated.

Despite the real and potential financial benefits of casino gambling, however, these have not yet proved to be the economic salvation of American Indians. The popular perception of newfound wealth does not characterize the vast majority of the Indian population. Clearly, the benefits have not been felt for all Indians, even of those tribes that operate flourishing casinos. In fact, among the 130 tribes with casino gambling, only a few near major population centers have thrived. For the remainder, gambling revenues have not been great enough to appreciably upgrade the quality of life (Pace, 2000). Nor has the unemployment rate on most reservations been significantly affected, since so many of the casino jobs are occupied by non-Indians. Above all, it must be remembered that two-thirds of all American Indians belong to tribes that do not own Las Vegas–style casinos. There is also fear among some Indian leaders that the attraction of tourists and the outside professional and managerial personnel who help operate the casinos will result in a further deterioration of tribal traditions and values (Pulley, 1999).

Whether casino gambling or other enterprises prove to be long-term sources of tribal revenue is not at all certain. The immense success of a few tribes has distorted the real condition of most, which have not benefited from profitable enterprises. The fact remains obvious that as a part of the American ethnic hierarchy, Native Americans are still in a very low position with unusually high unemployment and poverty rates (McAuliffe, 1996).

PREJUDICE AND DISCRIMINATION

The unique history of Indians as they encountered Europeans, compared to that of other groups who entered mainly as immigrants, shaped the nature of prejudice and discrimination to which American Indians have been subjected. The idea of Indians as "savages" corresponded to the threat that Indians posed to early settlers. Further, in many cases it served as justification for divesting them of lands. But once their defeat was accomplished, Indians ceased to pose an economic or political threat to Euro-Americans. Thus the need to create an elaborate racist ideology to justify the continuation of their subordinate status was not needed. Moreover, following the Indians' defeat, the bulk of their population was separated from the mainstream society on isolated reservations. The unique political status of Indian tribes and their placement on distant reservations resulted in the creation of a set of stereotypes, and their accompanying attitudes and actions, that relied largely on media depictions.

Stereotyping

From the start of contact between Europeans and Natives in North America, ethnocentric views prevailed on both sides. For Europeans, the distinctness of specific tribes was blurred by a perception of uniform inferiority. "That almost no account in the sixteenth century portrays systematically or completely the customs and beliefs of any one tribe," notes Berkhofer, "probably results from the newness of the encounter and the feeling that all Indians possessed the same basic qualities" (1979:26). This tendency to see Indians collectively rather than as separate and culturally varied units carried forward to the modern period and has been challenged only in recent times.

The view of American Indians as inferior served as a justification for expropriating Indian lands as settlers pushed westward. As noted earlier, in the perspective of Euro-Americans, land was to be settled and developed. Indians, therefore, stood in the way of American expansion. Those who fought Indians became folk heroes; those whom they fought were subhuman savages. "Over the course of the nineteenth century," notes John Coward, "the story of western expansion became part of a celebrated national myth where heroes rose to their civilizing task and Indians were worthy—and exotic—impediments to progress and empire" (Coward, 1999:6).

One unusual aspect of Indian stereotyping is that the "savage" characterization has, throughout American history, often been accompanied by admiration, yielding the "noble savage." Thus, on the one hand, Indians could be brave, innocent, and industrious; and on the other, cruel, treacherous, and lazy. But as Berkhofer has explained, "Beneath both the good and bad images used by explorer, settler, missionary, and policy maker alike lay the idea of Indian deficiency that assumed—even demanded—that Whites do something to or for Indians to raise them to European standards, whether for crass or idealistic motives" (1979:119).

However positive some characterizations of Indians were, the dominant white view was always one of inferiority. Indians lacked those practices and institutions that, in the Euro-American view, constituted civilization: a work ethic, effective government, a defensible culture, and almost all kinds of technology (Coward, 1999). Thus, in the confrontation between European and Native, the ultimate outcome in the European view was the triumph of "civilization" over "savagery." As Berkhofer notes, from the middle of the eighteenth century to very recent times, for whites, "the only good Indian was indeed a dead Indian—whether through warfare or through assimilation" (1979:30).

Native-American Stereotypes and the Mass Media The use of stereotypes in describing Native Americans proved critical in the production of a rationale for stripping them of their lands and for efforts at dissolving their cultures. From the early nineteenth century, the mass media played the major role in creating and communicating those stereotypes. As depicted in newspapers, which were aligned with both government and business interests, Indians were obstacles to economic growth and national expansion. Newspapers were also eager to offer their readers "colorful and exciting stories that confirmed the correctness of American values and goals" (Coward, 1999:39). Journalists, in their stories, conformed to a standardized, simplified view of Indians and their place in American society that could be easily understood by readers. Newspapers sustained the dual image of Native Americans as romantic on the one hand and savage on the other, but in all cases as fundamentally different from Euro-Americans and thus not in accord with Western civilization (Coward, 1999; Weston, 1996).

In the twentieth century, negative stereotypes endured largely through the portrayal of American Indians in motion pictures. Most Americans—indeed, most people of the world—have acquired their beliefs about North American Indians through films. Though created as entertainment, they nonetheless have shaped a perception of Indian culture that has been stubbornly resistant to change. Even American Indians themselves have drawn from these films in constructing their own views of their cultural heritage (Price, 1973).

The savage, hostile Indian, in confrontation with white settlers pushing westward, has been a common movie theme since the introduction of films in the early part of the twentieth century. From the beginning, films have consistently depicted Indians with a set of degrading stereotypes. "The prototype of the Hollywood Indian," notes Rita Keshena, "was treacherous, vicious, cruel, lazy, stupid, dirty, speaking in ughs and grunts, and often quite drunk" (1980:107). In the 1960s, films began to depict a more sympathetic view of American Indians, but one no less founded on inaccurate and simplistic stereotypes.

Not only have American films typically portrayed Indians as brutal and primitive, but they have blurred the cultural distinctions among them. Historically, films have presented Indians not as individuals or even as members of different tribes but as faceless parts of the collectivity labeled "Indians." Moreover, the Hollywood image of Native Americans has been derived mainly from those Indian societies with a historical reputation for violence, particularly the horsemen of the Plains and Southwest. Tribes with more passive and nonviolent reputations have been ignored. As Price (1973) notes, the emphasis on Plains Indians like Apaches and Cheyennes is ironic, because they did not evolve until after whites introduced the horse and were not similar to most other American Indians. North American Indians did not ride horses before Columbus, and most still did not hunt from horseback in the mid-1800s. Further, most were not hunters at all, but agricultural people.

In recent years, the use of Indian names by sports teams has raised additional issues with Indian stereotypes. Team names such as "Braves," "Redskins," and "Chiefs" are seen by many as demeaning, contributing to the perpetuation of the Indians' image as savage and warlike. A so-called Indian war chant—used by fans of the Florida State University athletic teams (called "Seminoles") as well as the Atlanta Braves baseball team—for example, evokes images of belligerence and brutality. The sports mascot issue seemed to resonate with the mainstream American media in 1990s, when several prominent newspapers editorialized sympathetically with those protesting the use of these names (Weston, 1996). Some college teams voluntarily changed their names and symbols. Stanford University changed its athletic teams' name from "Indians" to "Cardinal," St. John's University teams changed from "Redmen" to "The Red Storm," and the Miami of Ohio "Redskins" became the "Red Hawks." One newspaper, the *Portland Oregonian*, announced a policy of no longer using nicknames of sports teams that were deemed racially offensive.

ASSIMILATION AND AMERICAN INDIANS

Cultural Assimilation

Throughout the nineteenth century and the first half of the twentieth, U.S. policy toward American Indians was designed to destroy native cultures and political entities in hopes of creating assimilation into the dominant Euro-American culture and society. A major effort in this regard was the General Allotment Act of 1887. As noted earlier, the thrust of this legislation was to break up Indian tribes into individual farming units, hoping to accelerate assimilation. The driving force of the policy was the belief that once Indians were fully assimilated, the "Indian problem" would disappear.

From 1887 to 1934, the efforts of the BIA, supplemented by those of Christian churches, which were active on most reservations, were directed energetically to assimilation. Although the annihilation of Indian cultures and of Indian self-governing institutions was not fully realized, the effects on Native-American culture and social life were nonetheless very great. Moreover, during this period, the policy of forced assimilation engendered the deepest distrust of the federal government and the maximum hostility toward whites (Spicer, 1980).

The assimilation objective changed in 1934 with the appointment of John Collier to head the BIA. Collier, as mentioned earlier, championed the Indian Reorganization Act (IRA), which proved to be a watershed, bringing with it an entirely new approach and direction to Indian-white relations. Collier, it should be remembered, encouraged the preservation of Native cultures. But as Cornell explains, despite Collier's efforts and what became known as "the Indian New Deal," the basic assumption of Indian policy—eventual assimilation—was not essentially altered. Rather, this new approach was merely "a means of easing the process of change and giving Native Americans some measure of control over it" (Cornell, 1988:93). By the early 1950s, blatant assimilation sentiments again prevailed with enactment of the policy of termination.

Current Patterns For most of American history, then, forced cultural assimilation—that is, the disintegration of Native cultures—has been the primary objective of dominant political and social institutions. That objective has never been fully accomplished, but the continual assault on Native cultures has resulted generally in a significant dissolution of Indian cultural influences on individuals, especially regarding language. Dozens of tribal languages continue to exist and function, but their use is common mostly on reservations, and the proportion of Indians not also speaking English is low. Quite clearly, English today is the prevailing language among Native Americans. Many of those who are younger and have migrated to urban areas have been cut off from tribal cultural influences in addition to language, such as religion; hence, in some ways, their ethnic distinctness becomes increasingly difficult to sharply define.

Despite the steady dilution of Native American culture, the emergence of a strong Pan-Indian movement in recent years has emphasized common Indian values and practices, which may serve as a shield against more complete cultural assimilation. In the past three decades, American Indian art and other manifestations of Indian culture have seemed to proliferate. Many new tribal museums have been established, and numerous tribal and urban Indian powwows and arts festivals are held annually. Also evident is an explosion of Native-American literature, music, and film, as well as the founding of the National Museum of the American Indian in Washington, D.C. (Josephy et al., 1999).

There is, then, a somewhat contradictory pattern of cultural assimilation that has been occurring among American Indians. On the one hand, as Native Americans have increasingly moved off the reservations into urban areas and have intermarried with non-Indians, they have shed many tribal traditions. However, as Josephy et al. (1999) have explained, a countervailing pattern has emerged, bringing about a kind of "retraditionalization." The same urban Indians who enter the mainstream society return to the reservation periodically to take part in traditional practices as a means of reconnecting with past roots and Indian identity. These opposing trends have led to internal debates among Indians, raising questions of "Who is an 'authentic' Indian?" "Who is entitled to tribal membership?" "Who should represent Indians before the federal government?" and "Who should be eligible for affirmative action benefits?" (Glaberson, 2001; Tobar, 2001). These are questions that will continue to plague efforts at building a more cohesive American Indian ethnic population.

Structural Assimilation

In looking at the structural assimilation of Native Americans, it is important to keep in mind that this is an ethnic population divided into two demographic parts: those living on and those living outside the reservations. In either case, as already noted, Native Americans today remain at or near the bottom of the society's occupational, educational, and income hierarchies. This collective position has stifled rapid movement into mainstream economic institutions.

Secondary Structural Assimilation Regarding various measures of secondary structural assimilation, the distinction between reservation and nonreservation Indians is an important consideration. This is particularly apparent in looking at the participation of Indians in the political system. Those on reservations are subject to laws made by the tribal government; as a result, politics there are intratribal. Those off reservations do not represent a large enough group in any area to be able to carry much political weight. Thus American Indian representation in government at all levels has been minimal. One exception in recent years has been Ben Nighthorse Campbell, who twice has been elected to the U.S. Senate from Colorado. The economic success of the gambling industry for some American Indians, however, has led to more concerted and systematic lobbying to protect and enhance Indian needs (Dao, 1998). This has necessarily pulled Native Americans into some mainstream political issues.

Although substantial secondary structural assimilation has not occurred, one aspect of Indian life that seems to augur increasing assimilation in the future is the expansion of education. As noted earlier, a growing number of young Native Americans attend college and, as this occurs, more will inevitably take their place in the business and political worlds.

In considering secondary structural assimilation it is important to keep in mind that American Indians remain in a somewhat peculiar legal limbo. In some regards they have regained a semblance of political sovereignty, but in others they are still beholden to the federal government (Kehoe, 1999).

Primary Structural Assimilation   At the primary level of assimilation, those Native Americans who have migrated to the cities appear to have begun to integrate into the larger society. Perhaps the most telling indication of this is the high, and increasing, rate of intermarriage among American Indians and whites (Eschbach, 1995). Well over half of married Indians are married to non-Indians, a percentage considerably higher than for any other nonwhite racial-ethnic group (Snipp, 1997). This extremely high rate of intermarriage is particularly strong among those in urban areas.

What this high and, apparently, unflagging rate of intermarriage means for the future of American Indians can only be speculated on. In combination with the assimilating forces of urban life, it may indicate a decline of Indian ethnicity. Relatively undeterred by racial identity, Indians can more easily move into the mainstream—unlike African Americans, Asian Americans, and some Latinos. Intermarriage may also reduce the number of persons who, in the future, will identify themselves ethnically as Indians. However, the existence of tribes and of relatively cohesive Indian communities on reservations, where intermarriage rates remain lower, would seem to preclude the complete disappearance of American Indians as part of the American ethnic system (Eschbach, 1995). At the same time, it is unlikely that steady movement in the direction of assimilation will be seriously deterred as Native American ethnicity is increasingly diluted.

Indianization   The idea of an American-Indian ethnic group has, from the outset of European-Native contact, been a construct of whites. The extreme cultural and social diversity of the Native population was dissolved into a collective notion that created a single category. Among Indians, supratribal consciousness, though emergent in the early part of the twentieth century, did not become a well-developed movement until the 1950s and 1960s. Cornell (1988, 1990) attributes this in large measure to Indian migration to the cities. In the urban environment, Indians found that tribal identities and interests no longer held much usefulness as they found themselves in political and economic competition with other ethnic groups. This Indian ethnic consciousness, developed in the urban communities, reached back to the reservations through returning migrants.

In recent years, then, American Indians have begun to coalesce into a coherent ethnic group, in a kind of "cross-tribalism" or "Pan-Indianism." Tribal identities remain, but increasingly Indians have begun to develop an Indian consciousness that transcends tribal divisions. In a sense, what is emerging is an "American Indian" ethnic group (Nagel, 1995). This unity may result in greater political effectiveness and a stronger public awareness of Indian issues and problems.

SUMMARY

The conquest by European settlers of the Native Americans took the form of contact, ethnocentrism, competition over land, and the imposition of white dominance through superior force. Indians were exploited and largely annihilated in the nineteenth century. By the start of the twentieth century, American Indians were essentially wards of the state.

Though the Indian population was decimated in the nineteenth century through war, disease, and the breakup of tribal cultures, in recent decades it has been on the rise, due largely to the tendency of more people to claim Indian ancestry. About half of all Indians live in urban areas rather than reservations. Urban Indians are somewhat better off economically, but all Indians rank near the bottom of the economic class system. On various measures of socioeconomic status—income, education, occupation, health—the gap between American Indians and Euro-Americans is extremely wide. Some tribes have benefited from the establishment of gambling casinos, but the long-term benefits of these economic projects are not yet evident.

Native Americans have been stereotyped since the early nineteenth century as both savage as well as noble. Indian stereotypes, which have been promulgated in the twentieth century primarily through films, have been strongly resistant to change.

Attempts have been made repeatedly throughout various historical periods to force assimilation of Native Americans into the dominant culture. Although these attempts were never completely successful, Indian cultures have been diluted over many decades. As to structural assimilation, American Indians have intermarried with Euro-Americans at very high rates. This, as well as the assimilating forces of urban life, may lead to a further decline of Indian cultures and institutions. However, in the past two decades, a Pan-Indian movement has emerged, which has led to a renewed sense of Indian ethnic identity and efforts to revitalize Indian cultures.

Suggested Readings

Brown, Dee A. 1970. *Bury My Heart at Wounded Knee: An Indian History of the American West*. New York: Henry Holt. A classic work that chronicles the decline of Native Americans during the period of Western expansion between 1860 and 1890.

Cornell, Stephen. 1988. *The Return of the Native: American Indian Political Resurgence*. New York: Oxford University Press. Traces the historical development of Indian-white relations and concludes with a discussion of the emergence of a Pan-Indian ethnic group, one transcending particular tribal identities.

Coward, John M. 1999. *The Newspaper Indian: Native American Identity in the Press, 1820–90*. Urbana: University of Illinois Press. Describes how common negative stereotypes of American Indians were created and communicated to the public by newspapers in the nineteenth century.

Nagel, Joane. 1996. *American Indian Ethnic Renewal: Red Power and the Resurgence of Identity and Culture*. New York: Oxford University Press. An analysis of the social and historical forces that account for the increasing number of Americans who choose to identify as Indians.

Snipp, C. Matthew. 1989. *American Indians: The First of This Land*. New York: Russell Sage Foundation. A detailed collection of demographic and social statistics on Native Americans.

Weatherford, Jack. 1988. *Indian Givers: How the Indians of the Americas Transformed the World*. New York: Ballantine. Examines a number of New World Indian technologies, particularly agricultural, that had a lasting effect on the Western world.

Wright, Ronald. 1992. *Stolen Continents: The Americas Through Indian Eyes Since 1492*. Boston: Houghton Mifflin. Analyzes five Indian cultures: two North American (Cherokee and Iroquois), two Mesoamerican (Maya and Aztec), and one South American (Inca). Describes how each culture dealt with the European colonial assault and how each developed afterward.

ITALIAN AMERICANS

IN THIS CHAPTER, we look closely at Italian Americans, a group whose origins are chiefly part of the classic, or new, immigration period. Like all American ethnic groups, Italian Americans are a sociological case with many unique attributes. Yet in a number of ways this is the prototypical white ethnic group, and it illustrates well much of the general experience of that part of the American ethnic system.

It is important to look carefully at white (or Euro-American) ethnic groups, specifically those described in Chapter 5 as part of the second tier of the American ethnic hierarchy, for several reasons. First, they are a substantial element of ethnic America. Together, they represent the ethnic origins of perhaps as much as 20 percent of the total American population.[1] Second, although it is true that today these groups are increasingly less visible as distinct ethnic units and are being assimilated unwaveringly into the mainstream society, it is erroneous to assume that ethnicity for them is no longer of significance. Many observers maintain that ethnicity, though important to these groups in the past, no longer holds great meaning, either culturally or structurally. This view, however, seems premature. None of the groups have reached parity with the dominant Anglo group in economic status and

---

[1]This estimate is based on census reports of both single- and multiple-ancestry persons; that is, those who have claimed either one specific ethnic ancestry or a combination.

political power, none have entered fully into primary relations with the Anglo group, and none have completely lost their cultural forms through assimilation. Hence, although their conflict with the dominant group is muted by comparison with that of racial-ethnic groups—and as cultural and structural assimilation continue unabated, they are becoming less clearly defined—they remain discernible ethnic populations in American society and, to some degree, ethnic minorities.

Finally, it is important to study the white ethnic experience for comparative purposes. If white ethnics today remain minorities, the fact is they are minorities in a far less subordinate position than racial-ethnic groups. Recall the model of the American ethnic hierarchy in which groups of the second tier—including white ethnics—are increasingly merging socially and economically with those of the first tier—the dominant group, creating a dichotomy between primarily Euro-American and racial-ethnic groups. The social and economic disparities between them are a troubling issue of American society and have generated much debate among social scientists, policymakers, and the general public. These disparities in success have been interpreted by some as a product of internal group differences and by others as the result of inequitable and qualitatively different social contexts within which Euro-American and racial-ethnic groups have labored. To shed light on this issue, it is necessary to understand how the characteristics and social experiences of white ethnic groups have been similar to and different from those of the country's racial-ethnic groups. What cultural systems and social and economic skills did each bring with them? What was the nature of the opportunity structure open to each? What obstacles to social integration for each were dictated by the dominant group? To what extent did each seek assimilation or strive to preserve ethnic uniqueness? What was the nature and level of prejudice and discrimination directed at each? As we will see, although there are similarities between white ethnic groups and racial-ethnic groups on all these questions, there are also enormously consequential differences.

IMMIGRATION AND SETTLEMENT

Small numbers of Italians were part of American society almost from its founding, and a relatively small-scale immigration proceeded throughout the first three-quarters of the nineteenth century. The bulk of Italian immigration, however, occurred as part of the new immigration, from the 1880s until the mid-1920s.

The large-scale movement to the United States after 1880 represented the second great emigration from Italy. Earlier in the nineteenth century, Italians had immigrated mainly to Argentina and Brazil and only secondarily to the

United States. Many of these early immigrants, to the United States as well as to South America, were from northern Italy (Foerster, 1919).[2]

Beginning in the 1880s, both the origins and destinations of Italian emigrants changed. With the depletion of coffee plantations in Brazil and political turmoil in Argentina, most Italians now found the United States more attractive. And whereas the northern Italian provinces had previously supplied most of the immigrants, southern Italy now became the major region of emigration. This change in regional origin fundamentally altered the class and cultural character of the Italian population in the United States. By comparison with the more prosperous and literate immigrants from the North, those from the South were poor and uneducated. Further, whereas the northerners had been mostly artisans, shopkeepers, and professionals, those from the South were mainly agricultural peasants.

Southern Italy had historically been an impoverished, backward area, subjected to constant economic and political exploitation by the region's ruling groups; but unusually severe conditions were created in the latter nineteenth century. Depressed wages, little or no opportunity to improve a low agricultural output, little industrialization, poor health conditions, neglect by the central Italian government, and heavy indirect taxation all served to create a strong impetus for emigration. These factors, combined with the call for unskilled labor by a burgeoning American industrialism, resulted in a veritable exodus of southern Italian peasants. Between 1880 and 1915, four million Italians went to the United States, half of them in the single decade from 1900 to 1910. Between 80 and 90 percent were from the south, that area known to Italians as the Mezzogiorno, and about one-quarter of these were from Sicily (Foerster, 1919; Gambino, 1975; Lopreato, 1970; U.S. Census Bureau, 1975).

One extremely important characteristic of the Italian emigration of this time was that most who went to the United States did not expect to stay permanently. Most were men unaccompanied by their families, intent on earning money to purchase land after returning to their native villages. Many did go back. In fact no other sizable immigrant group displayed such a high rate of return migration. The comparative rates of repatriation of Italian and Jewish immigrants of this period are striking. Between 1908 and 1914, almost one-third of those who arrived from Italy eventually returned or migrated to another society; for Jews, the return rate was only 7 percent. Between 1915 and 1920, the corresponding figures

---

[2]Italian immigration to Argentina and the United States did overlap somewhat during the late nineteenth and early twentieth centuries. Although Italian immigrants to both societies were motivated primarily by the expectation of earning more money than they could at home, a greater percentage of those destined for Argentina were skilled workers. This proved to be an important factor in facilitating their more rapid and successful economic and social adjustment than occurred among Italian immigrants to the United States. See Baily (1996).

were 56.6 and 4.3 percent for the respective groups; and between 1925 and 1937, the figures were 40 and 3.8 percent (Learsi, 1954).

Urban Adjustment

The Italian immigrants of the late nineteenth and early twentieth centuries were in many ways unprepared for life in an industrial, urban society. Most lacked the skills and education that could be converted into rapid occupational mobility and economic success in the United States. As a result, most entered the labor force in unskilled jobs, where they and their children remained. Most settled in the cities of the Northeast, especially New York, where they found work in manual laboring occupations corresponding to their low skill and educational levels. Because cities like New York were at this time establishing their industrial infrastructures—roads, rail lines, subways, factories, housing—opportunities abounded for unskilled labor. It was there that Italian immigrants found their occupational niche. At the turn of the century, Italians made up three-quarters of the building laborers in New York City, for example, and almost the entire labor force constructing the city's subway system (Glazer and Moynihan, 1970).

Italian immigrants in the cities tended to settle in depressed areas, inhabiting run-down tenements that had been vacated by earlier ethnic groups such as the Irish. The cluster of Italians in these areas was partially the result of simple economic necessity; such housing was available and affordable. However, as with other groups of the new immigration, the creation of ethnic concentrations in New York and other cities was also the product of the natural attraction of immigrants to areas inhabited by those from their regions in Italy who had preceded them to the United States. Italian immigration to the United States was very much a chain phenomenon, in which those from one village gravitated to locations where previous emigrants from the village had settled. There was little identity with the Italian nation per se. Rather, immigrants identified with their province and, even more precisely, their village. As a result, Italian sections of American cities were internally subdivided into "Little Calabrias," "Little Sicilys," and the like, and neighborhoods were often settled by those from a common village or town in the province (Vecoli, 1964). Even outside the large cities, settlement patterns were chain-like. In one town on Long Island, for example, the nucleus of an Italian colony—along with a flourishing horticulture business—was established by Italian immigrants from a few small towns and their surrounding villages in the vicinity of Naples (LaGumina, 1988).

The Italian-American Population Today

Italian Americans remain one of the largest white ethnic groups. In 1990, almost 15 million people claimed either single or multiple Italian ancestry. Although the exact number of Italian Americans (or any other white ethnic group) cannot be pinpointed owing to the dilution of the group through ethnic intermarriage, there is no question that this is not only one of the largest but also one of the most heavily urbanized American ethnic groups. As can be seen in Tables 7.1 and 7.2, although it is far more dispersed than in earlier times, the Italian-American population continues to reside most heavily in the states and urban areas of the Northeast. Italian Americans in the New York–New Jersey metropolitan region, for example, constitute one of the largest elements of its ethnically diverse population.

ITALIAN AMERICANS IN THE CLASS SYSTEM

Italian Americans are well into the third generation; and many, especially those whose ancestors were northern Italian, are even into the fourth generation. Given the length of time since the group's large-scale entry, some evaluation can be made of Italian-American progress on various socioeconomic measures, both in absolute terms and in relation to other American ethnic groups. How far have Italian Americans come, what means have they employed in seeking upward mobility, and how might we account for their current status in the overall stratification system? It is important to keep in mind that most of the patterns displayed by Italian Americans are comparable to those of other white ethnic groups. The following sections, therefore, are generally descriptive of those groups as well.

Table 7.1 States with Highest Percentages of Italian Americans, 2000

STATE	ITALIAN AMERICANS	TOTAL	PERCENT OF TOTAL
Rhode Island	203,302	1,009,503	20.1
Connecticut	649,496	3,297,626	20.0
New Jersey	1,512,139	8,219,529	18.4
New York	2,724,039	18,395,994	14.8
Massachusetts	886,057	6,127,881	14.5
Pennsylvania	1,485,418	11,847,752	12.5

SOURCE: U.S. Census Bureau, 2001i.

Table 7.2 Metropolitan Areas with More Than 100,000 Italian-Americans, 1990

METROPOLITAN AREA	ITALIAN-AMERICAN POPULATION
New York–Northern New Jersey–Long Island, NY–NJ–CT CMSA	2,669,693
Philadelphia–Wilmington–Trenton, PA–NJ–DE–MD CMSA	665,626
Boston–Lawrence–Salem, MA–NH CMSA	516,445
Chicago–Gary–Lake County, IL–IN–WI CMSA	472,401
Los Angeles–Anaheim–Riverside, CA CMSA	439,257
San Francisco–Oakland–San Jose, CA CMSA	328,717
Pittsburgh–Beaver Valley, PA CMSA	266,604
Detroit–Ann Arbor, MI CMSA	219,325
Cleveland–Akron–Lorain, OH CMSA	193,587
Providence–Pawtucket–Fall River, RI–MA CMSA	163,326
Miami–Fort Lauderdale, FL CMSA	159,936
Buffalo–Niagara Falls, NY CMSA	146,626
Hartford–New Britain–Middletown, CT CMSA	142,749
Rochester, NY MSA	139,998
Washington, DC–MD–VA MSA	133,717
Tampa–St. Petersburg–Clearwater, FL MSA	126,333
New Haven–Meriden, CT MSA	116,580

Source: U.S. Census Bureau, 1990.

Social Class

In looking at the place of ethnic groups in the class structure, it is important to remember that these are collective pictures reflecting the general positions of groups, not particular individuals within them. Obviously, Italian Americans, for example, will be found in the entire range of the American class structure, from the very poor to the very affluent. Our concern, however, is with the group as a whole and how it compares generally with others.

By all indications, Italians have "made it" in American society to a greater degree than many ethnic groups and to a far greater degree than African Americans, Latinos, and American Indians. Although upward mobility was minimal for the first two generations, Italian Americans of the third and fourth generations are now exhibiting solid evidence of middle-class status in terms of income, occupation, and education. These trends are characteristic of white ethnics in general.

Income On the first measure of social class, income, Italian Americans are above the national average. Especially significant improvement occurred during the 1950s, with only Jews showing a greater increase among white ethnic groups. As is seen in Table 7.3, the median family income of Italian Americans is comparable to that of other Euro-American groups.

Clearly what has occurred in recent years is a convergence in average socioeconomic status among Euro-American ethnic groups. Southern and eastern European ancestry groups, like Italians, have reached a point where they are indistinguishable class-wise not only from each other, but more importantly, from northwestern Europeans—the society's dominant ethnic group (Alba, 1995).

Occupation Though above average in income, Italian Americans do not show a commensurate position in the occupational hierarchy. This is because little occupational mobility was evident for this group between the first and second generations (Greeley, 1977; Yans-McLaughlin, 1982). That is, most of the children of immigrant parents remained in working-class jobs. Because children in Italian families were expected to contribute to the family's support, they entered the

Table 7.3 Socioeconomic Standing of Euro-American Ethnic Groups, 1990

GROUP	MEDIAN FAMILY INCOME	PERCENT COLLEGE GRADUATES
Russian	$58,826	49
Dutch	43,415	18
Greek	43,330	28
Scottish	43,293	34
Hungarian	42,778	27
Italian	42,242	21
Polish	41,700	23
English	40,875	28
Swedish	40,459	27
Slovak	40,072	22
Scotch-Irish	39,816	28
Portuguese	38,370	12
German	38,216	22
Irish	38,101	21
French	36,237	18

Source: U.S. Bureau of the Census, 1993c.

labor force early, precluding the more extensive education needed for higher-status jobs. Most Italian children thus followed their fathers into manual laboring occupations. But those of the third and fourth generations, with greater education, have moved in significant numbers into all occupational areas—including the top professions like medicine and law, where they had been severely underrepresented (Alba, 1985; Glazer and Moynihan, 1970).

Most Italian Americans still remain more solidly working class than Jews, Irish, and several other white ethnic groups. But the extent of their mobility is nonetheless quite extraordinary considering how relatively ill equipped they were at the time of their arrival to succeed in the American economic system. Some have argued, however, that Italian-American mobility is a reflection not so much of the group's progress per se as of a change in the society's occupational structure. As more white-collar and professional jobs have been created, more Italian Americans have filled them (Alba, 1985; Gallo, 1974).

Education The relationship between occupation and education is ordinarily very strong; a group's position on one measure of socioeconomic status should therefore closely reflect the other. This is quite evident in the case of Italian Americans. Those of the first two generations lagged in educational attainment, but the third and fourth generations show educational levels that equal and in some cases even exceed those of other white ethnic groups (see Table 7.3).

Clearly, as with other indicators of socioeconomic status, in educational attainment Euro-American ethnic groups have in recent years converged, with marginal differences among them. Alba (1995) has shown that for men of southern and eastern European ancestry born between 1916 and 1925, there remain some apparent disadvantages when compared to the educational levels of those of northwestern European ancestry. But when those born between 1956 and 1965 are looked at (these would be third- or fourth-generation ethnics), southern and eastern European origin men are at or above the average of northwestern European origin men.

In sum, the class position of Italian Americans is on most measures at or above the average for white ethnic groups. Moreover, for the third and later generations and for those of part-Italian ancestry, parity with the society's core ethnic group, northwestern Europeans, has nearly been reached (Alba, 1985). Overall, the differentials between various European ethnic groups in the stratification system seem to be shrinking.

Political and Economic Power

Having looked at the place of Italian Americans in the economic realm, let's look briefly at the power hierarchy. Although there is commonly a close relationship between economic class and societal power, these are two distinct dimensions of stratification. As explained in Chapter 2, the power dimension entails the extent to which people are able to affect others in the society or community through their actions. Ordinarily in American society such power is attached to positions of authority in two key institutions: government and the corporation. The best indicators of societal power, then, are the number and level of top-ranking positions in these institutions held by members of the ethnic group. Where do Italian Americans currently stand, and what has been the extent of their mobility over several generations?

Government Patterns of Italian-American participation at the highest levels of government parallel group patterns in the economic realm: The first two generations made little progress, but in recent years, Italian Americans have begun to achieve important positions in modest numbers. This is a pattern generally duplicated by other groups of the new immigration.

Prior to the 1960s, the Italian-American presence at the highest levels of the federal government was limited. Not until 1950 was an Italian American elected to the Senate, and not until John F. Kennedy's administration in 1962 was an Italian American named to a cabinet post. Since then, Italian-American representation in the Congress and in the higher levels of the executive branch has steadily increased. By 1980, at least twenty-five Italian Americans were serving in the Congress, compared with four in the 1930s and eight in the 1940s (Rolle, 1980). Today, five U.S. Senators and twenty-seven members of the U.S. House of Representatives are of Italian ancestry (NIAF, 2001a). And the choice of a second-generation Italian American as a vice-presidential candidate in 1984 marked a watershed of sorts in national politics for Italian Americans. Moreover, in 1986, for the first time, an Italian American was appointed to the Supreme Court. This marked another ethnic watershed because not only Italians but southern and eastern European Catholics in general had historically been absent from the Court (Lieberson, 1980; Schmidhauser, 1960).[3]

The underrepresentation of all Catholic ethnic groups, not simply Italians, in top positions of the federal government has been a historically consistent pattern since the founding of the society. Although a Catholic was elected president for the first time in 1960, Catholics in general—and southern and eastern European Catholics in particular—have traditionally been underrepresented in the highest offices of the executive, legislative, and judicial branches (Alba and Moore, 1982; Mills, 1956; Stanley, Mann, and Doig, 1967). This historical pattern continues to change, however, with each succeeding generation. Catholics today are represented in Congress approximately in proportion to their share of

---

[3]However, five Catholics, including three Irish Catholics, had been appointed to the Court before World War II (Baltzell, 1994).

Table 7.4 Religious Makeup of the U.S. Congress[*]

	RELIGION	SENATE	HOUSE	U.S. POPULATION
1961	Protestant	87%	77%	68%
	Catholic	12	21	25
	Jewish	1	3	3
1979	Protestant	76%	66%	64%
	Catholic	15	27	25
	Jewish	7	5	2
2001	Protestant	64%	62%	56%
	Catholic	26	30	25
	Jewish	10	6	2
	Other, unspecified, or none	1	2	17

SOURCES: Hirschfeld, 2001; *Congressional Quarterly*, 1979; Gallup Organization, 2001b; Stanley, Mann, and Doig, 1967.
[*]Based on those who expressed a religious preference.

the total population (Table 7.4). Moreover, the choice of Michael Dukakis, the son of Greek-American immigrants, as the Democratic nominee for president in 1988 seemed to mark a culmination in the national political fortunes of southern and eastern European ethnic groups, indicating their fullest entrance into American politics.

In state and local governments, Italian Americans have attained substantial power in locales where they constitute a significant proportion of the population, such as New York, New Jersey, Massachusetts, Connecticut, and Rhode Island (Dinnerstein et al., 1996; LaGumina, 1988). Italian-American governors and lower officeholders in those states are today commonplace.

Part of the explanation for the slower development of Italian-American political power lies in the traditionally lower rates of political participation for this group. The tendency to ignore the political world can be seen as partially a product of the distrust of government that so characterized the contadini (peasants) of southern Italy. As Lopreato explains, the southern Italian, more than other Europeans, did not participate in politics. "Matters political," he notes, "had been done for him—or more precisely to him" (1970:113). Government, therefore, was an institution to be resisted, not embraced. The relationship between successive governments and the southern Italian peasant was characterized most frequently by exploitation or neglect.

Lower rates of political participation in the past among Italian Americans can also be attributed in some measure to class factors. The relationship between social class and political participation is one of the firmest assumptions of political sociology, backed by a wealth of empirical evidence. Put simply, the higher one's class, the greater the scope, intensity, and level of political participation (Lipset, 1960; Milbrath, 1982; Verba and Nie, 1972). Because they had been predominantly working class, lower participation among Italian Americans generally should have been expected, regardless of ethnic factors such as family and regional influences (Greeley, 1974). With progressive upward mobility, however, Italian Americans appear to be behaving politically in much the same manner as is typical of Americans generally (Barone, 1985).

The Corporate World Power in American society is by no means limited to the arena of government. In fact, an equally significant locus of societal power lies in the realm of economic activity—specifically, the giant corporation, where decisions are made that shape the society's workforce, consumer market, and general economic condition. How extensively have Italian Americans penetrated the highest positions of the corporate world?

Like most white ethnics in American society, Italians have displayed a slow entrance into the top echelons of economic power. Increasing political influence has generally not led to power in other institutional areas at the same pace, especially in the economy. Clearly, the ethnic origins of American business leaders have been and even today remain disproportionately northwestern European Protestant; Catholics and Jews, especially those of southern and eastern European origin, have been underrepresented (Davidson et al., 1995).

In the past, the number of Italian Americans in the executive suites of America's largest corporations had been well below their proportion of the general population. A study of Chicago's largest corporations, for instance, revealed that although Italians made up 4.8 percent of the city's population, they made up only 1.9 percent of these corporations' directors and 2.9 percent of their officers (Barta, 1979). In New Haven, Connecticut, Wolfinger (1966) found that although Italians in 1959 made up 31 percent of the registered voters and held 34 percent of major political offices, they made up only 4 percent of the city's economic elite.

Similar patterns have been evident for other white ethnics. In the past, few entered into the top ranks of economic power. For example, in the previously cited study of Chicago's top corporations, Poles, that city's largest white ethnic group, were shown to be even more severely underrepresented than Italians. Although they made up almost 7 percent of the city's population, they made up only 0.3 percent of these corporations' directors and 0.7 percent of their top officers. Numerous studies have shown the historically consistent domination of the American business elite by those of northwestern European and Protestant origins (Davidson et al., 1995; Marger, 1974; Mills, 1956; Newcomer, 1955). Sociologist Suzanne Keller (1953), for example, found that 89 percent of top business leaders in 1900 and 85 percent in 1950 were Protestants, most of British descent; 7 percent in each year were Catholics; and 3 and 5 percent, respectively, were Jews. By the 1970s, little had changed—Catholics made up 14 percent and Jews 6 percent of the corporate elite (Burck, 1976). Alba and Moore in 1982 found a continuation of Anglo dominance of economic power positions. However, as Anglo overrepresentation in power elites (including those of the corporate world) declines, Catholics are steadily moving toward parity (Davidson et al., 1995).

PREJUDICE AND DISCRIMINATION

Italians have been the objects of long-standing and often severe prejudice and discrimination in American society. The period of large-scale immigration beginning in the 1880s produced negative stereotypes and blatant anti-Italian actions. Though greatly diminished today in form and intensity, these attitudes and acts continue to affect the image and to some extent the treatment of Italian Americans.

It is important to understand that prejudice and discrimination leveled at white ethnic groups have historically never been as intense, widespread, or institutionalized as that aimed at racial-ethnic groups—African Americans, Latinos, American Indians, and Asian Americans. Though at times violent in form, discriminatory actions have ordinarily been limited to derogation and denial of access to certain jobs, schools, and neighborhoods. Never were white ethnics denied legal rights, as were blacks and, to a lesser degree, Asians; nor were they ever so thoroughly excluded from the larger society.

Images and Attitudes

As discussed in Chapter 5, anti-Catholic beliefs and actions in America became common beginning in the 1830s, when Irish immigrants began to arrive in large numbers. The anti-Catholic sentiments of Protestant Americans, then, were well established and widely proclaimed several decades before the large-scale influx of southern and eastern European Catholic groups began in the 1880s.

Italians suffered the usual antagonisms directed at Catholics generally, but anti-Italian feelings ran especially high as part of the newly inspired nativism and jingoism that arose in response to the new immigrants of the late nineteenth century. Indeed, with the arrival of hundreds of thousands of southern Italian peasants, Italians now replaced the Irish as the chief target group of the country's well-worn anti-Catholicism.

Much of the prejudice and discrimination was founded on racist ideologies. This was a time when ideas of social Darwinism prevailed among scholars and policymakers alike, and new racial theories were widely espoused and endorsed. Those groups composing the new immigration—overwhelmingly Catholics and Jews—were therefore described in racial terms, that is, as morally and even physiologically inferior to Americans of northwestern European origin. The new immigrants were perceived not simply as different from those who had preceded them but as threatening to the very social and biological fabric of the society (Higham, 1963). Italians were seen as the most degraded of all the entering groups.

The poor and illiterate peasants from southern Italy were received quite differently from the earlier northern Italian immigrants. Whereas the latter had been seen as honest, thrifty, law abiding, and generally "molto simpatici to the American character," the southern Italians were viewed as degenerate, ignorant, lazy, dirty, destitute, violent, superstitious, and criminally oriented (Train, 1912, cited in Moquin and Van Doren, 1974). Northern Italians, too, seemed to share the prevalent American view of their southern counterparts as a generally repugnant people (Gambino, 1975). One journalist of the time put the matter quite bluntly: "What shall we do with the 'dago'?" Note his description of the Italian as almost animalistic in nature:

> This "dago," it seems, not only herds, but fights. The knife with which he cuts his bread he also uses to lop off another "dago's" finger or ear, or to slash another's cheek. He quarrels over his meals; and his game, whatever it is, which he plays with pennies after his meal is over, is carried on knife at hand. More even than this, he sleeps in herds; and if a "dago" in his sleep rolls up against another "dago," the two whip out their knives and settle it there and then; and except a grunt at being disturbed, perhaps, no notice is taken by the twenty or fifty other "dagoes" in the apartment. He is quite as familiar with the sight of human blood as with the sight of the food he eats (Moquin and Van Doren, 1974:261).

The most serious elements in the negative image of Italians in America concerned their "criminal nature" and the existence of a widespread Italian criminal community, known as the Mafia, or the Black Hand. Few accounts of Italian immigrants during the late nineteenth century neglected the presumed criminal organizations among them. By the 1890s, the image of the criminally prone Italians and their association with the Mafia was well established (LaRuffa, 1982).

The strong anti-Catholic sentiments of a large segment of the American populace as well as government leaders aroused fear of Italian immigrants and not infrequently incited hostile actions against them. Anti-Italian antagonism was also partially a response of established workers, who saw the immigrants as a cheap labor force and thereby a threat to their jobs. Killings and lynchings of Italians, mostly in the South but in other regions as well, were numerous between 1890 and 1915. The most serious of these incidents occurred in New Orleans in 1891. Following the murder of the city's police superintendent, it was assumed that Sicilian immigrants were responsible, and hundreds of Italians were arrested. Nine were subsequently brought to trial, but none was found guilty. Anti-Italian sentiment was so feverish, however, that eleven of the suspects were taken from their jail cells by a mob and murdered; officials made no attempt to intervene. Afterward, local newspapers and business leaders expressed approval of the action (Gambino, 1977; Higham, 1963; Nelli, 1970).

Another especially notorious incident occurred in 1927. This was the famous trial of Sacco and Vanzetti. Two Italian immigrants, Nicola Sacco and Bartolomeo Vanzetti, were implicated in an armed robbery in Braintree, Massachusetts. Numerous witnesses attested to their lack of involvement in the crime, but despite the dubious evidence against them, they were found guilty. After numerous appeals, they were executed in 1927. The case created a storm of protest among civil rights groups throughout the country and remained a cause célèbre for many decades afterward. It emerged not only as an ethnic controversy, but an issue of symbolic class warfare. Those defending the two men's innocence saw their trial as a conflict between the oppressed working classes and the wealthy WASP elite (Baltzell, 1964).

The guilt or innocence of Sacco and Vanzetti is still debated, but historians generally agree that their trial was considerably biased and that the judicial process was seriously abused. Indeed, the perversion of justice in the case was finally acknowledged in 1977, when the state of Massachusetts officially proclaimed the trial unfair. In addition to the strong anti-immigrant sentiments of the time, the fact that the two defendants were avowed anarchists undoubtedly contributed heavily to their conviction. This was the era of the Red Scare, when government officials fanatically denounced and harassed union and socialist activities, in which Italians often took part. To many, Sacco and Vanzetti represented an affirmation of the perceptions held by much of the American public that Italian Americans were radical, conspiratorial, and criminal. These perceptions seriously colored the two men's arrest, trial, conviction, and finally, execution (LaRuffa, 1982).

This case had an enormous and lasting effect on Italian Americans, and for many years it influenced their view of the state and the judicial process. As Gambino has put it, "The outcome of the affair was simply another confirmation of the ancient belief of the Italian immigrants that justice, a very important part of their value system, had little to do with the laws and institutions of the state" (1975:122).

The Durability of Italian Stereotypes

The negative stereotypes of Italian Americans have been troublesomely persistent. Although they are rarely expressed today as virulently and hostilely as they were during the period of immigration, many of the early images remain dynamic elements of American ethnic mythology. Few would be hard put to describe the "typical" Italian American. Gambino observed the consistency of responses of non-Italians to his query about the Italian-American image: "the Mafia; pizza and other food; hard hats; blue collar; emotional, jealous people; dusky, sexy girls; overweight mammas; frightening, rough, tough men; pop singers; law and order; pastel-colored houses; racists; nice, quiet people" (1975:352).

These stereotypes are reinforced by media depictions. Lichter and Lichter's study (1982) showed that negative portrayals of Italian Americans in television outnumbered positive ones by two to one, and a disproportionate number of Italian characters were engaged in criminal activity. Difficulty in speaking English was also a commonly projected Italian-American trait. Comical or foolish characters accounted for a very large proportion of all Italian Americans portrayed on television. These stereotypes continue to be prevalent in Hollywood films and in other media depictions (Gambino, 1998). They are also frequently used in marketing products on television and even in packaging, despite the fact that they reflect little accuracy in portraying contemporary Italian-American life (Mitrano and Mitrano, 1996).

The most enduring negative image Italian Americans have had to deal with concerns the connection to organized crime. Because the Mafia is synonymous with organized crime and Italian Americans are identified with the Mafia, the logic, to many, becomes simple: "Italians = organized crime" (Iorizzo and Mondello, 1980). Although few deny the considerable involvement of some Italian Americans with organized crime, Italians have held no monopoly on such activities. As many have explained, crime has been a traditional avenue of upward mobility for members of most ethnic groups in American society (Albini, 1971; Bell, 1962; Ianni, 1974). Italian domination of organized crime did not emerge until the late 1920s, when Prohibition made trafficking in liquor profitable. Before that time, much of organized crime was controlled by the Irish. As illicit drugs replaced liquor as the major underworld-supplied item, Italian organized crime turned to more legitimate business areas, leaving the drug traffic to newer urban ethnic groups like blacks, Puerto Ricans, and Cubans (Cook, 1971; Ianni, 1974).

Even though Italians did not introduce organized crime to the United States and, in any case, the percentage of the Italian-American community engaged in such activities now or in the past has been small, the connection of Italian Americans generally with organized crime remains steadfast in the public imagery. Movies and television have been important purveyors of this stereotype (Carilli, 1998; Cortés, 1994). Dating from the 1930s, the association of Italians and organized crime remains a common Hollywood theme. Movies such as *The Godfather*, *Married to the Mob*, and *GoodFellas* and television programs such as "The Sopranos" are reflective of this genre. A survey in 2000 questioned teenagers of different ethnic backgrounds on their perception of movie and television stereotyping. When asked specifically to identify the role a character of Italian background would be most likely to play in a movie or on television, most (including Italian-Americans) responded "crime boss" (NIAF, 2001b). Recall the relationship of media depictions of ethnic groups with the creation and persistence of ethnic stereotypes, discussed in Chapter 3.

Prejudice and Discrimination Today

Surveys have regularly recorded the periodic changes in prejudice toward blacks, Jews, Latinos, and Asians. By contrast, there has been little study of shifting public attitudes toward Italian Americans. As we have seen, certain negative stereotypes remain well entrenched, but it is difficult to determine the nature of other aspects of prejudice as they affect this group. What little evidence exists, however, indicates a significant decline in prejudice against Italian Americans in recent decades.

Discriminatory actions are also difficult to measure. Although overt discrimination against Italian Americans has largely disappeared, subtle rejection of Italians in certain occupations and neighborhoods may become evident from time to time. It was revealed in 1960, for instance, that homeowners in the exclusive Detroit suburb of Grosse Pointe had maintained a rigid point system for screening potential home buyers, based on personal traits such as accent, name, swarthiness, and especially ethnicity (*Detroit News*, 1990; *Time*, 1960). Out of 100 points, Jews were required to score a minimum of 85, Italians 75, Greeks 65, and Poles 55. Blacks and Asians did not count at all! Antidiscriminatory legislation has made such cases today uncommon, but ethnic discrimination in subtle forms, even for white ethnic groups like Italian Americans, is not entirely something of the distant past (LaGumina, 1996).

Although objective measures of anti-Italian attitudes and actions are problematic, we might turn to the perspective of the group members themselves. How do Italian Americans perceive the levels of prejudice and discrimination against them? In his study of Italian Americans in Albany, New York, Alba found that very few had actually experienced discrimination within the five years preceding his survey, though a much larger percentage had heard about discrimination against Italian Americans and had encountered stereotypes about them (Alba, 1994). Crispino's study provides further illumination of the views of Italian Americans. He found that although they see declining discrimination, a substantial percentage feel there is still much prejudice against them. More than one-quarter of his third-generation sample reported such feelings. As to discrimination, 86 percent reported never having been refused housing, a job, or a promotion on the basis of ethnicity. The highest rates of discrimination were reported among those in higher-class positions, indicating that the occupational framework itself may affect the level of differential treatment. Crispino suggests that as long as Italian Americans remained within the working class, there was no need for discriminatory measures. Now that they have begun to enter higher occupational positions and to compete with the dominant group for power and privilege, discrimination "has become an important weapon in the fight to exclude them" (1980:118).

It is notable that in her study of ethnic identity among Euro-Americans, sociologist Mary Waters (1990) found that most of her interviewees, white ethnics of a variety of national origins, had themselves not experienced any type of discrimination or prejudice based on their ethnicity. There was some awareness, however, of past discrimination and prejudice against their particular groups.

Ironically, in recent decades Italian Americans themselves have sometimes been accused of racist attitudes and actions against African Americans. This animosity has been evident mainly in the urban environment, where the two groups have often found themselves competing for the same housing, jobs, and education. Working-class Italian neighborhoods have been more stable than those of most other ethnic groups, and fewer residents have fled to the suburbs with the advance of blacks into the core cities. Thus it is Italian Americans, particularly in cities of the Northeast, who have perhaps borne more of the effects of urban change than other ethnic groups. And it is African Americans who are seen as largely responsible for this social disruption (Rieder, 1985). Significant cultural differences between Italian Americans and African Americans have also contributed to intolerance. Gambino (1975) suggests that although racism is partially the basis of hostility between the two groups, Italian Americans are likely to respond vehemently to any group whose cultural patterns are interpreted as threatening to their family-oriented lifestyle. African Americans in central cities are seen in that light. Despite the apparent urban conflict between working-class Italian Americans and African Americans, however, studies have shown that when the racial attitudes of Italians are compared with those of other white ethnic groups, they are not different (Greeley, 1977; Taylor, Sheatsley, and Greeley, 1978).

ASSIMILATION AMONG ITALIAN AMERICANS

The applicability of either assimilationist or pluralist models of ethnic relations in the cases of African Americans and Hispanic Americans has often been debated. Whereas some see these groups as basically on a trajectory toward assimilation, others see them as excluded from the mainstream of American life; an inequalitarian pluralistic model (specifically, internal colonialism) is, the latter contend, more appropriate. For Italian Americans, however, as well as for other Euro-American groups, there can be little argument that assimilation has been the preeminent pattern of their adaptation to American society and their relations with the dominant group.

Although Italian Americans illustrate quite well the assimilation model of ethnic relations, their collective rate and extent of assimilation have differed in the cultural and structural dimensions. They have substantially adopted the dominant cultural ways but have not entered into equal relations with the dominant group to a corresponding degree at either the primary or secondary level. Nonetheless, movement in this direction is unequivocal.

Cultural Assimilation

Italian Americans have, with each succeeding generation, shed Italian cultural ways increasingly in favor of those of the core culture. In this they have exhibited a pattern common to most ethnic groups in American society, regardless of national origin or even physical distinctness. Quite simply, the social history of ethnic groups in the United States has been the adoption of major Anglo-American cultural patterns. As noted in Chapter 4, though some reciprocal changes are always evident in the interface of minority ethnic and dominant cultures, for the most part the exchange is one-sided.

Indications of cultural assimilation among Italian Americans are apparent in such areas as family patterns, language usage, and diet. The extent of these changes, however, varies among the generations. Whereas those of the third generation and beyond have substantially disposed of traditional ethnic cultural ways, members of the first and second generations have continued to adhere to them more closely. In his study of Italian Americans in Boston's West End, however, Gans (1982) found that even by the second generation most no longer remained attached to the major features of the southern Italian culture. This generation had retained only traditional cooking and foods as well as some use of the Italian language. At the conclusion of his study in 1962, Gans predicted that with each following generation the traditional Italian culture would be further diluted; subsequent research has confirmed his prediction. In a study of Italian Americans in Bridgeport, Connecticut, Crispino (1980) found that even in such areas as food and cooking, acculturation had taken place to a very great extent in later generations. Whereas 44 percent of first-generation respondents continued the tradition of making their own pasta (which, Crispino asserts, is "a more discriminating test of Italianness than cooking habits in general"), only 15 percent of the fourth generation did so.

Even intensely strong family ties and loyalties—the sine qua non of the Italian-American cultural system—are weakened by the third generation or have begun to be replaced by more common American patterns (Alba, 1985; Crispino, 1980; Gans, 1982). For example, whereas divorce is either taboo or limited in the first two generations, it becomes more accepted in the third. As Lopreato (1970) notes, those aspects of family behavior peculiar to Italian Americans are present to some degree, but they are overshadowed by those aspects they hold in common with other Americans in corresponding social classes.

Perhaps the most critical factor in the cultural assimilation of European ethnic groups in American society is the loss of the mother tongue. Because language is the means by which group ways are transmitted from one generation to the next, presumably as the language of ethnic origin is discarded and replaced by the host society's language, cultural assimilation will proceed apace. The relationship between loss of mother tongue and degree of cultural assimilation seems very strong among Italian Americans. In Crispino's study (1980), 72 percent of second-generation respondents (both parents born in Italy) spoke Italian, but only 14 percent of the third generation (both parents born in the United States) and 10 percent of the fourth generation (those most thoroughly assimilated culturally) did so. This pattern of language conversion is typical of all Euro-American groups, including those of southern and eastern European ancestry. While many older members of these groups still speak a European mother tongue in the home, among younger people, usually by the third generation, hardly any do (Alba, 1995).

Observers of the Italian-American ethnic group have uniformly noted the relationship between social class and cultural assimilation (Crispino, 1980; Gambino, 1975; Gans, 1982; Lopreato, 1970; Roche, 1982). As class rises, the level of cultural assimilation rises correspondingly. Today the Italian-American ethnic culture remains essentially a working-class phenomenon. As they continue to leave the working class, moving into middle-class occupations and neighborhoods, Italian Americans tend to become indistinguishable from other middle-class Americans. This is especially so among those of the third generation and beyond. These patterns are generally characteristic of all white ethnic groups in American society.

Structural Assimilation

At the secondary, or formal, level of structural assimilation, we have seen that Italian Americans have begun a large-scale movement into the mainstream of American society. Their occupational, income, and educational levels are at or above the average for the society as a whole. Most important, they have begun to enter positions of power within most institutional realms, though slowly and still in only small numbers at the top echelons. Clearly the process of structural assimilation at the secondary level is proceeding irrevocably.

Within the context of personal social settings (that is, at the primary group level), however, the pace and degree of structural assimilation for Italian

Americans are not so high. As with cultural assimilation, much of the difference among individuals in rate and extent of primary structural assimilation is related to social class and generation. For example, Gans (1982) found that second-generation West Enders in Boston maintained social relationships almost wholly within the ethnic community; for later generations, however, particularly those that had experienced upward mobility, close friendships were founded on common class interests and lifestyles, not ethnic or kinship ties (see also Alba, 1994; Crispino, 1980; Roche, 1982).

Residential Patterns A key indicator of primary structural assimilation is residence. The greater the residential integration of members of a minority ethnic group with those of the dominant group, the more completely they are assimilated structurally. As for Italian Americans in this regard, observers have often noted the long-lasting cohesiveness and stability of their neighborhoods. Indeed, some point to the neighborhood as one of the key elements, along with the family, in the chemistry of the Italian-American ethnic group (Glazer and Moynihan, 1970; LaRuffa, 1988).

But this evident stability of residential areas may be mostly a class-related phenomenon, rather than something uniquely Italian American. Working-class people are generally more stable residentially primarily because their opportunities for spatial mobility are more limited than those of middle-class people. Therefore, because most first- and second-generation Italian Americans were working class, they had been inclined to remain in their original central-city neighborhoods. When mobility into the middle class occurs, however, there appears to be as much propensity among Italian Americans as among other ethnic groups to leave the ethnic neighborhood for the more heterogeneous suburbs (Crispino, 1980; Gans, 1982). Alba (1995) found that in the greater New York metropolitan region, where they constitute one of the largest ethnic categories, Italian Americans in the 1980s were somewhat less likely than the average white to reside in suburbs rather than the central city, but the continued suburban movement among them was nonetheless evident. By 1990, 70 percent lived in suburbs.

Moreover, the residential solidarity of Italian Americans has often been exaggerated. Studies have shown that in Chicago and Philadelphia, for example, Italians from the outset moved frequently from their original areas of settlement, regardless of their social class. The "Little Italys" of various cities were in any case apparently never as homogeneous as has commonly been assumed (Hershberg et al., 1979; Lopreato, 1970; Nelli, 1970). At an early point of the new immigration, some of the European groups, including Italians, were highly segregated in northern cities, to be sure. But this pattern changed quickly and radically in the early years of the twentieth century (Lieberson, 1980). Today, even where Italian-American ethnic communities are evident, their homogeneity is evaporating. In the greater New York region, Alba found that "The Italians, some of whose ethnic communities are still conspicuous, reside mostly in non-ethnic areas, and their continuing suburbanization is eroding the most ethnic Italian neighborhoods" (1995:12).

Intermarriage Perhaps the most critical indicator of primary structural assimilation is the extent to which members of an ethnic group have intermarried with those of the dominant group or other out-groups. For Euro-American ethnic groups, there appears to be an increasing tendency toward marriage outside the group with each successive generation. Indeed, among whites, most marriages today involve some degree of ethnic intermixing (Alba, 1995). As Waters has noted, "For white ethnics, the longer a group has been in the United States and the greater the percentage of its members in later generations, the lower the in-marriage ratios" (1990:104).

Among Italian Americans specifically, Alba in 1981 found that whereas older, mostly first-generation individuals showed virtually no mixed ancestry, two-fifths of younger Italian Americans, mostly third- and fourth-generation, reported mixed ancestry. In this, Italian Americans were in line with other American Catholics of European descent. By the 1990s, an even stronger tendency toward interethnic marriage was apparent among Italian Americans as well as other southern and eastern European ethnic groups. Almost three-quarters of younger Italian Americans had spouses without Italian ancestry, and for Polish Americans the out-marriage figure was even higher (Alba, 1995).

The Future of Ethnicity Among Italian Americans and Other White Ethnics

It is clear that the strength of ethnicity for Italian Americans and other white ethnic groups like them diminishes in most ways as they move increasingly from the working class to the middle class. With upward mobility, social class rather than ethnicity becomes the most significant determinant of job, school, neighborhood, lifestyle, friendship circle, and even marital partner. For these groups, there is also strong evidence to support what has been called "straight-line theory," which posits that movement toward more complete assimilation proceeds irreversibly with each successive generation (Gans, 1979; Sandberg, 1974).[4]

Does this spell the eventual disappearance of the Italian-American ethnic community and others like it? This development may lie in the distant future, but it is unlikely in the lifetimes of contemporary white ethnics or their

---

[4]Gans, who first suggested the straight-line theory, has suggested that it be renamed "bumpy-line theory" to account for the unexpected movements that the assimilation process may take, sometimes even reverting to a more expressive ethnic identity. The direction of the movement—toward more thorough assimilation—does not change, however (Gans, 1992).

children. Whereas certain social forces are propelling assimilation relentlessly, others serve to retard the process.

In any case the form of ethnicity for current and future generations of Italian Americans and other white ethnic groups in the United States will not be the same as it was for their parents or grandparents. Although they will probably continue to assert an ethnic identity and retain a few major elements of the ethnic culture, their acknowledgment of ethnicity will be largely expressive rather than instrumental. Gans (1979) has suggested the idea of "symbolic ethnicity" to describe the continued and at times even accentuated expression of ethnicity among those of the third and fourth generations. Ethnicity, he explains, becomes a matter of voluntary personal identity, but it is largely devoid of vitality as a determinant of social behavior. It is, as Richard Lambert (1981) has put it, vestigial, something of the past that people can identify with but that has a very limited effect on their daily lives.[5]

What seemed to be a revitalization of ethnicity among whites during the late 1960s and early 1970s was interpreted by some as evidence of the enduring function of the ethnic group in providing a referent of personal identity and a sense of rootedness in an increasingly bureaucratized and depersonalized society (Greeley, 1971; Novak, 1972). Sporting buttons and bumper stickers proclaiming "I'm Polish and proud" or "Kiss me, I'm Irish," third- and fourth-generation whites seemed to develop a newfound ethnic consciousness. This "new ethnicity," as it was called, demonstrated to some social observers that continuing ethnicity was more than symbolic.

By the late 1970s, however, the resurgence of ethnic identity and the prominence of ethnic-related issues among whites had waned considerably. Given the movement toward assimilation of several generations, the ethnic revival was, as Steinberg (1989) contends, "doomed from the outset." Indeed, Alba suggests that ethnicity could be celebrated during the 1970s "precisely because assimilation had proceeded far enough that ethnicity no longer seemed so threatening and divisive" (1989:96).

Its transitory nature also seemed to indicate that the new ethnicity was more a political movement—specifically, a working-class political movement—than a revival of ethnic cultures and communities. Because social class has never served in the United States as an effective political mobilizing force, ethnicity has often functioned in this role. This is in large measure what occurred in the late 1960s and early 1970s in cities where white working-class people—mostly southern and eastern European ethnics—remained in significant numbers (Gans, 1974, 1979). As they perceived the economic and political concessions gained by African Americans and other minorities from the federal government and other

---

[5]For a rejoinder to Gans's thesis, particularly as it applies to Italian Americans, see Vecoli, 1996.

societal institutions, working-class white ethnics responded with a counter-movement, seeking to enhance and protect their interests—jobs, neighbor-hoods, schools, and so forth. Hence, what were in actuality class issues and conflicts were manifested and interpreted as ethnic ones. In a sense, ethnicity served as a substitute for class-based actions.

This kind of political ethnicity, in which ethnic groups serve as competing in-terest groups rather than as cultural islands, may be the chief form and function of ethnicity today for those few segments of white ethnic groups that still have substantial core-city communities (Bell, 1975; Glazer and Moynihan, 1970). For those members who with upward mobility continue their residential dispersion, however, ethnicity seems destined to function largely in a symbolic fashion.

Euro-Americans: A New Ethnic Group? As explained in Chapter 5, the sharpest lines of ethnic division in American society today are between Euro-Americans and racial-ethnic groups. As ethnicity for the former continues to de-cline as a consequential factor in various areas of social life, the boundaries of white ethnic groups become less and less distinct. In fact, Lieberson and Waters (1988) have found that a substantial and growing portion of the white popula-tion is apparently unaware of its ethnic origins, does not identify with any of the commonly understood ethnic groups, or simply refuses to identify its ethnic an-cestry. They refer to this grouping as "unhyphenated whites."

One of the forces contributing to this blurring of ethnic lines and the tendency for Euro-Americans to deemphasize ethnicity, at least in an instrumental fashion, is the enormous intermixing of ethnic groups that has been occurring continu-ously for the past 150 years. Today, it is increasingly difficult for people to declare their ethnic ancestry without recording two and often several additional cate-gories. For example, they may declare themselves "Irish," but "Italian" as well. And ethnic intermarriage continues unabated (Lieberson and Waters, 1988). The re-sult is that ethnic identity for Euro-Americans has become largely a matter of choice. And as ethnicity becomes more voluntary, resulting in fading ethnic boundaries, "in a sense," writes Richard Alba, "a new ethnic group is forming—one based on ancestry from anywhere on the European continent" (1990:3).

Waters (1990) has studied these fading ethnic divisions and finds that, in-creasingly, white ethnics, like Italian Americans, make conscious choices about their ethnic identity. Not only can they choose their ethnic affiliation, but they can choose whether to emphasize ethnicity in their lives at all. As she explains, "Ethnicity is increasingly a matter of personal preference" (1990:89). Given the strong rate of intermarriage among white ethnics, each generation's ethnic background becomes more complex and difficult to clearly define. Individuals—if they think of ethnicity at all—therefore begin to identify themselves in ethnic terms that reflect only a part of what may be their very elaborate ethnic heritage. People tend to stress only one part of their complex ethnic ancestry, for any number of reasons. They may know more about one parent's ancestry than another, or their surname or physical characteristics may conform to stereotypical images of the chosen group, or they may perceive certain of their ancestral lines as more socially desirable than others (Waters, 1990). In any case, for white ethnics, ethnicity is increasingly voluntary.

Yet, the fact remains that most people do choose and continue to affiliate with some ethnic group. Thus ethnic identity, which may be in large part uncertain or even fictive, continues to play a symbolic function for many people. However, it has little or no real bearing on their social behavior or economic standing. They may carefully pick and choose their ethnic affiliation, but, as Waters explains, "It is clear that for most of them ethnicity is not a very big part of their lives" (1990:89). Ethnicity becomes mostly something that provides a feeling of community as well as an identification that may set one off as unique and interesting. It has no impact on vital life decisions like where to live, whom to associate with, or even whom to marry.

The tendency for individuals to make choices about their ethnic identity demonstrates that ethnicity is fluid and can change in significance within one's lifetime or from one generation to the next. It is not fixed.

In considering these "ethnic options," as Waters refers to them, it is important to understand that the kind of freedom to select the content and strength of one's ethnic identity is not the same for racial-ethnic groups. Whereas for Euro-Americans the question, "What are you?" can be answered as the individual chooses, for members of racial-ethnic groups the answer has already been given by out-group members. African Americans, Latinos, Asians, and Native Americans do not, therefore, enjoy the same kind of options in the realm of ethnicity. "The reality," notes Waters, "is that white ethnics have a lot more choice and room for maneuver than they themselves think they do. The situation is very different for members of racial minorities, whose lives are strongly influenced by their race or national origin regardless of how much they may choose not to identify themselves in ethnic or racial terms" (1990:157). Failure on the part of Euro-Americans to acknowledge that difference often leads to their belief that ethnicity need play no greater role for African Americans or Hispanic Americans if they so choose. This creates misconceptions about the continuing strength of ethnicity and its impact on the lives of members of racial-ethnic groups.

SUMMARY

Italian Americans are representative generally of Euro-American ethnic groups that derive from the classic (or new) immigration period from the late nineteenth to the early twentieth centuries. Entering the society at the bottom of the class hierarchy, Italian Americans by the third generation had displayed significant upward mobility. Today they are at or above the national average in income, occupation, and education, and they have begun to enter positions of political and economic power, though slowly and still in small numbers.

Beginning with large-scale immigration in the late nineteenth century, Italians in the United States were the objects of prejudice and discrimination, sometimes of a particularly virulent nature. Perhaps the most persistent negative stereotype that Italian Americans have had to deal with involves their identification with organized crime.

Italian Americans today display a relatively high level of cultural assimilation, having substantially adopted the ways of the core culture. Although it is less thorough and rapid, structural assimilation is also occurring steadily. As this process continues, for Italian Americans as for other white ethnic groups, ethnicity is increasingly reduced to a symbolic function and becomes less a determinant of social behavior.

Suggested Readings

Alba, Richard D. 1990. *Ethnic Identity: The Transformation of White America*. New Haven, Conn.: Yale University Press. Examines the waning strength of ethnic identity among Euro-Americans, suggesting that assimilation has reduced ethnicity for these groups to a largely symbolic form.

Gambino, Richard. 1975. *Blood of My Blood: The Dilemma of the Italian-Americans*. Garden City, N.Y.: Anchor. Describes the culture and social institutions of Italian Americans.

Gans, Herbert J. 1982. *The Urban Villagers: Group and Class in the Life of Italian Americans*. Updated and expanded ed. New York: Free Press. A classic work that is required reading not only for its examination of the Italian-American experience but also for its presentation of issues regarding the retention of ethnic identity and the intersections of ethnicity and social class.

Lieberson, Stanley, and Mary C. Waters. 1988. *From Many Strands: Ethnic and Racial Groups in Contemporary America*. New York: Russell Sage Foundation. Analyzes the current meaning of ethnicity for Euro-American groups, like Italians, drawing on census data.

Lopata, Helena Znaniecki. 1976. *Polish Americans: Status Competition in an Ethnic Community*. Englewood Cliffs, N.J.: Prentice-Hall. A sociological analysis of another Euro-American ethnic group that, like Italian Americans, immigrated mostly in the late nineteenth and early twentieth centuries.

Mangione, Jerre, and Ben Morreale. 1992. *La Storia: Five Centuries of the Italian American Experience*. New York: HarperCollins. An account of the Italian-American experience from colonial times to the present.

Rieder, Jonathan. 1985. *Canarsie: The Jews and Italians of Brooklyn Against Liberalism*. Cambridge, Mass.: Harvard University Press. An ethnographic study of the response of a white ethnic community, largely Italian and Jewish, to the political and economic inroads of blacks.

Waters, Mary. 1993. *Ethnic Options: Choosing Identities in America*. Berkeley: University of California Press. Using census data as well as in-depth interviews, the author reveals how Euro-Americans choose their ethnic identities.

Yans-McLaughlin, Virginia. 1982. *Family and Community: Italian Immigrants in Buffalo, 1880–1930*. Urbana, Ill.: University of Illinois Press. An especially good analysis of the family as the key institution of Italian immigrants in their adaptation to American society.

8

JEWISH AMERICANS

IN MANY WAYS, as we will see, Jewish Americans are a unique case among American ethnic groups. As people who for centuries had suffered persecution and dislocation in almost all European societies, Jews found the United States a haven that provided for the first time a social atmosphere in which Jewish identity could be retained without fear of official repression and from which there was no need to contemplate flight. Jews in America also found their economic circumstances relatively unconstrained, in sharp contrast to the situation they had faced in Europe. So well nurtured were the opportunities offered them that Jews today represent the prototypical American ethnic success story. In only two generations, Jewish Americans collectively accomplished a truly phenomenal upward mobility that makes theirs a compulsory case for the analysis of American ethnic relations.

Unlike other American ethnic groups, Jews do not identify with a common homeland, having come from many nations. The broad geographical native region of the majority of Jewish Americans, however, is the same—eastern Europe—and a common language, Yiddish (a combination of mostly German and Hebrew elements), supersedes most national differences. Jews in the United States are best seen as an ethnic group because they clearly display the key features of such groups: a unique cultural heritage, a consciousness of community, and a group identity. Regardless of their degree of commitment to Jewish religious practices and beliefs (many are nonbelievers), and regardless of their national origins, Jews recognize themselves as a people with common roots.[1] They are seen in that fashion, as well, by the larger society.

That Jewish Americans lack a common homeland does not in itself constitute a significant difference from many white ethnic groups in the United States. In most cases, immigrants from southern and eastern Europe arrived with little national consciousness; regional or village boundaries generally defined their sense of peoplehood. Only in the United States, as a means of social adjustment and in response to out-group recognition, did they adopt a national identity.

IMMIGRATION AND SETTLEMENT

In some ways, Jewish patterns of immigration and settlement in the United States paralleled those of the Italians. Both ethnic groups were primarily part of the new immigration, and following large-scale entrance, both found themselves initially at the bottom of the class hierarchy. Like the Italians, Jews initially settled mostly in the large cities of the Northeast. Many of the occupations of the two groups during the major period of immigration also overlapped, as did their residential areas. As non-Protestants who had emigrated from relatively backward societies, both groups were made-to-order targets of the nativist and racist ideas of the turn of the century. Furthermore, because both groups remained concentrated primarily in eastern cities, they did not blend easily into the larger society. The prejudice and discrimination they faced, therefore, proved more pernicious and dogged than those aimed at other white ethnic groups. Despite the similarities in their mode of entrance and settlement patterns, however, important differences between these two groups were also apparent. These differences led, as we will see, to different rates of mobility in the second and third generations and to different positions in the class system.

Sephardic Jews

Jewish immigration to the Americas has a lengthy historical tradition. In fact, a few Jews supposedly accompanied Columbus on his first voyage. The earliest immigrants to the United States included Sephardic Jews, whose ancestors were Spaniards and Portuguese. When the Jews were expelled from Spain and Portugal in the late fifteenth century, many migrated to England, Holland,

---

[1] A national survey in 1990 found that most Jewish Americans define their Jewish identity in ethnic or cultural, not religious, terms. In fact, the survey found a low level of positive support for the religious group concept. Thus, most Jewish Americans consider themselves Jews primarily as members of an ethnic group (Kosmin et al., 1991).

Brazil, and the Caribbean islands. The first American Jews came from all these locations and even from the Iberian peninsula (Glazer, 1957). The Sephardim are to be distinguished from the Ashkenazic Jews of central and eastern Europe, who constituted almost all Jewish immigration into the United States after the colonial period.

German Jews

A second and more sizable Jewish immigration to the United States was made up of German Jews who came with the great wave of other German immigrants during the 1840s and 1850s. These Jews were socially and economically several notches below the earlier Sephardic Jews, who by that time were relatively prosperous and respected. Although most were merchants, their trade was at a level considerably less significant than that of their Sephardic predecessors. Many, in fact, were peddlers who moved westward with the country's expansion. Indeed, at the middle of the nineteenth century, most of America's 20,000 itinerant traders were German Jews (Lipset and Raab, 1995). German Jews dispersed widely to all parts of the country, in both large cities and small towns. Where they settled, they commonly established clothing and dry goods or general stores, the vestiges of which are seen today throughout the United States (Brockman, 2001; Pressley, 1999). A few developed into large national chains. Many familiar department store names, like Macy's, Bloomingdale's, Saks Fifth Avenue, Sears, and Neiman-Marcus, stem from German-Jewish founding families.

Collectively, the German Jews displayed rapid and substantial upward mobility; by the 1880s, most were middle-class entrepreneurs. Among the most successful of them developed a small, aristocratic upper class, which intermingled socially with the Gentile upper class and even intermarried among them (Baltzell, 1958, 1964). By the turn of the century, German Jews were so thoroughly assimilated that they had lost much sense of ethnic identity. Only their religious affiliation distinguished them from other Americans (Herberg, 1960).

East European Jews

By 1880, Jews were still only a tiny fraction of the American population, numbering no more than 250,000. In the next three-and-a-half decades, however, 2 million Jews would come to the United States. It is primarily from this wave of immigration that the contemporary Jewish-American population derives.

Jews coming to the United States after 1880 differed markedly from those who had settled earlier. To begin with, most were not from Germany but from eastern Europe, particularly Russia and Russian-occupied Poland. One of the key motivating factors behind the mass emigration of Jews from Russia was the severe oppression they were subjected to under tsarist policies. Following the assassination of Tsar Alexander II in 1881, bloody pogroms were sanctioned, and onerous restrictions were placed on Jews in all areas of social and economic life. America, therefore, was seen not only as a destination of economic opportunities but also as a political haven, a society that promised liberation and acceptance (Biale, 1998). Unlike the Italian emigration, which at first comprised mainly single men expecting to return, the Jewish emigration from Russia was a movement of family units that harbored few thoughts of returning to their homeland.

The East European Jews also differed from earlier immigrant Jews in their class origins: They were poorer and less educated. Jewish life in eastern Europe had been confined to all-Jewish villages and towns; as a result, these immigrants were extremely provincial in contrast to the relatively urbane and worldly German Jews. Striking differences between the two groups also lay in their attitudes toward religion. The East Europeans came with a version of Judaism far more traditional than the modernized and Americanized Reform Judaism of the Germans. To the German Jews, Judaism was simply a faith that was not to interfere with assimilation into the core American culture and society. To the East Europeans, coming from the shtetl, or Jewish village, where religion governed all aspects of life, Judaism was an entire social world. As Glazer and Moynihan explain, "In practice, tone, and theology, the Reform Judaism of the German Jews diverged from the Orthodoxy of the immigrants as much as the beliefs and practices of Southern Baptists differ from those of New England Unitarians" (1970:130).

In addition to differences in class and religious practice, East European Jews did not disperse geographically as had the earlier German Jewish immigrants. Instead, they remained in a few large cities, especially New York, where they were concentrated in lower-class ghettos. In this environment, the East European Jews, like the Italian immigrants of that time, experienced strikingly harsh conditions of work and residence. Many labored in garment factories, commonly called sweatshops, where they worked long hours for scant salaries. In 1911, one-third of Jewish heads of household earned less than $400 a year; and in 1914, the average hourly wage of male clothing workers was about 35 cents. Even by 1930, garment workers were still earning an average of only $24.51 a week—less than the pay of workers in stockyards and slaughterhouses (Howe, 1976). As a result, most families became work units in which women and children as well as men labored fifteen or eighteen hours a day, often taking piecework home with them.

Like the Italians who streamed into New York and other large eastern cities at the same time, the East European Jews accepted housing conditions that were frequently brutal and demeaning. They, too, crowded into rundown tenements, abandoned by previous ethnic groups. On the famous Lower East Side, 350,000 people per square mile jammed into squalid rooms and apartments in which sanitation facilities were meager at best. These degraded conditions of life only further confirmed the racist views of those who saw immigrant Jews of that time, along with other southern and eastern Europeans, as innately degenerate.

In sum, the urban environment into which immigrant Jews entered the United States in the late nineteenth and early twentieth centuries required struggle and stamina, which could be sustained only by the expectation that life conditions would improve—possibly for them but surely for their children. "The generation that entered the immigrant ghetto," writes Ben Halpern, "was confronted by one overwhelming task: to get out or enable the next generation to get out" (1958:36). This goal proved attainable for most. One important factor in accounting for the determination of immigrant Jews to succeed in American society was their understanding that there was no return to the homeland; for most Jews, the commitment to the United States was complete.

Given their striking differences in class, religion, and culture, it is little wonder that the German and East European Jews were initially antagonistic. The Germans viewed the new immigrants as illiterate, uncouth, and provincial greenhorns, who could only cause embarrassment to the American Jewish community and produce a backlash of anti-Semitism. By the 1890s, however, the divisions between the two groups were evaporating. Inspired partially by humanitarian motives and partially by the concern that they might be lumped with these poor and religiously traditional Jews in the eyes of their Christian neighbors, German Jews now accepted responsibility for uplifting and assisting their East European cohorts (Wirth, 1956). By the end of large-scale immigration in the early 1920s, the subdivision of the American Jewish community into German and East European groups was no longer apparent (Glazer and Moynihan, 1970; Yaffee, 1968).

Jewish Demography Today

Jewish Americans number about 6 million, little more than 2 percent of the total U.S. population (American Jewish Committee, 2000). This represents a declining Jewish population in both absolute and relative terms. Demographers have estimated that Jews in the United States will number less than 5 million by the middle of this century (Della Pergola et al., 2000). The decrease is due to an unusually low birthrate among Jews, as well as an increasing rate of intermarriage. Even though they are only a small part of the American population, and a declining one as well, Jews in the United States constitute over 40 percent of the world's Jewish population. Augmenting the Jewish-American population in the last three decades have been immigrant Jews from the former Soviet Union (most entering as political refugees) and from Israel. Together, however, these groups constitute only a small portion of Jews in the United States (Gold, 1999).

Jews, like Italians, are among the most urbanized of American ethnic groups. In fact, three-quarters of the entire Jewish population in the United States is found in just ten large urban areas (Table 8.1).[2] Geographically, the American Jewish population is relatively dispersed, though East and West Coast cities predominate. Similarly, at the local level, densely Jewish neighborhoods of the past are largely gone.

Table 8.1 Metropolitan Regions or Areas with Largest Jewish Populations

METROPOLITAN AREA	ESTIMATED JEWISH POPULATION
New York (metro region)	1,957,000
Los Angeles (metro region)	635,000
Miami–Ft. Lauderdale (Southeast FL)	514,000
Philadelphia–Wilmington (metro region)	276,000
Chicago	261,000
Boston (metro region)	227,000
San Francisco–Oakland–San Jose	210,000
Washington, D.C.–MD–VA	165,000
Detroit–Ann Arbor–Flint	101,000
Baltimore	95,000
Cleveland–Akron–Lorain	87,000
Atlanta	80,000
San Diego	70,000
Denver–Boulder	63,000
Phoenix	60,000
Las Vegas	56,000
St. Louis	54,000
Dallas–Fort Worth	50,000
Tampa–St. Petersburg–Clearwater	44,000
Houston	42,000
Minneapolis–St. Paul	41,000
Pittsburgh	40,000

SOURCE: American Jewish Committee, 2000.

---

[2]Studies of Jews in small towns and rural areas include Rogoff (2001) and Rose (1977).

That most Jews of the new immigration entered the society in relative poverty made their rapid upward mobility quite remarkable. As a group, Jews rose virtually from the bottom to the top of the stratification system in two generations. On all measures of social class—income, occupation, and education—Jews collectively rank higher than most other American ethnic groups.

The high socioeconomic position of Jewish Americans is well evident in family income. The median family income of Jewish Americans in the early 1970s was $12,630, more than $2,000 higher than the median for all U.S. families (Massarik and Chenkin, 1973; U.S. Census Bureau, 1973). By 1986, 77 percent of Jewish-American households earned more than $20,000, compared to 61 percent of American households in general (Cohen, 1987; U.S. Census Bureau, 1988). By 1989, the median Jewish-American household income was $39,000, about $10,000 higher than for the society as a whole (Kosmin et al., 1991; U.S. Census Bureau, 1995b). Table 8.2 shows the contrast between the median household income of Jewish baby boomers (those born between 1946 and 1964) and U.S. baby boomers as a whole. As can be seen, a far greater percentage of Jewish baby boomers fall into the higher income brackets.

Commensurate with their relatively high income, Jews display a disproportionate concentration in higher-status occupations; namely, the professions, business ownership, and managerial positions. Although the German Jews had, for the most part, established themselves as part of the middle class by the late nineteenth century, the rise of the East European Jews since the cessation of large-scale immigration in the 1920s has been even more striking (Chiswick, 1984). Only in New York City, containing the largest concentration of Jews in the United States, does there remain a substantial Jewish working class (Glazer and Moynihan, 1970; Goldstein, 1980; Sklare, 1971). Quite simply, as Calvin

Table 8.2 Household Income of U.S. and Jewish Baby Boomers, 1989

HOUSEHOLD INCOME	U.S. BABY BOOMERS	JEWISH BABY BOOMERS
Less than $50,000	72.8%	51.4%
$50,000–$149,999	25.8	44.3
$150,000 or greater	1.4	4.9

SOURCE: Waxman, 2000.

Goldscheider has stated, "poverty and unskilled occupations are largely uncharacteristic of contemporary American Jews" (1986:4). Table 8.3 vividly demonstrates the concentration of Jews in high-status occupations. Notice the extremely high percentage of Jewish professionals and the comparatively low percentage of workers.

Finally, the extremely high level of educational attainment among American Jews is well recognized, having been noted repeatedly in sociological studies (Goldscheider, 1986). Indeed, the high income and occupational levels of Jews can be accounted for in large measure by their similarly high educational level. As shown in Table 8.4, over 50 percent of Jews, compared to 20 percent of adult Americans in general, were college graduates in 1992. If education is the key to upward mobility in American society, Jews are the classic case of an ethnic group having converted educational opportunities into economic betterment in a relatively short time.

Ethnic and Class Factors in Upward Mobility

The saga of Jewish upward mobility is often proclaimed as the great success story of American ethnic history. Indeed, no other ethnic group has seemed to rise so far so quickly. As a collective rags-to-riches case, Jewish Americans are often held up as an example for other groups to emulate. What has provided for this extensive collective mobility in only two generations? Here we might compare the Jews with other white ethnic groups, such as the Italians, that entered the United States on a large scale at precisely the same time yet have not exhibited a similar degree of upward mobility. This comparison may give us strong clues to why different ethnic groups experience varying rates of success in the United States and in other multiethnic societies as well.

Table 8.3 Occupational Distribution of Jewish and U.S. Males

OCCUPATIONS	JEWISH MALES	U.S. MALES (WHITE)
Professionals	39.0%	15.8%
Managers	16.7	14.3
Clerical and sales	24.4	17.5
Workers	19.9	52.5
Total	100.0	100.1

SOURCE: Goldstein, 1992.

Table 8.4 Educational Attainment of Jewish and U.S. Adults

EDUCATIONAL LEVEL	JEWISH ADULTS	U.S. ADULTS
Less than college degree	28%	62%
Some college	19	17
College degree	53	21

SOURCE: Goldstein, 1992.

Herbert Gans has explained that in looking at the mobility patterns of ethnic groups, it is essential to try to understand whether what happens to them is "ultimately more or less a function of their characteristics and culture than of the economic and political opportunities which are open to them when they arrived and subsequently" (1967b:8). That is, patterns of mobility among Jews, Italians, or any ethnic group can be seen as the result of either the group's cultural traits or the opportunities to which the group is exposed, or perhaps a combination of those factors.

Group Culture Some contend that differences in ethnic achievement are attributable mainly to internal characteristics of the groups themselves (Fauman, 1958; Hurvitz, 1958; Lipset and Raab, 1995; Rosen, 1959; Schooler, 1976; Sowell, 1981; Strodtbeck, 1958). The crux of this argument is that some groups are better enabled to succeed occupationally by virtue of certain cultural values and behaviors. Presumably, those groups that have come to the United States with values similar to or compatible with those of the core group—individual achievement, competition, pragmatism, thrift—are likely to rise more rapidly.

The substance of this view is contained in what Rosen (1959) has termed the "achievement syndrome." Groups differ, he explains, in their orientation toward achievement. Some, like Jews, encouraged independence and stressed achievement values and high educational aspirations in children, whereas others, like Italians, did not. Italian Americans traditionally stressed accomplishment only as it benefited the family in some concrete fashion. Individual achievement, however, was deemphasized because it destabilized the structure of the family unit and threatened its authority. Glazer and Moynihan write that "that form of individuality and ambition which is identified with Protestant and Anglo-Saxon culture, and for which the criteria of success are abstract and impersonal, is rare among American Italians" (1970:194–95). Among Jews, however, it was preeminent. Thus, instead of plying their father's trade, as did most second-generation Italians, Jews aspired to higher occupational positions. As Philip Gleason explains, "Unlike Catholic peasant immigrants, the Jewish immigrants did not regard intensive education and professional careers as something beyond their experience or capacity; they saw them rather as their natural lines of aspiration" (1964:168).

Perhaps the most frequently cited cultural value in explaining Jewish mobility, particularly as it contrasts with the more limited mobility of other ethnic groups of the new immigration, is the group's attitude toward education. The value of learning has a rich tradition in Jewish culture extending as far back as biblical times, and it permeates all aspects of Jewish life (Fauman, 1958; Gleason, 1964; Strodtbeck, 1958). Because of this key cultural trait, it is held, Jews have quite naturally achieved proportionately greater success in business, academia, science, and the arts than have other ethnic groups.

The Jewish penchant for education is contrasted sharply with the traditional southern Italian apprehension about this institution. Rather than a means to social and economic advancement, education was viewed by immigrant Italians as having little practical value and in some ways being antithetical to family and group interests (Covello, 1967). Hence, they did not avail themselves of the educational opportunities in the United States as quickly and exhaustively as did the Jews. Some have observed that even later generations of Italian-American parents, including those of the middle class, have not overly pressured their children for achievement. "By their norms," concludes one study, "motivation as well as ability should come from the child, not the parent" (Johnson, 1985:189).

Intense family loyalty as part of the southern Italian culture was especially consequential for immigrants and their children. With an allegiance to family above all else, individual Italian Americans found it difficult to pursue occupations that did not concretely improve the family's welfare. The most important work objective was to find a steady, secure job that would contribute to the family's immediate well-being and that would not take the individual outside the realm of the family's influence. Such values restrained upward job mobility (Gambino, 1975; Gans, 1982).

Obviously the choice of fulfilling short-range work goals instead of seeking the long-range advantages of continuing education was not unique to Italian immigrants and their children. But the norm was perhaps more firmly rooted among them. There is little doubt, then, that their skeptical view of formal education contributed heavily to their slower and more limited upward mobility compared to Jewish Americans.

Another culturally based argument holds that, in comparison with other ethnic groups, Jews invest more heavily in their children. Because Jewish families are, on average, smaller than those of most other ethnic groups, Jewish parents, it is argued, are able to devote more time and other resources to child rearing. In turn, Jews are better able to translate their education into prestigious occupations and high labor market earnings. Moreover, more stable family living arrangements (lower rates of divorce, separation, out-of-wedlock births), in this view, additionally improve child quality (Chiswick, 1984).

Group Opportunities Disputing the contention that ethnic groups differ in achievement primarily because of internal group values, some maintain that differences depend mostly on the opportunity structure that groups encounter on entering the society and the skills they bring to that structure (Gans, 1982; Steinberg, 1989). The essence of this view is that class factors in general explain more about mobility than do ethnic cultures.

Although not denying differences in group culture, sociologist Stephen Steinberg (1989) argues that Jewish immigrants came to the United States with an advantageous occupational background that enabled them to cultivate the American economic system better than other ethnic groups could. Russian Jews entering the United States were not peasants but were urban people who had been engaged in manufacturing and commerce. Consequently, although they were certainly poor, they were equipped with skills that would serve them well in the expanding industrial economy. Italians and most other groups of the new immigration were, by comparison, mainly agricultural peasants. Moreover, Jewish occupational skills complemented almost perfectly the needs of the rapidly growing garment industry, where so many Jewish immigrants found employment. "Jews," notes economist Eli Ginsberg, "had the good fortune to be in the right place at the right time" (1978:115). Moreover, as the economy later shifted from agriculture and manufacturing to services, Jews benefited from their background in trade and commerce.

Glazer (1958) also points out that East European Jews who came to the United States had a long heritage of middle-class occupations, including the professions and buying and selling, by contrast with other immigrant groups. Although tsarist oppression had driven Jews from their traditional trades, their link to them was not broken. They came to the United States with little capital, then, but not with the limited horizons of the working class. In effect, the East European Jews were part of the working class for only one generation. They were neither the sons nor the fathers of workers (Herberg, 1960).

Steinberg also questions the attribution of Jewish occupational success to the high value attached to learning. Although he does not deny that the higher rate of literacy among immigrant Jews compared, for example, to that of southern Italians was instrumental in their more rapid mobility, he maintains that substantial educational advancement followed economic mobility rather than the converse. After looking at historical data, Steinberg concludes that for Jews, "economic success was a precondition, rather than a consequence, of extensive schooling" (1989:136). Sowell (1981), too, denies that education was the cause of mobility for Jews and, like Steinberg, explains it as the result of economic success. By the time Jews became a substantial part of New York's college population, he notes, mobility was already well under way.

In a similar fashion, the presumably lower educational and occupational aspirations of immigrant Italians can be accounted for by class as much as by ethnic factors. That is, their attitudes toward education and work, particularly among the first and second generations, may have reflected their primarily working-class status, not simply their southern Italian heritage. Gans (1982) explains that many of the behaviors and values he observed among second-generation Italian Americans in Boston's West End were not so much ethnic traits as working-class traits. Thus, the limited mobility aspirations among working-class Italians might be expected as well among, for example, working-class Poles, working-class blacks, or even working-class Jews. As Gans has observed, "I continue to be impressed by how similarly members of different ethnic groups think and act when they are of the same socioeconomic level, are alike in age and other characteristics, and must deal with the same conditions" (1982:277–78).

Looking at Italian-American achievement and attitudes toward education as the products primarily of a unique ethnic trait also fails to take into account the opportunity structure in the old country, which dictated limited aspirations. "To whatever extent Southern Italians exhibited negative attitudes toward education," writes Steinberg, "in the final analysis these attitudes only reflected economic and social realities, including a dearth of educational opportunities" (1989:142).

Historian John Briggs (1978) also discounts the idea of a predisposed negative view of education among southern Italians. He attributes their rejection of schooling in the United States, instead, to their expectation of an imminent return to Italy. It was not a lack of interest in education but the lack of relevance of American schooling that led to its widespread repudiation, he contends. Schooling in the new country could be of little value if one's stay was only temporary.

The Italian pattern also seems to verify the thesis that generally only after economic gains have been secured is education ordinarily extended. As we have seen, Italian Americans are only now beginning to attain educational levels in line with those of other ethnic groups, even though they earlier attained an above-average level of income. As educational levels rise, of course, the effect is circular, and Italian Americans will no doubt display subsequent rises in occupation and income. Data indicate that although Jews have progressed further and faster, Italians and other Catholic ethnic groups are now catching up (Alba, 1981; Greeley, 1977; U.S. Census Bureau, 1993c).

To put the matter simply, those who interpret different patterns of mobility among Jews and Italians in class rather than ethnic terms stress that, whatever the extent of a group's occupational and educational background or tradition, only when it is combined with a favorable opportunity structure can it enhance mobility. For Jews, there was a fortunate compatibility between the skills and experience they brought with them and the needs and opportunities of the American economic structure. And as Steinberg suggests, "It is this remarkable convergence of factors that resulted in an unusual record of success" (1989:103). Valid comparisons among ethnic groups can be made, then, only when consideration is taken of the society's constantly changing opportunity structure.

Evidence can be marshaled to support both the group culture and opportunity structure arguments regarding ethnic mobility; but as with most issues in sociology, the truth more likely lies somewhere between the two. On the one hand, had Jews not possessed at least some propitious cultural traits, including among them a reverence for learning, it is not likely that such rapid mobility would have been possible even within the context of a favorable opportunity structure. Educational opportunities in the United States existed for other European ethnic groups, but it seems evident that immigrant Jews availed themselves of these opportunities more thoroughly and avidly than did others. Moreover, Jews in other immigrant societies such as Canada, Argentina, and South Africa have displayed, as a group, economic and educational success comparable to that attained in the United States and have pursued similar occupations. On the other hand, there is little question that immigrant Jews brought with them a wider range of occupational skills that automatically placed them a notch ahead of other ethnic groups in the opportunity structure. As we will see in Chapter 10, the rapid economic success of Cuban immigrants in the United States in the past two decades, compared to that of other contemporary Hispanic groups, poses a quite similar case, as does the upward mobility of many Asian groups.

Political and Economic Power

Jewish participation at the higher levels of government has been relatively greater than that of Italians and most other southern and eastern European groups (Alba and Moore, 1982). Jewish entrance into the realm of American political power is a relatively recent phenomenon, however. Although Jews have historically been in advance of Catholics in their appointment to positions in the federal cabinet and judiciary, this is due to the prominence of a few earlier German Jews on the American political scene. Indeed, Jews did not serve in the federal establishment in a really significant way until Franklin D. Roosevelt's administration (Isaacs, 1974).

A few Jews were elected to Congress as early as the nineteenth century, but only in recent years have they been elected in meaningful numbers. Before 1945, only fifty-nine Jews had ever served in either house. The number of Jews in the U.S. Congress increased significantly after 1974 (Goldberg, 1996; Zweigenhaft and Domhoff, 1998). Today, in light of their minor proportion of the general populace, Jews are overrepresented in the federal legislature (see Table 7.4). Jewish Americans make up 10 percent of the Senate and over 6 percent of the House (Hirschfeld, 2001). They are also strongly represented as federal political appointees and civil servants (Alba and Moore, 1982) and have occupied a number of cabinet positions, particularly in Democratic administrations (Zweigenhaft and Domhoff, 1998). Moreover, Jews play an important role in political campaign financing at the national level, particularly for the Democratic Party (Goldberg, 1996).

In a few cases, Jews served as state governors in the late nineteenth century, but these were unusual occurrences (Lieberson, 1980). In recent years, Jews have been elected governor only in a few eastern states where their electoral power is of some consequence. Similarly, at the local level, Jews, like Italians, tend to be prominent mainly in those areas where they represent a substantial constituency. But this, too, is a relatively recent development. Consider that a Jew was not elected mayor of New York, the city with the largest Jewish-American population, until 1974.

It is important to consider that Jews today experience few impediments to elective office at any level of government. In fact, a 1987 Gallup Poll revealed that 89 percent of Americans would vote for a well-qualified Jewish presidential candidate; only 46 percent felt that way in 1937 (Gallup, 1982, 1988; see also Quinley and Glock, 1979). The acceptance of Jews in higher politics is demonstrated as well by the fact that Jewish candidates are today commonly elected to offices by overwhelmingly non-Jewish constituencies (Wyman, 2000). Both current U.S. senators from Wisconsin, for example, are Jewish, despite the fact that the state's Jewish population is one-half of 1 percent. The selection of Joseph Lieberman, an Orthodox Jew, as the vice-presidential candidate on the Democratic ticket in 2000 symbolically shattered any remaining barriers to Jewish-American penetration of the very highest levels of government. Jewish Americans have consistently shown a greater propensity than other ethnic groups for political participation, and this may account in some measure for their increasing prominence in elective offices.

With regard to economic power, despite their collectively high occupational status as well as their significant numbers among the society's affluent, Jews have historically been underrepresented in the American corporate elite (Alba and Moore, 1982; Korman, 1988; Zweigenhaft and Domhoff, 1982). The strong Jewish participation in the business world traditionally was in small or medium-sized firms (most of them self-owned), not the larger, more powerful corporations that dominate the American economy (Institute for Social Research, 1964; Kiester, 1968; Marger, 1974; Ward, 1965). Furthermore, where they had been successful in the corporate world, Jews more commonly entered the elite through the growth of their own companies rather than by climbing the organizational ladder of well-established concerns.

The historical underrepresentation of Jewish Americans in the top echelons of the corporate world can be explained partly as discrimination and partly as a propensity for Jews to prefer independence in business and the professions. The two are related. Faced with discriminatory hiring practices, which did not decline significantly until after World War II, Jews naturally turned to self-owned businesses or the independent professions, like law and medicine.

The traditional underrepresentation of Jews in the corporate elite is ironic in light of the commonly held assumption among most Americans that Jews wield inordinate economic power. Studies beginning in the 1930s consistently showed this to be a misconception (Burck, 1976; Fortune, 1936; Keller, 1953; Newcomer, 1955; Warner and Abegglen, 1963). To the extent that the public illusion of excessive Jewish economic power persisted, it may be explained to some extent by the conspicuous nature of certain industries in which Jews did exert a dominant influence, most notably movies and television, retail sales, and real estate (Glazer and Moynihan, 1970; McWilliams, 1948; Selznick and Steinberg, 1969).

Some hold that traditional patterns of discrimination in the corporate world have not changed measurably. Investigating evidence of corporate recruiting, Korman (1988) found that even as late as the 1980s, firms in many of the most important segments of the corporate economy continued to recruit more commonly at schools with few Jewish undergraduates. He concluded that at the managerial/executive level, Jews remain "outsiders" in the corporate world.

There is stronger evidence, however, that the traditional pattern of Jewish absence from the corporate elite has changed markedly, with Jews increasingly entering the top executive positions of major corporations (Klausner, 1988; Silberman, 1985; Zweigenhaft, 1984, 1987). In their study of the American power elite, Zweigenhaft and Domhoff (1998) conclude that, despite underrepresentation in some business sectors, Jews are actually overrepresented in the corporate elite. Whatever discrimination in large corporations and law firms continues to exist may be simply the lingering effect of past occupational barriers against Jews that, for the most part, have been formally taken down. Indeed, a study of the late 1980s concluded plainly that "the disadvantage of Jews in the executive suite has all but disappeared" (Klausner, 1988:33).

PREJUDICE AND DISCRIMINATION

For centuries, Jews have been a favored target of ethnic bigotry in Western societies. As non-Christians they were particularly vulnerable to scapegoating during times of social adversity and were generally relegated to the role of economic and political outcast. Jews were made especially ready targets for oppression after the thirteenth century, when they were forced into separate communities in almost every European society in which they resided. The term *ghetto* was first used to describe the area of Venice in which Jews were compelled by custom and law to live. It was subsequently applied to the Jewish quarter of all European cities.[3] Jews were forbidden to own land or join craft guilds and maintained an ever-tenuous civil status. The capricious nature of persecution and the lack of political security forced them to remain constantly vigilant and prepared to flee. Because of the need to remain mobile, they turned to occupations such as tradesman, moneylender, or professional, which provided for easy movement of capital and skills (Fauman, 1958; Glazer, 1958). These occupational roles often carried social stigmas, which further provoked resentment and hostility from those who were required to deal with them.

The Elements of Anti-Semitism

A set of distinct and consistent negative stereotypes, some of which can be traced as far back as the Middle Ages in Europe, has been applied to Jews. Among the most common of these are that Jews are monied, dishonest, and unethical; clannish, prideful, and conceited; and power hungry, pushy, and intrusive (Quinley and Glock, 1979). It is the connection of Jews with money, however, that appears to be the sine qua non of anti-Semitism.[4] As Glock and Stark put it, "Perhaps the most constant theme in anti-Semitism from medieval times down to the present is of the Jew as a cheap, miserly manipulator of money, forever preoccupied with materialism, and consequently possessing virtually unlimited economic power" (1966:109). As late as 1995, over a quarter of non-Jewish whites supported the assertion that "When it comes to choosing between people and money, Jews will choose money"; even larger percentages of African Americans, Latinos, and Asian Americans agreed (National Conference, 1995).

That Jews have more money than most people is, as Selznick and Steinberg (1969) explain, accepted as fact among a majority of non-Jews, though this in itself does not necessarily imply anti-Semitism. Nonetheless, even unprejudiced non-Jews exaggerate the extent of Jewish wealth, and it is, of course, only a short step to interpreting this belief in a negative manner. Although Jews in American society are, as we have seen, collectively higher in socioeconomic status than most other ethnic groups, this does not mean that all Jews are therefore wealthy. Like members of all ethnic groups, Jews will be found at every class level, including the very bottom.[5]

Closely akin to the beliefs regarding the connection of Jews with money are those concerning Jews and business. "Of all the crimes and misdemeanors charged to the Jews," notes Stephen Whitfield, "the worst has been an uncanny or even supernatural capacity to make money, perverted into avarice"

---

[3]On the development of European ghettos, see Wirth (1956).
[4]*Anti-Semitism* is the term most commonly used to describe prejudice and hostility toward Jews.
[5]For a description of the Jewish poor, most of whom are elderly, see Levine and Hochbaum (1974).

(1995:90). The shrewd and unscrupulous businessperson has been a tenacious negative stereotype of Jews for ages. Derivative of this image is the pejorative American phrase, to "Jew down," which means, roughly, to bargain craftily or to buy something more cheaply than the seller initially offers it. The beliefs regarding Jews and business deal not only with ethics, however, but also with power and manipulation. Jews have been commonly seen as maintaining excessive influence in the business world and as controlling large segments of the economy of the United States and of other societies. The fallaciousness of this image has been confirmed repeatedly; and as noted earlier, in the past Jews have been underrepresented in the higher executive positions of the most powerful American and multinational corporations.

The attributions of Jewish power are often blatantly contradictory. One of the staples of crude anti-Semitism concerns the notion of world conspiracy. Curiously, Jews have been portrayed on the one hand as leaders of a world capitalist conspiracy, and on the other as leaders of a world communist conspiracy. This inconsistency has never deterred anti-Semites from expressing both beliefs when the situation calls for it. Indeed, as McWilliams so aptly suggested, there is a tone of "demagogic genius" in this contradictory charge, "for it permits an appeal to the dispossessed and a threat to the rich to be voiced in a single sentence" (1948:92).

Anti-Semitism in the United States

Historically, three general periods of anti-Semitism in the United States can be defined, each characterized by a different level of intensity and content. During the early decades of Jewish settlement, before 1880, anti-Jewish prejudice and discrimination were minimal. With the influx of the East European Jews in the late nineteenth and early twentieth centuries, harsh and overt antagonism developed in response to the much larger number and very different character of these immigrants. The anti-Semitism that arose during this period peaked in the late 1920s and 1930s and did not abate until after World War II. Since the 1950s, prejudice and discrimination toward Jews have declined considerably, though they remain evident in subtle and subdued forms.

The Early Period Although there were anti-Jewish groups in colonial America, they were not excessively vocal or influential (Blau and Baron, 1963). During the era of the old immigration, between 1820 and 1880, German Jews were not essentially distinguished from their Christian counterparts except in religion; and as noted earlier, they mostly assimilated into the mainstream society. Like the early northern Italians, German Jews (and the earlier Sephardic Jews) were too insignificant numerically to evoke intense hostility. Furthermore, they were geographically dispersed and established themselves in economically se-cure positions, thereby avoiding the attention of nativist groups, which targeted the much larger, city-concentrated, and destitute Irish Catholic population.

This relatively concordant relationship with the Anglo-Protestant majority began to change, however, with the entrance of the East European Jews in the 1880s. A harbinger of the changed attitude toward all Jews occurred in 1877 when Joseph Seligman, a prominent financier and political adviser, was denied entrance to the Grand Union Hotel in Saratoga, New York, a noted upper-class resort (Baltzell, 1964; McWilliams, 1948). From that point on, Jews were com-monly barred from certain resorts, clubs, and college fraternities.

The Development of Popular Anti-Semitism Like the southern Italians, im-migrant East European Jews became prime subjects of the racist ideologies of the late nineteenth and early twentieth centuries. Just as the criminal tendency attributed to Italians was simply part of "their nature," the association of Jews with money was seen as deeply rooted in this group's heritage and perhaps even as genetic. Consider this brief passage from Jacob Riis's influential essay of 1890, *How the Other Half Lives*:

> As scholars, the children of the most ignorant Polish Jew keep fairly abreast of their more favored playmates, until it comes to mental arith-metic, when they leave them behind with a bound. It is surprising to see how strong the instinct of dollars and cents is in them. They can count, and correctly, almost before they can talk ([1890] 1957:84).

This description, it should be noted, was not necessarily written in a deroga-tory vein. Indeed, the essay was essentially sympathetic to the plight of both im-migrant Italians and Jews. Yet its racist assumptions are obvious. The racist ele-ment of anti-Semitism was, of course, carried to its ultimate by Hitler and the Nazis during the 1930s and 1940s, but its genesis can be traced to much earlier ideas and their exponents in the United States as well as Europe.

During the 1920s, American anti-Semitism was strongly abetted by the ac-tivities of the Ku Klux Klan and of Henry Ford, the noted industrialist. The Klan's anti-Semitic and anti-Catholic propaganda and actions during this time were as widespread and severe as its antiblack campaigns. As for Ford, his anti-Semitic program lasted from 1920 to 1927, during which time his newspaper, the *Dearborn Independent*, issued virulent attacks on Jews.[6] The "world Jewish conspiracy" idea was a regular component of the *Independent*, which used the no-

---

[6]McWilliams (1948) writes that the rise of the Ku Klux Klan and Ford's anti-Semitic campaign were not unrelated. The Klan began to attract a mass following in 1920 when Ford launched his attack.

torious *Protocols of the Elders of Zion* as its basis. This was a well-known anti-Semitic tract circulated originally in the early 1900s by the Russian secret police. Its thesis was a Jewish conspiracy to achieve world domination through control of finance and banking, and it was used by the Russian tsars as a rationale for their oppressive policies and actions against the Jews. Though its contentions were repeatedly disproved, this manifesto was subsequently circulated widely, and its theme became a standard item of the litany of anti-Semitism. Ford later apologized to the American Jewish community for his attacks, saying that he had been misled, but many of the ideas espoused by the Independent remained in wide circulation during the 1930s.

Although the anti-Semitism of this era was extremely vituperative, it rarely generated violence. Most of the discrimination against Jews involved their expulsion or exclusion from various areas of social life, including clubs and resorts, neighborhoods, schools and colleges, and certain occupations (McWilliams, 1948). As Jews began to occupy a disproportionate number of seats in elite colleges and professional schools in the early 1920s, restrictive policies were enacted, limiting their entrance. Strict Jewish quotas were established at schools such as Harvard and other Ivy League institutions as well as medical and law schools throughout the country (Baltzell, 1964; Bloomgarden, 1957; Korman, 1988). Jews were also severely discriminated against in the professions, especially medicine, law, and academia. Jewish doctors, for example, were often refused appointments to hospital staffs and were denied clinical privileges. Law firms and university faculties imposed similar restrictions.

The rising level of anti-Jewish prejudice and discrimination in the 1920s and 1930s can be seen as a part of the popular anti-immigrant and isolationist sentiments that so characterized that period. But it can also be seen as a response to Jewish upward mobility. Before the early years of the twentieth century, Jews were fairly well removed from the general occupational structure, concentrating mainly in a few industries, like the garment trade, or operating their own small businesses. By the second decade of the century, however, second-generation Jews were beginning to break out of the insular Jewish community and challenge—most often, successfully—members of the dominant group, particularly in the more prestigious occupations and educational settings. Artificial barriers, in the form of quotas and other discriminatory policies, were therefore thrown up. The anti-Jewish attitudes that emerged in the 1870s among the well-to-do have been interpreted in much the same manner, as social barriers that arose in response to competition for privilege (Higham, 1963; McWilliams, 1948).

Declining Anti-Semitism In the 1950s and, particularly, the 1960s, anti-Semitism in the United States declined precipitously. The anti-Jewish prejudice and discrimination that persisted took on more camouflaged and restrained forms. This downtrend has been unremitting.

The decline of negative Jewish stereotypes among the general American public has been demonstrated by several comprehensive studies and opinion polls (Anti-Defamation League, 1998; Chanes, 1994; Gallup, 1982; Glock and Stark, 1966; Quinley and Glock, 1979; Selznick and Steinberg, 1969; Stember, 1966; Williams, 1977). The unscrupulous businessperson, for example, is no longer a foremost Jewish image among most non-Jews. Nearly half the respondents in polls taken in 1938 and 1939 believed that Jewish businesspeople were less honest than others. A 1952 poll reported a similar percentage. By 1962, however, that stereotype had diminished considerably; 70 percent felt that Jewish businesspeople were no different from others, although the notion that they were shrewd in business remained strong (Stember, 1966). By the early 1980s, the percentage of those adhering to the stereotype had further declined (Raab, 1989).

A significant change was also registered in the perception of Jewish economic and political power. Public belief in the extensiveness of Jewish influence seemed to peak in the 1930s and 1940s. Forty-one percent of Americans in 1938 thought that Jews had too much power, and 58 percent thought so in 1945 (Quinley and Glock, 1979). In 1952, 35 percent of Protestants and 33 percent of Catholics still answered yes to the question, "Do you think the Jews are trying to get too much power in the United States or not?" But in 1979, the percentages of Protestants and Catholics answering yes were only 12 and 13, respectively (Gallup, 1982). Opinion polls throughout the 1980s indicated similarly that only a small percentage viewed Jews as having inordinate power and influence (Smith, 1991).[7] A 1998 national poll found that only 11 percent of Americans still embraced the stereotype of excessive Jewish power (Anti-Defamation League, 1998).

In the realm of personal relations, negative attitudes toward Jews have also declined markedly. In a 1948 survey, 22 percent expressed the view that they would prefer not to have Jewish neighbors. By 1964, only 8 percent answered similarly, a figure that did not essentially change by 1989. Even intermarriage with Jews, perhaps the most sensitive area of interethnic relations, was viewed negatively by only 34 percent of non-Jews in 1981 (*Public Opinion*, 1987; Smith, 1991).

What can account for this noticeable decline in anti-Semitic thought and action during the past few decades? Though such social trends are never attribut-

---

[7]There are noticeable differences, however, among respondents of different ethnic groups. The National Conference reported that 22 percent of non-Jewish whites, 37 percent of Latinos, 35 percent of Asian Americans, and 45 percent of African Americans held the belief that Jews "have too much control over business and the media." Similar percentages of each group felt that Jews "are more loyal to Israel than to America" (National Conference, 1995).

able to a single factor, the general rise in educational level among the American populace does stand out as an extremely important agent. Research and public opinion polls indicate consistently that anti-Semitism is related to educational level. To put the matter simply, the lower the educational attainment, the greater the degree of anti-Semitism. For example, surveys of attitudes toward Jews have repeatedly found college-educated people more tolerant than those with less formal education (Anti-Defamation League, 1998; Gallup, 1982; Gilbert, 1988). Thus, as the level of education has increased for Americans generally, anti-Jewish attitudes and actions have waned correspondingly (Glock and Stark, 1966; Lipset, 1987; Quinley and Glock, 1979; Selznick and Steinberg, 1969).

Too much significance should not be given to educational and class levels in explaining the causes of anti-Semitism, however. More correctly, the forms of anti-Semitism are different, depending on education and class. Some researchers have shown that those who are economically most secure and most educated are not necessarily more supportive of minority rights than other social classes (Hamilton, 1972; Huber and Form, 1973). Prejudiced members of higher-class groups may be more sophisticated in publicly proclaiming tolerant attitudes while continuing to deny Jews and other minorities full participation in various institutions.

Moreover, despite the clear decline of prejudice and discrimination, Jews continue to be denied entrance into certain fraternal groups, recreational clubs, and similar social organizations. In 1991, for example, the exclusion of Jews by the Kansas City Country Club prompted Tom Watson, the professional golfer, to resign from the club in protest. Such restrictive policies constitute an aspect of anti-Semitism that has been relatively untouched by both antidiscriminatory laws and the general tendency toward more ethnically open social organizations (Braverman and Kaplan, 1967; Selznick and Steinberg, 1969; Slawson and Bloomgarden, 1965). A study of Jewish life in Los Angeles (Sandberg, 1986) showed that Jews themselves perceive more anti-Jewish discrimination in private clubs than in any other area of social life.

Jews and Blacks In recent years, anti-Semitism has been more noticeable among blacks in American society than among other population elements (Anti-Defamation League, 1998; Golub, 1990). This is paradoxical in that Jews have traditionally maintained a more sympathetic attitude than other segments of the white population toward black interests and have been consistently supportive of black efforts to secure civil rights (Smith, 1990b). The apparent rise in black anti-Semitism is accounted for by several factors. First, the emergence of a more militant ideology within the black movement in the 1960s led to a wide-scale purge of whites from black civil rights organizations, within which many Jews had been strongly active. This created some distance between the two groups. Also, militant black rhetoric in the 1960s encouraged blacks to begin to assert control over their neighborhoods and communities. Because Jews in many large cities owned and operated a substantial number of small businesses and apartment houses in black areas, it was very often Jews with whom blacks were forced to deal in economic matters. As a result, Jews served as the closest and most visible symbols of white oppression and the most vulnerable targets of black rage.

A second major factor concerns the implementation of affirmative action programs. The past use of quotas to limit Jews in colleges and professional schools, occupations, and neighborhoods weighs heavily in the resistance by many Jewish leaders and organizations to the establishment of racial quotas today in these same social areas. Even though quotas today are intended to uplift the status of ethnic minorities, many Jews, having themselves experienced their adverse effects, see them as discrimination (albeit a kind of reverse discrimination) against which they have labored for several decades (Glazer, 1975).

Third, Black Muslim leader Louis Farrakhan, along with some black scholars, have in recent years rekindled anti-Semitic rhetoric in their speeches and writings. Jews are portrayed as having been especially instrumental in the slave trade and generally in the oppression of blacks in America; such claims have been strongly denounced and refuted by mainstream historians and other scholars (Jordan, 1995).

The Tenaciousness of Anti-Semitism Today, though prejudice and discrimination against Jews remain persistent among certain segments of the American populace, they are mainly random in occurrence, not organized and widespread as in the 1920s and 1930s. Racist anti-Jewish notions are almost nonexistent except among those who are part of the right-wing lunatic fringe of American sociopolitical life. Even black anti-Semitism does not represent more than what Robin Williams calls "situational antagonisms rather than deep-seated traditionalized hostilities" (1977:6). The anti-Semitism of most Americans today does not extend much beyond supporting mild forms of discrimination and accepting certain traditional Jewish stereotypes.

It is important to note that even during those times when it was intense and wide ranging, anti-Semitism in the United States differed from European forms in that it was never institutionalized, and no official anti-Jewish measures were enforced (Konvitz, 1978). Had this not been so, it is unlikely that Jewish Americans as a group could have attained their current high position in the society's stratification system.

Furthermore, despite the persistence of negative images of Jews, positive images of this group are also common in American society (Martire and Clark, 1982; Quinley and Glock, 1979). Jews are ordinarily seen as intelligent, religious, and family oriented and are often admired for their accomplishments in science and the arts. Indeed, several years ago, C. P. Snow, the famous British scientist and author, suggested that Jews might very well be genetically superior to other ethnic groups. Recognizing this as a thesis of reverse racism, many Jewish leaders were quick to denounce Snow's suggestion. Of course, some of the admiration of Jews expressed by non-Jews may be a manifestation of the tendency to look at the Jewish ethnic group in a negative light while expressing positive ideas about some of its individual members. As explained in Chapter 3, this is a common pattern of thought among the prejudiced, regardless of the group in question, perhaps best reflected in the familiar expression, "some of my best friends are . . . "[8]

Although anti-Semitism in the United States today is probably as mild and subdued as it has been at any time or place in modern Western history, this does not mean that more malignant and intense anti-Jewish thought and action might not be evoked under certain circumstances. As one researcher has explained, "Jews are recognized as an ethnic and religious out-group and are judged and treated in a distinctive manner accordingly" (Smith, 1991:27). Prejudice and discrimination are, it should be remembered, tools applied at varying times and degrees to protect the social, economic, and political advantages of some groups over others. For example, the overrepresentation of Jews in leadership positions of the mass media, particularly television and the movies, has produced frequent charges with anti-Semitic intimations of Jewish "control" of the media. Even international events may spark latent anti-Semitism—for instance, an Arab oil embargo or Israeli military actions in the Middle East.

In the 1980s and 1990s, the emergence of right-wing extremist groups, rabidly anti-Semitic, created increasing concern as their propaganda and violent actions seemed to have a public impact that went well beyond their tiny numbers. The proliferation of these groups' hate sites on the Internet also presented a new medium for exposing a potential pool of young people to their messages. Moralistic campaigns among leading political figures against the mass media, particularly films, also seemed to take on anti-Semitic overtones, given the strong influence of Jews in that industry. Even if these currents prove only transitory, they demonstrate how venerable are anti-Jewish stereotypes and how confounding is the phenomenon of anti-Semitism. Jews themselves perceive the continued potential of serious outbreaks of anti-Semitism. Surveys indicate that most Jewish Americans feel that anti-Semitism in the United States either is already or might become a serious problem (Cohen, 1985; Kosmin et al., 1991). In his study of Jewish baby boomers, Waxman found that 90 percent said they had not personally experienced discrimination. Nonetheless 80 percent agreed

---

[8]Although ethnic prejudice is a negative group evaluation, not all the traits alleged to typify a group are necessarily negative. Thus, in the case of the Jews, they may be seen by those prejudiced against them as intelligent, clever, and especially adept at business; but these traits are applied in such a way as to describe Jews as people who nefariously use their intelligence or business acumen to cheat and deceive non-Jews. Glock and Stark (1966:120) point out that a theme in anti-Semitism "is that the Jew constitutes a grave threat because he is such a worthy opponent."

somewhat or strongly that anti-Semitism "is a serious problem n the United States" (2001:97)

ASSIMILATION PATTERNS OF JEWISH AMERICANS

Jewish-American patterns of assimilation, like other features of ethnic life and relations, have to a large degree resembled those of other white ethnic groups. Cultural assimilation has taken place increasingly and irrevocably with each succeeding generation. Structural assimilation, however, has been slower and less profound, especially within the realm of primary relations.

Cultural Assimilation

Each successive generation of Jewish Americans has displayed cultural characteristics less uniquely Jewish than the previous one. And as with other white ethnic groups, much of this assimilation is related to upward class mobility. As they have moved from the working-class occupations and urban ghettos of the first generation to the middle-class occupations and suburbs of the second and third, Jews have steadily adopted lifestyles and values that characterize middle-class Americans generally. Like Italians of the second generation, the children of Jewish immigrants to some extent found themselves wavering between the ghetto and the larger society (Kramer and Leventman, 1961). Those beyond the second generation, however, have clearly chosen the dominant culture.

Perhaps the most significant indicator of cultural assimilation among Jewish Americans is the gradual movement, beginning with the second generation, away from the Orthodox branch of Judaism, encompassing traditional and rigid forms of worship and belief, to the Conservative and Reform branches, each reflecting more Americanized practices and doctrines. Orthodox Judaism continues to survive, however, and in fact has attracted increasing numbers of younger Jews in recent years (Goldberg, 1996; Heilman and Cohen, 1989; Niebuhr, 1997). It is of note that the division of Jews into three denominational subgroups is, in the main, an American phenomenon. In non-American Jewish communities, people are either Orthodox or nonreligious.[9]

---

[9]In this regard, Konvitz (1978) notes that although the Reform movement began in Germany, it was most fully developed by German Jews in the United States. Wertheimer (1993) describes a fourth branch of American Judaism, Reconstructionism, subscribed to by only 1 percent of adult Jews. It is also important to note that the Orthodox branch of American Judaism is itself split between those who believe that observance of Jewish religious ritual is compatible with the modern world (the "modern Orthodox") and those who separate themselves from the secular society (the "ultra-Orthodox," or "haredim").

Not only has there been a movement toward less rigorous forms of worship, but religious observance in general has diminished with each generation (Gans, 1958, 1967a; Kleiman, 1983; Sklare, 1969; Wertheimer, 1989, 1993). Although synagogue affiliation (not attendance) remains strong among even those of the third and fourth generations, researchers have cautioned against interpreting this as a resurgence of Jewish religious observance (Gans, 1967a; Sharot, 1973; Sklare, 1969). Rather, the synagogue today serves as the key institution by which Jews can express their ethnic identity, acquaint their children with their Jewish heritage, and interact with others of like ethnicity. Its purely religious aspects hold less importance. Thus, assimilation with the dominant culture, in a broad sense, continues, while Jews retain their ethnic identity.

Structural Assimilation

Jewish Americans have exhibited steady and in some ways substantial progress in the direction of secondary structural assimilation. We have seen that they have begun to participate fully in almost all areas of American society, collectively exceeding almost all other ethnic groups in levels of income, occupation, and education. Jews have also begun to enter elites in most institutional realms, though to a greater extent in some than others.

Structural assimilation at the primary level, however, is a more complicated matter. In general, Jewish Americans have traditionally exhibited strong in-group cohesion in the realm of primary relations (Gans, 1958; Lenski, 1963; Ringer, 1967; Sklare and Greenblum, 1967). This stems from attitudes and actions of the larger society and from the in-group preferences of Jews themselves. Sklare (1978) explains the ambiguity among out-groups regarding Jewish-Gentile social relations. Jews have commonly been seen as "cliquish" when they retreat to their own communities but as "pushy" when they make serious efforts to enter Gentile primary groups. Fearing rejection or at best a marginal acceptance, Jews more often than not were reluctant to leave the safety and psychological comfort of the ethnic group. As a result, concentrated Jewish residential areas were unlike black ghettos—which developed primarily because of overt discriminatory actions and policies—in that they evolved through a combination of out-group discrimination and in-group choice. In addition Jews, like other ethnic groups, initially congregated in relatively homogeneous areas where ethnic institutions were close at hand.

Today, discrimination and the attraction of ethnic institutions no longer play major roles in determining residential patterns, but strongly Jewish neighborhoods are still evident in cities with large Jewish populations. These seem to be a product mostly of housing markets, age, and socioeconomic factors, however, rather than the desire to live in Jewish areas (Goldscheider, 1986). Where they do not constitute a relatively sizable group, of course, Jews tend to be residentially dispersed (Rose, 1977). Moreover, Jewish Americans have generally followed the pattern of other white ethnics in removing to suburbs, where ethnic homogeneity is much less apparent. Waxman (2001) reports that 65 percent of Jewish baby boomers described their neighborhoods as little or not at all Jewish in character. Even in the suburbs, however, assimilation is not necessarily an unavoidable result. In his study of Boston, Goldscheider found that Jews living in neighborhoods of low Jewish density were not alienated from other Jews and that such individuals were not "on the edge of ethnic survival" (1986:40).

Intermarriage It is in the realm of intermarriage that Jewish Americans today give the greatest indication of accelerating primary structural assimilation. Like Italians and other southern and eastern European groups, Jews are increasingly intermarrying with members of the dominant group as well as other ethnic outgroups.

Until the 1960s, intermarriage among Jewish Americans was very low. Indeed, the rate of out-group marriage among Jews was lower than for other white ethnic groups. Change in this traditional pattern, however, is now evident. Data clearly point to a very substantial increase in Jewish intermarriage in recent decades (Goldstein, 1980; Kleiman, 1983; Massarik and Chenkin, 1973). From 1970 to 1990 the rate of intermarriage more than quintupled, from 9 percent for Jews married before 1965 to 52 percent for those married after 1985 (Steinfels, 1992). Thus more Jews now marry outside the ethnic group than within.

Increasing Jewish intermarriage is attributed to several factors. Jews in recent years have entered occupational areas and social groups previously closed to them, placing them in social settings where there is increased interaction with non-Jews. Only a third of Jewish baby boomers reported that all or most of their closest friends were Jewish (Waxman, 2001). Moreover, cultural assimilation has weakened ethnic traditions among younger Jews. Attitudes toward marriage outside the ethnic community are therefore not so uncompromisingly negative as they were for older generations. Further, as noted earlier, there is an increasingly tolerant view of the general populace toward intermarriage.

It is important to bear in mind that regarding assimilation, the Jewish-American population, like all other ethnic groups in American society, is internally variegated. Some individuals remain solidly attached to the ethnic group and retain a strong ethnic identity, while others have shed completely their ethnic attachments. Between these two extremes, a wide range of assimilation is evident. The Jewish-American population today seems to comprise three general components (Cohen, 1995; Liebman, 1973).

The "involved" or "affiliated" are immersed in Jewish life and are religiously observant. They belong to a synagogue, have close Jewish friends, and are self-consciously Jewish despite their adoption of non-Jewish values and behaviors. The "moderately affiliated" or "associated" engage in some Jewish practices and maintain close Jewish friendships but are not deeply involved in organized Jewish life or ritual. These are mainly younger, third- and fourth-generation persons. Finally, the "peripheral" or "nonassociated" are at best nominally Jewish. If they maintain Jewish friends, it is only accidental. Such people may be in the arts, professions, or other occupational areas in which a freer, more open social atmosphere prevails.[10] Cohen (1995) estimates the breakdown as "involved," 36 percent; "moderately affiliated," 41 percent; and "peripheral," 23 percent.

Recent immigrant Jewish populations from Russia and Israel have further complicated the range of Jewish assimilation. Israeli immigrants have generally been from among Israel's secular Jewish population, those not committed to religious orthodoxy. As to the Russians, most bring little if any Jewish background with them, and as one researcher has observed, "the challenge of integrating them into a Jewish American community has provided a degree of revitalization for American Jewry" (Sorin, 1997:244). Immigrant Russian Jews, as well as Israelis, have established institutionally complete enclaves that have further enabled them to confound rapid assimilation (Gold, 1999, 2000; Orleck, 1987).

The Future of Ethnicity among Jewish Americans

Patterns of assimilation and pluralism among Jewish Americans, although in some ways similar to those of other Euro-American groups, have been and will probably continue to be unique in other ways.

The Strength of Assimilation The forces of assimilation, barring the unlikely resurgence of overt forms of prejudice and discrimination, seem no less irresistible for Jews than for other white ethnics in American society. In considering the status of Jewish Americans today, Seymour Martin Lipset and Earl Raab have written that although they are not on the verge of disappearance, nonetheless, "current evidence suggests that group identity and cohesiveness are severely eroding for the large majority" (1995:47). In his study of Jewish baby boomers, Waxman, too, concludes that "in terms of social structural significance, as indicated by residential patterns, socioeconomics, language, and mate selection, there is a waning of Judaism and Jewishness for the group as a whole" (2001:117). With upward mobility, geographic dispersion, and increased intermarriage, what is left for most individuals is essentially symbolic ethnicity, having dwindling impact on social life. Indeed, many Jewish leaders have in recent years expressed concern over the threat to the continued existence of the

---

[10]On this point, see Gordon (1964).

Jewish-American community given the accelerated pace of acceptance of Jews into the mainstream society and the decline of blatant anti-Semitism. Such fears have been reinforced by declining birthrates and increasing intermarriage among the Jewish-American population. One survey indicated that among Jewish Americans in general, over 60 percent recognized serious dangers to the survival of the ethnic group in light of current rates of assimilation and inter-marriage (Cohen, 1987).

Some have questioned whether Judaism in America can be sustained without a return to a stronger expression of traditional Jewish beliefs and rituals. Jewish leaders have responded to this issue with opposing strategies: On the one hand, some prescribe an adaptation to make it easier for interfaith couples to become part of the Jewish community and for secular Jews to reaffirm their faith. Others, in contrast, argue that only by a reinvigoration of traditional Judaism can the Jewish-American community survive in the long term (Leiter, 2000).

Assimilation and Jewish Identity Despite objective indications of continued cultural and structural assimilation, ethnic identity among Jewish Americans has in some ways been strengthened in the last several decades. Some, in fact, strongly challenge the view that full assimilation is an eventuality. They see not an erosion of Jewish ethnicity but an adaptation to new conditions, which will assure the group's future integrity despite foreboding demographic trends (Cohen, 1988). Two events of the modern era account in largest measure for the revived sense of Jewish community and identity: the Nazi Holocaust and the founding of Israel as a nation-state. Both events served to make Jews in the United States more conscious of their heritage and instilled a sense of collective commitment to the preservation of the Jewish people.[11]

This reinvigorated Jewish identity, however, must be seen in much the same way as the resurgent ethnic identity among Italian Americans and other white ethnic groups: It is mostly symbolic. Jews continue to identify as Jews and to abide by some of the major Judaic rituals, such as attending synagogue during high holy days and having their sons (and increasingly their daughters) bar (bas) mitzvahed. But for most, the expression of their Jewishness does not extend much beyond these few practices. Put simply, a majority of American Jews have drifted away from religious participation (Goldberg, 1996; Wertheimer, 1989, 1993).

Although the decline of religious orthodoxy among most Jewish Americans is very evident, equally evident in recent years is a return to more traditional re-ligious practices by a small but growing minority. Ritterband, in fact, suggests a dichotomy in which the American Jewish community is becoming bimodal:

---

[11]By the late 1990s, there were indications that Jewish Americans were beginning to move away from their historic close ties to Israel (Heilman, 1999).

". . . The more committed Jews have increased their commitment while less committed Jews have more and more opened themselves up to the forces of assimilation" (1995:378). No one has better described the deep divisions that have arisen among Jewish Americans regarding this divergence than Samuel Freedman (2000). The split between those who desire a secularized Judaism and those who call for a reaffirmation of traditional beliefs and practices is, in his estimation, irreconcilable. So profound and abiding are the differences, writes Freedman, that Judaism may very well fragment into divergent faiths sharing only "a common deity and a common ancestry" (354). Clearly, then, American Jewry is moving simultaneously in two divergent directions. At the same time that Jewish identity is becoming more tenuous as assimilation proceeds unabashed, a resurgence in Jewish culture and a return to more traditional practices by many are unmistakable trends (Leiter, 2000). The former, however, appears to constitute the stronger force.

Even with a decline of religiosity, however, there remains a strong impetus for the continuation of Jewish identity. As observers have pointed out, the strength of Jewish communalism more than compensates for the weakness of Jewish religious associations (Lenski, 1963; Sandberg, 1986; Waxman, 1981). The commonality of social class and high educational achievement creates common lifestyles and, therefore, social bonds. "The concentration of Jews in particular jobs and in college and post-graduate educational categories," notes Calvin Goldscheider, "links Jews to institutions, networks, families, neighbourhoods and political interests and therefore has become a powerful basis of ethnic continuity" (1995:136). Moreover, it should be noted that, despite their peripheral commitment to Jewish religious ritual and even belief, very few Jews convert to Christianity.

Contemporary Jewish identity is sustained as well by an awareness that, though their place in American society is probably as secure as it has been in any society in modern history, Jews remain the potential object of differential treatment. The historical roots of anti-Semitism are deep and abiding. Ethnic identity, therefore, is a product of eternal vigilance, instilled by centuries of persecution. Jewish identity is reinforced also by the perception of exclusion. Despite their positions of prominence in academia, the arts, business, and politics, Jewish Americans may still feel they are outsiders, and thus not entirely part of the centers of secure influence (Cohen, 1985).

The marginality of Jews in American society has been substantially reduced in recent decades, but there nonetheless remains a fundamental cultural difference from their Christian neighbors—religion. This has created a kind of collective schizophrenia for Jewish Americans. As one writer has put it, they live "in a state of tension between the two values of integration and survival, and, regardless of where they stand on a survival-integration continuum, they find themselves pulled in both directions" (Liebman, 1973:134). Another has described their position in the American ethnic system as doubly marginal: "marginal to the majority culture, but also marginal among minorities" (Biale, 1998:27). Jewish Americans will therefore continue to be unique among American ethnic groups in their path toward assimilation.

SUMMARY

In many ways, Jews collectively represent the prototypical American ethnic success story. Paralleling the immigration and settlement patterns of other groups of the new immigration, the East European Jews subsequently outpaced all others in upward social mobility. Today, on various measures of socioeconomic status (income, occupation, education), Jewish Americans rank at or near the top. The high skill level of Jewish immigrants entering an industrializing economy provided broad opportunities for them, but certain aspects of the Jewish cultural heritage also helped to propel them quickly upward in occupational standing.

Anti-Jewish attitudes and actions (anti-Semitism) in the United States have included not only negative stereotyping but also widespread exclusion from certain educational institutions, business areas, and social organizations. Although certain forms of anti-Jewish prejudice and discrimination remain apparent, these are not organized and popular efforts, as they were in earlier periods.

As with Italian Americans, assimilation for Jewish Americans has proceeded with each successive generation, though structural assimilation has been slower and less extensive than cultural assimilation. At the secondary level, Jews have begun to participate fully in almost all major institutions of the society; they have also begun to penetrate the leadership echelons of these institutions. At the primary level, Jews continue to exhibit strong in-group cohesion. Sharply increasing rates of intermarriage, however, signify a growing change in this pattern. Although Jews continue to assimilate into American society, their religious uniqueness remains a source of ethnic identity and out-group recognition.

Suggested Readings

American Jewish Committee. (annually). *American Jewish Year Book*. A basic reference work for understanding the contemporary status of Jews in American society. Each edition contains up-to-date demographic statistics as well as research articles by social scientists.

Dinnerstein, Leonard. 1994. *Antisemitism in America*. New York: Oxford University Press. A history of anti-Semitism in the United States from the founding of the society to the present, concluding that it is a declining phenomenon, a trend not likely to reverse itself in the future.

Freedman, Samuel G. 2000. *Jew vs. Jew: The Struggle for the Soul of American Jewry*. New York: Touchstone. In novel-like fashion, Freedman describes the vast divisions that have arisen among the various branches of contemporary Judaism. Predicts that the Jewish-American community is heading toward an irreconcilable split into several separate faiths.

Goldscheider, Calvin. 1986. *Jewish Continuity and Change: Emerging Patterns in America*. Bloomington: Indiana University Press. A sociological analysis of the contemporary Jewish-American ethnic group, showing demographic and behavioral trends.

Hertzberg, Arthur. 1997. *The Jews in America: Four Centuries of an Uneasy Encounter: A History*. New York: Columbia University Press. A basic history of Jewish Americans, concluding that they have assimilated thoroughly into all major institutional realms.

Howe, Irving. 1976. *World of Our Fathers*. New York: Simon & Schuster. Richly describes the immigration, urban settlement, and emergence of ethnic institutions among East European Jews of the late nineteenth and early twentieth centuries, focusing particularly on New York City.

Liebman, Charles S., and Steven M. Cohen. 1990. *Two Worlds of Judaism: The Israeli and American Experiences*. New Haven, Conn.: Yale University Press. Compares the meaning of Jewish life and identity in the United States and Israel.

Lipset, Seymour Martin, and Earl Raab. 1995. *Jews and the New American Scene*. Cambridge, Mass.: Harvard University Press. Describes the current status of Jews in American society and the social, economic, and political forces that have brought them there. The authors see strong indications of an eventual complete assimilation of Jewish Americans.

Silberman, Charles E. 1985. *A Certain People: American Jews and Their Lives Today*. New York: Summit. Discusses the current position of Jews in American society, stressing their economic, political, and social achievements, as well as the declining potency of anti-Semitism.

AFRICAN AMERICANS

THE EXPERIENCE OF African Americans[1] is unique among American ethnic minorities. No other group entered the society as involuntary immigrants, and no other group was subsequently victimized by two centuries of slavery. The vestiges of these social facts account for the uninterrupted, if vacillating, conflict between whites and blacks throughout American history and the agonizing nature of the adjustment of blacks to a predominantly white society.

During the twentieth century, the study of race and ethnicity in America focused primarily on the relations between blacks and whites. Throughout most of its history, the American racial/ethnic system was, essentially, a binary system, in which one was either black or white. Asians and other groups distinctly non-European or non-African in origin were simply too small in population to significantly complicate this system. Obviously that is no longer the case. With the society's growing ethnic diversity in the past several decades, race and ethnic relations have taken on a much broader scope and become more complex.

---

[1]Some controversy concerns whether *African American* rather than *black* is a more appropriate and acceptable term for the American black population. A survey in 2001 by the National Urban League found that three-quarters of blacks preferred the term *African American* (Stafford, 2001), but recent Gallup polls indicate that most people do not have a strong preference for either (Gallup, 2001e). In this book, the terms are used interchangeably. Technically, however, whereas *black* stresses the racial distinction that sets off this group from others, *African American* emphasizes the group's ethnic identity, that is, its culture and historical origins (Patterson, 1997).

Nonetheless, despite the emergence of a more heterogeneous ethnic configuration, it is the place of blacks in the racial/ethnic system that continues to perplex the society more than any other domestic issue. African Americans remain the most visible ethnic group, a fact that must be combined with the heritage of slavery in accounting for the extraordinary character of the black experience. More than other ethnic groups and categories, therefore, African Americans—and their relations with Euro-Americans as well as other ethnic groups—continue to draw the intense scrutiny of sociologists.

THE DEVELOPMENT OF THE BLACK MINORITY

The history of blacks in American society is complex and subject to a variety of interpretations. There is no disagreement, however, that it is a history marked above all by the continued stress of accommodation to the culture and institutions of a white-dominated society. Tension between whites and blacks has varied in intensity at different times, but it has never been far from eruption into overt and violent hostility. With the understanding that conflict has been a constant underlying theme of black-white relations, we can subdivide the history of those relations into three broad epochs, each characterized by a somewhat different pattern: slavery, the Jim Crow period, and the modern era.[2] During these three periods, relations were shaped most fundamentally by social and economic trends and structures of the larger society. As those trends and structures changed, black-white relations, in turn, took on new form and meaning.

Slavery: Paternalistic Domination

Blacks first entered American society when they landed in Virginia in 1619. Their legal status, however, was undetermined for at least forty years. Until the 1660s, the status of blacks as servants was not essentially different from that of many others, including some whites.[3] Various forms of bondage had been prevalent in all the colonies almost from the founding of the society (Handlin, 1957). Perhaps the most common form was voluntary servitude, or indenture, in which a person was bound by contract to serve a master for a certain length of time, usually four to seven years (Jordan, 1969). In exchange, the servant's passage to the colonies was paid. A great number of people, particularly from Scotland and Ireland, entered the country in this manner.

---

[2]This broad historical subdivision of the black experience conforms generally to that delineated by William Julius Wilson (1980).

[3]Jordan (1969), however, maintains that some blacks were enslaved even before the 1660s.

Although no specific date marks the establishment of chattel slavery, slave codes and statutes evolved piecemeal, so that by the 1660s most southern states had enacted laws defining blacks as slaves rather than as indentured servants. Most of the features of American slavery were now settled, and at least in the slaveholding states, the ambiguity of black status was resolved.

The Choice of Blacks Other groups in the colonies seemingly could have served as slaves; only blacks, however, were inevitably to qualify. Other groups were treated harshly, and even the enslavement of some was seriously considered. The Irish seem to have been dealt with particularly severely, and attempts to enslave Indians were also made. But why, in the end, were only blacks chosen for indeterminate bondage, that is, slave status? Historian Oscar Handlin gives us some clues. In the mid-seventeenth century, the colonies were in dire need of labor; immigration from Europe therefore had to be stimulated. As a result, the colonies adopted legislation that promised better conditions for servants and expanded the opportunities for release from servitude. To encourage immigration, then, the master-servant relationship gradually evolved into a contractual one. But blacks did not benefit from the liberalization of the bondage system. Because they were involuntary immigrants, the application of these policies could have no effect on their numbers. In short, there was nothing to be gained by placing blacks under this legislation. "To raise the status of Europeans by shortening their terms would ultimately increase the available hands by inducing their compatriots to emigrate; to reduce the Negro's term would produce an immediate loss and no ultimate gain" (Handlin, 1957:13).

The choice of blacks was dictated as well by white ethnocentrism. Most simply, their physical and cultural traits were more obviously distinct than those of other groups. Especially significant to the English settlers was what they interpreted as the heathenism of blacks. This served as a vital part of the rationalization for enslavement. Indeed, as historian Winthrop Jordan (1969) points out, the religious difference between whites and blacks was of greater importance, at least initially, than the physical distinction. As Christians, white indentured servants could not be dealt with as nonhumans. Yet the religious factor is not in itself a sufficient explanation for black enslavement, because the colonists made no distinction between black nonbelievers and those who had been converted to Christianity; both groups qualified as slaves. Jordan concludes that it was an aggregate of qualities, the total of which set blacks sufficiently apart from others, that made them the most likely candidates for slavery. Their lack of Christianity and unique physical appearance were vital, but other traits that the English interpreted as savage and bestial "were major components in that sense of difference which provided the mental margin absolutely requisite for placing the European on the deck of the slave ship and the Negro in the hold" (1969:97). The prevalent white images of blacks, explains Jordan, can be traced to the first contacts between the English and Africans in the sixteenth century. From the first, "Englishmen found blackness in human beings a peculiar and important point of difference" (1969:20). Black skin became a key visible mark for identifying people who were believed to be defective in religion, savage in their behavior, and sexually wanton.

But what of Indians? Were not they, like blacks, nonbelievers and even physically distinct? On these bases, they were as subject to white ethnocentrism as were blacks and thus, logically, could have been eligible for slavery. But Indians maintained certain advantages. First, they were unused to settled agriculture and therefore not easily suited to plantation labor. Moreover, unlike blacks, Indians were familiar with the terrain and could more easily escape to their own people (Stampp, 1956). The most important difference between blacks and Indians as regards their ultimate status, however, was the power resources each could command. Put simply, Indians could offer greater resistance to enslavement. As William Wilson (1973) notes, blacks were forced to live in a foreign land, lacked organization, and were scattered about the countryside. In no way did Indians face a similar situation. Indians remained culturally and politically organized in nations, whereas blacks, following enslavement in Africa and the brutal middle passage, were dispersed and purged of their native cultures. As a result, they remained more dependent on white settlers in the way of culture and sheer survival and posed little collective threat (Jordan, 1969).

In sum, the cultural and physical differences of blacks were important rationales for their enslavement; but in the end, the most critical factor was differential power (Geschwender, 1978; Noel, 1968). Although black skin and "heathenism" gave rise to white ethnocentrism, this did not predetermine the choice of blacks. Rather, blacks simply lacked a viable community from which they might have mustered sufficient counterforce to resist slavery. Color became the visible symbol of slave status, and gradually a racist ideology was developed to buttress the system.

The American form of slavery that evolved was unique, having had no precedent in seventeenth-century England (Elkins, 1976; Handlin, 1957). Indeed, it was a system different from any earlier forms of slavery. The Portuguese and Spanish had already enslaved blacks in their colonies and had maintained slaves at home since the early fifteenth century. But the status of slaves in these societies differed somewhat from that of their American counterparts. As we will see in Chapter 14, in Brazil and other colonies of Portugal and Spain, slaves maintained certain property and family rights and were often freed. In the American form, however, the slave was essentially an object not to be afforded common human privileges.

In the final analysis, the evolution of slavery in the United States was primarily a consequence of economic rationality, prompted above all by the demand for cheap labor in the underpopulated colonies. "The use of slaves in southern agriculture," explains historian Kenneth Stampp, "was a deliberate choice (among several alternatives) made by men who sought greater returns than they could obtain from their own labor alone, and who found other types of labor more expensive" (1956:5). Although it had existed in all the colonies at one time or another, slavery by the middle of the eighteenth century was confined to the South, where plantation agriculture had become the foundation of the economy. In the North, political and moral factors had been influential in the decision to abandon slavery; but more important, the system was seen there as a hindrance to continued industrialization.

The Development of a Racist Ideology As explained in Chapter 3, negative belief systems regarding ethnic groups are more likely to follow than precede the development of systems of discrimination. This is clearly seen in the case of American slavery. The pseudoscientific form of racism—the belief that blacks are innately and permanently inferior to whites—did not arise forcefully until well into the nineteenth century. Racial prejudice was evident before that time, of course, but black inferiority was seen as a product of slavery itself, not of inherent deficiencies (Fredrickson, 1971). Before the nineteenth century, beliefs in cultural, not biological, inferiority were in effect (Wilson, 1973). Blacks were seen as heathens and as savages, but not necessarily innately so.

During the 1830s, slavery came under serious attack. Drawing on the environmentalist thought of the eighteenth-century Enlightenment, abolitionists argued that blacks should be freed to develop their capacities to the fullest. During the three decades between 1830 and 1860, proslavery theorists, in response to abolitionist thought, developed a racist belief system replete with ideas pertaining to irreversible physiological differences, such as cranial size and shape, which allegedly explained mental and physical inferiority. Also stressed in these new racist theories were the failure of blacks to develop what Westerners considered a "civilized" life in Africa and the dangers of miscegenation, leading to racial degeneracy (Fredrickson, 1971).

Racist ideology was not limited to the South, however. Most northerners were likewise unprepared to accept blacks as equals and subscribed to basically similar ideas. Fredrickson notes that "a common Northern dream depicted an all-white America where the full promise of equality could be realized because there was no black population to be relegated to a special and anomalous status" (1971:323). Even antislavery forces included at least one school of thought (whose supporters included Abraham Lincoln) holding that only by establishing a colony for blacks in the Caribbean or in Africa would the issue of slavery and the racist fears of white Americans be solved.

After the 1830s, almost all whites of whatever persuasion agreed that blacks were inferior to whites in certain fundamental qualities, especially intelligence and initiative, and that those differences were essentially unchangeable. Because of these differences, an integrated society, it was maintained, was not feasible. Barring their elimination, therefore, continued subordination of blacks was natural and necessary (Fredrickson, 1971). These ideas did not change fundamentally until well into the twentieth century.

The development of a racist ideology, then, is best seen as a response of southern slaveholders to the threat of abolition of the slave system, but a response that gained national acceptance. Perhaps most significant in contributing to the general acceptance of biological racism were the ideas of social Darwinism, which gained widespread recognition in the late nineteenth and early twentieth centuries. Indeed, the fullest flowering of the racist ideology did not occur until after slavery had been abolished (Woodward, 1974).

Master-Slave Relations The American brand of slavery was a system of paternalistic domination, creating, in some degree, a father-child relationship between master and slave (Elkins, 1976). The total control of the slave's existence by the plantation master was explicitly legitimized by the slave codes of the various slaveholding states. Slaves lacked virtually all legal rights. They could not own or inherit property, testify in court, hire themselves out, or make contracts. Slave laws clearly provided that marriage between slaves held none of the rights of marriage between free people. As a result, families could be broken up in trade with no consideration given to keeping husband, wife, and children intact as a single unit. Finally, laws forbade teaching slaves to read and write.

Enforcement and interpretation of the slave laws were left almost entirely to the master, not the courts (Fogel and Engerman, 1974). With such wide discretion, utter cruelty was not uncommon; but it was felt that brutal treatment would be controlled through public opinion, by the master's sense of decency, and above all, by economic interest (Elkins, 1976). Slaves, representing a significant financial investment, would not be physically or emotionally abused to the point of incapacity. Historians Robert Fogel and Stanley Engerman (1974) explain, for example, that although slave marriages were not legally recognized, masters nonetheless encouraged their slaves to marry and establish families in order to ensure an atmosphere of stability. Thus, slave families were not separated if this was avoidable (see also Gutman, 1976).

As noted in Chapter 4, systems of paternalistic ethnic relations do not require the maintenance of great physical distance between groups. Rather, it is the social distance that is of greatest importance. In the pre–Civil War South, a rigid stratification system was enforced through a well-understood racial etiquette, in which physical contact was not uncommon but the social places of whites and blacks were not violated. Social intimacy might often be seen between master and slave in the form of concubinage between white men and black women or the raising of white children by black "mammies." It was out of this apparent physical closeness that emerged the myth of racial harmony in the Old South. But such intimate personal contacts in a paternalistic system are always relationships between unequals. Moreover, the degree of physical intimacy between the two groups should not be overstated. On large plantations, where most black slaves were held, personal relations with masters were limited to household servants. The majority of slaves were field hands who rarely came into physical contact with the white master or his family.

Although most blacks in the South remained in rural slavery, a few hundred thousand free or quasi-free blacks, most concentrated in cities, lived within the slavery regime. But these free blacks were scarcely better protected by law than slaves, and the degree of discrimination mounted against them was not significantly less. They could make contracts and own property, but in most other respects their civil rights were not much greater than slaves' (Stampp, 1956; Woodward, 1974). Further, blacks living in the North, though not subjected to slavery, were by no means accepted by whites into the mainstream society. As Wilson describes it, "Whites rejected slavery as an acceptable institution in the North but were unwilling to endorse the view that blacks should receive social, economic, and political equality" (1973:94).

The immediate effects of slavery on blacks were obvious, but the impact of the system on whites must be considered more carefully. Most whites in the South did not accrue direct benefits from the slave system. In fact, most whites were small farmers who owned no slaves. In 1860 there were 385,000 owners of slaves among 1,516,000 free families in the South. Moreover, the typical slaveowner held only a few slaves. In 1860, 88 percent held fewer than twenty, 72 percent held fewer than ten, and 50 percent held fewer than five (Stampp, 1956). Obviously, large slaveowners—the planter class—represented only a tiny fraction of the southern white population. Most of the slaves, however, were concentrated in these few large plantations.

For the elite planter class, slavery provided substantial economic and social rewards; for smaller slaveholders, it offered correspondingly more modest benefits. But what did slavery provide for nonslaveholders, and why did they so universally accept it? Clearly, it could provide them no direct economic payoff; but it could serve as a means of controlling potential economic competition from blacks. In addition, it furnished to poor white farmers a kind of prestige that could be derived from the thought of belonging to a superior "race" (Stampp, 1956). In short, both fear of competition from blacks and social-psychological rewards account for the acceptability of slavery by nonslaveholders. Moreover, the power of the ruling planter class to impose its racist ideology on all whites, regardless of social class, was substantial. As explained in Chapter 2, nonruling groups ordinarily accept for the most part the prevailing social system in the belief that it works in their interests and is justifiable, even though it may work primarily in the interests of the powerful few.

In sum, the 200 years of slavery were a period in which the subservience of blacks to whites was legitimized in the South and informally recognized in the North. In addition, a racist ideology emerged out of the slave system that continued to influence the perception of blacks by whites well into the twentieth century.

Racial Caste: The Jim Crow Era

For a brief historical moment following the Civil War, it appeared that blacks might overcome the crippling social effects of slavery. Slavery had been abolished, and with the protection of the federal government, freed blacks began to acquire some of the power resources necessary to enter into the mainstream society with whites. They now exercised the vote and were even elected to high offices in state and federal governments. An educational system was also established that promised still greater potential opportunities. The decade of Reconstruction from 1865 to 1877, then, marked the end of paternalistic relations between blacks and whites and the beginning of competitive relations. But it was this competition that eventually led to the system of segregation as a means of controlling the black threat to white economic and social advantage. And to reinforce that control, a racist ideology was developed into a more sophisticated and generally acceptable belief system.

The process of denying blacks competitive power resources had begun well before 1877, but that year serves as a historical benchmark ending Reconstruction. The results of the presidential election of 1876 were disputed, and as a means of breaking the impasse, the Democrats and Republicans agreed to a compromise: Federal occupying troops, guaranteeing protection to blacks, would be removed from the South in exchange for the electoral votes of southern Democrats for the Republican candidate, Rutherford B. Hayes. This was not simply a symbolic action; it enabled whites to establish further control over the black population without fear of interference by the federal government. By that time, the planter class had already reasserted its traditional place of power, and the economic dependence of blacks on their former masters was revived with the advent of the sharecropping system. With the removal of the federal government's oversight, the drive to reinstitute white dominance was given greater impetus.

Perhaps the most significant mechanism in this process was the disfranchisement of blacks. When it became apparent that they might hold the balance of political power between competing white factions, efforts to deny voting rights to blacks began in earnest. This put an end to the incipient alliance that had begun to form within the Populist movement between poor blacks and poor whites. Through various constitutional provisions, blacks were effectively disfranchised throughout the South by the first decade of the twentieth century. In South

Carolina, for example, a clause was adopted that called for two years' residence, a poll tax of $1, the ability to read and write any section of the U.S. Constitution or to understand it when read aloud, the owning of property worth $300, and the disqualification of convicts (Franklin, 1980). Such measures, of course, automatically disqualified most blacks from the registration rolls.

The Growth of Segregation Following disfranchisement, the path was cleared for the enactment of the entire array of Jim Crow measures designed to separate blacks and whites in almost all areas of social life: housing, work, education, health care, transportation, leisure, and religion. Trains, waiting rooms, drinking fountains, parks, theaters, and other public facilities were now segregated. Access to hotels, restaurants, and barbershops was limited to those serving either blacks or whites, and by 1885, most southern states had enacted laws requiring racially separate schools. With the 1896 Supreme Court decision in the case of *Plessy v. Ferguson,* the "separate but equal" doctrine was upheld, and the system of segregation in the South was securely in place.

Segregation as it now appeared was not the same as what had been practiced during slavery. As van den Berghe (1978) explains, roles and institutions that had previously been racially complementary now became racially duplicatory. Rather than whites and blacks maintaining different statuses and roles within the biracial plantation, separate economic and social institutions now served each racial group in a castelike system. Although state laws requiring separate facilities for the two groups stipulated that these would be equal, the Supreme Court never defined *equality* while repeatedly upholding the "separate but equal" principle. As a result, the most consistent fact of this system was that separate facilities were inherently unequal; those reserved for blacks were grossly inferior or inadequate, usually by design.

In addition to physical separation, there was the need for a system of racial etiquette, for blacks and whites obviously could not be totally isolated from each other. The etiquette was intended to make clear the caste positions of dominant and subordinate groups. Whites could not shake hands with blacks, for example, nor would blacks be addressed as "mister" but rather as "boy" or by the person's first name. And most important, racial endogamy was rigidly enforced. Indeed, prohibition of interracial marriage was one of the first formal measures enacted by southern states in the formation of the Jim Crow system.

Racial segregation was maintained by both force and ideology. As to the former, blacks in the South could be dealt with in an almost totally unrestrained fashion. Where legal techniques were found inadequate, extralegal measures were used. With the rise of vigilante groups such as the Ku Klux Klan, violence and physical intimidation became the order of the day. Lynching, the most extreme technique employed to maintain black subservience, was resorted to with increasing frequency. From 1884 to the outbreak of World War I, over 3,600 lynchings occurred, most in the South and with most of the victims black (Franklin, 1980). Fear and humiliation were integral parts of everyday life for blacks in the southern states.[4]

Conditions in the North, though not supported by legal mechanisms as in the South, were not significantly different. Although black civil rights were acknowledged, segregation in virtually all areas of social life typified black-white relations. Indeed, as historian C. Vann Woodward (1974) points out, segregation actually emerged first in the North before the abolition of slavery. Despite the free status of blacks in the northern states, their inferiority and inability to assimilate had not been questioned in any quarter. Segregation was a means of assuring that blacks understood and kept "their place," and it was well established in all the free states by 1860. Moreover, blacks in most of these states had been effectively disfranchised.

The system of segregation in both the South and North was abetted by the fullest development of a racist ideology. Ideas of racial superiority and inferiority were now supported by scientific thought and were applied not only to blacks but also, as we saw in the case of Italians and Jews, to all non-Anglo-Saxon groups. Woodward (1974) notes that a further contributing factor to the vehement racism of this era was American imperialistic ventures at the turn of the century, involving nonwhite peoples in the Philippines and the Caribbean. The negative stereotypes of blacks that emerged from slavery—lazy, childish, irresponsible, uncouth—became even more potent. In short, ideas of white—specifically, Anglo-Saxon—supremacy were at a zenith at the turn of the century and for two decades afterward.

Early during the period of caste, the black leadership strongly debated whether to adapt to or challenge the system of segregation and denial of black rights. One school of thought, reflected most clearly by Booker T. Washington, contended that the most realistic course was to accept a form of separatism based on the idea of self-improvement. In his famous "Atlanta Compromise" address in 1895, Washington deemphasized the effects of racial discrimination and called for blacks to learn industrial skills and trades that would help strengthen the black community internally. Opposed to this strategy of accommodation and withdrawal were those led by W. E. B. Du Bois, who argued for the complete acceptance of blacks into all areas of the society and advocated militant resistance to white racism. Du Bois contended that it was not solely the responsibility of blacks, nor was it in their capacity, to alter their collective place in American society; but that it was primarily the responsibility of whites, who held the power to effect such change. His efforts were instrumental in the

---

[4]A poignant description of the constant terror and degradation to which blacks were subjected is provided in the works of Richard Wright, one of the great black writers of the twentieth century. See especially *Black Boy* (1945) and *Uncle Tom's Children* (1938).

founding in 1909 of the National Association for the Advancement of Colored People (NAACP), which eventually became the chief organization working for black civil rights.[5]

The alternative strategies of integration or separation were debated by succeeding generations of black leaders. The theme of separatism reemerged strongly in the 1920s when Marcus Garvey, a Jamaican immigrant, led a movement advocating the return of American blacks to Africa. Integrationist thought prevailed among most blacks, however, although the separatist issue, as we will see, surfaced again in the 1960s.

Northward Migration Relations between blacks and whites took on a more confrontational and violent tone during the early 1900s than at any previous time. This situation was brought about by demographic and economic changes in the society that placed the two groups in direct competition.

Industrial expansion in the North created an increased demand for labor, and with the cutoff of immigration from Europe, blacks from the South began to fill these places. If jobs were the key "pull" factor impelling black out-migration from the South, the declining cotton economy and the continued enforcement of Jim Crow were the chief "push" factors. The demographic changes involving the black populace were profound. In 1910 almost 90 percent of blacks were living in the South, but eighty years later, little more than half remained there. Streams of blacks from the South to the North and West reached epic proportions from 1940 to 1970, with almost 1.5 million leaving in each of these three decades. Moreover, this migration was almost wholly to the cities, making blacks an increasingly urbanized population.

With their movement to the cities, black workers began to pose a real labor threat to white workers in the North, and the violence and intimidation that had characterized southern race relations became a national phenomenon. Adding to the tension created by job competition was the discontent expressed by the many black World War I veterans who had been exposed to more liberal social conditions in Europe. The upshot of the rising competition for work and housing and the growing impatience of blacks was an increase in the frequency and severity of black-white hostilities. Serious race riots occurred in numerous cities throughout the first two decades of the century (Rudwick, 1964; Rudwick and Meier, 1969; Tuttle, 1970).

In sum, black-white relations from the end of Reconstruction in 1877 to the 1940s involved a system of restrictive competition (Wilson, 1973). The forces of industrialization created the potential for significant changes in the structure

---

[5]Washington's and Du Bois's major ideas concerning the place of blacks in American society and their approach to change are contained in Rudwick and Meier (1969).

of race relations, and blacks were now theoretically capable of competing with whites, in contrast to their previous slave status. But in fact, strict segregation in most areas of social life and the denial of political participation severely limited their accumulation of competitive resources—jobs, education, and housing. Wilson has noted that when groups lack the necessary political resources, whatever opportunities are created by industrialization are offset by new controls imposed by the dominant group. "Race relations become competitive, but only in the narrowest sense" (1973:64). Blacks could offer little resistance because of the strength and resolve of whites of all classes to maintain the discriminatory system through both legal and extralegal mechanisms. Furthermore, the system was buttressed by a now scientifically endorsed racist ideology, which posited the innate inferiority of black people.

Fluid Competition: The Modern Era

Whereas the Jim Crow period was one of restrictive competition between blacks and whites, World War II marked the emergence of more fluid competitive relations. In such a system, according to Wilson (1973), the minority group's power resources are increased to the point at which it can begin to challenge the dominant group's authority and invalidate racist stereotypes. Members of the minority begin to move into positions formerly occupied only by the dominant group; in the process, institutional racism is exposed and challenged, and individual racism is seriously undermined. As these changes occur, some members of the dominant group begin to question the legitimacy of racial discrimination and either refrain from practicing racist norms or openly attack the racist system. As Wilson explains, "The greatest threat to any form of racism, then, is the significant entry of minority members into upper-status positions within the larger society" (1973:59).

The mid-twentieth century was clearly that period in which the acquisition of power resources by blacks impelled the most forceful challenge to the American racist system. This challenge involved the efforts of organized pressure groups as well as unorganized forces. Together they made up what became the most compelling U.S. social movement of the twentieth century.

Several factors account for the emergence of the movement for black rights after World War II. Much of the impetus had been established in the 1930s, when the Roosevelt administration began to appoint blacks to nonpolicymaking government advisory positions (Franklin, 1980; Myrdal, 1944). Moreover, blacks benefited from the administration's relief programs even though these continued to be administered in a discriminatory fashion. Another significant inroad made by blacks in the 1930s was their entrance into the labor movement, specifically the Congress of Industrial Organizations (CIO). Unlike the American Federation of Labor (AFL), the CIO was relatively nonexclusionary in its organizing activities. Representing industries in which most black unskilled laborers were employed (steel, automobiles, mining, clothing), the CIO now afforded blacks labor union participation in significant numbers for the first time.

Wartime conditions in the 1940s opened new areas of work to blacks and provided an even greater push to the northward migration that had been occurring since the early part of the century. Jobs in the defense plants of cities like Detroit, Chicago, Philadelphia, Cleveland, and New York attracted a new surge of migrants from the South, whites as well as blacks. Blacks, however, continued to suffer the effects of discrimination in the distribution of these jobs. As a result, black labor leader A. Philip Randolph organized the March on Washington Movement and threatened to mobilize millions of blacks in protest. Hoping to squelch mass unrest, President Franklin D. Roosevelt in 1941 issued an executive order prohibiting racial discrimination in federal jobs and created the Fair Employment Practices Committee to monitor the order. Randolph's movement was an early sign of the effectiveness of black protest, which would reach much greater heights after the war.

In addition to the economic opportunities the war provided, it presented white political leaders with a dilemma: How could the United States support a racist system at home while it was at war abroad with the most flagrantly racist regime, Nazi Germany? This inconsistency was difficult to reconcile without a greater commitment to black rights. Moreover, following the war, racist policies proved an international liability in dealing with the emerging nations of non-white Africa and Asia. As a result, the federal government now seemed more amenable to efforts at ending patterns of discrimination. One of the first institutions to be desegregated was the armed forces.

Perhaps the most momentous single government action during the 1950s that stimulated the push for black advancement was the Supreme Court decision of 1954 in the case of *Brown v. Board of Education of Topeka*. In that case, the Court ruled invalid the separate but equal doctrine, which had been upheld in the famous *Plessy v. Ferguson* decision of 1896. The 1954 decision basically removed the legal foundation of segregation and thus marked the beginning of the end of the entire Jim Crow system. Influenced by the testimony of social scientists, the Court made clear in its ruling that separate schools were inherently unequal and imposed an inferior status on black children, causing irreparable psychological change. School segregation laws, the Court declared, therefore violated the equal protection clause of the Fourteenth Amendment of the Constitution. This clause stipulates that no state shall deprive people of life, liberty, or property without due process of law, or deny them the equal protection of the law. The *Brown* ruling was subsequently used as the legal basis for court decisions that struck down other segregated institutions.

The significance of the *Brown* decision cannot be overstated, for among blacks it provided the spark to ignite a surge of hope that had begun to build in the

1940s. Lewis Killian has described its psychological impact on blacks: "The authority of the law of the land was now cast on the side of change, not on the side of the status quo. Any Negro with even the vaguest idea of what the Supreme Court had said now had good reason to hope for an improvement in his life and, even more so, in the life of his children" (1975:43). The drive for integration and an end to societal discrimination may be said to have begun with *Brown*. One year after this decision was handed down, the civil rights movement was symbolically launched with the boycott by blacks of the segregated bus system of Montgomery, Alabama.

Nonviolent Protest From the late 1950s to about 1964, nonviolent protest and civil disobedience were the movement's chief tactics. The basic idea was that individuals had a moral duty to disobey the law when it was clearly unjust. This was a strategy that had been applied many times historically in a variety of social settings, its most notable success occurring in the 1940s in India, where Mahatma Gandhi led the movement for independence from Britain (Sibley, 1963). The technique was to actively oppose the law but in a peaceful fashion. The proponents of nonviolent protest, Martin Luther King Jr. in particular, emphasized that it was necessary to win over the opposition through friendship and understanding, not through defeat or humiliation. Those participating in sit-ins at segregated lunch counters, freedom marches, or freedom rides were therefore prepared to accept suffering without retaliating in kind (King, 1964).

Hostile white retaliation against demonstrators and civil rights workers, much of it in the South carried out by civil authorities, included shootings, beatings, firebombings, and even killings. A pivotal year was 1963, when several key events occurred. Police brutally suppressed black demonstrators in Birmingham, Alabama; a Birmingham church was firebombed, killing four black children; and National Guardsmen were used to carry out the court-ordered desegregation of the University of Alabama. These events were given prominent media attention and galvanized public and governmental support for black civil rights. This supportive mood climaxed with the march on Washington of over 200,000 people in August of that year. One year later, the most comprehensive civil rights measure since the Civil War was enacted, prohibiting discrimination in voting, public facilities, schools, courts, and employment. And in 1965 the Voting Rights Act was passed, ending the systematic disfranchisement of southern blacks.

In essence, the civil rights acts of 1964 and 1965 signified the demise of official segregation and discrimination in the United States. But the end of de facto racial separation and inequality was not yet in sight. Moreover, black leaders became increasingly aware of the discrepancy between what had been accomplished in a legal sense and what had not been accomplished in the actual improvement of life conditions for most blacks. This was sharply evident in northern cities, where blacks were concentrated in depressed ghettos. The black movement, accordingly, shifted its focus from the South to the urban areas of the Northeast and Midwest. This shift signaled a move not only in the geographical locus of protest but also in the goals and tactics of the movement itself.

Securing legal rights was not the primary aim of black protest in the North, as it clearly had been in the southern states. Technically, those rights had been in place for many decades. Instead, emphasis was placed on economic issues—jobs, housing, education, and all other life chances—which, through discriminatory actions, were being denied or poorly provided to most blacks. These issues, however, did not lend themselves to the kind of direct amelioration that had been the case with previously state-approved segregation and denial of civil rights. Legal maneuvering and sit-ins were inapplicable to socioeconomic problems.

As a result, the black movement passed temporarily through a more militant phase, in which the black leadership now proclaimed that ending racism required systemic changes in economic and political institutions, changes that purely nonviolent protest had failed to address. The shift away from defensive and nonviolent tactics produced not only more militant rhetoric but also serious urban violence throughout the late 1960s (Bergesen, 1980; Jacoby, 1998; National Advisory Commission on Civil Disorders, 1968).

Black Power Complementing the substantially changed strategy of black protest during this period was a reformulation of the movement's ideological goals. Until the mid-1960s, the chief aims of black protest were the abolition of segregated public facilities, the achievement of legal equality, and the encouragement of eventual racial integration at all societal levels. It was the third of these objectives that now came under attack from black leaders like Malcolm X and Stokely Carmichael, who pronounced the need for independence of black action and the establishment of black control of black community institutions. This was the essence of the idea of black power. Integration was rejected in favor of a separatist position, coalitions with white liberal organizations were severed, and whites were generally purged from the movement. Racial prejudice and discrimination were seen not simply as the attitudes and actions of individuals but as structural features of the American social order. The notion of institutional racism now became popularized (Killian, 1975). The ideology of black power (or black nationalism) was espoused most vehemently by the Student Nonviolent Coordinating Committee (SNCC) and, later, by the Congress of Racial Equality (CORE). Less militant black protest organizations, such as the National Association for the Advancement of Colored People (NAACP) and King's Southern Christian Leadership Conference (SLC), however, remained steadfastly committed to integration.

Black power was widely interpreted as the advocacy of militant tactics against the white power structure, and the stridency of its proponents represented what seemed to many whites a kind of racism in reverse. Black power, however, comprised a twofold campaign: to assert control over the development of the black community with minimal interference by whites and to instill self-pride among blacks, who, in the view of black power advocates, had for generations internalized the white stereotyped image of black people.

The black community in the United States was seen essentially as a colonial people dominated by white institutions—hence the rejection of integrationist objectives. Carmichael and Charles V. Hamilton stated this position: "The fact is that integration, as traditionally articulated, would abolish the black community. The fact is that what must be abolished is not the black community, but the dependent colonial status that has been inflicted upon it" (1967:55). Blacks were urged to build genuinely black communities by taking control of school boards, businesses, and political bodies in their areas of the central cities. Only by closing ranks and asserting power within their own communities, it was felt, could blacks emerge as a viable political force, able to compete within a pluralistic system.

The development of a black communal identity and pride that had long been stifled by white oppression prompted an acceptance of African cultural roots that had previously been consciously shrouded. As Isaacs contends, the slogan for which the black radicalism of this period is likely to be remembered is "Black is beautiful." "In both its literal and its symbolic meanings," he notes, "it became the password to a measure of self-acceptance by black Americans that generations of earlier leaders and tribunes of the people had sought in vain to achieve" (1989:85–86).

The militant phase of black advocacy was short-lived. In the late 1960s, public concern with black civil rights in general was increasingly displaced by preoccupation with the Vietnam War. Most important, with the advent of the Nixon administration, it became clear that the federal government's response to black nationalist goals and tactics would be severely repressive (Button, 1978). The militant stage of the movement came to an end, according to Killian, not because the racial crisis had passed but because the white power structure effectively demonstrated the dangers of such measures by imprisoning, assassinating, or forcing into exile the most "dramatically defiant" black leaders (1975:155).

Though ephemeral and seemingly unsuccessful in achieving its long-range goals, the black power phase of the civil rights movement may, in retrospect, be judged more positively. First, black nationalism in the 1960s created a degree of self-pride among blacks that had not been achieved by previous nationalist movements such as Garvey's in the 1920s. Second, short-run profits in the form of increased government assistance followed the urban turbulence. Although it is commonly believed that violence is counterproductive to groups employing such techniques in political protest, historical evidence seems to reveal the contrary. When ruling elites make concessions to challenging groups—or conversely, when challenging movements are effectively stifled—violence or the threat of violence on the part of the successful group looms large in the picture (Gamson, 1975). Piven and Cloward (1971) maintain that protest movements like the one for black civil rights do not gain establishment concessions because of the successes of their organization and leadership but because of the social disturbances they create. Relief programs such as welfare and unemployment payments are set up not so much because of moral incentives as because of the need to contain the social disorder created by mass protests. This conclusion seems to be borne out by events of the 1960s. Following the riots of that decade, the federal government extended numerous programs to the cities designed to address those social conditions that were believed to be at the core of black discontent. As with the aid provided to the unemployed of the Great Depression, write Piven and Cloward, "The expansion of the welfare rolls was a political response to political disorder" (1971:198).

Of course, when concessions are granted under these circumstances, they may be only superficial, momentary, or symbolic; that is, their long-range effect may not be great. As Piven and Cloward (1971) explain, government relief to the poor and unemployed is usually withdrawn once social order has been restored. Such was clearly the case with aid to the urban ghettos. In his study of government responses to the riots of the 1960s, James W. Button (1978) shows that once the level of hostilities subsided and the areas were relatively pacified, the programs were either discontinued or emasculated by underfunding.

Assessing the Civil Rights Era By the early 1970s the "Second Reconstruction" had ended, and blacks began to take stock of their advances and failures of the previous two decades. Most important, the legal infrastructure of the system of segregation in the South had been dismantled. No longer could state and local ordinances prevent blacks from entering hotels, theaters, restaurants, trains, buses, waiting rooms, and all other public facilities previously reserved for whites. Unlike the first Reconstruction following the Civil War, white vigilante groups like the Ku Klux Klan were limited in their influence and were not supported either materially or morally by most whites. Most important, the federal government accepted the commitment to black civil rights and effectively countered state and local governments that sought to sustain segregation in some form. This was in sharp contrast to the abandonment of southern blacks by the federal government during the post–Civil War period.

A second major achievement of the civil rights movement of the post–World War II era was the expansion of black economic opportunities and the consequent emergence of a sizable black middle class. But it is also in the economic

Part II: Ethnicity in the United States

realm that the movement's most glaring failures were to be found. As we will see in the following section, blacks accumulated much greater competitive resources, such as white-collar jobs, middle-level income, and higher education, than at any previous time. These were not distributed equitably among the black population, however. Thus, although the black middle class expanded considerably, the conditions of a significant black underclass, submerged mainly in urban ghettos, were not seriously affected.

Finally, if the decline of biological racist ideas was clearly on the horizon as early as the 1930s, the postwar period of black-white relations, despite its volatility, seemed to further diminish the persuasiveness of this ideology. The notion of innate black inferiority was no longer given credence by the vast majority of white Americans.

Black Demography Today

Population Comprising about 13 percent of the population—36.5 million persons—African Americans were until recently the largest ethnic minority in the United States, a distinction now shared with Hispanic Americans, also roughly 13 percent. This figure includes all those who, in the 2000 U.S. Census, identified themselves as either entirely or partially black. The latter, however, made up only a tiny portion (less than 5 percent) of the total. The black population grew by over 20 percent from 1990, and much of this numerical increase stems from the entrance of newer ethnic elements in recent years—immigrants from the Caribbean and from African nations.

Black Ethnic Diversity It is becoming increasingly difficult to describe the American black population not only in broad class terms, but in ethnic terms as well. Though the overwhelming majority of American blacks trace their heritage to slavery, an increasing minority are voluntary immigrants or their descendants, most from the West Indies but some from the African continent as well. A significant element of the newest immigration to the United States has comprised people from Caribbean countries—such as Haiti, Jamaica, Trinidad and Tobago, and the Dominican Republic—many of whom are, by American standards, "black." It is estimated that by 2010, over 9 percent and by 2020 almost 10 percent of the U.S. black population will be foreign-born (U.S. Census Bureau 2000b). These black immigrants bring cultural, and in some cases linguistic, features that are both varied and profound. While they may evince racial similarities, culturally they have little in common with African Americans.

As a result of the newest immigration, the American black population is evolving into a somewhat broad, multicultural ethnic category similar to that of

Asian Americans and Hispanic Americans, though certainly to a lesser degree. As the American black population becomes more culturally diverse, one can no longer make assumptions about group characteristics simply on the basis of racial designation.

Geographical Patterns As explained earlier, in the early and middle decades of the twentieth century, blacks moved in large numbers from the rural South to the urban North. In 1910 almost 90 percent of blacks lived in the South; by 1960 only 60 percent did so, and little more than half remained in the South by 1990. In the past decade, however, this pattern has begun to reverse itself as more blacks remigrate to southern states (Frey, 1998; Stack, 1996). Many are choosing to live in urban areas of the South, however, rather than traditional rural areas (Frey, 2001).

Today the South contains the largest black population and has the highest proportion of blacks in its total population. This can be seen in Table 9.1. In eight states, all of them located in the South, blacks represent 20 percent or more of the total population.

Blacks in the United States, whether in the North or the South, are today a largely urban population. The largest black populations are in New York, Chicago, Detroit, Philadelphia, and Houston (U.S. Census Bureau, 2001b). In fact, New York and Chicago together accounted for 9 percent of the total black population. Table 9.2 shows the ten cities with the largest black populations.

Table 9.1 States with Largest Percentages of Black Population, 2000

STATE	TOTAL POPULATION	BLACK POPULATION	BLACK PERCENTAGE OF TOTAL POPULATION
Mississippi	2,845,658	1,041,708	37
Louisiana	4,468,976	1,468,317	33
South Carolina	4,012,012	1,200,901	30
Georgia	8,186,453	2,393,425	29
Maryland	5,296,486	1,525,036	29
Alabama	4,447,100	1,168,998	26
North Carolina	8,049,313	1,776,283	22
Virginia	7,078,515	1,441,207	20

SOURCE: U.S. Census Bureau, 2001b.

Table 9.2 Cities with the Largest Black Populations, 2000

CITY	TOTAL POPULATION	BLACK POPULATION	PERCENTAGE OF TOTAL POPULATION
New York	8,008,278	2,274,049	28.4
Los Angeles	3,694,820	444,635	12.0
Chicago	2,896,016	1,084,221	37.4
Houston	1,953,631	505,101	25.9
Philadelphia	1,517,550	672,162	44.3
Detroit	951,270	787,687	82.8
Baltimore	651,154	424,449	65.2
Memphis	650,100	402,367	61.9
Washington, D.C.	572,059	350,455	61.3
New Orleans	484,674	329,171	67.9

SOURCE: U.S. Census Bureau, 2001b.

AFRICAN AMERICANS IN THE CLASS SYSTEM

In the post-civil-rights era, attention has been focused on the speed and extent of African Americans' movement toward parity with whites in various dimensions of social, economic, and political life. Where do African Americans stand in the society's class and power hierarchies, and how has that position changed during the group's sociohistorical development, particularly the last few decades? Answers to these questions must be approached cautiously because there are various measures of stratification, and data can be marshaled to support a variety of often contradictory views. As with Italian Americans and Jewish Americans, analyzed in the two previous chapters, we will concentrate on the effects of ethnicity on the place of African Americans in the occupational and political worlds. As explained in Chapter 2, minority status does not necessarily indicate the position of a group at the time its members enter the society in large numbers but rather the patterns that develop to keep them in a disadvantaged position generation after generation. Such patterns are clearer for African Americans than for any other ethnic group in American society.

The Economic Status of African Americans

The fact of slavery weighs so heavily on the black experience that its lingering effects continue to make African Americans a special group in the American ethnic hierarchy. Obviously, blacks' collective status under slavery was at the bottom of the economic class system, even for those who were free in the North or South. Following the abolition of the slave system, this status changed little, as we have seen. Indeed, the economic standing of blacks was even more tenuous following slavery because they were denied the opportunity to compete on an equal basis with whites. Clearly, they did not fare well in resisting a system of blatant discrimination and repression that kept them in a subservient position.

Even after large-scale migration to northern cities starting in the second decade of the twentieth century, a system of direct and intentional discrimination effectively thwarted serious competition with the white population. For the most part, blacks were shut out of higher-skilled, more prestigious, and higher-paying jobs. Recall the split labor market theory of discrimination, discussed in Chapter 3 (Bonacich, 1972, 1976). Not until the 1950s and 1960s did blacks begin to move out of the lowest-level jobs in significant numbers. In effect, then, meaningful comparisons with other ethnic groups can be made only for the post–World War II period.

African-American economic progress in the last five decades is debatable; social scientists as well as policymakers, both black and white, have reached no consensus on either its extent or substance. There is, however, relatively broad agreement regarding several general patterns: (1) the 1960s was a decade of substantial improvement in the economic status of blacks collectively, as measured specifically by income, occupation, and education; (2) by comparison, in the 1970s and 1980s the rate of progress was curtailed except for a few specific subsets of the African-American population; (3) in the 1990s and early 2000s, the economic status of blacks rose but, despite black upgrading, the economic gap between blacks and whites collectively today remains wide; and (4) the economic progress of the modern period has created a substantial black middle class but has had little impact on blacks at the bottom of the economic scale.

Let's look at some specific variables of economic class, comparing the position of blacks and whites.

Income and Wealth Depending on how the data are interpreted, there are grounds for both positive and negative views regarding black income. The real income of black families increased substantially after World War II, especially during the 1960s (Farley and Hermalin, 1972; Levitan, Johnston, and Taggart, 1975). And the enormous expansion of the black middle class is undeniable. Today, a majority of the black population have middle-class incomes (Smith, 2001). This is the positive picture. But the negative view arises when we look at the gap between median black income and median white income. As shown in Table 9.3, this gap has changed only slightly since 1950. In 2000, median black family income was 64 percent of median white family income—the same as it was in 1970.

Table 9.3 Median Family Income of Blacks and Whites

YEAR	BLACK	WHITE	RATIO B-W
1950	$ 1,869[*]	$ 3,445	.54
1960	3,230[*]	5,835	.55
1970	6,279	10,236	.64
1980	12,674	21,904	.58
1990	21,423	36,915	.58
2000	34,204	53,256	.64

SOURCE: U.S. Census Bureau, 1981, 1995b, 2001g.
[*]Includes other nonwhites.

The income gap between blacks and whites narrowed slightly in the late 1960s and early 1970s, but in the 1980s and early 1990s blacks seemed to fall further behind whites. In the late 1990s, however, black median income grew faster than white median income and the gap began to close again (Holzer, 2001; Smith, 2001). Nonetheless, income inequality between blacks and whites remains evident throughout the class hierarchy. Thus, poor blacks are generally poorer than poor whites, and affluent blacks are less well-off than affluent whites.

This general picture should not obscure the fact, however, that comparisons between black and white income are subject to a number of factors, including region and urban location. In a few communities, black median income has actually surpassed that of whites (New York Times, 1994). More important, the black population is polarized between relatively stable middle- and working-class persons on the one hand and those in poverty on the other. The substantially greater percentage of blacks below the poverty line in comparison with whites accounts in large measure for the general differences between black and white income patterns. Among blacks with higher education, the racial gap has diminished; for black women specifically it has vanished (Farley, 1993). And, married-couple black families are approaching income parity with married-couple white families.

Another mixed indicator of the current economic status of African Americans is the poverty rate. The discrepancy between white and black poverty has declined, but in 2000, blacks were still more than twice as likely to be in poverty than whites (Table 9.4).

An important factor in accounting for these statistics, both positive and negative, is changes in the African-American family structure. Those families that have experienced income increases have been dual-breadwinner families.

Table 9.4 Persons Below the Poverty Level

	BLACK		WHITE	
YEAR	*Number (1,000s)*	*Percent*	*Number (1,000s)*	*Percent*
1959	9,927	55.1	28,336	18.1
1971	7,396	32.5	17,780	9.9
1980	8,050	31.0	17,214	9.0
1990	9,302	30.7	20,785	10.0
2000	8,360	23.6	21,922	9.8

SOURCE: U.S. Census Bureau, 1979, 2002.

Among such families there are only slight differences between blacks and whites. Similarly, much of the halt in the relative rise of black income from the late 1970s through the early 1990s is attributable to the increasing number of female-headed, single-parent families (Farley, 1993). In 1970, 28 percent of black families were headed by a woman; by 1989, that figure had risen to over 43 percent and by 1997 to 47 percent; in 2000, it was still over 45 percent (U.S. Census Bureau, 1998a; 2002). Single-parent families earn considerably less than husband-wife families. In fact, over half of all families headed by women are below the poverty line.

Wealth, or property, refers to the economic assets that families possess— homes, bank accounts, stocks and bonds, automobiles, and so on. Wealth is extremely important as a determinant of class position because it can be invested and thus earn income, or it can be accumulated and passed down to other family members. On this measure of socioeconomic status, differences between blacks and whites are significantly greater than differences in income. Blacks are less likely than whites to own any type of wealth, and the holdings of those blacks who do own wealth are far smaller than for white wealthowners. For example, less than 7 percent of black households own stocks compared to 23 percent of white households (Oliver and Shapiro, 1995). In fact, the typical white household has assets about ten times those of the typical black household. Most black income, then, even within the middle class, does not originate from property or self-employment.

Occupation If income is the chief determinant of life chances, work is, for most people, the medium through which income is acquired. Occupationally, African Americans as a group have improved their place over the past several decades, progressively moving into better-paying and higher-status jobs. In 1890 most black males were in agricultural jobs, and females were in domestic and personal service occupations. By 1930 the structure of black employment had changed substantially, with an enormous drop in agriculture and an increase in manufacturing. This was due, of course, mainly to the out-migration of rural southern blacks to industrial cities of the North. But the most significant occupational changes occurred between 1940 and 1970. Whereas in 1940 about one-third of all employed blacks were farmworkers, by 1960 only 8 percent were, and by 1970, 3 percent. By contrast, in 1940 only 6 percent of all employed blacks were in white-collar occupations. By 1970, 24 percent were white-collar workers, and by 2000, well over half were (Table 9.5).

Although it is clear that blacks have been increasingly integrated into the mainstream workforce, black-white disparities remain evident. Blacks are underrepresented in jobs at the top of the occupational hierarchy and overrepresented at the bottom. As shown in Table 9.5, a far higher percentage of whites are in managerial and professional occupations, while blacks predominate in service jobs such as cleaners, cooks, security guards, and hospital attendants. Indeed, the only subset of the black population that has attained occupational parity with similar whites is college-educated women (U.S. Census Bureau, 2001h). Moreover, the apparent occupational advancement of blacks need not be seen only as a product of the decline of discrimination in employment. Much of it is due to changes

Table 9.5 Occupational Distribution of Employed Persons Age 16 or Older, 2000

	OCCUPATION	BLACK	WHITE
Males	Managerial, professional	20.4%	32.5%
	Technical, sales	17.4	19.0
	Service	15.7	7.3
	Farming	2.1	3.5
	Skilled blue-collar	15.4	20.6
	Unskilled blue-collar	28.9	17.1
Females	Managerial, professional	27.7	37.3
	Technical, sales	35.4	38.8
	Service	24.7	14.0
	Farming	0.2	1.2
	Skilled blue-collar	2.2	2.2
	Unskilled blue-collar	9.7	6.4

SOURCE: U.S. Census Bureau, 2002.

in the occupational structure itself as the economy shifted from an agricultural to an industrial base and subsequently to a service base.

Further, the discrepancy between blacks and whites in rate of unemployment is wide (Table 9.6). Since the end of World War II, black unemployment has been, except for a few years, at least double that for whites. When one considers that government unemployment statistics do not account for those discouraged persons who have given up the search for a job, the unofficial number and percentage of unemployed is much higher.

Education As with income and occupation, there is room for interpreting educational data for African Americans both positively and negatively. Measured by years of schooling, blacks have increased their educational achievement both absolutely and relative to whites during the last five decades. Especially noticeable gains were achieved in the number of high school graduates. As shown in Table 9.7, among the adult population, 78.5 percent of blacks in 2000 had completed high school or more, compared with 85 percent of whites. Thus, blacks have almost reached parity with whites in median years of schooling completed. Also, today there is only a slight difference in the high school dropout rate for blacks and whites (U.S. Census Bureau, 2001h).

Substantial gains have also been evident in black college attendance and graduation. In 1960 there were 250,000 blacks attending college; by 1990 there were 1.2 million (*Chronicle of Higher Education*, 1992a; Wilson, 1989). As shown in Table 9.7, whereas in 1980, 8 percent of adult blacks had earned at least a bachelor's degree, by 2000, 16.5 percent had done so. Black college attendance, however,

Table 9.6 Unemployment Rates for Persons Age 16 or Older

YEAR	BLACKS[a]	WHITES	RATIO B-W[b]
1950	9.0%	4.9%	1.8
1960	10.2	4.9	2.1
1970	8.2	4.5	1.8
1980	14.3	6.3	2.3
1990	11.4	4.8	2.4
2000	7.6	3.5	2.2

SOURCES: National Urban League, 1981; U.S. Census Bureau, 1979, 1999, 2002.
[a]Includes other races from 1950 through 1980. Over 90 percent of this category is black.
[b]This means that in 1950, 1.8 black workers were unemployed for each unemployed white worker.

Table 9.7 Educational Attainment of Persons 25 Years Old and Over, White and Black (percent)

EDUCATIONAL LEVEL	1960 W	1960 B	1970 W	1970 B	1980 W	1980 B	1990 W	1990 B	2000 W	2000 B
High school graduate	43.2	20.1	54.5	31.4	68.8	51.2	79.1	66.2	84.9	78.5
Four or more years of college	8.1	3.1	11.3	4.4	17.1	8.4	22.0	11.3	26.1	16.5

SOURCE: U.S. Census Bureau, 1979, 1998a, 2002.

has vacillated in the past two decades, rising in some years and declining in others. A similar trend is apparent in the number of blacks earning graduate degrees.

In any case, the quantity of education is not necessarily commensurate with its quality. Despite the narrowing differential in years of schooling, black achievement continues to lag behind white achievement at all school levels—elementary, secondary, and college (Coleman et al., 1966; Commission on Minority Participation, 1988; Farley and Allen, 1987; Jencks and Phillips, 1998; U.S. Department of Education, 1992). Moreover, it seems to hold across class lines as well. Lower educational achievements are, of course, translated into fewer opportunities for occupational advancement.

Explanations vary for racial differentials in educational achievement. At one extreme, some argue that much of the difference between blacks and whites is due to genetic factors (Herrnstein and Murray, 1994; Jensen, 1969, 1973). This view remains highly controversial and is not accepted by the vast majority of social scientists and educators. At another extreme, many maintain that the schools themselves are inherently racist in both content and technique and thus guarantee the failure of blacks (Bowles and Gintis, 1976). A third position emphasizes social and environmental factors. The essential argument is that family background and individual and peer-group attitudes are more important determinants of academic performance than either genetic inheritance or the characteristics of the schools (Coleman et al., 1966; Jencks et al., 1972; McWhorter, 2000).

In sum, black educational advances in enrollment in the past four decades have not resulted in comparable academic attainment. Blacks have approximated equality with whites in years of schooling, but what that means beyond a mere time increase is difficult to evaluate.

Class or Race? It is clear that on all economic measures, blacks as a group—despite their advances—continue to lag behind whites. The questions that such statistics prompt are whether the persistent gap between blacks and whites is attributable to lingering racism, whether deeper structural forces impede the attainment of economic equality, or whether something unique about the black experience contributes to the group's continued plight. The third issue is covered later in this chapter when comparing African Americans with other American ethnic groups. The first two questions have stirred a lively debate among sociologists for several years. The issue is essentially this: Does racial discrimination continue to affect the economic status of blacks more than their socioeconomic background? That is, do blacks succeed or fail mainly on the basis of their skills and education, or does the racial factor still play a major role in their economic destiny?

The issue of race versus class is addressed most directly in the work of sociologist William Julius Wilson (1980, 1987, 1996). His basic position is that although discrimination based on race accounted for the denial of necessary economic resources to blacks in earlier times, the traditional patterns of black-white interaction have been fundamentally altered so that class factors are today the dominant influence on the life chances of black people: "In short, whereas the old barriers portrayed the pervasive features of racial oppression, the new barriers indicate an important and emerging form of class subordination" (1980:2). Wilson does not maintain that prejudice and discrimination no longer exist in American society, but that so far as access to power and privilege are concerned, they are no longer of paramount significance.

Wilson explains that to understand the current plight of the black populace, it is necessary to carefully subdivide it into at least three distinct classes: a black middle class consisting of white-collar and skilled blue-collar workers, which has expanded measurably in the past few decades; a working class of semiskilled operatives; and a lower class made up of unskilled laborers and service workers. At the very bottom of this last group is an underclass that includes those lower-class workers whose income falls below the poverty level, the long-term unemployed, discouraged workers who have dropped out of the labor market, and the more or less permanent welfare recipients. It is the disjunction between the black middle class and working class, on the one hand, and the lower class (particularly its most destitute segment), on the other, that is most important in explaining why economic statistics seem to indicate a halt to economic progress for the black populace.

During the post–World War II decades, writes Wilson, middle-class blacks were able to take advantage of the breakdown of racial barriers and as a result experienced substantial upward mobility. Those in the lower class, however, were unable to benefit from the expanded opportunities because of their lack of competitive skills and education. In fact, their position deteriorated. Thus, according to Wilson, a growing bifurcation has arisen within the black community between the "haves" and the "have-nots." This growing gap is a product of changes in the society's economic structure that have made lower-class, unskilled blacks an excess labor force that is becoming a permanent underclass. The labor market today calls for jobs requiring technical and educational skills, for which such persons cannot qualify. Moreover, industry has moved to areas on the urban fringe, so that those low-level jobs still available are outside the range of central-city blacks. The result is a semipermanent state of unemployment. The lack of job opportunities creates a vicious circle in which blacks are forced to remain in inner-city ghettos, attending inferior schools and thereby reinforcing their disadvantaged position in the labor market.

Wilson further explains that the problems of the black urban under-class—high rates of joblessness, school dropouts, crime, teenage motherhood, and so on—have been exacerbated by the exodus of middle- and working-class families from ghetto neighborhoods, removing vital role models and a support system of basic institutions. He points out that poverty in black urban areas in the past was not characterized by the degree of social isolation that is such a striking feature of contemporary inner-city ghettos. Residents of these areas become concentrated in disadvantaged areas of central cities and thus are increasingly cut off from the mainstream society. As Wilson puts it, "the communities of the underclass are plagued by massive joblessness, flagrant and open lawlessness, and low-achieving schools, and therefore tend to be avoided by outsiders. Consequently, the residents of these areas, whether women and children of welfare families or aggressive street criminals, have become increasingly socially isolated from mainstream patterns of behavior" (1987:58). Observing the situation of inner-city blacks twenty years after the Kerner Commission Report warned of increasing isolation, Gary Orfield also has noted that the scale of isolation is much greater than it was in the 1960s: "To a considerable extent the residents of city ghettos are now living in separate and deteriorating societies, with separate economies, diverging family structures and basic institutions, and even growing linguistic separation within the core ghettos" (1988:103). Sociologists Douglas Massey and Nancy Denton (1993), in their study of residential segregation, reached a similar conclusion.

Wilson believes that significant improvement in the economic condition of the black urban poor is unlikely within the framework of current economic institutions. It is important to reiterate Wilson's key point, that it is not racial discrimination but the present occupational structure that sustains the black underclass. Programs aimed at alleviating traditional forms of racial exclusion and encouraging black competition with whites, such as affirmative action, are not relevant to the key problem: the surplus of unskilled labor in a modern industrial economy that calls for an increasingly better-trained workforce. Catastrophically high unemployment rates among black youth and the burgeoning of low-income, female-headed families unable to support themselves independently threaten to perpetuate the black underclass, by some estimates as much as one-third of the black populace.

Civil rights leaders as early as the 1960s recognized that the problems of the black underclass were not the same as those of the black middle class when they warned that, despite its political successes, the movement had not basically affected the life chances of ghetto blacks. What is controversial in Wilson's thesis is that race is no longer seen as the key factor in the perpetuation of these conditions. Although some agree fundamentally with his position (Gershman, 1980) or have largely corroborated his findings (Featherman and Hauser, 1976), others have criticized him for too easily dismissing the continued influence of racial discrimination in the economic sphere or, through their research, have offered findings that do not entirely support his conclusion (Cancio et al., 1996; Clark, 1980; D'Amico and Maxwell, 1995; Landry, 1987; Pinkney, 1993; Willie, 1978).

Some, for example, have pointed out that unemployment rates are higher for blacks than for whites with the same educational attainment and sometimes even higher than for whites with less education (Hill, 1981; Kasarda, 1985). Critics view this as evidence of continued racial discrimination in employment. Moreover, criticism is made that Wilson and others who attribute the lower occupational status of blacks to lack of education and skills ignore the extent to which failure to attain those qualifications may itself be attributable to racial discrimination (Steinberg, 1995).

Another aspect of Wilson's thesis concerns the question of whether middle-class blacks have attained the equivalent status of middle-class whites, or whether their middle-class status is marginal, making them "second-class" Americans despite their personal achievements (Landry, 1987; Pattillo-McCoy, 1999). Between 1960 and 1970 the percentage of middle-class blacks doubled, from about one in eight to one in four black workers. This gain, notes Bart Landry (1987), exceeded their total increase during the previous fifty years. Landry explains, however, that the decade of the 1960s was unique to black upward mobility because it was a period that combined both newly enacted civil rights laws with a general economic prosperity, thereby giving not only legal incentive to blacks but the promise of real opportunities in middle-class occupations. This propitious combination had not occurred in the 1950s, nor was it to become apparent in the 1970s, thus creating a slowdown in occupational gains for blacks.

Landry further maintains that the black middle class remains significantly different from its white counterpart. Its path to middle-class status has been much more tortuous than for whites, its living standard is well below that of the white middle class, and its middle-class position remains far more tenuous than for whites. Moreover, Landry notes the continuation of subtle forms of discrimination against the new black middle class, including the application of timeworn stereotypes and unconscious expressions of racism. Oliver and Shapiro (1995) also show the insecure perch of the black middle class by comparison with its white counterpart. The differences in assets that middle-class blacks and whites hold are striking, and they are critical to understanding that "an accurate and realistic appraisal of the economic footing of the black middle class reveals its precariousness, marginality, and fragility" (1995:92–93).

Whether the black middle class will stabilize, expand, or even constrict will rest heavily on the state of the U.S. economy. The recessions of the 1970s, 1980s, and early 1990s took a greater toll on the living standards of the black middle class than on the white, demonstrating that economic vacillations continue to affect the black population more severely than the white.

African Americans and Societal Power

As noted in Chapter 2, whatever changes occur in the distribution of the society's resources—education, income, jobs—necessarily depend on access to important decision-making positions in the polity and economy, those two institutions in which power lies most heavily. To what extent have African Americans begun to move into such positions?

Political Power The gains made by African Americans in the realm of politics in the past four decades have been largely a product of increased electoral participation. The black vote has become an important swing factor in many elections as well as a significant electoral force in its own right in certain localities. The post–World War II economic advances of blacks were very much tied to the changed role of the federal government in securing and protecting black rights. This was even more clearly the case in the striking increase in black political participation during the 1960s and 1970s, especially in the South. Before that time, various measures kept blacks off the registration rolls in the southern states, and in the North, gerrymandering of black areas neutralized their vote.

Although an increasing number of blacks had begun to register and vote in the South during the 1940s, legal impediments as well as threats of violence still discouraged most from trying to exercise the ballot. As Gunnar Myrdal explained, "It is no test of the franchise that some Negroes are permitted to vote in a given community, for what is permitted to a few would never be permitted to the many" (1944:489). With the advent of the civil rights movement in the 1960s, however, enormous gains were achieved in black voter registration in the South, aided by the civil rights legislation of 1964 and 1965. Over two million blacks were newly registered in the southern states between 1964 and 1972.

In the North, the expansion of black electoral power was not a product of the civil rights movement but rather of the demographic changes that had occurred during the three decades from 1940 to 1970. It was the transformed racial composition of cities of the Northeast and Midwest that accounted for the rise in black electoral power more than any increase in black voting or concessions of power by whites. As the central cities became increasingly black and the suburbs increasingly white, blacks in the central cities found themselves with the numerical power to influence the outcome of local elections.

The upshot of increased political participation in the South and altered urban demographics in the North was a sharp rise in the number of black elected officials, particularly at the local and state government levels. Whereas in 1970, some 1,500 blacks had been elected to political office, that number had swelled to almost 9,000 by the late 1990s (U.S. Census Bureau, 2001h).

The substantial increase in the number of black elected officials in the past two decades is, on the surface, an encouraging indication of increasing black political power. This is particularly so when compared with the virtual absence of blacks in politics during the early part of the century. In the 1990s, the black presence in national politics was very apparent. Fifty-three African Americans had been elected to the U.S. House and one to the Senate (Zweigenhaft and Domhoff, 1998). African Americans had served as governor of Virginia, as national chairman of the Democratic Party, and as chairman of the Joint Chiefs of Staff, the nation's top military post. Perhaps most significant, six African Americans had served in President Clinton's cabinet: secretaries of commerce, agriculture, energy, labor, transportation, and veterans affairs. For the first time, a black had even been given serious consideration as a potential presidential candidate. At the local level, the election of African-American mayors of major U.S. cities had become commonplace. Between 1967 and 1993, African Americans had been elected mayor in eighty-seven cities with a population of 50,000 or more, two-thirds in cities where blacks were a minority of the population (Thernstrom and Thernstrom, 1997).

The presence of blacks on the American political scene was no less apparent at the beginning of the twenty-first century. Black politicians had been elected mayor in cities with relatively small black populations, and a few congressional districts with white majorities had elected black representatives. These developments indicated that race may be declining as the primary factor in the election of African Americans to political office. As well, several blacks were serving in cabinet and advisory positions of the George W. Bush administration, including, most significantly, secretary of state.

Closer analysis, however, indicates that there are mixed results of African-American efforts at political equity. To begin with, despite the increased number of African Americans elected to office, they do not represent the numerical proportion of the U.S. black population or that of any individual state.

Second, the offices held by blacks remain predominantly at the local level. Although the election of black mayors and other local officials is symbolically significant, their power to affect income and jobs in their communities is severely limited. Local economic conditions are shaped mainly by corporate officials, not local political leaders, and mainly on the basis of short-term profit, not community needs. Moreover, the increasing black control of central cities has come at a time when political and economic power has shifted to the suburban areas. Industries as well as white middle-class residents have continued their flight to the suburbs, leaving the core cities with a declining tax base with which to finance public needs for those who remain—increasingly, the elderly, the poor, and disproportionately, the black poor. This has left the cities largely dependent on public funding. As they have become more dependent on state and federal governments to finance basic services, central cities have become captives of the decisions of government agencies and legislatures that may have no political interests or constituencies in these communities. Ironically, then, at a time when the difficulties of cities are vast—industrial flight, deteriorating schools, poverty, crime—they lack the financial means to deal with them and must increasingly appeal to often unresponsive and unsympathetic sources. Black control, therefore, may be a hollow victory. Moreover, starting in the 1990s the concentration of blacks in central cities, which had translated into a political power base, was beginning to disperse as blacks themselves increasingly moved to the suburbs. The result has been a weakening of black voting power (McWhirter, 2000).

Third, even in local areas where blacks constitute a significant potential electorate, their traditionally lower electoral participation has further diluted their political power. This tendency is in some part a function of the lower collective class position of African Americans (Verba et al., 1995). Those lower in occupation, income, and education generally participate in the electoral process to a lesser degree than those of the middle class and above. This trend, however, may be changing as the socioeconomic condition of the black population rises. In 1998, for example, whereas 47 percent of whites voted in the congressional election, 42 percent of blacks did so. Moreover, this represented the first time that the percentage of blacks voting increased from the previous congressional election, whereas the percentage of whites voting decreased (U.S. Census Bureau, 2000a). There is no question that in certain states, and at the national level as well, the black vote has proved critical and will continue to be given serious consideration by political candidates.

Finally, and perhaps most important, an increase in numbers may mean little if the positions to which blacks are elected or appointed do not entail broad policymaking powers that provide opportunities for changing the distribution of economic and political resources. Despite their growing presence in political offices at the local and state levels of government and in the U.S. Congress, the black presence at the highest ranks of the executive branch of the federal government—precisely where the power to effect changes in the collective life chances of African Americans is most concentrated—has not kept pace. The appointment by President Clinton of six African Americans to cabinet positions, however, spelled significant change in this regard. Also, the serious attention given Colin Powell in 1995 as a possible presidential candidate, as well as his later appointment by President Bush as secretary of state, must be seen as very positive signs. It is of some note that in 1999, 97 percent of whites said they were willing to vote for a black person as president; in 1958, only 38 percent had expressed that view (Newport, Moore, and Saad, 1999).

Economic Power The assertion by political scientist Thomas R. Dye that "there are very few blacks in positions of power in America" (1995:175) is clearly borne out in the corporate world. Few African Americans serve on the board of directors or in a top-level executive post of the largest corporations, and these few are often token rather than bona fide decision-making appointments. A 1995 study found that only 0.6 percent of senior executive posts were held by African Americans (Glass Ceiling Commission, 1995). Even those who have achieved middle-management positions are often assigned to race-related areas or to special marketing (that is, minority) groups rather than in decision-making positions dealing with production, planning, or the supervision of large numbers of employees (Clark, 1980; Collins, 1997; Jones, 1986). This is an area of authority in American society, of course, that is not amenable to change through the electoral process; African Americans and other ethnic minorities, therefore, are dependent on the inclinations of the corporate elite to grant them a share of power. An encouraging pattern is emerging, however, in which blacks continue to increase their presence in executive and managerial positions, small though their numbers are (Hammonds, 1998; Williams, 1995).

African Americans also control only a tiny fraction of the nation's income-producing assets, and they seldom own their own businesses (Swinton, 1992; U.S. Census Bureau, 1979). The meager share of business wealth controlled by blacks is reflected in a few comparative figures. In 2000, the 100 largest primarily black-owned industrial and service enterprises had combined revenues of $10.9 billion. This figure was equivalent to the revenues of the 176th largest U.S. corporation alone. As for banks, insurance companies, and other financial institutions, the marginal place of black enterprise is even more striking. For example, in 2000 the combined assets of the 25 largest primarily black-owned banks—$4.1 billion—were less than one-third of the yearly *profits* of the single largest American banking corporation (McKinney, 2001; Scott, 2001; *Fortune*, 2001).

PREJUDICE AND DISCRIMINATION

The modern period of black-white relations has seen profound changes in patterns of racial prejudice and discrimination. Whereas some form of biological racism was a solidly entrenched ideology in all parts of the nation well into the 1930s, by the 1960s a majority of Americans had rejected the notion of innate black inferiority. This shift in attitude accompanied the demolition of the legitimated system of racial segregation in which overt and direct forms of discrimination, both individual and institutional, were approved and defended by the state. But as these traditional forms of discrimination were abolished, they were replaced by covert and indirect forms—ones less amenable to alleviation.

Changes in Patterns of Discrimination

Before the civil rights movement of the 1960s, discrimination was de jure, that is, legitimized and enforced by local, state, and even federal governments. Today, however, prescribed discrimination has given way to de facto forms, which are the result not of direct antiblack measures but of spatial arrangements, demographic trends, and housing patterns (Pettigrew, 1979). Recall the notion of indirect institutional discrimination, discussed in Chapter 3.

The most evident, persistent, and consequential forms of racial discrimination in American society are found in the area of housing. Moreover, the patterns of discrimination here are direct and deliberate to a far greater extent than in the areas of work and education. The effects of lingering racial discrimination in employment, for example, are, as we have seen, debatable. Employers may continue to discriminate against blacks, but the extent and impact of those actions as well as their intent are difficult to precisely ascertain. In education, too, although many urban school systems are as segregated today as they were before the 1954 Supreme Court decision barring segregated schools, these patterns are maintained more by residential segregation and the past drawing of racially distinct school districts than by the conscious design of whites. Discrimination in housing, however, is much clearer, and its intent is not easily disguised. It is well recognized that less progress in breaking down racial barriers has been made in this sphere of social life than in any other (Levitan, Johnston, and Taggart, 1975; Massey and Denton, 1993; Pettigrew, 1979).

Because the patterns are so clear-cut, focusing on housing is essential in looking at current forms of discrimination leveled at blacks. However, there are additional important reasons for concentrating more intently on residential discrimination. Its ripple effects touch every other area of social life (Farley, Bianchi, and Colasanto, 1979; Holzer, 2001; Massey and Denton, 1993; Sigelman et al., 1996; Steger, 1973). Where people live determines in largest part the schools they attend (and thereby the quality of their education); the jobs they have access to (and thus their occupational destiny); the benefits they receive from public institutions like hospitals, recreational facilities, and transportation systems (and, as a result, their health and well-being); and the commercial establishments they have access to (and thus their style of life).

Moreover, the society's residential arrangements determine the essential character of interethnic relations. With continued segregation in housing, whites and blacks are prevented from interacting at personal levels, thus reinforcing racial attitudes and perpetuating structural pluralism. Though it is difficult to select one element of the system of American ethnic relations and call it the "key" to the maintenance of conflict and misunderstanding, residential segregation is unquestionably among the most fundamental. It would hardly be an exaggeration to proclaim that the present pattern of housing in the United States is the chief obstacle to progress in all other aspects of race and ethnic relations. Pettigrew has put the matter succinctly: "The residential segregation of blacks and whites has emerged as a functional equivalent for the explicit state segregation laws of the past in that it effectively acts to limit the life chances and choices of black people generally" (1979:124).

The Formation of Black Ghettos Wide-scale residential segregation has been a continuous fact of urban life in the United States since blacks began to move to cities in large numbers in the early twentieth century. As with European immigrants before them, they were concentrated in urban ghettos; but unlike those of the European groups, black ghettos did not progressively break up with each succeeding generation.

Indeed, what is most significant regarding the formation of urban black ghettos is that they have remained so intractably persistent. Patterns of residential dispersion exhibited by white ethnic groups as they have experienced upward mobility (the Italians and Jews described in previous chapters, for example) simply do not hold for blacks (Hershberg et al., 1979; Kantrowitz, 1973, 1979; Massey and Denton, 1987, 1993; Roof, 1978, 1979; Spear, 1967; Taeuber, 1975). The black ghetto today is an integral facet of almost every large or medium-sized city in the United States. In their study of urban residential segregation, Karl and Alma Taeuber concluded that "a high degree of racial residential segregation is universal in American cities. Whether a city is a metropolitan center or a suburb; whether it is in the North or South; whether the Negro population is large or small—in every case, white and Negro households are highly segregated from each other" (1965:2). Their study showed that not only had segregation in housing not lessened since the early 1900s, but it had actually intensified in the 1940s and 1950s.

These trends changed little throughout the 1960s and 1970s despite greater government activity in promoting desegregation, more liberal social attitudes, and more middle-class blacks with sufficient incomes to afford better housing (Farley, 1985; Levitan, Johnston, and Taggart, 1975; Massey and Denton, 1987, 1993; Orfield, 1985; Roof, 1979; Van Valey, Roof, and Wilcox, 1977; White, 1987). Table 9.8 shows the levels of black-white residential segregation in the largest metropolitan areas. Although these levels declined somewhat in the

1990s, blacks remain "a uniquely segregated group" (Farley and Frey, 1992:28). The simple fact is that most African Americans, regardless of their social class or the geographic area in which they live, continue to reside in predominantly black neighborhoods.

Migration to the suburbs, a movement followed heavily by whites since the end of World War II, has been replicated to some degree by blacks in recent years. The share of blacks living in suburbs rose from 21 percent in 1980 to 27 percent in 1990 and 39 percent in 2000. This is still well below the rate of suburban residence for whites, however, which is more than 70 percent (DeVita, 1996; Mumford Center, 2001). Moreover, rising socioeconomic status has not appreciably affected the overall level of black segregation (Massey and Denton, 1987, 1993). Hence, when blacks do move to suburban areas, they again find themselves for the most part in racially segregated communities (Dent, 1992; DeVita, 1996; Logan, 1988; Massey and Denton, 1987, 1993; Pattillo-McCoy, 1999).

Might these patterns be more a result of class than of race? That is, might the persistence of black ghettos merely reflect the lower economic status of blacks generally? If this were so, we could expect to find blacks and whites of similar social class residing in racially mixed neighborhoods. Studies have shown, however, that even where average rents paid by blacks and whites are equivalent, little racial integration is evident (Farley, 1985; Farley and Allen, 1987; Massey and Denton, 1993; Orfield, 1988; Taeuber, 1975). Blacks of every economic and educational rank are highly segregated from whites of the same economic and educational levels.

Class factors, however, do play a role in determining patterns of residential segregation. For one, middle-class blacks are more likely to live in integrated neighborhoods (Alba et al., 2000). Even in the most highly segregated urban areas, middle-class blacks are less segregated from whites than are poorer blacks. Also, the resistance of whites to black incursions is reduced when blacks are perceived as higher in social class. As Alba et al. explain, "Whites are reluctant to accept blacks as neighbors, but that reluctance is eased when the socio-economic characteristics of African Americans are superior to those of whites themselves" (Alba et al., 2000:557). That is, there is evidence to support the notion that whites avoid black neighbors for reasons related more to class than to race (Alba et al., 2000; Harris, 1999). However, stereotyping still plays a strong role in the makeup of neighborhoods and limits black choices. As Alba et al. point out, because whites are more willing to accept black neighbors who are perceived as superior to themselves in social class, blacks live in less affluent communities than do whites of the same socioeconomic status. That is, whites seek out neighborhoods that are relatively safe, clean, and have good public facilities (especially schools). These conditions are more prevalent in middle-class neighborhoods. Because blacks are commonly stereotyped as poor, however, this leads to a form of statistical discrimination

Table 9.8 Black-White Residential Segregation for Metropolitan Areas with the Largest Black Populations

REGION/ METROPOLITAN AREA	WHITE-BLACK RESIDENTIAL SEGREGATION SCORE*	
	1990	2000
Northeast		
New York	77.8	81.8
Philadelphia	81.5	72.3
Newark	82.8	80.4
Boston	70.4	65.7
Midwest		
Chicago	87.1	80.8
Detroit	88.8	84.7
St. Louis	81.1	74.3
Cleveland	86.1	77.3
Kansas City	75.5	69.1
South		
Washington, D.C.	67.5	63.1
Atlanta	72.6	65.6
Baltimore	75.4	67.9
Houston	69.0	67.5
New Orleans	73.8	69.3
Dallas	66.2	59.4

(discussed in Chapter 3). For whites who have had no experience in an integrated environment, this stereotype serves as the guiding mechanism for the choice of their neighborhood's racial mix.

Might a part of the continuation of residential segregation be the preference of blacks themselves to remain in racially homogeneous areas? There is evidence that racial segregation in housing is today in some measure voluntary (Thernstrom and Thernstrom, 1997). But this observation must be carefully qualified. For more than two decades, studies have shown that blacks favor residential integration and prefer mixed neighborhoods by a wide majority (Farley, 1985; Farley, Bianchi, and Colasanto, 1979; Massey, 2001; Sigelman et al., 1996).

REGION/ METROPOLITAN AREA	WHITE-BLACK RESIDENTIAL SEGREGATION SCORE*	
	1990	2000
South *(continued)*		
Memphis	75.6	68.7
Norfolk	56.7	46.2
Miami	74.6	73.6
Richmond	64.1	57.0
Birmingham	79.3	72.9
Charlotte	64.6	55.2
West		
Los Angeles	71.4	67.5
Oakland	68.7	62.8

SOURCES: Farley and Frey, 1992; Mumford Center, 2001.

*The segregation score or index of dissimilarity is a measure of the extent of residential segregation between two groups. It indicates the percentage of one group that would have to move in order to achieve total integration, that is, a population mix that corresponds to each group's proportion of the population of a designated area. For example, if a city were made up of 80 percent whites and 20 percent blacks, the two groups would be represented in each neighborhood in that same proportion. The index can range from 100, indicating total segregation, to 0, indicating total integration. Thus, the higher the index, the greater the degree of segregation. See Taeuber and Taeuber (1965).

Surveys show that whites, on the other hand, are willing to accept residential integration only if the number of blacks is minimal (Schuman et al., 1997).[6]

Studies indicate that Latinos and Asians, groups entering the society much later, have already begun to disperse throughout metropolitan areas to a far higher degree than African Americans (Alba and Logan, 1993; Farley, 1985;

---

[6]Farley et al. (1994), however, have found that black preferences for racially mixed neighborhoods have weakened in recent years. Whether this trend is the result of a genuine preference to live in black communities, apprehensions about white hostility, or the view that integration yields few benefits is not clear.

Farley and Allen, 1987; Farley and Frey, 1992; Langberg and Farley, 1985; Massey, 2001; Massey and Denton, 1987, 1989b, 1993; White, 1987). Segregation scores, or indexes of dissimilarity (see Table 9.8), indicate that residential segregation between Latinos and non-Hispanic whites as well as between Asians and whites is much lower than segregation between blacks and whites.

It is apparent, then, that the intensity and intractability of black-white residential segregation are unequaled in American society. Housing patterns between blacks and whites are, as Pettigrew (1979) has noted, qualitatively different from those between other groups. Massey and Denton have described this uniquely disadvantaged black urban environment as "hypersegregation," that is, segregation across several dimensions. "Black Americans in these metropolitan areas live within large, contiguous settlements of densely inhabited neighborhoods that are packed tightly around the urban core. In plain terms, they live in ghettos" (1993:77). Roughly one-quarter of the African-American population, in ten large metropolitan areas, lives in such an urban area. No other racial-ethnic group, Massey and Denton explain, experiences such multidimensional segregation. This environment shuts out most contact with whites except through participation in the labor force. Thus, African Americans living within the largest urban black communities remain, in comparison with other ethnic groups, highly segregated and spatially isolated (Massey, 2001). Moreover, these extraordinarily high levels of residential segregation almost assure their continuation into the foreseeable future.

The Maintenance of Residential Segregation   Essentially, then, there exist in the United States two housing markets: one white and one black. Despite the enactment of many state and federal antidiscrimination statutes since the 1940s, residential segregation is maintained by a set of discriminatory practices, both individual and institutional. Moreover, some are direct and intentional.

Past government policies created much of the framework for the current patterns of residential segregation. Beginning in the 1930s, the federal government encouraged home ownership among middle- and working-class people by establishing the Federal Housing Administration (FHA) as well as other housing-related agencies. The idea was to furnish cheap financing by providing government-backed mortgage insurance. From the outset, the FHA officially discouraged integrated residential areas by refusing to guarantee loans for homes that were not in racially homogeneous areas (Abrams, 1966; Feagin and Feagin, 1978; Massey and Denton, 1993). This policy was not changed until 1962, when nondiscriminatory pledges were required from loan applicants.

The real estate system has been another key element in the perpetuation of segregated housing. Until the 1950s, real estate boards, almost entirely white in membership, followed a strict code that restrained the renting or selling of property in white areas to blacks. Using various techniques, white real estate agents would steer blacks to black areas and whites to white areas in order to enforce this code. Banks and other lending institutions cooperated in the system by "redlining"; that is, designating certain areas within which real estate loans would not be made. Another device used by real estate brokers was "block-busting." By spreading word through a white neighborhood of an impending black influx, agents would frighten whites into selling their homes cheaply. These homes were subsequently sold to blacks at inflated prices. In the process, all-white areas were transformed quickly into all-black areas.

Where such tactics were unsuccessful, zoning regulations were established, specifying certain types of housing that could be erected in particular neighborhoods; these were designed to exclude low-income, mainly black, units. Restrictive covenants, agreements made by homeowners not to sell to members of particular groups, were also used to bar blacks and other minorities from white residential areas. Until the Civil Rights Act of 1968 prohibited discrimination in housing, such covenants had been widely applied and were even supported by law until 1948. When all else failed, violence and intimidation were frequently used to deter blacks from buying or renting in all-white neighborhoods.

Although these practices have been curtailed by antidiscriminatory legislation, they have been replaced by more subtle measures that sustain segregation. These include resistance to public housing desegregation, zoning and annexation policies, redlining by lending institutions, and various informal practices of the real estate industry such as steering people toward predominantly white or black neighborhoods (Massey and Denton, 1993, 2001; Taeuber, 1975; Yinger, 1996). Despite antidiscriminatory statutes, most of these techniques are difficult to detect and cannot be effectively policed. In addition to these institutional forms of residential discrimination, individual homeowners and landlords may discriminate in selling or renting their property with even greater impunity from legal restraints.

Other racially related problems of American cities, including school segregation and chronic unemployment, are in large part directly traceable to the perpetuation of black ghettos. School segregation, for example, is sustained chiefly by the maintenance of segregated neighborhoods (Farley, 1984; Santiago, 1991). As a result, legal efforts to integrate schools have been nullified in countless cases (Abrams, 1966; Cataldo, Giles, and Gatlin, 1975; Farley and Taeuber, 1974; Kantrowitz, 1979; Taeuber, 1979). As previously noted, spatial segregation creates numerous types of "side effect discrimination" (Feagin and Feagin, 1978). Residential discrimination, which is most often overt, direct, and intentional, produces, naturally and often unintentionally, indirect discriminatory effects in other areas. In sum, although racial discrimination in other institutional realms of American society has broken down to a great extent, it remains tenaciously persistent in residential patterns.[7]

Following the urban riots of 1967, the Kerner Commission issued its report on civil disorders, warning that the United States was "moving toward two societies, one black, one white—separate and unequal" (National Advisory Commission on Civil Disorders, 1968:1). In no area of social life does this prognosis continue to be verified better than in housing.

Pertinent to this issue is Joel Kovel's distinction between "dominative" and "aversive" racism. In the former, actions are taken to oppress racial minorities and keep them subservient. In the latter, inaction is characteristic. Racial minorities are simply ignored and avoided when possible; relations at best become "polite, correct and cold in whatever dealings are necessary between the races" (1970:54). Continued residential segregation provides the backdrop for such aversive discrimination in various areas of social life. While obvious and blatant forms of discrimination are no longer prevalent, more benign black-white relations are in large measure founded on intergroup avoidance (Pettigrew, 1980).

Changes in Racial Attitudes

As the nature of racial discrimination has changed in the past several decades, so have the attitudes of whites toward blacks. Although there is much room for optimism regarding these attitudinal shifts, there are also indications that racist beliefs are still firmly rooted in American thought, albeit in modified form. Moreover, there is ample evidence that blacks and whites do not see the same reality regarding the current status of blacks in the society or the nature and causes of black problems.

Attitudinal changes during the past forty years have been measured by extensive and frequent surveys among blacks and whites. At the outset, it should be noted that survey data regarding attitudes must be accepted cautiously as representative of people's real beliefs and feelings. At best they comprise rough descriptions, not precise readings. Because they are designed to draw collective or general pictures, they will not necessarily reflect the views of particular individuals within the groups. Obviously, blacks and whites as collectivities are internally variegated, and views among individuals will differ on the basis of age, gender, education, and numerous other social variables. Moreover, as explained in Chapter 3, surveys, particularly those dealing with socially sensitive issues like race and ethnic relations, may reflect mainly what people believe interviewers want to hear or what they feel to be the most socially acceptable response, not

---

[7]Marston and Van Valey (1979) argue that there are some positive effects of residential segregation for blacks, namely, that it creates and sustains an institutional structure that caters to blacks, induces ethnic group consciousness, and in some cities, produces local political clout.

necessarily their true feelings. Finally, attitudes may appear inconsistent from item to item or from survey to survey. For example, public opinion studies have shown a strong acceptance of integrated schools but an equally strong rejection of school busing (Gilbert, 1988; Greeley and Sheatsley, 1974). Interpretations of such inconsistencies are often problematic. In this case, the antibusing sentiment may be an indication of covert racism; or it may indicate other, nonracist attitudes such as commitment to the neighborhood school, fear of the dangers of busing, or resentment over the inconvenience caused by busing (Greeley and Sheatsley, 1974). It may also reflect a particular view of what government should or should not do to accelerate racial equality (Sniderman and Piazza, 1993). With these caveats in mind, let's look at some specific patterns.

Stereotypes and Social Distance First, it is clear that negative stereotypes of blacks have diminished and that less support is given to blatant forms of discrimination (Brink and Harris, 1963, 1967; Campbell, 1971; Greeley and Sheatsley, 1974; Pettigrew, 1975, 1981; Schwartz, 1967; Schuman et al., 1997; Sheatsley, 1966; Williams, 1977). As shown in Table 9.9, a relatively steady decline in white notions of black inferiority has been measured since 1963. In the fifteen years between 1963 and 1978, the percentage of whites who believed blacks to be innately less intelligent fell from 39 to 25. By the mid-1990s, only 12 percent

Table 9.9   White Americans' Stereotypes of Blacks

	PERCENT OF WHITES WHO AGREED				
STEREOTYPE	1963	1967	1971	1978	1995
Blacks tend to have less ambition than whites.	66%	70%	52%	49%	29%
Blacks want to live off handouts.	41	52	39	36	22
Blacks are more violent than whites.	—	42	36	34	38[a]
Blacks breed crime.	35	32	27	29	—
Blacks have less native intelligence than whites.	39	46	37	25	12
Blacks care less for the family than whites.	31	34	26	18	40[b]

Sources: Anti-Defamation League, 1993; National Conference, 1995; National Conference of Christians and Jews, 1978.
[a]1992
[b]Statement was "Blacks have less family unity."

maintained that view (National Conference, 1995; Sniderman and Piazza, 1993). It is true nonetheless that traditional negative stereotypes of blacks—poor, violent, irresponsible—are still potent, if diminished (Waters, 1999).

White attitudes toward racial integration have also changed in the past three decades (Table 9.10). Most whites now seem committed (at least verbally) to integration in most social areas entailing secondary relations (Greeley and Sheatsley, 1974; Schuman et al., 1997; Steeh and Schuman, 1992; Taylor, Sheatsley, and Greeley, 1978). Even the more negative policies and attitudes of the Reagan era of the 1980s did not seem to reverse the steady liberalization of racial attitudes among whites. This rise in racial liberalism may be attributed to higher levels of education, an increasing liberal perspective on social issues generally, and increased contact between blacks and whites.

As for social distance, only a minority of whites nationwide object to contacts with blacks on the job or in public facilities. But the percentage rises dramati-

Table 9.10 White Americans' Attitudes on Integration

| ITEM | RESPONSE | PERCENT RESPONDING | | | | | |
		1963	1970	1977	1982	1989	1996
"Do you think white students and black students should go to the same schools or to separate schools?"	"Same schools."	63%	74%	86%	90%	93%	96%*
"How strongly would you object if a member of your family wanted to bring a black friend home to dinner?"	"Not at all."	50	63	71	78	77	—
"White people have a right to keep blacks out of their neighborhoods if they want to, and blacks should respect that right."	"Disagree slightly" or "disagree strongly."	45	51	56	71	78	86
"Do you think there should be laws against marriages between blacks and whites?"	"No."	37	48	72	66	77	87

SOURCES: *Public Opinion*, 1987; Schuman et al., 1997; Smith, 1980; Wood, 1990.
*1995

cally when intimate contacts are involved. Few blacks today, regardless of social class, are likely to socialize with their white fellow employees outside the formal work setting (Sigelman et al., 1996). Attitudes differ on specific issues, however. An overwhelming majority of whites no longer object to interracial dating and marriage; but most, as we have seen, are also opposed to living in areas with a high percentage of blacks (Farley et al., 1978; Gallup, 1995; Schuman et al., 1997).

Attitudes and Actions Although white attitudes concerning discrimination and integration are apparently increasingly liberal, ambivalence arises when translation of these attitudes into concrete actions is suggested. This is obvious in various ambits of social life, including politics, jobs, housing, and education. Whites agree that blacks should have equal rights in all these areas but are less committed to approving measures necessary to guarantee them (Pettigrew, 1979; Schuman et al., 1997). For example, white attitudes favoring the principle of school integration have continually risen over the past 35 years, but support for school busing to achieve integration has not followed suit (Schuman et al., 1997).[8]

A similar inconsistency between attitude and action lies in the realm of residence. Although most whites favor racially integrated neighborhoods, white preferences for the degree of integration are quite different from those of blacks. Whereas blacks seem to prefer a relatively even mix, whites prefer a mix in which whites are clearly in the majority (Farley, Bianchi, and Colasanto, 1979; Schuman et al., 1997; Sigelman, 1996). In a sense, there is a circular movement in which prejudice feeds on segregation, and vice versa. As one observer has noted, "The most well-documented and well-replicated finding in research on the effect of interracial housing and neighborhood integration is that white prejudice declines with interracial contact, particularly interracial contact that occurs in a situation where whites and blacks are of equal social status and are aware that they share similar values" (D. Taylor, 1979:35). Increased contact, in other words, seems to moderate racial prejudice, especially if the groups are of the same social class. As mentioned earlier, a key reason for whites' fears of black penetration of their neighborhoods is the lower social class of blacks, not necessarily negative racial attitudes alone (Farley, Bianchi, and Colasanto, 1979). Whites are concerned that their neighbors will be poorer and will have lifestyles very much at odds

---

[8]The issue of school busing is complicated. In 1994, 86 percent of whites and 94 percent of blacks agreed with the Supreme Court's 1954 decision barring school segregation. But most blacks as well as whites preferred that students attend their local schools, even if it meant most of the students would be of the same race. Only 9 percent of whites and 33 percent of blacks preferred to assign students to schools outside their local communities as a way to achieve more integration (McAneny and Saad, 1994).

with their own. Although blacks moving into white neighborhoods are generally similar to their white neighbors in socioeconomic status, fixed negative images of blacks (as "poor," "lazy," "squalid") often prevent whites from perceiving the class distinctions among them—hence the continued resistance to integrated residential areas. Without social contact, then, prejudicial attitudes and images persist, in turn leading to intransigence on matters of neighborhood desegregation.

In his classic work dealing with black-white relations in the United States, Swedish social scientist Gunnar Myrdal (1944) described the disjuncture between white attitudes and actions vis-à-vis blacks as "an American dilemma." The dilemma, according to Myrdal, is the conflict between, on the one hand, adherence to the "American Creed," entailing a belief in equality and Christian precepts, and on the other, group prejudice and discrimination based on material and status interests. The dilemma remains evident today in many ways.

Views of the Socioeconomic Gap   Although whites generally recognize the disadvantaged status of a large proportion of the black population, there are striking differences in black and white explanations for that depressed condition. Most whites continue to view the problems as the result of blacks' own shortcomings, not social conditions or institutional forms of discrimination (Lipset, 1987; Schuman et al., 1997; Sniderman and Piazza, 1993). The majority, however, no longer attribute these deficiencies to biological inferiority (Schuman, 1982; Schwartz, 1967; Sniderman and Piazza, 1993). Most often, whites see black disadvantages as a product of "lack of ambition," "laziness," "irresponsibility," or "failure to take advantage of opportunities." The prevalent attitude is, "If we did it, so can they." Indeed, the image of blacks as lacking ambition and being socially irresponsible is today the most widespread and common negative stereotype (Sniderman and Piazza, 1993). Blacks, by contrast, see their plight as deriving primarily from discrimination or a lack of the same opportunities as whites. For instance, a 1997 national poll found that 80 percent of whites but only 46 percent of blacks believed that blacks had as good a chance as whites to "get any kind of job" (Gallup, 1997).[9]

Structural forces that strongly affect people's place in the stratification system (such as the society's changed occupational order or its basically altered demographic patterns) are not easily perceived. Thus, social success or failure is apt to be interpreted as primarily the result of individual effort. Moreover, the belief

---

[9]There is some ambivalence on the part of blacks, however. Although they are more likely to recognize structural factors as important in accounting for lower economic and social achievement, blacks no less than whites subscribe to the view that blacks fail to try hard (Sniderman and Piazza, 1993).

that each member of society is personally responsible for his or her social lot is a basic component of American ideology. The society's opportunity structure is pictured as open, providing equal chances for all to achieve material success or political power regardless of their class or ethnic origin. As a result, it seems only natural that whites would be more likely to attribute the lower social and economic achievement of blacks to individual and group characteristics than to the workings of a class and ethnic system that automatically favors the wellborn, especially those who are also white, and disadvantages the poor, especially those who are black. In addition, the more blatant forms of discrimination, which long held blacks in subservience—slavery, the Jim Crow system—are easier to perceive and understand by whites than are the indirect and largely covert forms of institutional discrimination that continue to render debilitating effects. Moreover, even at the individual level, discrimination in less easily measured forms—poor service, racial slurs, fearful or defensive behavior, lack of respect—is a commonplace experience for most blacks (Morin and Cottman, 2001) and renders a further injurious impact.

Problems and Solutions Whites and blacks do not share a common view of the problems facing African Americans. Whites today tend to view blacks as better off economically and socially than blacks view themselves; they also see the extent and pace of racial change in the United States as far greater and more rapid than do blacks. Most whites today believe African Americans enjoy equal opportunities in education, jobs, housing, and access to credit. African Americans have diametrically opposite views (Gallup, 1997; Gallup Organization, 2001; Morin, 1995b; National Conference, 1995).

Surveys indicate that whereas whites see problems as more of a "mopping up," with basic changes having been made, blacks perceive deeply rooted discrimination continuing to affect their chances in work and education. In a 1995 survey, middle-class whites and blacks were asked if they agreed with the statement, "Past and present discrimination is the major reason for the economic and social ills blacks face." Eighty-four percent of blacks agreed, but only 30 percent of whites did so (*Washington Post*, 1995; see also Schuman et al., 1997). Whites define racial progress mostly in political terms. If laws have been changed that previously denied blacks access to critical institutions, then it is assumed that an equal opportunity structure has been created. In this view, if blacks succeed or fail it is their own doing. Blacks, on the other hand, see not simply a few remnants of the past that need attention, but a system that continues to block their advancement. Moreover, even middle-class blacks, those who have seemingly "made it" in an economic sense, continue to experience what they perceive as indignities based on race (Cose, 1993) and express frustration regarding future opportunities (Hochschild, 1995).

Also, blacks see a justice system that treats blacks differently from whites (Gallup, 1997; National Conference, 1995; National Urban League, 2001). A strikingly contrasting view of the O. J. Simpson case provided stark evidence of this. While most whites saw Simpson as guilty and as having evaded justice, most blacks agreed with the "not guilty" verdict and were inclined to believe that the jury was fair and impartial (Whitaker, 1995). Many blacks saw the decision as a kind of symbolic retribution for what they viewed as a criminal justice system that more commonly victimizes blacks. The gap between whites and blacks in their view of how blacks are dealt with by the police in their communities remains particularly wide (Gallup Organization, 2001).

Black views of black progress are dramatically different from those held by whites. A majority of whites believe, for example, that blacks have an equal chance to succeed and that blacks on average hold jobs of equal or better quality to those of whites. Whites are also more likely to believe that racial injustices have been ameliorated and that discrimination therefore does not affect the pace and extent of black progress (Gallup, 1997; Hochschild, 1995; Gallup Organization, 2001; Morin, 1995). In his interviews of middle-class Americans, sociologist Alan Wolfe (1998) discovered that whites did not naively believe that discrimination against blacks no longer existed; but they maintained a much higher degree of optimism regarding the rate of change than did blacks, whose anger stemmed from a perception of change that was not occurring quickly enough. Wolfe also concluded that differences were based largely on conflicting definitions of fairness, equality, and other terms of the racial debate: "When discussing issues of racial equality, most Americans, black or white, rely on the same words: their disagreements are about what the words mean" (214). Nonetheless, a 1999 poll indicated that despite the perceived need to resolve serious problems, on many social issues, African Americans are even more positive about the future than white Americans (*Newsweek*, 1999).

AFRICAN AMERICANS: ASSIMILATION OR PLURALISM?

Chapter 4 pointed out that assimilation and pluralism are processes and outcomes of interethnic relations that are not necessarily mutually exclusive for any particular group in a multiethnic society. Rather, a group may display aspects of both simultaneously. In the case of African Americans, certain patterns of assimilation seem apparent even though pluralism, specifically the colonial form, has generally been more characteristic of their historical and even contemporary experience. African-American cultural assimilation has been strong, and secondary structural assimilation has gained momentum in the past several decades; primary structural assimilation, however, remains relatively weak.

Cultural Assimilation

Because the African cultural heritage was largely destroyed under slavery, blacks had little choice about the adoption of the dominant group's major cultural traits, especially religion and language.[10] Some hold that in light of this historical fact, cultural assimilation has been as fully accomplished for blacks as for any American ethnic group (Gordon, 1964; Pinkney, 2000). However, although it is true that blacks have adopted the dominant group's major cultural forms, it is equally evident that a black cultural variation has evolved in the United States, one that clearly sets blacks apart from Anglo-Americans and other ethnic groups (Hannerz, 1969; R. Taylor, 1979). African Americans have developed a linguistic style, have molded a unique version of Protestantism, and have created black art forms, especially in music.

As explained in Chapter 4, cultural assimilation is commonly a reciprocal process in which minority ethnic groups not only absorb the dominant culture, but transform it to some degree with their own contributions. This has clearly been the case for African Americans in recent decades. Indeed, sociologist Orlando Patterson maintains that the black influence on the prevailing American culture is today not simply evident, but pervasive (1997). Blacks, he suggests, dominate American popular culture—including music, dance, language, sports, and youths' fashion. "So powerful and unavoidable is the black popular influence," he writes, "that it is now not uncommon to find persons who, while remaining racists in personal relations and attitudes, nonetheless have surrendered their tastes, and their viewing and listening habits, to black entertainers, talk-show hosts and sit-com stars" (1995b:24). As a reflection of this cultural drift, in 1998 Oprah Winfrey was found to be America's favorite TV personality (Taylor, 1998).

Structural Assimilation

Structural assimilation at the secondary level has begun in earnest for African Americans as they increasingly acquire better jobs, higher incomes, and more education. For the black underclass, however, secondary structural assimilation has been stymied, and the prognosis for future movement in this direction is bleak. Even for the black middle class, of course, it is still only in the incipient stages.

Structural assimilation at the primary level—as measured by residential patterns, club memberships, friendship cliques, and intermarriage—remains low.

---

[10]There is an unresolved debate among sociologists and historians concerning the degree to which blacks retained elements of their native African cultures under American slavery. Some, like Frazier (1949), assert that blacks were essentially stripped of African cultural ways; others, like Herskovitz (1941), maintain that remnants remained intact.

Whites today do have more personal contact with blacks than they did in the past, but that contact continues to be minimal and casual (Sigelman et al., 1996). Residential integration is a very basic prerequisite to other aspects of primary and even secondary assimilation, but in this area, as we have seen, African Americans have not followed the usual path of American ethnic groups. Though residential dispersion accompanied upward class mobility for other groups, this has not occurred for African Americans, who remain highly segregated despite advances in income, occupation, and education.

As might be expected, interracial marriage also remains uncommon (Alba and Golden, 1986; Lieberson and Waters, 1988). Although the number of black-white marriages increased from 65,000 in 1970 to 363,000 in 2000, it is still an unusual phenomenon considering that in 2000 there were over 56 million married couples in the United States (U.S. Census Bureau, 1998a; 2002). Blacks and whites may increasingly work, shop, vote, and attend school together, but the formal interrelations in these areas of social life do not extend very far into more personal relations as neighbors, church members, close friends, or marital partners. Clearly, primary structural assimilation has progressed to a lesser extent for African Americans than for any other American ethnic group.

Assimilation: African Americans and Other Ethnic Groups

Are African Americans simply one ethnic group in American society, subject to basically the same patterns of ethnic relations as others? Or is their experience so exceptional that they cannot be compared using the same analytical models and criteria applied to other ethnic groups? This is an issue of some debate in sociology. The question is essentially this: Is the assimilation, or "ethnic" model—so clearly applicable to white ethnic groups, such as the Italians and Jews examined in previous chapters—relevant to African Americans, or are they best viewed in terms of some form of inequalitarian pluralism? Let's briefly examine the two sides of this issue.

The Assimilation Model Social scientists who have applied the assimilation model to African Americans see their social evolution as generally comparable to that of white ethnic groups. In this view, African Americans, by virtue of their migration from the rural South to the cities of the North, face problems similar to those faced by earlier urban immigrants. Just as Italians, Jews, and other southern and eastern Europeans arriving in the late nineteenth and early twentieth century had to compete for jobs and political power with the earlier-arriving Irish and others, so, too, African Americans must now compete with those who preceded them to the city. Changed urban conditions and group cultural differences are acknowledged, but the most important difference between African Americans and other urban ethnic groups, in this view, is simply the time that they have entered the city: African Americans are latecomers (Glazer, 1971; Glazer and Moynihan, 1970; Handlin, 1962; Kristol, 1970; Sowell, 1981). The process, therefore, may be longer and more painstaking, but eventually, African Americans, like other ethnic groups, will be assimilated into the mainstream of American society.

In this view, the steady economic and political progress of the black middle class is evidence of continuing assimilation much in the fashion of other ethnic groups. The low level of societal integration of the African-American underclass, by contrast, is explained in large measure as a product of their failure to adopt conforming, middle-class social values (Banfield, 1968, 1990 [1974]; Sowell, 1978, 1981). Discrimination, it is argued, is not the major impediment to further assimilation for lower-status African Americans; instead, it is largely a matter of internal group norms and values. Only if and when those norms and values reflect the mainstream culture will the black underclass advance economically and socially—and subsequently be integrated into the larger society. Thomas Sowell compared American blacks of slave ancestry with those whose families originally migrated from the West Indies and found striking occupational, income, and educational differences: West Indians ranked higher on all counts. Because the groups are racially the same, Sowell concluded, the development of different attitudinal traits, not racism, has been the major factor in the different rate of group advancement (1978, 1981).[11]

Assimilationists have found the contemporary black family to be a particularly critical hindrance to African Americans in following the adjustment patterns of earlier ethnic groups. Daniel P. Moynihan, in a controversial report issued in 1965, asserted that the black family's comparatively high instability, its disproportionate rate of illegitimacy, and its matriarchal structure were the chief sources of black social and economic problems. He traced these destabilizing family characteristics to slavery, in which, it was presumed, the black family was sapped of its vitality and a female-dominated unit was created (see Frazier, 1939). Moynihan's thesis stirred a great deal of controversy among social scientists and policymakers. Most criticism held that he had confused cause with effect: The unstable black family was not the cause of such circumstances as widespread unemployment and discrimination in the labor and housing markets but the result of those conditions (Billingsley, 1968; Farmer, 1967; Ryan, 1967). Some also pointed out that Moynihan and others who took his position had devoted most of their attention to problem-plagued lower-class families and had largely ignored stable middle- and working-class black families—hence the perception of the black family as "pathological" (Lewis, 1967; Staples, 1981).

---

[11]Farley and Allen (1987), however, suggest that the socioeconomic differences between blacks of West Indian origin and other black Americans may not be so great as has often been assumed.

Moreover, the issue of slavery and its effects on the black family was opened to rigorous debate shortly after Moynihan's report was published. Some historians maintained that the slave system had not had the devastating effects traditionally attributed to it. Instead, they suggested, the slave family had been a haven of psychological security for its members and a social mechanism that aided in survival (Fogel and Engerman, 1974; Gutman, 1976). The prevalence of matriarchy and instability was therefore seen as a product not of slavery but of twentieth-century occupational and residential discrimination.

The role of the family structure in the perpetuation of the black underclass remains disputable. There is little question, however, about the deleterious impact of the female-headed family on income, occupation, and education. Moreover, the percentage of black female-headed families has risen sharply since the mid-1960s, when the Moynihan report was issued. Then, about a quarter of black families were headed by a female; in 2000, almost half were (U.S. Census Bureau, 2002).

The Inequalitarian Pluralistic Model Some aspects of the assimilation model continue to be applied to African Americans. But a more prevalent view today regards the black experience as unique among American ethnic groups. Even exponents of the assimilation model have conceded that the path of societal integration followed by African Americans is exceptional vis-à-vis not only past immigrants, but even new immigrant groups (Glazer, 1997).

In explaining the noticeably different patterns of assimilation of African Americans compared with other ethnic groups, emphasis is placed on the fundamentally divergent African-American ethnic experience; no other group arrived so nearly completely as involuntary immigrants, none was subjected to two centuries of slavery, none experienced the depth and persistence of racial prejudice and discrimination, none was so totally divested of its native culture, and none was so indelibly marked.

Many social scientists contend that the African-American experience fulfills the basic components of what was referred to in Chapter 4 as the "colonization complex"—forced entry of the group, destruction of its indigenous culture, repression by the dominant group, and the application of a racist ideology (Blauner, 1969, 1972; Franklin and Resnik, 1973). Other ethnic groups did not face similar circumstances. In this view, then, African Americans clearly fit the internal colonial model, not the assimilation, or "ethnic," model.

Those who see African Americans as singular among American ethnic groups argue that the vestiges of slavery have immeasurably handicapped them in their efforts to compete with whites. The abolition of slavery did not change established relations of dominance and subordination. Blacks continued to be relegated to the lowest occupational levels and were denied fundamental political rights well into the 1960s. Furthermore, the economic and political structures of American society, they argue, have changed so basically from what they were for entering European immigrant groups that comparisons with them are rendered vacuous (Hershberg et al., 1979). African Americans simply cannot utilize economic and political institutions as European immigrants earlier did to secure better jobs, housing, and political clout and, thus, to effectuate assimilation.

Assimilation and Visibility As noted in Chapter 4, visibility is usually a decisive factor in the rate and degree of structural assimilation for ethnic groups in multiethnic societies. The greater the visibility, the slower and less intense is the assimilation process. This would seem to be of overriding importance in the case of African Americans.

Despite its abatement in the past two decades, the racial stigma of African Americans remains an ever-present fact in the United States, making black-white interaction at all social levels the most obdurate problem of American ethnic relations. As Hughes and Hughes put it five decades ago, "Few are the situations which both Negro and white Americans enter, side by side, unlabelled and undifferentiated. Even a monk or a hero must be named by his color unless he is white" (1952:28). Today, even those African Americans who have attained middle-class status have not been able to reverse the order of their social identity as "black" first and "middle-class" second. Isaacs has poignantly expressed the significance of visibility as it affects the lives of black people in America:

> For black Americans, more clearly perhaps than for any other group, the element of color and physical characteristics lies at the very center of the cluster that makes up their basic group identity. It is the one element around which everything else in their lives has been made to revolve, the heart of the identity crisis that is with them every hour of every day and which they need more than anything else to resolve. (1989:66)

In sum, although ethnicity may be declining in significance for most whites in American society, for blacks it continues to be the most crucial social characteristic, influencing almost all aspects of social life. Though the relations between blacks and whites are no longer the almost exclusive focus of ethnic analysts and policymakers—as they had been for most of the twentieth century—they will likely remain the major issue of interethnic relations in the United States for many generations.

SUMMARY

The black experience is unique among American ethnic groups. No other group entered the society so completely as involuntary immigrants, and no other group was subjected to such fully institutionalized degradation. The aftereffects of slavery continue to influence the patterns of black-white relations and the place of blacks in the social hierarchy.

Black-white relations can be divided into three major periods. Slavery, lasting from the 1600s until 1865, was characterized by paternalistic domination in which blacks were totally subservient, by law in the South and by informal processes in the North. Following the brief interlude of Reconstruction after the Civil War, black-white relations entered the Jim Crow era. Although no longer slaves, blacks were forced back into a subservient economic role and, through legal and extralegal measures, were denied the vote as well as other basic civil liberties. Moreover, a fully legitimized system of segregation was developed in the South, abetted by the application of a racist ideology. An era of fluid competition between blacks and whites commenced in the 1930s, and in the 1960s the Jim Crow system was demolished in the wake of the black civil rights movement.

As a collectivity, blacks in the past three decades have made significant strides in income, occupation, and education. However, they continue to lag behind whites in overall employment and income, and the poverty gap between blacks and whites persists. Furthermore, the gains of the past thirty years have not been experienced in a balanced fashion within the African-American community. One segment has taken its economic place in the working and middle classes; but the other segment comprises an underclass—chronically unemployed, undereducated, and acutely indigent. A debate has been waged between those who see the economic problems of poor blacks mostly as a function of their lack of skills and education and those who see these problems as a function of continued racial discrimination.

The levels of prejudice and discrimination against blacks have historically been more severe and persistent than those against any other ethnic group in American society. In the past three decades, however, these patterns have changed measurably. White attitudes are more tolerant, and discrimination today is confined mostly to the institutional variety. Housing, however, is an area still largely unaffected by changes in black-white dynamics. Although African Americans have reached a high level of cultural assimilation and are in the incipient stages of secondary structural assimilation, primary structural assimilation remains minimal. Little black-white interaction occurs outside formal social settings such as work and education.

There is some debate concerning the relevance to African Americans of either the assimilation model or the internal colonial model. Some maintain that African Americans are simply latecomers to the urban environment and will eventually display assimilation patterns much like those of white ethnic groups. Others, however, see African Americans as a group whose historical experience and current conditions are so distinctive that comparisons with white ethnic groups are difficult at best.

Suggested Readings

Hacker, Andrew. 1992. *Two Nations*. New York: Scribners. Examines the continuing divisions and inequalities in various spheres of social and economic life between blacks and whites in the United States.

Jacoby, Tamar. 1998. *Someone Else's House: America's Unfinished Struggle for Integration*. New York: Free Press. A finely written historical account of the growing divisions between blacks and whites during and after the civil rights era, using three cities as a backdrop: New York, Detroit, and Atlanta.

Killian, Lewis M. 1975. *The Impossible Revolution, Phase II: Black Power and the American Dream*. New York: Random House. One of the best accounts of the origins, strategies, and objectives of the black civil rights movement and the brief period of the black power movement. A rather pessimistic prognosis for American race relations that seems to have been borne out.

Lemann, Nicholas. 1991. *The Promised Land: The Great Black Migration and How It Changed America*. New York: Vintage. Chronicles the migration of blacks from the rural South to the urban North, interweaving personal narratives with an analysis of public policy.

Massey, Douglas S., and Nancy A. Denton. 1993. *American Apartheid: Segregation and the Making of the Underclass*. Cambridge, Mass.: Harvard University Press. Traces the development of residential segregation in U.S. urban areas. Argues that the maintenance of two housing markets, black and white, is the key to understanding the continuation of the economic gap and the social divisions between the two racial categories.

McWhorter, John H. 2000. *Losing the Race: Self-Sabotage in Black America*. New York: Free Press. Presents a controversial argument that the consistently lower academic performance of African American students is the result of a cultural attitude of anti-intellectualism and the embracing of victimhood as an identity rather than a problem.

Myrdal, Gunnar. 1944. *An American Dilemma: The Negro Problem and Modern Democracy*. New York: Harper & Row. One of the most influential books written about race relations in the United States. Myrdal, a Swedish sociologist, challenged Americans to recognize the inconsistency between their society's commitment to social democracy and its institutional discrimination against blacks.

Oliver, Melvin L., and Thomas M. Shapiro. 1995. *Black Wealth / White Wealth: A New Perspective on Racial Inequality*. New York: Routledge. Focuses on the differences in wealth held by whites and blacks, demonstrating the precarious position of the black middle class.

Patterson, Orlando. 1997. *The Ordeal of Integration: Progress and Resentment in America's "Racial" Crisis*. Washington, D.C.: Civitas / Counterpoint. A refreshing analysis of black-white relations in the United States, pointing out the significant changes of the past several decades in ethnic attitudes and the improved socioeconomic conditions of African Americans. Includes an innovative essay on the uses of affirmative action.

Stampp, Kenneth M. 1956. *The Peculiar Institution: Slavery in the Ante-Bellum South*. New York: Random House. Describes, in a very readable account, the nature of the American slave system as it developed and functioned in the eighteenth and nineteenth centuries.

Thernstrom, Stephan, and Abigail Thernstrom. 1997. *America in Black and White: One Nation, Indivisible*. New York: Simon & Schuster. Examines the current economic, political, and social status of African Americans by tracing changes from the end of Jim Crow and then since the civil rights era. Takes an optimistic view, in contrast to Hacker's *Two Nations*.

Waters, Mary. 1999. *Black Identities: West Indian Immigrant Dreams and American Realities*. Cambridge, Mass.: Harvard University Press. Describes issues of identity and adaptation of black immigrants to the United States, who often find themselves involuntarily caught in the maelstrom of the American racial system.

Wilson, William J. 1996. *When Work Disappears: The World of the New Urban Poor*. New York: Knopf. An update of Wilson's earlier works on the black underclass, arguing that the continuation of black inner-city poverty is the result not of racial discrimination but of a restructured urban economy.

HISPANIC AMERICANS

THERE ARE MORE people of Mexican origin in Los Angeles than in all but one city in Mexico; in New York City, there are more Puerto Ricans than in San Juan; and only in Havana are there more Cubans than in Miami. Together, Hispanic Americans number more than 35 million, 12.5 percent of the total American population. And the Hispanic-American ethnic category is growing rapidly and substantially. So rapid has the increase been, in fact, that today it has surpassed by a small margin African Americans as the country's largest ethnic minority.

Hispanic Americans[1] constitute several distinct ethnic groups, linked by a shared language and a cultural heritage derived, in the main, from Spanish colonialism. The most sizable groups among them are Mexican Americans—the largest—followed by Puerto Ricans and Cubans (Figure 10.1). The rest are people from Central and South America as well as a small group whose origins are

---

[1]Although the term *Hispanic* has been most frequently used in the past two decades, *Latino* is today also commonly used to describe those of various Latin American origins. In this chapter, the two are used interchangeably. Mexican Americans, specifically, are often referred to as *Chicanos*. The U.S. Census has, at different times, used a variety of terms to describe the Hispanic-American population. A 1995 government study found that a majority favored Hispanic (Tucker and Kojetin, 1996), similar to a Gallup poll finding in 2001 (Gallup, 2001e). A survey of the Hispanic-American population in 1992 (de la Garza et al., 1992), however, found that most Hispanics prefer to identify themselves in terms of their society of origin (e.g., Mexican, Cuban, etc.).

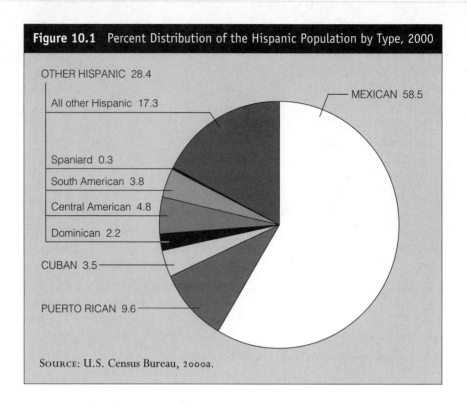

Figure 10.1 Percent Distribution of the Hispanic Population by Type, 2000

OTHER HISPANIC 28.4

All other Hispanic 17.3

MEXICAN 58.5

Spaniard 0.3

South American 3.8

Central American 4.8

Dominican 2.2

CUBAN 3.5

PUERTO RICAN 9.6

SOURCE: U.S. Census Bureau, 2000a.

traced to Spain. Because they are the largest elements of the Hispanic-American population, our concern in this chapter is mainly with Mexican Americans, Puerto Ricans, and Cubans.

THE "IN-BETWEEN" STATUS OF HISPANICS

If there is a common theme that runs through the unique histories and experiences of the several Hispanic groups in the United States, it is their intermediate ethnic status between Euro-American groups, on the one hand, and African Americans, on the other. In several important respects, Hispanics are an ethnic minority "in between."

Manner of Entry

Hispanic Americans have entered the society in a variety of ways. Although Mexicans have come to the United States in the twentieth century as voluntary immigrants, they retain a heritage of having been absorbed into American society originally as a conquered group. The area that today is the American Southwest was part of Mexico before being annexed by the United States in

1848 following the Mexican War. Mexicans living on those lands were subsequently incorporated into the United States, much in the fashion of classic European colonialism. Unlike any other ethnic group, then, Mexican Americans today inhabit an area that was once part of their native society.

Puerto Ricans are a group not clearly part of either voluntary immigration or conquest. Puerto Rico became a territory of the United States in 1898 following the Spanish-American War, and in 1917, the inhabitants of the island were given the status of American citizenship. This enabled them to immigrate freely to the mainland with no restrictions. Puerto Ricans, then, are not technically immigrants even though they come to the mainland from a distinctly foreign culture.

The movement of Cubans to the United States in the past four decades has been a voluntary immigration. Yet it has been different from those of most past groups in that it was initially impelled mainly by political rather than economic motives. Moreover, the class characteristics of immigrant Cubans and the reception given them by the host society were unusual. As sociologist Alejandro Portes has put it, "Seldom has a foreign group come to the United States so well prepared educationally and occupationally and seldom has this country received one so well" (1969:508).

Hispanic Americans and Race

Perhaps the most obvious and consequential in-between feature of Hispanic Americans is their racial status. All three major groups include a range of physical types.

Mexico is racially amalgamated to a greater degree than perhaps any modern society (van den Berghe, 1978). Over the past two centuries, its population has been transformed into one dominated by mestizos, a physical type combining European and Indian traits. Although many are either basically European, or at the other extreme, purely Indian, the vast majority of Mexicans today are racial hybrids. Most Mexican immigrants to the United States are mestizo, making them physically distinct from the Anglo majority but still "the least divergent racially of all non-Caucasian groups" (Murguía, 1975:51). So ambiguous is the racial status of Mexican Americans that the U.S. Census Bureau has in the past classified them both as a separate "race" and as part of the white population. In any case, most Mexican Americans are physically distinct enough to be perceived by many Anglos in racial terms. (The term *Anglo* is used in the Southwest and in other areas with a large Hispanic population to distinguish any non-Hispanic white.)

Puerto Ricans' in-between racial status is even more problematic. Whereas Mexicans are mostly mestizo, Puerto Ricans inherit a racial background that is a combination primarily of European and African but also some Indian elements. The large African population originally brought to the island as slaves intermarried with whites over several generations, and today, although they cover the entire color spectrum, many Puerto Ricans are racially "intermediate" (Glazer and Moynihan, 1970). On the island, racial distinctions are not so acute, and one's color is seen along a scale or continuum (Duany, 1998; Fitzpatrick, 1987; Padilla, 1958; Rodríguez, 1996). But problems arise in the United States, where there is a racial dichotomy, allowing for no intermediate categories. Many who are "white" on the island may therefore find themselves classified as "black" when they come to the mainland. The differential treatment they subsequently receive on the basis of their darker skin (by American standards) is frustrating and problematic because interracial marriages are commonplace in Puerto Rico, and social treatment of darker groups is more equitable than on the mainland. Similar problems are encountered by Dominicans who immigrate to the United States. They may consider themselves white, but in the American context are often perceived as black (Duany, 1998).

In both Mexico and Puerto Rico, racial distinctions are more often a function of class and cultural differences than of physical type. Those who rise in socioeconomic status or who adopt Western cultural ways are accordingly "whitened." As we will see in Chapter 14, this is also the case in Brazil. In the United States, however, racial distinctions are based on skin color or other physical characteristics, in addition to ancestry, and are not modified by changes in class or culture.

Among the Hispanic groups in the United States, Cubans are the least racially heterogeneous, most falling into the American category of "white" (Diaz, 1980). Cuba itself is a racially variegated society, however, suggesting that the Cuban immigration has been racially selective (Portes, Clark, and Bach, 1977). Most early Cuban immigrants were of higher social status in Cuba and hence whiter in color, whereas many of the more recent immigrants have been of lower class origins and of darker color (Pedraza-Bailey, 1985).

THE DEVELOPMENT OF THE LATINO MINORITY

Despite broad similarities, Mexican Americans, Puerto Ricans, and Cuban Americans each display a unique history of immigration as well as unique patterns of settlement and adaptation.

Mexican Americans

As they have with blacks and American Indians, Anglos have come in conflict deeply and consistently with Mexicans throughout the history of the Southwest. Early in the contact between these groups, a pattern of Anglo domination and Mexican subordination emerged. And as with American Indians, the vehicle through which domination was initially established was physical conquest.

Conquest of Mexico Mexico gained its independence from Spain in 1810, and at that time its territory encompassed an area as far north as what is today Colorado. Most of this northern Mexican region, far removed from the administrative capital of Mexico City, was inhabited by Indians and a smaller number of Mexicans. As a means of populating the area, the Mexican government in 1821 granted permission for foreigners—mainly Americans—to settle in the area that today is Texas. By 1830, about 20,000 Anglos were living in the Texas region, constituting a numerical majority. Culturally and physically distinct from the indigenous Mexicans and Indians, the Anglo settlers from the outset translated their differences into contempt and hostility toward these groups.

The Mexican government could exercise only limited control over the Texas colony, and pressures from Anglos for independence grew steadily. In 1835–1836, the colony revolted and established itself as an independent republic. Mexico never recognized the independence of Texas; and when it became clear that the United States favored annexation of the territory, much friction was created between the American and Mexican governments, leading eventually to war in 1848.

The Mexican War proved disastrous for Mexico, which lost more than half its territory. The United States acquired what are today the states of California, Colorado, New Mexico, Nevada, Utah, and most of Arizona in addition to Texas, which had been annexed previously. Under the Treaty of Guadalupe Hidalgo, ending the war, Mexicans living in what was now American territory were guaranteed the rights of American citizenship if they chose to remain. Most did, and it was from this nucleus that the Mexican-American ethnic group evolved.

The significance of the Texas rebellion and the Mexican War cannot be overstated as factors in the shaping of subsequent relations between Mexican Americans and Anglos in the Southwest. The borderlands remained an area of overt conflict long after the conclusion of the war, and this protracted hostility intensified the negative perceptions that Anglos and Mexican Americans had of each other (Grebler, Moore, and Guzman, 1970). Anglo-Mexican conflict did not characterize all of the Southwest to the same degree. In New Mexico, for example, relations between the two groups were relatively peaceful by comparison with Texas, where the strife reached its most violent and bloody form. But in all cases, the war left a residue of mutually antagonistic feelings and group images among Anglos and Mexicans that continued to color interethnic relations for many generations.

Although their property rights and political rights were guaranteed by the Treaty of Guadalupe Hidalgo, Mexicans in the newly acquired American territory found themselves increasingly displaced by Anglos. Their property was taken through official and unofficial force and fraud, and they were gradually transformed into a colonized workforce serving the area's labor needs. Mexicans were also dispossessed politically, as Anglos firmly established themselves in power. The scenario of heavy immigration of Anglos followed by the divesting of Mexicans' economic and political power was played out somewhat differently from area to area. In New Mexico, for example, the indigenous Mexican population, called Hispanos, did retain some political and economic power. These people were descendants of the original Spanish settlers dating back to the 1600s and did not identify themselves with the later-immigrating Mexicans (Nostrand, 1973). In general, however, Anglos perceived Mexicans as a conquered—and therefore inferior—people who quite naturally were relegated to subordinate ethnic status (Alvarez, 1973).

The development of ethnic stratification in the Southwest seems clearly in accord with Noel's (1968) theory, discussed in Chapter 2. All three factors necessary to the emergence of an ethnic hierarchy—ethnocentrism, competition, and above all, differential power—came into play after 1821 in the confrontation of Anglos and Mexicans (McLemore, 1973). Anglos settling the region were most intent on securing land held by Mexicans—hence the ensuing competition. With them, they also brought their racial attitudes, which led them to see Mexicans as an inferior people whose exploitation could therefore be justified. Finally, with their numerical power—and following annexation by the United States, their political power—Anglos became the dominant group, displacing Mexicans both legally and by force.

Mexican Immigration As a result of the Texas and Mexican wars, the first Mexicans were incorporated into American society through conquest. But subsequent generations of Mexicans entered the United States as voluntary immigrants. Indeed, it must be remembered that the original Mexican Americans, those living on annexed lands, were a relatively small group, perhaps no more than 75,000 (McWilliams, 1968). There were over twice that number of Indians in the same territory. The overwhelming majority of Mexican Americans, then, historically entered as voluntary immigrants, motivated by the same push-pull factors as other immigrant groups.

The flow of Mexican immigration during the nineteenth and twentieth centuries was dictated primarily by the labor needs of the American Southwest (Camarillo, 1979; Tyler, 1975). Most simply, movement across the border was heaviest during periods when the demand for cheap labor was high. Although immigration was evident during the remainder of the nineteenth century following the Mexican War, it reached enormous proportions during the twentieth century with the expansion of the American Southwest and the corresponding inadequacies of the Mexican economy. The pressures of a continually growing population and the political destabilization brought about by the Mexican Revolution in 1910 further impelled migration northward.

Between 1910 and 1930 the development of railroads, mining, and agriculture, all labor-intensive industries, spurred the first great wave of Mexican immigration into the United States (McWilliams, 1968). As the chief source of unskilled labor in the Southwest, Mexicans became a subordinate labor force. They were commonly kept in debt by their employers, paid less than others for the same work, concentrated in the least desirable occupations, often used as strikebreakers, and laid off most easily in distressful times (Barrera, 1979).

When they represented an excess labor supply, however, Mexicans were driven back across the border. During the 1930s, for example, not only were their jobs abolished, but they served as convenient scapegoats for the area's unemployment and growing welfare rolls. Deportations were launched, in which illegal immigrants were rounded up and sent back to Mexico. Repatriations also took place, in which Mexicans, many of whom had been born in the United States, were induced to leave the country. This pattern was repeated in the 1940s and 1950s. During World War II, a severe labor shortage once again prompted the importation of Mexican workers. But by the early 1950s, the deportations and repatriations of the 1930s were grimly repeated in a policy called Operation Wetback, designed to stem the flow of illegal Mexican immigrants. In five years, 3.8 million persons were sent back to Mexico (Moore and Pachon, 1976). Relatively few were actually deported in formal proceedings; most were simply rounded up and expelled. In the process, Mexicans were once again reminded of their tenuous status in the United States. Persons who simply "looked Mexican" were stopped and required to present evidence of their legal status. Although there is some question about the official number actually apprehended and deported (Garcia, 1980), Operation Wetback in any case created fear and suspicion among the Mexican-American populace.

Today Mexicans are the country's largest immigrant group, and their numbers are swelled by a continual flow of illegal entrants. During the past three decades, Mexicans have constituted about a quarter of all legal immigrants to the United States. As in the past, economic conditions are the chief motive of this migration.

The Chicano Movement and the Development of Ethnic Consciousness As with other ethnic groups, in the 1960s and 1970s, Mexican Americans engaged in significant political activity as part of a period of growing ethnic consciousness. Perhaps the two most dramatic episodes of Mexican-American political activism during this period were the movements led by Cesar Chavez and Reies Lopez Tijerina. Chavez's objective was to organize migrant farmworkers, and Tijerina focused on regaining communal lands that he charged had been illegally taken from Hispanos in New Mexico.

Chavez's well-publicized movement of this period consisted of efforts to force growers in California and the Southwest to recognize his United Farm

Workers Union (UFW) as the representative of thousands of migrant laborers, most of them Mexican Americans. Past efforts had been made to organize farm-workers in the Southwest, some even dating back to the early years of the twentieth century, but Chavez was the first to succeed. Through his organizing activities, Chavez became something of a folk hero to Mexican Americans, and his charisma thrust him into the forefront of the larger Chicano movement (Solis-Garza, 1972; Steiner, 1970; Stoddard, 1973).

The movement led by Tijerina and his organization, the Alianza de las Mercedes, was more openly militant, and its objectives were less pragmatic than Chavez's. Tijerina argued that several million acres of land in New Mexico had been fraudulently taken from Hispanos following annexation of the territory at the conclusion of the Mexican War, and his aim was to re-cover them. To draw attention to his cause, he provoked several confronta-tions with federal and state officials, leading to his arrest and subsequent imprisonment (Knowlton, 1972). Although the Alianza helped in creating a national awareness of the Chicano movement, its goals were too visionary to yield concrete benefits.

Chavez's and Tijerina's movements were the two most prominent elements of the larger Chicano movement of the 1960s and early 1970s, which extended well beyond specific personalities and organizations. *Chicanismo* was very similar to the black nationalist movement and was spawned in the same general atmosphere of minority activism in the 1960s. Its major goal was to galvanize Mexican-American political action in order to assert group rights and expand economic and social opportunities. Many of the same tactics employed by the black movement, including legal cases, protests, boycotts, and direct encounters with authorities, were adopted.

Like its black nationalist counterpart, the Chicano movement also repre-sented an assertion of militant cultural pluralism. Rather than assimilation, the idea was to affirm the unique Mexican-American culture and to instill in Mexican Americans a sense of ethnic pride and awareness (Macías, 1972; Steiner, 1970). Younger activists began to refer to themselves as Chicanos rather than Mexican Americans and to proclaim a more militant and antiestablishment position. The term *Chicano* had in the past referred to lower-class Mexicans, but it now began to represent an ethnic viewpoint (Macías, 1972; Moore and Pachon, 1976). Just as *black* had been substituted for *Negro*, *Chicano* came to symbolize an ethnic identity that repudiated the negative group image that had been imposed for generations by Anglos.

That *Chicanismo* seemed to coincide with black nationalism was not acciden-tal; indirectly, the latter contributed to its emergence. Black power demon-strated that securing civil rights did not in itself assure meaningful social and economic change. Mexican Americans observed aid programs to black inner cities and began to petition for their own "fair share" of assistance. Their problems, they felt, were equally pressing; and like black nationalism, Chicano ideology placed the onus for minority deprivation on the larger society.

Just as black militancy subsided in the 1970s, so, too, did the Chicano movement. But as with black power, the long-range effects of *Chicanismo* were more substantial than they first appeared. Undeniably, concrete gains were not broad; governmental programs aimed at depressed Mexican-American communities were short-lived. Intangible benefits, however, were more meaningful and lasting. A positive group identity emerged among Mexican Americans, and ethnic unity was instilled through the idea of *La Raza*.[2] Most important, national attention was drawn for the first time to Mexican Americans as a sizable minority group.

Puerto Ricans in the United States

The formation of the Puerto Rican minority in the United States differed considerably from the development of the Mexican-American community. To begin with, Puerto Ricans are a relatively recent immigrant group. Even though they had maintained a presence on the mainland for most of the twentieth century, their numbers were quite small until after World War II. The greatest influx occurred during the 1950s, when nearly 20 percent of the island's population moved to the mainland. Whereas in 1940, only 1,000 Puerto Ricans migrated to the mainland, in 1953, almost 75,000 did so (U.S. Commission on Civil Rights, 1976). In the 1970s, as the U.S. economy stagnated and low-skilled jobs disappeared, the migration to the mainland ceased and, in fact, a net return migration was sustained throughout the decade.[3]

The push-pull factors impelling Puerto Ricans to migrate have been similar to those for other voluntary immigrant groups: A surplus population in an economically depressed society seeks economic betterment in a society that promises improved conditions. Puerto Rican migration has therefore fluctuated, depending on economic conditions both on the island and on the mainland. An unusually high birthrate coupled with a seriously underdeveloped economy created great pressures for out-migration from the island after World War II. Two factors increased the appeal of the mainland United States. First, with

---

[2]*La Raza* (the race) is not similar in meaning to the North American idea of race, but is a somewhat vague concept denoting a common heritage or peoplehood.

[3]Fitzpatrick points out that immigration statistics involving Puerto Ricans are subject to serious difficulties in interpretation. "There is no way of telling the characteristics of the person traveling, that is, how many are Puerto Ricans coming to the mainland for the first time and how many are coming with the intention of returning to the island" (1987:17). Calculating net migration as the difference between the number arriving on the mainland and the number leaving Puerto Rico (as has customarily been done) may therefore tell us little about the actual size of the Puerto Rican migration.

citizenship rights, Puerto Ricans could enter the country with no restrictions, quotas, or other legal steps. Second, with the establishment of airline service, transportation between the mainland and the island became relatively cheap and rapid. The migration process for Puerto Ricans, then, was considerably less complicated than that for other groups.

Although the absolute size of the Puerto Rican migration to the mainland is, compared with past groups, small, it is numerically significant in relation to the island's population. By 1960 there were one-third as many Puerto Ricans on the mainland as in Puerto Rico, and by 1970 one-half as many (Bahr, Chadwick, and Stauss, 1979). Moreover, migration statistics do not accurately indicate the frequent back-and-forth movement between island and mainland, so it is quite reasonable to assume that a much larger number of islanders have, at one time or another, migrated. Indeed, during the three decades between 1940 and 1970, it is unlikely that a single Puerto Rican family was not affected by this movement (U.S. Commission on Civil Rights, 1976).

The Puerto Rican population is increasingly a second- and third-generation ethnic group. This is especially evident in New York, where the largest number of Puerto Ricans continues to reside (Fitzpatrick, 1995).

Cubans in the United States

The conditions of immigration for Cubans and their subsequent settlement patterns have been quite different from those experienced by either Mexicans or Puerto Ricans.

Given its geographical proximity to Cuba, the United States often served as a haven for Cuban exiles during times of political turmoil before and after the Cuban movement for independence from Spain in the late nineteenth and early twentieth centuries. In fact, one of the key figures of Cuban political history, José Martí, sometimes called Cuba's George Washington, spent much of his exile during the 1880s in the United States. Small colonies of Cubans had for many years been present in Key West, Tampa, Miami, and New York. But only in the 1960s, following the revolution led by Fidel Castro, did Cubans begin to come to the United States in significant numbers.

Whereas the pull factors of migration were stronger for Mexicans and Puerto Ricans, the push factors seemed more crucial for early Cuban immigrants. Their chief motivation was political refuge rather than economic betterment. Nonetheless, the underlying economic objectives of the émigrés should not be overlooked. Because the revolution imposed a socialist system in Cuba, the power and economic opportunities for most of the early immigrants—who were mainly middle-, upper-middle-, or upper-class people—had been severely constricted. As Portes has explained, "The wealthy, the successful, the powerful, the educated saw their status challenged, their influence radically curtailed, and their economic position continuously menaced" (1969:506). Departure from Cuba, then, was not simply politically inspired but was calculated in the hope of reestablishing a relatively high socioeconomic status in the United States and returning to Cuba when political conditions were favorable. For later Cuban immigrants, economic factors in the decision to migrate were even more clearly apparent (Portes, Clark, and Bach, 1977; Wong, 1974).

Cuban immigration from 1960 onward was aided immeasurably by American policies that in effect permitted the entrance of Cuban refugees without restriction. The political objectives of this policy are not difficult to understand. The chilled relations between the Cuban and American governments soon after the Cuban Revolution were followed with strongly belligerent actions by the United States, including an American-supported invasion of Cuba at the Bay of Pigs in 1961 and the imposing of a trade embargo against Cuba in the same year. The Cuban missile crisis of 1962 firmly established the Castro regime as a basic element in the Cold War, and from that point forward, Cuba became a bête noire for successive American administrations. Cuba was portrayed in the United States as a "communist stronghold" and the Cuban people as "politically enslaved." The emigration of large numbers of Cubans, therefore, was seen as confirmation of this perspective, and the refugees were welcomed enthusiastically.[4]

Large-scale Cuban immigration ceased in 1973 but resumed for a brief period in 1980, when more than 125,000 people came to the United States. The conditions under which this second immigrant wave left Cuba were graphically described by the American media, and the manner in which they arrived—most in small boats—further dramatized their entrance. They were dubbed *Marielitos*, in reference to the port of Mariel, from which most of them departed Cuba. These immigrants were generally lower in social class than the original Cuban refugees, and many were black or mulatto (a mixture of black and white). As a result, they did not find the adjustment to American society as smooth or their reception as enthusiastic. They experienced higher rates of unemployment, low-paid work, and dependence on charity and public welfare (Portes, Clark, and Manning, 1985). Moreover, the presence of a criminal element among them served to taint the entire group.

Except for part of the second wave, most Cuban immigrants entered the United States with a strong family and kin support system that aided their adjustment considerably (Portes, Clark, and Bach, 1977). Indeed, following the initial immigration in the early 1960s, succeeding immigrants were encouraged

---

[4]For thirty-five years, the U.S. federal government maintained an open-door policy for Cubans who fled to the United States. Under the policy, they were given immediate permanent resident status—terms afforded no other immigrant or refugee group. The policy was effectively terminated in 1995 with the elimination of Cubans' automatic acceptance as political refugees. See Nackerud et al., 1999.

to come to the United States by relatives or friends who could offer them physical and economic assistance (Fagan, Brody, and O'Leary, 1968). Although the presence of friends and relatives has also characterized the Mexican and Puerto Rican migrations, the support system of Cubans has been stronger and more complete. Moreover, Cubans have experienced substantial economic success and have therefore been enabled to offer greater assistance to those who have followed them. Mexicans and Puerto Ricans, by contrast, have remained collectively in a far more insecure economic position.

Hispanic-American Demographics Today

Population Trends Hispanic Americans as a whole are a relatively young and, by comparison with others, rapidly growing minority. Between 1990 and 2000 the Hispanic population increased by 58 percent, more than four times as fast as the U.S. population in total. Hispanic growth is accounted for by a combination of continued heavy immigration and a comparatively high birthrate.

Geographic Trends The Hispanic population is rapidly dispersing to all regions of the United States. Consider that in Atlanta and Memphis, Hispanics, never in the past more than a minor element of the population in those cities, now dominate the construction and landscaping trades (Schmitt, 2001); they have become the main source of labor for meatpacking plants in Omaha, Nebraska; and they make up the bulk of the workforce in poultry plants in Delaware and North Carolina (Viglucci, 2000). Despite this dispersion, however, the vast majority of Hispanic Americans are concentrated in the Southwest and particularly within just a few states. Over three-quarters live in the seven states with Hispanic populations of 1 million or more (Table 10.1). Half of all Hispanics live in just two states alone—California and Texas. Looking at the proportional breakdown of states, New Mexico has the largest percentage of Hispanics that make up its total population: over 42 percent. Other states with large percentages of Hispanics are California, 32 percent; Texas, 32 percent; Arizona, 25 percent; Nevada, 20 percent; and Florida, 17 percent (U.S. Census Bureau, 2001a).

Like African Americans, Latinos are primarily urban dwellers, whether Mexicans in the Southwest, Puerto Ricans in the East, or Cubans in Florida. In fact, all three groups are more urbanized than the general U.S. population. Eighty percent of Mexican Americans and virtually all Puerto Ricans and Cubans live in urban areas (Bean and Tienda, 1987). Table 10.2 shows the major metropolitan areas in which Latinos represent a significant population.

Mexican Americans In recent decades there has been a dispersion of Mexicans throughout the United States, particularly to the industrial cities of the Midwest

Table 10.1 States with the Largest Hispanic Populations, 2000

STATE	NUMBER OF HISPANICS (1,000s)	PERCENT OF STATE POPULATION
California	10,967	32.4
Texas	6,670	32.0
New York	2,868	15.1
Florida	2,683	16.8
Illinois	1,530	12.3
Arizona	1,296	25.3
New Jersey	1,117	13.3
New Mexico	765	42.1
Colorado	736	17.1

Source: U.S. Census Bureau, 2002.

(Aponte and Siles, 1994). As shown in Table 10.2, among major urban areas, Chicago has the fifth highest number of Hispanics and most of these are Mexicans. In the past ten years, Mexicans have also become a significant portion of New York's large Latino population (Choi, 2001). Nonetheless, the vast majority of Mexican Americans remain in the Southwest, where they are numerically the most significant ethnic minority. Southwest cities with the largest Mexican populations are Los Angeles, San Antonio, Houston, and Phoenix. Within the Southwest, however, there are regional differences in the size and importance of the Mexican-origin population. In some southern Texas cities and towns, for example, Mexicans are a huge numerical majority, whereas in other areas of Texas, they may be a minuscule element.

Puerto Ricans Throughout the 1950s and 1960s, the Puerto Rican population was heavily concentrated in New York City. In 1950 over 81 percent of all Puerto Ricans on the mainland resided there, and in 1970, 62 percent. During the 1970s and 1980s Puerto Ricans began to disperse, though almost half remained in New York City. As Joseph Fitzpatrick explains, "the Puerto Rican experience is no longer a predominantly New York experience, as it once was" (1987:10). Other significant Puerto Rican communities are found in Chicago, Philadelphia, and northern New Jersey, especially Newark and Jersey City. The remainder are primarily in cities of the Northeast. Nonetheless, as economic opportunities shift, Puerto Ricans are becoming a more geographically dispersed population (Torrecilha et al., 1999).

Table 10.2 Hispanic Population in Metropolitan Areas with More than 200,000 Hispanic Persons

METROPOLITAN AREA	HISPANIC POPULATION (1,000s)	PERCENT OF TOTAL METRO AREA
Los Angeles–Riverside–Orange County, CA	6,081	38.5
New York–Northern NJ–Long Island	3,490	17.4
Miami–Ft. Lauderdale, FL	1,407	38.5
San Francisco–Oakland–San Jose, CA	1,314	19.3
Chicago–Gary–Kenosha, IL–IN–WI	1,213	13.8
Houston–Galveston–Brazoria, TX	1,084	24.6
San Antonio, TX	823	53.5
Dallas–Fort Worth, TX	757	15.8
San Diego, CA	722	25.9
Phoenix–Mesa, AZ	601	20.5
El Paso, TX	524	74.5
McAllen–Edinburg–Mission, TX	460	88.1
Washington–Baltimore, DC–MD–VA–WV	387	5.3
Fresno, CA	369	42.4
Denver–Boulder–Greeley, CO	344	14.6
Boston–Worcester–Lawrence–Lowell–Brockton, MA	323	5.5
Philadelphia–Wilmington–Atlantic City, PA–NJ–DE–MD	300	5.0
Austin–San Marcos, TX	283	25.6
Brownsville–Harlingen–San Benito, TX	278	85.2
Albuquerque, NM	266	39.2
Sacramento–Yolo, CA	252	15.0
Tucson, AZ	229	28.9
Corpus Christi, TX	225	58.2
Bakersfield, CA	222	35.1

SOURCE: U.S. Census Bureau, 2001h.

The decline of New York City's Puerto Rican population has become more evident in recent years. Being Hispanic in New York once meant being Puerto Rican. In 1950, almost 80 percent of all Latinos in the City were Puerto Rican. But today, with a far more diverse and substantial population of people from Latin America and the Caribbean, Puerto Ricans—while still the City's largest Hispanic group—account for only 37 percent of its Hispanic population. Not only have New York's Puerto Ricans dispersed to the suburbs and to other parts of the United States, but many—even those born on the mainland—have returned to Puerto Rico (Navarro, 2000).

Cuban Americans Cubans are the most concentrated of the three major Hispanic groups, with about two-thirds residing in the metropolitan Miami area. Other relatively sizable Cuban communities are found in New York, northern New Jersey, Tampa, and Los Angeles.

Central and South American and Caribbean Groups In the past two decades the Hispanic-American population has been infused with newer groups, from Central and South America and from the Caribbean, whose numbers in the United States previously were very tiny. Among the Hispanics who are not part of the Mexican, Puerto Rican, or Cuban groups, 1.7 million are Central American, 1.4 million South American, and 765,000 Dominican (U.S. Census Bureau, 2001a). For the most part, these groups have clustered in a few large urban areas.

Most Dominicans arrived in the 1970s and, especially, the 1980s and are today the largest immigrant group in New York City. Dominicans have fared more poorly than most other ethnic groups in the City. Most of them work in low-paying service or factory jobs and are handicapped by low levels of education. Roughly half of Dominican families are headed by women and suffer a poverty rate almost double the overall poverty rate in the City (Ojito, 1997). The racial factor has also affected this group since most identify themselves as "mulatto" or "black" (Pessar, 1995).

New York City has been the major destination point as well for those from South American countries, especially Colombia, Brazil, Ecuador, and Peru. Other large South American populations live in Los Angeles, Chicago, and Miami. It is difficult to estimate the size of these groups because so many have come to the United States as illegal immigrants. What is certain is that they are substantially larger than census figures would indicate (Margolis, 1998). Like others who preceded them, these are immigrants who come in pursuit of economic betterment; many harbor expectations of returning to their origin societies after working for a period in the United States. Many are highly skilled workers who are fleeing political instability and civil conflict in their home countries (Forero, 2001).

Central Americans immigrated to the United States in large numbers starting in the 1980s, mostly because of political unrest in the region. Many arrived as refugees and many were undocumented. The largest among them are from El Salvador, Guatemala, Honduras, and Nicaragua. By far the largest concentration of Salvadorans and Guatemalans is in Los Angeles, where about one-third of all Central American immigrants reside (Rodriguez and Hagan, 1999). The largest cluster of Nicaraguans is in Miami, where a community of 125,000 makes up the second-largest Hispanic group (Colon, 1999). While in the past it had been a primarily Cuban-American city, Miami in recent years has become a magnet for all Hispanic groups, including tens of thousands from virtually every country in Central and South America and the Caribbean. In addition to Los Angeles and Miami, other large Central American populations are found in New York, Houston, and San Francisco (U.S. Census Bureau, 2001a).

LATINOS IN THE CLASS SYSTEM

Socioeconomic Status

As in other societal dimensions, Hispanic Americans seem to occupy an intermediate position—below Euro-Americans, but in several ways above African Americans and American Indians—in the degree of economic success they have achieved and in the paths to upward mobility they are following. However, keep in mind that "Hispanic American" is an umbrella category encompassing several distinct groups. Thus, important differences will be evident among them. Specifically, there is a class range among the three major Latino groups, with Cubans at the top, Mexicans in the middle, and Puerto Ricans at the bottom.

Income As shown in Table 10.3, the discrepancy between Hispanic and non-Hispanic white median family income is substantial. Despite moderate gains during the past two decades, Hispanic income is still only about 60 percent of white income. Also evident in Table 10.3, however, is a range within the Hispanic category. Whereas the median family income of Mexicans in 2000 was $31,123, for Puerto Ricans it was $30,129, and for Cubans, $38,312. All three, however, were well behind the median income of non-Hispanic whites.

With incomes well below average, Latinos—especially Mexicans and Puerto Ricans—are, predictably, disproportionately among America's poor. Over one-fifth of all Hispanic families are below the poverty line, a rate almost four times higher than for non-Hispanic white families. As with income, however, there is a poverty range within the Hispanic-American category. As shown in Table 10.3, Mexicans and Puerto Ricans below the poverty line are more numerous than Cubans. Indeed, among a significant part of the Puerto Rican population, economic conditions are much akin to those of the black underclass (Fitzpatrick, 1995; Lemann, 1991).

Occupation The low incomes and high poverty rates of Latinos, particularly Mexicans and Puerto Ricans, reflect their generally lower occupational levels. Except for Cubans, Latinos are underrepresented in the higher-status occupational categories and overrepresented in the lower ones. They also exhibit high

Table 10.3 Hispanic-American Socioeconomic Status, 2000

GROUP	MEDIAN FAMILY INCOME	PERCENT FAMILIES IN POVERTY	PERCENT UMEMPLOYED
Hispanic	$31,663	20.2	6.8
Mexican	31,123	21.2	7.0
Puerto Rican	30,129	23.0	8.1
Cuban	38,312	15.0	5.8
Central and South American	33,105	16.3	5.1
Other Hispanic	34,935	18.1	7.8
Non-Hispanic White	56,442	5.3	3.4

SOURCE: U.S. Census Bureau, 2001g, 2001k, 2002.

unemployment rates (Table 10.3). And, as with African Americans, official statistics understate the extent of this problem.

What accounts for the relatively low levels of income and occupation of Hispanic Americans? While employment discrimination against Latinos has historically been a factor for particular groups, especially Mexicans in the Southwest and Puerto Ricans in New York, today low levels of education and their relatively recent entry into the labor force seem most critical (Camarillo and Bonilla, 2001). Large proportions of all Latino groups have immigrated to the United States since 1950 (Massey, 1981a). As latecomers to the American postindustrial economy, they have necessarily occupied the least skilled and lowest-paying jobs. Like African Americans, they have become disproportionately part of the most expendable element of the workforce, generally lacking the requirements for positions above the semiskilled or unskilled levels. Moreover, as an unusually young population, Mexican Americans and Puerto Ricans are most likely to lack marketable skills and work experience. Education, however, appears to be an even more important factor in explaining the lower income and occupational positions of the Hispanic-American population. As the economic gap grows between those with higher education and those without, Latinos have been especially hard hit, given their relatively lower educational levels generally and the low proportion with college education specifically (Carnoy et al., 1993).

The effects of lingering discrimination can in some cases present additional difficulties even after age, education, and lack of skills are accounted for (Meléndez, 1993). Here it is important to consider differences among specific Latino groups and the regions where they live. While discrimination against Mexicans in the Southwest or Puerto Ricans or Dominicans in New York may be a continuing problem in attaining higher-level jobs, Cubans in Miami, by contrast, have faced no such obstacles. Indeed, regarding employment, for them ethnicity may serve as a positive rather than negative influence.

Education As noted, perhaps the most serious factor in accounting for the lower occupational status and earnings of Hispanic Americans in general is their lower educational level. As shown in Table 10.4, only 57 percent of Latinos 25 years of age and over have completed high school, compared with 88 percent of non-Hispanic whites. Dropout rates among Latinos also are extraordinarily high, compared to those of both whites and blacks (Fletcher, 1998; U.S. Census Bureau, 2001h). Advances in recent years have begun to close this educational gap, however. The proportion of younger Latinos who have completed high school is substantially higher compared to older Latinos, and the proportion of Latinos over age 25 who completed four years or more of college doubled between 1970 and 1990. Also, as with income, it is important to consider the differences in educational attainment within the Latino category. As is seen in Table 10.4, the discrepancy between Mexicans and Cubans, for example, is very great.

Reasons for the lower educational attainment of Latinos, particularly Mexican Americans and Puerto Ricans, have been debated. Some contend that inappropriate cultural values fail to encourage successful school performance; others maintain that schools themselves, Anglo-dominated as they are, do not address the unique needs of Latino students. One point about which there is general agreement is that language difficulties intensify the academic problems of these students (Celis, 1992; Nava, 1975; U.S. Commission on Civil Rights, 1972,

Table 10.4 Hispanic-American Educational Level
Persons 25 Years Old and Over, 2000

GROUP	HIGH SCHOOL GRADUATE OR HIGHER	BACHELOR'S DEGREE OR HIGHER
Hispanic	57.0%	10.6%
Mexican	51.0	6.9
Puerto Rican	64.3	13.0
Cuban	73.0	23.0
Central and South American	64.3	17.4
Other Hispanic	71.6	14.5
Non-Hispanic White	88.4	28.1

SOURCE: U.S. Census Bureau, 2001k.

1974, 1976). Many Mexican-American students in the Southwest are from families with non-English-speaking backgrounds, making their school experience trying, at best. Language problems create or exacerbate much of the high dropout rate and poor academic performance among Puerto Ricans as well. In New York in 1990, 66 percent of Puerto Ricans born on the mainland had finished high school, but only 37 percent of the first generation (Puerto Rican–born) had done so. Dropout rates were still high, however, even for the second generation, due in large measure to poor language proficiency (Fitzpatrick, 1995).

In sum, Mexican Americans and Puerto Ricans are underrepresented at the higher levels of the mainstream economy and overrepresented among the poor. In looking at the socioeconomic levels of these two groups, it should be noted that these are aggregate figures reflecting the groups' general position in the American stratification system. There will obviously be an entire range of occupational and income positions within each group despite their disproportionate concentration at the lower levels. Moreover, for Mexicans there are regional differences of some importance. In the Southwest, where they have traditionally played a subservient economic role, they clearly make up the bulk of the poor. But in northern industrial cities, they are a more stable and well-entrenched part of the blue-collar workforce (Cardenas, 1976; Estrada, 1976; Nodín Valdés, 1989; Aponte and Siles, 1994). For Puerto Ricans, too, analysis of New York City data reveals that there has been a gradual increase in professional and managerial positions (Fitzpatrick, 1995). In general, however, Mexican Americans and Puerto Ricans have entered the U.S. workforce at an inopportune time. Because they face the same increasingly technical and high-skilled job market as African Americans with a correspondingly low skill level, they will continue to make up a large segment of the poor in the foreseeable future. And, like the black underclass, many will continue to face the intractable problems of chronic unemployment and a large percentage of female-headed families.

Cubans: A Special Case Cubans are unlike Mexican Americans and Puerto Ricans in their adaptation to the American economic system. Indeed, few groups in American ethnic history have equaled the phenomenal socioeconomic rise of the contemporary Cuban-American community in so short a time. On all measures of socioeconomic status, Cubans are clearly well in advance of Mexicans and Puerto Ricans despite their more recent arrival. Their income and educational levels are higher, their unemployment rates are lower, and they are well represented across the occupational spectrum. What accounts for these striking differences?

Cuban immigrants to the United States brought with them a class background markedly different from that of other Latinos. A disproportionate number derived from the middle and upper strata of prerevolutionary Cuban society. In their study of early Cuban refugees, Fagan, Brody, and O'Leary

(1968) found that professional and semiprofessional occupations were heavily overrepresented among this group, whereas fishing and agricultural jobs—those predominating in Cuba—were hardly represented at all. The occupational skills Cubans brought with them enabled them to work themselves more easily into the mainstream economy and climb rapidly even though most of them experienced downward mobility when they first entered the United States. Indeed, research has shown that among younger Cubans in the United States, the rate of upward mobility has been similar to that of immigrants from English-speaking countries (Massey, 1981b). In addition to their high occupational status in Cuba, the refugees were far better educated than most Cubans, and most were urban residents. Succeeding immigrant cohorts displayed a somewhat lower class background but still represented a privileged sector of Cuban society (Portes, Clark, and Bach, 1977).

Cubans in the United States, then, differ radically from other Hispanics in their social origins. Unlike Mexicans and Puerto Ricans, early Cuban immigrants were likely to have been occupationally successful and well educated before emigrating. These higher skill and educational levels have translated into rapid upward mobility in the United States. Also, most immigrant Cubans have been white and therefore have not been exposed to the added handicap of racial discrimination in the labor market, with which many Mexicans and Puerto Ricans have had to deal. These same advantages—higher educational and skill level and the absence of a racial disadvantage—have characterized much of the new Latino population from South American countries.

Another advantage Cubans have had over other Hispanic groups has been the creation of an institutionally complete community in Miami, where most Cuban Americans reside. As described by Wilson and Portes (1980), an ethnic enclave economy was established by early Cuban immigrants in Miami, comprised of numerous small businesses serving the ethnic community. In addition, a small manufacturing sector emerged that subsequently employed other Cuban immigrants who arrived with fewer skills and financial capital. The presence of an ethnic subeconomy not only created jobs for these immigrants but also enabled them to avoid the secondary labor market of the mainstream economy, where jobs ordinarily are temporary and do not provide for occupational advancement. Many of the Cuban immigrant workers who were employed by Cuban employers eventually acquired the skills and capital needed to open their own establishments, which created an expanding Cuban subeconomy (Portes and Bach, 1985). Neither Mexicans nor Puerto Ricans have been as demographically concentrated as Cubans, thus making the development of similar economic arrangements impracticable. Moreover, neither group contained a substantial entrepreneurial class, possessing both business capital and skills, that could re-create in the United States the position it held in its country of origin and thereby serve as the nucleus of an ethnic subeconomy. Pedraza-Bailey (1985) further asserts that Cuban economic success in the United States has been very much a product of the fact that Cubans were received as political migrants—that is, as refugees—rather than as economic migrants. As a result, the U.S. government provided aid to early-arriving Cuban immigrants that helped them in reestablishing their class position. Mexicans, by contrast, have traditionally been received as economic migrants and have never been given similar governmental assistance.

Despite the extraordinary achievement of Cubans relative to other Hispanic Americans, their economic success in the United States should not be overstated. Though they are certainly in a very different category from Mexicans and Puerto Ricans, their median income is still below Anglos', and their poverty rate and unemployment rate are higher. Also, even though they are well represented in the higher-level white-collar occupations, most are still in clerical, semi-skilled, or service jobs (Pérez-Stable and Uriarte, 1993). Moreover, the economic gap between Cubans and other Hispanics is closing (Betancur et al., 1993; Diaz, 1980; Perez, 1986).

Latinos and Political Power

Latino political power in the United States is still in its infancy. Because Latinos are regionally concentrated, their political gains have necessarily been limited mostly to the local and state levels. Even at these levels, however, Latinos have not yet mobilized as effectively as have African Americans in recent years. Nonetheless, Hispanic Americans represent a growing and increasingly powerful political force.

Mexican Americans As the largest and geographically most widespread group, Mexicans constitute potentially the most significant political force among Hispanic Americans. This potential, however, has not yet been realized. Mexican Americans have maintained a low rate of political participation (Garcia and Arce, 1988; McClain and Stewart, 1998; Verba et al., 1995). Several factors may account for this. First and perhaps most important, a large percentage of Mexican Americans are poor, and as noted previously, lower-income people in general display lower rates of political activity. Second, the fluid nature of Mexican immigration, in which people constantly move back and forth across the border, has militated against the development of long-term political loyalties and interests. Third, a vast number of Mexicans in the United States either are not naturalized citizens or are illegal immigrants, further diluting the Mexican constituency (Kirschten, 1992). Finally, in the Southwest, tactics designed to prevent Mexican Americans from voting, such as the poll tax, gerrymandering of ethnic districts, literacy tests, intimidation, and even violence, were in the past commonly applied (Camarillo and Bonilla, 2001; Pachon and Moore, 1981; U.S. Commission on Civil Rights, 1970). Though these impediments have been removed by means of voting rights legislation in recent decades, their aftereffects are still felt in many areas.

Participation of Mexican Americans in local and state politics has varied in different locations. In some cities, political underrepresentation has been very evident. In Los Angeles, for example, with the largest Mexican-American community in the country, not until 1991 was a Mexican American elected to the county board of supervisors, an extremely powerful local governing body. By contrast, San Antonio, also with a very large Mexican population, achieved proportional ethnic representation on its city council by the late 1970s, and in 1981 a Chicano was elected mayor. Increasingly, Mexican Americans have had success in electing individuals to statewide offices, especially in states with large Latino populations (McClain and Stewart, 1998).

In New Mexico, Mexican Americans (Hispanos) have held a relatively equitable share of political power with Anglos since the original annexation of the territory in the nineteenth century (Garcia and de la Garza, 1977; Knowlton, 1975; Padilla and Ramirez, 1974). The American conquest of New Mexico was not violent, as it had been in Texas and other parts of the Southwest, resulting in less historical division between the two groups. Moreover, after the Mexican War, Hispanos in New Mexico were a numerical majority. By contrast, in Texas, even before annexation Mexicans were outnumbered five to one, and Anglo political and economic dominance were almost complete. The upshot was that Mexicans there had to struggle for local and state political power (Moore, 1970).

Today, Mexican Americans are a growing political force in those areas where they make up a significant portion of the population, namely southern California and south Texas. At the national level, too, Mexican-American political influence appears to be growing. In 2001, fifteen members of Congress were Mexican American, though most were from locales with large Mexican-American populations (Hirschfeld, 2001). Moreover, two Mexican Americans had served in President Clinton's cabinet and several have played a prominent role in the Bush administration.

Puerto Ricans To a limited extent, Puerto Ricans participated in the ethnic militancy of the 1960s and early 1970s. Several groups emerged at that time, advocating control by Puerto Ricans of their communities and emphasizing ethnic identity and pride. The Young Lords, for example, a militant group in Chicago and New York, imitated the activities of the Black Panthers in organizing Puerto Rican ghettos and calling for a redress of squalid conditions (Lopez, 1980). As with black and Chicano political activism, Puerto Rican militancy resulted in drawing the serious attention of national authorities to the Puerto Rican community for the first time. But just as the attention paid other minority ghettos dwindled with the passing of militancy, so, too, most of the programs of federal and community agencies directed at Puerto Ricans disappeared in the 1970s.

Today, the political activity and influence of Puerto Ricans in the United States are limited primarily to New York City, because in no other area do they represent a group large enough to make a serious impact on local politics. Political participation among Puerto Ricans, however, as among other Latinos, is lower than average. The same factors at work for Mexicans—heavy back-and-forth movement between the United States and the society of origin, language difficulties, a young population, and above all, lower-class status—have militated against significant Puerto Rican political participation. Moreover, Puerto Ricans in the New York metropolitan area are relatively dispersed residentially, so that in few districts are they alone able to carry an election (Fitzpatrick, 1987). Also, their numbers are declining in the area (Navarro, 2000). Despite their relatively low political profile in New York City, Puerto Ricans have become more politically active in recent years and have recorded some electoral successes. In 2001, three Puerto Ricans had been elected to the U.S. Congress and several to state and city offices.

Cuban Americans Cuban Americans have emerged as the dominant political force in Miami with its large Hispanic population. Almost all high-ranking local political positions are held by Cubans. Moreover, nearly half of Miami-Dade County's 27-member delegation in the state legislature and two of its six members of the U.S. Congress are Cubans (Navarro, 1999). Also, Cuban Americans participate in the political system at a significantly higher rate than other Latino groups (Verba et al., 1995). Quite simply, the political influence of Cubans in Miami is unequaled by other Hispanic groups nationwide (Morgan, 1998a).[5]

The political leanings of Cubans in general tend to be more conservative than those of either Mexican Americans or Puerto Ricans, which may be a reflection of their higher-class position. When they do participate in the electoral process, they tend toward the Republican Party, in sharp contrast to the overwhelmingly Democratic leanings of Mexican Americans and Puerto Ricans. Cuban conservatism is also the product of a very strong anticommunist sentiment, particularly among the first-generation refugees (Portes and Mozo, 1985). Within the Cuban community, much attention remains focused on the political scene in Cuba, and political issues are often heavily tinged with ideology (Rieff, 1993). Younger Cuban Americans, however, are likely to view political relations between Cuba and the United States in less passionate terms than their parents (Canedy, 2001; de los Angeles Torres, 1988; Perez, 1980). Moreover, politics at the local level among Cuban Americans is increasingly concerned with local, not foreign policy, issues (Portes and Truelove, 1987).

---

[5]Despite the strong influence of Hispanics, and Cubans specifically, a 2000 survey showed that a majority of key public and private power positions in Miami were still held by non-Hispanic whites (Branch-Brioso, Henderson, and Chardy (2000).

The Outlook for Latino Political Power Unlike African Americans, Latinos have not yet been able to exert significant political influence at the national level. This is due primarily to two factors. One is the relative youth of the Hispanic-American populace. This is a young population with a large element under age 18 and thus not eligible to vote. Another substantial portion of the population is among the age categories (18 to 24 years) that are least active politically. A second, and more important, factor is the relatively low percentage of citizenship among Hispanic Americans. Thus, although the Hispanic-American population has increased dramatically, its political clout has not followed suit.

Although Latino political power has not yet realized its potential, there are growing signs of change. If a good measure of a group's position in the political system is the presence of that group's members in elective offices, Latinos have made significant progress in the past decade. By 1994, almost 5,500 Latinos held publicly elected offices at various government levels—local, state, and federal—an increase of almost 50 percent from 1984 (*National Roster*, 1994). Nineteen Latinos were members of the U.S. Congress in 2001 (Hirschfeld, 2001). Although most of these officeholders were elected by majority Latino constituencies, they represented a variety of geographic areas.

Indications of increasing Latino political power are especially evident in states and communities where Latinos constitute a numerically substantial part of the population. In California, for example, where the largest part of the Hispanic-American population resides, four Latinos were elected to the state Senate in 1996 and fourteen to the Assembly, representing the largest number of Latinos in either body in the state's history. Moreover, for the first time, a Latino was elected speaker of the assembly, one of the most powerful state offices (McClain and Stewart, 1998). Most significant, in 1998 a Mexican American was elected lieutenant governor, the first Hispanic official to be elected statewide in more than 120 years (Purdam, 1998). Although the Hispanic electorate in any state does not yet correspond to the Hispanic percentage of the population, it can be a determining swing vote in many elections.

The long-range outlook for Hispanic political power, then, is mixed. Although Hispanic voting rates will lag behind those of whites and blacks for some time to come, an expanding population base and continued assimilation will probably stimulate increased political activism. Between 1994 and 1998, for example, the Latino vote in Los Angeles County grew by almost 50 percent, compared to 10 percent for the non-Latino vote (Tomás Rivera Institute, 2001). That Latinos are a growing element of the population and a potential electoral force has not been ignored by the two major parties at both local and national levels. Both presidential candidates in 2000, for example, made appeals to the Latino vote. Moreover, President George W. Bush has courted the Hispanic electorate and given several high-level appointments in his administration to Hispanic Americans.

Today it is possible to speak of a Latino vote, and as the Hispanic-American population increases, it will grow in significance.

Economic Power As with African Americans, the role of Latinos in leadership positions of the American economy is limited. Fewer than 150 of the 1,000 largest U.S. corporations have Hispanics on their boards of directors (*Hispanic Business*, 2001). Despite their underrepresentation in executive positions, an increasing number of Latinos are moving through the corporate elite pipeline (Zweigenhaft and Domhoff, 1998). Also, an increasing number of Hispanic-owned firms are growing in economic significance, though most enterprises owned by Mexican Americans are small, family-operated businesses, and the same is true of Puerto Rican businesses (Fitzpatrick, 1980; *Hispanic Business*, 1995; O'Hare, 1987).

The majority of highly successful Latino entrepreneurs and corporate executives come from Cuban-American backgrounds (Whitefield, 2000; Zweigenhaft and Domhoff, 1998). The more substantial economic success of Cuban Americans by comparison with other Latinos is the result primarily of their generally higher-class background and the entrepreneurial skills and resources many brought with them from Cuba. In Miami, Cuban Americans are the economically dominant ethnic group, both in large and small enterprises. Cubans are president of the city's largest bank, owner of the largest real estate developer, managing partner of the largest law firm, and publisher of the largest newspaper (Cardwell, 1998; Navarro, 1999).

PREJUDICE AND DISCRIMINATION

Because of their "in-between" minority status, Latinos have not been subjected to the dogged prejudice and discrimination aimed at African Americans; but neither have they been dealt with as European immigrant groups were. John R. Howard refers to such intermediate groups as "partial minorities" and calls attention to their paradoxical position: "By being somewhat below the threshold of public consciousness with regard to 'minority problems' they escape some of the pervasive indignities visited upon blacks, but, precisely because of this it becomes difficult to arouse public indignation at the depredations they do suffer" (1970:8–9).

More precisely, an entire range of attitudes and actions has characterized the relations between Anglos and the several Hispanic groups. Variations have turned on each group's historical context, its geographical location, and its racial makeup. Those close to the dominant Anglo group, like most Cubans, have experienced little prejudice and discrimination; at the other extreme, dark-skinned

Puerto Ricans and Dominicans have often encountered ethnic antagonism equal in form and degree to that suffered by African Americans.

Hispanic Stereotypes

For as long as they have been part of the American ethnic mosaic, Mexicans and Puerto Ricans have labored under strongly negative stereotypes. These have arisen in response to the manner in which both groups became part of the society (more so in the case of Mexicans than Puerto Ricans) and as manifestations of American racist ideology.

Mexican Americans Feelings of racial superiority on the part of Anglos and policies reflecting those beliefs have been obvious since the American conquest of the Southwest in the nineteenth century. More than anything, the historical circumstances that originally brought Mexican and Anglo into confrontation fashioned a set of group images that have remained relatively fixed since that time.

As noted in Chapter 3, negative group images are ordinarily formed as a result of some form of interethnic competition, usually economic. In the Southwest, land and labor were the critical economic components in the development of mutual antagonism between Anglos and Mexicans. "To the early American settlers," writes Carey McWilliams, "the Mexicans were lazy, shiftless, jealous, cowardly, bigoted, superstitious, backward, and immoral" (1968:99). These images subsequently served to rationalize the dispossession of Mexicans' lands and the exploitation of their labor. The negative stereotyping of Mexicans was abetted by the fact that they were a conquered people. To the Anglos this only confirmed even more positively their inferior nature. Correspondingly negative stereotypes of Anglos naturally arose in this situation. To the Mexicans, Anglos were "arrogant, overbearing, aggressive, conniving, rude, unreliable, and dishonest" (McWilliams, 1968:99).

Also critical in the development of negative stereotypes of Mexicans were the racial attitudes Anglo settlers brought with them to Texas. From the beginning, Anglos viewed Mexicans as innately inferior on the basis of their racial characteristics (Moore and Pachon, 1976). Mexicans were different in language, culture, and religion, but they were also physically distinct from the Anglos. As a mestizo people, they were not blacks, but neither were they whites. This in-between racial status led to the classification of Mexicans as a separate racial group, one perhaps a notch above blacks but clearly inferior to the Anglos.

The poverty and illiteracy of those Mexicans with whom Anglo settlers had the most dealings further strengthened these negative views. McWilliams notes that most of the Anglos who settled in the Southwest borderlands in the mid-nineteenth century were middle-class people who failed to find a Mexican middle-class counterpart. "If a larger middle-class element had existed," he explains, "the adjustment between the two cultures might have been facilitated and the amount of intermarriage might have been greater" (1968:75). Although upper-class Hispanos were initially seen in a different light, they too were eventually absorbed into the adverse image.

The negative stereotypes of Mexicans that developed during and after the period of conquest were maintained and even strengthened not only in the Southwest but also in the society generally. Mexicans were included among the "undesirable" groups denoted by the Dillingham Commission in the early 1900s, along with southern and eastern Europeans, falling somewhere between blacks, on the one hand, and "white" Americans, on the other. It is instructive to note that despite their undesirable status, Mexicans were so vital as a cheap labor supply in the Southwest that they were exempted from the restrictive immigration laws that were drafted at that time largely on the basis of the commission's report.

The stereotypes developed in the nineteenth century continued to influence the views of Anglos in the Southwest and elsewhere throughout most of the twentieth century. In his study of a south Texas community in the early 1970s, Simmons (1971) found that Anglos believed Mexicans to be unclean, prone to drunkenness and criminality, deceitful in dealings with Anglos, hostile, and unpredictable. Among favorable images of Mexicans were their "romantic" rather than "realistic" attitude and their fun-loving nature. But even the positive traits implied a childlike and irresponsible character, making the Mexicans appear deserving of their subordinate status.

Television and movies have played a large role in perpetuating negative stereotypes of Mexican Americans, much as they have for African Americans and American Indians. In Hollywood films, Mexicans have traditionally been portrayed as villains in confrontation with Anglos; and even when they are allied with the Anglo heroes, their roles are ordinarily subservient and often ludicrous (Petit, 1980; Wilson and Gutiérrez, 1995). Television commercials have also perpetrated the villainous image of Mexicans (Martinez, 1972; Wagner and Haug, 1971). One of the better known of these ads, the Frito Bandito commercial, provoked a strong protest by Mexican Americans, and eventually the Frito-Lay Company was forced to remove it. The ad featured a cartoon character who appeared as a Mexican bandit sneaking Fritos corn chips from the kitchens of unsuspecting housewives. Because he was cunning and clever, they were advised to buy two bags (Westerman, 1989; Wilson and Gutiérrez, 1995).

Latinos' visibility in general on American television does not correspond to their proportion of the population. Moreover, although criminality among Hispanics is a declining theme, it is still more evident than for whites and blacks (Lichter and Amundson, 1997). As TV producers and mass media advertisers have become more attuned to a growing Latino market, however, such stereotypical images are subject to change. Ten years after it was forced to remove its

Frito Bandito ad, for example, the Frito-Lay Company advertised Tostitos, another corn chip product, this time with a character who spoke with a lilting Spanish accent and presented a romantic image of a distinguished, cultured Latino (Wilson and Gutiérrez, 1995).

Puerto Ricans Although they have not experienced a history of confrontation and conflict with Anglos, as Mexicans have in the Southwest, Puerto Ricans have nonetheless commonly been stereotyped negatively. This is mainly a product of their racial distinction and their generally lower-class position. Because much of the Puerto Rican population in the United States is dark-skinned, they often experience the same forms of prejudice as non–Puerto Rican blacks. As a group whose members are disproportionately among the poor, Puerto Ricans are also often characterized as lazy, irresponsible, and welfare dependent. This is no better illustrated than in the remark of businessman J. Peter Grace, who, as head of the Reagan administration's committee on governmental waste, derisively described Puerto Ricans living in New York City as "all on food stamps," thus making the federal food stamp program "basically a Puerto Rican program" (*New York Times*, 1982).

Cubans As a group heavily concentrated in a single city and having arrived more recently, Cubans have not had the kind of national exposure that Mexican Americans or even Puerto Ricans have had. Moreover, as a group that is for the most part physically indistinguishable from the Euro-American population, Cubans automatically enjoy a status that precludes racially based stereotypes. Further, their generally higher class position has made for a more positive image among Anglos.

Although the attitudes of Anglos toward Cubans in Miami were initially benign, certain events have created an increasing bifurcation between the two groups. One area that has fueled Anglo resentment is language. Increasingly, Miami has become a bilingual city, and this has met with Anglo resistance. In 1980 a referendum was held to bar Dade County (Miami) from using public funds to conduct official business in any language but English and from promoting any but "American" culture. The measure, which narrowly passed, polarized the community along ethnic lines (Burkholz, 1980; Castro, 1992). The county commission in 1993 voted unanimously to repeal the 1980 anti-bilingualism ordinance.

The referendum on bilingualism disclosed the depth of the anti-Cuban sentiment among Anglos. A survey conducted immediately following the vote revealed that nearly half of those who had voted for the ordinance did so as an opportunity to express their protest, not because they saw it as a good idea. Further, more than half the Anglos who supported the measure said they would be pleased if it "would make Miami a less attractive place for Cubans and other Spanish-speaking people" (Tasker, 1980). In recent years Miami has become a predominantly Hispanic city, but language continues to draw the strongest divide between Hispanics and Anglos. As one Anglo expressed it, "Sometimes, I feel like I'm a visitor in my own country" (Morgan, 1998b).

An additional source of animosity has been Cubans' rabid and sometimes violent politics, specifically as they relate to Cuba. Nothing better illustrated this than the Elián González affair in 2000. It was perhaps the most polarizing single event separating the Cuban and Anglo populations in the Miami urban area, and even more broadly, American society as a whole. Following his rescue at sea after his mother, along with several others, had drowned in an attempt to sail to Miami from Cuba, six-year old Elián was temporarily placed in the custody of his great-uncle's family. The family subsequently refused to release the child for return to his father's custody in Cuba, which set off a five-month standoff with the U.S. Justice Department. The incident created a national drama—most Cuban Americans felt the child should not be returned to his father in Cuba, while most Anglos in Miami, as well as in the rest of the country, felt otherwise (Viglucci and Marrero, 2000). Many held that Elián was being used by the Miami Cuban community as a political pawn in their long-standing ideological war with Fidel Castro. The incident left Miami bitterly divided.

Discriminatory Actions

The in-between status of Hispanic Americans—between Euro-American ethnic groups, on the one hand, and African Americans, on the other—is quite evident in looking at patterns of discrimination. Hispanic Americans have lacked the heritage of a legal system that, in the case of African Americans, made discrimination legitimate. As a result, such actions have ordinarily been extralegal rather than formally endorsed by the state. Moreover, although discrimination against Latinos has been widespread and often intense, it has not been so profound and intransigent as that experienced by African Americans.

This is illustrated by Pinkney's study (1963) of one community with sizable Mexican-American and black populations. General patterns of discrimination for both groups, he found, were similar but were much stronger in degree for blacks. For example, greater acceptance of Mexican Americans was evident in public accommodations like hotels and restaurants as well as in residential areas and social clubs. Anglos also expressed greater approval for granting equal rights in employment and residence to Mexican Americans. Pinkney attributed these differences in part to differences in skin color: Mexican Americans, closer to the dominant Anglo group, were more acceptable.

That it has not been as rigid, intense, and formalized as that aimed at African Americans should not disguise the severity of discrimination to which Latinos in some circumstances have been exposed, particularly Mexican Americans in the Southwest. Many of the actions against them in areas such as south Texas have been similar in degree and scope to those experienced by blacks, including segregated restaurants, churches, schools, and other public facilities (Burma, 1954; Grebler, Moore, and Guzman, 1970). Roberto Suro describes this social environment:

> In South Texas in the 1950s, discrimination against Latinos was different because it was not based on the single immutable characteristic of skin color and because it was not primarily a matter of law. Culture and convenience dictated that poor, dark-skinned Mexicans should not mix with whites, that they should go to inferior schools, and that they would always work at the bottom of the labor force. And while segregation applied to all blacks, the discriminatory culture of South Texas distinguished among Latinos, treating some like Mexicans and others not (Suro, 1998:81).

Physical violence against Mexicans in the Southwest also has a long heritage. McWilliams notes that in the mining camps of the nineteenth century, any crime committed was immediately attributed to Mexicans, for whom lynching was the accepted penalty. "Throughout the 1860's," he writes, "the lynching of Mexicans was such a common occurrence in Los Angeles that the newspapers scarcely bothered to report the details" (1968:130).

Police Relations A particularly harsh and often violent relationship has traditionally existed between Mexican Americans and various law enforcement agencies in the Southwest. The police and courts have seemed to consistently mete out a different standard of justice to Mexican Americans (Moore and Pachon, 1975; Morales, 1972; U.S. Commission on Civil Rights, 1970). Although the level of discrimination in the criminal justice system has historically been unusually high in Texas, several of the most serious large-scale incidents of the modern era have taken place in Los Angeles. These include the notorious pachuco, or zoot suit, riots of 1943 and the East Los Angeles riot of 1970. In both cases, the police played major roles in either precipitating or sustaining the hostilities.

The events surrounding the zoot suit riots have been well documented (Garcia and de la Garza, 1977; McWilliams, 1968). During the 1940s, the Los Angeles press waged a virulent anti-Mexican campaign, creating a rancorous mood among Anglos. Sensational accounts were presented of Mexican-American youth gangs whose members took to wearing bizarre outfits called zoot suits. Following an incident involving Mexican-American youths and sailors stationed in Los Angeles, 200 servicemen began a four-day rampage, in which they drove through the Mexican area in taxis, randomly attacking those on the streets. The Los Angeles police did virtually nothing to halt the violence and in fact arrested many of the victims. Serious rioting continued for almost a week, ending finally with the intervention of the military police. This incident permanently damaged Mexican-Anglo relations in Los Angeles, and it confirmed the ever-precarious status of the Mexican-American community in an Anglo-dominated environment. Although other such incidents have rarely garnered as much public attention, commonplace police brutality (or indifference, as the case may be) in dealing with Mexican Americans has been repeatedly affirmed.

Residential Segregation For Latinos, discrimination in housing has been generally less severe than for African Americans (Grebler, Moore, and Guzman, 1970; Massey, 1981a, 1981b; 2001; Massey and Denton, 1993; U.S. Commission on Civil Rights, 1976; Camarillo and Bonilla, 2001). Recall the index of dissimilarity, a measure of residential segregation, explained in Chapter 9. Studies using this measure have shown that Hispanic-Anglo segregation is markedly lower than black-white (Anglo) segregation (Massey, 1979; 2001; Massey and Denton, 1989a, 1989b, 1993; Mumford Center, 2001a; White, 1987).

There are specific differences, however, in spatial patterns among the three major Hispanic-American groups. Whereas both Mexicans and Cubans are highly segregated from blacks, Puerto Ricans are less segregated from blacks and more segregated from whites (Bean and Tienda, 1987; Camarillo and Bonilla, 2001; Massey and Bitterman, 1985; Massey and Denton, 1989a). In housing patterns, therefore, Mexicans and Cubans are increasingly similar to the general white population, whereas Puerto Ricans are much like African Americans. Massey and Denton suggest that given their continued spatial isolation, Puerto Ricans are, like many African Americans, in danger of "becoming part of a permanent urban underclass" (1989a:75).

In sum, discrimination against Latinos has been ongoing throughout American history, but it has differed in scope and intensity for each Latino group and has differed as well by region. In any case, its severity has not been similar to that experienced by African Americans, and it appears to be declining. It is of note that a majority of Mexicans, Puerto Ricans, and Cubans reported in a 1992 national survey that they had not been discriminated against because of their ethnicity (de la Garza et al., 1992). Today, anti-Latino prejudice and discrimination seem focused mostly on large immigrant populations in specific locations like southern California and south Florida, particularly on those who have entered the United States illegally. At times, of course, this has had a spillover effect on all Hispanic Americans.

HISPANIC-AMERICAN ASSIMILATION

In Chapters 7 and 8 we saw that Italians, Jews, and other European ethnic groups in American society are best analyzed within the framework of the assimilation model of ethnic relations, wherein ethnicity becomes less intense with each succeeding generation. African Americans, by comparison, were seen in Chapter 9 as a case that more closely fits the inequalitarian pluralistic model; rather than decreasing in significance, ethnicity continues to shape intergroup relations and to affect in a basic manner most life activities of group members. Where do Hispanic Americans fit on the continuum of ethnic relations, and which model is best suited to their experience in American society? As with African Americans, there is some disagreement over whether they are basically like previous immigrant groups who have gone through stages of assimilation or whether they are essentially a semicolonial people. Here it is important to distinguish among the three major Latino groups since patterns of assimilation and pluralism for Mexicans, Puerto Ricans, and Cubans have not been entirely similar.

Patterns of Assimilation

Those who see Hispanics as simply the latest among a continuous stream of American immigrant groups emphasize the slow but steady upward mobility of Mexicans, Puerto Ricans, and Cubans; the resemblance of their social adjustment to that of previous groups; and their increasing social interaction with non-Hispanics (Chavez, 1991; Handlin, 1962; Jaffe, Cullen, and Boswell, 1980). There is evidence to support the view that Latinos in general have experienced a relatively low but increasing level of both cultural and structural assimilation. More specifically, however, Hispanic Americans exhibit a wide range of assimilation, extending from Hispanos in New Mexico—who have strongly assimilated both culturally and structurally into the dominant group—to black Puerto Ricans, who have displayed only minimal structural assimilation.

Cultural Assimilation Unlike African Americans, Hispanic Americans were not stripped of their native cultures on entering the society and have retained many ethnic traits, especially the Spanish language. Nonetheless, Latinos are moving toward cultural assimilation, though more slowly than Euro-American groups of the past.

Latino cultural assimilation is strongly evident in language usage. Although Spanish is the most widely used non-English language in the United States, only a minority of Hispanic Americans speak only Spanish, and most of these are first-generation immigrants. As Latinos move to the second and third generations, English becomes their primary language (Cafferty, 1985; Crawford, 1992; Portes and Schauffler, 1996; Veltman, 1983). Moreover, even among those who continue to use Spanish in the home, the majority use English in other social contexts, like work and friendship groups. De la Garza et al. (1992) found that not only did most U.S. citizens from each of the three major groups use English or a mixture of English and Spanish at home, but they preferred English-language media to either bilingual or Spanish-language media. Further, more than 90 percent of Latinos supported the proposition that citizens and residents of the United States should learn English. Clearly, then, contrary to a common public perception about the maintenance of Spanish, most Latinos prefer English and support its usage.

The persistence of Spanish in some degree, however, has contributed to difficulties in educational and occupational adaptation (Bean and Tienda, 1987). Moreover, the tendency for Latinos to concentrate in particular cities and regions and in some cases to maintain relatively tightly bound ethnic enclaves (Cubans in Miami, for example) retards language assimilation (Portes and Truelove, 1987). Some also believe that bilingual education, rather than accelerating adaptation to English, has done the opposite (Chavez, 1998).

Another factor contributing to slower cultural assimilation among Latinos is the proximity of their society of origin. This is especially evident for Mexican Americans. Nearly four decades ago, McWilliams observed that "In migrating to the borderlands, Mexicans have not found immigrant colonies so much as they have 'moved in with their relatives'" (1968:38). Today that observation seems even more compelling. In a sense, the Mexican immigrant in the Southwest never really leaves Mexico. The frequency and ease of travel between the United States and Latin America have created a similar phenomenon for other Latino groups. Immigrant networks have made back-and-forth movement so commonplace that family units may be said to reside simultaneously in two societies.

Language and closeness of the native culture have also affected the assimilation patterns of Puerto Ricans and, less so, of Cubans. In the case of Puerto Ricans, the easy access to and from the island acts as a disincentive to assimilate culturally into dominant American ways (Garcia-Passalacqua, 1994; Lewis, 1965). For this group, a racial factor also complicates matters. For those darker in skin color, the incentive to retain the Spanish language and culture is a product of their perceived need to separate themselves from the African-American population. It is more advantageous, they feel, to be identified with the Latino populace than to be lumped with non-Hispanic blacks. For those who are neither clearly white nor black by American racial standards, the problem is even more acute, for they find themselves marginal to both categories (Mills, Senior, and Goldsen, 1950). Among Cubans, a relatively substantial portion who arrived as part of the 1980 refugee wave are, in the American sense, racially defined as black. They, too, find themselves in a marginal position, peripheral to the larger Cuban community and unable to assimilate into the African-American community (Nordheimer, 1987; Ojito, 2000).

As for most Cubans, because they originally came not expecting to stay and take on the life of the new society, they have perhaps clung to the old culture more strongly than most other ethnic groups. This pattern is not uncommon in the case of political exiles, whose commitment to their adopted country is never complete. Resistance to cultural assimilation is particularly evident among first-generation

Cuban Americans, and this will no doubt change with the second and third generations, for whom there are no thoughts of return. Further, political realities have prevented frequent back-and-forth movement between Cuba and the United States, which has had a stabilizing effect on the Cuban-American community.

Another factor that has contributed to a retention of the ethnic culture among Cubans is their concentration in a single community. Unlike Mexicans and Puerto Ricans, Cubans in the United States, especially those in Miami, have established a well-knit institutional structure that caters exclusively to the unique social and cultural needs of the group. Raymond Breton (1964) has suggested the idea of "institutional completeness" to describe the extent to which the needs of an ethnic group are met by institutions within the ethnic community—businesses, churches, newspapers, schools, and so on. Ethnic communities may range from those that are institutionally complete, in which members need to make no use of the host society's institutions, to those that are institutionally incomplete, in which the network of interpersonal relations is almost totally within the context of the host society. Cubans in Miami appear to be at the extreme of institutional completeness. There is not even a need to learn English, because Miami is very much a bilingual city. Television stations broadcast and newspapers publish in Spanish; bilingual public schools are common; and food stores, restaurants, and businesses of all kinds cater to the unique ethnic needs of the Cuban community.

Despite the problems of language and proximity to the origin societies, Latinos do appear to be moving along a trajectory of cultural assimilation. Peñalosa, more than three decades ago, maintained that Mexican Americans were "clearly moving further away from lower-class Mexican traditional culture and toward Anglo-American middle-class culture" (1967:43). He believed that Mexican Americans were moving increasingly into the mainstream of American life, just as previous immigrant groups had done. The Latino National Political Survey revealed much evidence that Latinos in general, not only Mexican Americans, are in fact moving in that direction (de la Garza et al., 1992). Regarding social networks and social distance, for example, although members of each specific national-origin group felt closer to and interacted more frequently with coethnics (that is, Mexicans with Mexicans, Cubans with Cubans, and so on), most felt closer to Anglos than to any of the other Hispanic national-origin groups.

The survey found that most Latinos do not identify themselves as members of a panethnic group—Hispanic or Latino—but as members of their specific national group (Mexicans, Puerto Ricans, Cubans) or simply as Americans. Although members of the three groups expressed common views on some issues and attitudes, on many they also expressed marked differences. Hence, "to the extent that the Hispanic political community exists, there is scant evidence that it is rooted in alleged distinctive cultural traditions such as Spanish-language maintenance, religiosity, or shared identity" (de la Garza et al., 1992:13–14). In sum, the survey found that Latinos were well within the mainstream of American political thought and action. "There is no evidence here," the study concluded, "of values, demands, or behaviors that threaten the nation's cultural or political identity" (de la Garza et al., 1992:16).

Moreover, as with African Americans, cultural assimilation is not entirely a one-way process. Rather, there is a growing interchange of Latino and American core cultures. Latino cultural influences are increasingly evident in American music, language, fashion, and cuisine: Salsa and Latin jazz are now standard musical genres; pop artists like Jennifer Lopez, Ricky Martin, and Marc Anthony sell millions of records; Spanish is today the most popular foreign language, taught nationwide in primary and secondary schools and in every U.S. college; and Americans prefer salsas and picantes as often as ketchup (Fabricant, 1993; Garcia, 1999; La Ferla, 2001). In a sense, what we are witnessing is continued assimilation alongside ethnic cultural affirmation—a phenomenon similar to that experienced by earlier generations of immigrants.

Secondary Structural Assimilation As we have already seen, all three major Hispanic-American groups continue to lag behind Anglos in income, occupation, education, and political power. Despite these shortcomings, steady—if slow—progress has been made in closing the gap. Each succeeding generation has improved its status over that of the previous one, and it appears that Latinos are moving into greater participation in the society's mainstream economic and political systems. The pace and extent of secondary structural assimilation, however, have been uneven for the three major groups. Puerto Ricans have proceeded more slowly and less effectively than Mexicans, who in turn have not progressed as rapidly as Cubans.

Entering the urban environment later than other ethnic groups, Mexicans and Puerto Ricans have not been able to use the economic and political systems as leverage in their pursuit of upward mobility. By the time they began to enter the cities in great numbers, the unskilled factory jobs that prior groups had used as economic springboards were already dwindling. And the local political machines, which had doled out many benefits to immigrant groups, were also no longer of much impact. The economic and political conditions faced by these groups, then, are closely parallel to those encountered by African Americans.

The factor of race in the case of Puerto Ricans and, less so, of Mexican Americans may also account for some of the lag in secondary structural assimilation (Rodríguez, 1996). As noted earlier, prejudice and discrimination, though not as well rooted and intense as those experienced by African Americans, are continuing facts for Latinos, particularly in areas where they constitute a numerically significant part of the population.

Cubans in the United States have clearly exceeded Mexicans and Puerto Ricans in both the pace and degree of economic success. However, their upward economic mobility in the past four decades has been not so much within the mainstream economic system as within a parallel Cuban economy, particularly in the Miami area. Nonetheless, given their class advantage—and, for most, their racial indistinctness—assimilation at the secondary level should be a less painstaking and lengthy process than it has been for either Mexican Americans or Puerto Ricans.

Primary Structural Assimilation At the primary level of structural assimilation, Latinos have clearly advanced further than African Americans. We have already seen that on one important measure of primary structural assimilation, residential integration, Latinos have begun to show strong signs of dispersion. Nearly half of Hispanics now live in suburbs, for example (Mumford Center, 2001b). There are similar indications in other areas.

As explained earlier, the rate at which members of an ethnic minority group marry those of the dominant group is perhaps the best evidence of how far the assimilation process has advanced. We have seen that marriage across ethnic lines has occurred increasingly among Euro-American ethnic groups, like Italians and Jews, but has not occurred significantly among African Americans. As in other aspects of comparison, Latinos seem to fall somewhere between the two.

Past studies have shown a trend toward increasing marital assimilation for Hispanics (Fitzpatrick, 1987; Grebler, Moore, and Guzman, 1970; Jaffe, Cullen, and Boswell, 1980; Mittelbach and Moore, 1968; Murguía and Frisbie, 1977; U.S. Commission on Civil Rights, 1976; Valdez, 1983). Later data indicate a continuation of that trend. In 1999 there were more than 1.6 million couples in the United States in which one spouse was of Hispanic origin and one not. These constituted almost 27 percent of the 6 million couples in which the husband or wife were of Hispanic origin (U.S. Census Bureau, 2001h).

Internal Colonialism and Hispanic Americans

Although there is evidence that Latinos are assimilating into the mainstream society, albeit at a slow and uneven pace, there are nonetheless aspects of group history and interrelations with Anglos that conform to the internal colonial model.

The Mexican-American experience, in particular, seems to closely fulfill several features of internal colonialism: forced entry of the dominant group, alteration of the indigenous culture, administration by the dominant group, and the application of a racist ideology. Following the U.S. conquest, Mexican culture was denigrated by Anglos, indigenous Mexicans were stripped of political and economic power, and the system of dominant-subordinate relations that emerged was justified by a racist ideology: Mexicans, it was believed, were innately inferior to Anglos.

Some have maintained that the conditions established at the initial encounter of Mexicans and Anglos in the Southwest have been upheld over the years so that Mexicans, like African Americans, remain a colonized people who lack the fundamental political and economic power to control their destiny. Schools, police, businesses, and other vital institutions, it is argued, are controlled by Anglos and do not serve the needs and interests of Latino communities (Acuña, 1972; Alvarez, 1973; Garcia and de la Garza, 1977; Murguía, 1975). Unlike European immigrants who were able to utilize unskilled jobs to sponsor either their own or their children's upward mobility, Latinos, in this view, have been held at the lowest occupational levels not only because of the society's changing work structure but also because of discrimination. The depressed barrios of southern Texas and other parts of the Southwest are seen as similar to urban black ghettos, and in such cases, the internal colonial model is deemed relevant.

Although they have entered the society as voluntary immigrants, Puerto Ricans also seem to evince several aspects of the internal colonial model, notably minimal control of their communities on the mainland. Puerto Rico itself has been an American colony for most of the twentieth century, and though its official status was changed in 1952 to "commonwealth," the colonial-like relationship between island and mainland has remained essentially unchanged (Christopulos, 1980). The status of Puerto Ricans in this situation is somewhat in limbo. Although they are officially American citizens, they are not permitted to vote in presidential elections and do not have voting representatives in the U.S. Congress. The issue of commonwealth status, statehood, or full independence for the island has been an overriding and often emotional issue in Puerto Rican politics and remains unsettled among both Puerto Ricans and American policymakers (Rodríguez-Vecchini, 1994).

Most important, Puerto Ricans have traditionally been viewed with a colonial mentality. This is largely a product of their racial background, which in the American racial system often places them among the darker-skinned categories. As noted earlier, racial factors have seriously impeded Puerto Rican assimilation. The dilemma of Puerto Ricans regarding their racial status in the United States is expressed dramatically by Piri Thomas, a dark-skinned Puerto Rican who has written of his New York experience. Thomas tells of an early incident that thrust on him the American race ethic. After applying for a door-to-door sales job, he is told by the employment manager that he will be called when "the new territory is opened up." His white friend applies immediately afterward and is told to start "on Monday." Realizing what has happened, Thomas relates the story to his black buddy, who tells him that "a Negro faces that all the time." "I know that," Thomas replies, "but I wasn't a Negro then. I was still only a Puerto Rican" (1967:104).[6]

Assimilation versus Pluralism: An Assessment

Clearly, neither the assimilation model nor the internal colonial model of ethnic relations alone can account for the patterns of adjustment of Hispanic Americans and their relations with Anglos. Again we return to the theme of intermediacy, or in-betweenness, of Latinos. Unlike Euro-Americans, who obviously fit the assimilation model, or African Americans, who in large measure do not, Hispanic groups fall somewhere between these two extremes.

Murguía has written that Mexican Americans "seem to lie in that area between those groups that are so culturally and racially similar to the dominant society that they will assimilate completely in time, and those groups that are so racially and culturally divergent from the majority society that, given present attitudes and behavior, they will never assimilate with the dominant society, but instead will proceed with a decolonization movement of cultural nationalism and separatism" (1975:57). This seems equally true in the case of Puerto Ricans. At present, only for Cubans among the three major Hispanic-American groups does the traditional ethnic, or assimilation, model seem largely relevant (Arias, 2001). For Mexicans and Puerto Ricans, however, the nature of their entry into the society, their racial characteristics, and their class handicap make the assimilation model less clear-cut and impose at least some aspects of the internal colonial model.

Some analysts maintain that Hispanics, for the most part, may be following the general path of assimilation followed earlier by Euro-American groups, though their slower and more tortuous movement in that direction is occurring with little notice. Linda Chavez (1991), for example, argues that ongoing Hispanic assimilation is concealed by the continued entrance of new Hispanic immigrants in great numbers who remain more attuned to the ethnic culture and language. She also contends that Hispanic leaders, strongly committed to multiculturalism, are responsible for generating heightened ethnic consciousness in Latino communities in their attempts to retard assimilation. We will revisit the issues of immigration and multiculturalism in Chapter 12.

In considering the future of Hispanic Americans it is well to reiterate a point emphasized at the outset of this chapter: There are extremely diverse class, ethnic, and racial origins among those who make up this panethnic category, as well

---

[6]Although the problem of racial identity remains unsolved, New York Puerto Ricans often identify themselves as "New Yoricans," "New-Ricans," or "Ricans," terms implying recognition of and pride in the synthesis of the two cultures (Duany, 1998; Rodríguez, 1989).

Part II: Ethnicity in the United States

as great differences among its generations. As with Euro-Americans and, as we will see in the following chapter, Asian Americans, it is misleading to describe the broad composite of groups that make up this category in holistic terms. To speak of "Hispanic Americans" as if this were a term with real meaning rather than a convenient Census Bureau category is, as we have seen, grossly simplistic and inaccurate. Hence, the kind of adaptation Hispanic Americans make in their absorption into American society—as well as the imprint they make on the society and its culture—will undoubtedly be disparate, if uniformly strong.

SUMMARY

As part of the American ethnic system, Hispanic Americans occupy an intermediate status between Euro-Americans and African Americans. Three major groups make up the Hispanic-American minority—Mexicans, Puerto Ricans, and Cubans—but groups from other Latin American and Caribbean societies have, in the past two decades, become a large and significant component of this panethnic category. Each group is concentrated regionally, and each is primarily urban. Although Mexicans, the largest Hispanic group, have entered American society mostly as voluntary immigrants, their roots in the United States reach back to Spanish colonial times before the American conquest of the Southwest. Puerto Ricans are technically American citizens, yet they have arrived from a foreign culture, much as other immigrant groups have. Cubans are more recent arrivals and differ from Mexicans and Puerto Ricans in class origins and racial features. Newer groups from Central and South America and the Caribbean have entered as voluntary immigrants, though some, particularly from Central American countries, arrived as political refugees.

Hispanic Americans generally are below average on all measures of socioeconomic status and are a large element of the American poor. There are differences among the major groups, however, with Cubans in a relatively higher position than Mexicans, who are somewhat higher than Puerto Ricans. Latinos are in a less developed stage of political participation than are African Americans, and like the latter, are underrepresented in top positions in all institutional areas.

The level of prejudice and discrimination directed at Hispanic Americans has not been as intense as that experienced by African Americans but has been more severe than that suffered by Euro-American ethnic groups. Mexican Americans have commonly been the chief targets of ethnic hostility in the Southwest, where traditionally most have lived.

As with African Americans, there is some disagreement over whether Hispanic Americans more appropriately fit the assimilation model or the internal colonial model. Although in some regards Cubans seem to conform to the immigrant experience of previous white ethnic groups, in others they are more

insular. Mexicans and Puerto Ricans exhibit certain features of internal colonialism. Compared with both African American and Euro-American ethnic
groups, Hispanics have displayed a slower pace of cultural assimilation, though
movement in that direction is clearly evident. The speed and extent of secondary
structural assimilation for Latinos have generally resembled African-American
patterns, but at the primary level, Latinos have outpaced African Americans.

Suggested Readings

Chavez, Linda, 1991. *Out of the Barrio: Toward a New Politics of Hispanic Assimilation*.
New York: Basic Books. Argues that the economic and social assimilation of
Hispanic Americans is concealed by the flood of new Hispanic immigrants and
by Hispanic leaders with an interest in fueling separatism.

Connor, Walker (ed.). 1985. *Mexican-Americans in Comparative Perspective*.
Washington, D.C.: Urban Institute Press. Essays by social scientists that deal
with social, economic, and political issues of the largest Hispanic-American
group.

Fitzpatrick, Joseph P. 1987. *Puerto Rican Americans: The Meaning of Migration to the
Mainland*. 2d ed. Englewood Cliffs, N.J.: Prentice-Hall. A comprehensive study
of the Puerto Rican experience in the United States, especially New York,
where the bulk of the Puerto Rican population has traditionally resided.

Grasmuck, Sherri, and Patricia R. Pessar. 1991. *Between Two Islands: Dominican
International Migration*. Berkeley: University of California Press. Examines one
of the newer, and growing, Hispanic groups in the United States, particularly
the process of their immigration from the Dominican Republic.

Grenier, Guillermo J., and Alex Stepick III (eds.). 1992. *Miami Now! Immigration,
Ethnicity, and Social Change*. Gainesville, Fla.: University Press of Florida. A
collection of articles describing the social composition of the most thoroughly
Hispanic city in the United States and the political, economic, and social issues
that center on its various ethnic groups.

Massey, Douglas, Rafael Alarcón, Jorge Durand, and Humberto González. 1987.
Return to Aztlan: The Social Process of International Migration from Western Mexico.
Berkeley: University of California Press. An analysis of the factors that motivate
Mexicans to immigrate to the United States. Shows that immigration is driven
not only by economic circumstances but also by the establishment of social
networks.

Portes, Alejandro, and Robert L. Bach. 1985. *Latin Journey: Cuban and Mexican
Immigrants in the United States*. Berkeley: University of California Press. Deals
with the economic and social adaptation of immigrants of two of the major
Hispanic groups in the United States. Also contains an important theoretical
discussion of immigration and of different modes of immigrant adaptation.

Rieff, David. 1993. *The Exile: Cuba in the Heart of Miami*. New York: Touchstone.
Describes the Cuban community of Miami—how it has affected the Anglo

community and, conversely, how it has increasingly been transformed by American culture.

Rodriguez, Richard. 1983. *Hunger of Memory: The Education of Richard Rodriguez.* New York: Bantam. In this autobiography, Rodriguez tells of his Mexican-American origins and his path toward assimilation into mainstream American society. Argues strongly against bilingual education.

Suro, Roberto. 1998. *Strangers Among Us: How Latino Immigration Is Transforming America.* New York: Alfred A. Knopf. Weaving personal accounts into his narrative, Suro portrays the wide class and ethnic discrepancies among the specific groups that make up the Hispanic-American population. Describes how Latinos are changing American society at the same time they are being changed by it.

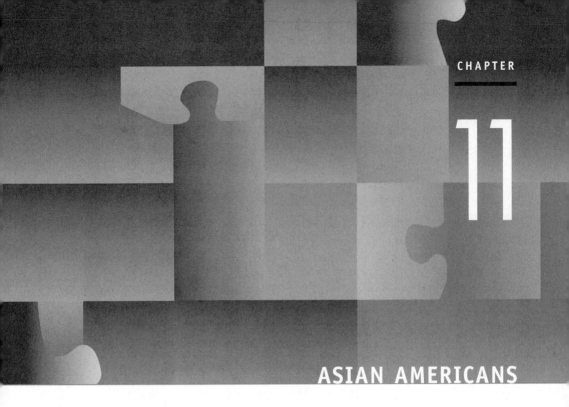

ASIAN AMERICANS

ASIANS HAVE BEEN in the United States since the early part of the nineteenth century. Only in recent decades, however, have they emerged as a sizable element of the American ethnic configuration. Like other ethnic populations, Asians have displayed variant patterns of settlement and adaptation and have been received by the dominant group in different ways, depending on social, political, and economic circumstances. As it has been with other groups, our objective is to trace the historical development of Asians in the United States and then to examine their place in the stratification system, the forms of prejudice and discrimination they have encountered, and their patterns of assimilation into the larger society.

At the outset, it is important to stress that Asian Americans constitute an extremely diverse panethnic category. In some ways it is even more disparate than the Hispanic-American category, most of whose groups share a common language and religious tradition. To lump together those from numerous Asian societies as "Asian Americans" is therefore a serious oversimplification. In fact, Asian Americans consist of at least a dozen distinct groups, with extremely diverse cultural and even physical features. Currently the largest groups are Chinese, Filipino, and Asian Indian. But substantial populations of Japanese, Koreans, and Vietnamese, as well as small numbers from other Asian societies, are included. The term *Asian American* must be understood as nothing more than a convenient category that enables us to look at the general characteristics of the various Asian groups together, in comparison with other ethnic populations in the United States.

IMMIGRATION AND SETTLEMENT

The Asian experience in the United States can be divided into two distinct and divergent parts. The first consists of the old Asian immigration, occurring roughly from the middle of the nineteenth to the early years of the twentieth centuries. The Chinese were the first to arrive, followed by the Japanese and, in much smaller numbers, Koreans and Filipinos. These first Asian immigrants were mostly unskilled laborers, recruited for construction or agricultural work. Following restrictive measures, very few additional Asians entered the society until the revision of immigration laws in 1965. That marked the onset of the second period of immigration to the United States from Asia, which continues to the present.

The second, and current, immigration has been markedly different from the earlier movement in that most of the new immigrants have been noticeably higher in class origin, and many have been well educated and occupationally skilled. Moreover, the new Asian immigration has been considerably more diverse in national origin, made up of people from almost every contemporary Asian society. Today, except for the Japanese, Asians in the United States are predominantly first-generation immigrants (Lee, 1998).

Chinese Americans

The Asian-American experience may be said to begin in the mid-nineteenth century. Unfavorable social and economic conditions and political unrest at that time in China, combined with the lure of labor opportunities in Spanish, Portuguese, Dutch, and British colonies, created a swell of Chinese emigration, mostly from the country's southeastern coastal provinces.[1] The United States was also among the destinations of these emigrants, mostly single males, some of whom were lured by the discovery of gold in California in the 1840s (Dinnerstein and Reimers, 1988; Lyman, 1970). In the 1860s, the construction of the transcontinental railroads drew many additional Chinese laborers to the United States. Like other immigrants before them, most intended to return after bettering themselves economically.

---

[1] Chinese immigration over many generations has constituted one of the great diasporas. Today Chinese communities can be found in almost every country of the world (see Poston and Yu, 1990). Ethnic Chinese minorities have played an especially important economic role in most Southeast Asian societies (Minority Rights Group, 1992).

In a very short time, the influx of Chinese produced one of the most hostile movements in American ethnic history. The cast of an anti-Chinese movement had been set even before the Chinese landed in the United States, notes Stanford Lyman, having been "preceded by a richly embellished but almost entirely negative stereotype" (1974:56). Perceived by white workers as a labor threat, the Chinese were forced out of their jobs and commonly driven from one town to another. In addition to the antagonism they encountered from hostile white workers, the Chinese were severely restricted by legal measures used to limit their occupational and residential movement. These culminated in the Chinese Exclusion Act, passed by the U.S. Congress in 1882 and designed to preclude Chinese immigration for a ten-year period. The act was extended for another ten years in 1892, however, and the ban against Chinese was made permanent in 1907. These measures marked a watershed in American ethnic history: For the first time, a specific group was formally barred from entrance. In fact, the Chinese Exclusion Act was the precursor of numerous anti-immigration laws that would be enacted in the following decades.

Although the exclusionary laws never fully suspended Chinese immigration, they effectively checked any significant population growth of the Chinese in America. When the initial act was passed in 1882, there were around 125,000 Chinese in the United States. For the next six decades, the Chinese population not only failed to grow but actually declined. By 1910 it had dropped to around 70,000, a figure that did not change radically until after 1965. This decline was the result not only of the restrictive legislation itself but also of the overwhelmingly male composition of the Chinese population that had made up the mid-nineteenth-century immigration. Before 1882, more than 100,000 men but fewer than 9,000 women had immigrated to the United States from China (Lyman, 1974).

Because the Chinese immigrants were almost totally male, there existed no possibility of a natural population increase and thus no creation of a stable and thriving ethnic community. The severely unbalanced sex ratio, in addition to the relentless discrimination they experienced, forced the Chinese at the turn of the century into urban ghettos, the familiar "Chinatowns," from which they did not begin to disperse until well after World War II.

By the early 1960s the Chinese population had begun to grow somewhat, due to the repeal of the Chinese Exclusion Act in 1943 as well as to the entry of war brides, refugees, and some scientific personnel after World War II. This growth also helped correct the unbalanced sex ratio. As with other non-European groups, however, it was the revised immigration legislation of 1965 that created a large-scale Chinese population increase in the United States. Not only were more immigrants admitted, especially from Hong Kong and Taiwan, but the

Chinese population was augmented as well by a large number of Vietnamese refugees who, though having lived in Vietnam, were ethnically Chinese. We look more closely at these new immigrants, along with others from Asia, in later sections of this chapter.

Japanese Americans

The formal exclusion of the Chinese provided for the entrance of Japanese immigrants beginning in the latter years of the nineteenth century. In their early immigration and settlement patterns, the Japanese in some ways resembled the Chinese who preceded them. Most of the first Japanese immigrants were males who were confined to lower-status occupational positions. And, like the Chinese, they soon came to be seen as a labor threat in California, where most had settled. The result was a movement of severe antagonism, culminating in formal legislation that prohibited their further entrance. In 1907–1908, the United States and Japan entered into a Gentlemen's Agreement, which effectively restricted further Japanese immigration to the United States. Under the terms of the agreement, only nonlaborers and relatives of resident Japanese would be permitted to enter. The Oriental Exclusion Act of 1924 carried the restriction a step further, effectively barring *all* subsequent Japanese immigration.

Despite some similarities, in a number of important ways the early Japanese immigrant experience differed from that of the Chinese. To begin with, although they initially worked in many of the same jobs as the Chinese, a large percentage of the early Japanese immigrants gradually turned to farming, where they carved out a niche for themselves in the California economy. More important, the sex ratio of the Japanese immigrant population, though heavily male in the early years, eventually was balanced with the introduction of a substantial female immigration. The Gentlemen's Agreement of 1907–1908 had permitted wives and families of men already in the United States to join them. As a result, the social difficulties created by the demographic imbalance in the case of the Chinese did not materialize. With families intact, the base was provided for a second generation and thus a natural increase in the Japanese-American population. By 1940 nearly two-thirds of the Japanese in the United States were native born (Kitano and Daniels, 1995). Today, unlike other Asian-American groups, the Japanese have not increased their numbers through large-scale immigration. As a result, the Japanese-American population has stabilized and is rapidly being surpassed by other, more recent, Asian groups.

The demographic and historical characteristics of the Japanese-American group today set it apart in some ways from others that make up the Asian-American population. Perhaps most important, because most are native born, Japanese Americans have a longer history in the United States and are generationally divergent from others.[2] As a result, Japanese-American assimilation is far more advanced than for other Asian groups. Also, there is a strong anti-Japanese sentiment among many of the recent Asian immigrants, stemming from the harsh experiences of their home countries under Japanese colonization, particularly during World War II.

Other Asian-American Groups

The Chinese and Japanese were the first and largest of the Asian groups to enter the United States. Not until the liberalized immigration law of 1965 did a new stream of Asian immigrants augment these two communities. The number and variety of Asian societies represented by the post-1965 stream were considerably greater than any previous immigration, and it is these groups that today make up the major components of the Asian population in the United States. Although small numbers of all these groups had been in the society since the late nineteenth century, none could be said to represent a sizable community comparable to those of the Chinese and Japanese. Other than the Japanese, no more than half the total population of any Asian-American group is native born. The history of these new Asian-American groups, therefore, is essentially of the last three decades.

Koreans In the early years of this century, a few thousand Koreans were recruited to work the sugar plantations in Hawaii. Ironically, they were needed to replace Chinese workers who had been barred by the 1882 Chinese Exclusion Act. Further Korean migration to the United States in any significant numbers did not occur again until the late 1950s, following the Korean War, when many came either as refugees or war brides. Prior to this time, the Korean population on the U.S. mainland never exceeded 2,000 (Light and Bonacich, 1988).

It was in the post-1965 period, however, that the formidable growth of a Korean ethnic group in the United States occurred. Between 1970 and 1980 the Korean-American population increased five times—from 70,000 to 355,000. By 2000, Korean Americans numbered over 1 million. Like other Asian groups, Koreans have concentrated in California, where almost one-third of the Korean population is found—17 percent in Los Angeles alone (Light and Bonacich, 1988). Contemporary Korean immigrants have tended to be from the urban middle class of Korean society, including many college-trained professionals. What's more, most have come as family units, and most are Christians (Kitano and Daniels, 1995; Min, 1991). Koreans in the United States have been

---

[2]Although the Chinese have an even longer history in the United States, most of the current Chinese-American population is foreign born.

especially prominent as owners of small businesses, usually operated as family enterprises (Light and Bonacich, 1988; Kim, 1981; Min, 1988, 1995).

The flow of immigration between 1970 and 1990, which had created such a substantial increase in the Korean-American group, declined considerably during the 1990s; and many chose to return to Korea. Economic opportunities in Korea were seen by many as more promising, and the political instability that had characterized Korea in the past had been reduced. These factors diminished the appeal of immigration to the United States. They also created an incentive for some to return to Korea. For those who had been successful in the United States, having earned a university degree or having run a profitable business, return represented an opportunity to enjoy even greater success in their country of origin, where they may have felt more culturally at ease. Return was also appealing for those who had experienced increasing difficulties in adapting to American culture or who found that operating a small business, so common among Korean Americans, was simply too risky in terms of physical safety as well as economic security (Belluck, 1995; Min, 1995a). The image of Korean shops being burned and looted during the Los Angeles riots of 1992 and the frequent harassment of Korean shopowners in other cities made Korean Americans more conscious of ethnic conflict, a condition not encountered in a homogeneous Korea.

Filipinos Few people think of Filipinos[3] when they consider the makeup of America's Asian population. Yet, Filipinos are now the second-largest Asian-American group. As with other Asian groups, the liberalization of immigration laws in 1965 accounts for their rapid and substantial increase. During the past three decades, only Mexicans have outnumbered Filipinos as immigrants to the United States.

The Philippines, before becoming an American possession in 1899, was a colony of Spain for over three centuries. As a result, most Filipinos are Roman Catholics and have Spanish surnames. After the Spanish-American War, the United States took possession of the Philippines, and the American cultural influence became dominant. Following independence in 1946, political ties between the two countries remained close, and American cultural and economic influences continued to be strong. English is one of the country's major languages and is spoken by most educated people. Historically and culturally, then, the Philippines has had little in common with other Asian countries. Geographically, however, it is part of the Pacific Rim, and thus Filipinos are considered Asian people.

---

[3]There is no "F" sound in Tagalog—the language, other than English, that is most widely spoken in the Philippines. Hence, rather than Filipino, Pilipino is sometimes used as a preferred term of reference.

No group better illustrates the link between immigration and labor needs than do the Filipinos. Like the early Koreans, a few thousand Filipinos came to Hawaii in the early 1900s, recruited as agricultural workers to replace the excluded Chinese. Many more immigrated to the mainland in the 1920s as California farm producers faced a labor shortage created by newly restrictive quotas limiting cheap labor from Mexico. By 1930, almost 50,000 Filipinos were living on the mainland United States, most in California (Mangiafico, 1988). Because the Philippines at this time was a U.S. territory, Filipino immigrants were actually considered American nationals though not U.S. citizens. Because of this, they were not subject to the same kind of quota restrictions as other Asians until 1935, when the Philippines was granted deferred independence.

Like the Chinese before them, early Filipino immigrants were almost entirely single males. And with the general anti-Asian social and political climate on the West Coast, where most settled, Filipinos found themselves subjected to the same kinds of discriminatory actions as the Chinese and Japanese. Following the outbreak of World War II, the Philippines was a critical ally in the Pacific war against Japan, and as a result the social climate in the United States reflected a more tolerant view of Filipinos.

Along with discrimination, the low level of education of Filipino immigrants produced a restricted range of occupational opportunities. Most worked as seasonal agricultural laborers and in the service sector in low-status positions such as restaurant workers, hospital attendants, and hotel workers.

Between 1960 and 1970 the number of Filipinos in the United States nearly doubled. Cultural compatibility—especially language—and political ties account for the attraction of the United States for Filipinos. Also, the strong U.S. military presence in the Philippines for over four decades following World War II enabled many Filipino women to immigrate as wives of U.S. servicemen. It is important to consider that the social characteristics of the new immigrants were considerably different from those of the earlier period. Most were now families with children, seeking the promise of better economic opportunities. Recent Filipino immigrants have also been better educated than those of earlier decades, and many come with professional credentials, enabling them to move into highly skilled occupational fields like medicine and education (Cariño, 1987). Their generally higher occupational positions, along with English proficiency, have dictated against the formation of cohesive and distinguishable ethnic enclaves comparable to Chinatowns (Agbayani-Siewert and Revilla, 1995).

Vietnamese The entrance of Vietnamese into the United States has been unlike that of any other Asian group. Most have come as political refugees since the end of the Vietnam War. Prior to that time, very few Vietnamese had been resident in the United States.

Between 1966 and 1975, some 20,000 Vietnamese arrived in the United States, but the Vietnamese immigration reached epic proportions following the fall of the South Vietnamese government in 1975. In the nine years between 1975 and 1984, over 700,000 Southeast Asian refugees came to the United States, most of them Vietnamese (Gardner, Robey, and Smith, 1985). Perhaps the most dramatic entrance of these refugees occurred in 1979 and 1980 when thousands fled Vietnam, Laos, and Cambodia in crude boats. The United States absorbed over 200,000 of these so-called boat people in a period of fifteen months. Most of the refugees coming in the second wave were unlike the earlier Vietnamese in that they were relatively unskilled occupationally, less educated, and spoke little English (Wong, 1986). Regarding differences in education, for example, almost 80 percent coming to the United States between 1965 and 1969 were college graduates, as opposed to less than 16 percent of those coming between 1975 and 1980 (U.S. Commission on Civil Rights, 1988). Their adjustment to American society, as a result, was far more difficult, and today they remain the most economically depressed element of the Asian-American population. The Vietnamese, then, display a bipolar class breakdown—those who have successfully adapted to American society and are at a comfortable economic level, and those at the opposite end of the class hierarchy, disproportionately unemployed, among the poor, and socially alienated (Weiss, 1994).[4] Other Asian-American groups show a somewhat similar class division.

Though the heaviest concentration of Vietnamese is in California, the group has dispersed widely throughout the United States. Substantial Vietnamese populations can be found today in Houston, New Orleans, and Arlington, Virginia, among others. The largest single community, however, is in Orange County, California. There, in the town of Westminster, a thriving population of 80,000 has created a Little Saigon, made up of about 1,000 stores, offices, and restaurants serving the ethnic enclave.

Other Southeast Asians In addition to the Vietnamese, several hundred thousand refugees from other Southeast Asian countries arrived in the United States following the end of the Vietnam War in 1975. The largest among these were Laotians (mostly Hmong people) and Cambodians.

These groups, like many Vietnamese, have not easily adapted to American society. The intention of U.S. policy when they entered as refugees was to resettle them in various regions and cities. The expectation was that this would accelerate their adaptation and would avoid any negative impact on the labor market in communities that might absorb a large concentration of the refugees.

---

[4]As a result of subsequent immigration waves, the Vietnamese in America today constitute not only a class-diverse group, but one that reflects a cross section of Vietnamese society in terms of language, religion, and region of origin.

The resettlement of the Hmong, in particular, has been fraught with serious problems (Beck, 1994). Culturally, the group had not been exposed to Western ways before emigrating from a relatively isolated region of Laos. The class and educational characteristics of most of the refugees have been significantly lower than those of other Asian groups, further contributing to their problems of adaptation.

Asian Indians Like the Chinese, Indians have a long heritage of immigration to societies in all parts of the world. Until recently, however, the United States had not been one of their principal destinations. In the nineteenth century, most had gone as indentured servants to various parts of the British colonial empire, including South Africa and a number of Caribbean societies. In the early years of this century, a few thousand workers from India were recruited as agricultural laborers in California and Washington; but because of immigration restrictions, the Indian population did not expand. Like other Asian immigrants at this time, they experienced extremely virulent forms of discrimination. In this racially hostile atmosphere, the relatively small Asian-Indian population remained isolated in agricultural communities, mainly in California.

As with other Asian groups, the liberalized immigration measure of 1965 represented the point at which Indians began to enter the United States in significant numbers. And like the others, the new immigrants were sociologically quite distinct from their earlier counterparts. Those who have come in recent decades represent, for the most part, a highly educated and occupationally very skilled population. Indeed, their educational level, compared with that of other ethnic groups, is quite remarkable. Almost 60 percent of the adult Asian-Indian population are college graduates, and fully one-third have earned graduate or professional degrees (U.S. Census Bureau, 1993a). Clearly, Asian Indians are among the most highly educated segment of the American population. What's more, their income levels are far above average.

Most Indians coming to the United States, then, represent a very select subset of the Indian population—more educated, more skilled, and more generally privileged. They are part of the worldwide migration of those with educational and occupational skills leaving developing societies in favor of the industrialized societies of the West, where economic and professional opportunities are more plentiful. In fact, many of the Asian-Indian immigrants of the recent period have been students at American universities who, after completing their studies, find employment and remain. Asian-Indian software engineers, scientists, and other highly skilled technical workers constitute a huge segment of the workforce of the American computer industry (Boxall, 2001). The relatively high occupational status and high level of education of Indian immigrants in the past two decades have enhanced the speed and extent of their economic and social adaptation to American society. Indians have also benefited from the fact that on arrival most are fluent in English and have already been exposed to Western values.

In the past decade, the Asian-Indian community in the United States has become more diverse, with a larger commercial class augmenting the predominantly professional and technical character of the first post-1965 immigrants.[5] Hence, in addition to their place in skilled occupational areas, Asian Indians have played a significant role as entrepreneurs in the small-business sector; for example, more than half of all motels in the United States today are owned by Asian Indians (Varadarajan, 1999). This entrepreneurship is comparable to the economic path followed by the recent generation of Korean immigrants. In California's Silicon Valley, where their presence in the computer industry is so evident, immigrants who are not high-tech workers—taxi drivers, restaurant workers, truckers, and so on—have begun to create a more class-diverse Asian-Indian community (Boxall, 2001).

It is important to understand that those classified as "Asian Indian" represent a group that is internally quite diverse culturally. India itself is an ethnic potpourri, with virtually dozens of languages and cultural communities. The religious division among Hindus, Muslims, and Sikhs is only the most apparent and profound variance. Moreover, the national and geographical origins of Asian Indians in the United States are varied. While most have come from India, sizable numbers have come from Pakistan, Bangladesh, and Sri Lanka, as well as other countries where East Indians had settled earlier in the nineteenth century as part of the Indian diaspora. Among the latter are Indo-Caribbeans, from Caribbean countries like Trinidad and Tobago and Guyana (Sengupta and Toy, 1998), and those from Kenya and Uganda, who had been forced to flee these African countries when they were expelled by oppressive governments.

Asian-American Demographics Today

Today, Asian Americans remain a relatively small part of the total American population—3.6 percent—but one that is rapidly growing not only in numbers but in economic and cultural significance.[6] In fact, the Asian-American population is today the fastest-growing ethnic minority in the United States. Between 1970 and 1980 it grew by 128 percent, and between 1980 and 1990, by 108 percent. It is projected that the current Asian-American population of 10.2 million will double by 2020, and by 2050 it will be 38 million (U.S. Census Bureau, 2000b). This rapid and sizable population growth is attributable mainly to the fact that Asians now account for almost a third of all immigration to the United States.

---

[5]It is interesting to consider that the post–World War II Indian immigration to Britain, by contrast, comprised not only a much larger but also a more heterogeneous population, with many unskilled and poorly educated among them (see Watson, 1977).

[6]If those who claim Asian ancestry in combination with one or more other racial categories are included, the figure jumps to 4.2 percent (U.S. Census Bureau, 2001d).

Figure 11.1 shows the breakdown of the Asian-American population today. In 1970, the Chinese and Japanese composed the bulk of the Asian-American population, but today these two groups are less than one-third. The Chinese category remains the largest component (almost 24 percent), with Filipinos the second largest (18 percent). The Asian-Indian population has grown more rapidly than any of the other Asian-American ethnic groups during the past decade—more than 100 percent—and is now the third largest, with almost 1.7 million. The Vietnamese and Korean populations have also continued to grow steadily, each with over 1 million. The Japanese-American population is growing much more slowly because of very limited immigration from Japan and a low fertility rate; with fewer than 800,000, it is today one of the smallest Asian-American groups.

During the 1980s, the Asian-American population expanded not only in size but in diversity as well. Hmong tribespeople from the mountains of Laos, for example, increased their numbers substantially, as did Cambodians (Worsnop, 1992).

The rapid increase in the Asian-American population is evident in the numbers of those who arrived in the past twenty years. In 1990, two of every three Asian Americans were foreign born, including 80 percent of Vietnamese, 75 percent of Asian Indians, and 73 percent of Koreans (Lee, 1998). The majority entered the United States after 1980. Among the Asian-American groups, only the Japanese are not predominantly foreign born.

Although they have begun to settle in all areas of the United States, Asian Americans are most heavily concentrated in five states: California, Hawaii, New York, Illinois, and Washington. More than one-third reside in California alone. Among Asian Americans, only Asian Indians have not clustered in

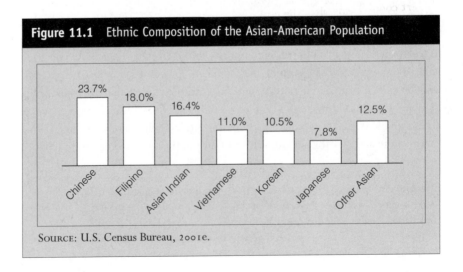

Figure 11.1 Ethnic Composition of the Asian-American Population

SOURCE: U.S. Census Bureau, 2001e.

California, having dispersed more widely. Also, like other ethnic populations that have arrived primarily in the post-1965 era, Asian Americans dwell mostly in large metropolitan areas. Almost half the Asian-American population is found in just four urban areas—Los Angeles, New York, San Francisco, and Honolulu. As shown in Table 11.1, Asians make up more than 11 percent of the total metropolitan population of Los Angeles, and in San Francisco, they make up almost 19 percent.

One of the more intriguing demographic patterns exhibited by Asians, compared to other nonwhite groups, is their rapid rate of suburbanization. Unlike blacks and Hispanics, Asians, especially those coming since 1970, have been leaving the central cities in ever greater numbers (Alba et al., 1999). In the Los Angeles metropolitan area, many of the Asian enclaves are now found in suburbs rather than the central city. Monterey Park is an interesting example. In the early 1970s, it was a Los Angeles suburban town of mainly Anglos and Hispanics. Now the population is about 50 percent Asian, mainly middle class, and most of its businesses are Asian-owned (Horton, 1995; Fong, 1994).

Table 11.1 Metropolitan Areas with the Largest Asian Populations

METROPOLITAN AREA	ASIAN POPULATION (1,000s)	PERCENT OF TOTAL METRO AREA
Los Angeles–Riverside–Orange County, CA	1,799	11.4
New York–Northern NJ–Long Island	1,343	6.7
San Francisco–Oakland–San Jose, CA	1,279	18.8
Honolulu, HI	566	64.9
Washington–Baltimore, DC–MD–VA–WV	373	5.1
Chicago–Gary–Kenosha, IL–IN–WI	367	4.2
San Diego, CA	297	10.7
Seattle–Tacoma–Bremerton, WA	280	8.2
Houston–Galveston–Brazoria, TX	229	5.2
Boston–Worcester–Lawrence–Lowell–Brockton, MA–NH	217	3.7
Philadelphia–Wilmington–Atlantic City, PA–NJ–DE–MD	179	3.0
Dallas–Fort Worth, TX	175	3.7
Sacramento–Yolo, CA	171	10.1
Detroit–Ann Arbor–Flint, MI	110	2.0

SOURCE: U.S. Census Bureau, 2001h.

ASIAN AMERICANS IN THE CLASS SYSTEM

Socioeconomic Status

What is perhaps most striking about the Asian-American population today is its unusually high ranking on most measures of socioeconomic status. Put simply, Asian Americans rank higher than most ethnic groups in family income, occupational prestige, and level of education.

The socioeconomic success of Asians in the United States, however, must be looked at with some caution. There are differences among specific groups as well as among members of the same group. For example, although nearly half of employed Asian Indians are professionals or managers, less than 20 percent of Vietnamese are in such occupations (U.S. Census Bureau, 1993a). The income and educational levels of the latter are also considerably lower than those for other Asian-American groups. Another internal difference is the tendency toward economic polarization. That is, members of specific Asian-American groups tend to cluster at the two extremes of the economic hierarchy, either at or near the top (such as professionals and business owners) or at or near the bottom (service workers and other low-status job classifications).

Income The family income of Asian Americans far exceeds that of most other minority ethnic categories and, with the exception of Koreans and Vietnamese, surpasses most Euro-American groups as well. Median family income for Asian Americans in 1999 was $56,316, compared to $54,121 for non-Hispanic whites and $48,950 for the U.S. population as a whole (U.S. Census Bureau, 1999a).

Here again the internal differences within the Asian category need to be carefully considered. Table 11.2 shows these differences in 1993. As can be seen, the median income for Laotians, for example, is less than half of what it is for Filipinos or Asian Indians. Moreover, although most Asian groups have higher family incomes than whites, they also have more family members in the labor force than whites.

The same internal discrepancy is evident regarding poverty. While the Asian-American poverty rate (10.7 percent in 1999) is only slightly higher than that for non-Hispanic whites (7.7 percent), it is extraordinarily high for some Southeast Asian groups, like Cambodians and Hmong.

Occupation The rate of labor-force participation for Asian Americans is higher than that of any other ethnic category, and the range of occupations held by Asian Americans is broad, extending from working-class jobs to professions such as medicine and engineering.

Here the split between those of previous generations and the more recent immigrants is very evident. In striking contrast to the earlier Asian immigrants,

Table 11.2 Asian-American Median Family Income

	ALL	NATIVE-BORN	FOREIGN-BORN	PERCENT IN POVERTY
All Asians	$41,583	$50,513	$40,011	11.4
Chinese	41,316	56,762	37,966	11.1
Filipino	46,698	42,114	47,595	5.2
Japanese	51,550	52,728	47,836	3.4
Asian Indian	49,309	38,256	49,567	7.2
Korean	33,909	42,427	33,401	14.7
Vietnamese	30,550	23,819	30,621	23.8
Cambodian	18,126	21,974	18,063	42.1
Hmong	14,327	22,171	14,272	61.8
Laotian	23,101	20,795	23,144	32.2
Non-Asians	35,108	35,451	30,354	9.9

SOURCE: U.S. Census Bureau, 1993a.

most of whom were unskilled laborers, Asians today are disproportionately among the most highly skilled. A key feature of the immigration reform of 1965 was the preferential treatment given to family members and to those with valuable occupational training. Because Asians had been virtually excluded for several decades, most of the early post-1965 immigrants entered not as family members but as skilled workers. As a result, many of the recent arrivals are found in professional, technical, and managerial jobs. Between 1976 and 1980, almost half of working Asian immigrants reported professional occupations, whereas no more than 10 percent of any Asian group reported laborer occupations (U.S. Commission on Civil Rights, 1988). For the past twenty years, Asian immigrants to the United States have been helping to fill shortages in a number of professional and technical areas. For example, 16 percent of the nurses and 18 percent of the lab technicians in the Los Angeles area are Filipino (Jacoby, 2000).

High occupational rank is characteristic of native-born and foreign-born Asian Americans alike. Except for Filipinos and Vietnamese, Asian-American men are more heavily represented in white-collar occupations and less represented in blue-collar occupations than are non-Hispanic white men. The percentage within the managerial and professional category is especially noteworthy. More than one-third of Asian-Americans are in managerial and professional occupations, about the same as for non-Hispanic white males and twice as high as for other ethnic minorities. The sharp stratification among Asian Americans is, again, very evident in the occupational world. At the same time that Asian Indians, Filipinos,

Japanese, Koreans, and Chinese are disproportionately in high-ranking professional and technical fields, a high percentage of Vietnamese and other Southeast Asians are in low-paying, unskilled occupations. The percentage of other Asian groups in unskilled jobs is about the same as for non-Hispanic whites and well below the percentage for blacks and Hispanics (Lee, 1998).

Asian Americans are rising faster in the corporate world than any other minority group and are increasingly occupying high-powered jobs in finance and advertising. One should not misread such trends as evidence of the arrival of Asian Americans into the corporate world on a par with Euro-Americans, however. They are underrepresented in managerial positions despite their comparatively large numbers in professional and technical positions. And at the very top of the corporate ladder, their numbers are as small as for African Americans and Hispanic Americans. In 1990, 0.3 percent of senior corporate executives were Asian Americans (U.S. Commission on Civil Rights, 1992). Asian Americans have often spoken of a glass ceiling, referring to the limits to which they are able to rise in the corporate world. For example, in California's Silicon Valley, the major center of the American computer industry, Asians make up as much as one-third of the engineering workforce. But they are not well represented in upper management at the larger companies. Some, as a result, have begun to start up high-technology enterprises of their own (Pollack, 1992). This was a route followed earlier, in the 1950s, by the late An Wang, Chinese-American founder of Wang Laboratories, for many years one of the country's largest computer companies.

Several explanations have been offered to account for the underrepresentation of Asian Americans in executive posts. Some see them at a disadvantage in vying for administrative jobs because of a lack of communication skills. Also, Asian Americans are often stereotyped as docile and thus not prepared for leadership positions (Min, 1995b). Despite their scant numbers in the executive suites of the American corporate economy, however, the extraordinarily high level of education of most Asian Americans will likely accelerate their movement into top management positions at a faster rate than that of other minorities. Furthermore, as the number of Asian Americans in the general population grows, corporations will find it in their interests to seek out and place more Asian Americans into higher-ranking positions, much as they have done with African Americans and Hispanic Americans.

A salient feature of the economic role played by Asian Americans today is the unusually large number who choose to own small businesses. Indeed, Asian Americans have the highest rate of business ownership of all U.S. minority groups. Thirty percent of all minority-owned firms and over half of all sales of such firms are Asian (U.S. Census Bureau, 2001j).

Specifically among Asian-American groups, the rate of business ownership among Koreans, Asian Indians, Japanese, and Chinese exceeds the rate for all Americans. Koreans have been especially prominent in small business (Kim,

1987; Light and Bonacich, 1988; Min, 1988, 1991; Yoon, 1995). Indeed, among all U.S. ethnic groups, a greater proportion of the Korean population is active in small-business enterprises (Hoffman and Marger, 1991). In New York and Los Angeles, with the largest Korean-American populations, Korean enterprises number in the thousands (Kim, 1981, 1987; Light and Bonacich, 1988).

Several factors seem to account for the high rate of participation of Asians in the small-business sector. Most of these are family-operated businesses, in which several or even all members of the family participate. As a result, labor costs are minimal, and relatively small amounts of capital are needed. Many immigrants thus find readily available opportunities in small enterprises. Many also fear that opportunities for advancement within the mainstream sector of the economy, dominated by big corporations, are limited for nonwhites, especially those who do not speak English fluently. Owning a small business thereby gives immigrants some independence, and enables them to avoid having to deal with what are seen as unreceptive dominant economic institutions.

Education The differences between Asian Americans and other minority ethnic groups regarding education are striking. All Asian-American groups are either close to or exceed the average level of schooling of the white population. Eighty-eight percent of white non-Hispanic adults in the United States are high school graduates, compared to 85 percent of Asians. The rate of college graduation is even more dramatic. More than 42 percent of adult Asian Americans in 1999 had graduated from college, compared to 28 percent of non-Hispanic whites (U.S. Census Bureau, 1999b). Many Asian immigrants, of course, are highly educated before they enter the United States, making them distinctly different from most contemporary Hispanic immigrants as well as Asian immigrants of past eras. Table 11.3 shows the percentage of college graduates among native-born and foreign-born Asians. Only the Southeast Asian groups (Vietnamese, Cambodian, Hmong, Laotian) do not have higher rates than non-Asians.

In academic achievement, Asian-American students are well above average on almost every measure. For example, Asian-American high school students score higher than students from any other ethnic group in mathematics scholastic aptitude tests and are second only to whites on verbal aptitude tests (U.S. Department of Education, 1992). Predictably, Asian students enter top universities far out of proportion to their numbers. At Harvard and Yale, for example, Asian Americans constitute 15–20 percent of U.S. students, and at elite University of California campuses they outnumber all other groups, including whites (Jacoby, 2000).[7] Moreover, Asian Americans have the highest graduation

---

[7]It is noteworthy that in 1990, an Asian American was named chancellor for the prestigious Berkeley campus. This represented the first appointment of an Asian American to head a major research university on the U.S. mainland.

Table 11.3 Asian-American Educational Level,
Persons 25 Years Old and Over

	PERCENT HIGH SCHOOL GRADUATES		PERCENT COLLEGE GRADUATES	
GROUP	Foreign-Born	Native-Born	Foreign-Born	Native-Born
All Asians	74.8	86.4	37.0	32.0
Chinese	70.1	91.1	38.7	51.0
Filipino	82.2	85.1	42.3	22.3
Japanese	86.0	88.3	35.1	34.2
Asian Indian	84.8	82.2	58.5	44.2
Korean	80.0	82.7	34.4	36.5
Vietnamese	61.0	67.7	17.4	16.9
Cambodian	34.6	55.4	5.5	12.4
Hmong	31.0	41.3	4.8	9.4
Laotian	39.9	43.5	5.2	13.1
Non-Asians	54.1	76.9	15.4	20.3

SOURCE: U.S. Census Bureau, 1993a.

rate among all racial and ethnic groups at Division I colleges and universities (American Council on Education, 2000).

Science, medicine, and other highly technical fields are academic areas in which the excellence of Asian-American students is especially apparent. But they are entering the humanities and the arts in increasingly greater numbers and are excelling in those fields as well. A quarter of the student body of the Juilliard School of Music in New York, perhaps the best-known and most prestigious institution of its kind, is Asian and Asian American. The representation of Asian Americans on the faculties and in the administration of American universities is also growing (Tien, 1995).

Ethnic and Class Factors in Upward Mobility

The economic and educational record of Asian Americans is truly extraordinary compared with averages for other ethnic groups. The now well-recognized success of Asian Americans has prompted a number of questions. Why have they moved up the economic ladder more quickly than other ethnic groups, and why do they seem to outdistance others in the realm of education? Is it because of a unique cultural heritage, or are class factors more important? Might Asian Americans even be innately different from other racial and ethnic groups?

In looking at the Asian-American experience, two caveats are in order. First, we need to remember the diversity of this ethnic category. Not only is the Asian-American population made up of many different groups, but each group itself is internally varied, stretching over an entire range of social class. Second, the Asian-American population consists of two historical components—those who came as early as the mid-nineteenth century, and those who have come since 1965. As we have now seen, many of the more recent Asian Americans have come with class and educational backgrounds very different from those of earlier ones.

Attempts to explain the unusually high rates of Asian-American mobility, educational achievement, and social adaptation have employed several different lines of thought. First, some believe Asians are innately superior. An eminent pediatrician at Harvard, for example, suggests that Chinese and Japanese babies are more alert and sensitive at birth, making them faster learners (Butterfield, 1986). As explained in Chapter 1, most social scientists do not recognize the validity of such biological arguments, whether applied to seemingly high-achieving or low-achieving groups. Moreover, this only prompts the nature-nurture debate, which as suggested earlier is seemingly irresolvable. The intelligence factor is one that nonetheless confounds observers because there is no denying that on certain measures of intelligence, Asian Americans consistently outperform other ethnic groups.

As noted in Chapter 8, social scientists have emphasized either structural opportunities or cultural factors in explaining why some ethnic groups seem to display greater collective achievement in American society than others. In the Asian-American case, many sociologists and psychologists point to cultural factors in accounting for high achievement. Stressed in this explanation are values that emphasize the importance of education and the cohesiveness of the family. Almost all studies that have addressed this issue seem to converge on one point: Asian Americans are prepared to work arduously and, in turn, are able to instill in their children a great motivation to work hard. Hence, Asian-American business owners, it is often observed, work longer hours than their competitors, and time spent on homework or other school activities by Asian-American students considerably exceeds that spent by non-Asian students (Butterfield, 1986; Kim, 1981; Light and Bonacich, 1988). In his study of Koreans in New York, Illsoo Kim traces their business success largely to cultural factors. "An emphasis on economic motivation and mobility," he writes, "has contributed to acculturation in the new land because it was already a deep part of contemporary Korean culture and values" (Kim, 1981:295). He claims that values such as industriousness and the control of impulses and emotions, which correspond closely to dominant American values, were carried by Korean immigrants to the United States. A similar argument has been advanced regarding the economic success of Japanese Americans, who it is claimed brought values from Japan highly compatible with white, Protestant, middle-class American values (Caudill and De Vos, 1956).

That cultural factors are of key importance in explaining Asian-American educational and economic success is suggested by a study, conducted by University of Michigan researchers, investigating the adaptation of Vietnamese refugees. At the time of their arrival in the United States, most of the subjects of this study (part of the boat people of the late 1970s) had completed little formal education, few had any marketable skills, and only one in a hundred spoke fluent English. The researchers found that despite these handicaps, many had achieved economic independence after only a few years. Furthermore, whatever success they had realized was almost entirely a product of their own efforts; the assistance of outside agencies had played a very limited role (Caplan, Whitmore, and Choy, 1989).

The academic achievement of school-age children from these families was more remarkable still. Despite their severe language deficiencies and family poverty, after an average of only three years, they were experiencing success in American schools. On national standardized tests of academic achievement, 27 percent of the refugee children scored in the ninetieth percentile on math achievement—almost three times better than the national average. And despite their somewhat lower scores in English-language proficiency, they earned higher grade point averages than their peers (Caplan, Choy, and Whitmore, 1991, 1992). The researchers found that this rapid educational success was closely linked to a cluster of core values deriving from the Confucian and Buddhist traditions of East and Southeast Asia: education and achievement, a cohesive family, and hard work. These, they found, "are, if not identical, congruent with what have been viewed as mainstream middle-class American beliefs about the role of education in getting ahead" (1991:172).[8]

Other researchers have also emphasized the importance of Confucian values in explaining the rigid work ethic that Asian Americans and their children seem to exhibit. Kim (1981), for example, views Confucianism as a significant factor in accounting for Korean business success. Zhou and Bankston (1998) also find this to be an important part of the explanation for the inordinate educational performance of Vietnamese children. In addition to the sanctified place of education and the family, Confucian thought stresses that people can always be improved with effort and instruction, values well suited to the competitive worlds of American business and schooling.

---

[8]The researchers did find a critical difference between the two value systems, however, in their orientation to achievement: "American mores encourage independence and individual achievement, whereas Indochinese values foster interdependence and a family-based orientation to achievement" (Caplan, Choy, and Whitmore, 1992:41).

Although most observers today recognize the importance of unique cultural factors in explaining Asian-American success, many also see a fortuitous fit between the opportunity structure of contemporary American society and the class background of these immigrants. We have already noted the disproportionate number of highly skilled and educated persons who make up the Asian immigrant population. Hence, the existence of many doctors, engineers, and other highly trained professionals among them is largely a function of their preimmigration class position. In some ways, then, today's Asian immigrants represent a select subset of the population of their origin societies. What has been referred to as a "brain drain" consists of the out-migration of highly skilled and educated persons from developing societies. Such persons recognize the more abundant opportunities to employ their skills and training in the United States as well as the possibility of earning substantially greater income. Hence, many are not poor, underprivileged immigrants but are already imbued with achievement-oriented values along with the class qualifications to implement those values. The importance of a propitious opportunity structure, therefore, cannot be overstated in explaining the rapid and substantial upward mobility of Asian Americans, especially the more recent arrivals. In brief, many are people who bring their skills and training with them, enabling them to adapt more easily than others with fewer skills and lower educational backgrounds. Structural opportunities, then, in addition to a cultural heritage compatible with American achievement values, are critical in accounting for the relatively rapid success of so many Asian Americans.

Asian Americans and Political Power

Even though their accomplishments in the economic and educational spheres of American society are increasingly evident, Asians have not exhibited a similar prominence in the society's political institutions. In the recent past, Asian-American representation in the political world was limited mainly to Hawaii's congressional and senatorial seats and a few congressional seats from California. There is much evidence to indicate that Asian Americans are on the brink of much fuller participation in the political world, both as an electorate as well as occupying positions of leadership (Nakanishi, 2001).

The traditionally low level of participation of Asian Americans in political life has been attributed to a number of factors. The nationally diverse and factionalized nature of the Asian-American population seems to have been a critical hindrance to effective political power. Asians also are not as concentrated spatially as Latinos or African Americans, making it difficult to form cohesive voting blocs. Moreover, because such a large proportion of the Asian-American population is foreign born, many are not U.S. citizens and thus not eligible to vote. Also, among older Asian Americans, especially those who have been victimized by past discrimination, the tendency to remain as invisible as possible continues to militate against strong political activism (Gross, 1989). Finally, many of the newest Asian immigrants come from societies where citizen participation in politics is not institutionalized and where there exists a profound distrust of government generally.

As their population base widens and as the newer immigrants experience progressive assimilation, Asian Americans are beginning to play a more active political role, and not only in those areas where they constitute a sizable proportion of the population or where they are a significant minority. Between 1996 and 1998, the number of Asians elected to political office grew by 10 percent, totaling 2,000 in thirty-three states (Ratnesar, 1998). Asian-American elected officials are unique compared to African-American and Latino elected officials in that they tend to emerge from political districts where their group does not account for a substantial portion of voters (Lai, 2000). One of the most visible indications of an increasing Asian-American presence on the American political scene was the 1996 election of Gary Locke as governor of Washington. Locke is a second-generation Chinese-American whose election marked the first time an Asian American had been elected to a governorship outside of Hawaii. Also in 2001, Japanese American Norman Mineta, a former congressman, was appointed secretary of transportation in the Bush administration.

Although Asian Americans still represent only a small percentage of the national population, their increasing visibility, numbers, and financial power will necessarily force politicians to take notice of them, particularly in states like California, where they are strongly concentrated. Already, for example, Asian Americans have established themselves as significant campaign donors (Kirschten, 1992; Stokes, 1988). Perhaps most important, future issues may arise that will have a mobilizing effect on the Asian-American population. Widespread occurrences of discrimination, for example, real or perceived, are apt to solidify the disparate Asian-American ethnic groups and to create a more potent political base. Interethnic competition and hostilities, such as the boycotts against Korean shops by blacks or the attacks on Korean businesses during the Los Angeles riots of 1992, may also stimulate Asian-American political participation. A government investigation of the contributions of Asian nationals to the Democratic Party in 1996, causing investigators to look primarily at contributors with Asian-sounding names, for example, as well as Republican proposals to curtail immigrant family reunification and deny welfare benefits to legal immigrants prompted many Asians to become citizens and thus eligible to vote (Rodriguez, 1998). Moreover, the growing importance of U.S. economic and political relations with Asian nations may create issues that will redound to the Asian-American population, pushing them increasingly into the domestic political arena.

Among voluntary immigrants to American society, none suffered more severe forms of prejudice and discrimination in their settlement experiences than Asians. Although they were not enslaved, like African Americans, nor were genocidal measures taken against them, as with Native Americans, they were constant targets of all forms of social and physical abuse. Indeed, if there is a common thread running through the early history of Asian-American groups, it is the experience of rampant prejudice and discrimination of the most vehement and often violent nature.

The Anti-Asian Heritage

Anti-Asian sentiments have lengthy historical roots, reaching back, as we have seen, to the entrance of the Chinese into the American West in the mid-nineteenth century. The intensity of early anti-Asian actions traditionally seemed to hinge on the perceived threat of Asian immigrants to the jobs of native workers. The anti-Chinese movement, which began in the mines of California, spread to other West Coast areas and eastward to other parts of the country (Lyman, 1974). Everywhere, Chinese were met with derision and hostility. They enjoyed hardly any legal rights and were common targets of assault and even murder. "The bewildered Chinese," note tenBroek, Barnhart, and Matson, "speaking little or no English and ignorant of Western customs, became the victims of every variety of fraud and chicanery, abuses encouraged by the absence of an active public opinion which might have alerted the police and the courts, and by the law prohibiting the testimony of Chinese in cases involving whites" (1968:15).

The "Yellow Peril" The standard panoply of negative ethnic stereotypes—dirty, immoral, unassimilable—was applied to the Chinese. In addition, they were seen as sly, untrustworthy, and inscrutable. Much of the negative stereotyping of the Chinese, and later the Japanese, stemmed from their labor role and their alleged impact on the job market. The term *coolie labor,* for example, denoting undignified work, derives from this period. That the Chinese were inferior to whites was beyond question in most people's minds. Perhaps the most devastating aspect of Chinese stereotyping, however, involved the notion of a "yellow peril." Chinese immigration, in this view, was an invasion of people who were loyal only to their country of origin and who, if not stopped, would eventually take over the United States (tenBroek, Barnhart, and Matson, 1968).

This sinophobia, fueled especially by labor leaders, culminated in the previously mentioned Chinese Exclusion Act of 1882. By 1910, the anti-Chinese movement had achieved its primary objectives: Chinese workers had been all but eliminated from the labor market, and restrictive legislation was in place that effectively barred further Chinese immigration. Moreover, the ban on future immigration, particularly of women, seemed to assure that the Chinese-American population would, in the long run, disappear.

Many of the same attitudes and actions were cast on the later Japanese immigration. At the time of their first substantial entrance in the 1880s, the Japanese were received enthusiastically by large farm owners and manufacturers, who saw them as a natural substitute for the Chinese laborers who had been excluded by the restrictive measure of 1882. So long as they occupied only menial or unskilled jobs, for which there was a shortage of workers, they were seen as filling a need. But like the Chinese before them, they soon found themselves the targets of labor groups who viewed them as potential competitors. Anti-Chinese agitation easily shifted to the Japanese, with many of the same stereotypes now applied to them, including the yellow peril (tenBroek, Barnhart, and Matson, 1968). Despite the Gentlemen's Agreement between the United States and Japan in 1907–1908 limiting Japanese immigration, as well as a California measure restricting Japanese ownership of land, anti-Japanese ferment never fully subsided.

In many ways, the anti-Japanese movement of the 1910s and 1920s was merely a continuation of the earlier anti-Chinese movement. Despite their modest group size (in California, where most were concentrated, they numbered but 70,000 in 1919, little more than 2 percent of the state's population), the Japanese were seen by many as a threat to white dominance. As already noted, the Japanese in California had become successful farmers, and although they controlled only about 1 percent of California's farmland, their highly efficient agricultural practices produced more than 10 percent of the value of the state's crops (Kitano and Daniels, 1995). Many whites were alarmed at the growing Japanese economic influence and began to agitate for restrictions not only on further immigration but on the right of the Japanese to own land in the state. California newspaper publishers William Randolph Hearst and V. S. McClatchy emerged as two of the leading figures in the anti-Japanese movement, outlining in their newspapers the themes of the Japanese "threat." Their propagandistic attacks claimed that the Japanese refused to assimilate, that their birthrate was so great they would eventually outnumber whites, and that their "low standard of living" presented an unfair advantage in economic competition with whites (McClatchy, 1978). All these charges, of course, had been applied to southern and eastern European immigrants as well, but the anti-Japanese campaign brandished a unique fear of an immigrant "invasion." Following World War I, a full-blown exclusionist movement developed in California, designed to isolate those Japanese already in the United States and, as one California politician described it, "to influence Congress and the administration at Washington to enact such legislation, even if the amendment of the Constitution be necessary, as will protect the white race against the economic menace of the unassimilable Japanese" (quoted in Daniels, 1977:84).

The vehemence of the anti-Japanese movement in California was unmatched against any other American ethnic group outside the South. Much of this hostility was the result of the racial distinctiveness of the Japanese. Southern and eastern European groups, though the targets of virulent racist attacks, were not so indelibly marked. Perhaps of greater importance, however, the Japanese were economically successful in both agriculture and business; therefore, they were resented and easily targeted as competitive threats to the majority population. The Japanese fear was further fueled by the growing disrepute of Japan as an aggressive, militaristic international power during the early decades of the century. The Japanese military threat, covered sensationally by the press, served as cause for suspicion of Japanese Americans. Finally, that the Japanese were concentrated in California provided an opportune setting for a movement aimed at an Asian group. As one historian describes it, "California, by virtue of its anti-Chinese tradition and frontier psychology, was already conditioned to anti-Orientalism before the Japanese arrived" (Daniels, 1977:106).

Japanese Internment Fueled by the shock of Pearl Harbor and reinforced by the stereotype of Japanese treacherousness, the anti-Japanese movement culminated after the start of World War II with the internment of the Japanese population living on the West Coast. No other series of events in American ethnic history, other than the enslavement of African Americans and the genocidal measures employed against Native Americans, was comparable to this action. Two months after the Japanese attack on Pearl Harbor in 1941, President Franklin D. Roosevelt, acting on the recommendation of Secretary of War Henry L. Stimson, issued Executive Order 9066, authorizing the U.S. Army to remove any group viewed as a security risk. Almost all those of Japanese ancestry living in California, Oregon, Washington, and Arizona—120,000—were affected. They were subsequently rounded up and sent to internment camps (called "relocation centers") in several Rocky Mountain states and in Arkansas. Having no time to prepare, most were forced to quickly liquidate their businesses; many abandoned their homes and possessions, losing their life savings. They were permitted to carry with them only a single suitcase of personal belongings.

The Japanese Americans were held in the detention camps almost until the end of the war. In an ironic twist, thousands of incarcerated Japanese men were permitted to enter the U.S. Army as a special unit; in the European theater, where they fought the Nazis, they became one of the most highly decorated battle units in American military history.

Several aspects of the Japanese incarceration stand out. To begin with, most of the detained had been born in the United States and were American citizens. Hence, people were denied their fundamental civil rights solely on the basis of their ethnicity. In his final report, the commanding U.S. Army general responsible for carrying out the Japanese removal put the matter bluntly: "The Japanese race is an enemy race and while many second and third generation Japanese born on United States soil, possessed of United States citizenship, have become 'Americanized,' the racial strains are undiluted" (quoted in tenBroek, Barnhart, and Matson, 1968). Second, that the removal of the Japanese was racially motivated was demonstrated by the fact that Americans of neither German nor Italian ancestry were similarly treated despite the war against Germany and Italy. Moreover, after the war it was acknowledged that the Japanese on the West Coast had never presented a military threat, and in fact their incarceration may actually have retarded the war effort. The emptiness of the security issue is reflected in the fact that the larger Japanese population in Hawaii was never removed despite its more strategic location. In the end, the exclusionist movement that had begun in California in the earlier years of the century had accomplished its objective— removal of the Japanese. The fear, suspicion, misperception, and envy bred by decades of negative stereotypes were released in a grievous deed, responsibility for which was shared by nativist groups, farmers, entrepreneurs, the military, the courts, and state and federal politicians.

Not until 1983 did a congressional committee formally acknowledge the political injustice that had been dealt the Japanese Americans, recommending a formal apology and compensation of $20,000 in damages to each survivor of the incarceration. Legislation to accomplish this was enacted in 1988.[9] The long-term effects of the internment camps, of course, could not be erased so easily, particularly for first- and second-generation Japanese Americans. For them, American democracy had failed to provide its promise of protection for all citizens. Even for the third and fourth generations, the World War II internment remains the paramount event of the history of the Japanese in America.

Although the Japanese internment is the most infamous case of direct institutional discrimination, all Asian groups have been subjected to various legal forms of discrimination at one time or another. Indeed, during the early years of the century, anti-Chinese and anti-Japanese measures were usually designed to apply to all Asians. In California, intermarriage of whites and Asians, for example, was barred, and residential segregation was enforced by restrictive covenants aimed specifically at Asians. Even federal laws were designed to handicap Asians. Perhaps most revealing, it was not until 1952 that federal legislation

---

[9]A similar internment was carried out by the Canadian government at the start of World War II against the 23,000 people of Japanese origin living in the West Coast province of British Columbia, 75 percent of whom were Canadian citizens (see Adachi, 1976; Daniels, 1981; Sunahara, 1980). In much the same fashion as the U.S. Congress, the Canadian parliament in 1988 moved to formally apologize to Japanese Canadians and to make a reparation payment of $19,325 to each survivor (Burns, 1988).

was passed making all Asian immigrants eligible for U.S. citizenship (U.S. Commission on Civil Rights, 1988).

Prejudice and Discrimination in the Modern Era

Expressions of prejudice and acts of discrimination against Asian Americans in recent years are in no way comparable to the widespread and vehement forms these took in the nineteenth and first half of the twentieth centuries. Nonetheless, negative Asian stereotypes as well as disturbing acts of hostility have been an integral part of the Asian-American experience in recent years.

Prejudice   A recent national survey sponsored by the Committee of 100, a national organization of leaders of the Chinese-American community, revealed a mixed view of Americans, both positive and negative, toward Asian Americans generally and Chinese Americans specifically (Yankelovich, 2001). The great majority of Americans believed Chinese Americans to have strong family values, to be honest in business, to be patriotic Americans, and to place a higher value on education than most other U.S. groups. At the same time, however, about one-third of Americans felt that Chinese Americans were more loyal to China than to the United States and had too much influence in the U.S. high-tech sector. Moreover, nearly half believed that Chinese Americans passing secret information to China was a problem. As with other studies of ethnic prejudice, it was concluded that prejudice toward Chinese Americans was most closely correlated with social class—those at lower educational and income levels were more commonly prejudiced.

Regarding attitudes toward all Asian Americans, the study indicated that almost one-quarter of Americans are uncomfortable with the idea of an Asian American as president of the United States; only 15 percent expressed a similar feeling about an African-American president, and only 11 percent about a Jewish-American president. One-quarter of Americans indicated that they would disapprove to some degree if a member of their family married an Asian American, as opposed to 34 percent expressing disapproval of marriage to an African American and 24 percent to a Hispanic American. One in six (17 percent) also expressed at least some displeasure with a substantial number of Asian Americans moving into their neighborhood.

Discrimination   Many acts of discrimination against Asian Americans in recent years are simple cases of verbal harassment or of indirect forms of institutional discrimination, but a number have been more overt and even violent (Noble, 1995; U.S. Commission on Civil Rights, 1986, 1992). As with Jewish Americans in past eras, the very success of Asian Americans in education, business, and the professions has created a potential backlash of resentment from majority whites, who see themselves being surpassed by a nonwhite minority in competition for the society's rewards. One of the most serious charges of discrimination emanating from Asian-American success has been leveled against some of the most prestigious colleges and universities in the country, which have been accused of limiting Asian enrollment (Biemiller, 1986; Johnson, 1989; Salholy, 1987; U.S. Commission on Civil Rights, 1992). Such measures are markedly similar to the quotas used by Ivy League schools in the 1920s to restrict the admission of Jews, who qualified in disproportionate numbers.

In addition to the success of Asian Americans relative to other ethnic groups, anti-Asian sentiments in the 1980s and early 1990s were fueled by the economic prowess of Asian societies, especially Japan, and their ability to compete with the United States in the American as well as the global market. Business leaders, labor officials, and politicians accused Japan of unfair trade practices, which they claimed had created many of the ills of American industry. Advertisements encouraging people to buy American-made products were tinged with anti-Asian sentiments, and many politicians used economic nationalism as a primary component of their campaigns. These actions redounded to Asian Americans (Mydans, 1992a).

Perhaps the most compelling and dramatic illustration of the impact of economic nationalism on Asian Americans was the killing in Detroit of Vincent Chin in 1982. While celebrating his impending wedding with friends at a bar, Chin, a young Chinese-American engineer, was verbally accosted by two white autoworkers, one of them unemployed at the time. Assuming that Chin was Japanese, they told him that he and one of his friends, also an Asian American, were "the reason we're all laid off." A fight ensued, after which the two chased Chin and clubbed him to death on the street. The attackers were fined $3,780 and given three years probation after one pleaded guilty and the other pleaded no contest to charges of manslaughter. The leniency of the sentence outraged the Asian-American community, and the U.S. Justice Department subsequently filed civil rights charges against one of the attackers. He was convicted by a federal jury of violating Chin's civil rights and sentenced to twenty-five years in prison. That conviction was later overturned on appeal, however, and he was acquitted of all charges, ending any further criminal prosecution (Moore, 1987).

In addition to the Chin case, which strongly aroused the ire of Asian Americans, several other incidents in the 1980s provoked similar reactions. Each involved violent attacks which resulted in only nominal punishment by the courts (U.S. Commission on Civil Rights, 1986). Although these cases received much publicity, other less prominent but no less severe anti-Asian actions, many motivated by unadulterated racist sentiments, were reported in the 1980s and early 1990s (Noble, 1995; U.S. Commission on Civil Rights, 1992). In Jersey City, New Jersey, for example, a community of several thousand Asian Indians had been subjected to constant harassment by white, black, and Hispanic youths. In 1987, the violence took on a far more serious cast as one Indian was beaten to death and another found beaten unconscious on a busy street corner (Marriott, 1987). Anti-Indian incidents continued to occur frequently in the Jersey City area for at least a year after the killing.

A 1995 study indicated a 35 percent increase in anti-Asian hate crimes nationwide over the previous year, involving a range of acts from racist graffiti to beatings. In one case, the home of a Chinese city councilman in Sacramento, California, was firebombed (Noble, 1995). In fact, the extent of anti-Asian hate crimes may be greater than is actually reported. Asian Americans, especially recent immigrants, are reluctant to report crimes for a variety of reasons, including language problems, distrust of police, the desire not to cause problems, and shame at becoming a victim of crime (U.S. Commission on Civil Rights, 1992).

Two events of the late 1990s seemed to galvanize Asian Americans around issues of perceived discrimination. In one case, some contributions given to the Democratic National Committee (DNC) during the 1996 presidential election raised by a Chinese-American former official of the DNC were later found to be illegal, having been channeled by overseas Asian political influence seekers. Subsequent investigation of illegal contributions cast suspicion on Asian American contributors, who had done nothing wrong (Tajitsu Nash and Wu, 1997). In another incident, Wen Ho Lee, a Taiwanese-born U.S. citizen and former government-employed nuclear scientist, was accused of espionage in 1999. He was indicted on several dozen felony counts for downloading classified nuclear design and testing information at the Los Alamos National Laboratory in New Mexico. Subsequently all but one charge was dropped in what was seen as a major embarrassment for federal prosecutors. Both of these cases caused much chagrin among Asian Americans, many of whom saw them as blatant instances of anti-Asian stereotyping and discrimination.

Asians and Other Ethnic Minorities In addition to harassment and hostilities issuing from the white majority, Asian Americans have sometimes been targets of violence by inner-city blacks and Latinos. As previously noted, many Asian immigrants turn to small business, and they often operate groceries, restaurants, liquor stores, and laundries in minority areas of large cities. In this, they are playing a middleman minority role. And as we saw in Chapter 2, such middleman business owners frequently become the targets of hostility from their customers. Tensions between Asian business owners and blacks in inner-city areas have been especially high in recent years. Animosities have arisen from the perception of blacks that Asian shopkeepers often deal with them disrespectfully and have no interest in the black communities in which they operate, other than earning a profit. Hence, they see Asians as exploiters. Shopkeepers, on the other hand, complain of the frequency of harassment, theft, and vandalism of their businesses. In a widespread pattern, black boycotts of Asian-owned businesses occurred in several cities in the 1980s and 1990s including New York, Philadelphia, Washington, and Los Angeles. During the Los Angeles riots of May 1992, shops owned by Koreans in the South Central district, where most of the disturbances occurred, seemed to be particular targets of looters and arsonists. Before the riots, relations between Korean shopkeepers and their black customers had been volatile for years and seemed to reach a high point when a Korean-born grocer shot to death a fifteen-year-old black girl whom the merchant had accused of stealing a bottle of orange juice. Black resentment grew even stronger when the woman grocer, who was convicted of voluntary manslaughter, was given only a probationary sentence (Mydans, 1992b; Ong et al., 1992). While discordant relations between Koreans and blacks were most dramatic in Los Angeles as a result of the 1992 riots, a similar under-current was evident in the inner cities of other urban areas where Koreans operated small businesses serving minority residents (Chesley and Gilchrist, 1993; Goldberg, 1995; Wilkerson, 1993).

Friction between Koreans and blacks in business relationships should not be seen as unique, however. Similar animosities have arisen among blacks and mer-chants of other ethnic groups operating their enterprises in inner-city areas. In Detroit, for example, most inner-city neighborhood groceries and liquor stores are owned by Arab Americans or Chaldean Americans, who are seen in much the same way as Koreans in other cities (Dawsey, 1992; Gerdes, 1993). Moreover, conflict between neighborhood residents and shopowners of another ethnic group has a long tradition in many countries and has been evident in diverse parts of the world in recent times (White, 1993).

Moderating Levels of Prejudice and Discrimination Despite evidence of anti-Asian sentiments as well as disturbing acts of violence in recent years, there seems little chance of a return to the extreme levels of prejudice and discrimi-nation that characterized the Asian-American experience of earlier decades. One of the more apparent differences today is that negative stereotypes of the past have given way to a generally positive collective image. Indeed, Asian Americans are increasingly seen as groups to which others should look as role models. As sociologist Peter Rose has put it, "The pariahs have become paragons, lauded for their ingenuity and industry and for embodying the truest fulfillment of the 'American Dream'" (1985:182). Moreover, despite cases of blatant and direct discrimination, such as those described, most instances of discrimination today are more subtle and indirect.

One key indication of a lower rate of discrimination against Asian Americans than against other nonwhite groups today is their more rapid rate of residential integration, exceeding that for both blacks and Latinos. Even though their presence in American society is more recent than either of these two groups, the level of Asian-white residential segregation is comparable to that of Hispanics from whites and much lower than the segregation of blacks from whites (Alba and Logan, 1993; Langberg and Farley, 1985; Massey, 2001; Mumford Center, 2001a).[10]

There is some evidence of a declining perception of discrimination among Asian Americans themselves. A 1997 poll of Chinese Americans in the Los Angeles and San Diego metropolitan areas, for example, found that only 12 percent of respondents felt that racism was holding back Chinese people in Southern California. Only 10 percent said that they had experienced a great deal or fair amount of discrimination (*Los Angeles Times,* 1997). That compared to 14 percent for Asians overall in a 1993 poll (Kang, 1997).

Asian Americans as a "Model Minority"

The clear movement toward assimilation, demonstrated most powerfully by the extraordinary success they have achieved in the realms of education and the economy and by their comparatively less problematic adaptation to American society, has led in recent years to the popular characterization of Asian Americans as a model minority. Although the label is seemingly complimentary, many Asian Americans today see it as condescending in a number of ways and, paradoxically, a case of damaging stereotyping. Some have pointed out that this new, positive image tends to gloss over the shabby treatment of Asians in the past and disguises the serious social and economic problems that continue to confront some elements of the Asian-American population (Hurh and Kim, 1989). Obscured is the fact that some, particularly Southeast Asian refugees, remain poor, uneducated, and relatively unsuccessful.

In addition, the model minority image has become so well entrenched that it places tremendous social and psychological pressures on Asian Americans, especially young people. Students, for example, must do exceedingly well in their studies and in other pursuits in order to uphold this image and may even feel that they must outperform their non-Asian peers. This becomes especially problematic for recent Asian immigrants whose command of English may be minimal and who therefore suffer added burdens in an educational environment that is not attuned to the unique cultural needs of Asian students (Bernstein, 1988).

The model minority image has also seemed to create a new set of stereotypes, which though generally positive, nonetheless create the same kinds of group effects as other, negative, stereotypes: They induce the perception of

---

[10]An apparent exception are Vietnamese. Eighteen percent of the respondents in a 1989 national survey indicated their aversion to having Vietnamese as neighbors. This was a higher negative response than for any other ethnic group, though not significantly different from those given for Hispanics (16 percent), Indians or Pakistanis (15 percent), or Koreans (14 percent) (*Gallup Report,* 1989).

commonality of traits among all individual ethnic group members. A Chinese-American schoolteacher expresses the frustration produced by the new stereotypes: "Years ago they used to think you were Fu Manchu or Charlie Chan. Then they thought you must own a laundry or restaurant. Now they think all we know how to do is sit in front of a computer" (quoted in *Time*, 1987). Moreover, as explained in Chapter 3, positive traits attributed to ethnic groups may easily be manipulated so that they become negative in content and thus serve as the rationales of prejudicial thought and discriminatory behavior. If Asian Americans today are seen as "brainier" and "more ambitious" than others, such seemingly positive traits may easily be transformed into negative ones like "shrewd" and "pushy," should Asians come to be seen as outcompeting the dominant group. Recall the similar circumstances regarding the rise of anti-Semitism in the United States in the late nineteenth and early twentieth centuries.

The characterization of Asian Americans as a model minority may also work to deflect attention from structural problems of other minority ethnic groups in the society. If Asian Americans can attain success despite their racial visibility, it is argued, even race cannot be viewed as an overwhelming handicap in American society. As Asian Americans are seen as hardworking, persevering, and highly motivated, the failure of other minority racial and ethnic groups to reach comparable levels of achievement is attributed to a lack of determination and hard work rather than to structural problems created by changing economic and social institutions (see, for example, D'Souza, 1995).

ASSIMILATION AMONG ASIAN AMERICANS

As with economic stratification, in a number of ways, assimilation for Asians in the United States takes on a dual pattern. On the one hand, within all Asian American groups, those who are highly educated and who hold prestigious occupations are increasingly moving toward both cultural and structural assimilation. On the other hand, those less affluent and educated retain much of their traditional ways and have had more difficulty in moving beyond the ethnic enclave.

Chinese and Japanese Patterns

In examining assimilation among Asian Americans, it is important to consider that the bulk of the Asian-American population today is foreign born. It is thus too early to marshal solid evidence regarding the rate and extent of assimilation for those who have not gone beyond the first generation. Most evidence must necessarily come from Chinese and Japanese Americans, both groups well into the third and fourth generations. With these limitations in mind, let's look at the patterns and prospects of Asian-American assimilation.

Chinese-American Assimilation As in other societies to which they have migrated, the Chinese in America traditionally exhibited a resistance to assimilation. The characteristic features of Chinese ethnic communities worldwide seem to be social and cultural exclusiveness and a low level of absorption into the larger society (Lyman, 1968b, 1974; Purcell, 1980). Much of this self-imposed isolation is the product of Chinese community organizations, which in the new society reflect extended family patterns, known as lineages. Traditionally, much of Chinese social life in the United States was organized around various kin, clan, and secret societies.

Additional reasons have been suggested for the historical lag in assimilation among Chinese Americans. For one, the early demographics of this group—almost all males without wives and families—prohibited a second generation from establishing stable families and moving along an assimilation path comparable to that of other immigrant groups (Kitano and Daniels, 1995; Lyman, 1974). Moreover, the Chinese were prohibited from acquiring citizenship and thus remained legally marginal. Finally, the levels of prejudice and discrimination were unusually harsh, barring Chinese from closer relationships with members of the dominant society or from entering into mainstream institutions. For the first 100 years, then, the Chinese were a relatively isolated ethnic group in American society.

Despite past patterns of insularity, it appears that Chinese Americans are moving rapidly in the direction of assimilation, both cultural and structural. Regarding cultural assimilation, a 1997 study of Chinese Americans in six Southern California counties found that although 79 percent spoke mostly Chinese at home, 68 percent spoke English when conducting personal business; among those between the ages of 18 and 29, 88 percent spoke English outside the home. Only 16 percent of the respondents' children spoke only Chinese (*Los Angeles Times,* 1997).

Structural assimilation is evident at the secondary level in occupational and residential patterns (Jiobu, 1988) and at the primary level in the substantial rise in intermarriage. One study (Wong, 1989) indicates that over 30 percent of all Chinese marriages involve non-Chinese, most of them whites. For Asian Americans generally, intermarriage is substantially higher than for African Americans and Hispanic Americans (Lee, 1998). Asian groups with the lowest level of intermarriage are those with the newest immigration cohorts, such as Indians, Vietnamese, and recent Chinese (Lee and Yamanaka, 1990). This may be seen as solid evidence of the diminishing of social boundaries between Asian Americans and the larger society with each generation.

Japanese-American Assimilation Among Asian-American groups, the Japanese have experienced a more profound assimilation than others. This is partly because they have resided in the United States for several generations and their numbers have not been swelled in recent years by new immigrants. Patterns of assimilation among Japanese Americans, however, differ on the basis of generation. Japanese Americans refer to each generation with a particular name. The first, or immigrant, generation is called *Issei*; the children of the Issei, or second generation, are referred to as *Nisei*; and their children, the third generation, are called *Sansei*. Japanese Americans are now entering the fourth generation, referred to as *Yonsei*.

The Issei represent a generation that did not assimilate. In this, they did not differ essentially from other immigrant groups. However, the much more severe prejudice and discrimination directed against them produced a greater disincentive to venture out of the ethnic milieu and attempt to interact with the larger society.

By contrast, Nisei, and especially Sansei and Yonsei, are highly assimilated culturally, as indicated by the very thorough adoption of dominant cultural traits (Levine and Rhodes, 1981; Onishi, 1995). This is not surprising in light of the high level of education among Japanese Americans of the second generation and beyond and the great strides each generation has made in socioeconomic advancement.

Regarding secondary structural assimilation, we have already seen the marked success of Japanese Americans in economic and educational realms of the society and their increasing movement into various institutional areas. Residential patterns of Japanese Americans also suggest a relatively fluid movement into the mainstream society (Kitano, 1976; Levine and Rhodes, 1981).

At the primary level of structural assimilation, there is strong evidence that Japanese Americans have begun to leave the ethnic community and forge close associations with non-Japanese. This is particularly apparent among those beyond the second generation (Levine and Rhodes, 1981; Onishi, 1995; Qian et al., 2001). The most telling indicator of primary structural assimilation is intermarriage, and in this, several studies have found significant increases, especially for the Sansei (Kikumura and Kitano, 1973; Kitano and Daniels, 1995; Levine and Rhodes, 1981; Montero, 1980; Tinker, 1973). Among his sample of Japanese Americans, Montero (1980) found that 40 percent among the Sansei had married a non-Japanese. This compared to only 10 percent of the Nisei and 1 percent of the Issei. Kikumura and Kitano (1973) found a similar pattern and concluded that Japanese Americans now choose marital partners as frequently from outside the ethnic group as from within. Data on out-marriage of Asian Americans in Los Angeles between 1975 and 1989 show a rate of over 50 percent for Japanese Americans (Kitano and Daniels, 1995). A recent study confirms the strong trend toward exogamy among Japanese Americans, indicating that, among Asian-American ethnic groups, they are most likely to intermarry with whites (Qian et al., 2001).

In general, these studies indicate a progressive trend toward primary structural assimilation among Japanese Americans, and an increasing dilution of the ethnic community with each succeeding generation. In light of their rapid and extensive assimilation, some have even begun to question the ability of Japanese Americans to maintain a viable ethnic community into the next (Yonsei) generation (Montero, 1980; Onishi, 1995).[11] Others, however, maintain that despite very strong assimilation tendencies, Japanese Americans do not appear to be losing all cultural distinctiveness and identification (Kitano, 1976; Woodrum, 1981). One study, in fact, concludes that Japanese Americans appear to have been able to successfully combine high levels of structural assimilation with a strong retention of ethnic identity and group cohesion (Fugita and O'Brien, 1991).

The Future Path of Asian-American Assimilation

In examining issues of stratification, the internal variability of the Asian-American population was emphasized. In looking ahead to future assimilation patterns, these differences once again must be kept in mind. As with stratification, the rate of assimilation among Asian Americans will differ from group to group. Vietnamese and other Southeast Asian groups, most recent and least prosperous, currently appear to be that part of the Asian-American population experiencing the most tortuous path toward assimilation. At the other extreme, Japanese Americans, one of the oldest and most prosperous among Asian-American groups, continue to move most directly and rapidly in that direction (Jiobu, 1988). For the new immigrants the occupational and educational achievements of many will enhance the prospects for both cultural and structural assimilation, particularly for those beyond the first generation. Already it is evident, for example, that Asian Americans are dispersing throughout the urban areas in which they are concentrated. Almost 60 percent of Asians lived in suburbs in 2000, up from 53 percent in 1990 (Mumford Center, 2001b). This is higher than for any other ethnic minority, despite their relatively recent arrival in large numbers.

Chapter 8 noted that the assimilation of Jews in the United States had proceeded as far as in any society where Jews had lived in large numbers. Yet, the religious divergence between Jews and non-Jews, it was suggested, remains a factor that continues to impede full assimilation of Jewish Americans. The same can be said for Asians, whose physical distinctiveness makes them even more indelibly conspicuous within the larger society. Because of this visibility, Americans continue to see Asian minorities as somehow not fully American, as

---

[11]Montero (1980) notes that ironically, the remarkable upward economic mobility of the Japanese in America may lead to a diminution of the traditional ethnic values that helped to stimulate that advancement in the first place; hence, the loss of those values may create a leveling off of the achievement of future generations.

"outsiders"—even though their ancestors may have been in the United States for several generations, they may speak only English, and they have no more ties to Asian societies than Swedish Americans have to Sweden or Italian Americans to Italy. Because of their racial visibility, as Peter Rose explains, "No matter how adaptive in values and aspirations, no matter how similar to whites in mannerisms and actions, Asian Americans cannot be members of the majority" (1985:212). Ethnic identity and consciousness among Asian Americans, therefore, regardless of the extent of their assimilation, is not likely to fully erode.

As will be discussed in Chapter 12, the preeminence of multicultural ideology in schools, universities, and the media has made it easier for ethnic minorities to preserve or perhaps reaffirm aspects of the ethnic culture. This is no less the case for Asian Americans. Multiculturalism is also helping to create a consciousness of what may have been a quite inchoate Asian-American panethnicity. Tamar Jacoby (2000) maintains that young college-educated Asian Americans appear to be experiencing a kind of ethnic reaffirmation in large measure as a product of the oppositional multiculturalism to which they are exposed in their university experience. Moreover, the creation of institutionally complete ethnic enclaves in areas where Asian Americans constitute a significant portion of the population—California in particular—provides a further defense against assimilation.

Despite these impediments to assimilation, however, Asian Americans, like Hispanic Americans, appear to be reconciling the competing demands of ethnic retention and assimilation. Asian Americans seem determined to be both Asian and American, seeing no inherent incompatibility between the two and no conflict in balancing them. Jacoby quotes a recent Stanford graduate, who asserts that "I'm determined to hold onto my Chinese heritage and my Chinese culture. But there's absolutely nothing about America that I reject because of my ethnicity." And a Chinese business owner echoes this view, claiming that assimilation and ethnic retention are "two parallel tracks, and there's no reason to choose between them" (Jacoby, 2000:25).

Asian-American Panethnicity The term *Asian Americans* is something of a misnomer given the cultural and physical variety of the population to which it refers.[12] Yet, a social and political coming together of the various elements of this broad and extremely diverse ethnic category may become increasingly apparent in the future. The continued use of the collective *Asian American* by government, academic, and media institutions already reflects the perception of a more coherent ethnic population than actually exists. This is not unlike the use

---

[12] Jacoby points out that "the very term 'Asian-American' traces back to the 1960's, when it was coined, in the spirit of "Black Power" and "Chicano Studies," as a kind of battle cry" (2000:22).

of the term *Hispanic American* or *Latino* to collectively describe the various groups of Latin American origins. And it is not unlike the process of ethnic group formation that characterized the American experience of many European groups of the earlier immigration periods. Only in the American context were Italians from various provinces in Italy, for example, designated and dealt with collectively as "Italians." Through this common treatment by Americans, an Italian-American group consciousness was engendered, and eventually Italians in America identified themselves primarily as "Italian," rather than "Calabrian," "Sicilian," or "Neapolitan." Though surely a long-term process for the various Asian-American groups, such a collective identity may emerge more clearly with each successive generation. Moreover, Asian Americans of diverse ethnic origins themselves may increasingly recognize this collective designation as a useful political strategy, just as Latinos are beginning to do (Espiritu, 1992; Shinagawa and Pang, 1996). As separate and disperse groups, none can exert the kind of political leverage at the national or even local level that a more consolidated position may produce.

SUMMARY

Asian Americans are a diverse collection of distinct ethnic groups, the largest among them Chinese, Filipinos, and Asian Indians. Asian immigration to the United States consists of two periods. The first began in the mid-nineteenth century with the immigration of Chinese manual laborers. Exclusionary measures beginning in 1882 halted further Chinese immigration and provided an opening for Japanese workers to replace them in the early part of the twentieth century. Discriminatory laws brought further Asian immigration to a virtual halt in the 1920s. The second period extends from 1965, when immigration laws were liberalized, to the present. The vast majority of Asian Americans today are first-generation immigrants who stem from this second wave. Many of those of the second wave have come to the United States with much higher occupational skills and more education than earlier immigrants. Both immigrant waves settled primarily on the West Coast, especially California, where Asian Americans remain heavily concentrated.

Asian Americans, as a whole, are at or near the top of the stratification system in terms of income, education, and occupation. There are important differences among the various Asian groups, however. Whereas Japanese and Asian Indians rank among the highest income earners and most highly skilled workers, the Vietnamese and other Southeast Asian groups, most of whom have come as political refugees, are among the most impoverished groups in the United States. There are also internal differences in class within each Asian-American group.

High levels of prejudice and discrimination have marked the Asian-American experience from the outset of immigration. Exclusionary measures designed to bar the entry of Asians into the United States were enacted in the nineteenth and twentieth centuries, and basic civil rights were denied Asians in Western states until well into the twentieth century. In recent years, Asian Americans have been portrayed as a model minority because of their exceptional economic and educational attainments. Nonetheless, remnants of anti-Asian prejudice and discrimination remain evident. More serious is the potential backlash of anti-Asian sentiment that may emerge from these groups' economic success.

It is still too early to gauge the extent of assimilation among Asian Americans because most are first-generation immigrants. However, high levels of cultural assimilation characterize those groups who have entered the third and fourth generations—specifically, the Japanese and Chinese—and there is also evidence of increasing rates of both primary and secondary structural assimilation among them.

Suggested Readings

Bonacich, Edna, and John Modell. 1980. *The Economic Basis of Ethnic Solidarity: Small Business in the Japanese American Community.* Berkeley: University of California Press. A good account of how Japanese Americans carved out a place for themselves in the California economy. Also contains a valuable explication of the role of small business among ethnic minorities.

Caplan, Nathan, John K. Whitmore, and Marcella H. Choy. 1989. *The Boat People and Achievement in America.* Ann Arbor: University of Michigan Press. Along with its companion work, *Children of the Boat People: A Study of Educational Success* (1991), explores why Southeast Asian refugees have excelled educationally and occupationally despite their language and other handicaps on arrival in the United States.

Espiritu, Yen Le. 1992. *Asian American Panethnicity: Bridging Institutions and Identities.* Philadelphia: Temple University Press. Describes the process by which an inclusive Asian-American ethnic group may be forming, one that transcends its members' specific national origins.

Horton, John, and Jose Calderon. 1995. *The Politics of Diversity: Immigration, Resistance, and Change in Monterey Park, California.* Philadelphia: Temple University Press. Traces the social and political transformation of this Los Angeles suburb from a primarily white to majority Asian-American community.

Kim, Illsoo. 1981. *New Urban Immigrants: The Korean Community in New York.* Princeton, N.J.: Princeton University Press. Examines the social and economic institutions of the Korean ethnic group in the United States and the entrepreneurial role Koreans play in many large cities.

Kwong, Peter. 1987. *The New Chinatown*. New York: Noonday Press. The author describes the changing makeup and social institutions of New York's Chinatown by drawing on his observations as a long-time resident.

Lyman, Stanford M. 1974. *Chinese Americans*. New York: Random House. An examination of the Chinese-American experience, with good discussions of the development of Chinese community organizations and the historical patterns of discrimination, both formal and informal, suffered by the Chinese in the United States.

Min, Pyong Gap (ed.). 1995. *Asian Americans: Contemporary Trends and Issues*. Thousand Oaks, Calif.: Sage. Describes the history, current demography, sociological features, and patterns of assimilation of the major Asian-American groups.

Rangaswamy, Padma. 2000. *Nemasté America: Indian Immigrants in an American Metropolis*. University Park: The Pennsylvania State University Press. Using Chicago as her research site, the author draws a picture of a burgeoning—and heterogeneous—Asian Indian community, similar to those emerging in other major U.S. urban areas.

Root, Maria P. P. (ed.). 1997. *Filipino Americans: Transformation and Identity*. Thousand Oaks: Sage. A collection of articles by Filipino/a-American scholars that explore various social, political, and economic issues of the second largest Asian-American ethnic group.

Takaki, Ronald. 1989. *Strangers from a Different Shore: A History of Asian Americans*. New York: Penguin. Traces the history of Asian Americans beginning with the first immigration of workers in the nineteenth century up to the most recent wave of immigrants.

12

CURRENT AND FUTURE ISSUES
OF RACE AND ETHNICITY
IN THE UNITED STATES

AMONG MULTIETHNIC SOCIETIES of the modern world, the United States seems to fall at a midpoint along a scale of ethnic conflict and inequality. As we will see in Part III, in some societies, ethnic divisions and stratification are far more severe. But understanding that other societies may suffer even greater problems that derive from ethnicity should not cause us to lose sight of the fact that the United States has been and remains a society in which racial and ethnic inequality is a fundamental characteristic. That inequality, in turn, produces ethnic conflict—ordinarily subdued and contained, but at times blatant and explosive.

Currently, three interrelated issues seem to clearly stand out regarding American race and ethnic relations: (1) How should public policies address the continued economic gap between Euro-American and most Asian-American groups on the one hand, and African-American, Hispanic-American, and Native-American groups on the other? (2) How is the influx of large numbers of new immigrants and the creation of a more pluralistic ethnic mix affecting the society today, and how will it affect the future of race and ethnic relations? (3) Should the end product of interethnic relations be some form of assimilation, in which diverse groups become more culturally alike and socially integrated, or some form of pluralism, in which they maintain or perhaps increase their cultural differences and social boundaries?

THE CONTINUING GAP BETWEEN EURO-AMERICANS AND RACIAL-ETHNIC MINORITIES

Let's briefly review the American ethnic hierarchy that has been explored in the previous six chapters. Roughly, the hierarchy is divided into three comprehensive tiers: a top tier of white Protestants, essentially the dominant ethnic group; a second, intermediate tier made up of white Catholics of various national origins, Jews, and many Asians, for whom ethnicity remains of significance but decreasingly so; and a bottom tier, comprised of racial-ethnic minorities, for whom ethnicity continues to render great effect in the distribution of societal rewards and in shaping patterns of social interaction. The most important division within this hierarchy is between the bottom tier and the other two.

The third tier of groups continues to maintain a collective place at the bottom of the society's economic and political hierarchies. This does not mean that many individuals in each of these groups have not demonstrated substantial upward mobility, particularly in the past three decades. Indeed, evidence is unmistakable that the resistance of whites to nonwhite economic, political, and social advancement has greatly diminished, and as a result, the economic and political status of large numbers of racial-ethnic minorities has risen significantly. But in looking at their collective status vis-à-vis Euro-Americans and some segments of the Asian-American population, members of these groups remain disproportionately among the poor and the powerless. Moreover, the social boundaries separating the third tier of groups from the first and second, though dissolving slowly, remain unmistakably in place. Members of third-tier groups continue to encounter higher levels of prejudice and discrimination, and the extent to which they have been afforded entry into the full range of societal institutions remains lower.

Efforts to deal with this persistent economic gap and the social distance between majority and minority ethnic groups have, in recent decades, given rise to a number of controversial public policies.

Compensatory Policies

Out of the atmosphere of the civil rights movement and the new pluralism of the 1960s and 1970s, the federal government undertook sweeping measures to foster political and economic equity among ethnic groups. Specifically, these policies were originally aimed at improving the social standing of African Americans, against whom official discriminatory measures had been historically applied. Other long-term victims of discrimination—Latinos, Asian Americans, and American Indians—were subsequently brought under the compensatory umbrella and, later, women were added as a target group. These new public policies, designed as they were to raise group positions, marked a significant departure from the traditionally understood role of government in the area of civil rights.

In the past, government's function had been to ensure that everyone was afforded equal opportunities in work and education, regardless of ethnicity. Because this guarantee had been violated for nonwhites, legislation was enacted beginning in the 1960s that was designed to protect members of minority groups from discrimination in schools, workplaces, and other societal institutions. But the question now raised was whether eliminating discriminatory practices alone could counteract the effects of past discrimination. Given the generations of denied opportunities, was it fair to expect African Americans and other racial-ethnic minorities to compete on an equal basis with the white majority? How could these groups ever catch up if they entered the competition burdened by decades of imposed disabilities? President Lyndon Johnson, in proclaiming the government's intention to address this issue, put the matter squarely:

> You do not wipe away the scars of centuries by saying: Now you are free to go where you want, do as you desire, and choose the leaders you please. You do not take a man, who for years has been hobbled by chains, liberate him, bring him to the starting line of the race, saying 'you are free to compete with all the others,' and still justly believe you have been completely fair. Thus it is not enough to open the gates of opportunity. All our citizens must have the ability to walk through those gates. This is the next and more profound stage of the battle for civil rights." (Johnson, 1965:635)

In response to this dilemma, a series of compensatory measures and programs evolved, grouped under the rubric of *affirmative action*, that was intended to advance the economic and educational achievement of the minorities that had been most severely and consistently victimized by past discrimination. That government was now acting to guarantee not simply equality of opportunity but also equality of result evoked a strong public debate, which continues to generate much controversy.

As affirmative action policies unfolded in the 1960s, they stipulated that those doing business with the federal government (universities as well as businesses) were required to take steps to increase their minority representation and to establish goals and timetables to meet that objective. The aim was basically to affect minority employment and student admissions. With more widespread and stringent application in the 1970s, affirmative action policies grew increasingly unpopular and created a backlash among majority whites. The arguments on both sides of this issue have not basically changed in the past thirty years.

Affirmative Action: The Debate Proponents of affirmative action have argued that advancing the position of minorities through the use of goals and preferential hiring practices is necessary if the victims of past discrimination are eventually to attain equity with the majority (Kennedy, 1994; Livingston, 1979; W. Taylor, 1995). Even if direct forms of discrimination no longer prevail, they point out, the indirect and institutional forms continue to perpetuate nonwhite and female disadvantages in the labor market and in higher education. Simply protecting minority individuals against ethnic discrimination, therefore, is inadequate by itself. Hiring workers or admitting students without regard to ethnicity (a "color-blind" process) will automatically preserve the disproportionate representation of whites because they enter the competition with background advantages accumulated over many generations. Artificial incentives for minorities are needed temporarily, therefore, until the opportunity structure is made more truly equitable. Most important, proponents of affirmative action argue that without such policies, there is a risk of returning to a norm of discrimination.

Some maintain that the critical importance of affirmative action lies in providing access for minorities to social networks that aid in securing better jobs and educational opportunities. In this view, what is most important is being enmeshed in key networks, through which individuals find jobs, meet influential people, and learn bargaining skills. Individuals may possess formal qualifications, but unless they are able to move into the inner circles of the educational and work worlds, their chances of acquiring top jobs are diminished. African Americans have lacked the connections—social capital—ordinarily developed through ties made in universities, in the corporate world, or through intermarriage, that Euro-Americans have availed themselves of more easily. Affirmative action, it is argued, can therefore cast African Americans and other minorities into networks that they otherwise would have little chance of penetrating (Loury, 1998; Patterson, 1997, 1998).

Those who oppose affirmative action or who strongly criticize the way it has been carried out argue that, in effect, these programs have become a kind of reverse discrimination in which those previously discriminated against are given preference over others merely on the basis of ethnicity (Glazer, 1975; Yates, 1994). Hence, they maintain, the very objective intended—reducing ethnic discrimination—has been undermined, the victims now being Euro-Americans. Government is seen as re-creating racial and ethnic categories that had rightfully been abandoned. The effect of affirmative action in many cases has been to create quotas favoring minorities, a particularly sensitive issue for certain groups, such as Jews, who themselves were the past victims of quotas limiting their entrance into prestigious colleges and professional schools.

Opponents of affirmative action have pointed out that preferential measures shifting emphasis from equality of opportunity to equality of result create aspects of corporate pluralism in which social benefits are distributed on the basis of group membership, not individual merit.[1] Hence, less qualified minority persons may be promoted over better qualified majority persons. Moreover, the targeted groups of affirmative action, they argue, are stigmatized because of their special treatment, thus producing negative social and psychological effects on those who are supposedly the beneficiaries of these programs (Sowell, 1990). Shelby Steele (1991), for example, has suggested that affirmative action programs create a kind of implied inferiority. Whites commonly view blacks as having acquired their positions on the basis of special preference and thus as less than qualified or competent. This not only fuels negative stereotypes by whites, argues Steele, but also creates self-doubt among blacks and other minorities who are the beneficiaries of affirmative action (see also Heilman, 1996; McWhorter, 2000).

In addition to questions of reverse discrimination and effects on employment and educational qualifications, affirmative action programs have been criticized as too sweeping in application and therefore unable to distinguish from among the various targeted minorities those who are truly the past or present victims of discrimination. For instance, although Latinos are covered under the principles of affirmative action, we have seen in Chapter 10 that this broad ethnic category comprises several disparate components, each with different American experiences. Should Cubans, who have not encountered discrimination in work and education, be entitled to the benefits of these programs in the same way as Mexicans in the Southwest, who have been severely discriminated against? Such problems suggest that minority group membership is no longer unambiguous. Moreover, many have pointed out that the major recipients of affirmative action benefits have been middle-class racial and ethnic minorities, not the "truly needy"; that is, those in disadvantaged class positions. Filtering out those truly deserving of compensatory benefits has therefore become more complicated not only between various groups, but within them as well.

The issue of affirmative action continues to generate heated debate, with advocates of these programs countering their detractors on various points. As to the lack of qualifications of those employed or admitted through preferential policies, for example, proponents argue that "qualifications" can be variously interpreted, no matter how seemingly objective and valid tests or other sorting mechanisms may seem. Placement tests are often biased in favor of those with a white middle-class background, and in any case are never wholly adequate in

---

[1] Ethnic pluralism as either "liberal pluralism," in which ethnicity does not enter into legal or political matters and does not define a person's rights or duties, or as "corporate pluralism," in which group affiliation is very much a determinant of rights and benefits, was discussed in Chapter 4. See also Gordon (1975, 1981).

measuring one's potential on the job or in the classroom. Moreover, merit, they point out, has never really been the sole criterion used in filling occupational and educational positions. On the matter of quotas, it is argued that the objective of affirmative action is to facilitate the entrance of minorities into various institutions, not to keep them out, as was the purpose of earlier discriminatory quotas. Furthermore, they point out, preferential treatment is already given certain groups, such as veterans or athletes, in employment or education. And, while opponents see racial minorities as being stigmatized by affirmative action, defenders argue that their absence in jobs and schools would create an even greater stigma (Kennedy, 1994).

The Legal Issues of Affirmative Action

The legal questions of affirmative action programs have involved their scope and intent as well as whether they are racially discriminatory and therefore in violation of the Fourteenth Amendment of the Constitution. They have been tested in the courts on numerous occasions, but judicial decisions have been inconsistent and have not fully clarified these issues.

Major Court Cases As to their scope, an early case of great significance was *Griggs v. Duke Power*, heard by the Supreme Court in 1971. The specific question addressed in this case was whether tests given by companies to potential employees could be required if it was believed that they had adverse effects on minority applicants. This went beyond the Fifth and Fourteenth Amendments stating that only *intentional* harmful actions against people on the basis of race or ethnicity violated their constitutional rights. The Court ruled that company tests as a qualification for employment could be invalidated unless they could be shown to be directly related to job performance or were a business necessity. That is, the use of practices such as testing was now severely limited if their *effect*—not simply their intent—was to eliminate minority applicants. If the number of minorities in a company's workforce, for example, did not reflect the proportion of the minority population in the area where the company was located, that fact might be sufficient to demonstrate the effect of such indirect discrimination or, as the term came to be known, "disparate impact." Thus, in this case the Court extended affirmative action to address instances not only of overt, intentional discrimination, but institutional discrimination as well. In 1989 in another ruling, however, the Court weakened the business-necessity requirement and shifted the burden of proof for demonstrating such indirect discrimination from the employer to the applicant (Ezorsky, 1991).

As to whether affirmative action programs are themselves discriminatory, the *Bakke* case is perhaps the most significant, particularly as affirmative action relates to higher education. Allan Bakke was denied admission to the medical school of the University of California at Davis even though his entrance qualifications exceeded those of minority applicants who had been admitted under a special program. As a means of increasing the number of minority students, 16 of 100 places in each entering class had been reserved for minority applicants. Bakke maintained that he had been discriminated against on the basis of ethnicity and sued the university for admission. The case eventually reached the U.S. Supreme Court, and in 1978, by a five-to-four decision, the justices ruled in favor of Bakke. Their decision was ambivalent, however, leaving the way open for schools to establish goals for meeting an ethnic balance. In effect, the Court ruled that quotas (which in this case the university was found to have used) were illegal, but that the use of race as a criterion of preference was legitimate as long as it was one among many criteria and its purpose was to create a more ethnically balanced student body.

In another case, *Weber v. Kaiser Aluminum*, affirmative action policies that affected employment were tested in the Supreme Court one year after the *Bakke* decision. Brian Weber, a white worker, contended that he had been passed over in a company training program in favor of blacks with less seniority. As in Bakke's case, a fixed number of places had been reserved for underrepresented blacks. Here the Court ruled with less ambiguity than in *Bakke*, declaring that employers seeking to increase minority representation in certain occupational areas not only could consider race as a criterion of hiring and promotion but also could do so even if such efforts involved the use of quotas. A similar decision was rendered in 1984 in a case involving the promotion of blacks as officers in the Detroit Police Department. Both of these decisions were read as victories for affirmative action. In another 1984 case, however, the Court seemed to move in the opposite direction, ruling that employer layoffs had to be made according to applicable seniority rules, even if increases in minority employment from past affirmative action were wiped out in the process.

Public officials in the 1990s voiced uncertainty about how, and whether, affirmative action could be sustained. A 1995 Supreme Court decision reflected a movement toward reversing the scope and impact of these measures. In June of that year, the Court issued an important ruling in the case of *Adarand Constructors v. Peña* that placed tight limits on affirmative action. Though not going so far as declaring them unconstitutional (two of the justices did, in fact, support that position), it struck down federal contracting programs that favored firms owned by racial minorities (so-called set-asides), ruling that such programs had to be narrowly tailored to address identifiable past discrimination.

A number of other legal rulings dealt further blows to affirmative action. A federal appeals court in 1994 found that the University of Maryland's scholarship program for black students violated the rights of students of other races (a ruling that the U.S. Supreme Court later upheld), and in 1995 the governor of California declared an end to the use of preferential measures in admissions to the University of California system. In 1996, a U.S. Court of Appeals rendered a decision that appeared to rescind the *Bakke* decision allowing public universities to consider race as one criterion in choosing students. In what became known as the *Hopwood* case, judges ruled that the University of Texas Law School could no longer use race as a factor in its admissions decisions simply to increase ethnic diversity. In addition to these legal decisions, voters in California and Washington State passed ballot initiatives that ended consideration of race by public agencies and universities in hiring and admissions. All of these measures seemed to indicate a growing trend toward severely limiting or perhaps even abolishing affirmative action policies altogether.

Legal rulings in 2000 and 2001, as well as others pending, continue to reveal an ambivalence on the part of public officials and the courts toward affirmative action policies. A federal judge in December 2000 upheld the undergraduate admissions policy of the University of Michigan, which clearly considered the race of applicants. This followed another similar ruling that same month involving the University of Washington Law School. While these cases seemed to uphold the legality of affirmative action, several other cases produced rulings that augured the very opposite trend. In a 2001 case involving the University of Georgia, a federal court ruled unconstitutional an admissions policy that gave a slight preference in bonus points to nonwhite applicants. The policy's rationale was to guarantee a diverse student body. The court's ruling suggested, however, that diversity could not be defined only by race but that other factors needed to be considered, ones that could apply to white as well as nonwhite students. A white applicant from Appalachia, for example, might contribute more to the creation of a diverse campus than a black applicant from suburban Atlanta, the judges said (Firestone, 2001). Earlier in the year a similar ruling was handed down in a case involving the law school of the University of Michigan (Wilgoren, 2001). Both cases resembled the 1996 *Hopwood* case, and one of them seemed likely to eventually be decided by the U.S. Supreme Court.

The Politics of Affirmative Action

As affirmative action policies have evolved and been applied during the past forty years, political debate has crystallized around their legitimacy and fairness. As a result, affirmative action has been a key issue separating liberal and conservative sides of the American political spectrum.

The Political Contest Starting in the late 1960s and throughout the 1970s, affirmative action policies evolved through various administrative decisions of both Democratic and Republican administrations (Skrentny, 1996; Swain, 2001). The Reagan administration, however, unlike previous administrations, declared its opposition to the use of affirmative action measures and throughout the 1980s sought to subvert them through actions (or inactions) of the Justice Department.

"Opposition to race-based affirmative action," notes Thomas Edsall, "became for the Reagan regime a matter not only of principle and of policy, but of partisan strategy" (Edsall and Edsall, 1992:187). During the early 1980s the Supreme Court, despite its inconsistent rulings, generally resisted these destructive efforts, upholding affirmative action programs in a number of decisions. With the appointment of additional conservative justices by Reagan, however, the Court by 1989 was philosophically poised to reduce the impact of affirmative action and to weaken civil rights measures. In several decisions, it clearly moved in that direction.

The Reagan administration's position on affirmative action seemed to indicate a general shift in presidential policy regarding ethnic minorities and civil rights issues (Edsall and Edsall, 1992; Miller, 1984). Though its attempts to undermine federal policy were not entirely upheld either in the courts or by Congress, the administration seemed intent on rescinding the federal government's role in advancing and protecting minority rights. George Bush Sr.'s presidency seemed to sustain the efforts of Ronald Reagan to pull back from the application of affirmative action. Although not driven by the ideological fervor of the Reagan presidency, Bush, like Reagan, employed opposition to affirmative action as part of a strategy through which Republicans could galvanize white electoral support.

Since the inception of affirmative action in the 1960s, Democratic and Republican parties had generally aligned themselves on opposite sides of the issue (Edsall and Edsall, 1992; Hacker, 1995). The election of a Democratic president in 1992 therefore augured a more sympathetic presidential view of minority concerns. While Bill Clinton did in fact seem to uphold a commitment to the principle of affirmative action, a growing public disaffection with these policies and the election in 1994 of a deeply conservative Republican Congress, most of whose members strongly opposed affirmative action, combined to create the most serious movement to significantly modify or entirely abolish these policies.

Public sentiment regarding affirmative action measures had always been ambivalent. In the 1990s, however, the antipathy to affirmative action among whites grew increasingly popular and vocal. A 1995 national poll revealed that 79 percent of white Americans opposed racial preferences in employment or college admissions (Fineman, 1995).[2] Quite simply, whites had come to see such measures as "reverse racism" and in violation of the American creed. So negatively had they come to see affirmative action that, as political scientists Paul Sniderman and Thomas Piazza concluded, it had "led some whites to dislike blacks—an ironic example of a policy meant to put the divide of race behind us in fact further widening it" (1993:109). Furthermore, a widespread public perception, particularly on the part of whites, was that by the 1990s an equal opportunity structure had been created, obviating the need for compensatory measures.

---

[2] The poll revealed that although racial minorities were not as strongly opposed to such preferences (46 percent disapproved), they were by no means overwhelmingly supportive either.

It is important to consider that the public response to affirmative action is not so clear-cut as one might expect. To see the issue as a fundamental split between whites and racial minorities—whites oppose affirmative action, racial minorities support it—is too simple. As with racial attitudes generally (discussed in Chapter 9), whites today seem to generally accept the appropriateness of compensatory policies for racial minorities in principle, but their support wanes when specific measures are applied that are interpreted as special preferences. National surveys have consistently indicated that a majority of Americans, regardless of race or ethnicity, support affirmative action policies as long as they are not perceived as "rigid quotas" (Harris, 1991a, 1991b; Morin, 1995; Plous, 1996). Even a significant percentage of blacks are opposed to affirmative action when it takes such form. As Sniderman and Piazza found in their surveys, substantial numbers of whites are prepared to support policies that improve the status of blacks, but "The idea of quotas and preferential treatment is the reef on which affirmative action founders" (1993:130). Racial attitudes, then, are not necessarily the critical—much less the only—factor comprising this issue. People may oppose or feel resentment toward affirmative action policies not because they are antiblack or wish to keep things as they are (surely many *do* oppose them for these reasons), but because they see them as fundamentally in violation of the value of fairness.

Fairness and Objectives How compensatory measures can be made fairer has been subject to similar debate. Some maintain that instead of race or ethnicity, social class should be the major criterion by which people are deemed eligible for compensatory advantages (Kahlenberg, 1996; Schrag, 1995), or that policies should be applied in a "race neutral" fashion (Wilson, 1994). Others, however, maintain that such measures bring with them the same kinds of inherent problems as those based on race and ethnicity; that is, issues of determining who should be eligible and how they are to be chosen (Hacker, 1994; Kinsley, 1991).

Another confounding aspect of affirmative action concerns the very objective of such policies. Originally, they were intended to provide a temporary advantage to those racial minorities who had been the historic victims of systematic discrimination. As they evolved, however, their purpose was transformed into an effort to create schools and workplaces that resembled proportionately the racial and ethnic makeup of the society, state, or community. The question, then, is whether their purpose is to help bring about greater equity among the society's diverse groups or to create more ethnic diversity within the society's major institutions. Efforts at attaining one of these objectives do not necessarily correspond to attainment of the other.[3] Moreover, if it can be shown that creating diversity is in the society's interests, affirmative action to achieve this end can

---

[3] A national survey indicated that a majority of Americans recognize the value of diversity in higher education and support efforts to ensure a diverse student body (Morin, 1998).

be defensible for an indefinite period. This is unlike affirmative action policies that seek to correct past wrongs, upon which time limitations can be imposed; once it can be demonstrated that the playing field for all groups is in fact level, such measures would no longer be called for.

The Future of Affirmative Action

Whether affirmative action policies will continue to be applied by governments, corporations, schools, and the media is no longer certain. It is unlikely, however, that preferential measures in some form will be completely abandoned in the near future, whether mandated by government policy or not. Corporations, for example, understand that recruiting more minority managers is simply good business in an increasingly ethnically diverse society as well as a global economy (Glater and Hamilton, 1995). Moreover, they recognize that minority groups will supply much of their labor force in future. Three out of four *Fortune* 500 firms have diversity programs aimed at attracting and retaining minority workers (Cloud, 1998). And as already noted, most Americans seem to support some form of affirmative action so long as it does not appear to involve special treatment to particular groups (Sniderman and Carmines, 1997). But if compensatory policies in some form are likely to remain in place, rising political and social pressures seem to make it even more likely that the use of affirmative action, particularly in government and higher education, will be reduced sharply and its shape reformed in a basic way.

Affirmative action, in sum, has proven itself one of the most vexing issues of race and ethnicity in America. Yet, it is important to consider the widespread use of such measures in other multiethnic societies. Preferential policies designed to bring about more equity among diverse groups have been adopted in countries as divergent as Australia, India, and Malaysia. Thus, we should not think that the problems revolving around this issue are unique to the United States. As countries everywhere are transformed into multiethnic societies, mitigating the inequitable distribution of societal resources among various ethnic groups has become an inescapable predicament.

ISSUES OF THE NEWEST IMMIGRATION

The prominence of ethnic issues in the United States today stems not only from the continuing socioeconomic gap between whites and nonwhites but also from the large-scale immigration of diverse peoples in the past three decades. These latest immigrants have begun to have a significant impact on the ethnic flavor of the society and have introduced new problems and questions of ethnic relations.

Economic Issues of the Newest Immigration

The economic impact of the newest immigration is widely debated. What effects are these immigrants having on the American economy? Does immigration produce net benefits for the labor force, or does it negatively affect native workers? Do immigrants become self-supporting members of the society, or do they become a drain on public services like schools and hospitals?

Immigration and the Labor Force Whereas some argue that immigrants constitute an added burden to an already swollen labor pool, others contend that the jobs they typically hold are those that native workers shun—and furthermore, that they create as many jobs as they take. Similarly, while some maintain that immigrants overburden the social welfare system, others hold that they pay in taxes far more than they collect in benefits (Borjas, 1994; Espenshade, 1998; Simon, 1991).[4]

The economic impact of the newest immigration is not uniform. Instead, it appears to favor some sectors of the economy and harm others. Most of the immigrants from Mexico (the largest single group) and the Caribbean are unskilled and take their place at the lowest employment levels. They benefit employers in labor-intensive industries, such as clothing manufacturing or other work areas calling for cheap labor; but they depress the job opportunities of native low-status workers, especially African Americans (Borjas, 1998; Smith and Edmonston, 1997). As explained in Chapter 5, not all the new immigrants are impoverished and unskilled, however. A large segment of some new groups, particularly Asian Indians, Chinese, and Filipinos, are highly trained professionals and managers whose economic impact is far different from that of those entering with few occupational resources. Many immigrant doctors and nurses, for example, now staff inner city and rural hospitals, which would find it difficult to operate without them. The United States has also lagged in producing engineers and other highly trained scientific workers, and many of the newest immigrants are filling these needs (Muller, 1993; Suro, 1989; Szabo, 1989).

Those most profoundly affected by immigration are apt to express strongest support or opposition to its rate and extent. Industrialists and business owners ordinarily favor an open-door policy with few restrictions on continued immigration. Immigrants, in this view, serve as a needed workforce that helps to control wages. Those at the other end of the class spectrum, however, see immigration from an entirely different perspective. In the early 1990s, with the U.S.

---

[4]Another argument for restricting the number of immigrants to American society is that they constitute a growing environmental problem. The United States, it is held, has limits on what it can provide while still maintaining a high standard of living. Greater numbers of newcomers strain the society's resources and create environmental problems (Beck, 1996; Bouvier, 1992).

economy still in recession, immigration provoked negative reactions not only from whites but also from racial and ethnic minorities, who perceived the new groups as a threat to either their jobs or as an increasing pool of welfare recipients. This was especially apparent in areas like Miami, Los Angeles, and New York, which had absorbed large numbers of Hispanic immigrants (Beck, 1996; *Business Week*, 1992; Portes and Stepick, 1993; Warren and Stack, 1986). As we saw in Chapter 11, anti-Asian incidents also became more common with the continuation of Asian immigration.

It is especially illegal immigration that has come under public attack, since it is popularly perceived that illegals put an undue burden on public services, like schools and hospitals, at taxpayers' expense. Following the 2000 U.S. Census, estimates of the size of the illegal immigrant population rose considerably (Sachs, 2001). Because such a large proportion of illegal immigrants originate in Mexico, this is an issue with particular currency in states like Texas and California, which have large Mexican-American populations. In a 1994 referendum, California voters approved Proposition 187, which denied undocumented immigrants vital public services in education, welfare, and health care. Although the measure was subsequently declared unconstitutional, thereby preventing it from ever taking effect, the debate surrounding the referendum was extremely bitter and reflected the strong public sentiments regarding the entire illegal immigration issue.

Regarding the economic impact of the newest immigration, the debate waxes and wanes, hinging in large measure on the state of the U.S. economy. When immigrants enter during times of economic dislocation, workers are fearful of losing their jobs; immigrants, from this perspective, are seen as a potential threat, prepared to work at lower wages and under poorer conditions. Immigration is thus viewed negatively. When unemployment is low and jobs plentiful, however, arguments against immigration are muted and support wanes for reductions or restrictions on immigration levels.

These shifting trends were demonstrated in the 1990s. In the early years of the decade the U.S. economy was in recession; the rate of unemployment was relatively high and wages were stagnant or falling. Accordingly, anti-immigration attitudes seemed to be growing. By the latter part of the decade, however, a huge economic recovery was underway, with a tight labor market and rising wages. Immigrants were now seen as a valued commodity, filling labor gaps and revitalizing urban neighborhoods that the middle classes had abandoned. Many industries—hotels, restaurants, poultry-processing plants, garment factories—found themselves heavily dependent on immigrant labor, including illegal workers (Schmitt, 2001b). At the high end of the occupational hierarchy, too, a shortage of skilled workers, such as software programmers and computer engineers, gave rise to calls for the liberalization of immigration laws to permit the easier entrance of critically needed technical workers. As jobs remained plentiful, even some groups that had been staunchly critical of an open-door immigration policy were no longer pressing the issue. In 2000 the AFL-CIO, for example—long opposed to high levels of immigration, viewing immigrants as a threat to jobs and wages—called for amnesty for all illegal workers, viewing immigrants as potential union members.

In addition to the robust economy, a changed view of immigration was prompted by the increasing realization of elected officials that immigrants represented a potentially powerful political force. As a result, in the 2000 election and afterward, both Democratic and Republican parties engaged in efforts to woo immigrants. This was especially evident with the emergence of Hispanics as the largest of America's minority ethnic groups.

These vacillating attitudes toward immigration can be clearly seen in national surveys of the 1990s. In 1993, when the U.S. economy was still in a recession, 65 percent of Americans wanted a decrease in the level of immigration to the United States. In 2001, only 43 percent wanted immigration reduced (Gallup, 2001d). A downturn in the early 2000s, however, again raised the possibility of an anti-immigration backlash.

Social Issues of the Newest Immigration

The current immigrants are highly visible, bringing to American society cultural and physical features distant from those of the white majority. This makes them considerably different from previous immigrants who, though always viewed ethnocentrically by those already present, were nonetheless primarily European. The debate, therefore, concerns whether these new cultural and racial strains constitute a positive or negative social influence.

Some believe that currently high levels of immigration should be continued or made even higher to allow for additional immigrants. Their view is that immigrants not only create economic activity, but are a fresh cultural influence on the society. American society is in fact pluralistic, they hold, and should further develop its multicultural character, creating a more equitable society in the process (Isbister, 1996). Others, however, support highly restrictionist measures, maintaining that the new immigrants are causing a radical and unprecedented change in the social and cultural makeup of the society, which can only lead to more racial and ethnic conflict. Some see this change as undesirable in itself, suggesting that the United States remain a primarily European-origin society (Brimelow, 1995). The new groups are seen as unassimilable and thus a continual drag on the society. These issues have a familiar ring, recalling those that raged at the turn of the century with regard to southern and eastern European immigrants.

The mixed feelings of Americans regarding the social and cultural impact of immigrants is evident in national surveys. In 1993, 55 percent of Americans felt that the increased diversity of immigrants was a threat to American culture. In 2001, 45 percent said that increasing diversity in the United States as a result of immigration mostly improves American culture; but this was still only slightly higher than the 38 percent who said it was threatening to the culture (Gallup, 2001d). These shifting views may reflect changing economic conditions, as mentioned earlier.

Language Questions and attitudes regarding the impact of immigrants on American culture have come together most clearly around the issue of language. In Canada, as we will see in Chapter 15, the nucleus of ethnic conflict is the preservation of one group's language. A somewhat similar situation has begun to surface in certain regions of the United States, where large numbers of immigrants, particularly the Spanish speaking, use their native tongue. That these immigrants are able to maintain close links with their origin societies through easy communication and travel and through a constant influx of fellow immigrants from the homeland means that there is no pressing need to quickly relinquish the native language (Chavez, 1991; Heath, 1985; Muller, 1993). This makes their situation somewhat different from that of past immigrants. Although the latter also came speaking languages other than English, it was assumed that through the use of English in the school and other institutions, those languages would gradually be abandoned, if not by the first then by the second generation. Ethnic groups, of course, were not prohibited from maintaining their own institutions (private schools, newspapers, and so on), which often served as preservers of the native language for a time. Today, however, the assumption of language assimilation is being challenged. A greater tolerance of ethnic pluralism has led to efforts to provide public educational and other services in the language of the new groups, especially Spanish. This has created controversy among those who favor or oppose such measures (Baron, 1990; Cafferty, 1983; Chavez, 1991; Crawford, 1992).

Bilingual education has emerged as a particularly divisive political issue, especially in states with large Hispanic populations like California, Texas, New York, and Florida. In New York City, classes are now taught not only in Spanish, but in Chinese, Haitian Creole, Russian, Korean, Arabic, Vietnamese, Polish, Bengali, and French. In some of the city's neighborhoods, eight in ten school-age children live in a home where the family communicates in a non-English language (Sachs, 1999). And in Los Angeles, almost half of all students have limited English proficiency (Lopez, 1996). In Miami-Dade County, almost 60 percent of public school students speak a language other than English at home; and many schools at all levels are bilingual institutions, teaching in both English and Spanish (Nazareno, 2000). The issue is not limited to these high-profile immigrant-receiving cities, however. Almost half of all U.S. school districts have at least some non-English-speaking children (Hornblower, 1995). The controversy swirls around whether such students should be instructed in their native languages or in English, and in what proportion.

Some opposing greater language diversity have sought to have English legally declared the "official" American language. By the mid-1990s, more than twenty states had passed legislation to that effect. Proponents see such legislation as a means to reverse a perceived decline in English language use and to force immigrants to more quickly adopt the dominant culture. They advocate strict limits on bilingual education, the elimination of voting ballots in languages other than English, and raised language-proficiency standards for prospective citizens. Supporters of the movement to make English the country's official language maintain that without a common language, new immigrants will resist assimilation, and the United States risks becoming linguistically divided like Canada. In response, critics of the English-only movement charge that efforts to declare English the official language of the United States are a backlash against the new immigrants and essentially a mask for ethnic antagonism. Moreover, they claim that such legislation is useless because it does nothing to help promote the learning of English (Braverman, 1988; Crawford, 1992). By the late 1990s, public opinion seemed to favor the "official English" side of the controversy. A ballot initiative in California outlawing most bilingual classes was approved in 1998, and a national poll indicated that most Americans were in agreement with California voters in their opposition to bilingual education (Newport, 1998).

The issue may be moot in any case because studies indicate that, as in the past, the children of recent immigrants adopt English as their major language and gradually drift away from using their native language outside the home (Portes, 2002; Portes and Rumbaut, 1996; Saad, 1995). This shift to the English language, notes sociologist Calvin Veltman (1983), has been characteristic of every previous immigrant group and is evident among the newer groups today. "There is no language group in any region," Veltman concludes, "which possesses the retentive characteristics which would remotely sustain the theory that continued immigration is laying the groundwork for linguistic nationalism" (1983:217). Research among Latinos, for example, shows that first- and second-generation immigrants are usually bilingual, but by the third generation, English has become the primary—and for most, the only—language spoken (Fishman, 1987: Portes and Schauffler, 1996). The Latino National Political Survey revealed that among U.S.-born Latinos, 62 percent of Mexicans, 50 percent of Puerto Ricans, and 31 percent of Cubans used English predominantly or exclusively at home. More than two-thirds, regardless of national origin, were better at English or spoke no Spanish (de la Garza et al., 1992). An earlier study carried out in Miami revealed that 98 percent of Latino parents felt it was essential for their children to become competent in English (Braverman, 1988). And as we saw in Chapter 10, the vast majority of Latinos believe that citizens and residents of the United States should learn English (de la Garza et al., 1992). What keeps Spanish alive in the United States, therefore, is not a determination of Latinos to retain their native language but the continuation of large-scale Hispanic immigration.

Issues of Immigrant Adaptation and Integration

As explained in Chapter 5, immigration to the United States has reached epic proportions in the past two decades. Of no little significance is the fact that most of the newest immigrants are from non-European societies. How they will be absorbed into the society and the place they will eventually take in the ethnic hierarchy, as well as the ways in which they are changing the society, are issues of profound importance.

For several reasons, the typical patterns of adaptation to American society for past immigrants are not likely to be followed in quite the same way for the newest immigrants. European immigrants of the nineteenth and early twentieth centuries generally followed a path of eventual assimilation into the mainstream society, leading to upward mobility, over two or three generations. The newest immigrants may follow a more tortuous route and less certain outcome of social and economic adaptation, referred to in Chapter 4 as *segmented assimilation*. Some may follow the traditional assimilation pattern, but others may retain their ethnic cultures and social structures for many generations. Still others may find themselves stuck in a low-income status, with stifling job and educational opportunities that confine them to a condition outside the mainstream society. Most of the newest immigrants are nonwhite, which automatically places a racial barrier before them that past immigrant waves, mostly European, were able to overcome relatively quickly. However, the newest immigrants face significantly different social and economic circumstances as well.

First, the labor market encountered by the newest immigrants is one in which solid industrial working-class jobs, which in the past provided immigrants with opportunities for upward mobility, have disappeared. What has taken its place is a labor market more clearly divided into two segments—high-status (highly skilled professional and technical workers) and low-status (unskilled, mostly service-sector jobs). The latter are poorly paid and essentially dead-end jobs that have been disproportionately occupied by newly entering immigrants. The fear is that a semipermanent immigrant underclass will be created, impeding integration of these groups into the mainstream society for many generations.

Second, recent immigrants enjoy communications and transportation technologies unimagined by past newcomers. To fly from New York to the Dominican Republic, for example, involves a few hours. In the mid-nineteenth century it took about 5 weeks to sail from Britain to America, though bad weather could make the trip twice as long (Van Vugt, 1999). Similarly, immigrants today can remain in constant contact with relatives and friends in the origin country by telephone and electronic mail. This ease of communication enables immigrants to retain links with the ethnic culture and to resist rapid cultural and structural assimilation. For many, the proximity of the origin country also contributes to the retention of ethnic cultures and social ties. Immigrants from Mexico, for example, make frequent trips back and forth across the border, and a similar pattern of movement is common among immigrants from the Caribbean.

The newest immigrants are likely to see a continuation of immigration from their origin societies well into the twenty-first century. This is unlike the earlier European immigrant waves, which for the most part ended in the 1920s. The older immigrant groups were therefore not constantly infused with new members. Cut off more completely as they were from their societies of origin, they were faced with a greater incentive to assimilate quickly into the mainstream society. Current immigrants and their children, by contrast, are able to continually interact with coethnics, which will likely make for longer-lasting ethnic communities and cultures (Min, 1999). On the other hand, however, the mass media have a way of homogenizing people quickly and with great force. American popular culture is a global phenomenon; most immigrants therefore have absorbed a good deal of it before they actually migrate. Gans (1999) refers to this as "anticipatory acculturation." Roberto Suro explains that it is impossible to grow up in Latin America, for example, without being saturated with American images and information from birth. As a result, most immigrants can feel quite comfortable with U.S. culture when they arrive, at the same time feeling no need to discard their "Latino sensibilities" (Suro, 1998:71).

Finally, in recent years major institutions in the society, particularly education and the media, have moved toward an acceptance of ethnic pluralism, or multiculturalism, and away from the Anglo-conformity ideology that typified the dominant response to previous immigrant waves. The economy, too, in various ways has begun to cater to the unique needs and tastes of ethnic groups. In short, American institutions have today made it easier to retain one's ethnicity.

At the same time that social forces seem to present unique issues of social and economic adaptation, there are strong indications that the newest immigrants are, in a number of ways, following an assimilation path not radically different from past groups. As we have seen in previous chapters, language assimilation for both Latino and Asian groups is proceeding apace, as are residential integration and interethnic and interracial marriage.

Richard Alba has pointed out that assimilation has been incorrectly interpreted by some as a one-way process, in which immigrants surrender their culture and social structure to the dominant society and become carbon copies of the dominant ethnic group (1999:7). But in fact assimilation, as noted in Chapter 4, is a reciprocal process involving a cultural exchange between dominant ethnic groups and immigrant minorities. Thus, immigrants themselves inject change into the society's sociocultural system. As Alba has put it, "immigrant ethnicity has affected American society as much as American society has affected it" (1999:7). Moreover, a distorted conception of assimilation assumes a homogeneity of American culture that simply does not exist. Thus assimilation as it worked for past immigrants will in some regards probably work for current immigrants in a similar fashion, though not as clearly and directly. The remnants of ethnic cultures and social structures therefore will remain evident. As Alba explains, "assimilation most often occurs in the form of a series of small shifts that takes place over generations; those undergoing assimilation still carry ethnic markers in a number of ways" (1999:21). Assimilation, in other words, entails a decline of ethnic distinctions, not their absolute disappearance.

A Revitalized Nativism?

During the past decade, public views of immigration have been ambivalent. This is revealed in national surveys of the 1990s and early 2000s (*American Enterprise*, 1994; *Business Week*/Harris Poll, 1992; Gallup, 2001d, 2002; Jones, 2001; *Newsweek*, 1993, 1995b; *USA Today*, 1995). [5] Most Americans recognize the great historical role played by immigration in shaping their society but do not necessarily see present-day immigration in the same light. As explained earlier, tolerance for immigration seems to rise and fall with economic conditions. Even during economically good times, however, immigration is not seen positively by a substantial portion of the population. For example, in 1996, 50 percent said they would favor a law that would stop almost all legal immigration to the United States for the next five years; three years later, with the economy more robust than it had been in many years, almost 40 percent still stated their preference for a cut-off of immigration (Gallup, 2001d). Even though the newest immigrants were seen by most as very hardworking, they were also viewed by most as taking jobs away from American workers, driving down wages, and using too many government services. Although anti-immigrant sentiment is by no means as virulent as in the earlier part of the twentieth century, when it eventually led to a halt to the large-scale immigration of southern and eastern European groups, the racial and ethnic character of the newest immigrants, when combined with fluctuations in the state of the domestic economy, continually threatens to create a revitalized nativism.

---

[5]This was true of ethnic minorities as well as Euro-Americans. The Latino National Political Survey showed, for example, that more than 65 percent of U.S.-born Mexicans, Puerto Ricans, and Cubans believed that there were currently too many immigrants coming to the United States (de la Garza et al., 1992).

The flow of international events can also affect public attitudes toward immigration. In the wake of the terrorist attacks on the United States in 2001, for example, new thinking about immigration issues surfaced quickly. A national debate ensued regarding whether adequate security measures were in place to sift out undesirable immigrants, how illegal immigration could be better controlled, and whether U.S. policy was too lenient toward immigrants generally. It is highly probable that U.S. immigration policy will be affected fundamentally by the war against terrorism, as will public attitudes toward immigration.

Whatever changes are enacted in immigration laws are unlikely to bring to an end the debate regarding the number and character of immigrants that should be permitted to enter the United States. Those seeking more stringent limits will continue to argue that immigrants have a negative economic and social influence, whereas those advocating more liberal immigration laws will maintain that immigrants ultimately have a positive impact, both economically and culturally. These are hardly new points of debate, however. Historically, the absorption of immigrants has been a persistent theme of American political argument. In one sense, the United States has always been regarded as a "golden door," open to all seeking economic opportunity or political refuge. But the acceptance of new groups has been countered with a tradition of protectionism, which has manifested itself repeatedly in efforts to limit or exclude newcomers. The current public controversy is, therefore, only the latest in a long tradition.

It is important to note that the current public debate in the United States regarding immigration and the generally negative public view of immigration are not without parallel in other societies. In an analysis of public opinion toward immigrants and immigration issues in Australia, Canada, France, Germany, Great Britain, Japan, and the United States, Simon and Lynch (1999) show that despite these countries' different current and historical immigration policies, a majority of the public in each express relatively similar views: They want fewer immigrants admitted, especially immigrants of color; they feel that priority should be given to immigrants with special skills rather than to those seeking family unification; and they believe that their country has done more than its share in accepting refugees.

CULTURAL ASSIMILATION OR PLURALISM? COMPETING GOALS

As we have now seen, the ongoing population shift in American society has provoked fundamental economic, political, and cultural issues. But it has also prompted an ideological issue: Given America's increasing ethnic diversity, should the end product of interethnic relations be some form of cultural assimilation, in which diverse groups become more alike, or some form of pluralism, in which they maintain or perhaps increase their differences? The question of whether the proper course for disparate ethnic groups in American society should be aimed at eventual assimilation and, concomitantly, whether ethnic policies should be directed toward assimilationist goals, has arisen many times throughout American history. The political and social conditions of the last three decades, however, have given this issue renewed substance and made it especially salient.

Multiculturalism: The New Pluralism

In the past, ethnic issues in the United States were generally reduced, by sociologists, policymakers, and laypeople alike, to problems of assimilation (Metzger, 1971). How, it was asked, could diverse peoples achieve a maximum of social integration with a minimum of social travail? Put differently, the direction of ethnic relations was rarely questioned; it was assumed that groups would move toward some form of assimilation. Today that assumption is very much in dispute. In the 1960s the black movement, especially its nationalistic phase, served as a catalyst in prompting feelings of pluralism across the entire spectrum of ethnic groups. Heightened collective consciousness among African Americans and black political activism stirred similar movements among Latinos, Asians, American Indians, and even white ethnics. A new ideology of cultural pluralism seemed to materialize in which ethnic differences were not only to be tolerated and respected but perhaps even to be encouraged. In the past three decades the failure of group boundaries—particularly between whites and blacks—to dissipate, despite higher levels of racial tolerance and the end of blatant forms of discrimination, along with the influx of millions of new immigrants radically different in culture from the European norm, has given added incentive to stronger forms of cultural pluralism. As explained in Chapter 5, this ideology has increasingly been referred to as "multiculturalism." The turn toward a multicultural alternative represents a fundamental shift in the direction of American ethnic attitudes and ideology.

Multicultural ideas and policies are reflected in almost all major societal institutions, including education, business, government, and the mass media. The impact has been especially great in education, where non-European cultures are increasingly acknowledged in school and university curricula, and educational institutions generally conform to the multicultural ideology. The essential objective is to give greater recognition to the contributions of non-European groups in U.S. and global historical development and to study those groups' literature and art. These efforts have been met with heated debate among educators and public officials. One side views it as an attempt to demean European culture, and in the process to create sharper group divisions; the other side views it as necessary to counterbalance the traditional emphasis on "white" (that is, European) cultural and historical traditions (Goldberg, 1994; Ravitch, 2002; Schlesinger, 1992).

Language, as noted earlier, has been another flashpoint of ethnic ideological and policy conflict, pitting proponents of bilingualism in education and public communication against those who view language concessions as not only contrary to traditional American principles and practices, but as a dangerous opening to a corporate pluralistic system. It is a vivid illustration of the clash of assimilationist and pluralist perspectives.

A New "Melting Pot?" In a larger sense, however, this issue may be hollow. Despite the accentuated cultural diversity of the United States, acculturative forces are at work that seem to have homogenizing effects on all groups. In this new social environment, it becomes increasingly difficult to clearly identify the "dominant" ethnic culture.[6] Mass forms of communication and transportation increasingly break down cultural singularities and compress them into common forms, spiced with only slight ethnic variations. Billboards that in Des Moines picture white people and are written in English may in Detroit feature blacks or in Miami be written in Spanish, but they advertise the same products for all. Moreover, as the society becomes more ethnically varied, cultural and social influences begin to cross in bewildering combinations. Journalist David Rieff describes, for example, the ethnic fusion occurring in contemporary Los Angeles:

> Just as the Irish, Poles, Jews and Italians, who had rubbed shoulders and more in the wake of the European immigration of 1900, had, by the 1950s and '60s, begun to intermarry en masse, so the process is beginning to take place among the recent arrivals in L.A. One can find every sort of nonwhite combination in the city now: Hmong and Salvadoran, Ethiopian and Taiwanese, Mexican and Filipino. (1991:20)

In ethnic cauldrons like California, New York, and South Florida, the mix is so intense that ethnicity itself begins to lose significance. As one Asian Indian entrepreneur put it, "California is so cosmopolitan, you don't have to think about ethnicity here. It just doesn't matter—to anyone" (quoted in Jacoby, 2000:25). These centers of ethnic heterogeneity, of course, do not typify the United States; and the likelihood that such demographic conditions and their attendant

---

[6]Some suggest that in this new multicultural environment, even WASP Americans are evolving into simply one more ethnic group among many. William Greenbaum may have overstated their declining influence only somewhat when he declared that "after having been the society, the Protestants have been relegated to a place within the society, and increasingly they experience a bewildering sense of themselves as a new minority" (1974:412).

attitudes will soon follow in the American heartland is at best very long-term. Nonetheless, the seed has been planted.

It is in this sociocultural sense that, ironically, the "melting pot"—an inaccurate metaphor of American ethnic relations in previous generations—may be closer than ever to realization today. Steinberg (1989) suggests that although pluralist principles have become much stronger in recent years, the fact remains that ethnic differences have been diminishing. Thus, though groups extol the need to retain an ethnic culture and encounter declining resistance to its retention from the dominant group, societal trends continue to erode those cultural differences. Moreover, the insistence by ethnic leaders that ethnic cultures must be preserved does not seem to be shared with the same intensity by immigrants themselves. Contrary to conventional thought, immigrants seem most committed to assimilation, not retention of their ethnic cultures. A national survey in 1995 found that only 27 percent of immigrants favored the so-called multiculturalism effort to preserve ethnic identity, whereas 59 percent favored assimilation (Saad, 1995).

This homogenized culture is brought home strongly when Americans of any ethnic origin travel to the society of their forebears. Most quickly realize how little they have in common with the people of Ireland, if they are Irish Americans; of Poland, if they are Polish Americans; or of African nations, if they are African Americans. It is in such foreign contexts that a unique "Americanness" is most fully revealed. The same acculturative process will no doubt have a similar effect on the newest immigrants to the United States.

Looking Ahead

Although ethnic strife has been a historically constant theme in the United States, there appears to be a direness about the state of ethnic relations today—given the disturbing regularity of occurrences of interethnic hostility, the tenaciousness of ethnic inequality, and the emergence of a population more sharply divided by race and culture. Moreover, national surveys in recent years have consistently revealed public pessimism regarding current and future ethnic relations. Following the Los Angeles riots in mid-1992, for example, only 25 percent of Americans thought that race relations were "good," whereas 68 percent thought they were "bad" (*Newsweek*, 1992b). But even before the riots, surveys had revealed a similar public perception of the deterioration of ethnic relations in the United States (*Black Enterprise*, 1990; Hugick, 1992; *Newsweek*, 1992a). Three years after the riots, the same bleak picture seemed evident—only 23 percent of whites and 12 percent of blacks said that race relations were "good" or "excellent" (*Newsweek*, 1995a). By 2001, little had changed (Gallup, 2001a).

In all likelihood, then, ethnic discord will continue at a high level well into the twenty-first century. Still, there is cause for at least some cautious optimism.

First, we should remember that there have been few times in American history when ethnic issues have not been in the forefront of public consciousness and the focus of public policy. We should therefore not think of ours as a unique era.

Second, although the United States' pluralistic tendency and the accommodative responses to it by its major institutions are the source of continued public debate, there is unarguable evidence of a growing ethnic tolerance in the society as a whole. Table 12.1 shows the results of a poll in which a national sample of Americans in 1964 and 1989 was asked to rank different American ethnic groups. On a scale of 1 to 9, 1 was the lowest standing and 9 was the highest. The figures shown are averages. (Note the inclusion of a fictitious group, "Wisians," in 1989.) The rank order of the groups corresponds closely to the social distance hierarchy that emerged from the earlier Bogardus studies of social distance (described in Chapter 3), indicating much continuity in the social standing of American ethnic groups—the groups closest culturally and racially to the dominant group occupy the highest positions. Perhaps equally significant, however, is that the score for almost all groups rose during the twenty-five years between 1964 and 1989. Especially large increases are noticeable for blacks and Asian-American groups. Recall as well the demise of crude racist ideas among whites vis-à-vis racial minorities, noted in Chapter 9.

Another indication of positive change in the character of American ethnic relations can be detected in the response to the attack on the World Trade Center in New York in 2001. In the weeks following the catastrophe, anti-Arab sentiment ran very high and, as noted earlier, Arab Americans were the targets of verbal and in some cases physical abuse. Public opinion polls, moreover, confirmed the readiness of Americans to subject Arab Americans to special security measures, much in the fashion of racial profiling. In the wake of the society's war atmosphere, Arab Americans' fear of being targets for severe discrimination, comparable to the way German Americans were dealt with during World War I and Japanese Americans during World War II, was well grounded. But unlike the tragic attitudes and actions of those past cases, government officials—including the president—as well as media, education, and business leaders made direct and forceful appeals to the American public for tolerance, specifically to disassociate the terrorist acts from Arab Americans and Muslim communities in the United States.

Third, despite what at times seems a hopelessly racially and ethnically divided society, the United States may in fact be more socially and politically cohesive than is often assumed. Closer analysis reveals that on basic values regarding issues such as teaching a common American heritage and values, there is overwhelming approval by all groups; in fact, more minorities than whites endorse this position (Etzioni, 1998). Moreover, ethnic leaders often convey the view that race or ethnicity is the sole or major determining factor in setting peoples' perspectives on social and political issues. But this view ignores more

Table 12.1 Changes in the Social Standing of American Ethnic Groups

ETHNIC GROUP	1964	1989	CHANGE
Native white Americans	7.25	7.03	−0.22
People of my own ethnic background	6.16	6.57	0.41
British	6.37	6.46	0.09
Protestants	6.59	6.39	−0.20
Catholics	6.36	6.33	−0.03
French	5.73	6.07	0.34
Irish	5.94	6.05	0.11
Swiss	5.50	6.03	0.53
Swedes	5.41	5.99	0.58
Austrians	5.06	5.94	0.88
Dutch	5.60	5.90	0.30
Norwegians	5.48	5.87	0.39
Scotch	5.73	5.85	0.12
Germans	5.63	5.78	0.15
Southerners	5.25	5.77	0.52
Italians	5.03	5.69	0.66
Danes	5.20	5.63	0.43
French Canadians	5.08	5.62	0.54
Japanese	3.95	5.56	1.61
Jews	4.71	5.55	0.84
People of foreign ancestry	4.84	5.38	0.54
Finns	5.08	5.34	0.26
Greeks	4.31	5.09	0.78
Lithuanians	4.42	4.96	0.54
Spanish-Americans	4.81	4.79	−0.02
Chinese	3.44	4.76	1.32
Hungarians	4.57	4.70	0.13
Czechs	4.40	4.64	0.24
Poles	4.54	4.63	0.09
Russians	3.88	4.58	0.70
Latin Americans	4.27	4.42	0.15
American Indians	4.04	4.27	0.23
Negroes[*]	2.75	4.17	1.42
"Wisians"		4.12	
Mexicans	3.00	3.52	0.52
Puerto Ricans	2.91	3.32	0.41
Gypsies	2.29	2.65	0.36

SOURCE: Copyright © 1992 by The New York Times Company. Reprinted by permission.

[*]Blacks were referred to as Negroes in the 1989 survey to conform with the wording in the 1964 survey.

fundamental class differences that, as we have seen in the preceding few chapters, internally divide popularly conceived panethnic categories.

Fourth, intermarriage across racial and ethnic lines, as we have seen, continues to rise unabated. Thus the social walls separating racial and ethnic groups continue to crumble, and as they do, the arbitrary division of the society into conveniently demarcated racial and ethnic categories continues to lose meaning.

Finally, as we will see in the next five chapters, the United States is not alone in its ethnic problems. Indeed, few societies of the contemporary world do not face equally daunting problems stemming from an expanding ethnic diversity and its attendant intergroup conflict. And, as will be revealed, the level of American conflict is relatively mild compared to that of most other ethnically diverse societies.

SUMMARY

Three interrelated issues of race and ethnic relations in the United States are today of pressing importance. One concerns the continuing socioeconomic gap between Euro-American groups and racial-ethnic groups, specifically African Americans, Latinos, and Native Americans. To address this problem, during the past three decades government and other institutions have introduced preferential policies aimed at creating greater ethnic equity. These programs have been the subject of much controversy among politicians, scholars, and the general public. An intense debate concerns the fairness of these programs and whether they do, in fact, accomplish their ostensible objectives.

A second major issue concerns the large-scale immigration to the United States that has occurred since the mid-1960s. During this period, more immigrants have entered the society than at any time since the classic period of immigration in the early years of the twentieth century. Most important, the vast majority of the current immigrants are non-European, coming mostly from Asia and Latin America. Their presence has prompted much social debate regarding their economic impact and their social integration. Who and how many should be allowed into the United States as permanent residents? What effect is continued large-scale immigration having on the economy, education, and other institutions? Are social divisions becoming more severe with the expansion of ethnic diversity? Proponents of liberal immigration policies argue that immigrants bring many benefits to the society and have a generally positive economic and social impact, while those who advocate more stringent policies contend that immigrants have a negative economic impact and exert a potentially divisive social influence.

Immigration has become one of the most virulent issues of national politics. Public opinion for the past decade has seemed to favor more restrictionist measures, though the intensity of public reaction has hinged mostly on the state of the economy. And in response to periodic public fervor, politicians at the national level and in states and locales where the impact of large-scale immigration is most acute have exploited this issue repeatedly.

The final issue of American race and ethnic relations concerns the question of whether the society is—or should be—moving toward greater cultural assimilation or pluralism. Unquestionably, the United States today is a more ethnically diverse society than at any time in its history. Moreover, there appears to be greater tolerance for that diversity, with various societal institutions acknowledging and even sanctioning the expression of ethnic differences. Whether in the long term these differences will dissipate or be sustained is not yet clear.

Although ethnic conflict is an inherent feature of the United States as an ethnically diverse society, there are clear indications that the level of conflict is lower than in past times and that ethnic tolerance is at a high point. Moreover, among the world's ethnically divided societies, the intensity of American ethnic conflict is comparatively low.

Suggested Readings

Beck, Roy. 1996. *The Case Against Immigration: The Moral, Economic, Social, and Environmental Reasons for Reducing U.S. Immigration Back to Traditional Levels*. New York: Norton. Argues that large-scale immigration to the United States during the past three decades has contributed significantly to American economic, social, and environmental problems.

Crawford, James. 1992. *Hold Your Tongue: Bilingualism and the Politics of "English Only."* Reading, Mass.: Addison-Wesley. Traces the controversies surrounding the issue of bilingualism, prompted by the surge of new immigrants, especially Latinos, whose native language is not English.

Ezorsky, Gertrude. 1991. *Racism and Justice: The Case for Affirmative Action*. Ithaca, N.Y.: Cornell University Press. Presents moral and practical arguments for affirmative action policies and assesses their impact, particularly on African Americans.

Glazer, Nathan. 1997. *We Are All Multiculturalists Now*. Cambridge, Mass.: Harvard University Press. A prominent sociologist offers his views on the decline of the assimilation model and the increasing acceptance of ethnic pluralism as a central value of American society.

Hartman, Chester (ed.). 1997. *Double Exposure: Poverty and Race in America*. Armonk, N.Y.: M. E. Sharpe. Social scientists present their views on the major issues of race and ethnicity in the United States today, particularly the continuing gap between Euro-Americans and racial-ethnic minorities.

Kahlenberg, Richard D. 1997. *The Remedy: Class, Race, and Affirmative Action*. New York: Basic Books. Argues for the use of class-based factors as the rationale for affirmative action, rather than racial and ethnic factors.

Mills, Nicolaus. 1994. *Debating Affirmative Action: Race, Gender, Ethnicity, and the Politics of Inclusion*. New York: Delta. A collection of essays by social scientists, policymakers, and political commentators that explores all sides of the affirmative action debate.

Muller, Thomas. 1993. *Immigrants and the American City*. New York: New York University Press. Explores issues pertaining to the newest immigrants to American cities. Maintains that they have had a generally positive, sometimes transformative, economic and social impact on urban areas.

Schlesinger, Arthur M., Jr. 1992. *The Disuniting of America: Reflections on a Multicultural Society*. New York: Norton. A prominent historian analyzes the upsurge in ethnic consciousness and its impact on education, especially the teaching of history as part of a multicultural curriculum.

Skrentny, John David. 1996. *The Ironies of Affirmative Action: Politics, Culture, and Justice in America*. Chicago: University of Chicago Press. Meticulously traces the origins and development of affirmative action policies, stressing the political process which involved little public debate.

Steele, Shelby. 1991. *The Content of Our Character: A New Vision of Race in America*. New York: HarperPerennial. A black scholar argues that affirmative action policies no longer function beneficially for African Americans but in fact act as a group handicap and accentuate perceived racial differences.

PART

III

ETHNIC RELATIONS IN COMPARATIVE PERSPECTIVE

HAVING INVESTIGATED PATTERNS of race and ethnic relations in the United States, we now examine some non-American cases as a basis for comparison. People often assume that what occurs in their own society is unique and that others are simply not like them. Conversely, they sometimes assume that what happens in their own society is characteristic of human societies in general. In part, both of these suppositions are valid. To ascertain in what ways U.S. society is similar to and in what ways it is different from others requires that Americans transcend the limited perspective imposed by the social system with which they are most familiar and comfortable. That is, they need to adopt a broader, less parochial approach to human affairs.

The societies examined in this part lend themselves readily to comparison with the United States. It will not be difficult to detect similarities, nor will the differences be obscure. Each is part of a different geographic region, is a product of a distinct historical tradition, and displays quite different bases of ethnic stratification and conflict.

The patterns of race and ethnic relations in each of these societies represent one of the major types described in Chapter 4. South Africa, the focus of Chapter 13, resembles in its recent history the *colonial* (or extreme inequalitarian pluralistic) type; and Brazil, examined in Chapter 14, is an example of the *assimilationist* type. In Chapter 15, Canada will be seen as a society with many features of a *corporate pluralistic* system. Northern Ireland, described in Chapter 16, exhibits aspects of both colonial and corporate pluralistic systems. These societies also seem to fall roughly on a scale, or continuum, based on the severity of ethnic stratification and conflict in each. South Africa is at one extreme, characterized throughout most of its history by gross ethnic inequality and, until recently, institutional separation. At the opposite pole is Brazil, where the blending of groups culturally, structurally, and even physically is officially endorsed, if not realized in fact. As in the United States, it is the disparity between groups of different color that represents the thrust of ethnic relations in both South Africa and Brazil. In Canada and Northern Ireland, however, ethnic relations revolve primarily around the juxtaposition of two cultural groups, English and French in Canada and Protestant and Catholic in Northern Ireland.

In Chapter 17, we explore the rising significance of ethnicity across the globe as nations everywhere, through immigration and political turmoil, become more ethnically diverse. We look first at the impact of this new heterogeneity on western European countries. This will provide us an opportunity to compare their relatively fresh encounter with ethnic conflict and change to the American experience, in which ethnic issues have been a wellspring of social friction and strife for many generations. Focus then falls on the former Yugoslavia. There, as we will see, conflict among ethnic groups in the 1990s sank to a level of brutality and hate unseen in Europe since World War II. Although comparison with the United States is dubious, the Yugoslav case nonetheless forces us to consider what can happen to a society in which fear and distrust of ethnic out-groups become the preeminent characteristics of interethnic relations.

13

SOUTH AFRICA:
SOCIETY IN TRANSITION

UNTIL VERY RECENTLY South Africa, in its system of ethnic relations, was unique in the modern world. As an inequalitarian pluralistic society, it once had no peer. Despite persistent intergroup conflict and hostility, almost all contemporary multiethnic societies made attempts at ethnic unification; but for most of the past fifty years, South Africa moved in the opposite direction, compelling greater divisions among its ethnic populations. What made South Africa most singular among modern societies was that its racist system was fully legitimized and enforced by the state. South Africa, as we will see, has been a congeries of societal inconsistencies and contradictions. Throughout most of its history, however, one thing was clear and unambiguous—the dominance by whites, the numerical minority, over blacks, the numerical majority.[1] No area of South African life was untouched by the racial policies and ideology of the dominant white group.

---

[1] The ethnic terminology used in describing South Africa's population has not been consistent, and as in the United States, it has political connotations. Traditionally, black Africans were called *Bantu*, a name corresponding to their languages. During the apartheid regime, they were officially called *blacks* but referred to themselves as *Africans*, a term used by most scholars as well. Some use the term black to describe the entire nonwhite population, including Coloreds and Asians, as well as Africans. Today *black African* or *African/black* is used commonly to refer specifically to Africans, and that usage is adopted in this chapter. As the new South Africa evolves, ethnic terms will likely take on new meanings (see Adam, 1995; Horowitz, 1991; Lever, 1978; Thompson, 1990; and van den Berghe, 1967).

Today South Africa has embarked on what appears to be an irreversible course of change, moving toward a more equalitarian system. Indeed, the scope and depth of change this society has experienced in the past decade would have been unimaginable just a few years earlier. South Africa therefore provides an instructive case of the fluidity of ethnic relations and how, no matter how rigid and seemingly implacable, they can change profoundly in a relatively short period of time. Despite the magnitude of change, however, the ultimate shape of the new South Africa as a multiethnic society remains very much uncertain. Its racial and ethnic problems, therefore, will engage the attention of the world for many years. Furthermore, although its system of ethnic separatism has been dismantled officially, enough of it endures informally to make this society still an extraordinary case in its extremes of ethnic division and inequality.

As we look at South Africa it is important to keep in mind that events of change have moved so swiftly in recent years that any description of its prevailing institutions risks almost immediate obsolescence. South Africa's significance as a case study of race and ethnic relations, however, lies as much in its history as in its current affairs. Furthermore, its system of extreme ethnic separation and stratification is not something of the distant past, but has only recently been successfully challenged. It is fresh enough, therefore, to warrant a careful examination.

THE DEVELOPMENT OF ETHNIC INEQUALITY

Few who read newspapers or watch television are unaware of at least the basic outlines of the ethnic drama that began to unfold in South Africa in the 1980s. However, the extraordinary social and political changes that took place—and continue to take place—and the new South Africa that is emerging as a result can be understood only if they are positioned alongside this society's long heritage of intergroup conflict. As in other multiethnic societies, ethnic stratification in South Africa has been the result of a historical sequence of diverse groups coming together, viewing one another ethnocentrically, competing for scarce resources, and ultimately forming a hierarchy in which one imposes its superior power on others.

White Settlement

Historical evidence indicates that what is today South Africa was inhabited by Bantu-speaking peoples, the society's major African linguistic group, as early as the sixteenth century and probably much earlier (Wilson and Thompson, 1969). Modern South African history, however, involving the confrontation of European and native African peoples, begins with Dutch settlement in 1652. At that time, the Dutch East India Company established a colony in Cape Town as a refreshment station for its ships traveling to and from India. The indigenous people of the area, the Khoikhoi (Hottentots) and the San (Bushmen), were unable to satisfy the labor needs of the colonists, who therefore began to import slaves from other parts of Africa and from the Dutch East Indies only a few years after their original settlement. The colonists' negative view of these slaves did not prevent widespread miscegenation; hence, a mixed racial group was created whose descendants are today called Coloreds.

As the colony grew, many of the Dutch, German, and French Huguenot colonists chose permanent settlement and started at the beginning of the eighteenth century to leave Cape Town, establishing farms on land farther into the interior. Thus began the penetration of the northern and eastern frontiers, a movement called trekking. These settlers became known as Boers, a Dutch word meaning "farmers." The trekking movement lasted 150 years, with the Boers continually pushing farther and farther inland. As they moved into the interior, the major patterns developed that would basically affect the shape of South African society—conflict with the indigenous peoples and the emergence of a unique white South African ethnic group.

The Boers came into conflict not only with native African peoples occupying these lands but also with migrating African tribes moving across the continent, westward and southward. These were Bantu-speaking tribes who were well organized and who presented serious resistance, unlike the Khoikhoi and the San, who had been more easily subdued. The clashes over land and cattle that developed between the Boers and the Bantu, called Kaffir wars, lasted for many decades before the dominance of the Boers was finally established. The disdain for the natives based on their racial features was exacerbated by the bitterness that resulted from these encounters. Additionally, because they greatly outnumbered the Boers, the Bantu were seen as a constant threat whose suppression had to be ensured. Many of the racial attitudes of white South Africans in recent times can be traced to these initial contacts with Africans in the interior.

The second major consequence of the trekking movement was that the Boers increasingly distanced themselves from their European roots. As the settlers pushed inland, Dutch control over them waned. The farther they removed themselves from Cape Town, the more they became "a law unto themselves" (Thompson, 1964:186). As subsistence farmers, they disengaged themselves even further from intercourse with the economic and political institutions of the colony and its civil administration.

With their economic and political isolation, the Boers developed a culture thoroughly distinct from their European heritage. Differences were most apparent in language and religion. A unique Dutch dialect evolved over several generations, producing eventually the Afrikaans language, today only remotely related to the original Dutch. Divergent religious beliefs also contributed to the cultural distance between the Boers and their European ancestors. The Boers resisted the enlightened Protestantism that was emerging in Europe in the late seventeenth and early eighteenth centuries, clinging instead to a primitive Calvinism whose ideas of predestination and rugged individualism seemed a fitting complement to the pioneering conditions of life in rural South Africa. They increasingly saw themselves in a biblical light as "chosen people," destined to conquer the frontier against the numerically superior Bantu and later the colonial British (Fredrickson, 1981). The chosen people theme continued to run strongly through the ideology of the Boers (today called Afrikaners) and was reflected in their stubborn resistance to changes that threatened to reduce their power.

By the late eighteenth century, the Boers were a unique social entity with a culture and language distinct from any other. They were geographically and intellectually isolated. Their European roots, notes historian Leonard Thompson, "were almost completely severed" (1964:187). Thus was born the Afrikaner ethnic group of South Africa. The Afrikaners' knowledge of and ties to Europe were so minimal that they had become, in effect, "white Africans." Unlike other European settlers on the African continent, then, the Afrikaners did not see themselves, nor did they even function, as colonials representing and maintaining strong cultural and political ties to a motherland.

This sense of isolation and permanence explains much about the determination of Afrikaners in recent decades to defend their way of life and what they came to see as their territory against the threat of black rule. "The clue to South African history," wrote Douglas Brown over thirty-five years ago, "is that the Boer, though early forced into an intimate local relationship with other races, brown, white, and black, has since the first decade of the eighteenth century had no backing, no allies, no means of reinforcement, no possibility of retreat. His nationalism soon became absolute, and so it remains in the jet age, to the discomfiture of the rest of the world" (1966:19). Not until the late 1980s were the Afrikaners compelled to alter that view. As we will see, Afrikaner nationalism has been tempered and has given way to a recognition that only by sharing societal power with those groups it dominated so completely for most of South Africa's history could it survive in the modern world.

The British Entrance

Whereas the first century and a half of modern South African history is largely a chronicle of confrontation between white and black—Afrikaner and African—the following 150 years are marked principally by the political rivalry between the Afrikaners and the colonial British. The British entered the South African scene at the start of the nineteenth century when, in a move designed to protect their colonial interests in the East, they took possession of the Cape colony. The next fifty years saw the development of a competitive and often bitterly hostile relationship between the British and the Afrikaners that, although greatly moderated, has been a continuing element of South Africa's social dynamics.

The British were not a large settler group like the Boers, and they brought a strikingly different political and social temperament to South Africa, more liberal and cosmopolitan. Their values clashed with the stern parochialism of the Afrikaners, and a wide gulf opened between them. This gap has remained relatively broad to the present time.

Soon after their arrival in South Africa, the British instituted liberalized policies regarding race relations, which exacerbated the already strained relations with the Afrikaners. The British attitude toward the native Africans was no different from the Afrikaners' in its substance—white dominance. But the British saw black Africans in a more paternalistic frame, whereas the Afrikaners saw them as a threat to their ultimate survival as a people. "The abstract difference between British imperial and Afrikaner republican conceptions of how to rule Africans," explains historian George Fredrickson, "might be described as a conflict between the trusteeship ideal and what the Boers called *baaskap*—which in essence meant direct domination in the interest of white settlers without any pretense that the subordinate race was being shielded from exploitation or guided toward civilization" (1981:193). The practice of white supremacy by the Afrikaners was thus consistent with their fundamental belief that force and exploitation were perfectly legitimate means to protect their group interests.

In 1834, slavery was abolished throughout the British empire. This act was followed during the next two decades by the lifting of other legal restrictions against nonwhites. These radically new racial policies acted as a major incentive for the second and far more substantial migration of Boers into the interior. In 1836, a sweeping movement was launched in which one-quarter of the entire Afrikaans-speaking population would eventually set out on their own into the interior (Robertson and Whitten, 1978). This was the commencement of the Great Trek, comparable in many ways to the westward movement of Americans in the nineteenth century. As the Western frontier had served as a kind of social safety valve in the United States, the Boer trek northward and eastward after 1836 provided a means of escape for those who found the British presence a constant source of anguish.

These Voortrekkers, as they were called, were doggedly pursued by British colonial advances, however. After fierce battles with the Zulu tribe in 1838, Boers founded the republic of Natal, which was subsequently annexed by Britain in 1843. Pushing farther northward, the Voortrekkers founded the Orange Free State and the Transvaal, which were eventually recognized as independent states by the British. These states became the core of the Afrikaner culture, and within them the sense of autonomy and group solidarity was developed to its fullest. In both, the notion of white supremacy was a fundamental principle.

The discovery of diamonds around Kimberley in 1867 and of gold in the Transvaal in 1886 once again spurred British advances. Britain had earlier attempted to extend its influence into the Transvaal, resulting in the first Boer War.

Although the Boers successfully resisted at that time, the influx of British railroads, mining interests, and workers following the gold discovery led to a more serious challenge to Boer autonomy. The British finally intervened with military force, provoking the second Boer War in 1899, a conflict that would leave permanent scars of bitterness between the two groups. The Boers used guerrilla tactics against the larger and superior British forces; in response, the British interned the Boers in concentration camps where 26,000, including women and children, died of disease. This is an event of enormous significance in Afrikaner history, symbolizing the martyrdom of the Boer people and their determination to survive against all odds. This second encounter with the British was a key factor in the subsequent development of militant Afrikaner nationalism. As we will see, though they were defeated in war, the Afrikaners eventually asserted political ascendancy and emerged as South Africa's dominant ethnic group.

Unlike the Afrikaners, the British never coalesced into a nationlike unit in South Africa. Their loyalties were fragmented between South Africa and the British Commonwealth (Thompson, 1964). As a result, Afrikaner dominance—though repeatedly diluted by British incursions in education, immigration, and especially the economy throughout the nineteenth century and part of the twentieth—was eventually entrenched and solidified. Although they had been militarily successful in the second Boer War, the British began to lose influence afterward and reached a tenuous political compromise with the numerically superior Afrikaners.

In 1910 the four territories of South Africa—the Cape, Natal, the Orange Free State, and the Transvaal—were merged into a union. This marked the beginning of a period in which the British and moderate Afrikaner political forces were sufficiently consolidated to frustrate the efforts of the more extreme Afrikaner-dominated National Party to excise the British influence. Following the formal union of South Africa, the foundation of the society's racial laws was also secured. Although the subordination of blacks had always been informally understood and enforced, during this period a succession of government policies institutionalized many segregationist and inequalitarian practices, particularly those regarding participation of Africans in government and the economy.

The Nationalist Ascendancy

The modern period of South African history can be said to have begun in 1948 with the political victory of the National Party. This ended the almost half century of delicate balance between British and Afrikaner interests and established the Afrikaner political ascendancy. The National Party represented the most extreme element of Afrikaner nationalism, viewing itself as the spearhead of a God-ordained mission to shape South Africa in the interests of Afrikaners and to throw off the yoke of British influence. With the withdrawal of the Union of

South Africa from the British Commonwealth in 1961, South Africa's British ties were severed once and for all.

Most important, the Afrikaner-dominated government began to formally erect an institutionalized system of white supremacy, called *apartheid*, the dynamics of which we will look at shortly. While other African nations of the post–World War II era were moving toward national unification and a breakdown of tribalism, South Africa was moving in the opposite direction, creating a social order of strict separation between whites and blacks and encouraging the retention of tribal cultures among the African populace. In the modern period of South African history, the Afrikaner-led whites presented to the world a system of formalized ethnic separatism unmatched in rigidity and scope.

ETHNIC STRATIFICATION

In no other modern multiethnic society was ethnic stratification so clear-cut as in South Africa, and in no other was it so inflexible. Until the official demise of apartheid in the early 1990s, the boundaries between ethnic strata were, by law, relatively impermeable. South Africa's racial system was castelike, sustained by endogamy and by the maintenance of separate institutional systems for each ethnic group. Only in the larger economic system did the different ethnic groups come together, and there, as in all other societal areas, the intergroup relations were grossly inequitable. Indeed, along with its rigidity, the other major feature of the South African system of ethnic stratification was the enormous gap between the dominant white group and the several subordinate black groups—a gap that, as we will see, remains very much in evidence today. South Africa, then, was the prototypical inequalitarian pluralistic society, in which only the economic system served as a kind of social glue holding the various groups together.

Undergirding and enforcing the system of ethnic inequality was the South African state—dominated by whites, specifically, the Afrikaners. So complete was white control of the society's political apparatus that blacks were allowed no meaningful participation and were accorded few legal rights. Pierre van den Berghe (1978) referred to the South African sociopolitical system as a Herrenvolk democracy, that is, a state that provides most democratic features of political rule to whites while ruling blacks dictatorially. Others described the system as a "race-oligarchy" (Adam, 1971b) or a "pigmentocracy" (Thompson and Prior, 1982). All these appellations denote the fact that skin color was the single overriding criterion of societal power in South Africa.

In almost all multiethnic societies, ethnic status is a significant factor in the distribution of life chances. The allocation of justice, health, occupation, education, living quarters, and so on depends, to some degree or another, on one's ethnic classification. But in no modern society has that classification been so critical in determining people's social place as it was in South Africa under apartheid. Today the ethnic hierarchy continues to closely overlap the class hierarchy, so that class stratification in South Africa remains very much a function of ethnic status.

To ensure white dominance in all areas of social life, the castelike system of apartheid required a formal division of the population into specific ethnic categories—whites, Coloreds, Asians (mostly Indians), and Africans. Though no longer officially recognized, this ethnic division continues to strongly shape South Africa's social, political, and economic dynamics.

The Whites

Whites (or Europeans) constitute about 10 percent of the population (Table 13.1) but, until 1994, completely dominated government. As the only group with effective political rights, whites for decades filtered all social rewards, first siphoning off their disproportionate share before distributing the remainder to the three subordinate groups. Although they no longer dominate the political system, whites continue to monopolize positions of power in the economy and own most of the society's wealth. In an economic sense, then, they remain the dominant group, despite the fact that they have been displaced as the political ruling group.

As already noted, however, whites are by no means a culturally or even politically unified group. Rather, they comprise two major elements: Afrikaners and the English-speaking, most of the latter of British origin. The two major white subgroups differ not only in language and culture but also, to some degree, in

Table 13.1 Ethnic Population of South Africa

YEAR	TOTAL (1,000s)	AFRICAN (1,000s)	%	COLORED (1,000s)	%	INDIAN/ ASIAN (1,000s)	%	WHITE (1,000s)	%
1911	6,044	4,091	67.7	525	8.7	152	2.5	1,276	21.1
1936	9,591	6,597	68.8	770	8.0	220	2.3	2,004	20.9
1951	12,672	8,560	67.5	1,103	8.7	367	2.9	2,642	20.8
1970*	21,794	15,340	70.4	2,051	9.4	630	2.9	3,773	17.3
1991	36,198	27,400	75.7	3,254	9.0	960	2.6	4,238	11.7
2000	43,686	33,880	77.5	3,797	8.7	1,092	2.5	4,522	10.3

SOURCE: South African Institute of Race Relations, 1992, 2001.
*Includes all ten homelands.

class status. With the victory of the National Party in 1948, Afrikaners for almost four decades dominated all aspects of the political system. The English-speaking whites traditionally maintained the strongest influence in the society's economic institutions—trade, finance, mining, and manufacturing. The English, compared with the Afrikaners, traditionally occupied higher-ranking jobs, were better educated, and were in general wealthier (Adam, 1971b). These discrepancies, however, have diminished in recent years.

The English, whose South African roots do not extend as far back as the Afrikaners', are less cohesive and do not display the same sense of ethnic solidarity. Moreover, though most are of British origin, among the English-speaking whites there are several ethnic communities, including 115,000 Jews and other smaller groups of European origin.[2]

The Afrikaner-British conflicts of the nineteenth century left a residue of intergroup animosity that continued to manifest itself for much of the twentieth century. On the one hand, Afrikaners resented the economic role of the English, and anti-British sentiment was a staple of Afrikaner nationalism for many generations. On the other hand, the English-speaking whites viewed the Afrikaners as parochial and unenlightened. Other than business and politics, social contacts between the two groups were minimal (Brown, 1966; Lemon, 1987; Thompson and Prior, 1982).

Language was a particularly important factor in separating the two white groups. Afrikaans, the language of the Afrikaners, traditionally was established throughout the state school system, and when Afrikaners dominated South Africa politically, it was the language of government and politics. But in business and industry, dominated as they were by the English, the predominant language was English. The language division, however, has transcended politics and commerce. For Afrikaners, it has been a vital component of Afrikaner nationalism (Lemon, 1987).

The class, status, attitudinal, political, and linguistic divisions between the Afrikaners and the English, however, should not be overstated. First, with the ascendancy of the Nationalists in 1948, a significant closing of the gap in wealth and economic influence between the two groups gradually occurred (van zyl Slabbert, 1975). Class differences, though real, were minimized by the primary importance of one's racial classification. Second, both groups benefited from the castelike system and were united in their desire to preserve white rule. This common interest was strong enough to override the political differences between them. Though the English-speaking whites traditionally displayed greater liberalism regarding ethnic stratification, their differences with the Afrikaners

---

[2]On the Jewish community in South Africa, see Della Pergola and Dubb (1988) and Shimoni (1988).

revolved mainly around the most efficient and politically feasible means of en-suring white dominance, not necessarily changing South Africa into a fully dem-ocratic society. Thus, their opposition to apartheid stemmed mainly from the view that it was economically impractical and inefficient, not that it was morally objectionable (Banton, 1967; Lewin, 1963; van den Berghe, 1967). Although the social gap between the English and Afrikaners remains evident, the political dif-ferences between them have diminished considerably (Horowitz, 1991).

With the end of apartheid, in various ways, the political and cultural power of the Afrikaners is in decline. Afrikaners no longer play a major role in govern-ment, and the Afrikaans language is under attack in the schools and media, where English is now preferred (Vestergaard, 2001).[3] The civil service, once a stronghold of Afrikaners, now hires mostly blacks (Daley, 1998a). Obviously, this former dominant group has been transformed into a minority. And, that mi-nority status has fundamentally affected the very self-identity of Afrikaners, many of whom no longer even define themselves in this ethnic fashion. In his fieldwork in South Africa, Mads Vestergaard found that "When asked, individu-als who formerly might have called themselves 'Afrikaner' now increasingly define themselves according to their profession, their geographical location, or simply as 'South African'" (2001:30).

The Coloreds

Out of the sexual and sometimes marital unions of whites and blacks during the early settlement of the Cape was created an intermediate racial group called Coloreds. During the nineteenth century, the Cape continued to evolve along different social and political lines from the frontier settled by the Voortrekkers. As a result, for a long period following the abolition of slavery in 1834, the le-gal system of the Cape was essentially color-blind, allowing for a great deal of racial intermixing (Fredrickson, 1981; Mason, 1970). Today, almost all of South Africa's three and a half million Coloreds live in the Western Cape province, the majority in or around Cape Town.

The Coloreds are perhaps the most peculiar element within the South African ethnic system because in all ways except their physical features, they are thoroughly Europeanized. Most speak Afrikaans, are Christian (mostly Protestant), and are in other cultural ways barely distinguishable from the Afrikaners. The sole, and obviously critical, distinction between the two groups is skin color. Many Coloreds, of course, are even physically indistin-guishable from the Afrikaners.

---

[3] With the replacement of Afrikaners as the dominant political group by blacks in recent years, many believe that Afrikaans will increasingly be displaced by English in all areas of social and political life (McLarin, 1995).

The position of the Coloreds has been marginal to both whites and Africans not only physically but also politically, economically, and psychologically. During the apartheid regime, Coloreds maintained a few political privileges that—limited though they were—placed them apart from the Africans. The economic position of the Coloreds has also been somewhat privileged by comparison with that of Africans, though well below the white standard. Most are part of the unskilled proletariat, both in agriculture and industry, though there is a substantial middle class.

Finally, the Coloreds are psychologically marginal to both the white and African groups. That they are thoroughly Europeanized, and under apartheid held at least a few token privileges, created feelings of superiority to and distance from the Africans. Their attitudes toward apartheid, therefore, were somewhat ambivalent. As Dickie-Clark (1972) has noted, through apartheid they lost their earlier legal equality with whites but gained a somewhat more advantageous position over Africans and Indians. This accounts in some measure for their reluctance to identify or interact more closely with the Africans (Gevisser, 1994).

The Asians

The bulk of South Africa's Asian population is made up of Indians, mostly descendants of indentured workers brought from India in the late nineteenth century to work in the sugar-cane fields of Natal. Following their term of service, many remained and became traders, shopkeepers, and workers at various skill levels. Most still live today in Natal, in and around Durban, the province's largest city. Within the South African system they are, like the Coloreds, a marginal group—part of neither the white nor the African populations.

Ethnic Indians in South Africa are a complex group in both culture and class. They are of several linguistic origins and are further divided along religious lines. Within the Indian population is an entire range of classes, from wealthy merchants to unskilled agricultural and industrial workers. All, however, exist in the context of an institutionally separate community. Indians generally occupy a more privileged economic position than either Coloreds or Africans (in most occupational sectors, they earn more than Coloreds and far more than Africans), but most rank well below whites.

South Africa's Indians have functioned as a classic middleman minority. In Chapter 2, such minorities were described as occupying an intermediate economic status between dominant and subordinate ethnic groups. Anthony Lemon has written that "European attitudes to Indians have been characteristically ambivalent; employers found their labour useful, but traders and businessmen feared competition" (1987:245). For a long time the permanence of Indians in South Africa was questioned by the white power structure. Moreover, prohibitions on their movement in South Africa as well as against further immigration limited their size. At the same time that Indians have been viewed suspiciously and with derision by the whites, they have been seen by Africans as exploiters. One scholar, in fact, asserts that in general "Indians are more resented by most Africans than are Whites" (Moodley, 1980:226).

The Africans

As part of the South African ethnic breakdown, the African population is by far the largest (over 75 percent) and traditionally had been the most subordinate. Although until recently Africans shared a common societal powerlessness, they are not a single culturally unified group. Rather, they derive from several Bantu tribes, the largest among them the Zulu, Xhosa, and Sotho. Even though almost all Africans still speak a native tongue and retain certain other uniquely Bantu traits, the tribal culture for most is residual. Except for those who remain in rural areas, they are assimilated culturally into Western ways. They are at least nominally Christian and in urban areas speak either English or Afrikaans in addition to their Bantu language. Within the urban environment, tribal or subethnic boundaries become indistinct, and the common political and economic oppression they have suffered has impelled the formation of an encompassing African ethnic group.[4]

A crude economic hierarchy is evident among Africans, just as it is among the other three ethnic populations. Occupational differences are particularly acute between those living in urban areas—mostly industrial workers—and those living in rural areas—mainly subsistence farmers or agricultural workers. In addition, a substantial number of migrant laborers from neighboring African nation-states work in the gold and diamond mines.

Most Africans have occupied the lowest-ranking jobs in the labor structure, those requiring little or no skill. With the end of the official policy of racial discrimination, however, this traditional pattern is changing. Job opportunities for blacks have opened in significant numbers, and blacks have been afforded increasing access to higher education. As a result, the black middle class, growing even before the end of apartheid, is expanding rapidly.

These marked improvements in the status of blacks, however, should not disguise the fact that the economic differences between blacks and whites remain vast. As one observer of South Africa has described it, "perhaps no country on earth has cultivated such a wide gulf between the country-club, luxury-car, boutique life style of white privilege and the wretchedness of black denial" (Keller, 1995:A6). The rate at which the socioeconomic differences between blacks and

---

[4]A particularly sharp political division within the African population has emerged in the last several years, however, which will be discussed in a later section of this chapter.

whites can be reduced will determine whether South Africa will move through its social transformation relatively peacefully or with renewed conflict.

PREJUDICE AND DISCRIMINATION

In Chapter 4, inequalitarian pluralistic systems were seen as characterized by several key traits—separation and duplication of institutions among the society's various ethnic groups, extreme polarization between dominant and subordinate groups, high levels of prejudice and discrimination, and a vastly inequitable distribution of wealth. The coordinating mechanism of such systems is the absolute authority and power of the dominant group. Under apartheid, the South African system was characterized by all these traits. A system was created by which white supremacy was assured and separation between the racial categories rigorously maintained.

Starting in 1948 and for the next four decades, political policies of the Afrikaner-dominated National Party were designed to create and sustain inequality among South Africa's four ethnic populations. These policies were grounded in two overriding principles: (1) the subordination of nonwhites in all areas of social, political, and economic life; and (2) the maximum physical and social separation of the four groups, particularly the two major groups, white and African. The latter principle was justified by the premise that peace and order could be ensured "only if people of the various colours mix as little as possible" (Wiechers, 1989:8). So great was the incompatibility of black and white ethnic groups, it was felt, that contact between them could lead only to conflict; hence, minimizing contact would be the logical solution to this perceived problem.

The Dynamics of Apartheid

Apartheid provided specific legislation mandating the social separation of the racial groups, the maintenance of separate social institutions for each, and eventually the formal division of South Africa into separate and independent white and black nations. Indeed, so blatant and uncompromising was the South African system that the very term *apartheid* has, in recent times, come to be applied commonly outside the South African context, used to signify any rigid and durable aspect of ethnic separation and inequality.

In 1994, the Nationalist regime relinquished power in a peaceful transfer to a democratically elected black-dominated government. This officially ended apartheid. Although it is no longer formally in effect, examining this system of ethnic separation today is instructive in that we may see how extreme inequalitarian pluralism operated in a modern society. Based on an ideology of ethnic incompatibility, this was a complete system of ethnic separatism that shaped all aspects of social life. There have been few comparable cases in the contemporary world. Although, as we will see, with the advent of black political dominance apartheid was officially dismantled, its basic principles and underlying framework continue, in large measure, to guide interethnic relations unofficially and informally. Let's look more closely at several of the most important measures that shaped the apartheid system.

Racial Classification Where physical distinctions among people are not clear-cut, ethnic boundaries are difficult to define and enforce. In South Africa the physical dissimilarities between whites and blacks are usually apparent, but in some cases they are not. To establish and enforce a classification system that could be used to distinguish the racially dominant and subordinate groups, the Nationalist government soon after it came to power in 1948 installed several formal measures.

The first cornerstone of apartheid was the Population Registration Act of 1950, providing for the official racial categorization of all persons. Questionable cases were decided by race classification boards, made up of whites, who ruled on the identity of such persons. Although the registration system was technically applied to everyone in South Africa, it was designed primarily to monitor the movement of Africans and was enforced mainly against them. A "passbook" comprising detailed identification papers was required of all Africans, and failure to produce it on demand was an automatic assumption of criminal status. Over the years, eighteen million arrests were made for passbook violations (Lipton, 1987). The passbook system was a particularly odious element of apartheid for Africans and was a constant reminder of their almost total subordination (Klaaste, 1984). The pass laws were abolished in 1986. Most important, the Population Registration Act was repealed in 1991, removing in effect one of the most fundamental legal props of the apartheid system.

Petty Apartheid Physical contact between blacks and whites was minimized through a system of petty apartheid, or what van den Berghe (1967, 1979a) called "micro segregation," consisting of dozens of laws and mandates created to maintain ethnic separation in almost every conceivable area of social relations. Restaurants, hotels, buses, trains, public toilets, waiting rooms, hospitals, schools, parks, beaches, theaters, and the like were for the most part segregated. Moreover, these separate facilities were invariably unequal in quality and quantity. Crowning the rules of petty apartheid was a policy of strictly enforced endogamy, prohibiting interracial marriage and even sexual intercourse between whites and blacks. In the realms of social life where people come into physical contact with each other, then, the South African apartheid system was strikingly similar to the Jim Crow system of segregation enforced in the pre-1960s American South. Unlike the Jim Crow system, however, there were no pretensions of "separate but equal."

Most of the measures of petty apartheid were officially repealed beginning in the 1980s. Neighborhoods, hospitals, public transportation, parks, beaches, and other public facilities were desegregated. But these official changes have not been implemented uniformly, nor have social attitudes, cast by decades of socialization to apartheid, been easily reversed. As a result, petty apartheid continues to guide many areas of social life. "To the casual visitor," noted a study group to South Africa in 1986, "apartheid may appear to be on the way out. In its essential elements, it remains very much intact" (Commonwealth Group, 1986:33). Almost a decade later, the observation still seemed valid. A former Canadian member of parliament visiting South Africa in 1995 related an incident that poignantly illustrates apartheid's legacy. Riding on the airport bus, he struck up a conversation with a black politician whom he recognized. Two other whites on the bus immediately stopped speaking. Although he didn't think much of it at the time, later the black politician chatted with him about the shock that many South African whites still feel about "normal conversations between a black man and a white man" (Langdon, 1995).

Residential Segregation A more fundamental and significant level of apartheid concerned the maintenance of separate living areas for whites and nonwhites. Officially mandated residential segregation, then, became another cornerstone of apartheid. Through the Group Areas Act of 1950, each racial group was assigned specifically demarcated living areas. Any area might be proclaimed "reserved" for a particular group, after which residents not belonging to that group had to move elsewhere. As a result, depending on their skin color, people might be arbitrarily removed and sent to their "appropriate" area. Because this measure was designed to assure racial integrity of white areas and to displace nonwhites from them, over the years its effects were felt primarily by the latter.

The designation of specific areas for each ethnic category was closely related to the South African labor system. With the industrialization of the society, Africans functioned as the unskilled industrial proletariat. South African industry was located in traditionally white areas, and this spurred the migration of African workers into these areas. Because the industrial economy could not function without them, some provision had to be made to accommodate Africans without upsetting the ethnic purity of white neighborhoods. The solution was a migrant laboring system in which Africans were permitted to enter white areas to work but would be housed in separate communities, called townships, adjacent to the white cities. These townships, then, served essentially as dormitories for Africans who labored in white areas. Africans who lived in the townships were officially only "temporary" residents or migratory workers, even if they were born in the urban areas and their ancestors had lived there for generations.

The township of Soweto, adjacent to Johannesburg, is perhaps the epitome of this purposefully designed system of residential segregation. Constructed to house Johannesburg's African migrant laborers, Soweto was actually larger than the white population of Johannesburg itself! Because it was seen by the government as simply a housing area for temporary workers, the physical conditions often were squalid and extremely dense.

Years of restricted movement of Africans created severe social disruption among the urban African populace. In a classic illustration of the self-fulfilling prophecy, social conditions were produced among urban Africans that inevitably led to the decadent behavior commonly attributed to them. Forced to migrate to the cities to find employment, African men were forbidden to bring their families with them. Demoralization and alienation consequently thrived among them, in turn leading to high rates of crime, violence, and alcoholism in the townships.

Restricting the movement of blacks into white areas was referred to by the Nationalist government as "influx control," and the passbook system was a key device in enforcing this policy. The system, however, failed to stem the urban flow of Africans, and the policy was scrapped in 1986. Nonetheless, residential segregation today remains the norm, in large measure because blacks in general lack the means to acquire housing in what have been traditional white areas (Sachs, 1992).

The Homelands As the Nationalist government shaped apartheid, it envisioned the culmination of this racially divided system as the complete separation of the ethnic populations in independent nations. This policy of ultimate exclusion, or what van den Berghe (1967, 1979a) called "macro segregation," entailed the division of the society into geographic areas corresponding to race and ethnicity. As a result, certain areas, variously called "homelands," "Bantustans," or "black states," were set aside for Africans, based on the Nationalist version of historic tribal areas. Each homeland roughly corresponded, in this view, to an area originally inhabited by a particular Bantu tribe. The areas set aside for Africans were less than 14 percent of the land area of South Africa, leaving more than 85 percent for whites. All Africans, regardless of where they resided or where they were born, were declared "citizens" of one of the homelands, based on their ethnic or tribal origins. This included those living in officially designated white areas, the majority of the African populace.

When the homelands policy was formulated, the Nationalists expected that eventually all Africans in South Africa would live in one of these areas, each of which would be granted independence. Once this grand design was fully realized, it would be possible for the South African government to declare that there were no Africans in South Africa, only "guest workers" temporarily living in the remaining (white) areas of the country. In short, until all Africans could be "repatriated to their homelands," those remaining in the white areas were to be considered transient laborers, migrants entitled to few legal rights or privileges.

Sebastian Mallaby refers to the homelands as "the most ambitious injustice that apartheid ever devised" (1992:26).

From the outset, the homelands were fictional creations, constructed essentially to provide South Africa with a continued flow of cheap, exploitable labor and to serve as a convenient dumping ground for unemployed, aged, and infirm Africans (Carter, 1980). None of these areas was economically viable, capable of sustaining itself independently of the South African economy. Moreover, Africans living in South Africa's cities had no cultural or even social ties to their official homeland. The Nationalist government acknowledged these facts by encouraging and even subsidizing industries to locate in adjacent border areas or in the homelands themselves so that jobs could be created for Africans living there (Stultz, 1980). None was ever recognized by any international body or nation outside South Africa. With the advent of a black-dominated government in 1994, the homelands as political entities were officially abolished. Today these areas are extremely poor, even by South African standards (Daley, 1999).

Apartheid and Economic Inequality

One of the cruel ironies of apartheid is that it was financed primarily by its victims. All nonwhites—Africans, Coloreds, and Indians—were discriminated against, yet they had to underwrite the oppressive system by accepting artificially low wages and seriously deprived working and living arrangements. Plainly, apartheid enabled whites to live at a level of comfort they would not otherwise have enjoyed.

The class of poor whites, a substantial part of the Afrikaner population before World War II, was essentially eliminated following the advent of the Nationalist regime. This was accomplished through discriminatory occupational and wage policies that guaranteed to whites the higher-ranking jobs and dictated grossly discrepant wages between blacks and whites.[5] As a means of ensuring the most desirable occupations for whites, a system of job reservation was established whereby no black could advance above a white in the same occupational area. In certain areas of work, blacks were totally excluded. To complete the system of labor exploitation, trade union activity and strikes by African workers were forbidden.

The requirements of South African industry forced the circumvention of these measures, however, and dictated official policy changes. Whites were unable to fill all higher-skilled jobs, and the movement of blacks into these positions therefore was necessitated. In 1969, for example, only 3 percent of professional and managerial positions were occupied by blacks; but by 1989, blacks accounted for 14

---

[5] With the end of apartheid, the protection afforded poor, unskilled whites no longer prevailed. Today many of these people are economically desperate (Arnold, 2000).

percent of them (South African Institute of Race Relations, 1992). With the end of apartheid, many companies have adopted affirmative action policies and are making efforts to employ more blacks in managerial positions.

In the post-apartheid period, the economic gap between whites and blacks has declined, but huge discrepancies remain. These are illustrated in Table 13.2. Almost 70 percent of all Africans live below the poverty line, compared to just 8 percent of whites; the unemployment rate is seven times higher for Africans than for whites; and per capita income of whites is almost seven times what Africans receive. In all occupational areas, the wage gap between whites and blacks remains wide, though it continues to narrow.

The results of the huge discrepancy in resources allocated to blacks and whites during the apartheid era are most apparent in education. Under apartheid, white schools were funded many times higher than black schools. Not surprisingly, the illiteracy rate among blacks is four times higher than among whites (P. Taylor, 1995). And dropout rates for blacks are extraordinarily high (Lemon, 1994). Although in the post-apartheid era, funding no longer favors white schools, black schools continue to experience serious problems including poorly trained, underqualified teachers, lack of resources, and lack of student discipline (Asmal and James, 2001; Nattrass and Seekings, 2001). South Africa no longer compiles education statistics by race, but exam results of recent years seem to indicate that predominantly white schools continue to flourish while black schools languish (Duke, 1998). Although schools are increasingly integrated, many whites have abandoned the public educational system in favor of private schools. In 1991 there were 93 private schools in South Africa; in 1995, 550; and in 1999 about 1,300 (Daley, 1998b; Swarns, 2000).

At the university level, institutions that previously had barred blacks now are increasingly racially integrated. A majority of the 1995 freshman class at the

Table 13.2 South African Socioeconomic Differences, by Race

SOCIOECONOMIC INDICATOR	AFRICAN	COLORED	ASIAN	WHITE
Percentage of people in poverty (1996)	67.8%	35.3%	14.3%	8.4%
Rate of unemployment (1999)	44.0%	23.6%	20.2%	6.8%
Disposable per capita income (2000)*	7,567	12,960	25,541	50,804

SOURCE: South African Institute of Race Relations, 2001.
*Figures are in South African Rands. In 2000, one Rand was the equivalent of approximately 15 cents U.S.

University of Witwatersrand in Johannesburg, one of South Africa's most prestigious universities, was black (Keller, 1995). Below the university level, however, schools remain de facto segregated. Nonwhites make up less than 1 percent of students in traditionally white schools, and these are mainly Colored children in the more liberal Cape Town area (Berger, 1992; Keller, 1992). Schools in the black townships have also been disrupted in recent years by the sporadic violence that has gripped these communities. Racial parity in education, then, remains a distant goal.

Enforcement of Apartheid: Force

As noted in Chapter 4, all societies are characterized by elements of order and conflict. Social control is therefore always a combination of consent and coercion. In a society in which the majority of the population does not accept the legitimacy of the prevailing system, however, coercion becomes the more common of the two. In South Africa, where four-fifths of the population rejected the system of white domination, "the element of consent," wrote South African legal expert Albie Sachs, "must be extremely low and the element of coercion extremely high" (1975:224).

The South African criminal justice and legal institutions traditionally provided for suppression of any challenges to white rule. Indeed, the use of terror against the African population was routine. Writer Mark Mathabane recalls his own experience: "I remembered the brutal midnight police raids launched into the ghetto to enforce apartheid; the searing images of my father's emasculation as he was repeatedly arrested for the crime of being unemployed; my parents constantly fleeing their own home in the dead of night to escape arrest for living together as husband and wife under the same roof" (1994:38). Police violence against Africans was so pervasive and routine, explains political scientist Gwendolyn Carter, as to be almost unnoticeable except when a well-known person was affected or the circumstances "so unusual as to attract attention" (1980:12).

The preparedness of the Nationalist government to use force in averting change in the distribution of power was demonstrated repeatedly. At Sharpeville in 1960, the police fired on a crowd of Africans peacefully protesting the pass laws, killing sixty-nine. Sachs has pointed out that the Sharpeville massacre, though the best-reported incident of its kind, was "merely one in a series of characteristically South African episodes in which police fired on crowds of blacks that were either totally unarmed or armed only with sticks and stones" (1975:227). A second well-reported incident occurred in Soweto in 1976. What began as a student protest against inequalities in the educational system culminated in the slaughter of over 600 Africans. Thousands were injured in the disturbances that followed, and schools and universities were closed and industries shut down by strikes. This event represented a spontaneous protest against the entire apartheid system and afterward took on great symbolic importance among the African populace.

A Truth and Reconciliation Commission, created in 1996, revealed the systematic nature of human rights violations that were committed by the Nationalist government in enforcing apartheid. Testimony given by perpetrators as well as victims demonstrated that during the apartheid regime the police considered brutality, torture, and even murder as legitimate tools in stifling political opponents (Daley, 1997; Handley and Herbst, 1997:223).

Socialization to Apartheid: Ideology

South Africa under apartheid is a marvelous illustration of a dominant group's efforts to uphold a system of ethnic supremacy using not only the power of raw force and political manipulation but also material, psychological, and ideological commitment. As in any multiethnic society, most members of the dominant group in South Africa were effectively socialized to accept and justify the prevailing system of ethnic inequality. Socialization to apartheid was especially pronounced among Afrikaners. Traditionally, in every institution—family, school, church, government—the values of Afrikaner nationalism and white dominance were reinforced.

South African whites, particularly Afrikaners, believed themselves to be natural overseers of primitive Africans, whose shortcomings were not likely to change significantly. White rule, therefore, was seen as a kind of paternalistic relationship in which more intelligent and civilized peoples asserted their domination over inferior peoples in the interests of both (Mulder, 1972). In this view, notes Leonard Thompson, "It was the duty of the Whites, constituting the civilized, Christian race, to use their control of the state to prevent racial friction and racial bastardization by ensuring that the races would be separated from one another" (1985:190).

This essentially racist ideology underwent alteration during the apartheid regime. Although the crude belief in biological superiority had been by no means unvoiced, the white resistance to changes in the system of racial inequality was based mainly on the notion of cultural inferiority and ethnic incompatibility (Adam, 1971a, 1971b; Thompson, 1985; Welsh, 1975). The cultural gap between the races was so vast, it was claimed, that blacks were simply incapable of operating the institutions of a modern industrial society. As Lawrence Schlemmer explained,

> The perception, and to a very great extent the reality, of a conflict of interests is not simply between classes and ethnic or racial interests, but between a first world and a third world social order. In the collective fears

of whites, the third world within the country is associated with a lack of sophistication, disorder and lower standards. The perverted vision of a future under majority rule most frequently conjured up among rank-and-file whites is one of irregularities of administration, loiterers in parks, dirt in the street, a lack of public discipline and power-hungry politicians with a fateful sway over slavishly adoring masses. The fact that none of these images is confined to the third world, and that all are gross over-simplifications, does not alter the power of the perception (1988:42).

Hence, the continued domination of the society by the white numerical minority was, in this view, essential.

The view that it was necessary to maintain white rule, based on the understood cultural superiority of whites, was thoroughly incorporated into all institutions, especially the school and the mass media. Through the Bantu Education Act of 1953, complete control of black education was given to the central government, thereby assuring that the history and social science taught would reflect the Nationalist ideology. Blacks, it was thought, would be trained to accept their subordinate position in the society. Prime Minister Hendrik Verwoerd articulated this view:

> The school must equip the Bantu to meet the demands which the economic life of South Africa will impose on him. . . . There is no place for him in the European community above the level of certain forms of labour. Within his own community, however, all doors are open. . . . Until now he has been subject to a school system which drew him away from his own community and misled him by showing him the green pastures of European society in which he is not allowed to graze. . . . What is the use of teaching a Bantu child mathematics when it cannot use it in practice? (quoted in Harrison, 1981:191)

The separation of white and black was deemed necessary to protect whites from cultural pollution by an obviously inferior people. A schoolbook, for example, described the differences between white and black: "The White stands on a much higher plane of civilisation and is more developed. Whites must so live, learn and work that we shall not sink to the cultural level of the non-White" (quoted in Harrison, 1981:204).

In addition to the rationale based on economic and political efficiency, an element of religious mission infused the determination of the Afrikaners, particularly the zealously nationalistic element among them, to maintain white supremacy. In this, the Afrikaners diverged from the English-speaking whites, whose motivations to support the system of ethnic inequality rested mostly on the desire to protect their economic interests. As already noted, the Afrikaners have historically invoked a biblical analogue to portray their place in South Africa, and their self-image of a chosen, superior people is strongly expressed in various aspects of Afrikaner culture and social life. The Dutch Reformed Church has served as an integral part of Afrikaner nationalism. For many, the church is the focal point of Afrikaner culture and the religious expression of its nationalism. Unlike most English-language churches, it maintained a close link with the Nationalist state and abided apartheid.

To better comprehend the determination of Afrikaners to preserve their power and, in so doing, their ethnic culture, one needs to keep in mind the unique status of the Afrikaners on the African continent. Unlike traditional European colonial groups in Africa, Afrikaners have no cultural or political ties to a European homeland. Generations of Afrikaners have known nothing but South Africa, and there is no thought of returning to a society of origin, for there no longer is one. The Afrikaners are, in all ways, a group sui generis. The prevailing belief has been that the Afrikaner way of life would be submerged if the numerically dominant nonwhite populace were afforded equal access to social and political institutions. Black political and cultural dominance, therefore, would represent to them the dissolution of their ethnic group.

Although these beliefs were the ideological bases of the apartheid system for a large element of the Afrikaner populace, the practical foundations of the system lay mostly in the resolve of the dominant whites, most Afrikaners and English alike, to sustain their supremacy. Whites well understood that a truly democratic political system would inevitably produce an African-dominated government, given the numerical imbalance between the two groups. This, in turn, would spell the end of white economic and social privilege.

Whatever their psychological and ideological commitment to white supremacy may have been, for most whites, social pressures ordinarily compelled conformity. As noted in Chapter 3, prejudice and discrimination are most effective and self-sustaining where such thought and behavior are socially expected and rewarded. The economic benefits that flowed to whites from black exploitation were, as we have seen, enormous. Even those who may have been bothered by feelings of moral contradiction would therefore have been reluctant to call for the total dismantling of a system from which they greatly profited. Within the South African context, prejudicial attitudes were therefore easily converted into degrading and humiliating forms of behavior toward nonwhites, enacted through a whole etiquette of intergroup relations, similar to that of the pre-1960s American South. Thus, even after the abolition of the formal apartheid system, the informal system of white supremacy remains evident.

Even nonresidents of South Africa, used to more fluid social relations, might easily be caught up in the everyday workings of a system of extreme social inequality and separation that may at first have been repugnant. One American sociologist studying in South Africa in the 1970s reflected with astonishment on his own behavior: "One falls in with the culture and those with whom one must cooperate in assumptions and behavior. Even in my short stay I could observe the process occurring in myself: by the time I left I was beginning to unconsciously accept conditions that I found shocking when I first confronted them" (Mechanic, 1978:137).

Not only is it difficult to resist patterns of thought and behavior that reward conformity so generously, but it is especially hard to question a system of exploitation that, so far as it is recognized at all, is perceived from a distance. Because for whites the standard of living has been so comparatively high and the political order essentially democratic, there can arise among them only a limited understanding of black economic and, until recently, political deprivation. Adding to the incomprehension has been the essentially separate societies in which whites and blacks have lived. For the most part, whites have seen blacks only as workers and servants, not as neighbors and schoolmates. Few whites speak any of the various African languages, and few ever venture into black townships. Put simply, blacks and whites in South Africa have lived in different worlds.

FORCES OF CHANGE: BUILDING A NEW SOCIETY

More than three decades ago, one observer of South Africa declared that "the best way of describing South African society is to say that it consists of groups of peoples thrown together by history, all hating each other, but not enough to want to end their relationship. South Africa is, in other words, the same as everywhere else—only more so" (Feit, 1967:403). In the last decade, the ethnic makeup of South Africa has not basically changed, but the relationship among its groups has been altered in a fundamental way. Indeed, if during the apartheid era one were to have suggested the scope and depth of change in the ethnic system that has taken place in South Africa since 1989, that person might have been seen as hopelessly naive. The unimaginable, however, has occurred. Today the government is led by blacks, and the most egregious policies of racial separation have been repealed. Official apartheid has ended and the society is moving, however tortuously, toward ethnic democracy. What propelled this amazing change in a system so seemingly uncompromising in its commitment to ethnic inequality?

Changes in the Political System

The radical transformation of South Africa that began in the late 1980s did not emerge suddenly. Its roots reached back at least a decade when it became evident to the white power structure that changes in the apartheid system would have to be made if the society were to avoid a bloody confrontation. Former Prime Minister P. W. Botha publicly declared in 1978 that South Africa must "adapt or die," and reforms in petty apartheid and in the labor system instituted in the 1980s were manifestations of that view. As a desperate effort to preserve white rule, a new constitution was adopted in 1983, creating a tripartite parliamentary body, representing whites, Coloreds, and Asians. This new governing system, however, assured the continued dominance of whites and, most important, made no provision at all for the representation of Africans. The reformulated constitution was interpreted as essentially an attempt to co-opt the Colored and Indian groups, and it was generally recognized as a sham.

The exclusion of Africans from the constitution was one of the precipitating factors in the civil unrest that occurred in the 1980s (Adam and Moodley, 1986; Dugard, 1992; Mallaby, 1992; van zyl Slabbert, 1987). Widespread violent protests took place in black townships, the target of which was the system of black local authority. Black officials in the townships were seen as collaborators in the oppressive system and were driven from their communities, forced to resign, or in some cases even murdered. In addition to the violence in the townships, nationwide marches, rent strikes, consumer boycotts, student agitation, and labor protests became commonplace (Meredith, 1987; Murray, 1987). In attempting to control the protests, the government responded with force, using the South African Army as well as the police, who were given broad discretionary powers. Hundreds of blacks were killed in government actions to reassert control in the townships, and in 1985 a state of emergency was declared, providing for arrest and detention without trial and searches without warrant.

South Africa's panoply of internal contradictions brought the society to a watershed beginning in the late 1980s. A sociopolitical system enforced by coercion no longer seemed viable, and major reforms were now proposed. Much of the impetus behind the reforms was provided by the newly elected president, F. W. de Klerk, who acknowledged the failure of apartheid. As Mallaby describes him, "de Klerk had finally abandoned whites' hopes of extending their privileges beyond the colonial age" (1992:78). In October 1992, de Klerk basically declared an official end to the apartheid system. Referring to his National Party, he stated that "For too long we clung to a dream of separate nation states when it was already clear that it could not succeed. For that we are sorry. That is why we are working for a new dispensation" (South African Institute of Race Relations, 1993:25).

Pressures for change arose primarily from international opprobrium of the apartheid regime, the requirements of economic interdependence, and political movements within the black population.

International Pressures Throughout the Nationalist ascendancy during which apartheid was at its peak, most nations viewed South Africa as a kind of pariah state. Nonetheless, South Africa remained strategically important as a source of vital minerals as well as a major link in the worldwide production and distribution networks of multinational corporations, most of them based in Britain or the

United States. Thus, although the apartheid system was condemned, economic relations between South Africa and the Western industrial nations generally were unaffected.

The often brutal suppression of the black protests starting in 1984 led to a stepped-up international campaign that included calls on multinational corporations to divest their interests in South Africa and to force Western nations to impose trade sanctions. Although these measures seemed to only minimally affect the ability of the South African economy to function, they contributed to economic uncertainty and, combined with the civil unrest of the late 1980s, impelled the South African business community to call for greater and more rapid social reforms (Blumenfeld, 1987; Harari and Beaty, 1989).

Economic Forces More significant changes in the system of racial inequality were impelled by the demands of economic interdependence. The South African economy required an increasingly skilled workforce that could not be supplied by the white population alone. As noted earlier, this resulted in modifications in the job reservation system, permitting more blacks to fill higher positions. In effect the system of job reservation, for decades restricting black labor mobility, collapsed (Adam and Moodley, 1986).

Labor needs also compelled relaxation of prohibitions against black trade unions and the right to strike, creating a growing and increasingly significant black labor movement (Thompson and Prior, 1982). Whereas 292,000 Africans were members of registered unions in 1980, by 1991, African union membership was over 1 million (South African Institute of Race Relations, 1992). Strikes by black workers arose frequently starting in the mid-1980s in certain industries, most notably mining, a particularly critical sector of the South African economy.

Internal Political Forces After the Nationalist regime came to power in 1948, opposition black political organizations and movements were declared illegal, forcing them to operate underground or in neighboring African countries. The most important of these were the African National Congress (ANC), the Pan-Africanist Congress (PAC), and the Black Consciousness Movement. In 1990 the de Klerk government restored the legality of these political organizations, providing a new element in the chemistry of change in South Africa.

During apartheid the ANC, the most significant of the black political organizations, had assumed the status of "government in exile" for the African population. Its official leader, Nelson Mandela, was released from prison in 1990 after serving a sentence of twenty-eight years for antiregime political activity. Mandela's release was seen as a threshold. From that point forward, the end of white South African rule and the emergence of a black majority government were only a matter of time. Shortly thereafter, the South African regime entered into talks with the ANC and other black groups to negotiate the entrance of blacks into the political system and to dismantle the apartheid system.[6] The historic elections held in 1994, in which all South Africans—black and white—voted for the first time, were won by the ANC, bringing Mandela to power as the country's president. South Africa was now poised to create a nonracial state.

Future Prospects

The political upheaval in which South Africa has been engulfed during the past decade has produced changes of a truly revolutionary nature. As changes continue to unfold, the shape of the society that ultimately will be produced remains unsettled. All of the key issues facing South Africa today, however, in one way or another concern its movement toward greater ethnic equalitarianism.

Political Issues As we have now seen, the legal and structural roots of the apartheid system have been removed, and significant policy changes have muted the most blatant forms of racial discrimination. Most importantly, blacks have been brought into the political process. Awarding full and equitable voting rights to the black population was the major prerequisite to any peaceful solution of South Africa's ethnic conflict. The great fear of whites was precisely what was enacted: a one-man, one-vote, majority-rule system that, as they realized, spelled the end of white dominance. Astonishingly, this change was brought about peacefully. In light of the generations of racial hostility and the determination of the white regime to retain power, few could have predicted that the resistance to a black-majority political system would evaporate so relatively swiftly and nonviolently.

In 1999, the ANC was again victorious in South Africa's second national election. With an overwhelming majority, the ANC's political dominance is almost unchallenged, but a democratic national government now seems solidly embedded in South Africa's constitutional system (Mattes, 2002). Mandela himself, however, retired in 1999; and as the most widely supported and admired figure in South Africa—among whites as well as blacks—he will not be easily replaced as a unifying symbol. As one journalist described him, he served as the nation's "moral guide" and, given his history of personal sacrifice, was ideally suited to convince the country of the need for compromise, reconciliation, and change (Duke, 1999b).

The major political division among Africans lies between the ANC and Chief Mangosuthu G. Buthelezi's Inkatha Freedom Party, a political movement of mainly Zulus. Buthelezi has challenged the tactics and objectives of the ANC for years,

---

[6] An important stimulus to change was a 1992 referendum, in which white voters presented de Klerk with a mandate to forge ahead with negotiations with black leaders to ultimately construct a new multiracial constitution.

rallying his followers in opposition. Before the election of 1994, Inkatha and the ANC had engaged in a bitter and bloody conflict in many black townships, resulting in the deaths of 6,000 in only four years. The hostilities between the ANC and Inkatha have since declined; but this division, as well as other less profound ones, demonstrates the serious differences within the black population regarding future scenarios for South Africa and the strategies that are believed to be most effective in bringing about a transformation of the society.

Whites are divided between moderates who pragmatically recognize the inevitability of fundamental change in all dimensions of society, and conservative hard-liners who stubbornly refuse to accept the transfer of political power from whites to blacks (Schiff, 1996). Some among the latter envision the eventual formation of an autonomous, white territory of their own. These extreme right-wing elements, however, no longer seem capable of presenting serious disruptions to the current South African government (Bratton, 1998; Vestergaard, 2001).

Social and Economic Issues Securing political equality is only a first step toward the elimination of ethnically based social inequalities in jobs, education, health, and other life chances. As noted earlier, dramatically wide discrepancies remain in the socioeconomic status of whites and blacks. Perhaps most critical is South Africa's high unemployment—many times higher for Africans than for whites (South African Institute of Race Relations, 2001; Duke, 1999a).

Paradoxically, changes effected in the political status of the black population may inevitably provoke more, not less, social unrest. Now that blacks control the political system, expectations are very high that change in their economic status will follow rapidly. This places great pressure on the government to provide jobs, housing, and increased public services, especially education and health care, to an enormous part of the black population that remains impoverished. If expectations exceed the government's ability to meet them, however, blacks may become frustrated and place even greater pressure on leaders to produce economic betterment. As many have observed, it is, ironically, only after conditions have begun to improve that people may be drawn into political action (Hoffer, 1951; Runciman, 1966; Turner and Killian, 1972). Thus, failure to rapidly alleviate black-white economic disparities may foment social and political unrest and place increased stresses and strains on the government. In fact, dissatisfaction with economic policies of the ruling ANC has become increasingly evident within the party ranks, particularly among labor unions, traditionally the party's strongest allies (Swarns, 2001a).

To view the relationship between race and class in South Africa today as a simple dichotomy—black/poor; white/rich—is, however, a serious distortion. More accurately, the society is one in which there is declining *inter*racial inequality but rising *intra*racial inequality. Specifically, South Africa is divided into three broad classes: an increasingly multiracial class of well-off families, including corporate executives, professionals, and managers; a middle class of mostly urban, industrial, or white-collar workers; and a marginalized class of black unemployed (Nattrass and Seekings, 2001). Even before the end of apartheid, blacks had been experiencing upward occupational mobility. In the post-apartheid period, despite the still very sharp disparity, there has been a marked narrowing of the economic gap between white and black populations. This can be seen in Table 13.3. From 1970 to 1996, the share of total income received by Africans rose from 20 percent to almost 36 percent, an increase of 80 percent. At the same time, however, inequality *within* the black population has increased. The black middle class, able to take advantage of new opportunities arising with the end of racial discrimination, is rapidly growing, but in the process is leaving behind a depressed black underclass. The gap between well-off and middle-class blacks on the one hand and poor blacks on the other thus continues to widen. This is not unlike the internal division within the African-American population, discussed in Chapter 9, in which a growing middle class stands in sharp contrast to a marginalized underclass.

Although whites have been displaced in the political realm, they continue to control key elements of the economy. The ruling black regime desperately needs the capital and managerial and technical skills that are possessed primarily by the white community. Thus it cannot afford to alienate whites, who might threaten to leave and take with them their businesses and economic resources. The ANC has put forth a reconstruction and development plan that outlines its goals in transforming South Africa and bringing about a more equitable distribution of jobs, health care, housing, and land. Here the government faces the delicate task of raising public revenues—which must come primarily from white-controlled businesses and the wealthier white community—to finance ambitious public projects designed to lift the economic and social conditions of blacks.

As long as white South Africans retain control of the bulk of the society's economic resources, the wide differentials between whites and blacks, even after fundamental political changes have been effected, will not decline. The South African regime, then, must walk a fine line in trying on the one hand to allay

Table 13.3 Income Shares by Race, 1970–1996				
RACE	1970	1980	1991	1996
African	19.8%	24.9%	29.9%	35.7%
Colored	6.7	7.2	6.8	7.9
Indian	2.4	3.0	3.8	4.5
White	71.2	65.0	59.5	51.9

SOURCE: South African Institute of Race Relations, 2001.

white fears of loss of economic power and status, and on the other to meet the rising economic and social expectations of blacks now that political democracy is firmly in place.

Failure to seriously reduce socioeconomic inequality may contribute to another serious problem—a dramatically high incidence of crime. The murder rate in South Africa is today the highest in the world—ten times the U.S. rate—and violent crime in general has become rampant (Marais, 1999; South African Institute of Race Relations, 2001). Failure to arrest this soaring crime rate may have serious economic ramifications. Those with capital may be less inclined to invest in South Africa and those with highly needed occupational skills may be induced to emigrate (Daley, 1995; Shaw and Gastrow, 2001).[7]

Another issue that threatens serious social disruption in the new South Africa is the extraordinarily high rate of AIDS. One in nine South Africans and one in four adults are infected with the virus, more than in any other country (Swarns, 2001b). South Africa recorded a quarter of a million deaths from AIDS in 2000, and it is estimated that over 10.5 million may die from AIDS-related illnesses by 2015 (South African Institute of Race Relations, 2001).

The Continued Significance of Racial Thinking The different, racially defined, social worlds to which whites and blacks in South Africa were traditionally socialized continue, for the most part, to exist in post-apartheid South Africa. "Race," notes Vestergaard, "remains one of the main organizing principles in social life" (2001:30). Racial stereotypes, which formed such a basic part of the dominant ideology—that blacks simply were unable to maintain the institutions of a modern society—have not disappeared. As Vestergaard explains, more than once in his interviews of Afrikaners—after apartheid had been abolished and blacks were in control of the government—did he hear the complaint that the country's ongoing economic problems were due to the fact that blacks "simply don't have what it takes" to run a modern economy (2001). Although they are far more subtle than in the past, expressions of racial thinking continue to color whites' views of social problems like education and crime. Vestergaard cites the view of a fifty-seven-year-old Afrikaner who worked for the Department of Education on problems in black schools:

"The teachers cramp three classes together in one room, about eighty to ninety kids, in order for the two others to sit outside—they are like lizards, they like to sit in the sun. That is why there always are chairs outside. Africans just have a much more lazy way of doing things and they're

[7]Although popular perception would have it that violent crime in South Africa is mostly committed by Africans against whites, in fact most victims are African and poor (Shaw and Gastrow, 2001).

never in a hurry." When he described the white schools, his mood brightened: "There the kids WANT to be better and the teachers WANT to take it to a higher level! The black kids have no ambitions" (2001:30–31).

It would seem naïve to assume that attitudes and ideas so prevalent among whites during apartheid would be subject to fundamental change in just a few years. As one observer of South Africa today has put it, "generations of whites were brought up to be racists and they are unlikely to be able, even if theoretically willing, to change" (Arnold, 2000:18). This applies to blacks' views of whites as well, of course. The distrust nurtured during generations of apartheid is not likely to be overcome for many years. "Thus," observes Vestergaard, "when a white person makes a statement, it is perceived as a *white* person speaking—making a *white* statement" (2001:30).

Ethnic Diversity In addition to the overwhelming problems of trying to create a system that will provide for a stable government and thriving economy, South Africa is faced with ethnic diversity that was camouflaged for decades under the oppression of apartheid. Besides the enduring black-white schism, there remain other ethnic divisions in South Africa. The split of the African population along ethnic lines is most apparent in the conflict that has been ongoing between the ANC and the primarily Zulu-based Inkatha Freedom Party. Also, many Coloreds and Indians feel marginalized by a government that is now led primarily by Africans (Duke, 1999a).

To think of South Africa as a society neatly divided into black and white, then, is a glaring oversimplification. Nothing could better illustrate this than the fact that today South Africa recognizes eleven official languages.

Reconciliation De facto segregation continues to characterize most institutions of South Africa, sustaining the enormous discrepancies between whites and blacks. Removing the legal and political foundations of apartheid, therefore, cannot guarantee rapid changes in other institutions—particularly the economy—that have real bearing on the distribution of power and wealth. As one observer has put it, "South Africa today is a place where the white minority of 5 million owns the economy, the 32 million-strong black majority runs the politics, and both races have a living memory of three centuries of oppression" (P. Taylor, 1995:16). It is the latter issue that has created another vexing problem for the new South Africa: how to bring about a reconciliation of those who were the victims of apartheid with those who created and perpetrated that system.

In 1995 the Truth and Reconciliation Commission, under the supervision of Bishop Desmond Tutu, was established in order to bring to light human rights violations committed by the apartheid regime, to offer amnesty to those who admitted their complicity in those abuses and whose motives had been political, and to make reparations to the victims of apartheid's most heinous crimes. A less concrete goal was to change the thinking of white South Africans about the apartheid regime and to bring them to a greater awareness of past injustices. Although the latter goal apparently was not accomplished and its deliberations produced much bitterness and pain, the Truth Commission was a vital part of the policy of reconciliation and of the broader effort of the new regime to thrust South Africa beyond the apartheid era (Bratton, 1998; Duke, 1999b).

Clearly, the post-apartheid healing process will remain a long-term affair and will be much dependent on the success of efforts to bridge the economic gap between whites and blacks. A 1999 public opinion poll seemed to offer both positive and negative views on the future of South African ethnic relations: 63 percent of blacks and whites agreed with the statement, "It will take a long time, but we will eventually become a united nation"; but 44 percent also agreed that "blacks and whites will never trust one another" (*Washington Post*, 1999).

The kind of multiethnic society into which South Africa will evolve can only be speculated upon. Few observers, even the most informed, were able to predict the scope and speed of the changes that occurred in less than a decade. To forecast the specifics of South Africa's future, therefore, would surely be presumptuous. What is clear, however, is that South Africa is in the midst of basic change in its ethnic structures and relations. As that change unfolds, much of the world's attention will be focused on this society at the outset of the twenty-first century.

Despite its lengthy history of extreme ethnic division and conflict, South Africa presents itself today as, ironically, a potential model for other multiethnic societies. If the most oppressive system of ethnic relations in the contemporary world can be fundamentally altered in a mostly peaceful fashion, and if those changes can be sustained, there is strong reason to expect that comparable change can be achieved in other societies with severe ethnic divisions.

SUMMARY

Among contemporary multiethnic societies, South Africa has been the most extreme case of inequalitarian pluralism. The society is a racial dichotomy, black and white, but there are more specific group subdivisions. The whites, roughly 10 percent of the total population, comprise two major subgroups, Afrikaners and English. Blacks comprise three subgroups: Coloreds, a racially mixed group; Asians, mostly Indians; and Africans of several different tribal origins. Africans are by far the largest racial-ethnic category, making up over 75 percent of the population.

Ethnic stratification in South Africa has traditionally been a castelike system in which each ethnic group is institutionally separated by a customary and, until recently, legal code. The system of ethnic separation and inequality, called

Comparing South Africa and the United States

Points of Similarity

- In both South Africa and the United States, the dominant group emerged from a process of indigenous subordination, in which a settler group subdued the native population.
- Both societies supported the institution of slavery, though far more extensively and for a longer period in the United States.
- A castelike system of separation was constructed in both societies, with a well-understood racial etiquette and an official racial classification scheme.
- Both societies developed institutionalized systems of racial separation, apartheid in South Africa and Jim Crow in the southern United States.
- Both societies abolished their castelike systems but continue to struggle with efforts to fully integrate socially and economically the formerly subordinate black population.

Points of Difference

- For most of its history, South Africa's dominant ethnic group (whites) was a numerical minority vis-à-vis subordinate Africans, along with Coloreds and Asians. In the United States, by contrast, whites were always a numerical, as well as sociological, majority.
- The economic and social gap between whites and blacks is today much wider in South Africa than in the United States.
- In multiethnic South Africa, blacks today are the dominant group, at least politically. This is obviously not the case in the United States.
- The United States is an ethnically more varied society than South Africa.

---

apartheid, was designed to maintain white supremacy in all areas of social, economic, and political life. Africans were excluded from the political system and were protected by few civil rights. Coloreds and Asians were only slightly better off. On economic measures, the inequality between whites and blacks remains extreme, though the gap is beginning to narrow.

Starting in the late 1980s, the apartheid system began to be dismantled as black protest and economic uncertainty created strong pressures for reform on the South African regime. The national election of 1994 was the first in which blacks participated as equals, and a black-dominated government, led by Nelson Mandela and his ANC, came to power for the first time. This spelled the end of the apartheid system. Internal divisions within the black majority, however, as well as challenges from dissident whites, continue to create political unease.

Though the official system of racial inequality has been abolished and blacks now dominate the political system, the major components of white economic domination remain in place. The key issues of South Africa's immediate future, therefore, concern the manner in which the tremendous economic gap separating whites and blacks will be reduced. The end product of the radical changes that have occurred during the past decade, therefore, is still uncertain.

Suggested Readings

Beinart, William. 1994. *Twentieth-Century South Africa*. New York: Oxford University Press. A succinct history of modern South Africa through the end of apartheid and an analysis of its attempts to create a new national identity.

Daedalus. 2001. "Why South Africa Matters." Vol. 130. Cambridge, Mass.: The American Academy of Arts and Sciences. A special issue of this quarterly journal deals exclusively with South Africa, focusing on current social, political, and economic issues.

de Villiers, Marq. 1987. *White Tribe Dreaming: Apartheid's Bitter Roots as Witnessed by Eight Generations of an Afrikaner Family*. New York: Penguin. The author describes the development of ethnic identity among Afrikaners and their perspective on South Africa by tracing his own family's history.

Fredrickson, George M. 1981. *White Supremacy: A Comparative Study in American and South African History*. New York: Oxford University Press. Examines the historical development of South African race relations and American race relations, comparing their ideologies, policies, and actions.

Holland, Heidi. 1994. *Born in Soweto: Inside the Heart of South Africa*. London: Penguin. A poignant description of life in South Africa's largest and most infamous black city, created by the apartheid system.

Mallaby, Sebastian. 1992. *After Apartheid: The Future of South Africa*. New York: Times Books. A journalist explores the social and political forces that brought an end to apartheid. In predicting the ultimate outcome of South Africa's transition, the author draws on the experience of other African nations following the end of white rule.

Sampson, Anthony. 1999. *Mandela: The Authorized Biography*. New York: Alfred A. Knopf. Traces the life and times of modern South Africa's most important political figure, emphasizing the years of political struggle to defeat apartheid.

South African Institute of Race Relations. (published annually). *Race Relations Survey*. Johannesburg: South African Institute of Race Relations. An invaluable source of social and economic data pertinent to all aspects of South African ethnic relations.

Thompson, Leonard. 2001. *A History of South Africa*, 3d ed. New Haven: Yale University Press. A comprehensive history of this society, tracing the development of race relations through various periods, written by one of the outstanding historians of South Africa.

van den Berghe, Pierre L. 1967. *South Africa: A Study in Conflict*. Berkeley: University of California Press. Although dated, this work remains one of the few thorough sociological analyses of South Africa under apartheid.

14

BRAZIL: ETHNIC PARADISE?

MOVING FROM SOUTH AFRICA to Brazil is analogous to traveling from one planet to another. A more striking contrast in ethnic relations could not be found in the modern world. Instead of an ethnic purgatory, Brazil has often been praised, by Brazilians and outside observers alike, as a racial and ethnic paradise where people of varied physical features and with diverse cultural heritages live amicably together. Whereas the dominant ideology of South Africa, until recently, was ethnic pluralism in a most extreme form, in Brazil the predominant philosophy has been ethnic assimilation. Here the overriding belief is that the more quickly the society's various ethnic and racially defined groups can be merged into one, the better the society will be.

Whether Brazil is in fact an ethnic paradise is, as we will see, by no means universally agreed on either by scholars or by Brazilians themselves. Another, less laudatory, view of Brazil places it well within the bounds of other multiethnic societies characterized by intergroup conflict and division. In any case, it is clear that the direction of ethnic relations in Brazil is toward assimilation, in which ethnic convergence and the breakdown of distinct physical and cultural groups are the society's long-range goals. More significant, Brazil has advanced further in the realization of these objectives than almost any other major multiethnic society. This fact alone distinguishes Brazil not only from societies with inequalitarian pluralistic systems but also from the United States and other societies like it that have traditionally proclaimed an assimilationist ideology but have often exhibited contradictory patterns of action and policy. If ethnic relations are not so ideally harmonious as Brazilian ideologists would have the world believe, they are nonetheless far less combative than in most other multiethnic societies. In what ways and why Brazil's ethnic relations are more amicable are issues that deserve our careful attention.

MAJOR FEATURES OF BRAZILIAN SOCIETY

Brazil seems especially appropriate to compare with the United States because the two societies exhibit strongly parallel ethnic developmental patterns. Both were originally colonized by Europeans who overwhelmed the indigenous population; both imported vast numbers of Africans who were the mainstays of an institutionalized slave system that lasted until the late nineteenth century; and both were peopled by immigrants from a variety of European societies. There are, of course, important ethnic differences between the two societies as well. Africans taken to Brazil as slaves far exceeded in number those taken to the United States, and as a result, blacks composed the largest element of the Brazilian population until the nineteenth century. Although whites predominate today in both societies, in Brazil the African cultural influence has been more widespread and significant. More important, Brazil's ethnic and racial mix is far more amalgamated than that of the United States.

Although Brazil is as ethnically varied as any society of the modern world, its dynamics of race and ethnic relations revolve primarily around its major racially defined groups. Our discussion, therefore, will concentrate mainly on patterns of black-white relations and only secondarily on interrelations among European ethnic groups and indigenous Indians.

Brazil is the largest nation of Latin America, both in area and in population.[1] Yet it is clearly divergent from other societies of the region in at least two important respects. For one, its colonial heritage derives from Portugal rather than Spain. Its language and other major cultural elements are therefore Portuguese. As we will see, some hold that the Portuguese cultural legacy largely accounts for Brazil's seemingly more pacific patterns of ethnic relations.

Second, the ethnic composition of Brazil is different from that of almost all other Latin American societies. Except for a few Caribbean nations (most notably Cuba), Brazil is unique in that it comprises a combination of European, Indian, and African peoples. Specifically, it is the European-African ethnic synthesis that sets off Brazil so sharply from other Latin American societies. It is important to consider, however, that although European, African, and Indian racial strains are evi-

---

[1] Geographically, Brazil is about the size of the United States exclusive of Alaska. Its population is 170 million.

dent, the Brazilian population today is so thoroughly mixed physically that it is difficult to determine most people's racial derivation with any precision.

Brazil is a huge and regionally diverse nation. It covers almost half of the South American continent, and its population is almost as great as that of all the remainder of Latin America exclusive of Mexico. Given this vastness, social and economic conditions are, predictably, not the same in all regions of the country. Brazil's major ethnic groups are also regionally concentrated, and patterns of ethnic relations therefore vary accordingly.

In the Northeast, most of the population is either black or mulatto (a combination of European and African), and the African cultural heritage is very apparent. This is Brazil's poorest region, an agrarian and largely underdeveloped area that has never fully recovered from the decline of the slave-supported sugar economy of the seventeenth and eighteenth centuries, of which it was the center. The aftereffects of slavery are still felt, and paternalistic black-white relations are more evident than in other regions of Brazil.

The Southeast, dominated by the metropolis of São Paulo, is the most industrialized and urbanized region of Brazil and is the core of its wealth and power. Ethnically it is primarily European, especially Italian, though its industrial preeminence has served as a magnet for out-migrants from other areas, especially the depressed Northeast. This has contributed to an increasing ethnic heterogeneity in recent decades though whites remain an overwhelming majority. The other region of greatest European influence is the extreme South, where German, Polish, Russian, Italian, and Portuguese farming communities are numerous alongside the traditional *estancias*, or cattle ranches.

The Indian influence in Brazil's population is most significant in the extreme North and in the interior, especially the Amazon basin. This is an enormous but sparsely populated region where most of the remaining Indians (around 200,000) live, most in isolated and primitive settings. Indian cultural and physical influences, however, are strongly evident here, and most of the population derives from a combination European-Indian ancestry.

DEVELOPMENT OF THE BRAZILIAN ETHNIC MÉLANGE

The shaping of Brazil's ethnic mélange resembles that of the United States in one important respect: Almost all its peoples have been immigrants, either voluntary or involuntary. The indigenous population at the time of Portuguese settlement in the early sixteenth century, perhaps two and a half million, was reduced to virtual insignificance in less than three centuries (Hemming, 1987). Unlike the U.S. colonists, who made an unsuccessful attempt to enslave Indians, the Portuguese used the indigenous peoples as slave laborers almost from the outset of colonization. The Indians were devastated in a relatively short time, however, because of their susceptibility to European diseases such as smallpox and measles. Another factor contributing to their demise was the intensity of agricultural labor, with which they were unfamiliar.

In the late sixteenth century, sugar cane became Brazil's chief cash crop. As a labor-intensive industry, sugar required large plantations; and as the Indian slave population diminished, the introduction of Africans in their place proved vital. Marvin Harris (1964) explains that with the development of the world market for sugar, the importation of African slaves—a costly enterprise—became economically feasible. Black slaves were many times more valuable than their Indian predecessors, for they seemed better prepared for field labor and were more resistant to European diseases. The African slave population in Brazil grew rapidly, and by the end of the eighteenth century, it numbered around a million and a half. Estimates of the actual number of slaves imported into Brazil between 1550 and the 1850s, when the slave traffic was ended, are varied, ranging from 3 million to 18 million (Knight, 1974; Pierson, 1967; Poppino, 1968; Ramos, 1939; van den Berghe, 1978). Because all records of the slave trade were purposely destroyed following the abolition of slavery in 1888, there are no precise data concerning the number of slaves actually imported into Brazil. That far more were brought to Brazil than to the United States, however, is undisputed.

The Brazilian and American Slave Systems

To better understand the roots of Brazil's system of ethnic relations, it is necessary to look at its long experience with slavery. It is especially important to compare slavery as practiced by the Portuguese in Brazil with the American slave system because many have argued that contemporary differences in black-white relations in the two societies are traceable to differences in the practice of slavery in each. Indeed, the similarities and differences in the two slave systems and the subsequent effects of each on patterns of race relations together make up one of the most widely debated issues of comparative social history.

Brazilian Slavery: A Milder System? Gilberto Freyre, Brazil's most renowned social historian, contended that slavery in Brazil was a milder, more humane form than its North American counterpart. This he attributed to several factors, the most important of which pertain to differences between Anglo-Saxon and Portuguese culture and society. Different practices of slavery, in turn, produced divergent patterns of relations between blacks and whites once slavery was ended in each society.

Freyre held that the Portuguese came to Brazil with a more tolerant attitude toward people of color. Portugal had experienced contact with African peoples in its colonial ventures as early as the fifteenth century and had endured a long period of rule by the Moors even earlier. The Portuguese sense of racial difference, therefore, was not so acute and absolute as that of the Anglo-Saxons, who had had little contact with nonwhites. A greater racial tolerance, according to Freyre's thesis, accounts for the more harmonious, less conflict-ridden nature of race relations in Portuguese colonies and the greater tendency among the Portuguese to intermix with native and slave populations. Freyre considered especially significant the extremely high rate of miscegenation between Portuguese slave masters and African slaves in Brazil. This, he contended, is evidence of a more compassionate and essentially altruistic master-slave relationship (1956, 1963a).

The thesis that the Portuguese slave system in Brazil was basically more humane than its North American counterpart was supported by an American historian, Frank Tannenbaum (1947). Tannenbaum explained that a milder form of slavery was evident in Brazil and in the Spanish colonies of Latin America because of differences between the Portuguese (and Spanish) and the Anglo-Saxons in their religious and legal concepts of slavery. He, like Freyre, maintained that there was a preconceived notion among the Portuguese regarding their treatment of slaves—deriving from principles established by both church and state—that fostered a less harsh slave system. In Brazil, the African did not lack a "soul" but was perceived simply as an unfortunate human being, a slave by accident of fate. In the United States, however, the African was perceived as less than human and thus naturally enslaved. Slavery in Brazil was understood as an economic necessity, hardly a relationship whose rationale derived from religious—or later, biological—principles, as in the United States. In short, in Brazil the human character of slaves was never denied. In the Portuguese view, slavery was a necessary evil. Thus, it was not defended so firmly on the notions of racial inferiority and superiority.

The Brazilian and U.S. legal codes dealing with slaves also differed significantly. Portuguese (and Spanish) laws pertaining to slavery protected the slaves' human rights, at least technically. Thus, slaves could not be dealt with by their masters with total impunity. Whereas slaves in the United States were chattel, reduced to mere property, in Brazil they were recognized as humans with certain legal rights. They were entitled to own property, marry freely, seek out another master if they were dealt with too harshly, and even buy their freedom, which many apparently did. It is this last-mentioned feature of the slaves' status, their greater opportunities for manumission, that Tannenbaum explained as an especially critical difference. Compared with the United States, manumission in Brazil was both common and expected. Its much higher incidence, Tannenbaum held, is in itself evidence of a milder form of slavery. More important, however, the accomplishment of manumission brought the freed slave automatic entry into Brazilian society with full and equal rights. In the United States, by contrast, even if slaves were freed, they did not acquire full citizenship but continued to suffer the economic, social, and legal handicaps that attached to their racial status.

The ability of the crown or the church to enforce these protective slave laws, of course, was quite another matter. These were, in effect, paper rights whose effectiveness was not sufficient to materially alter the nature of life for Brazilian slaves. "In Brazil, as everywhere in the colonial world," notes Harris, "law and reality bore an equally small resemblance to each other" (1964:77). As to the specific role of the church, its position on the issue of slavery was ambivalent and its power to influence Brazilian slave codes, in any case, limited (Marx, 1998). Neither Freyre nor Tannenbaum contended that slavery in Brazil was not harsh and inherently cruel. The key difference between the Brazilian and American varieties, however, was that "in the Spanish and Portuguese colonies the cruelties and brutalities were against the law, that they were punishable, and that they were perhaps not so frequent as in the British West Indies and the North American colonies" (Tannenbaum, 1947:93). Most important, the openings for freedom were much greater in the Brazilian system, as was the advancement to full citizenship following manumission. Given the different legal and philosophical status of the slave and the freed black in Brazil as compared with the United States, more benign and assimilationist relations were engendered between whites and blacks following the formal abolition of slavery.

Brazilian Slavery: A Harsher System? Revisionist thinking on the matter of Brazilian slavery has seriously questioned whether it was in fact a distinctly more humane system than those established in the United States and the Caribbean. Some have even concluded that in many ways it was a system more dehumanizing than the Anglo-Saxon variety. That slaves in Brazil were acknowledged as human and that they possessed certain legal and religious rights did not seem to discourage the common application of brutal and inhumane forms of punishment by masters and the maintenance of severely debased living conditions by any standard. Historian Charles Boxer (1962) has described the brutish and often sadistic cruelty with which slave discipline was enforced by the Portuguese planters. To tie a slave to a cart; flog him; rub his wounds with salt, lemon juice, and urine; and then place him in chains for several days was not considered excessive. Runaways were dealt with, as described by Brazilian scholar Darcy Ribeiro (2000), by branding with a hot iron, having a tendon cut, living shackled to an iron ball, or being burned alive.

That such punishments did not prevent slaves from continuing to run away on a large scale bespeaks the desperation to which they were driven. "Every black bore an illusion of flight in his breast," writes Ribeiro, "and was sufficiently bold to attempt flight when there was an opportunity" (2000:77). Other features of Brazilian slavery provide further evidence of its unusually harsh nature. Slaves commonly committed suicide, and insurrections among them were far more frequent than in the United States. Indeed, compared with the comparatively few rebellious actions by American slaves, Brazilian slave insurrections, many of them well organized, appear to have been very common. In fact, organized settlements of fugitive slaves, called *quilombos*, sprang up in various locales (Vieira, 1995).

Historical evidence, then, seems to have proved beyond much doubt that "humane" slaveowners in Brazil were the exception rather than the rule (Boxer, 1962, 1963; Davis, 1966; Graham, 1970; Marx, 1998). Recent historians have also disputed Freyre's and Tannenbaum's assumption that more frequent miscegenation and manumission as features of the Brazilian slave system substantiate the greater humanity and tolerance of the Portuguese. As for miscegenation, Skidmore (1993) notes that this was a common occurrence within the American system as well. The early enactment of laws prohibiting interracial marriage in the United States confirms the strong sexual attraction between the two groups, as does the wide range of physical traits of American blacks. Moreover, in Portugal's African colonies like Mozambique, the extent of miscegenation was never as significant as it was in Brazil. Clearly, factors other than the natural attraction of Portuguese men to native and slave women and their greater racial tolerance account for the high rate of interracial sexual encounters in Brazil. A more compelling explanation lies in the simple fact that colonial Brazil lacked sufficient Portuguese women (Marx, 1998). Unlike North American colonists, who came mostly in family units, Portuguese settlers were mainly unaccompanied males who quite naturally turned to native and slave women to meet their sexual needs. Harris has put the matter succinctly: "In general, when human beings have the power, the opportunity and the need, they will mate with members of the opposite sex regardless of color or the identity of grandfather" (1964:68).

As to the question of more common manumission granted Brazilian slaves, historian Carl Degler (1971) has suggested that this may have been simply a method of relieving slaveowners of the responsibilities and costs of caring for elderly or physically unfit slaves. Moreover, slaves were often unable to save enough to buy their freedom. Marx (1998) suggests that, given these conditions within the context of a highly stratified society, manumission in Brazil "did not threaten the social order and freed blacks were often left worse off than slaves" (52). In any case, Harris (1964) notes that although manumission was more common in Brazil, it was not so much more common as to build a case on this fact alone for a presumably milder form of slavery.

Degler has pointed out that the American and Brazilian slave systems may not have differed so fundamentally even on the matter of slaves' legal and moral status. Though legal decisions were not consistent from time to time and place to place, both church and state, at least technically, recognized slaves as humans in the United States as well as in Brazil. In neither society, however, was the law, whether church or state, effective in protecting slaves.

The issue of whether the Brazilian system of slavery was more humane than the American may, in the final analysis, be meaningless. How can such an argument be verified in any case? Indeed, the very notion of a "humanized" or "mild" slave system would be interpreted by most as a grotesque contradiction. Whatever the circumstances, slavery was dehumanizing and was resisted by the slaves. Moreover, the tremendously high mortality rate among the slave population would seem to render the issue "inconsequential in light of the overriding fact that slavery in Brazil was astonishingly deadly" (Marx, 1998:52).

Abolition and Its Aftereffects Although the actual conditions of slavery may not account in large part for the subsequently more benign relations between blacks and whites in Brazil, the conditions leading to the abolition of slavery and the status of blacks following their freedom may tell us a great deal.

Slavery endured in Brazil even longer than it did in the United States, ending formally in 1888. Its demise had been assured for several decades, however, and occurred in a gradual, evolutionary manner. This piecemeal abolition of slavery resulted in less wrenching social aftereffects than occurred in the United States, where slavery was ended abruptly following a cataclysmic civil war.

With the more common manumission of slaves in Brazil, blacks had already been a part of free Brazilian society for many decades before the formal end of slavery. In 1872, sixteen years before abolition, freed blacks were almost three times as numerous as slaves. By comparison, in the United States at the start of the Civil War, less than half a million out of a total black population of four and a half million were free (Skidmore, 1972). This meant that blacks and whites had experienced a long period of social interaction exclusive of the master-servant relationship. As Tannenbaum explained, in Brazil, "The Negro achieved complete legal equality slowly, through manumission, over centuries, and after he had acquired a moral personality. In the United States he was given his freedom suddenly, and before the white community credited him with moral status" (1947:112). As a result, there was no question of the social standing of free blacks in Brazil and their access to full citizenship. In Brazil, freed slaves were in fact free; in the United States, they were only technically free.

Furthermore, blacks in colonial Brazil had played a variety of occupational roles—even as slaves—and their entrance into the competitive labor market was therefore not so precipitous as it was in the United States (Karasch, 1975; Klein, 1969). Harris (1964) argues, in fact, that the continual shortage of skilled and semiskilled labor in colonial Brazil created the need to fill numerous positions with blacks, which, in turn, contributed more than anything else to the ease and frequency of manumission.

Freed blacks in Brazil, then, were not seen as either a social or an economic threat; therefore, resistance from whites did not strongly materialize. Brazil thus avoided the development of a racial caste system designed to permanently handicap blacks, as was the eventual outcome of abolition in the United States.

Although Brazilian society was not so fundamentally dislocated by the abolition of slavery as the American South had been, the consequences faced by newly freed slaves were in many ways similar. Freed slaves either wandered back to the rural workforce, seeking out their former masters, or migrated to the cities with few urban labor skills. They were further handicapped because they were forced into competition with more skilled European immigrants, especially in the industrializing Southeast.

The release of former slaves into a competitive labor market for which they were ill prepared produced a debilitating effect on black mobility that has remained a basic feature of the Brazilian social structure to the present day. As Brazilian sociologist Florestan Fernandes explains, blacks emerged from slavery materially and psychologically ravaged and "lacked the means to assert themselves as a separate social group or to integrate rapidly into those social groups that were open to them" (1971:28). Further disabling the freed black was the whites' eager relinquishment of any legacy of the slave system and its consequences. The occupational benefits of an industrializing Brazil with its competitive social order was therefore reaped primarily by European immigrants.

European Immigration

Beginning in the late nineteenth century, the ethnic composition of Brazil was altered fundamentally by the large-scale immigration of Europeans. Under the colonial regime, immigration into Brazil had been severely restricted. As a result, the bulk of the population was descended from three major ethnic lines: the Portuguese colonials, the indigenous Indians, and the African slaves. Following Brazilian independence in 1822, immigration remained sporadic until the mid-nineteenth century. An upward trend turned into a veritable flood tide of immigrants in the century's last two decades.

With the realization that slavery was doomed, coffee planters increasingly turned to Europeans, especially Italians, as a labor force. The bulk of European immigration beginning in the late nineteenth century, therefore, was directed to the state of São Paulo, where the major coffee-growing estates were located. Growers saw Europeans as better and more reliable workers than native Brazilians, especially freed blacks from the declining Northeast. São Paulo also became the primary destination for European immigrants because it represented the nucleus of burgeoning Brazilian industry. Here were to be found the industrial jobs for which these immigrants were better prepared than were Brazilian blacks.

European immigration was encouraged for another reason having nothing to do with the needs of labor. Racially, these immigrants were white and therefore contributed to the "whitening" of the society, an ideal that was now prevalent among the Brazilian political leadership and intelligentsia (Skidmore, 1990). Since abolition, the notion of an increasingly whiter population had been a staple of Brazilian intellectual thought. Abetted by the widespread influence of social Darwinism at the turn of the century, whites felt that Brazil should seek to dilute as quickly and completely as possible the black element of its racial composition. Put simply, the idea was "the whiter Brazil, the better" (Skidmore, 1972). Rather than encouraging a return of blacks to Africa, as was the case in the United States, Brazil chose to import whites. The goal in each case was the same—a dilution of the black population (Marx, 1998).

The mass immigration from Europe gave Brazil's whitening process a giant infusion. So limited had immigration been previously and so substantial had been the importation of African slaves that by the end of the eighteenth century, Africans were a majority of the Brazilian population. European immigration not only increased racial mixing, but reversed this ratio.

In the century between 1850 and 1950, five million immigrants came to Brazil. The peak era of immigration, however, occurred between 1870 and 1920, when over 3 million Europeans entered the society (de Azevedo, 1950; Skidmore, 1974). Almost 2 million entered in the little more than two decades between 1881 and 1903. Most were absorbed by the state of São Paulo, whose foreign population rose from 3 percent in 1854 to 25 percent by 1886 (Dzidzienyo and Casal, 1979). More than half the population by 1934 were first- and second-generation immigrants (Mörner, 1985). Italians were by far the largest immigrant group, but Portuguese, Spanish, and Germans were also significant. By 1897, Italians outnumbered Brazilians in the city of São Paulo two to one (Morse, 1958). Today about one-third of São Paulo state's population is of Italian descent (Freyre, 1963a)[1]. Europeans were also attracted to Brazil's extreme southern states, where large numbers of Germans and Poles established cohesive ethnic communities (Luebke, 1987, 1990).

In addition to non-Portuguese Europeans, another distinct ethnic element was added to the Brazilian mosaic beginning in 1908, when Japanese immigrants entered the society in large numbers for the first time. By 1941 almost 190,000 Japanese had immigrated into Brazil, with another 50,000 arriving after World War II. Today the Japanese ethnic community of Brazil numbers more than a million. Most Japanese originally settled as agricultural workers in the states of São Paulo and Paraná, and they still remain heavily concentrated in those areas (Makabe, 1999).

ETHNIC STRATIFICATION

In the United States, ethnic stratification, particularly as it pertains to racial-ethnic groups, is clear-cut: Those classified as white are at the top of the ethnic hierarchy

---

[1] Until the early 1900s, Brazil and Argentina—not the United States—were the major destinations of Italian emigrants.

Part III: Ethnic Relations in Comparative Perspective

and disproportionately occupy most top positions of power and privilege, while those classified as nonwhite are at the lower end of the hierarchy and collectively possess less power and privilege. (Recall the three-tiered hierarchy outlined in Chapter 12.) Furthermore, in social relations, the lines dividing white and black are relatively rigid, with groups coalescing around separate social institutions except for the economy and polity. Brazil's system of ethnic stratification resembles that of the United States in at least one regard, the rank order of ethnic categories. Whites disproportionately hold the society's power, privilege, and prestige. But several unique features of the Brazilian system distinguish it from that of the United States, as well as other multiethnic societies, making it one of the most far-reaching attempts at racial and ethnic democracy in the modern world.

The Brazilian Multiracial System

Perhaps the key factor differentiating the Brazilian system of ethnic relations from those of other societies in which there are several distinct physical types is the very concept of race and the manner in which people are racially classified.

In the United States, one's racial status is a product of one's descent; that is, individuals, regardless of physical appearance, are assigned to a racial status on the basis of ancestry. Such policies are based on what Harris (1964) calls the rule of "hypo-descent"—a person with any known or recognized ancestry of the subordinate group is automatically classified as part of that group rather than the dominant group. This is the so-called one-drop rule, referred to in Chapter 5. There are no provisions for "intermediate" cases, so that the system of racial classification is essentially a dichotomy—white and black. Thus, in the United States one may exhibit generally white physical features yet still be classified as black. For example, in 1983 a Louisiana judge declared a woman who had been raised as a white legally black on the basis of a distant black ancestor. She and her siblings were the great-great-great-great grandchildren of a black slave and a white planter. The decision was based on a Louisiana law declaring anyone with one-thirty-second "Negro blood" legally black. The woman was listed as three-thirty-seconds black (*Miami Herald*, 1983).

In Brazil, however, this strict rule of descent is not operative. One is not automatically placed into a racial category on the basis of family of origin. Instead, one's total physical appearance—skin color, hair texture, facial features—are the "obvious" determinants of one's racial classification. But in addition to these strictly physical aspects of racial identity, in Brazil certain social factors are used to classify people racially. The most important of these is social class (Harris et al., 1993). Thus, as one experiences class mobility, one's racial identity may to some extent also change. Popular Brazilian expressions such as "money whitens" or "a rich Negro is a white man, and a poor white man is a Negro" connote that as people improve their class status, they are perceived as lighter racially. Ribeiro relates the story of a well-known Afro-Brazilian painter, speaking with a young black man struggling to rise in a diplomatic career. After hearing the young man's complaints of the barriers he faces because of his color, the painter responds, "I understand your case perfectly, my dear boy, I was black once too" (Ribeiro, 2000:157). To a degree, then, racial status in Brazil is subject to redefinition in different social circumstances.

The imprecise nature of racial classification in Brazil means that racial groups are not castelike as in the United States, where "black" describes anyone with traceable African heritage, including those of mixed origins. But this understanding does not hold in Brazil. *Branco* and *prêto*—literally, white and black—merely denote people who are predominantly white or black in appearance, regardless of actual racial origin (Nogueira, 1959; Pierson, 1967). Simply having some traces of black ancestry does not result in the classification of someone as black. Indeed, most people classified as white in Brazil exhibit some evidence of black descent. Many American blacks, then, would be considered white in the Brazilian context.

Racial and Ethnic Categories This perception of race, different from American racial perceptions, has produced a correspondingly different system of racial categorization. Whereas the United States maintains an essentially biracial system in which people are classified as either white or black,[2] Brazil's multiracial system provides for more than two categories. Three major racial groupings are recognized in Brazil, but the boundaries between them are neither rigid nor clear-cut. *Branco* (white), *prêto* (black), and *pardo* (mulatto) are the most encompassing terms, but Brazilians employ literally dozens of more precise terms to categorize people of various mixed racial origins, depending on their physical features. Racial terminology is strongly localized, and from region to region—and even community to community—standards of classification vary.

The most minute physical features are often used to sort out people racially. The complexity and exactness of Brazilian racial typologies can baffle the American familiar with only a two-part racial scheme. Consider the descriptive categories used by the people of Vila Recôncavo, a town near Salvador in the state of Bahia, studied by anthropologist Harry Hutchinson:

The *prêto* or *prêto retinto* (black) has black shiny skin, kinky, woolly hair, thick lips and a flat, broad nose.

---

[2]The increasing Asian population in the United States has, of course, forced a third major category into the racial classification scheme used by government and other institutions. The principal racial division, however, remains black and white.

The *cabra* (male) and *cabrocha* (female) are generally slightly lighter than the *prêto*, with hair growing somewhat longer, but still kinky and unmanageable, facial features somewhat less Negroid, although often with fairly thick lips and flat nose.

The *cabo verde* is slightly lighter than the *prêto*, but still very dark. The *cabo verde*, however, has long straight hair, and his facial features are apt to be very fine, with thin lips and a narrow straight nose. He is almost a "black white man."

The *escuro*, or simply "dark man," is darker than the usual run of mestiços, but the term is generally applied to a person who does not fit into one of the three types mentioned above. The *escuro* is almost a Negro with Caucasoid features.

The *mulato* is a category always divided into two types, the *mulato escuro* and *mulato claro* (dark and light mulattoes). The *mulato* has hair which grows perhaps to shoulder length, but which has a decided curl and even kink. . . . The *mulato's* facial features vary widely; thick lips with a narrow nose, or vice-versa.

The *sarará* . . . has a very light skin, and hair which is reddish or blondish but kinky or curled. . . . His facial features are extremely varied, even more so than the *mulato's*.

The *moreno* . . . is light-skinned but not white. He has dark hair, which is long and either wavy or curly. . . . His features are much more Caucasoid than Negroid. (1963:28–30)

These are only the major categories, explains Hutchinson, each of which is further broken down even more precisely. Even the whites are subclassified on the basis of skin, hair, and facial features.

These categories, however, even the major ones, do not denote discrete racial groupings. Rather, they blend into one another as on a scale (Figure 14.1). As already noted, miscegenation has, from the time of the initial Portuguese settlement, been a constant and widespread practice in Brazil among the society's three major population types—European, African, and Indian. As a result, the racial origins of people are thoroughly interwoven in various combinations. *Black*, *white*, and *mulatto*, therefore, represent very broad categories, each with a wide range of physical types. Although there is general agreement on the terms *branco* and *prêto*, it is the mulatto, or *pardo*,

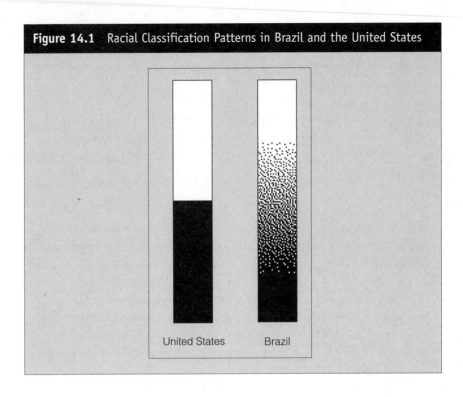

Figure 14.1 Racial Classification Patterns in Brazil and the United States

United States   Brazil

category, between extreme white and extreme black, that provides the almost endless variety of descriptive combinations throughout Brazil. As shown in Table 14.1, mulattoes officially composed 40 percent of the population in 1999; but unofficially, the percentage is much higher.[3] Again, it is important to emphasize the indistinct lines between these categories. Racial definition is largely subjective and arbitrary. Because, as we will see shortly, it is socially advantageous in Brazil to be considered white, the 1999 population figure of 54 percent white in Table 14.1 is greatly exaggerated in terms of the North American conceptualization of race (Harris et al., 1993). In any case, these census figures indicate a progressive decline of the white percentage, signaling that even when measured by people's self-identification, Brazil is gradually emerging as a racially mixed society.

Brazilians, then, are very conscious of color distinctions among people; indeed, they are far more conscious than North Americans or even South

---

[3]The *pardo* category consists not only of mulattoes (those of mixed European and African origins) but also those who are mixed European and Indian, usually referred to as *caboclos*, and those of mixed African and Indian origins, *cafusos. Cafusos*, however, are numerically a small element of the Brazilian population.

Part III: Ethnic Relations in Comparative Perspective

	1940	1960	1980	1999[*]
GROUP	Percent	Percent	Percent	Percent
Branca (white)	63	61	55	54
Parda (mulatto)	21	29	38	40
Prêta (black)	15	9	6	5
Amarela e Indígina (yellow and Indian)	1	1	1	1

Table 14.1 Brazilian Population by Color

SOURCE: Instituto Brasileiro de Geografia e Estatística, 1981, 1992, 2002.
[*]Does not include those who did not declare their color or race.

Africans, perceiving numerous physical types between black and white.[4] These gradations of color, however, make for a fluid system of ethnic stratification. It is the color continuum of Brazil as opposed to the rigid black-white dichotomy of the United States that explains much of the difference in the tone of ethnic relations in the two societies. Degler (1971) contends that the existence of the intermediate—mulatto—category in Brazil, more than anything, accounts for the diminished hostility between blacks and whites. The presence of what he calls "the mulatto escape hatch" makes infeasible the development of segregated institutions, which have so dominated most of the history of black-white relations in the United States. Where there are so many recognized racial combinations and where people's racial status is so arbitrary and personalized, it is impossible to erect castelike structures based on racial definitions. "The presence of the mulatto," explains Degler, "not only spreads people of color through the society, but it literally blurs and thereby softens the line between black and white" (1971:225). Furthermore, where there is such a varied biological mixture of the population, racial ideologies propounding the notion of superior and inferior races are difficult to sustain.

Skidmore (1993) casts doubt on the assumption that the presumed multiracial classification system of Brazil is today basically different from the U.S. biracial system. He contends that while there is an increasing awareness of the racial variability—and thus color gradations—of the U.S. population, at the

---

[4]When asked to describe their color in the 1980 census, Brazilians responded with 136 different labels, and in an earlier national household survey, with almost 200 different labels (Andrews, 1991).

same time there is a growing tendency among Brazilian scholars and others to describe Brazil in bipolar racial terms, that is, simply "white" and "nonwhite." Moreover, Marx (1998) points out that the Brazilian construction of a "triracial" system of categorization, though more fluid than the biracial U.S. system, still provided for the domination of whites over both blacks and mulattoes.

The Brazilian Ethnic Hierarchy

At first glance it might seem pointless even to envision an ethnic hierarchy—that is, a ranking structure of ethnic groups—in a society where hard and fast lines between ethnic groups are not recognized. Ethnicity would seemingly play no role in the allocation of social rewards. This, in fact, is "officially" how Brazil has been portrayed. Several aspects of the Brazilian system of ethnic relations, however, belie this ideal picture.

The Class-Ethnic Correlation By looking at the general class positions of Brazil's racial-ethnic categories, one can conclude without hesitation that there is indeed an ethnic rank order in this society. Although the boundaries of the three major groupings are not precise, there is little question that whites, mulattoes, and blacks represent a hierarchy of economic, political, and social standing. Whites are clearly at the top, followed in order by mulattoes and blacks. A stranger in Brazil could quickly confirm through simple observation the validity of an old Brazilian rule of thumb, "The darker the skin, the lower the class" (Wagley, 1971). Studies compiled by social scientists have repeatedly corroborated this visual impression. In his classic study of Salvador, Bahia, in the late 1930s, Donald Pierson (1967) found that Europeans were heavily concentrated at the top of the city's occupational structure even though they were a numerical minority. Though his findings were limited to one region, Pierson maintained that they were generally applicable to Brazilian society. More recent studies have confirmed that his conclusions generally remain valid for all regions of the country (Dzidzienyo and Casal, 1979; Fernandes, 1971; Garcia-Zamor, 1970; Hasenbalg, 1985; Lovell and Wood, 1998; Reeve, 1975, 1977; Silva, 1985; Twine, 1998; Wagley, 1971).

In looking at the class-ethnic relationship, one must be careful to keep in mind the considerably different class structure of Brazil compared with those of the United States and other advanced industrial societies. Brazil, like most Latin American societies, traditionally maintained a two-class system—an upper class, made up of landowners as well as wealthy merchants and professionals, and a huge lower class, comprising much of the remaining population. This traditional class system is gradually breaking down with the continued industrialization of the society, however, and class patterns are beginning to resemble those of advanced industrial societies. As this occurs, an expanding middle class of salaried professionals and white-collar workers is emerging, though it is a much smaller proportion of the population compared to the United States and other advanced industrial societies.

The ethnic hierarchy in Brazil has customarily paralleled the class system (Figure 14.2). At the pinnacle of the structure are whites, and "as one moves down the social hierarchy, the number of racially mixed or otherwise nonwhite individuals gradually increases" (Wagley, 1971:121). Although mulattoes and blacks are represented throughout the occupational structure, they are not found in significant numbers or anywhere near their proportion of the population in positions of authority or decision making. The elite levels of government, the economy, the military, and education are thoroughly dominated by whites. For example, the only black cabinet minister of the federal government in recent times has been Pelé, the retired soccer star. And it was not until 1996 that a black was elected mayor of São Paulo, the country's largest city. At the opposite end of the hierarchy, whether in rural or urban areas, the bulk of the poor and uneducated are black. Only in sports and entertainment do Brazilian blacks play a prominent role (Leite Lopes, 2000; Schneider, 1996).

Income and education statistics also demonstrate this racial hierarchy. Most simply, whites not only hold a disproportionate share of more prestigious jobs but

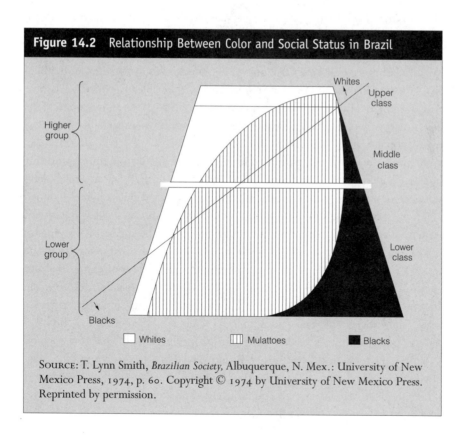

Figure 14.2 Relationship Between Color and Social Status in Brazil

Source: T. Lynn Smith, *Brazilian Society*, Albuquerque, N. Mex.: University of New Mexico Press, 1974, p. 60. Copyright © 1974 by University of New Mexico Press. Reprinted by permission.

also receive a disproportionate share of income (Hasenbalg, 1985; Silva, 1985). Nonwhites comprise almost two-thirds of the poorest Brazilians, those earning the minimum wage or less (Margolis, 1988). One study showed that nonwhites in all regions of the country earned about half of what whites earned. And these differentials were shown to be attributed not to differences in schooling or work experience but to racial discrimination (Webster and Dwyer, 1988). More recent studies have shown that differences between Afro-Brazilians and whites in wages and occupations and in the distribution of education and in literacy rate are evident even after controlling for race (Lovell and Wood, 1998; Schneider, 1996). Studies show that nonwhite children are disadvantaged in their access to education at all levels, as well as in the chances of completing their schooling, regardless of their socioeconomic status. At the university level in particular, whites are far more likely to matriculate than are blacks and mulattoes (Silva and Hasenbalg, 1999). In the early 1990s at the prestigious University of São Paulo, less than a dozen out of thousands of undergraduate and graduate students were Afro-Brazilian (Skidmore, 1999). Between racial groups, such discrepancies in income and education lead to other social inequalities, such as child mortality (Lovell and Wood, 1998; Schneider, 1996; Wood and Carvalho, 1988).

The "Whitening" Ideal Another clear-cut indication of an ethnic hierarchy in Brazil is the social preference for white over black. Despite the amalgam of skin colors and other physical features in its population, "white" has always been understood as more desirable than "black." The socialization process is quite effective in conveying the notion that to be white is to be socially favored. The previously quoted adages, implying that one gets "whiter" as one moves up in social class, are manifestations of this racial ideology. Brazil's white partiality has found expression at both individual and societal levels.

At the individual level, the white ideal is most evident in the preference for lighter-skinned marital partners. Such a marriage usually earns the darker-skinned partner a higher social status and in any case produces lighter-skinned children, who are assured of better social opportunities. Degler notes that "when a black or mulatto succeeds in marrying lighter, he speaks of 'purging his blood'—a phrase that in itself is revealing of what the process connotes in the minds of Brazilians" (1971:191–92). The white preference is often imparted to children in blunt fashion. Sociologist Oracy Nogueira explains that "from the start it is impressed upon the white child's mind that negroid characteristics make [the] bearer ugly and undesirable for marriage" (1959:171). Adults frequently tease a white boy that when he grows up he will "marry a Negro." Whereas marrying into a lighter family and having lighter offspring is the most effective form of whitening for Afro-Brazilians, others include upward economic mobility, associating with white friends and acquaintances, and adopting a white middle-class lifestyle (Andrews, 1991).

At the societal level, the "white is best" attitude has been evident in various governmental policies, particularly those pertaining to immigration. As earlier noted, European immigration was encouraged not only to expand the industrial labor force but also, as Roger Bastide puts it, "to submerge the descendants of Africans into a more prolific white population, and, in the last analysis, to change the ethnic composition of the population of the country" (1965:17).

The desire to create a whiter population stems from the response of Brazilian political and intellectual elites in the late nineteenth and early twentieth centuries to racist theories emanating from Europe and America, propounding notions of inferior and superior races. Without having to accept openly the idea of white superiority and black inferiority, they posited that the black (and presumably inferior) element of Brazil's population would be submerged through miscegenation and European immigration. These, they assumed, would progressively "whiten" the population, thus avoiding problems of race. The whitening process is still seen as desirable and is expressed, as we will see, in the tendency to discount racial inequalities and to squelch the development of racial consciousness. Brazilians, writes historian Thomas Skidmore, "are still implicit believers in a whiter Brazil, even though it may no longer be respectable to say so" (1990:28).

PREJUDICE AND DISCRIMINATION

It would be expected that in a society openly and ardently proclaiming the desirability of amalgamation of its various groups, ethnic prejudice and discrimination would be minimal. Brazilians do customarily assert that such prejudice and discrimination in Brazil are insignificant. As with ethnic stratification, however, there are glaring inconsistencies between the official accounts of racial democracy and the realities of individual and group relations.

Visitors to Brazil often comment on the seemingly placid and harmonious nature of interethnic relations. People of the entire range of color and of widely varied cultural origins appear to interact at all social levels with little or no difficulty arising from their ethnic differences. An American journalist stationed in Brazil described his surprise at witnessing one such social encounter, a dance at a neighborhood-based club:

> The crowd was interracial, with both blacks and whites in substantial numbers. At first glance, not much different from the crowd at the average Washington Bullets basketball game. But if you looked closer, you saw that blacks and whites were arriving together, dancing together and leaving together in couples and groups, not in separate cliques. This was different. And if you went down and made your way through the crowd, you saw that up close it was hard even to tell who was black and who was

white. . . . [L]ater . . . I pondered what I had found so striking about the evening. I realized that I had been in the midst of an interracial crowd that didn't really feel like an interracial crowd, at least not like one in the United States. (Robinson, 1996:23)

Such impressions extend back even to the period of slavery. Henry Koster, an Englishman traveling in Brazil at the beginning of the nineteenth century, remarked that "it is surprising, though extremely pleasing, to see how little difference is made between a white man, a mulatto and a creole Negro if all are equally poor and if all have been born free" ([1816] 1966:152).

Observant visitors have not been deceived. Relations between ethnic groups in Brazil *are* clearly more benign than those in most multiethnic societies of the modern world. Overt interethnic conflict is rare, and in no institutional area of the society have measures designed to discriminate against people on the basis of ethnicity ever been common or officially sanctioned. But such manifestations of ethnic harmony do not tell the entire story. Closer analysis reveals that prejudice toward nonwhites is a well-founded aspect of this society's culture. Moreover, these attitudes are translated into stringent rules of social distance, which are often disguised by the customary physical contact between ethnic groups in public places. Particularly at the primary levels of social interaction, the major color groups do not interact to a significant extent, and at the secondary levels, too, interaction is less cordial and acceptable than superficial observations indicate. Patterns of discrimination against nonwhites, though less overt and intensive than in other multiethnic societies, are also evident.

Prejudice

All who have carefully studied Brazilian ethnic relations have attested to the widespread prejudicial beliefs about and attitudes toward blacks and mulattoes. Specifically, ethnic prejudice is manifested in negative stereotyping of nonwhites, in patterns of social distance, and in the general disapproval of intermarriage between those of extremely divergent racial features.

Stereotyping Numerous studies have verified the prevalence of negative stereotypes of Brazilian people of color (Andrews, 1991; Bastide, 1965; Bastide and van den Berghe, 1957; Fernandes, 1971; Harris, 1963; Hutchinson, 1963; Pierson, 1967; Sanders, 1981; Saunders, 1972). Furthermore, similar stereotypes of blacks and mulattoes exist both in regions with great numbers of people of African origin and in areas with few (Bastide, 1965; Cardoso, 1965). Bastide and van den Berghe (1957) found, for example, that among white middle-class university students in mainly white São Paulo, blacks were commonly viewed as dirty, physically unattractive, superstitious, profligate, immoral, aggressive, lazy, lacking persistence at work, and sexually "perverse." Mulattoes were seen in much the same manner. Almost half of their sample also felt that blacks were intellectually inferior to whites. It is interesting to consider the resemblance of these stereotypes to those historically applied to blacks, Hispanics, and American Indians in the United States.

In Minas Velhas, a small town in the state of Bahia where most of the inhabitants were of mixed European-African origin, Marvin Harris reported generally similar beliefs and attitudes. "The superiority of the white man over the Negro," he wrote, "is considered to be a scientific fact as well as the incontrovertible lesson of daily experience" (1963:51). Harris noted the abundance of legends, folk stories, and pseudoscientific notions that supported the negative beliefs about blacks. A school textbook used in Minas Velhas, for example, stated that "of all races the white race is the most intelligent, persevering, and the most enterprising," whereas "the Negro race is much more retarded than the others," a view not disputed by any of the village's six teachers (Harris, 1963:51–52).

Harris explained that although blacks in the town were commonly derided as ignorant and ugly, such views were rarely expressed bitterly or hatefully.

> In Minas Velhas, to the white and to a certain extent to the Negro himself, the Negro is primarily a curious laughable anomaly. He is looked upon as a sport of nature, as a being with certain substandard and grotesque characteristics which make him amusing rather than disgusting. A white man will say: "*Negro desgraçado. Que bicho feio!*" (Miserable negro! What an ugly creature!) and smile broadly as though he were speaking of some rare amusing freak. (1963:53)

Moreover, the black stereotypes here (as well as in other regions) are often contradictory. Blacks can be seen at once as stupid and cunning, honest and dishonest, naive and shrewd, or lazy yet fit only for hard work. Three white attitudes, however, are constant and basic: (1) the black is inferior to the white; (2) the black does and should play a subservient role to the white; and (3) blacks' physical features are displeasing (Harris, 1963:56). "On the whole," noted Harris, "there is an ideal racial ranking gradient in which whites occupy the favorable extreme, Negroes the unfavorable extreme and mulattoes the various intermediate positions" (1964:60).

Social Distance Studies of social distance also reveal strongly prejudicial attitudes toward nonwhites (Bastide and van den Berghe, 1957; Garcia-Zamor, 1970; Zimmerman, 1963). In their study of white university students in São Paulo, Bastide and van den Berghe (1957) found that almost all accepted the idea of equality of opportunity for everyone, regardless of color. Better than half also accepted the desirability of casual relations between members of different racial groups. But beyond this level, social distance increased markedly. Sixty-two percent opposed intimacy with blacks beyond simple comradeship, and 77 percent opposed miscegenation with blacks (55 percent also opposed miscegenation with mulattoes). The most unwavering attitudes involved intermarriage; 95 percent indicated they would not marry a black, and 87 percent would not marry a light-skinned mulatto.

Endogamy The infrequency of marriages between those of noticeably different color is perhaps the clearest evidence of negative attitudes toward nonwhites. The disapproval of interracial marriage may seem curious in a society that has countenanced miscegenation for its entire social history. Compared with the United States, of course, there is a great deal of intermarriage across color lines, and the racially mixed character of most Brazilians testifies to the persistence of this practice. But the degree of social tolerance for such unions drops sharply when the contrast in color of the partners is great (Cardoso, 1965; Degler, 1971; Fernandes, 1971; Hutchinson, 1963; Reeve, 1977). Most marriages are between persons of similar color, and most intermarriages are between blacks and mulattoes rather than between whites and either blacks or mulattoes. Moreover, most intermarriages occur at the lower end of the class hierarchy. Hutchinson (1963), for example, noted that in the small Bahian community he studied, the aristocratic white upper class was completely endogamous. He went on to explain, however, that "no one in any class likes a marriage in which the skin colours are too far apart. Any marriage of dark with light or white will be referred to as *mosca no leite*, or fly in the milk, and a certain repugnance is felt by all" (1963:40). In marriages of black and white, the union is considered fortunate for the former and a case of lowering oneself for the latter (Nogueira, 1959).

The disapproval of marriage or even romance between persons of extremely divergent colors is revealed by the storm of protest that followed the portrayal on a popular television series of a love affair between a black foreman of a plantation in Haiti and the boss's white daughter. The public outcry was so great that the story was rewritten. A prominent white producer stated that on Brazilian television, "A white man can kiss a black woman, but a black man cannot kiss a white woman" (Vidal, 1978).

Brazilians often cite the tradition of miscegenation as proof of their lack of prejudice toward people of different colors. Yet it has been pointed out that the commonness of this practice may simply reflect the accessibility of black and mulatto women to white men throughout Brazilian history. The mulatto woman is traditionally seen in Brazil as epitomizing female beauty, but this has been interpreted by some as simply a sexual attraction, indicative of an exploitative relationship. Indeed, Bastide (1961) contends that rather than

Part III: Ethnic Relations in Comparative Perspective

proof of a lack of prejudice, miscegenation is a manifestation of it because it occurred primarily in the context of slavery, not marriage, and certainly not marriage of social equals.

Discrimination

Although studies of Brazilian ethnic relations have rarely failed to note the prevalence of negative stereotypes of dark-skinned persons, they have often pointed out that these attitudes are not ordinarily translated into discriminatory behavior (Freyre, 1963b). Indeed, some hold that the key difference between ethnic relations in the United States and Brazil is that in the latter, prejudicial attitudes do not fundamentally affect people's behavior toward members of particular groups (Harris, 1964). That discrimination against nonwhites in Brazil is essentially unlike ethnic discrimination in the United States is clear enough. The forms of blatant derogation, denial, and physical violence historically so well rooted in American society are simply not present in the same manner or degree in Brazil, and the institutionalized forms of discrimination that characterized the United States before the 1960s are not found at all. Whatever the extent of ethnic discrimination in Brazil, then, it is hardly comparable to discrimination American style. But its less overt, less socially acceptable, and far less virulent style does not mean that ethnic discrimination is alien to Brazil.

Quite clearly, we have already seen that color plays a fundamental role in the determination of people's life chances. Simple observation as well as careful analysis reveals that Brazil is a society in which those of darker skin are consistently relegated to the lower rungs of the social hierarchy. Although white skin does not guarantee social success, "it always improves one's position over that of a darker person" (Degler, 1971:191). Thus, as noted earlier, whites continue to occupy most of the higher-status, better-paying jobs in business and government.

Class or Race? Although the lower place of blacks and mulattoes in the occupational structure of Brazil is undeniable, many observers—as well as Brazilians themselves—have customarily attributed this situation not to discrimination but to the effects of class disadvantage (Freyre, 1963a; Pierson, 1967; Twine, 1998). The argument is essentially this: The lower place of blacks and mulattoes is the product of a heritage of slavery. Whites have enjoyed the historic advantage of being at the top of the social hierarchy and therefore continue to disproportionately occupy the society's better jobs and leadership positions. Unlike the situation in the United States, the disadvantaged place of nonwhites is not attributable to institutionalized and customary forms of racial discrimination. Differences in class, then, generally do reflect differences in color; but this relationship is due to historical accident, not patterns of discrimination. Because there is no castelike system in Brazil, as blacks and mulattoes rise in the class hierarchy, they are not confined to their own communities and institutions but are recognized by and accepted into the dominant (white) society. That is, the individual's class status takes precedence over his or her ethnic status, as opposed to the continued prominence of ethnicity in the United States. In Brazil, a successful black politician or doctor, for example, will not be identified invariably as a *black* politician or *black* doctor but only as a politician or doctor.

There is little question that class factors weigh heavily in the social placement of Brazilians. But the argument that color is simply an expression of class in Brazil has been challenged by many social scientists (Hasenbalg, 1985; Lovell and Wood, 1998; Morse, 1953; Russell-Wood, 1968; Silva, 1985; Skidmore, 1974; Twine, 1998; van den Berghe, 1978). Their studies have shown that although one is not denied opportunities automatically on the basis of racial ancestry, one's color is indisputably significant in the distribution of wealth and power.

In the past, color in Brazil did not seem as overwhelmingly decisive as in the United States. Whereas in the United States one's racial classification seemed to supersede all other social characteristics, in Brazil social class seemed more significant, particularly as one approached the lighter pole of the color continuum. That may no longer be the case, however. Through an analysis of census and other data, Andrews (1991) compared the degree of racial inequality in Brazil with that in the United States over several decades and found that in some ways there is a lower degree of inequality between black and white in the United States. Although in past decades racial disparities in life expectancy and occupation were not as great in Brazil, by 1980 those indicators had reversed themselves. Although Brazil showed less residential segregation and proportionately fewer Afro-Brazilians in poverty, these were counterbalanced by greater racial equality in the United States in school enrollment, median years of schooling, and earnings.

Andrews concluded that the United States has generally moved toward a reduction in racial inequality in the last several decades, while Brazil has moved in the opposite direction. He attributed this to several factors. First, blacks in Brazil did not move out of the country's most economically backward region (the Northeast) as quickly as African Americans moved from the rural South to the industrial North, where economic and educational opportunities favored upward mobility. Second, the benefits of economic growth in Brazil have tended to flow disproportionately to the upper and middle classes, overwhelmingly white; in the United States, a more equitable distribution of wealth generated by the post–World War II expansion created a more favorable climate for black mobility. Finally, an active role by the U.S. federal government in reducing discrimination through civil rights laws and other antidiscrimination measures has had no Brazilian counterpart.[5]

Overt Discrimination In addition to the indisputable relationship between skin tone and social class, blatant and direct forms of ethnic discrimination have frequently been revealed by social scientists and journalists in the past several decades. Blacks may be excluded from hotels and clubs and are commonly harassed by the police. Apartment house elevators in São Paulo, Brazil's largest city, commonly are marked "social," for residents and their guests, or "service," for maids and workmen; blacks, no matter their status, are routinely steered into the latter (Schemo, 1995). Discrimination is particularly evident in the area of employment (Sanders, 1981). In São Paulo, jobs requiring contact with the public, such as receptionists, salespeople, and waiters in better restaurants, are often closed off to Afro-Brazilians. Various subtle techniques are used in screening black job applicants. Employment want ads, for example, will often use the phrase "good appearance" (*boa aparência*), which is understood by all to mean "white." Such restrictions on Afro-Brazilians dealing with the public extend even to higher-ranking jobs, including management and the professions (Andrews, 1991).

There are, of course, regional differences in Brazil regarding the level of discrimination against blacks and mulattoes. In the South (the most heavily white region) discriminatory actions are more obvious, and the social lines between color groups are more rigidly drawn. At times, discrimination can manifest itself starkly. For example, in Rio Grande do Sul, a mostly white state, a mulatto woman won the right to represent her town in the state beauty contest as a preliminary to choosing Miss Brazil. A member of the jury, however, said it would not "look right" for a primarily white state to be represented by a "colored person." The jury, after several more votes, named the runner-up— white—to represent the town (Vidal, 1978). Even in the more socially and physically integrated Northeast, however, patterns of overt discrimination are not uncommon, and certain social and occupational areas are well understood to be off-limits to blacks (Degler, 1971).

Perhaps the most telling evidence of ethnic discrimination in Brazil came forth in 1951, when the Brazilian legislature was prompted to enact an antidiscrimination measure, the Afonso Arinos law. The significance of this legislation lay not so much in its subsequent effect but in the formal recognition for the first time in Brazilian history that discriminatory practices did in fact exist. Ironically, the measure was reportedly prompted by the refusal of a Rio de Janeiro hotel to accommodate a visiting black American entertainer and the clamor that

---

[5]Andrews notes, however, that these U.S. trends came to a halt in the 1980s.

followed. The ineffectiveness of this measure in alleviating discrimination was generally recognized, and in 1988 the Brazilian Congress passed a stronger antidiscrimination law that provided for increased penalties for discriminatory acts (Andrews, 1991).

STABILITY AND CHANGE

In looking at the forces of stability and change in Brazilian ethnic relations, we might focus on the two most prominent features of this society's ethnic ideology—the commitment to full assimilation of diverse peoples and the denunciation of racial prejudice and discrimination. How fully has Brazil attained these objectives, and will they continue to guide the course of ethnic relations in the future? In no society, of course, should we expect to find consistency between ideology and reality in all aspects of ethnic behavior and attitudes. The United States, as we have seen, has maintained for most of its history an equalitarian ideology even though official policies and individual actions have not always reflected that belief system. Brazil's affirmation of equalitarian and assimilationist principles, however, has been so unfailing that as a multiethnic society it promises to be a positive model for others.

The Goal of Assimilation

In some ways, Brazil seems to epitomize the assimilationist type of multiethnic society, described in Chapter 4.

Amalgamation Without question, the intermixing of diverse peoples— amalgamation—has advanced further in Brazil than in almost any other nation of the modern world. Several factors have contributed to this dilution of racial lines. First, immigration from Europe substantially increased the white element of the population. As we have seen, before the late nineteenth century whites were a numerical minority in most parts of Brazil, and European immigration played a large part in reversing that ratio. Second, a system of sexual exploitation had been in effect since the first Portuguese arrived in the early 1500s, a system whereby white men enjoyed relatively unlimited access to black and Indian women. Finally, as shown in Table 14.1, both black and white components of the Brazilian population have declined in percentage over the past few decades. Correspondingly, the racially mixed category (mulatto) has grown larger. This may simply be a result of the increasing tendency for people to identify themselves as neither white nor black, but it may also suggest a steady pattern of increasing intergroup marriage and thus a continuation of racial amalgamation (Sanders, 1982).

Cultural Assimilation The "official" view of assimilation in Brazil is that the society is fusing not only into a single racial group but also into a single Brazilian culture, comprising Portuguese, Indian, and African elements. "The Brazilian ideology of interracial and interethnic relations," wrote Nogueira, "is pro-amalgamation concerning physical traits and pro-assimilation as to cultural characteristics" (1959:172). The assimilationist goals of Brazil have been expressed most articulately by Freyre, who asserted that the society "has not attempted to be or become exclusively European or Christian in its styles" (1963a:156).

This view has largely been verified. African and, to a lesser extent, Indian elements have been blended into the Brazilian culture far more than have African or Native American traits in the United States. African cultural elements are especially evident in religion, food, music, and dance, particularly in those areas of Brazil where the population is heavily of African origin. For example, West African religious rituals have been fused with Catholicism in cults known as *candomblé* and *macumba* in Bahia and Rio de Janeiro (Rohter, 2000).

The fusion of Portuguese with African and Indian cultures, however, should not be overstated. As Wagley has noted, "The only cultural models for the African slave, the Indian, and their mixed descendants were their Portuguese, Catholic, and aristocratic masters" (1971:9). Thus, Brazil has evolved into a society in which African and Indian features are welded into what is fundamentally a Luso (Portuguese)-Brazilian culture.

As for the various European ethnic groups that have populated Brazil since the late nineteenth century, problems of assimilation have seemed less acute than in other multiethnic societies though separatist elements have at times evidenced themselves. For example, the German communities of the southern Brazilian states in the early twentieth century sought to perpetuate their language and to remain unassimilated into what they saw as an "inferior" Luso-Brazilian culture (Luebke, 1983). Even in more recent times they have remained relatively separated from other Brazilian groups, as have the Japanese (Hastings, 1969; Rodrigues, 1967; Suzuki, 1969). Moreover, the first two generations of other European groups have displayed the resistance to assimilation generally characteristic of early generations of voluntary immigrant groups. By the third generation, however, they have become an integral element of the Brazilian economic and political elites and have served as a prime segment of the expanding Brazilian middle class (Horowitz, 1964; Willems, 1960).

Brazil, then, has clearly achieved a level of physical and cultural assimilation surpassing that of most other multiethnic societies. Although there remain in Brazil strong regional patterns of culture and of ethnic composition, giving the society a pluralistic flavor, the commitment to an assimilationist ideology has been constant and has yielded an increasingly integrated society, physically and culturally. Two qualifications, however, bear mention. First, as we saw earlier, the amalgamation of Brazilian peoples has proceeded within the context of a white ideal. This essentially racist notion does not negate Brazil's integrative achievement relative to that of other societies, but it certainly reveals its less-than-democratic stimulus. The contradiction between pride in the development of a "racial democracy" and a social preference for whiteness is transparent. Brazilians officially perceive that they are moving increasingly toward a racially amalgamated society in which color distinctions are gradually eroding. Yet there is no denying that this is an amalgamation in which the darker hues are—desir-ably—being diluted by the lighter hues.

The continuation of the white ideal as part of Brazilian culture is no better il-lustrated than in the phenomenal popularity of Xuxa (pronounced "SHOO-sha"), a Brazilian media superstar. During the 1980s and 1990s she was the country's most celebrated entertainer, appealing in particular to children. In-deed, her popularity at times seemed better described as idolatry. "The star's presence can be so commanding," noted one observer, "that people burst into tears at the sight of her" (Simpson, 1993:2). What has made the Xuxa phenom-enon so intriguing is that within the racial continuum of Brazil, she is so thor-oughly atypical: blonde, blue-eyed, and fair-skinned. In short, she projects per-fectly the white ideal.

Second, although cultural assimilation is apparent, structural assimilation of blacks at both the primary and secondary levels has lagged far behind. As noted earlier, blacks and even mulattoes have not entered the higher echelons of Brazil's political and economic institutions at anywhere near a level commensu-rate with their percentage of the population, nor have they begun to interact personally in any significant way with whites. In short, regarding structural as-similation, Brazil is not essentially unlike the United States and other societies with contrasting racial-ethnic groups.

The Denunciation of Prejudice and Discrimination

The most glaring inconsistency between the ideology and the reality of ethnic relations in Brazil involves the irrefutable evidence of racial prejudice and dis-crimination, even though it is of a far milder and less intense form than in the United States. There is no doubt that the belief in equality of opportunity is widely accepted and that public opinion strongly opposes racial discrimination; but it is clear that there is an ethnic hierarchy in which blacks and mulattoes ex-perience differential treatment.

Brazil's nondiscriminatory ideology, however, has been so preeminent that revelations or accusations of discrimination are customarily met with denials or disbelief. There is a well-understood principle that any expression of racial dis-crimination should be attacked as "un-Brazilian" (Smith, 1963). Even speaking of a "racial problem" is met with disapproval. In the late 1970s, for example, cen-sors banned a program dealing with racial issues from Brazil's largest television network (Vidal, 1978). Official denials of discrimination tend to perpetuate the mystique of racial and ethnic democracy and make any attack on discriminatory practices more difficult (Hanchard, 1994). If officially "there is no racial discrimination," public opinion cannot easily be galvanized to confront it.

A further impediment to the elimination of ethnic discrimination in Brazil is the subtlety of its form. Although violations of the antidiscrimination law of 1951 were often prominently reported, most were dismissed by the authorities for lack of evidence. One observer notes that whites were rarely charged under the law, probably because discrimination is so common (Sanders, 1981). Moreover, the subtle and mild forms of prejudice and discrimination may cause blacks themselves to remain relatively unaware of them (Hanchard, 1994).[6] Even when they do express awareness of discrimination, blacks lack the economic and political resources to address such issues. Most are part of the lower class, who—like the poor and uneducated everywhere—are ordinarily pawns, not movers, in the society's political processes.[7]

That most blacks in Brazil are part of the lower class makes the issue of discrimination more complex because, as mentioned earlier, class and ethnic factors become closely intertwined and difficult to distinguish. Here it is instructive to recall the debate among American sociologists regarding William J. Wilson's thesis of the "declining significance of race" (Chapter 9). As noted earlier, Brazilian whites—and many blacks—prefer to explain prejudice and discrimination against darker-skinned people as a product of their class, not their racial status. Furthermore, the explanation that blacks are disproportionately occupants of the lower social positions in the society because of class factors is more readily acceptable because this places the blame on "natural" social and economic forces, not those purposely designed to limit the placement of people. Accusations of class prejudice and discrimination are therefore accepted, and accusations of racism are denied. Episodes of discrimination obviously based on race are treated as individual and exceptional cases, not as reflections of a customary pattern (Fontaine, 1981; Hanchard, 1994; Twine, 1998). Also, failure to improve one's social position is seen, as in the popular American ideology, as attributable to the individual's shortcomings, further obscuring the effects of structural disadvantages (Hasenbalg and Huntington, 1982/1983).

---

[6]Telles found a more moderate level of residential segregation among Brazil's racial groups compared to other countries with large populations of African and European origin. That blacks, mulattoes, and whites live in similar neighborhoods, he concluded, "may strengthen the perception that race has little or no effect on life chances, at least for individuals of the same social class" (1992:195).

[7]It is of note that the limited Brazilian black movement is led primarily by middle-class Afro-Brazilians with comparatively light skin (Burdick, 1995).

Ironically, the difficulties of attacking prejudice and discrimination in Brazil are due not only to the effectiveness of the society's assimilationist and equalitarian ideology but also to the presence of a color continuum, as opposed to the rigid color lines of the United States. Because racial lines are not inexorably fixed, there is little ground for organizing people on the basis of color. "In assimilationist circumstances," notes Ribeiro, "blackness is diluted on a broad scale of gradations, which breaks solidarity and reduces combativeness by insinuating the idea that the social order is a natural order, even a sacred one" (2000:157). Hence, the development of a strong civil rights movement, aimed at alleviating black deprivation and discrimination, is undermined. How can there emerge a "black movement" when there are varying perceptions and definitions of who is black? Blacks in the United States, by contrast, cannot escape racial prejudice and discrimination no matter what the shade of their skin and no matter what their class standing; all are part of the category "blacks." Creating solidarity among them therefore becomes easier despite internal group differences of class and status.[8] Although Brazilian civil rights organizations have emerged from time to time—such as the *Frente Negra Brasileira* (Brazilian Black Front), a movement of the 1930s whose designs were to integrate blacks into mainstream Brazilian society (Fernandes, 1971)—they have not displayed the endurance or the effectiveness of American organizations like the NAACP or the National Urban League (Hanchard, 1994).

What's more, that people may marry lighter and thus produce lighter children further detracts from building enthusiasm for black political activity. Bastide quotes a black woman who, looking at her lighter-skinned children, declares: "They're white already. What's the use of fighting, forming leagues for the defence of the Negro, and all that?" (1961:12). This low level of racial consciousness and militance is also a result, in large measure, of the feeling of political leaders and intellectuals that to heighten racial awareness, as occurred in the United States during the black civil rights movement, would be "racist" and thus a denial of the idea of Brazil as a racial democracy (Skidmore, 1990). Regardless of its inconsistencies, the idea of racial democracy has been accepted by most Brazilians, and it continues to undermine efforts to galvanize blacks into an effective social movement (Marx, 1998; Hanchard, 1994; Twine, 1998).

A combination of an effective antiracist ideology, more benign forms of prejudice and discrimination, the complex relationship between color and class, and

---

[8] As Toplin (1981) has pointed out, however, in the United States, too, lighter-skinned blacks have always fared better than darker-skinned blacks. This pattern began during slave days, when mulattoes were often given house duty and more frequently granted manumission. The continuity of this pattern is evident in the traditional predominance of lighter-skinned blacks as part of the black upper class.

a multiracial classification scheme has served to make the long-range improvement in the status of Brazilian blacks perhaps even more problematic than raising black status in the United States. In Brazil, as Degler (1971) has pointed out, there is still lacking a widespread acknowledgment that blacks are socially disabled by a heritage of slavery and discrimination. In the United States, this fact has been generally recognized, and compensatory measures, such as affirmative action programs, have been instituted. Comparable measures in Brazil, however, have only recently been suggested (Fry, 2000). This, of course, requires the admission of racial prejudice and discrimination, which, Degler notes, "the national myth explicitly denies" (1971:271). A São Paulo city council member, introducing an antidiscrimination measure, stated that "Everybody behaves as if inequality doesn't exist. It's therefore much worse, because unlike in the United States, we've never faced it directly. It's disguised" (Schemo, 1995). Some maintain, however, that today the myth of racial democracy in Brazil is waning because the long-standing depressed status of blacks in the society has become more difficult to ignore (Andrews, 1991; Vieira, 1995).

Changing Patterns of Ethnic Relations

In Chapter 4 we saw that in preindustrial or colonial societies, relations between ethnic groups are generally paternalistic. People "know their place" and do not question the ethnic hierarchy. As a result, the level of intergroup hostility may be low, despite gross inequities. As societies gradually shift to industrial capitalist systems, competitive ethnic relations emerge. The demands of economic rationality now require an open labor market, and in such settings, groups use whatever means are at their disposal to protect and improve their economic position. This commonly produces virulent forms of prejudice and discrimination that were previously unnecessary and brings to the fore racist ideologies that serve to rationalize the denial of opportunities to minorities.

Some have suggested that Brazilian race and ethnic relations may move in the direction of increasing competitiveness as Brazil continues to industrialize (Bastide, 1965; Fernandes, 1971). This means that ethnic conflict may intensify and that the ethnic consciousness of both blacks and whites may rise. For example, in the 1970s, a black-consciousness movement arose among young blacks in the *favelas* (slums) and working-class neighborhoods of Rio de Janeiro, much influenced in style and substance by its earlier American counterpart (Barcelos, 1999; Fontaine, 1981; Hanchard, 1994; Vidal, 1978). What's more, numerous political and cultural organizations developed, stressing black awareness (Sanders, 1981). Although this movement did not succeed in galvanizing poor and working-class blacks, it did force Brazil's major institutions, particularly government, to begin to acknowledge and seriously address issues of racial inequality for the first time (Andrews, 1991).

Compensatory Policies The increasing recognition of racial discrimination has become even more pronounced in the past several years, as policies and ideologies more in line with U.S. conceptualizations of race and affirmative action have emerged. In 1995 the government introduced a National Human Rights Program, aimed at enhancing the interests of the black population. It went beyond merely enforcing antidiscrimination measures, but called for a system of racial classification along U.S. lines and for U.S.-style affirmative action policies (Fry, 2000). Thus, rather than denying racial distinctions, race was formalized as a criterion for making and targeting policy. This American-style approach seems to be gaining support. Consider the statement of Brazil's president, Fernando Henrique Cardoso (also an eminent sociologist) in this regard in 1995:

> We want to affirm, and truly with considerable pride, our condition as a *multi-racial society* and that we have great satisfaction in being able to enjoy the privilege of having *distinct races* [*raças distintas*] and distinct cultural traditions also. In these days, such diversity makes for the wealth of a country. (cited in Fry, 2000)

The use of affirmative action to address issues of inequality has taken on many of the same controversial trappings as it has in the United States—whether such compensatory policies should reward people on the basis of group membership as opposed to individual merit, whether they can effectively help those most in need, and whether they solidify, rather than weaken, racial and ethnic boundaries. In addition, however, the controversy has demonstrated a unique Brazilian twist in that it throws into question the very notion of how race is conceived in Brazil. Classifying people racially using clear-cut lines of division, as in the United States, and subsequently using those classifications to apply affirmative action measures, also like the United States, contradicts the popular notion that racial discrimination does not occur in Brazil and challenges Brazil's self-proclaimed color blindness.

Affirmative action has therefore created a profound debate within Brazil (Guimarães, 1999). There are those who favor the adoption of a more American-like approach to racial inequality, using affirmative action to improve the position of blacks. This, of course, would necessitate the marking of explicit racial boundaries, something never done before. On the other side are those who assert that the system of racial classification in Brazil, along with its ambiguities and inconsistencies, is uniquely Brazilian and should be retained. Racial democracy in this view is, correctly, the society's goal—if not the reality—of race relations in Brazil. The movement for change in Brazil's racial politics seems intent on finding a way to reconcile these two positions—to promote greater racial equality without completely abandoning the unique Brazilian racial mixture and the ideal of racial democracy (Fry, 2000).

A New Movement Organizations have emerged in the past decade whose objectives are to improve the status of Afro-Brazilians and to fight racial discrimination. These have seemed to attract only a minority of the Afro-Brazilian population, however (Heringer, 1995). The ideology of racial democracy, the white ideal, and white political dominance have combined with the absence of formal segregation to frustrate the formation of racial consciousness and racial solidarity among Afro-Brazilians, which might lead to a more potent social movement (Skidmore, 1999). Furthermore, most of the population refuses to accept racial labels, particularly the terms *prêto* and *pardo*, preferring to call themselves *moreno* or any of the numerous terms traditionally used to describe people in Brazil (Schwartzman, 2000). Thus, despite the strong awareness of racial differences, racial boundaries remain indistinct. All of these factors, in turn, contribute to the perpetuation of a racial hierarchy (Hanchard, 1994). Nonetheless, the movement to formally recognize racial discrimination and to effect some form of affirmative action for Brazil's black population has become part of the public discourse. The paradox of this new movement, however, is that it has compelled Brazil to question the very notion of racial democracy and the declared irrelevance of race in the distribution of the society's resources.

Some have also noted that ethnic relations in Brazil and the United States seem to be on convergent paths (Degler, 1971; Toplin, 1981). In the United States, the deprived place of the black underclass is increasingly attributed to class factors as the more blatant and legitimized forms of racial discrimination are eradicated. In Brazil, however, the tendency seems to be just the reverse. Racial discrimination is becoming more obvious and acknowledged as a factor in black deprivation, and traditional class explanations have been seriously questioned.

Despite its growing awareness and acknowledgement of racial discrimination, Brazil can proclaim success in creating a common national identity that transcends the racial and ethnic components of its varied population. "One and the same culture takes in all," writes Ribeiro, "and a vigorous self-definition that is more and more Brazilian animates everyone" (2000:169). In this sense, Brazil can be considered an exemplar of an assimilationist society, having transformed its diverse ethnic elements into a culturally homogeneous social entity.

In sum, Brazil presents a complex and often paradoxical case of ethnic relations. In its commitment to racial amalgamation, it has few peers among current societies. Similarly, its embracing of the theme of ethnic equality has been unwavering since the abolition of slavery. Furthermore, the inconsistencies between ideology and reality cannot repudiate the fact that relations between blacks and whites in Brazil are more benign than in other societies with diverse racial-ethnic groups. If Brazil is not the racial and ethnic paradise its advocates have proclaimed, it is nonetheless a society in which interethnic

Points of Similarity
- The historical development of the racial/ethnic system in both societies was parallel: a combination of conquest of the indigenous (Indian) population by Europeans, importation of an African slave population, and immigration primarily from Europe.
- Slavery was institutionalized in both societies for more than two centuries.
- Blacks in both societies (as well as mulattoes in Brazil) are the primary minority ethnic group. In both societies blacks rank lowest on various social and economic indicators, including income, occupation, education, and political power.

Points of Difference
- Racial divisions are less fixed in Brazil. Numerous racial categories are used to designate people, based on observable physical traits, not on ancestry.
- Throughout its history, Brazil has encouraged amalgamation between whites and blacks, mainly as a way of diluting the black element of its population.
- The African influence, derived from slavery, has remained more vibrant and expressive a part of Brazilian culture.
- Race relations historically have been more benign in Brazil.

relations are less volatile than in most others and in which giant strides have been made toward ethnic assimilation. At the same time, however, Brazil demonstrates the difficulties of mitigating the intergroup conflict that, as noted in Chapter 4, seems endemic to multiethnic societies.

SUMMARY

Brazil presents perhaps the best case for comparison with the United States. Its ethnic composition and the historical development of its ethnic diversity closely parallel the American experience. As in the United States, the main focus of ethnic relations is the confrontation of blacks and whites. Other ethnic elements are present, however, including many European ethnic groups that entered as voluntary immigrants, and indigenous Indians.

Brazil was an integral part of the African slave trade, and its system of slavery endured even longer than its American counterpart. Debate among historians and social scientists concerns differences in the practice and aftereffects of slavery in the two societies. Some hold that the Portuguese brand of slavery in Brazil was more humane than the Anglo-American version and consequently produced more benign black-white relations following its abolition.

Ethnic stratification in Brazil features relatively fluid boundaries between groups. Racial classification is not a dichotomy but a continuum in which groups blend into one another. Racial categories are based not on ancestry but on physical appearance and social class. Despite blurred lines of division between groups, there is an ethnic hierarchy in which whites maintain a generally advantageous status over blacks and mulattoes, the latter a large intermediate racial grouping.

Although Brazilians proclaim themselves a society free of prejudice and discrimination based on race and ethnicity, these phenomena are present, though in forms milder than those found in the United States. Blacks are the main objects of this largely subtle ethnic antagonism.

Brazil is most clearly an assimilationist society whose long-range objective is ethnic convergence. And it has, in fact, seemed to progress further toward both physical and cultural assimilation than most other multiethnic societies of the world. Hence, although the historical conditions of race and ethnicity are similar to those of the United States, the outcome of ethnic relations in Brazil and the ideology on which they are founded have been quite different, particularly as they pertain to the relationship between blacks and whites.

Suggested Readings

Andrews, George Reid. 1991. *Blacks &Whites in São Paulo, Brazil, 1888–1988.* Madison: University of Wisconsin Press. Analyzes changes in Brazilian race relations over the past century, looking specifically at individual and institutional forms of prejudice and discrimination and patterns of racial inequality.

Degler, Carl. 1971. *Neither Black nor White: Slavery and Race Relations in Brazil and the United States.* New York: Macmillan. Explains how the emergence of a mulatto, or intermediate, racial category between black and white in Brazil produced a set of race relations markedly different from that in the United States, with its dichotomous system of racial classification.

Fernandes, Florestan. 1971. *The Negro in Brazilian Society.* New York: Atheneum. The major work of one of Brazil's leading sociologists, explaining the development of race relations following the abolition of slavery in 1888. Questions the notion of Brazil as a "racial democracy."

Fontaine, Pierre-Michel (ed.). 1985. *Race, Class, and Power in Brazil.* Los Angeles: Center for Afro-American Studies, UCLA. Essays by social scientists that document the critical role of race in the distribution of jobs, income, education, and other life chances in Brazil.

Freyre, Gilberto. 1963. *New World in the Tropics: The Culture of Modern Brazil.* New York: Vintage. A brief work by Brazil's most celebrated historian, with a valuable discussion of slavery and the evolution of Brazil's racial-ethnic system. Serves as an introduction to Freyre's lengthier and more detailed works, *The*

Mansions and the Shanties (1963) and *The Masters and the Slaves* (New York: Knopf, 1956).

Hellwig, David J. (ed.). 1992. *African-American Reflections on Brazil's Racial Paradise.* Philadelphia: Temple University Press. A unique collection of observations of Brazil by African-American scholars from the early 1900s through the 1980s.

Marx, Anthony W. 1998. *Making Race and Nation: A Comparison of South Africa, the United States, and Brazil.* Cambridge: Cambridge University Press. A sociohistorical study exploring the question of why patterns of race relations and the development of a racial system in Brazil evolved differently from those of the United States and South Africa.

Page, Joseph A. 1995. *The Brazilians.* Reading, Mass.: Addison-Wesley. A lively description of Brazil's culture and of the major racial and ethnic elements of its population.

Ribeiro, Darcy. 2000. *The Brazilian People: The Formation and Meaning of Brazil.* Trans. Gregory Rabassa. Gainesville: University Press of Florida. The work of one of Brazil's best-known intellectuals, describes the building of a multiethnic, multiracial society and the development of its assimilationist character.

Skidmore, Thomas E. 1999. Brazil: *Five Centuries of Change.* New York: Oxford University Press. A brief history of Brazilian society that considers the crucial issues of slavery and its abolition.

15

CANADA: ETHNIC MODEL OF THE FUTURE?

ASK AMERICANS TO NAME the capital of Russia, a society very distant from the United States both geographically and culturally, and most will probably answer correctly. Ask them to name the capital of Canada, a society bordering the United States for the entire length of the continent and culturally almost as close, and most will be hard-pressed. One reason Americans pay so little attention to and know so little about Canada is that its sociological features are assumed to be replicas of their own. In the American mind, Canada might easily pass for the fifty-first state.

Undeniably, the economic and cultural hegemony of the United States on the North American continent for two centuries has created great similarities in the two societies. But the fact that Americans can easily feel at home in Canada tends to disguise some fundamental differences. Common cultural preferences, consumer patterns, political alliances, and for two-thirds of Canada, language, do not make Canadian society simply a northern microcosm of the United States. These differences are nowhere more apparent than in the realm of ethnic relations.

Canada is a society whose ethnic structure is today extremely diverse. Moreover, it displays some features of the corporate pluralistic model described in Chapter 4. This state of affairs stems primarily from the presence of two founding groups, British and French, each of whose language and culture have been legally protected since the eighteenth century. The present language division in the society is shown in Table 15.1. Most of Canada's French-speaking people

Table 15.1 Mother Tongue of the Canadian Population

LANGUAGE	NUMBER	PERCENT
English	16,890,615	60.0
French	6,636,660	23.6
Chinese	715,640	2.5
Italian	484,500	1.7
German	450,140	1.6
Other	2,948,005	10.5
Total	28,125,560	100.0

SOURCE: Statistics Canada, 1996.

(Francophones) live in the province of Quebec, creating for them, in addition to their linguistic and cultural distinction, a territorial base. English-speaking Canadians (Anglophones), by contrast, are dispersed throughout Canada and include groups of various national origins. Those of British origin, however, are recognized as the cofounding group and remain the largest segment of English Canada.

In a sense, then, Canada is not one nation but two. As we saw in Chapter 1, when ethnic groups occupy a definable territory—as do French Canadians—they also commonly maintain or aspire to some degree of political autonomy. Hence, they are "nations within nations." The Canadian experiment in national dualism is still being conducted, and the issues separating the English and French fragments have by no means been resolved. Indeed, the historical discord between these two major cultural groups has in recent decades intensified, making the union of Canada more precarious than ever.

The relations between English and French Canadians constitute the first and most important dimension of ethnicity in Canada, cutting across other lines of ethnic diversity. As such, English-French relations will occupy most of our attention in this chapter. Two other dimensions of ethnicity in Canada, however, are extremely important, even if their manifestations have not been so dramatic. One concerns the various ethnic groups whose origins are neither British nor French, together composing one-third of the Canadian population. Canada is a society of virtually dozens of distinct ethnic groups varying in national origin, culture, and racial features. The important social issues pertaining to these groups involve the nature and extent of their cultural and structural assimilation. Because Canada is at least officially a bilingual society, however, these groups have fallen within the realm of either Anglophones or Francophones and are therefore part of the dual-nation issue as well.

The third ethnic dimension involves the relations of native (Aboriginal) peoples with the rest of Canadian society. Canadian Indians, usually referred to as "First Nations," and Inuit (Eskimos) are, along with their American counterparts, the original peoples of the continent. Today they are a relatively tiny proportion of the Canadian population. Their official status, their cultural systems, and the nature of their relations with other groups, however, largely set them apart from the rest of the society. Hence, they must be looked at as a unique component of the Canadian ethnic configuration.

These three dimensions create a complex ethnic picture in which the different elements are interwoven in complicated and, to American eyes, often bewildering combinations (Figure 15.1). Moreover, this ethnic diversity, particularly in recent decades, has yielded public policies more pluralistic in approach than those of the United States. Whereas the United States has, for most of its history, been committed to an assimilationist approach (at least for those groups

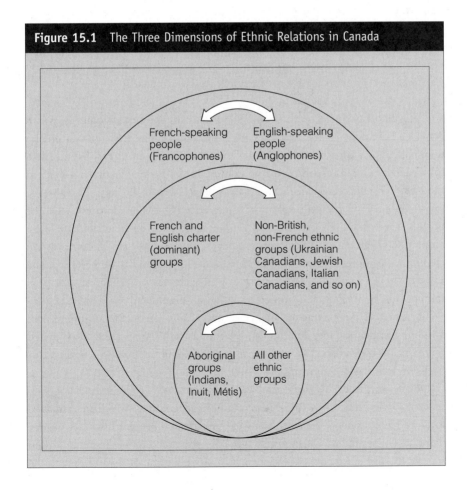

Figure 15.1   The Three Dimensions of Ethnic Relations in Canada

French-speaking people (Francophones)

English-speaking people (Anglophones)

French and English charter (dominant) groups

Non-British, non-French ethnic groups (Ukrainian Canadians, Jewish Canadians, Italian Canadians, and so on)

Aboriginal groups (Indians, Inuit, Métis)

All other ethnic groups

of European origin), the Canadian philosophy has traditionally been more tolerant of the continued expression of cultural differences among diverse groups.

FORMATION OF THE CANADIAN ETHNIC MOSAIC

We can begin to make sense of Canada's ethnic mosaic by looking first at the historical process by which ethnic diversity in this society has taken shape.

The Evolution of Two Nations: The English-French Schism

Most Americans have in recent years become at least vaguely aware of ethnic stirrings in Canada, specifically, the conflict in Quebec concerning French and English speakers. The roots of this conflict, however, reach back over 200 years. Although other ethnic groups would subsequently contribute to Canada's population, the confrontation of English and French groups consumed the affairs of state from the outset and continues to play the preeminent role in internal relations. French-speaking Canadians are about 24 percent of the Canadian population, and as earlier noted, they live for the most part in a single province, Quebec. The historical and contemporary relations between French and English Canadians form the major focus of ethnic conflict in Canada.

The English Conquest Both Britain and France established colonies in North America beginning in the early seventeenth century and vied for continental dominance for almost a century and a half. The victory in 1759 of the British forces led by General Wolfe over the French on the Plains of Abraham marked the end of French colonialism in North America, but it did not eliminate the presence of a French cultural group. New France (Quebec) was now made politically part of the British colonial empire, but its French inhabitants were granted the right to certain cultural privileges, including the retention of French civil law, the use of the French language, and the practice of Catholicism. The price of this cultural autonomy, however, was that French Canada was placed in the position of a permanent minority within an English milieu. From the moment of British ascendancy in Quebec, French Canadians were consumed with avoiding assimilation into the English-speaking North America that surrounded them. And it is this objective of cultural survival that remains at the heart of today's English-French schism.

During and after the American Revolutionary War, thousands of English-speaking emigrants from the thirteen American colonies who had remained loyal to Britain entered Canada. Many of these United Empire Loyalists, as they were called, settled in Quebec. This gave the province for the first time a substantial English-speaking population which quickly came into conflict with the

French majority over economic and political issues. In an attempt to separate the two groups, Britain established an Upper and a Lower Canada, the former composed mostly of English-speaking people, the latter of French-speaking. These would later become the provinces of Ontario and Quebec. The national duality was thus firmly established, with two peoples, differing essentially in language and culture, juxtaposed geographically and socially.

Although the French in Quebec were not threatened in a numerical sense by the influx of English-speaking people, this period of English-French relations marked the establishment of English dominance of the Quebec economy. With the exit of the French commercial elite following the British conquest, business and financial activities became the domain of the English, and the French Canadians remained mostly on the land as subsistence farmers. This cultural division of labor became a fixed arrangement that lasted into the modern era, making the French a numerical majority but an economic minority in their own province.

The disparities in power, wealth, and social position between English and French in Lower Canada led to the emergence of a French nationalist political element that sought the colony's independence. An open rebellion ensued in 1837, after which Britain consolidated Upper and Lower Canada into a single union. The underlying objective of this move was the eventual assimilation of the French into the English colonial society. To bring about this end, discriminatory measures were enacted assuring Anglo dominance.

In addition, with the failure of the rebellion, the conservative forces of French Canada were now ascendant. The clergy, in particular, reinforced its influence in the life of the French-speaking people and extended the doctrine that the French-Canadian culture could be sustained only through loyalty to the Catholic faith and the French language. This traditional nationalism, combining a rural-oriented way of life with a staunch Catholicism, held sway among the French-Canadian masses until the mid-twentieth century (Juteau Lee, 1979). The clergy, then, assumed the dominant institutional role in Quebec, promoting détente with the English colonialists as a means of assuring *la survivance*, the survival of the French culture.[1]

Confederation In 1867, the Canadian provinces were linked in a federal system. The effect of confederation was to further isolate Quebec from the rest of Canada and to heighten the minority status of French Canadians. Although the French and English languages were both protected by law, only in Quebec was the principle of bilingualism instituted, and there it was strictly one-sided— Francophones were forced to use English in dealings with the Anglo Quebecers, but the reverse was not the case. Outside Quebec, Francophones found

---

[1]A classic description of this traditional Quebec culture is found in Miner (1939). More recent essays are contained in Rioux and Martin (1978).

themselves defenseless against the Anglophone majority, which chose to ignore the need for French schools and other institutions. The powerlessness of French Canadians was driven home in numerous instances in which minority interests or wishes were swept aside by the dominance of English Canada.

For French Canadians perhaps the most humiliating aspect of the relationship with English Canada was the minority status to which they were relegated in their own province. English control of Quebec's commerce and finance produced an English-speaking business elite, assuring that higher-status positions would remain the reserve of Anglophones.

The Catholic church's control of education in the province also handicapped the upward mobility of the French-speaking masses. Expounding the philosophy of French-Canadian cultural survival through retreat to church and land, the clergy emphasized the humanities, classics, and religion in schools and colleges—not the commercial and technical skills that were appropriate to an industrial system. Hence, few French Canadians were prepared for skilled positions in business and science. As French Canada increasingly industrialized, Francophones therefore occupied the least-skilled and lowest-ranking jobs (Porter, 1965).

Emergence of a New French Nationalism Industrialization beginning in the early decades of the twentieth century set in motion social processes that would gradually spell the demise of the pastoral, clerical Quebec culture. Urbanization accompanied the growth of industry. Whereas Quebec had been 60 percent rural in 1900, by 1931 it was 63 percent urban (Legendre, 1980). The focus of this rural-urban movement was Montreal, which tripled in size from 1901 to 1921. Industrialization and urbanization conflicted with the traditional Catholic nationalism, which continued to stress agrarian and ecclesiastical values. The customary culture of Quebec society was clearly no longer compatible with the forces of modernization.

The industrialization of Quebec, however, was not sponsored by French-Canadian capital but by outsiders—specifically, English-Canadian, American, and British capitalists. Quebecers thus found themselves in a colonial-like situation. The skilled and more desirable jobs naturally went to the Anglophones, and the French speakers, though a numerical majority, were given the unskilled, menial positions. By the 1960s French Canadians were the most poorly paid workers in Quebec, below even the newly arrived immigrants from Europe, and their standard of living was below that of the average Canadian (Royal Commission on Bilingualism and Biculturalism, 1969a).

The English of Quebec, as members of an insular group with their own institutional structure—schools, businesses, churches, and neighborhoods—held an attitude toward the Francophones of arrogance and even disdain. They made little effort to learn French, for it was unnecessary. On the contrary, because so many jobs were controlled by the Anglophone business class, all the pressure was on the Francophones to learn English. In his classic account of a French-Canadian community of the 1930s, Everett C. Hughes noted that not only was there no need for the English residents of the town to speak French, but to do so in any case would have upset the subordinate-superordinate relationship:

> In fact, the English do not have to learn French to keep their position in industry. The housewife does not have to learn French to keep her housemaid. If they were to speak French in these relationships—except in a joking or patronizing spirit, as is occasionally done—they would be in some measure reversing roles. For they would then be making the greater effort, which generally falls to the subordinate; and they would speak French badly, whereas the subordinate generally speaks English pretty well. (1943:83)

Living in cohesive areas and in control of the most important commercial and financial institutions, Quebec's English community could, by and large, ignore the French around them. It was a situation described by novelist Hugh MacLennan (1945) as "two solitudes," French and English living side by side but essentially in different social worlds.

The Quiet Revolution The decade of the 1960s in Quebec is often referred to as the "Quiet Revolution," for it was then that the transformation of French-Canadian society that had begun in the early part of the century culminated in the emergence of a powerful nationalist ideology espoused by Quebec leaders determined to make Francophones *maîtres chez nous*—"masters in our own house." Though revolutionary only in a figurative sense, the changes accomplished during this period basically redefined both the role of government in Quebec and the identity and goals of French-Canadian society (Posgate, 1978).

The key instrument for change was the state. In the past, the provincial government had been viewed as a force in sustaining the status quo; but it now became the principal vehicle of change. Responsibility for health, welfare, and especially education became the concern of the Quebec state, whereas in the past these had been largely church-sponsored institutions. In the economic realm, the state pursued the objective of improving the position of Francophones and eventually ending Anglophone control of the important segments of the economy (McRoberts, 1988). In short, with the development of a dynamic state led by a new middle class, technologically skilled and ideologically committed to a strong and self-sufficient society, Quebec by 1970 was well on the way toward a social and political transformation.

The Rise of the Parti Québecois The new nationalism spawned strong sentiments favoring the separation of Quebec from the rest of Canada and the formation of an independent French-speaking nation. In the late 1960s, elements of the Francophone leadership that had emerged from the Quiet Revolution began to promote the idea that there was no longer any middle ground between partnership within the Canadian confederation and full sovereignty for Quebec; the latter was the only realistic option if Quebecers were to fully control their destiny. To consolidate and advance the achievements of the Quiet Revolution, Quebec, they felt, had to fully sever its ties with Canada (Posgate, 1978). Led by René Lévesque, they left the ruling Liberal Party (which, during the regime of Jean Lesage, had initiated the radical changes of the Quiet Revolution) and formed the Parti Québecois, declaring as its objective the political liberation of the Francophone community. Recognizing the need to retain an economic link with the rest of Canada, however, Lévesque created the notion of sovereignty-association, which proposed that a politically independent Quebec would retain close economic ties to Canada. Nothing short of the removal of Quebec from its historical place in the Canadian union, however, was the avowed aim of the Parti Québecois. In its view, the split with English Canada was absolute and irreversible. "First and foremost I am a Québecois," declared Lévesque, "and second—with a rather growing doubt—a Canadian" (Saywell, 1977:4).

By the early 1970s the growing support among Quebecers for the new, radically nationalist Parti Québecois was evident and alarming to English Canada. Even more alarming were the activities of a small extremist separatist group, the *Front de Libération du Québec* (FLQ), whose violent tactics and revolutionary ideology provoked a crisis in 1970 prompting the federal prime minister, Pierre Trudeau, to briefly invoke the War Measures Act, giving the police sweeping powers to arrest and detain suspected persons.

This was a wrenching experience for Quebec and all of Canada that in part forced the separatists to moderate their proposals for change. In response, Lévesque stressed the democratic processes by which Quebec's independence was to be attained and emphasized the "economic association" aspect of the envisioned sovereignty-association relationship with Canada. The Parti Québecois declared that if it were elected to power, it would put the issue of sovereignty-association before the people of Quebec as a referendum, to accept or reject.

In 1976, only eight years after its formation, the Parti Québecois won election as the government of Quebec. Its stunning victory sent shock waves throughout English Canada. For the first time, the contemporary Quebec nationalist movement drew the serious attention of even the United States. What had been feared since 1970 had occurred—the election of a provincial government ideologically committed to the separation of Quebec from Canada, led by the charismatic Lévesque. It now seemed that the division of Canada into two separate nations, in a real rather than a symbolic sense, was possible.[2]

Sovereignty-Association: The Recurrent Issue In 1980, the referendum on sovereignty-association that Lévesque had promised was held. Lévesque downplayed the independence aspects of the measure and emphasized that approval of sovereignty-association meant only that Quebec would enter into "a new partnership" with Canada. Moreover, the referendum was worded so that approval was only a first step toward political independence. It would give the Quebec government merely a mandate to negotiate a new arrangement with Canada, approval of which in any case would be left to the people through another referendum.

By a margin of three to two, Quebecers rejected the idea of sovereignty-association. Not unexpectedly, virtually all Anglophones voted no, but a slight majority of Francophones did as well. For the moment, then, the intensity of the Quebec nationalist movement had been quelled. In 1985 the Parti Québecois was defeated and replaced with a government led by Robert Bourassa—the same Bourassa whom, nine years earlier, it had defeated.

A decline in Quebec nationalism was induced as well by the economic changes that had occurred in the province during the previous decade. Business participation by Francophones was encouraged by the government, and managerial positions were increasingly made available to them. With the decline of English economic dominance, a growing number of Quebecers took control of much of the Quebec economy, forming a kind of Francophone economic elite, replacing the traditional Anglophone economic leadership.

The fundamental issues of the relationship between Quebec and the rest of Canada, however, had not been resolved. In the late 1980s, these issues surfaced again resoundingly. The catalyst was an effort to induce Quebec to sign the new Canadian constitution, adopted in 1982. To this end, an accord was reached in 1987 between Quebec and the federal government, giving the province broader independent powers. Most important, under the agreement, Quebec was to be given special recognition among Canadian provinces as a "distinct society." Some in English Canada, acknowledging Quebec's cultural uniqueness, believed that such exceptional political recognition was correct. Others held that Quebec should be accorded no special status. This pact, called the Meech Lake Accord, failed to be ratified by all the Canadian provinces in 1990 and thus was not adopted. This was widely interpreted in Quebec as a rebuff by English Canada, again igniting strong support for the nationalist cause of separation from Canada.

---

[2]Much of the support for the Parti Québecois was based not on its advocacy of separatism but on its promise to deliver "good government" and on public disaffection with the relationship between Quebec and the federal government (McRoberts, 1988).

Two years later, another attempt was made at bringing Quebec into the Canadian constitution. Much like the Meech Lake Accord, the 1992 proposal would have given Quebec special status in Canada as a distinct society and certain numerical guarantees of political and judicial representation in the federal government. In a national referendum, Canadians rejected the proposal by 60 to 40 percent. Temporarily, this seemed to put to rest the issue of Quebec sovereignty, but the forces in Quebec working for independence were given renewed strength and promised to put the issue before Quebecers again.

The opportunity for a revisit to the issue came in 1994 when the Parti Québecois was elected once again. The party leaders had made the sovereignty issue, as before, the core of their electoral program. As in the past, their message was familiar: Quebec would never be afforded the kind of unique place in the Canadian confederation that would allow it to protect and promote its language and culture; thus, political independence was the only logical alternative. To temper the uncertainties of separation, however, Quebecers were promised that after separation, the province would be able to continue to interact closely with the rest of Canada in economic matters such as trade and use of a common currency. Canadian federal leaders warned Quebecers against making such an assumption. The Parti Québecois promised to present the issue to Quebecers soon after the election, and in October 1995—fifteen years after the 1980 proposal was defeated—a referendum on sovereignty was once again held in the province.

As in 1980, the measure was defeated, but this time by the barest of margins—50.6 percent to 49.4 percent. A closer analysis of the results revealed that Francophones in the province had supported the separatist proposal by 60 percent to 40 percent, while Anglophones and Allophones (immigrants speaking neither French nor English) had roundly voted against the measure by over 90 percent. The closeness of the vote and the inability of the federal government to mollify the concerns of those supporting separatism seemed to augur another referendum on Quebec sovereignty in the near future.

Although leaders of the Parti Québecois have vowed to present to Quebecers another opportunity to vote on the issue of sovereignty through another referendum, the current political climate does not seem to favor it. Public opinion polls in 2001 indicated that the issue of sovereignty was a very low priority for Quebecers. Moreover, when asked how they would vote if a referendum were held on the question of separation, only 39 percent said they would vote yes; 62 percent would vote no, the lowest support for sovereignty in twenty years (Mackie, 2001). Quebecers even expressed a stronger sense of Canadian identity. Only 15 percent described themselves solely as Quebecers; most characterized themselves as both Quebecers and Canadians. Also it seems clear that even those who currently favor sovereignty are prepared only to accept a sovereign Quebec that would maintain a close association with Canada (Center for Research and Information on Canada [CRIC], 2002).

The sovereignty issue has for the moment been pushed into the background, but this by no means assures that it will not resurface with great force at some point in the future. Indeed, the intransigence of an element of Quebec society to accept nothing short of political independence, as well as the low-level animosity that seems so often to define the relations between Quebec and English Canada, promise to push the issue forward again eventually. As one observer has put it, the issue of Quebec separation may simply remain "a recurrent component of the Canadian agenda" (Banting, 1992:161).

Language Although Quebec remains, perhaps tenuously, part of Canada, the policies prompted by the ongoing separatist movement have created certain basic changes in the political and social status of French Canadians within Quebec and within the society as a whole. In one way or another, all of these policies revolve around the issue of language. In Quebec itself, the most far-reaching measures prompted by the movement have involved the primary use of French in all spheres of public life—including, most important, business and education. Because language continues to evoke such strong emotions on both sides and so thoroughly encompasses all facets of division and hostility between English-speaking Canadians and Quebecers, let's briefly review the issue as it has evolved in the past twenty-five years.

Language is the very foundation of any people's culture. Hence, it was long felt that assimilation into the dominant English-Canadian society would be inevitable for French Canadians if measures were not taken to assure the preservation of the French language. Moreover, language for the French Canadians, as for many ethnic collectivities in multiethnic societies, is the key symbolic marker setting them off from other groups. Quite simply, that they speak French is what most essentially defines the uniqueness of the French Canadians. As Lévesque put it, "Everything else depends on this one essential element and follows from it or leads us infallibly back to it" (1968:14). It would not be overstating the case to assert that the question of language was at the core of all historical divisions between English and French Canada from the time of the British conquest. It is no less so today.

In the 1970s, aggressive pro-French language policies were enacted by successive Quebec governments. These policies must be seen in the context of *la survivance*, the survival of the French-Canadian culture. French Canadians in the 1970s were faced with an overarching demographic problem: How could they remain a numerical majority in their own province? This problem was created by two population trends—a lowered birthrate and the influx of non-French-speaking immigrants. Quebec had historically maintained the highest birthrate in Canada. Thus, even though few immigrants from France were attracted to Quebec after the colonial period, the ratio of French to English in Canada held relatively constant for two centuries. This was derisively referred to as the "revenge of the cradle," the implication being that the complete English conquest was foiled by the high French birthrate. Quebec's extraordinarily high birthrate, however, dropped dramatically in the 1960s, becoming the lowest in all of Canada. In addition to this precipitous fall, the arrival of large numbers of European immigrants, most adopting English as their new language, threatened to further dilute the Francophone population.

In the late 1960s, the federal government under Trudeau began to promote the notion of a truly bilingual Canada. Responding to the recommendations of the Royal Commission on Bilingualism and Biculturalism, which had been charged with studying how a more equitable balance could be established between the two founding peoples, the government adopted and avidly supported a policy of official bilingualism. But this effort was seen as essentially meaningless to Quebecers, who in many cases were already bilingual—of necessity. Even today, it is only in Quebec where bilingualism is characteristic of a large percentage (over one-third) of the population. Although 17 percent of the total Canadian population is able to converse in both English and French, over half of these bilinguals live in Quebec (Statistics Canada, 1997). Moreover, it was the French language that served as the very basis of a distinct French-Canadian society and, as Dale Thomson notes, "the equality of French and English was incompatible with that goal" (1995:76).

The first move in the direction of French-language preeminence in Quebec came in 1974, when the Liberal provincial government of Robert Bourassa instituted a new language law, declaring French the official language of the province and making access to any but French-language schools difficult for those whose native tongue was not English. This created strong opposition from Montreal's English business community as well as the city's large number of European immigrants, who sought to maintain their right to choose the language of instruction for their children. Neither was the measure acceptable to the separatists, who felt that it did not go far enough in establishing the primacy of the French language. In 1977, therefore, Lévesque's government enacted a far more sweeping and radical language law, Bill 101, which mandated that French would be the prime, if not sole, language used not only in official matters but also in commerce and industry. Limitations on the rights of parents to educate their children in English were also made more stringent.

The aim of the new language law was to alter the structure of opportunity in favor of Francophones, who it was felt had been kept in a subordinate position by the Anglophone business elite. Legendre (1980) explains that the use of the language issue as the spearhead for Francophone control of the economy reflected a basic difference in philosophy between the Parti Québecois leadership of the 1970s and the leadership of the Quiet Revolution in the 1960s. Whereas the latter saw the past failure of Francophones to improve their economic status as the result of an outdated and ill-focused educational system as well as the continued adherence to traditional values, the new leaders placed the blame more fundamentally on the structure of Quebec's economy, which fostered institutional discrimination against Francophones. Requiring the use of French in business and industry, it was felt, might attack the Anglophone domination of economic institutions directly. No longer could managerial positions be denied Quebecers on the basis of language.

The response of the English-dominated business world of Montreal was indignation. Rather than acquiescing to the stringent new language regulations, many large corporations chose to move their offices to Toronto and other Canadian cities or at least to limit their Montreal operations. This proved economically costly to the province.

Even more controversial was the effect of the language law on Quebec's educational system. Restrictions on instruction in English prompted an out-migration of Anglophones. But the law's primary target, many believed, was the large European immigrant population in Montreal whose mother tongue was neither French nor English (Arnopoulos and Clift, 1980; Levine, 1990). Most of these groups had ordinarily chosen English-language schools for their children, believing that in the North American context, fluency in English would afford them greater occupational and status opportunities. Under the new law, however, almost all were now required to send their children to French schools. Because most immigrants as well as the English community in Quebec resided in Montreal, the conflict centered primarily in that city.

Montreal continues to be the focal point of the language issue (Thomson, 1995). However, the context of the issue has been fundamentally changed. Despite periodic controversies, clearly, the primacy of the French language in Quebec is no longer disputed.

French-Canadian Ethnic Identity French Canadians are not simply one more ethnic group within the Canadian mosaic. The French in Canada are a people with a linguistic and cultural autonomy formally recognized from the very founding of the society. Moreover, unlike other Canadian ethnic groups, French Canadians maintain a territorial base. For Canada's French-speaking people who live outside the province of Quebec, the tendencies toward assimilation into the dominant English-speaking society are very evident (Joy, 1972). More important, the contemporary nationalist movement in Quebec has jettisoned these French-speaking non-Quebecers. It is Quebec alone that has become, in the eyes of the separatists, the only meaningful base for a sovereign French-speaking society in North America.

Manifestations of ethnic identity among Quebecers also illustrate their uniqueness within the Canadian ethnic system. Unlike English Canadians, French Canadians—specifically, those in Quebec—are faced with a dual national identity. Whereas those outside Quebec think of themselves simply as "Canadians," French Canadians are both Canadiens and Québecois (Quebecers).[3] In recent decades, the latter identity has, for most, seemed to take precedence (CRIC, 2002). This dual identity, however, demonstrates well the idea of a unique French-Canadian nation in which people's allegiance and consciousness of kind do not focus necessarily on the same national unit perceived by other Canadians (Brunet, 1969). One must keep in mind, of course, that there are different degrees of nationalism among Quebecers. Some are vehemently and radically nationalistic and are unrelenting advocates of separatism, whereas others see themselves more as part of a unified Canada.

There is great irony in the desire today of many Quebecers for full independence from the Canadian federation, for there is no longer any question of the dominance of the French language in Quebec. Moreover, never before has Quebec achieved such great economic development, and at no other time have Quebecers played such a strong role in the federal government. The crux of the problem, however, is a view of Canada—and thus Quebec's place within it—that is very different from the view of English-speaking Canadians. The latter conceive of Canada as a nation of individual citizens who are equal before the law, regardless of their ancestors' language or ethnic origin, and who live in a federation of provinces with equal constitutional status. Quebecers, on the other hand, view Canada as a nation of collectivities defined primarily by language, specifically English and French (Bercuson, 1995). Because they are heavily outnumbered by English-speaking Canadians, most Quebecers feel that they are entitled to special status and not subject to the same institutions that are dominated by Anglophones. This leads to the view on the part of many that only independence can assure the survival of the French language and culture.

The consequences of the dual nationhood of Canada are not simply problems of identity and cultural maintenance. As we have already seen, there is a historical relationship in Canada between language and culture on the one hand and class, status, and power on the other. Indeed, it is precisely this relationship that in the past made French Canadians a minority group. We will look later at these patterns of ethnic stratification.

Canada's Other Ethnic Groups: The Third Force

Although the division between English and French Canadians overshadows all other ethnic issues, Canada today is far more ethnically diverse than is reflected

---

[3]The inhabitants of the original French colony in Canada called themselves *Canadiens*. Only later was the term *Canadian* adopted by the English. Hence, *Canadien* connotes, in the French-Canadian view, the idea of the original or true Canadians (see Brunet, 1969).

in this schism. Indeed, Canada is one of the most heterogeneous multiethnic societies in the world, and the noncharter groups stand out increasingly in the Canadian ethnic picture. As already noted, they affect the English-French schism on the issue of language; but perhaps more important, they comprise one-third of the total Canadian population and thus play an enormous role of their own in all aspects of ethnic relations and policy. They constitute an extremely diverse spectrum, including groups from all parts of Europe, East and Southeast Asia, the Indian subcontinent, the Middle East, the Caribbean, Africa, and Latin America (Table 15.2). Americans, too, are among these groups, but they are hardly distinguishable except in a formal sense.

It is obvious that the size of these groups as well as their cultural variety have basically altered Canada's ethnic composition. To speak any longer of "English Canada" is a misnomer. Those who see themselves as British in origin continue to decline, and those whose ethnic origins are neither British nor French continue to comprise a greater and greater share of Canada's population. Today Canada is very much an immigrant society, particularly in its largest cities—Toronto, Vancouver, and Montreal. The changed ethnic profile of Toronto is especially dramatic. Immigrants today account for over 40 percent of the city's population, a higher percentage than any city in North America. It is not simply numbers, however, that demonstrate the profound changes occurring, but also the sources of immigrants to Toronto. Immigrants arriving before 1961 were almost exclusively people born in the United Kingdom and other European countries, like Italy and Germany. Since the 1960s, the number of European-born immigrants in Toronto has steadily decreased . By the late 1990s, those born in non-European countries made up six out of ten of all immigrants living in the metropolitan area; the Asian-born alone were over 36 percent of immigrants (Statistics Canada, 2001b).

The shaping of Canada's ethnic diversity can be seen historically as a process encompassing two major periods of immigration—the influx of peoples during the late nineteenth and early twentieth centuries, and the post–World War II stream, which continues today. Each period differed in the origins of the major immigrant groups and in the groups' settlement patterns in Canada.

The Period of Western Settlement Although non-British and non-French groups had been present in Canada before the late nineteenth century, they were not large enough to make a significant impact on the society's ethnic composition. In 1871, of a total population of 3.7 million, the largest of these groups were the Germans, who numbered 200,000; others were considerably smaller (Hawkins, 1972). Throughout the nineteenth century, the British—English, Scots, Welsh, and Irish—dominated immigration and reinforced the Anglo-Canadian preeminence (exclusive of Quebec, of course), both numerically and culturally.

Table 15.2 Top 25 Ethnic Origins in Canada, Showing Single and Multiple Responses, 1996[a]

ETHNIC ORIGIN	TOTAL RESPONSES	SINGLE RESPONSES	MULTIPLE RESPONSES
1 Canadian[b]	8,806,275	5,326,995	3,479,285
2 English	6,832,095	2,048,275	4,783,820
3 French	5,597,845	2,665,250	2,932,595
4 Scottish	4,260,840	642,970	3,617,870
5 Irish	3,767,610	504,030	3,263,580
6 German	2,757,140	726,145	2,030,990
7 Italian	1,207,475	729,455	478,025
8 Aboriginal origins	1,101,955	477,630	624,330
9 Ukrainian	1,026,475	331,680	694,790
10 Chinese	921,585	800,470	121,115
11 Dutch (Netherlands)	916,215	313,880	602,335
12 Polish	786,735	265,930	520,805
13 South Asian origins	723,345	590,145	133,200
14 Jewish	351,705	195,810	155,900
15 Norwegian	346,310	47,805	298,500
16 Welsh	338,905	27,915	310,990
17 Portuguese	335,110	252,640	82,470
18 Swedish	278,975	31,200	247,775
19 Russian	272,335	46,885	225,450
20 Hungarian (Magyar)	250,525	94,185	156,340
20 Filipino	242,880	198,420	44,460
22 American	211,790	22,085	189,705
23 Spanish	204,360	72,470	131,895
24 Greek	203,345	144,940	58,405
25 Jamaican	188,770	128,570	60,200

SOURCE: Statistics Canada, 1996.

[a]Data in this table are based on a 20 percent sample. Respondents who reported one ethnic origin are included in the single-response column. Respondents who reported more than one ethnic origin are included in the multiple-response column for each ethnic group they reported. For example, a respondent who reported "English and Scottish" is included in the multiple-response count for "English" and the multiple-response count for "Scottish." The total-response column indicates the number of respondents who reported each ethnic origin.

[b]Respondents were given the option of writing in their ethnic origin(s). Many, therefore, chose to identify themselves simply as "Canadian."

In the latter years of the century, spurred by the promotion of western settlement, large numbers of European immigrants were attracted to the prairie provinces of Saskatchewan, Manitoba, and Alberta. Germans and, to a lesser extent, Dutch were the major immigrant groups until the early 1900s, when people from central and eastern Europe, especially Ukrainians, entered in large numbers. Americans, eager to acquire lands that were no longer available after the closing of the U.S. frontier, were also a significant immigrant group during this time (Troper, 1972).

Although immigrants did settle in Canada's cities, the objectives of nineteenth-century immigration policy were mainly to attract people who would develop the agricultural potential of the vast western domain. This did not mean an open-door policy, however. British peoples, or at least those from northwestern Europe, were deemed most desirable. Ukrainians and others from eastern and central Europe, though technically not among these, were recruited nonetheless because their agricultural background was considered more critical than their national origin. What the West needed above all were farmers; therefore, even nonconforming religious sects like Hutterites, Mennonites, and Doukhobors were promised parcels of land to settle, often as entire communities. Expressly rejected, however, were non-Europeans.

Post–World War II Immigration After World War II, the next great phase of immigration began, considerably changing the Canadian ethnic composition. Both their societies of origin as well as their Canadian destinations distinguished the new immigrant groups from those of previous eras.

Whereas British, northwestern European, and, selectively, central and eastern European groups had been the major immigrants of the past, the new arrivals were mainly from southern and eastern European countries, in addition to the continued large-scale entry of the British. The largest among the non-British groups were the Italians, almost a half million of whom had come by 1971 (Iacovetta, 1992).

The contemporary period of immigration also introduced an entirely new dimension to Canada's ethnic mosaic because it included for the first time nonwhite peoples in significant numbers. Although not officially stated, Canada's immigration policy before 1962 had been, in effect, "white only" (Hawkins, 1989; Richmond, 1976). Discriminatory measures favoring northwestern Europeans, particularly those from the British Isles, were now dropped, profoundly affecting the makeup of immigration. Whereas blacks and Asians before 1967 had constituted fewer than 4 percent of immigrants, by 1973 they were approximately one-third, and by 1981, one-half. As shown in Table 15.3, a broader immigration into Canada developed in the post–World War II era. Whereas European countries were the major source of immigrants in 1961, by

1971 East and South Asian countries were among the leading sources, a trend that was even more pronounced in the 1980s and 1990s.

Asian immigrants in the contemporary period have come from almost every country of the region, with especially large numbers of Chinese (from Hong Kong, Taiwan, and the mainland), Filipinos, and Vietnamese, as well as large numbers of Indians, Pakistanis, Sri Lankans, Lebanese, and Iranians. Blacks have come mostly from the Caribbean—especially Jamaica, Trinidad and Tobago, Guyana, and Haiti—and have settled mainly in Toronto and Montreal (McGill Consortium, 1997). Before this period, most of Canada's tiny black population was of American origin. Most Canadian blacks during the colonial period had come with the American loyalists during the Revolutionary War or during the War of 1812, settling mainly in Nova Scotia and New Brunswick (Clairmont and Magill, 1974; Winks, 1971). Many American slaves also sought refuge in Canada before the Civil War via the Underground Railroad.

In addition to their diversity, the immigrants of the post–World War II era differed from those of previous eras in that their destinations were almost entirely urban; specifically, Canada's three major metropolises of Toronto, Montreal, and Vancouver. More than a third of all the new immigrants entered Toronto alone, making that city the core of Canada's new ethnic character. What had been a staid, conservative, and relatively provincial British-dominated city was transformed into a cultural and linguistic panorama not unlike American cities of the East and Midwest at the turn of the century.

Table 15.3 The Top Ten Countries of Canadian Immigrants

IMMIGRATED BEFORE 1961	RECENT IMMIGRANTS (2000)
United Kingdom	China
Italy	India
Germany	Pakistan
Netherlands	Sri Lanka
Poland	Philippines
United States	Korea
Hungary	Iran
Ukraine	United Arab Emirates
Greece	Russia
China	Jamaica

SOURCES: Employment and Immigration Canada, 1988; Citizenship and Immigration Canada, 2000.

As we will see, Canada's groups that are neither British nor French in origin have established themselves as an important political and social element in their own right, a so-called Third Force. Their growing numbers and diversity more than anything else have contributed to the current philosophy and government policy of multiculturalism, which is discussed later in this chapter.

Visible Minorities Nonwhite groups in Canada are referred to as visible minorities. The immigration of the past thirty years has moved Canada much further along these lines of ethnic development than at any time in the past. Excluding Aboriginal groups, the visible minority population makes up over 11 percent of the total Canadian population. As shown in Table 15.4, the largest groups are Chinese, South Asians (mostly from India), and blacks (mostly from the West Indies). It is expected that by 2016, visible minorities will make up 20 percent of the total Canadian population (Statistics Canada, 1996).

Aboriginal Peoples

The third major element in the Canadian ethnic picture is the native, or Aboriginal, groups, who together make up about 4 percent of the total population. Among these peoples are Native Indians (or First Nations), Inuit, and Métis, the last of mixed racial origins (Table 15.5).

Culture and physical features no longer define Native Indians in Canada; rather, "Indianness" is legally defined (Frideres, 1998). Canadian Indians are

Table 15.4 Visible Minority Population		
	NUMBER	PERCENT
Total Canadian population	28,528,125	
Total visible minority population	3,197,480	100.0
Chinese	860,150	26.9
South Asian	670,590	21.0
Black	573,860	17.9
Arab/West Asian	244,665	7.7
Filipino	234,195	7.3
Latin American	176,975	5.5
Southeast Asian	172,765	5.4
Japanese	68,135	2.1
Korean	64,840	2.0
Multiple visible minority	61,575	1.9
Visible minority not included elsewhere	69,745	2.2

SOURCE: Statistics Canada, 2001A.

Table 15.5 Canadian Aboriginal Population

	SINGLE AND MULTIPLE ORIGINS
Indian (First Nations)	554,290
Métis	210,190
Inuit	41,080
TOTAL	799,010

SOURCE: Statistics Canada, 1996.

officially designated as "status" or "nonstatus." Status (or "registered") Indians, of whom there are about 600,000, have been classified as Indians under the Canadian Indian Act and are the direct responsibility of the federal government. About two-thirds live on reserves (the counterpart of reservations in the United States) established by the government. As in the United States, the Canadian First Nations are ethnically diverse and have long seen themselves as distinct peoples. It is their common relationship with the federal government, however, that links them together (Gibbins, 1997).

The Métis are marginal to both white and Native Indian societies. During the eighteenth and early nineteenth centuries, fur traders, mostly French, often lived with or married Indian women. The offspring of these mixed unions developed a unique culture, partly European and partly Indian. (*Métis* is a French word similar in meaning to the Spanish mestizo, "mixed.") In western Canada, where the intermixture occurred in a geographically and socially isolated environment, the Métis grew in number and developed a distinct ethnic identity, even declaring themselves a nation. Following an abortive rebellion in 1885, however, they became a highly stigmatized group. The Métis were recognized officially as a distinct people until 1940, when the Canadian government reversed this position. Today most Métis live in the Prairie provinces of Canada and a majority live in urban areas (Normand, 1996).

Canada's Inuit have remained relatively isolated geographically in the far North and historically have not been subject to colonization to the extent that Native Indians have. Thus, although they do qualify for special political status, technically they are not Indians.

Natives and Whites: Changing Relations Except for being markedly less violent, the historical relations between Canada's native peoples and white settlers have been, tragically, not essentially different from those that evolved in the United States. The technological superiority of the whites and their desire for land spawned policies and practices that led to deculturation, dependency, and impoverishment among the Aboriginal populace. The history of native-white relations in Canada can be divided into pre- and postconfederation eras.

Accommodation characterized the white attitude toward native peoples from the time of initial contact until 1867. The indigenous peoples were vital to the fur trade, and as enterprises like the Hudson's Bay Company expanded their operations, Indians and Inuit were engaged in trapping. As they were drawn into this trade, they became increasingly dependent on the same companies for their subsistence; in the process, their traditional cultures were basically overturned. During this period, however, there was little direct intent by whites to impose their cultural ways on the native peoples. Their prime concern was profit maximization. As Breton et al. have put it, "The traders and whalers wanted productive trappers and hunters, not North American versions of themselves" (1980:73). White domination, then, was real though not necessarily deliberate.

With confederation and the western expansion, this general attitude of accommodation changed to one of domination. The federal government entered into treaties with the native peoples for the acquisition of occupied lands, and Indians were increasingly relegated to reserves, isolated from the mainstream society. The government subsequently assumed the role of patron, regulating various aspects of Indian life. Moreover, the reserves were increasingly incapable of supporting hunting and fishing economies, forcing Indians into an even greater dependence on the state.

Concurrent with government jurisdiction of reserve lands were efforts by missionary schools to assimilate natives into either the English or French culture, resulting in the denigration of indigenous cultures. Thus, while they were being more and more isolated from white society, they were expected to assimilate into the white culture (Breton et al., 1980). Indeed, until the 1970s, the goal of assimilation, though largely unsuccessful, had been the foundation of government policies toward native peoples since the nineteenth century (Gibbins, 1997).

In recent years, the native peoples of Canada have developed a resurgent group consciousness and pride and have become more politically mobilized in movements that seem to parallel those among Indians in the United States, with whom they share generally common origins and cultures. Viewing themselves as a colonized minority in Canada, they have sought the renegotiation of land claims, and in recent years, policies have been instituted that have led to greater native control over their reserves (Asch, 1984; Frideres, 1990; Ponting, 1997a). In a sense, the political movement among Aboriginal peoples in Canada has displayed many basic similarities to the Quebec nationalist movement. Each has sought greater freedom from the control of the federal government, and each has espoused an assertive and sometimes militant nationalism.

Several events occurred in the early 1990s that significantly altered the relations between Native Canadians and non-natives. All involved land claims. In

1989, in the small Quebec town of Oka, about twenty miles west of Montreal, the town council announced plans to expand its golf course to eighteen holes. To do that, however, they needed part of a disputed land area that included a Mohawk cemetery. In March 1990, Mohawks resisted by occupying the area, and after Quebec provincial police moved on the occupying group, one policeman lay dead. The siege then spread as Mohawks blocked the Mercier Bridge, part of a freeway into Montreal used as a major commuter artery. Before a settlement was reached, the occupation by Mohawks lasted several months and involved the eventual deployment of 4,000 Canadian troops and hundreds of Quebec provincial police. The confrontation provoked strong protest across Canada not only among Indians but also among many prominent Canadians, who condemned the government's actions in the affair and its failure to negotiate with the Mohawks (Wright, 1992). This was an agonizing affair for Indian-white relations in Canada and sparked a public awareness of the land claims and political status of Native Indians. It also contributed strongly to the growing sense of solidarity among Indians not only in Canada but across North America. A series of confrontations between Native Indians and government authorities occurred again in the summer of 1995 across Canada, over issues ranging from fishing rights to land claims.

An important event in 1993 moved some of those claims to actualization. The federal government signed an agreement with Inuit leaders that provided for the creation of a new territory called Nunavut ("our land" in the Inuit language), to be carved out of the Northwest Territories. Under the terms of the agreement, beginning in 1999, Inuit were given effective control of 135,000 square miles (one-fifth of Canada's landmass) and the equivalent of about $1 billion over a fourteen-year period (Légaré, 1997; Stout, 1997). Although the creation of Nunavut had no bearing on the movement for self-government among Native Indians of southern Canada, it nonetheless marked an important step in giving back control of land to indigenous Canadians.

Two events occurred in the late 1990s that augur additional significant changes in the relations between Aboriginal peoples and the rest of Canada. First, the Canadian government in 1998 offered an historic apology to native peoples for 150 years of paternalistic and sometimes racist treatment. This represented an important symbolic gesture. More substantive, however, was an earlier Supreme Court ruling that validated Indian claims to land in British Columbia. The decision was based on oral histories presented by the Gitxsan tribe, not on written treaties. This ruling will likely have a profound effect on future treaty negotiations between First Nations and the federal and provincial governments and on decisions regarding land use and rights to timber and mining resources (DePalma, 1998).

Despite these positive signs in the relations between Aboriginal peoples and the rest of Canada, recent national surveys indicate a growing proportion of

Canadians who believe that those relations are deteriorating. Moreover, in much of the country there is a declining view of the validity of Aboriginal land claims. For example, outside of Quebec, only a third of Canadians feel that Aboriginal peoples should enjoy special hunting and fishing privileges in their traditional living areas (CRIC, 2002).

ETHNIC STRATIFICATION

Because Canada is one of the most diverse multiethnic societies of the modern world, it would be foolish to expect it not to exhibit a system of ethnic stratification. One looking superficially at Canadian society, however, might be hard put to perceive such a system. Looking at the ethnic composition of the political elite, for example, it is obvious that Quebecers commonly serve in the highest-ranking positions of the federal government, including prime minister. Or, looking at ethnic patterns of housing, it is clear that neighborhoods thoroughly segregated on the basis of race and ethnicity, so common to American cities, are not part of the Canadian urban scene. But to see these cases as proof of the absence of dominant-minority relations is misguided, for there is indeed an ethnic hierarchy in Canada. In its basic outline, it is not unlike the ethnic stratification system of the United States: an Anglo-Canadian group dominant, other white ethnic groups and Asians in intermediate positions, and blacks and Aboriginals at the bottom.

The Vertical Mosaic

In a study that has become a classic of modern Canadian sociology, John Porter (1965) described the system of ethnic stratification in Canada as a "vertical mosaic." As he explained it, a strong relationship was evident between ethnicity and various measures of social class. The general pattern that emerged from his analysis was a three-part structure made up of British and French charter groups, the former at the top of the income and occupational hierarchies, followed by later-arriving European ethnic groups and, finally, native peoples. The notion of a vertical mosaic suggested that Canadian society was not only ethnically differentiated but was ethnically stratified as well. As his work proceeded, Porter explained, "The hierarchical relationship between Canada's many cultural groups became a recurring theme in class and power" (1965:xiii).

The Royal Commission on Bilingualism and Biculturalism in 1969 reaffirmed Porter's basic findings regarding the class hierarchy and revealed the concentration of particular ethnic groups in certain occupational areas. British, Jewish, and Asian groups were overrepresented in the managerial, professional, and technical occupations. Ukrainian, Scandinavian, Dutch, German, and Russian groups were overrepresented among farmers (a reflection of these groups' significant presence in the agricultural provinces of Manitoba, Saskatchewan, and Alberta). And Italians, French, Polish, and Hungarians were noticeably over-represented among blue-collar workers (Porter, 1985; Royal Commission on Bilingualism and Biculturalism, 1969b).

The English-French Difference Perhaps the most glaring difference in class position, historically, was between the two charter groups—English and French. Whereas the English were traditionally at the top of the hierarchy, the French were no higher than most noncharter groups and actually lower than some.

Surprisingly, French Canadians traditionally occupied a lower collective economic position in Quebec, where they had always been a numerical majority, than in Canada as a whole. Porter found that in Quebec, "by and large the British run the industrial life" (1965:92). As occupational status rose, he noted, so did the proportion of English personnel. The tendency for French Canadians to fill the lower-level working-class positions in the Quebec economy had been vividly demonstrated over two decades earlier by Everett Hughes in his study of a Quebec town. In the town's major industries, Hughes found, the English held "all positions of great authority and perform[ed] all functions requiring advanced technical training" (1943:46). The French, by contrast, predominated in the lower occupational ranks and eventually disappeared in number as the degree of authority rose. "French Canadians as a group," noted Hughes, "do not enjoy that full confidence of industrial directors and executives that would admit them easily to the inner and higher circles of the fraternity—and fraternity it is—of men who run industry" (1943:53). Hughes concluded that the same situation prevailed throughout the province generally.

The English-French variance shown by Hughes and Porter was later corroborated by the Royal Commission on Bilingualism and Biculturalism. It was shown, for example, that in Quebec, over 30 percent of men of British origin were part of the professional and managerial occupational categories, compared with less than 15 percent of men of French origin.[4] This disparity was also apparent in wages. Canadians of French origin were shown to earn incomes that were on the average 20 percent lower than those of English Canadians. In Quebec, their incomes were 35 percent lower (Royal Commission on Bilingualism and Biculturalism, 1969a).

To see the historically lower place of the French Canadians in Quebec's economy and that of Canada as a whole as a product simply of discrimination by a powerful Anglo elite, however, is a gross oversimplification. The factors that created and sustained the lower occupational levels and incomes of French Canadians are complex and subject to much debate. Were they the result of the traditional

---

[4]Even as late as the early 1980s, only 26 of 165 enterprises in Quebec with an annual production in excess of $10 million were French-Canadian owned (McRoberts, 1988).

clerical emphasis on nontechnical and nonadministrative skills that permeated French-Canadian education until the 1960s? Were they the result of Francophone geographic concentration, making it difficult for French Canadians to move into English-speaking areas and to compete with English speakers? Was achievement motivation among the English and other ethnic groups different from that of the French Canadians? Did the basic power differential not permit the French to acquire the necessary capital to challenge English control of the economy? Elements of each of these explanations may be involved (Blishen, 1970; Fenwick, 1982; Levine, 1990; Porter, 1985; Richer and Laporte, 1979; Royal Commission on Bilingualism and Biculturalism, 1969a).

In the last three decades, these patterns have changed dramatically. The income discrepancies between Anglophones and Francophones have narrowed significantly, and occupational differences have declined. From an extensive study of mobility in Canadian society, Pineo and Porter (1985) concluded that there is little meaningful difference in the degree of equality of opportunity enjoyed by Francophones and Anglophones throughout Canada. But in Quebec specifically, the study found that Anglophones continue to be over-represented in higher-ranked occupations. Even this situation appears to be changing as more Francophones assume professional, technical, and managerial positions in all sectors of Quebec's economy (Levine, 1990). Particularly striking changes have occurred at the elite level. As French has become the primary language of business and commerce, a new Francophone economic elite has emerged, making Anglophone dominance largely a phenomenon of the past (Forcese, 1997). "There is now a genuine Francophone capitalist class headquartered in Montreal," notes Marc Levine, "strongly influencing not only the Quebec economy, but also emerging as powerful actors in the entire North American economic system" (1990:194). Most significant, the state-run hydroelectric power industry, Quebec's most important, is thoroughly Francophone at all managerial levels.

Ethnicity and Class in Canada Today Although the historically consistent pattern of English-Canadian dominance of the higher occupational ranks and the most important decision-making positions of Canada's major institutions remains apparent, it is not the dominance of the past. The requirements of a postindustrial economy have created a more ethnically varied workforce at all occupational levels. Thus, even though collective differences among the various ethnic groups remain evident, ethnic inequality in Canada is declining (Brym, 1989; Darroch 1979; Isajiw et al., 1993; Lautard and Guppy, 1990; Pineo and Porter, 1985). "The relationship between ethnic origin and class position is in flux," concludes one study, "and no ethnic group unequivocally dominates the Canadian class structure" (Nakhaie, 1995:187). Some, in fact, maintain that intergroup differences among white ethnic groups are now virtually nonexistent (Breton, 1989).

The large influx of non-British immigrants since World War II has provided further pressures to afford upward mobility and a greater share of power to minority Canadians. Some, like Jews and Asians, have experienced relatively rapid and substantial mobility into high occupational positions and incomes. Others, like southern and eastern European groups, have not exhibited such significant upward movement, though their overall occupational distribution has improved and become more like the national average (Breton et al., 1980; Brym, 1989; Nakhaie, 1995; Pineo and Porter, 1985). As with earlier European immigrants, the visible minority immigrant groups of recent years will likely experience upward social mobility at different rates (Breton, 1998).

Even at the elite level, the dominance of Anglo-Canadians appears to be eroding with the advent of a more ethnically diverse institutional leadership. In the past, non-British groups had been severely underrepresented in the power elites of major Canadian institutions (Clement, 1975; Kelner, 1970; Lautard and Guppy, 1990; Newman, 1975, 1979; Presthus, 1973). Today, elites in all major institutions—political, economic, educational, labor—are increasingly ethnically varied (Ogmundson and Fatels, 1994; Ogmundson and McLaughlin, 1992).[5]

Aboriginal groups merit a special note. Although their circumstances are much improved from earlier generations, they continue to occupy the lowest rung on all dimensions of social class and consequently have the poorest life chances of all Canadians. They exhibit the highest unemployment rates in the country, have the lowest levels of income and education, and disproportionately occupy low-level jobs. Unemployment among Aboriginals, for example, is three times higher than the national average (Frideres, 1998). The standard of living on Indian reserves is characterized by poor and overcrowded housing, extremely low incomes, and low participation in the labor force (Ponting, 1997b; Siggner, 1986). For Indians who migrate to the cities, a social pattern has developed that is similar to that of American urban Indians—high unemployment, high criminality, and high alcoholism (Frideres, 1993; Statistics Canada, 1993).

The social consequences of these conditions are predictable. Life expectancy rates are considerably lower, and infant mortality rates, though declining, remain higher among Canadian Indians than among the general Canadian populace (Barsh, 1994; Frideres, 1998; Norris, 1990). Standards of housing, education, nutrition, and general health remain below the national averages (Barsh, 1994; Blue, 1985; Frideres, 1998; Kendall, 2001; Siggner, 1986; Statistics Canada, 1993). Moreover, Aboriginal people make up a highly disproportionate number of inmates in Canadian prisons. Although less than 4 percent of Canada's population, they account for about 12 percent of male and 17 percent of female con-

---

[5]The class and ethnic origins of Canadian elites, however, remain a contested issue among social scientists. See Ogmundson (1990) and Clement (1990).

Part III: Ethnic Relations in Comparative Perspective

victs in federal penitentiaries. In provincial prisons, the figures are even higher. In Saskatchewan, where Aboriginals account for 13 percent of the population, they make up almost 70 percent of prisoners (Stackhouse, 2001).

Despite these chilling facts, in both income and education, Indians have clearly raised their status relative to the total Canadian population. In 1971, for example, only 3 percent of Indians had attained some postsecondary education; by 1991, almost 34 percent had done so (Ponting, 1997b; Siggner, 1986). In sum, the life chances of native peoples in Canada have improved, but they continue to represent a markedly deprived segment of the society and, in many ways, are divorced from mainstream institutions.

PREJUDICE AND DISCRIMINATION

When comparing their own country with their American neighbor in the conduct and character of ethnic relations, Canadians usually see themselves in a favorable light. The prevalent view is that their treatment of and attitudes toward minority ethnic groups have been and remain more tolerant and equitable. This popular Canadian perspective is in some regards valid. Though incidents of racial and ethnic violence are scattered throughout Canadian history, they have never reached the magnitude, frequency, or intensity of ethnic violence in the United States. Moreover, as we will see in the following section, an ethnic ideology more pluralistic in content has traditionally been proclaimed in which the society's various cultural groups have not been forced into a monolithic "Canadian" mold. That there have in essence been two Canadian nations, English and French, no doubt played an important role in establishing and sustaining this seemingly greater tolerance of ethnic diversity.

Despite its more subdued character, however, ethnic conflict in Canada has been revealed in patterns that are in some ways parallel to those of the United States. Like most other Western societies, Canada does not lack a racist tradition, though it has been more muted in expression and less malignant in consequence. Historically, Canadian racism has been evidenced in its immigration policies and in its treatment, both official and unofficial, of nonwhites. Indeed, prejudice and discrimination have traditionally been aimed primarily at the visible minorities and only secondarily at white ethnic minorities. It is those groups who have borne the brunt of negative attitudes as well as individual and institutional forms of discrimination.

Also, in its ethnic attitudes and actions, Canada at times takes its cue from the United States. For example, following the terrorist attacks on the World Trade Center on September 11, 2001, a number of anti-Muslim incidents, some of them violent, occurred in Canada, as they did in the United States. And on the issue of immigration, after the attacks, over 83 percent of the population held the view that Canadian immigration should be made more restrictive (Leger Marketing, 2001).

Immigration and Racism

Historian Howard Palmer has written that "the more one scratches the surface of the period up to 1920, the more difficult it becomes to differentiate between the immigration histories of Canada and the United States" (1976:499). Official government policies regarding who and how many would be admitted and what place they would take in the occupational structure, as well as nativist sentiments and attitudes toward immigration in general, were all basically similar.

One important aspect of the early history of immigration in both Canada and the United States was the decidedly racist character of the selection process. Like the dominant Anglo group in the United States, "English Canadians," notes Evelyn Kallen, "have, from the beginning, exercised control of federal immigration policies responsible for determining which ethnic groups would be allowed into Canada, where they would settle, what jobs they could assume, and what ranking and social position would be accorded them within the existing system of ethnic stratification" (1995:131). Those deemed most suitable for Canada were thus, not surprisingly, most like the English Canadians, culturally and physically.

After 1870, when large-scale immigration to Canada began, the preference for British, or at least northwestern European, immigrants was an outspoken national policy. The pretext for this selectivity was the matter of assimilation. Those from Britain and culturally similar societies, it was asserted, were more easily absorbed into the mainstream Canadian society. Others were not culturally or, as some believed, biologically fit.

That prejudice and discrimination against southern and eastern European groups are no longer significant should not imply that they have no heritage in Canada. In fact, Italians, Jews, and other such groups—especially those who immigrated into Canadian cities at the turn of the century—met with a brand of nativism very similar to the U.S. variety (Avery, 1979; Harney and Troper, 1977). But it was nonwhites, specifically Asians, who were least welcome and against whom the most blatantly racist policies and actions were directed.

The Chinese were the major target group. They had been permitted entry into Canada as workers in the construction of the Canadian Pacific Railway. Once this project was completed, however, they became a surplus labor force, seen by white workers and organized labor as an economic threat. During the late nineteenth and early twentieth centuries, discriminatory policies were enacted to discourage their entry, and in 1923 an exclusionary act totally barred their immigration (Elliott, 1979). Other restrictive measures severely limited entry into Canada of those deemed "unassimilable," understood to mean nonwhites. In 1911, for example, blacks seeking to emigrate from the United States were rejected because they were presumably unable to adapt to Canada's harsh winters (Palmer, 1976).

Discriminatory immigration policies restricting Asians and other nonwhites were not basically changed until the 1960s. Indeed, as recently as 1947, Prime Minister Mackenzie King stated bluntly Canada's intention to encourage population growth through selective immigration. Large-scale Asian immigration, he stated, "would change the fundamental composition of the Canadian population"; therefore, no changes in immigration regulations were to be made (Corbett, 1957:36).

Seeking to enhance its international image and to more realistically meet its need for human resources, Canada abolished its racially discriminatory immigration policy in legislative acts of 1962 and 1967. The selection of immigrants was no longer to be based on nationality or race but on a system of points that objectively evaluated each immigrant's potential economic and social contribution to Canadian society (Hawkins, 1989). As in the United States after it discarded restrictive immigration quotas in 1965, the ethnic origins of immigrants entering Canada changed radically—most were now non-European.

Ethnic Attitudes

As we have seen, Canada in the last three decades has become a society far more ethnically diverse than at any time in its history, with many of the newcomers among the visible minorities. This has prompted attitudes and actions that, though mild by American standards, indicate some degree of racism. Such patterns, of course, are not of recent vintage. In addition to the virulent anti-Asian feelings of an earlier period, whites' generally abusive interrelations with native groups as well as episodes of anti-Semitism have resembled American historical patterns (Henry and Tator, 1985). Studies of social distance, however, indicate that Canadian racial attitudes remain somewhat less intense than those of Americans.

In their national attitudinal study, Berry, Kalin, and Taylor (1977) found that respondents in general reacted very favorably to English and French Canadians but less favorably to non-British and non-French groups. Specifically, northwestern Europeans were judged most favorably, central and southern European groups next, and nonwhite groups least favorably, except for Japanese. This social distance scale has been demonstrated by other studies as well (Mackie, 1980; Pineo, 1987). It is interesting to observe its essential similarity to the ethnic scale that, as noted in Chapter 3, has been consistently reaffirmed in the United States, though the differences among groups are declining in both societies (Reitz and Breton, 1994).

Despite this clear acknowledgment among Canadians of an ethnic hierarchy, Berry and his colleagues found no evidence of extreme ethnic prejudice. Though the rank order of groups was very apparent, the differences among them were not exceedingly great. Race (that is, physical differences among groups) was found to be an important dimension of group perception among Canadians, but the researchers concluded that "Canadians reject explicit racism" (1977:206). Later studies reaffirmed this conclusion (Berry and Kalin, 2000; Kalin and Berry, 1994). Bibby's surveys (1995) indicate increasing ethnic tolerance on several measures, including intermarriage and perceptions of visible minorities. Acceptance of marriage between whites and blacks, for example, rose from 57 percent in 1975 to 81 percent in 1995.

Despite the evident lower level of racism among Canadians, at least by U.S. standards, some suggest that Canadian prejudice, especially toward nonwhites, may simply be more subtle and covert and may reveal itself when carefully probed (Frideres, 1976; Kallen, 1995).

Discriminatory Actions

Although the overt suppression of minority peoples is generally lacking in Canadian history, there are nonetheless some parallels between Canada and the United States in the treatment of ethnic minorities, particularly nonwhites. In Canada, these racial-ethnic groups have suffered various forms of discrimination including, at one time or another, restrictions in voting, employment, land ownership, housing, and public accommodations (Davis and Krauter, 1971; Kallen, 1995).

Arrant anti-Asian immigration measures were noted earlier. In addition to this official discrimination, violence against Chinese and Japanese workers in British Columbia was common in the late nineteenth and early twentieth centuries. The most blatant discriminatory action against Asians, however, occurred during World War II. As in the United States, those of Japanese origin were forcibly removed from their homes and businesses and placed in internment camps for the duration of the war (Adachi, 1976).

Serious racist actions have also been directed at people of East Indian origin. Along with the Chinese and Japanese, East Indians were barred from entry into Canada earlier in the century, and those remaining were subjected to discrimination in almost all areas of social life. East Indians in the past were seen as intrinsically dirty, sinister, immoral, prone to overcrowding, and generally inferior to whites. Buchignani suggests that East Indians in Canada before World War II were "an almost ideal type subordinate racial caste" (1980:129). They were denied entry into various occupational fields and could not vote or hold citizenship. Today there are almost 700,000 East Indians and other South Asians in Canada, and, while serious discrimination against them rarely surfaces, well-worn negative stereotypes do at times come to light (Berry, Kalin, and Taylor, 1977; Henry, 1978; Henry et al., 1995; Ramcharan, 1982; Statistics Canada, 1999).[6]

Blacks have never made up a significant element of Canada's ethnic population; today they are about 2 percent (Statistics Canada, 1999). Although their circumstances have in some ways conformed to those of American blacks, negative actions and attitudes toward blacks in Canada have never equaled in scope and intensity U.S. patterns. A study of urban housing, for example, revealed that blacks in Canadian cities do not experience similar levels of residential segregation, as is so common in the United States (Fong, 1996). As with other ethnic minorities, of course, immigration restrictions and discrimination in schools, housing, and public accommodations were once common features of the Canadian black experience (Walker, 1980). Today, however, the official and obtrusive forms of discrimination of the past have given way to milder and more subtle forms that are not easily verified (Bolaria and Li, 1988; Henry, 1994; Henry et al., 1995; Hill, 1977; Thompson and Weinfeld, 1995).

The 1990s produced a number of incidents that may augur more serious problems for Canada's visible minorities. These have been limited primarily to Toronto, Montreal, and Vancouver, where, because of immigration, nonwhites constitute an increasing presence. In Montreal, with a growing black population made up primarily of immigrants from the Caribbean, especially Haiti and Jamaica, charges of police brutality against blacks surfaced several times, and other racially inspired incidents arose with some frequency (Farnsworth, 1995). Police relations with blacks were a source of considerable tension in Toronto, and in Vancouver, where most immigrants have been Asian, a backlash against newcomers from Hong Kong was evident (Farnsworth, 1991; Katz, 1992; McGraw, 1990). The frequency of hate crimes, directed mostly at visible minorities and Jews, has also increased in recent years (Corelli, 1995; Wood, 1998).

Such episodes prompt serious consternation among those who have viewed ethnic discrimination as uncharacteristic of Canadians. In 1974 Hughes and Kallen explained that "Canadians, consciously or unconsciously, attempt to boost their feelings of self- and national identity by contraposing their 'peaceful and just' society to that of their 'violent and racist' neighbour to the south" (1974:213). A national survey taken twenty years later indicated the continuing strength of that attitude (Dwyer, 1994). Whether there are fundamental differences between Canadian and American beliefs and actions toward nonwhite peoples, however, remains a debated issue. Reitz and Breton (1994) suggest that when black-white relations are put aside, the differences between Canada and

---

[6]A number of physical attacks against East Indians occurred in the 1970s in some of Canada's largest cities (Ferrante, 1977; Hill, 1977; Pitman, 1977).

the United States regarding racial and ethnic tolerance are not great. They cite the cases of the historical treatment of native peoples as well as the treatment of Japanese Canadians during World War II, which they see as fairly similar in the two societies.[7]

STABILITY AND CHANGE

As a multiethnic society, Canada in some ways faces even greater problems in seeking societal unification among its various ethnic groups than does the United States. But the attempts to solve these problems do not necessarily resemble those of the United States, because Canada is more pluralistic in its ethnic ideology. The ethnic end product envisioned by Canadians is not a duplicate of that envisioned by Americans. Whereas the United States can be categorized as more assimilationist than pluralistic in orientation (more so, of course, for white ethnic groups than for nonwhites), Canada displays many elements of a corporate pluralistic society.

Melting Pot versus Mosaic

It has often been noted that one of the key factors differentiating the Canadian and American ethnic systems is how the two societies regard relations among diverse groups and the eventual absorption of these groups into the larger society. Canada has commonly been observed as a society in which ethnic group differences are tolerated more than in the United States. The popular phrases of comparison are *melting pot*, supposedly characteristic of the United States, and *mosaic*, supposedly characteristic of Canada.

The Melting Pot In the United States, as we saw in Chapter 5, the idea of the melting pot—the fusing of many immigrant groups into an American hybrid culture—became popular beginning in the early 1900s. Never, however, was it translated into public policy to any serious extent. In reality, the expectation was always that immigrants would conform to the dominant Anglo-Protestant culture. Even in recent years, the assumption that new groups will quickly learn English and generally adopt dominant norms and values has continued to guide social thinking despite increased pluralistic rhetoric.

The Mosaic The Canadian ideology, in contrast, has historically favored a more pluralistic outcome of the massing of various ethnic groups. There have been and

---

[7]It is of note that a national poll conducted in 1992 indicated a public perception of growing racial intolerance in Canada (Bozinoff and MacIntosh, 1992).

remain greater awareness and tolerance of ethnic separateness. A simplified view is "unity in diversity." Canada, in this ideal view, is a mosaic, the various pieces fitting together within a common political and economic framework. What has accounted for this mosaic ideology, and how valid is it in light of the reality of Canadian ethnic relations?

To begin with, the dual national character of Canada has made ideas of ethnic assimilation problematic. Given the historical fact of two founding groups, neither the melting pot nor the Anglo-conformity model could have the same meaning in the Canadian context as in the American. The question is, how can ethnic groups assimilate into the dominant culture and society when there are two dominant groups? The idea of "Canadianizing" people becomes an empty notion when there is no uniform "Canadian way of life" to serve as a societal reference point. As Hiller writes, "If one group had dominated, there would have been more accord about the specific nature of the dominant culture; but since the British and French were in conflict themselves, the society had a greater built-in tolerance for the perpetuation of ethnic identities" (1976:107–8).

Although it is also true that one would be hard put to clearly define an "American way of life," there is nonetheless no essential ambiguity with regard to language and other major elements of American culture. In Canada, however, the presence of two founding peoples with distinctly different cultural systems has allowed ethnic groups that entered the society after the British and French to parlay this basic schism into significant freedom to retain their culture and group structure.

Some have suggested that one explanation for Canada's greater tolerance of ethnic pluralism lies in the fact that Canada has historically retained a more aristocratic and particularistic social system, in contrast to the equalitarian and universalistic system of the United States (Clark, 1950; Lipset, 1968, 1990). A traditionally greater emphasis on hierarchy and status has restrained pressures to melt down group differences. The necessity for ethnic groups in the United States to "become like others" has therefore not found a strong counterpart in Canada.

Finally, the impact of immigrant groups on Canadian society has in a way been even more profound than their impact in the United States, making assimilation extremely challenging in any case. Though not as numerous in an absolute sense, immigrants in Canada have been a considerably higher percentage of the population than immigrants in the United States. In the early years of this century, the foreign-born in some Canadian provinces outnumbered the native-born by two to one (McKenna, 1969). Since World War II as well, the relative impact of large-scale immigration on Canada has been greater than on the United States. Today, whereas the foreign-born in the United States are about 10 percent of the total population, in Canada they are 16 percent.

In comparing the ethnic ideologies of Canada and the United States, it is important to recognize that just as the American melting pot has been more myth than reality, so, too, the Canadian ethnic mosaic is not a true reflection of public attitude and policy toward ethnic differences. For both societies, the reality of ethnic relations lies somewhere between these two ideals. As Porter points out, "In practice neither [ideal] has been practicable and neither has been particularly valued by the respective societies despite the rhetoric in prose and poetry that has been devoted to it" (1979:144). Just as Anglo-conformity was the dominant public policy with respect to ethnic groups in the United States after its emergence as a heterogeneous society in the late nineteenth century, so, too, in Canada the first period of heavy immigration at the turn of the century produced an ethnic policy that encouraged assimilation into the dominant Anglo-Canadian group. This basic policy did not change for the next three decades (Burnet, 1976, 1981; Palmer, 1976). Although immigrants entered a society in which assimilation could not be enforced as strongly as it was in the United States, there was never any question that Anglo-conformity was, except in Quebec, the guiding force of government policy toward the newcomers.

In a comparison of ethnicity in the two societies, Reitz and Breton (1994) suggest that the differences between Canada and the United States regarding assimilation, at least as it pertains to European-origin groups, are today not as great as has been commonly assumed. "The fact is," they note, "in both Canada and the United States, the public discourse on immigration reflects both a tolerance for diversity and a bias toward assimilation" (1994:10). Moreover, their analysis suggests there is no evidence to support the notion that ethnic minorities in Canada retain their ethnic identity and culture longer or more strongly than American ethnic minorities. In Canada, as in the United States, ethnicity declines in significance with each passing generation, taking on a largely symbolic function.

Multiculturalism in the Modern Era

The post–World War II changes in Canadian society, especially the influx of hundreds of thousands of immigrants from various nations, created the setting for a more robust ethnic mosaic than had existed at any previous time. In the 1960s, Canada experienced a flurry of ethnic activity and an upsurge in ethnic consciousness, much as did the United States and many other Western societies.

The chief catalyst of these tendencies was the Quiet Revolution in Quebec and the subsequent rise of Quebec nationalism. The appointment of the Royal Commission on Bilingualism and Biculturalism to study ways of reconciling the age-old but newly freshened rift between the two Canadas, English and French, reflected the acknowledgment of a society faced with a powerful and potentially destructive ethnic conflict. The commission supported ways of strengthening the bilingual and bicultural character of the society and of equalizing the place of English and French cultures.

But the non-British and non-French groups, the other major element in the Canadian ethnic amalgam, were not content with the commission's recommendation that Canada was and should remain a bicultural society. The special cultural and language privileges of the two founding groups were, they believed, being furthered in disregard of the historical and contemporary role of other Canadian ethnic groups. Also being ignored was that these groups now constituted almost one-third of the total population. In their view, they had been relegated to the status of second-class citizens. Hence, this so-called Third Force demanded not a *bi*cultural but a *multi*cultural Canada, in which the retention of a unique ethnic culture would be recognized and supported not only for the French but also for all other ethnic groups.

The Royal Commission subsequently added to its report a volume dealing with Canada's non-British and non-French groups (Royal Commission on Bilingualism and Biculturalism, 1969b). In it, these groups' contribution to the Canadian mosaic was clearly acknowledged. The commission recommended that ethnic groups need not surrender their cultures to either Anglo or French Canada, but on the contrary should be encouraged, if they wished, to maintain them strongly (Burnet, 1976). In response, the federal government of Prime Minister Trudeau announced in 1971 the policy of "multiculturalism within a bilingual framework." This appeared to be a compromise measure to satisfy the French concerns about language retention and the other ethnic groups' concerns about their cultural status. One important aspect of the policy was the awarding of grants to ethnic groups to help them preserve their cultures. An even stronger commitment to the policy was made in 1988 when Canada became the world's first country to enact a national multiculturalism law (Fleras and Elliott, 1992). Multiculturalism, then, was firmly established as a public doctrine and set of public policies that recognize, protect, and encourage the retention of ethnic cultures within the larger Canadian society (Elliott and Fleras, 1990; Hawkins, 1989).

Multiculturalism, however, has not been met with overwhelming enthusiasm. Many have seen it as merely part of a political strategy designed to gain the support of ethnic minorities. There also is concern that it is moving Canada away from its traditional open, individualistic political system toward one in which, as one observer has described it, "the cake has to be sliced very carefully among powerful and competing groups" (Hawkins, 1989:216). Moreover, the notion of multiculturalism within a bilingual framework has seemed to satisfy neither the non-English, non-French ethnic minorities nor the Quebec nationalists. The former argue that without protection and support of language rights (that is, multilingualism), multiculturalism is rendered meaningless because group culture cannot be sustained without the ethnic language (as the Quebecers have so effectively argued). The French in Quebec, on the other hand, argue that multiculturalism at the national level reduces them from charter-group status to simply one more ethnic group in Canada (Rocher, 1976).[8]

There is general agreement that multiculturalism as a government policy has been at least symbolically beneficial in that Canada now officially celebrates its ethnic diversity and encourages the expression of unique ethnic cultures. Minority ethnic groups have established themselves as part of the Canadian identity and are now recognized in the school curriculum and in national patriotic events. "Tolerance towards diversity and the acceptance of pluralism," notes political commentator Richard Gwyn, "have become the defining characteristic of the country and its citizens" (1995:203). But in the final analysis, multiculturalism touches only the most superficial aspects of ethnic relations in Canada and does not speak to the more significant issues of ethnic inequality (the vertical mosaic) and of prejudice and discrimination toward the visible minorities.

Moreover, whether succeeding generations of minority ethnic groups in Canada can (or will want to) resist the forces of assimilation into the dominant Anglo culture seems doubtful. Some have pointed out that immigrants and their children are usually determined to take advantage of the opportunities of their new society rather than preserve the same kind of life they left in their societies of origin. For them, as Freda Hawkins has claimed, "the preservation of cultural heritage is a lesser concern and . . . the whole concept of multiculturalism can be confusing" (1989:217). Here the experience of white ethnic groups in the United States may be compared. Although at times tortuous, a high level of assimilation, both cultural and structural, has been the fate of most groups by the third generation. This seems to be the pattern followed as well by Canadian groups (Reitz and Breton, 1994). Moreover, assimilation expectations seem solidly part of the view of most Canadians. In 1975, 85 percent of Canadians agreed with the statement that "immigrants to Canada have an obligation to learn Canadian ways." Twenty years later the percentage agreeing was 88. Furthermore, this view was no less evident among those who had come to Canada since the 1960s, that is, immigrants themselves (Bibby, 1995).

Finally, not all see the aims of multiculturalism as socially beneficial. Some view it as divisive in effect and stultifying as well for minority ethnic groups (Porter, 1975, 1979). Sustaining and enhancing ethnic pluralism, they maintain, can only hinder the movement of these groups into mainstream institutions, thus perpetuating the system of ethnic stratification. Moreover, as writer Neil Bissoondath has pointed out, rather than instilling cultural understanding among ethnic groups, the policy of multiculturalism may actually have the opposite effect by reinforcing and perpetuating stereotypes. "Multiculturalism," he writes,

---

[8]In recent years, however, Quebec, in light of its growing minority ethnic population, has adopted its own policy of multiculturalism, recognizing the unique needs of these ethnic subcultures while trying to integrate them into Quebec institutions (Levine, 1990).

"with all its festivals and celebrations, has done—and can do—nothing to foster a factual and clear-minded vision of our neighbours. Depending on stereotype, ensuring that ethnic groups will preserve their distinctiveness in a gentle and insidious form of cultural apartheid, multiculturalism has done little more than lead an already divided country down the path to further social divisiveness" (Bissoondath, 1994:89–90).

It is instructive to consider public opinion on this issue. In comparing mosaic and melting pot models, in 1985, 56 percent of Canadians said they preferred the mosaic, only 28 percent the melting pot. Ten years later, the preference for the mosaic had declined to 44 percent while the melting pot had risen to 40 percent (Bibby, 1995). The Canadian commitment to multiculturalism, then, is hardly firm and unequivocal.

The Future of Canadian Ethnic Relations

Canada is a North American experiment in ethnic pluralism. But the experiment is still very much in the testing stage, and its outcome is uncertain. In considering the future course of ethnic relations in Canada, we might return to the theme of the three ethnic dimensions denoted at this chapter's outset.

The overriding ethnic issue of Canadian society remains, of course, the French-English schism. This is the most encompassing dimension of ethnicity in Canada and continues to present the most vexing problems of intergroup relations. Whether Quebec will retain its place in the Canadian union or eventually go the way of independence is a question that at the beginning of the twenty-first century remains unresolved. More than any other, this ethnic conflict will dominate internal politics in Canada in the coming years and, in a real sense, will define the future of Canada as a nation-state. Whether a politically and economically centralized society can sustain what are in essence two nations is a question that has been addressed in other corporate pluralistic societies of territorially based ethnic groups divided along the lines of language and culture. Judging from those cases, a continuing separatist movement of some scope can be expected as long as Francophone Quebec remains part of Canada.

The roots of English-French discord in Canada are, as we have seen, historically deep and abiding. But this is a conflict that, compared with the ethnic strife in the former Yugoslavia or even black-white relations in the United States, is subdued. Whatever the nature of its resolution, it is not likely to degenerate into warfare or to erupt periodically in violence. The democratic context of Canada makes ethnic conflict amenable to political solution.

The second major issue concerns the place of the non-British, non-French groups in the Canadian ethnic order. Will multiculturalism become more than an appeasement to these groups and eventually translate into a real alteration of the ethnic hierarchy—affording them greater and eventually proportional power, privilege, and prestige? As we have seen, this appears to be unmistakably the trend. Whether these groups will be able, or will choose, to preserve their ethnic cultures in a more corporate pluralistic society, however, is questionable. In any case, it is now clear that these groups have fundamentally altered the ethnic composition and flavor of Canada. As journalist Peter Newman has written, the members of Canada's two founding groups are no longer in the ascendancy, which means that "a new and radically different country has been created" (1995:34).

Another ethnic issue that will absorb the attention of Canadians in the next decade involves the racial and ethnic characteristics of its new immigrants—mostly non-European and nonwhite. This, as we have seen, is a critical ethnic issue in the United States as well. Although it is clear that Canada, with a very low birthrate, must continue to attract new immigrants in order to maintain its population, there is a growing restiveness over the shift in immigration patterns. Already the change in the racial and ethnic makeup of new immigrants has called into question public support for the official policy of multiculturalism. A 1989 national poll indicated that though 63 percent of Canadians supported the policy, 61 percent also felt that immigrants should change their distinct culture in order to "blend with the larger society." This response was even higher than the 51 percent of American respondents who, in a parallel poll, felt the same about new immigrants to the United States (*Maclean's*, 1989). A later national survey (Edwards and Hughes, 1995) revealed that nearly half of the Canadian public felt that Canada should accept fewer immigrants. In 2001, in the wake of the September 11 terrorist attack on the United States, 83 percent of Canadians thought that Canada "should be stricter when it comes to immigration" (Leger 2001).[9] As in the United States, attitudes toward immigration in Canada tend to vacillate with changing economic conditions. When the economy is strong and unemployment low, people tend to see immigration in a more positive light; when the economy is in recession, immigrants are perceived as a labor threat and public views take a negative turn (CRIC, 2002; Esses et al., 1999).

Finally, the place of Aboriginal peoples has yet to be clearly determined. Whether they will participate with other Canadians as equals in all aspects of citizenship or whether they will be granted greater autonomy is still an unfolding issue.

Canada will hold the attention of students of ethnic relations in the coming decades, for it may provide an answer to the question of how much diversity can be accommodated by a society that remains a centralized nation-state. Factors irrespective of ethnicity have hindered the development of a definitive Canadian national identity. Great regional differences as well as the constant specter of the United States, with its enormous economic and cultural influences, continue to

---

[9]The same percentage, however, wanted Canada to remain open to people of Arab descent or Muslim faith.

Comparing Canada and the United States

Points of Similarity
- The ethnic development of both societies began with a process of migrant super-ordination, in which an indigenous population was overcome by a colonizing force. Following conquest, native populations in both societies were reduced to a state of dependence on the dominant group.
- Both societies were populated by successive immigrant waves. Current immigration in both societies is comparable in its societal impact. Both societies are among the major destinations of global immigration, and most immigrants are non-European, creating, in the process, a more heterogeneous society.
- Both societies are today among the most ethnically diverse in the world.

Points of Difference
- Canada is a two-nation society, with two distinct cultural groups having charter-group status. In effect, there are two dominant ethnic groups in Canada.
- Canada has adopted a more pluralistic approach to the absorption of various ethnic groups into its population, though in fact progressive assimilation of various ethnic groups has generally paralleled U.S. patterns.
- Canada did not abide slavery and thus did not experience the travails of integrating a former slave population into its mainstream.
- Race and ethnic relations in Canada have historically been more benign and ethnic tolerance higher than in the United States.

divide Canadian society and lead to divergent visions of the country's future. But above all, it is issues of ethnic diversity that today Canada must address in assuring its national survival.

SUMMARY

Canada is geographically and culturally close to the United States, but its ethnic structure is decidedly different in several respects. In its ethnic makeup, Canada is a dual-nation society, French and English, within which additional ethnic groups have taken their place. There are three dimensions of ethnicity in Canada: the French-English bifurcation; other ethnic groups, neither French nor English in origin; and Aboriginal peoples.

The overriding issue of ethnic relations is the relentless conflict between French and English Canadians. The division between the two derives from the French colonial defeat by the British in the eighteenth century. Conflict has revolved around the efforts of French Canadians to retain their language and culture within the context of a surrounding English majority. The rupture intensified in the 1970s with the emergence of a separatist movement in Quebec, where most French Canadians live, and remains wide today.

A particularly large and varied cluster of non-British, non-French ethnic groups from virtually all nations of the world immigrated to Canada beginning in the late 1950s, changing the very ethnic character of the society. Today, these groups make up over one-third of the Canadian population.

Aboriginal peoples—Indians, Inuit, and Métis—are a numerically tiny element of the society.

Ethnic stratification in Canada was, in the past, a vertical mosaic in which those of British origin disproportionately occupied the higher occupations and the most important decision-making positions of major institutions. Today French Canadians and, increasingly, those of neither British nor French origin have narrowed those discrepancies considerably and are playing a more prominent role in Canadian economic and political affairs.

Certain patterns of prejudice and discrimination in Canada, although milder in form and substance than those expressed in the United States, are evident in past immigration policies and in the treatment of nonwhite peoples.

Canada has maintained a greater tolerance toward ethnic diversity than has the United States and can be placed among those societies we have called corporate pluralistic. Rather than a melting pot, Canada historically has advanced a mosaic ethnic ideal, according to which ethnic minorities are not expected to fully adopt the cultural ways of the dominant group. In past eras this philosophy was mostly disregarded in favor of Anglo-conformity; today, with Canada's growing ethnic diversity, the mosaic has become the society's official policy, though questions regarding its full acceptability by the Canadian public remain evident.

Suggested Readings

Bissoondath, Neil. 1994. *Selling Illusions: The Cult of Multiculturalism in Canada*. Toronto: Penguin. The author, himself a first-generation immigrant, argues that the Canadian policy of multiculturalism is fundamentally flawed, leading only to greater societal divisiveness and the perpetuation of ethnic stereotypes.

Fleras, Augie, and Jean Leonard Elliott. 1992. *Multiculturalism in Canada: The Challenge of Diversity*. Toronto: Nelson Canada. Explains the various dimensions of multiculturalism, how it has been implemented in policy, and how Canadians have responded to it.

Fournier, Marcel, Michael Rosenberg, and Deena White (eds.). 1997. *Quebec Society: Critical Issues*. Scarborough, Ontario: Prentice Hall Canada. An anthology that explores various aspects of contemporary Quebec, including the issues of language and sovereignty.

Gwyn, Richard. 1995. *Nationalism without Walls: The Unbearable Lightness of Being Canadian*. Toronto: McClelland & Stewart. One of its most noted political commentators addresses Canada's ability to forge a national identity from its increasingly diverse population and its future as a multiethnic society.

Levine, Marc V. 1990. *The Reconquest of Montreal: Language Policy and Social Change in a Bilingual City*. Philadelphia: Temple University Press. A very thorough examination of the critical role of language in sustaining Quebec nationalism and the French-English schism.

McRoberts, Kenneth. 1988. *Quebec: Social Change and Political Crisis*, 3d ed. Toronto: McClelland and Stewart. A comprehensive analysis of the economic and political transformation of Quebec society, addressing the overriding issue of political sovereignty.

Reitz, Jeffrey G., and Raymond Breton. 1994. *The Illusion of Difference: Realities of Ethnicity in Canada and the United States*. Toronto: C. D. Howe Institute. Based on survey data, the authors suggest that the ethnic systems of the two societies, on such issues as ethnic tolerance, acceptance of assimilation, and the economic incorporation of immigrants, are not so different as social scientists have assumed.

Resnick, Philip. 1994. *Thinking English Canada*. Toronto: Stoddart. Argues that English Canada (that is, Canada exclusive of Quebecers and Aboriginal peoples) constitutes a coherent sociological nation, the distinguishable features of which are described.

Thomas, David M. (ed.). 2000. *Canada and the United States: Differences that Count*. 2d ed. Peterborough, Ontario: Broadview Press. A collection of essays comparing U.S. and Canadian social institutions, policies, and ideologies, including those that revolve around ethnicity and ethnic relations.

Tulchinsky, Gerald (ed.). 1994. *Immigration in Canada: Historical Perspectives*. Toronto: Copp Clark Longman. A collection of essays by historians, describing the arrival and adaptation of various groups that have contributed to the Canadian mosaic.

16

NORTHERN IRELAND: ENDURING ETHNIC CONFLICT?

As is the case with other ethnic conflicts that erupt in periodic violence, Americans are faintly aware of the ethnic strife that has plagued Northern Ireland for more than three decades. Their understanding of the strife derives primarily from descriptions by the news media. Media narratives of ethnic relations in Northern Ireland, however, give a simplistic and somewhat distorted view, for they are sensational and largely superficial. Indeed, little information about this society ever reaches the American public (or, for that matter, the rest of the world) other than accounts of a grisly shooting or a horrifying bombing. Such accounts create the impression of a society in perpetual turmoil. In reality, although strife is certainly closer to the surface than it is in most other multiethnic societies, it is not ceaseless and does not affect all people in the same manner or degree. Of greater importance than body counts or gruesome descriptions of the physical and psychological damage wrought by the recurrent terrorism of both sides are the historical antecedents and the current structural conditions that underlie the ethnic chasm between Protestants and Catholics. These are the constant features of life in Northern Ireland that need to be examined to gain a more thorough understanding of the ethnic relations of this small but highly polarized society.

Although the protagonists in Northern Ireland's tragic conflict are Protestants and Catholics, it must be understood at the outset that the struggle is not a religious one per se. Rather, it is an ethnic conflict in which religious identities mark off the boundaries of the two major ethnic groups. Although religion certainly impinges on group relations, matters of doctrine and dogma are not the basic issues. Regarding the role of religion in the conflict, Harold Jackson and Anne McHardy write that it is simply "the handiest identifying mark available to the two sides" (1984:6). Or, as John McGarry and Brendan O'Leary have explained, religion in Northern Ireland "is best seen as an ethno-national marker rather than as an important independent motivator of violent conflict" (1995b:852).

Political scientist Richard Rose (1971) has referred to Northern Ireland as a "biconfessional society" because almost all people identify themselves as either Protestant or Catholic. Those claiming to be atheists, it is joked, must say whether they are Protestant atheists or Catholic atheists. These identities, however, do not mean the same as they do in the United States or other Western nations. In those societies, *Protestant* and *Catholic* denote denominational affiliations that no longer carry with them significant social consequences. In Northern Ireland, these are ethnic categories that determine in a fundamental manner one's place in the society and the nature of one's relations with others. Religious affiliation is quite simply the most important social characteristic, taking precedence over one's social class and even one's sex (Harris, 1972). In some measure, it is not unlike the significance of racial identity for persons in the United States or South Africa. The relationship of religious affiliation to almost every other aspect of social life—occupation, education, politics, residence, leisure activity—makes Northern Ireland a case of ethnic pluralism in the extreme. Not only do separate institutional structures exist for each group, but each also maintains basically different perspectives on how and by whom the society should be governed.

The essence of the ethnic conflict in Northern Ireland is this: In 1920, six counties of the Irish province of Ulster, which today make up Northern Ireland, were adamant in retaining their tie to Britain following Ireland's successful movement for independence. Protestants outnumbered Catholics in this area, unlike the remainder of Ireland, and were resolved to remain apart from a sovereign, Catholic-dominated society. As a result, Ireland was partitioned into two states, the Irish Free State (later the Republic of Ireland), with a Catholic majority, and Northern Ireland, with a Protestant majority. Unfortunately, this left a substantial Catholic minority in the six counties.[1] The dispute since that time has concerned this artificially created boundary. Most Protestants seek to maintain the boundary as it is drawn and to remain linked with Britain, whereas most Catholics do not recognize the permanence of the dividing line and visualize an eventually united Ireland.

---

[1] The province of Ulster originally comprised nine counties, but with partition it was reduced to six to assure a more comfortable Protestant majority.

Today Protestants are a numerical majority within Northern Ireland by about 10 percent, but in a united Ireland they would be outnumbered by about three and a half to one. (Ninety-five percent of the Republic's 3 million people are Catholic.) Hence, the root of the Protestants' resistance to anything that threatens the current political status of Northern Ireland is the fear that they will be absorbed into a unified Ireland and thus become the minority within a Catholic-majority state. The nub of the matter, then, is that Protestants wish to remain the dominant group in Ulster. (In this chapter, *Ulster* and *Northern Ireland* are used interchangeably.)

Although these conditions are the immediate sources of the contemporary conflict, the Protestant-Catholic division has much deeper roots reaching back many centuries. Ethnic polarization well antedates the partition of 1920. Evidence of this long history of sectarian struggle is visually abundant on the streets and buildings of Belfast and Londonderry. Defiant anti-Catholic slogans scrawled on the walls and fences of Protestant areas are rejoined by anti-Protestant shibboleths in Catholic areas. Perhaps the most telling messages, however, are those with a historical content. Visitors to Northern Ireland may be baffled by the sight of graffiti exhorting people to "Remember 1690" or "Remember 1916." What conceivable relevance could events of over 300 or even 85 years ago have on current societal issues? As puzzling as it may seem to those unfamiliar with Northern Ireland, events of past ages in this society may just as well have occurred yesterday. They are constantly resurrected as a means of keeping the struggle alive.

HISTORICAL DEVELOPMENT OF ETHNIC RELATIONS

The conflicting ethnic identities and loyalties of Protestants and Catholics in Northern Ireland emerge out of what both groups see as very different and opposing heritages: British for Protestants, Irish for Catholics. Each, therefore, maintains a distinctly different view of history. Let's look briefly at the major events of Ulster's past that have created these contrasting perspectives.

The Plantation of Ulster

The beginnings of conflict between Protestants and Catholics in Northern Ireland are traceable to the early seventeenth century, when England embarked on the colonization of Ireland. The most comprehensive colonizing effort was made in Ulster, where Irish lands were confiscated and Scottish Presbyterians and English Anglicans were settled on them. The native Catholics were either driven out of Ulster or subdued into a totally subordinate position.

The period of English and Scottish colonization, known as the Plantation of Ulster, marked a group confrontation that clearly illustrates the process by which ethnic stratification is established: ethnocentrism on the part of each group, competition between them for scarce resources, and, finally, the establishment of dominance by one through superior power.

In the English view, the native Irish were uncivilized people against whom the most brutal tactics could be employed. Indeed, Fredrickson (1981) suggests that the experience of the English in Ireland foreshadowed their colonizing practices and policies toward indigenous peoples in North America. If the natives were uncooperative in ceding their lands, the "savage" image could be invoked, and they could then be either exterminated or confined to reservations. Unlike the settlers' experience with the Indians in North America, however, the Scottish and English settlers could not so easily dispose of the Catholics. For one, there were simply too few settlers to impose total control. Furthermore, Catholics resisted, and in 1641 they revolted in a massacre of Protestants. This was avenged with a countermassacre a few years later by the armies of the English dictator Oliver Cromwell, and most of the remaining land held by Catholics was confiscated. By 1703, Catholics possessed less than 14 percent of all land in Ireland, and in Ulster, land was almost entirely held by Protestants (Rose, 1971).

The colonization and suppression of the Catholics in Ulster produced antipathies and armed hostilities that continue, so evidently, to the present day. For Protestants, a siege mentality was created, not unlike that of the Afrikaners in South Africa, that remains the core of Protestant militance. Historian Liam de Paor vividly describes the trepidation of the English and Scottish settlers, who harbored a constant dread that the indigenous Catholics would someday wreak their revenge:

> Fear, contempt and hatred animated the opinion held by the planters of the ragged native Irish who worked on the land, or passed homeless on the roads, fear, and a deep-seated uneasiness compounded of cant, guilt, and a determination to hold what they were building where before there had been nothing. Like white farmers in Kenya watching their Kikuyu workers and thinking of the midnight advent of the Mau Mau, the English and Scottish planters watched and waited. (1970:23)

At the end of the seventeenth century, the Protestant ascendancy in Ulster was firmly secured as the result of several political and military events. Most important of these was the defeat of James II, the last Catholic English monarch, by William of Orange at the Battle of the Boyne in 1690. The commemoration of this battle each year on July 12 is the most important unifying ritual for Protestants. Elaborate parades take place and bonfires are lit throughout the

Protestant communities; these festivities are provocatively anti-Catholic in tone (Jarman, 1997; Ruane and Todd, 1996).

What followed William's victory at the Boyne was an even more sustained dominant-subordinate relationship between Protestants and Catholics. Penal laws were enacted, further suppressing the Catholic populace. The Catholic clergy was also restricted, as was the Irish educational system. Although the Penal Laws were applied most severely against the Catholics, many were also employed by the dominant English against the Scottish Presbyterians. This prompted the mass emigration of thousands of Ulster Presbyterians to the United States. These were the so-called Scotch-Irish, who made up such a major element of the colonial American population in the eighteenth century.

At the close of the eighteenth century, one of the few nonsectarian efforts by the Irish to secure independence from British rule emerged in an alliance of radical dissenting Presbyterians and Catholics. The movement was short-lived, however, and led subsequently to the abolition of the Irish parliament. Thereafter, Ireland was ruled directly from London.

Home Rule and Irish Rebellion

Late in the nineteenth century, sentiment grew in England for the establishment of home rule for Ireland. This was strongly resisted by the Protestants of Ulster, who saw themselves at the brink of absorption into a Catholic majority. The slogan "Home rule is Rome rule" expressed this fear, and an effort was begun to defeat any movement toward Irish autonomy. This campaign against home rule in the 1880s produced a set of institutions that united Protestant Ulster and that remain the political foundations of the Protestant community. Chief among these was the Orange Order, an organization combining political, military, religious, and fraternal elements, whose raison d'être was to defend the Protestant ascendancy. Originally formed as an anti-Catholic vigilante group, it emerged in the 1880s with a newfound respectability as the nucleus of the anti-home-rule movement, uniting disparate unionist factions (Darby, 1976). The Orange Order would remain one of the strongest forces in Northern Ireland's politics and social life. Even today it is the largest Protestant fraternal organization and still wields much social and political influence.

Closely allied to the Orange Order and developed as a means of parliamentary resistance to home rule was the unionist movement. With the victory of a Conservative government in Britain in 1886, home rule, having been favored by the Liberals but not by the Conservatives (with whom the unionists of Northern Ireland had aligned), was shelved until the advent of a new Liberal government in 1906. Once again, militant Protestant resistance arose. This time, Ulster Protestants were prepared to resist with arms, and they formed a paramilitary group, the Ulster Volunteer Force. Moreover, almost all adult

Protestants in Ulster (nearly half a million) signed a covenant declaring their refusal to recognize any home rule measure passed by the British Parliament.

A home rule bill was enacted in 1914, but with the onset of World War I, British attention was deflected from the Irish issue. In 1916, however, impatient Irish nationalists sparked an armed rebellion against British rule in what has become known as the Easter Rising. Like the Battle of the Boyne for Protestants, this event has assumed great symbolic meaning for Catholics. The rebellion eventually led to the Government of Ireland Act of 1920, providing for the partition of Ireland into North and South, with the latter granted full sovereignty from Britain a year later.

The partition was a compromise measure that fully satisfied neither the North nor the South. The Ulster Protestants wanted no form of autonomy for Ireland, preferring retention of the British union, and the South desired a fully unified Irish nation, including Ulster. Northern Ireland, then, was essentially created by the British to avoid bloodshed between Protestants and Catholics. As one historian has written, "No one in Ireland, of any political persuasion, wanted or welcomed it. Its creation was an expedient imposed on the country by a hard-pressed British cabinet, which sought only to hold two groups of Irishmen from each other's throats and to give them an opportunity of living peaceably apart, since they could not live peaceably together" (Beckett, 1972:14).

A recalcitrant nationalist element led by the Irish Republican Army (IRA) provoked a bitter civil war over the issue of partition, refusing to accept its legitimacy. Thus, the new Irish Free State as well as the politically questionable entity of Northern Ireland were founded amid the violence and bloodshed its creators had hoped to avoid. Not until 1925 did all parties—Britain, Ireland, and Northern Ireland—finally accept the arrangement as permanent. Nonetheless, Irish Catholics, in both the North and South, never fully relinquished their belief that Ireland would eventually be unified.

The Civil Rights Movement

From the late 1920s until 1969, Northern Ireland experienced four decades of relatively unperturbed ethnic relations. Although anti-Catholic violence flared in the depression years of the 1930s, World War II brought economic prosperity to Ulster with its important shipbuilding and aircraft industries. After the war, new industries were attracted, many of them international firms whose hiring practices did not conform to the common anti-Catholic discriminatory patterns of the past. The opening of new occupational fields to Catholics served as a hopeful sign of increasingly equitable economic status for the minority as well as the possibility of more harmonious ethnic relations in general.

The increased economic prosperity of the 1950s and 1960s was paralleled by a substantial government emphasis on education that greatly benefited the Catholic minority. Many for the first time went on to a university education, seeking upward mobility in jobs commensurate with their educational qualifications (McAllister, 1977). A political leadership emerged from this new Catholic middle class determined to arrest the inequities long suffered by Catholics in various institutional areas. Like civil and minority rights movements in other countries during the 1960s (including the United States and South Africa), a civil rights movement petitioned for equal rights for Catholics. Peaceful protest tactics similar to those employed by African Americans were adopted, including marches and sit-ins. The protesters even sang the American civil rights movement's anthem, "We Shall Overcome." Specifically, the movement sought to end discriminatory practices in voting procedures, housing distribution, and employment, those areas in which Catholics had been most severely and continually aggrieved.

The civil rights movement differed from previous political efforts of the Catholic minority in that its immediate objective was not the end of partition and the eventual unification of Ulster with the Republic. Instead, the movement's focus was on issues pertinent to Catholics in Northern Ireland; that is, treatment as equal citizens under British rule. Implicit in this focus was a recognition of the legitimacy of the Northern Irish state and a willingness to participate in its operation. Previously, most Catholics had felt that participation was meaningless within governmental institutions—especially Ulster's parliament—that were thoroughly dominated by Protestants who refused to yield an inch to Catholic interests.

The prospects for a rapprochement between the Protestant and Catholic communities in the 1960s were given further stimulus by the views and actions of Northern Ireland's prime minister, Terence O'Neill. Although not deviating basically from the standard loyalist policies and objectives of the Protestant-dominated Unionist Party, O'Neill seemed intent on creating an economically sound Ulster that would usher in an era of more cooperative relations between Protestants and Catholics (Arthur and Jeffery, 1996). As Rose has described him, in the context of Northern Ireland, O'Neill "was an innovator, even a revolutionary, for he sought nothing less than a fully legitimate regime in which Catholics would support the Constitution as well as comply with its basic laws" (1971:97). In 1965, O'Neill pulled off a startling political act of reconciliation by meeting publicly with the prime minister of the Irish Republic. Given the suspicion and animus of past unionist governments toward the Irish state in the South, this was an extraordinary symbolic act. It was an event that "had all the overtones and anxieties, in Irish terms, of a summit meeting between Russian and American leaders" (Rose, 1971:99).

Tragically, both the civil rights movement and O'Neill's reformist government fell victim to Protestant paranoia and extremism—the politics of the old. That the civil rights movement had been founded primarily on issues of minority rights in Northern Ireland, not on the issue of unification with the South, was ignored; Protestant extremists, led by the demagogic preacher Ian Paisley, interpreted the movement as a Republican effort and mobilized a militant Protestant countermovement. Furthermore, Paisley and other Protestants of the extreme right had even earlier accused O'Neill of appeasement and sought his political defeat. With the outbreak of Protestant violence against civil rights marchers, the highly volatile situation rapidly deteriorated, and O'Neill was forced to resign in 1969.[2] What had promised to be the onset of an era of peace and conciliation between Protestants and Catholics degenerated into the most stressful period of Northern Ireland's history.

The Troubles

People of Northern Ireland refer to the contemporary phase of the conflict, beginning in 1969, simply as "the Troubles." By August of that year the situation had become extremely tense, and civil rights marchers were attacked by Protestant extremists. Rioting began in Londonderry, in turn producing a brutal police response against Catholics. The riots spread to Belfast, where Protestant mobs invaded the Catholic Lower Falls area; seven people were killed, and over 3,000 were left homeless as entire neighborhoods were set afire (Darby, 1976). These events prompted the British government to send troops into both Belfast and Londonderry to protect Catholics from what seemed to be turning into a bloody pogrom. At this time, Catholics welcomed the British Army because there was no longer any assurance that the Royal Ulster Constabulary (Northern Ireland's police) could or would protect them. Indeed, in some cases the RUC appeared to be in collusion with Protestant attackers (Coogan, 1980).

With the formation of the Provisional IRA in 1970, the major protagonists in the contemporary period of conflict were in place. Although the IRA had been sporadically active in attacks on the unionist government since the 1920s, it had gained little support among the Catholic community, and by the 1960s it was seen as something of an anachronism (Holland, 1982). With the emergence of renewed violence, the group split into two factions, the Official IRA, committed to a primarily nonviolent political course, and the Provisional IRA, which favored the use of violent tactics to effect unification with the South. Eventually the Provisionals (or, as they are commonly known, "Provos") gained ascendancy and became the major paramilitary group among the Catholics.

The rebirth of the IRA was given considerable impetus by the actions of the British Army. Even though resistance to reforms and much provocation of the violence was coming from the Protestant side, in June 1970 the Catholic Falls Road area of Belfast was sealed off by the British Army, which proceeded to

---

[2]For a view of the Catholic civil rights movement and responses to it, expressed by one of its key participants, see Devlin (1977).

ransack homes in search of arms. The Army's foray into the Catholic area was unnecessarily severe and was afterward referred to as the "Rape of the Falls." Several people were killed in the operation, many were injured, and the search produced very few arms (Coogan, 1980). After that incident, the British Army was viewed by Catholics no longer as protectors but as oppressors. The Catholic ghettos of Belfast and Londonderry increasingly turned to the IRA for protection. Although the IRA's initial purpose had been essentially to defend Catholic areas, it later began an offensive campaign against the British forces in Ulster, often involving particularly callous and shocking actions.

The rise of the IRA was countered by Protestants with vigilantes of their own, and for the next two and a half decades these paramilitary groups on both sides would be the chief forces in a turbulent and violent political environment that often seemed to defy rational explanation.

Two other events hardened Catholic attitudes toward the British Army. One was the 1971 reactivation of the policy of internment, carried over from the time of the Irish rebellion in the earlier part of the century. It provided that anyone suspected of terrorist activities could be arrested without charge, jailed, and held without bond or legal aid. Although the reinstitution of internment was ostensibly intended to control the violence emanating from Protestant as well as Catholic groups, it was applied in a less than evenhanded manner. Further adding to the fallen image of the British Army among Catholics was the shooting of thirteen unarmed demonstrators in Londonderry in 1972, an event annually recalled by Catholics as "Bloody Sunday."

With the intensification of hostilities, the British government in 1972 took full control of Northern Ireland, suspending Ulster's parliament and establishing direct rule from London. This proved no antidote to the level of violence, however. An attempt was made a year later to establish a new Northern Ireland Assembly, which would have provided a form of power sharing among the major political parties, but a general strike led by intransigent Protestant workers brought a return to direct British rule.

Until 1994, little had changed in the basic situation. The British Army, though reduced in size, remained in Northern Ireland. The paramilitary groups on both sides continued to engage in a kind of urban guerrilla warfare, intent on protecting their communities and interests and promoting the causes of either unification on the one hand or the status quo on the other. From time to time, sensationalistic events, such as the hunger strike of imprisoned IRA men in 1981, provoked a round of demonstrations and counterdemonstrations, violence and counterviolence. Meanwhile, a political middle course, aimed at reason and reconciliation, was stymied. A cease-fire was effected in 1994, representing the first real break in the cycle of violence and holding out promise that the Troubles might end. We will return to the current politics of the Northern Ireland conflict in a later section.

Having briefly traced the history of intergroup relations in Northern Ireland, we now turn to a more detailed sociological look at this ethnically polarized society. In several ways, Northern Ireland illustrates almost classically many of the notions discussed in Chapters 2, 3, and 4. To begin with, there is a system of ethnic stratification among Protestants and Catholics in which the former have traditionally maintained their power not only through superior numbers but also through dominance of the society's vital institutions. Second, the extreme social division of the two groups is supported by a high level of prejudice and discrimination and the implementation of an ideology by the dominant group that rationalizes and justifies its actions toward the minority. Finally, Northern Ireland is a prototypical pluralistic society in which each group maintains a relatively cohesive and institutionally complete community.

The Protestants

Northern Ireland's population of about 1.6 million is made up of approximately 51 percent Protestants and 43 percent Catholics (Table 16.1). The Catholic community is growing more rapidly, and its proportion of the total population has increased considerably in the past several decades (Darby, 1995; McGarry and O'Leary, 1995a). Nonetheless, Protestants remain both a numerical as well as a sociological majority.

Although they have generally been undivided in opposition to changing the political status quo, the Protestant population should not be seen as a thoroughly unified community. There are not only class differences among them but also significant sectarian differences. The two major sects are Presbyterians, most of whom derive from the Scottish settlers of the Plantation era, and those belonging to the Church of Ireland, the Irish counterpart of the Anglican church. These two Protestant subgroups constitute a certain degree of cleavage within the dominant group though in no way as acute as the schism that existed between, for example, Afrikaner and English South Africans.

Religious practices and beliefs are attended to quite seriously by Ulster Protestants (as well as Catholics, as we will see). The intentness with which they approach religion is reflected in the fact that church attendance is higher among Northern Ireland's Protestants than among any other Protestant groups in Europe or even the United States (Livingstone et al., 1998; Moxon-Browne, 1983). Most important, religion for Protestants is the key organizing feature of social life. Little is unaffected by religion, including politics. Although the ethnic conflict in Northern Ireland is hardly reducible to religious matters, certain Protestant elements do see their cause as the defense of their religion against an all-powerful Catholic church (Darby, 1976). This view is epitomized by Paisley,

Table 16.1 Religious Distribution of the Population of Northern Ireland

RELIGIOUS PROFESSION	1926		1951		1961		1971		1981		1991	
	Number	%	Number	%	Number	%	Number	%	Number	%	Number	%
Roman Catholic	420,428	33.5	471,460	34.4	497,547	34.9	477,919	31.4*	414,532	28.0*	605,639	38.4
Presbyterian	393,374	31.3	410,215	29.9	413,113	29.0	405,719	26.7	339,818	23.0	336,891	21.4
Church of Ireland	338,724	27.0	353,245	25.8	344,800	24.2	334,318	22.0	281,472	19.0	279,280	17.7
Methodist	49,554	3.9	66,639	4.9	71,865	5.0	71,235	4.7	58,731	4.0	59,517	3.8
Others	52,177	4.1	63,497	4.6	69,299	4.9	87,938	5.8	113,244	7.5	122,448	7.8
None	—	—	—	—	—	—	—	—	—	—	59,234	3.8
Unstated	2,304	0.2	5,865	0.4	28,418	2.0	142,511	9.4	274,162	18.5	114,827	7.3
Total	1,256,561	100.0	1,370,921	100.0	1,425,042	100.0	1,519,640	100.0	1,481,959	100.0	1,577,836	100.0

SOURCES: Doherty, 1993; Northern Ireland, Department of Health and Social Services, 1984; *Ulster Year Book*, 1963–1965, 1983.

*The Catholic population in both 1971 and 1981 is seriously understated by these figures because of the large percentage of the general population who chose not to respond to the voluntary census question on religious affiliation (9.4 percent in 1971 and 18.5 percent in 1981). The actual Catholic population was estimated to be about 37 percent in 1971 and possibly higher in 1981 (see Compton, 1982). The problem of no response was not so serious in 1991, when only 7.3 percent failed to declare their religious preference (see Doherty, 1993). Most of the latter, however, are assumed to be Catholic, thus raising the actual Catholic percentage.

whose anti-Catholic rhetoric is replete with admonitions that the aim of the Catholic church is to engulf Protestantism. For most Protestants, however, the major fear of unification with the Republic of Ireland centers on the social and political influence of the Catholic church in that nation and becoming a minority in that context (Livingstone et al., 1998).

Although Protestants look to Britain for their historical identity, there is among them a curious ambivalence regarding their relationship with the British. There is, on the one hand, a resistance to certain aspects of English life as well as the prevalence of strongly negative English stereotypes; on the other hand, however, Protestants maintain an unusually strong British political identity— some would say an overidentity (Arthur, 1980; Harris, 1972).

The Catholics

As both the numerical and sociological minority, Catholics in Northern Ireland are a more homogeneous community than are Protestants. If religion is the key organizing ingredient of social life for Protestants, it is even more so for Catholics. Almost all Catholic children, for example, attend parochial schools. Whereas church attendance for Protestants is uncommonly high, it is even higher for Catholics (Livingstone et al., 1998; Moxon-Browne, 1983).

The most telling difference between Protestants and Catholics, aside from matters of faith, involves their national identities and aspirations. Catholics identify themselves overwhelmingly as "Irish," whereas Protestants more commonly see themselves as "British" or "Ulster" (Boal et al., 1997; Boyle and Hadden, 1994; Elliott, 1997; Moxon-Browne, 1983; Pollak, 1993). Many Catholics continue to recognize the Republic of Ireland as the rightful government of all of Ireland and thus do not view local unionist or British governments as entirely legitimate. Hence, they retain the hope that unification will someday be realized. The past tendency for Catholics to reject the legitimacy of the Ulster government is a key factor in explaining Protestant militancy.

In brief, Protestants and Catholics in Northern Ireland constitute two sharply divided ethnic communities whose differences transcend religion. Most important, they hold strongly opposing views on the society's key political issues and maintain distinctly different cultural and national identities. Although they share some common features, neither community, as Ruane and Todd have described them, "has a developed concept of a general interest which transcends its own particular interests" (1996:78).[3]

---

[3]Boyle and Hadden (1994) suggest that there is really a third community, made up of those who do not identify with either Protestant or Catholic communities or reject the hard boundaries between the two. The percentage of those who fall into this category may vary, depending on different social and political issues.

Ethnic Boundaries

In the United States, although there are no "official" ethnic groups, ethnic or racial status is clearly defined in a customary and rigid fashion; rarely are there questions about who, for example, is white and who is black. In Northern Ireland, however, there are no apparent physical distinctions between the two ethnic groups, and except for religion, even cultural differences are not readily evident. How, then, do interacting persons perceive ethnicity in a society in which it is the most critical social identity? Social interaction with members of the out-group calls for a very different set of behavioral responses than interaction with one's coreligionists. Some technique is therefore necessary by which religious affiliation can be determined in the course of social interaction.

With obvious physical marks such as skin color not available, cultural cues and symbols, to which the people of Northern Ireland are finely attuned, serve effectively in their place. Names, for example, often suffice as indications of ethnic identity. Certain surnames indicate an English or Scottish (and thus Protestant) or Irish (and thus Catholic) ancestry, and even Christian names are commonly associated with one side or the other. Other cues are accent, dress, demeanor, and—some would insist—even facial features (Fraser, 1973). Protestants sometimes assert that they can recognize Catholics because of their "generally dark Celtic features," and Protestants may be described as having "an Anglo-Saxon or Scottish 'look'" (Sheane, 1977:67).

The ethnic symbolism is so strong that even certain words or references can alert people to one's religious affiliation. For instance, Londonderry, Northern Ireland's second largest city, has great historic meaning for both groups. Catholics refer to it as Derry, its pre-Plantation name, whereas Protestants prefer Londonderry. Thus, maneuvering a conversation to include the city's name becomes a ploy to ascertain religion (Darby, 1976). Because residence and education are so rigidly segregated, however, the surest indicators of religious affiliation are one's address or school attended. Knowing either of these social facts is ordinarily sufficient to identify a person ethnically.

Burton (1978) refers to the process by which Protestants and Catholics distinguish each other in their day-to-day interactions as "telling." "Telling," he explains, "is a necessary social skill for triggering the cognitive store of sectarianism. It is a trigger which brings into play the typifications and stereotypes about the nature of the 'other side'" (1978:37). Although each cue is subject to various interpretations and is often unclear and imprecise, what remains constant is "the desire and necessity to tell." Within the rural village she studied, anthropologist Rosemary Harris (1972) also noted the need to determine the ethnicity of strangers in order to facilitate interaction. Whereas outsiders might explain as bigotry the tendency for people, on meeting others for the first time, to ask about religious affiliation, Harris points out that "in fact this question is not in itself an expression of religious prejudice but an obvious response to the necessities of social existence in this area" (1972:148).

These ethnic cues by which people attempt to identify others and thus avoid uncomfortable interactions (or perhaps even dangerous situations) reinforce the effects of physical segregation in restricting the knowledge each group has of the other. Interpreting these cues, notes Burton, "constitutes mental bricks and mortar. It is the conceptual and cognitive ghetto of Northern Irish ideological social relations" (1978:66). Northern Ireland demonstrates dramatically that visible physical differences or even vast cultural differences are not necessary to create and sustain fierce ethnic hostility. Groups need only perceive their differences as unbridgeable.

Class Divisions

Class differences between Protestants and Catholics are noticeable, though in light of the intensity of social conflict they are not as severe as might be expected. Discrimination in employment, a major grievance of Catholics in Northern Ireland, was an important factor in energizing the Catholic civil rights movement in the late 1960s. Traditionally, higher-status jobs were reserved for Protestants, and the workforce was highly segregated. Today those discriminatory patterns have been rectified to a great extent.

Table 16.2 Composition of the Northern Ireland Workforce[*]

OCCUPATIONS	PROTESTANT	CATHOLIC
Managers and administrators	61.7%	38.3%
Professional occupations	56.1	43.9
Associate professional and technical occupations	55.7	44.3
Clerical and secretarial occupations	59.8	40.2
Craft and skilled manual occupations	64.3	35.7
Personal and protective service occupations	66.5	33.5
Sales occupations	58.5	41.5
Plant and machine operatives	58.9	41.1
Other occupations	58.5	41.5

Source: Equality Commission for Northern Ireland, 2001.
[*]Percentages are based on those whose religion was determined. This was over 95 percent for almost all occupational categories.

As shown in Table 16.2, Protestants are more strongly represented in the higher occupations, and Catholics in the lower, but the disparity is not great. The percentages for each occupational area are fairly close to Protestant and Catholic proportions of the total population. The differences are in no way comparable to those between blacks and whites in South Africa or even between whites and racial-ethnic minorities in the United States. The most serious discrepancy is in the rate of unemployment, which is markedly higher among Catholics (Darby, 1995; Pollak, 1993; Smith and Chambers, 1991).

Studies indicate a continuation of the trend toward greater ethnic equity in the Northern Ireland class structure (Boyle and Hadden, 1994; Cormack and Osborne, 1994; Equality Commission for Northern Ireland, 2001; Osborne and Cormack, 1991). Catholics are moving into higher managerial and supervisory positions and into professional occupations in significant numbers, and the Catholic middle class in general is expanding. Moreover, Catholics have achieved educational parity with Protestants. Also apparent, however, is the relatively minor change between Protestants and Catholics among the unemployed. Catholics are twice as likely to be unemployed as Protestants (Darby, 1995). This sharply higher unemployment rate has become a focus of grievance for Catholics, who see it as the result of continued discrimination. This issue will be considered in a later section of this chapter.

Political Power

A key to understanding ethnic relations in Northern Ireland lies in the political sphere. As previously noted, the correlation between political preference and religious affiliation is extremely high. This is similar to other highly stratified multiethnic societies, where various forms of inequality stem from great differences in political power.

Essentially there have been two diametrically opposed sides to the political conflict in Northern Ireland, each identified with either the Protestant or Catholic communities. One is the *unionist* movement, whose basic position, since its emergence in the late nineteenth century, has been preservation of the union between Northern Ireland and Great Britain (McGarry and O'Leary, 1995a). The unionist movement—in recent years internally split among several political parties—is made up primarily of Protestants who, in line with their intention to keep Northern Ireland British, have traditionally opposed efforts at developing closer ties with the Republic of Ireland. Any such efforts have been interpreted as the first steps toward eventual unification. Though tempered somewhat in recent years, this position has remained essentially unchanged. Protestants, then, whether they are politically active or not, are seen as supporters of unionism. The *nationalist* movement, on the other hand, is made up primarily of Catholics whose objective is eventual unification of Northern Ireland with the Republic of Ireland. The Nationalist Party and, in recent years, its successor, the Social Democratic and Labour Party, have been identified with Catholic issues and interests, especially on the question of unification.[4]

Because Protestants were always a numerical majority, the power of the unionists was unchallengeable. In a sense, before 1972, when Britain took full control of Northern Ireland's central government, Northern Ireland had been a one-party state. This dominance was exacerbated by the tendency of Catholics not to participate in parliamentary politics because they found themselves a permanent minority.

As a result of the political dominance of the unionist Protestants, Catholics had been, in effect, excluded from the political process at the national level and seriously underrepresented at the local level (Aunger, 1981; Rose, 1971). Before direct British rule in 1972, Catholics had not been appointed to any cabinet positions, and senior civil service and police appointments had also been almost exclusively Protestant. Unionist control of local councils had a similar discriminatory impact, effectively excluding Catholics from public jobs and offices.

The wide discrepancy in political power between Protestants and Catholics resulted in severe inequalities in other areas of social life, especially jobs and housing. When Britain assumed control of the Northern Ireland government in 1972, however, Protestant domination was ended. Britain, not unionist Protestants, controlled the central government. In addition, Britain imposed political reforms at the local level, resulting in increased Catholic representation on local councils. Catholic employment in civil service jobs subsequently increased substantially, and Catholics now are represented in such occupations proportionate to their share of the general population (Cormack and Osborne, 1994; Fair Employment Commission for Northern Ireland, 1997). Changes also occurred in the distribution of housing. Political control of public housing was an especially significant power held by Protestants because a high percentage of people in Northern Ireland rent their dwellings from public agencies. A centralized housing agency replaced local agencies in 1972, resulting in a more equitable distribution of public housing (Darby, 1995).

---

[4]Protestants are commonly referred to as "unionists" or "loyalists," in accordance with their desire to maintain Northern Ireland's union with Great Britain. Catholics are described as "nationalists" or "republicans," reflecting their favor of an eventual unification with the Republic of Ireland. There are subtle political distinctions in these terms, however. A reference to "republicans" or "loyalists" is to imply a stronger militancy regarding issues of the conflict.

As explained in Chapter 3, in most multiethnic societies the dominance of one ethnic group over others is enforced through discriminatory practices and is rationalized through a set of negative beliefs regarding minority groups. In some societies, this system is officially sanctioned and rigidly administered. Negative beliefs about the subordinate groups may also be encompassed in an elaborate racist ideology. This characterized South Africa under apartheid. In other societies, like the United States, the system of ethnic dominance depends on custom rather than law, and the ideology on which such practices are based is subject to strong dispute. In still other societies, like Brazil, ethnic dominance may be denied entirely though its manifestations are subtly evident. Where, in this range of ethnic patterns, does Northern Ireland stand? Prejudice and discrimination in this society are perhaps most similar to patterns of black-white relations in the United States. They are neither officially legitimized nor enforced through the application of coercive power, but they are not difficult to detect.

Protestant and Catholic Prejudice

In explaining ethnic prejudice and discrimination in Northern Ireland, two views have been put forth. One is that Protestants harbor deep suspicions of Catholicism. The Catholic church is seen as wielding great influence on its parishioners, and, because it is the central institution for the Catholic population, it becomes the focus of Protestant hostility (Bruce, 1986; Elliott, 1997). "For many Protestants," notes Heskin, "it seems that Catholic Churches are mysterious and forbidding places and Catholicism is a powerful, sinister and potentially engulfing creed" (1980:27). For extremists, he explains, Catholicism symbolizes evil, much like some Americans once perceived Communism. A more prevalent view, however, rejects the notion that anti-Catholic antagonism derives from Protestants' disdain of Catholicism per se. Rather, it is seen as stemming mainly from the nationalist sentiments of Catholics: If Northern Ireland were to be united with the heavily Catholic Irish Republic, the cultural identity and political power of Ulster's Protestants would disappear and their economic position decline (Jackson and McHardy, 1984; McGarry and O'Leary, 1995a).

Because most Catholics continue to hope for the eventual unification of Northern Ireland with the Republic, they are commonly viewed suspiciously by Protestants as disloyal citizens whose reluctance to accept the British link justifies their minority position. The dynamics of prejudice and discrimination by Protestants against Catholics, then, are fueled in part by the perception of Catholics as being intent on destroying the present political arrangement and thus turning the Ulster Protestant majority into an Irish minority. Protestant fears also stem from the belief that Catholics may eventually become the numerical majority in Ulster, a belief that has a solid basis in demographic trends. Since the 1960s, the Catholic population has been steadily growing as a proportion of the total population, due in large measure to their higher birthrate (Doherty, 1993; McGarry and O'Leary, 1995a).

Anti-Catholic stereotypes buttress these beliefs. Many of the stereotypes commonly applied to Catholics are part of what seems to be an almost standard litany employed by dominant groups in Western multiethnic societies—lazy, dirty, shiftless, oversexed, ignorant, quarrelsome, and so on (Bell, 1996; Bell, 1976; Fraser, 1973). In the rural village of Ballybeg, Harris (1972) found that Protestants routinely regarded Catholics as poor and as intent on cheating the welfare system. She quotes one of the villagers as believing that "Catholics have their own people in the Civil Service in Belfast and they tell them how to get money they don't deserve." Catholic doctors, the villager thought, "were always ready to say a Catholic was sick even when he was healthy, if only it would get him government money" (1972:174).

Fraser writes that of all negative characteristics, the size of Catholic families "seems to provoke Protestant hostility more than anything else" (1973:94). This stems primarily from the fear of being displaced as the numerical majority but also from the resentment of larger Social Security benefits paid to Catholic parents. Fraser quotes one young Protestant boy of eleven, already well imbued with anti-Catholic conviction:

> The Catholics have big families and they won't work. Daddy says we're poor because the Government takes his money to give to the Catholics and all their children and their priests to keep them in luxury because they don't work. In the South they wouldn't get any money and would have to work for it. Here they keep us all poor and don't appreciate what they get. (1973:96)

For Catholics, anti-Protestant animosity stems primarily from a heritage of discriminatory treatment rather than from any fears of Protestantism per se. The objections to and suspicions of Protestants focus on their political power and the society's social and economic issues, which tend to be shaped along religious lines (Barritt and Carter, 1972; Bell, 1996; Pollak, 1993). The animosity of many Catholics toward Protestants can be as deep as that which flows in the opposite direction. O Connor (1993) has trenchantly described this mind-set:

> Every Catholic knows someone in their circle who has either reacted with horror when told their child intended to marry a Protestant, tried to prevent such a marriage, or been wounded by such a reaction from a relative. The objectors may say their concern is based on religious conviction, that they fear for their child's religion, their future grandchildren's religion,

even their child's safety. But the sheer distaste many show when confronted with the imminent arrival in their family circle of a Protestant cannot be explained as anything other than prejudice. (169–170)

In a study conducted among both Protestants and Catholics, O'Donnell found that out-group stereotypes were not entirely negative. Members of both groups, he reported, saw each other basically as "decent, fine, ordinary people" (1977:165). As in other multiethnic societies, negative stereotypes are ordinarily applied to out-groups as a whole rather than to specific individuals. One's out-group neighbor, therefore, may be seen as "different from the others." Negative group images, however, are clearly expressed. Protestants, on the one hand, see Catholics as being Irish nationalists who are brainwashed by priests and as having too many children. Catholics, on the other hand, view Protestants as holders of power who are unwilling to relinquish any of it, whatever the cost. The images each side has of the other, then, are manifestations of the ideological basis of the conflict—Protestants' fear of unification and Catholics' fear of Protestant political power.

It should be remembered, of course, that in any multiethnic society not all members of one group exhibit equal levels of negative beliefs about members of the other. In each ethnic community, there is always a range of thought among individuals, extending from those who are highly prejudiced to those who genuinely desire a mixed society. So segregated are the two ethnic groups in Northern Ireland, however, that little opportunity arises for even moderate thought to take hold among most of the population.

Discrimination

Because the two groups are so segregated in their own communities, it is difficult to ascertain the actual degree of discrimination applied by Protestants against Catholics, or vice versa. In a social environment in which each group maintains its own complete communities, it is natural for institutions to become ethnically homogeneous. Schools, pubs, recreational organizations, and so on all cater to a clientele made up primarily of members of one group or the other. The question, then, is how much of the separation of the two groups and the resultant social differences between them is the result of voluntary forces and how much the result of forces imposed on them.

There is also the question of institutional discrimination. How much of the deprived condition of the minority is the result of direct actions by the dominant group, and how much is the product of the workings of institutions that, through historical evolution and unchanged manner of operation, discriminate against the minority routinely?

Although these questions are certainly pertinent in the case of Northern Ireland, it is generally agreed that ethnic discrimination has been direct as well as institutional, particularly in employment and politics, and in both areas Catholics have been the chief victims.

Employment In the past, discrimination in the workforce on the basis of religion was widespread (Barritt and Carter, 1972; Bell, 1976; Boyle and Hadden, 1994; Hoare, 1982). In 1989 a fair employment act, designed to monitor the employment practices of companies with 250 or more workers, was instituted in Northern Ireland. A similar measure in 1976 seemed to have minimal effect on reducing ethnic discrimination in employment, but the new act provided for a kind of affirmative action as a means of promoting ethnic equality (Edwards, 1995; Smith and Chambers, 1991). No longer could ethnic discrimination in the workforce be viewed as acceptable or unavoidable (Boyle and Hadden, 1994). Such efforts at overcoming discrimination appear to have had a positive impact, as Catholics have increasingly moved into higher-status jobs and into occupational levels, such as professional and managerial, once closed to them (Edwards, 1995).

However, as noted earlier, unemployment in Northern Ireland remains much higher among Catholics, and it is here that questions of discrimination continue to arise. Although Catholics see chronically higher Catholic unemployment rates as a result of continued Protestant discrimination, structural factors unquestionably contribute as well. Residential segregation and differences in educational attainment, for example, play some part in explaining different rates of unemployment between the two groups. Regional location also has an effect. Northern Ireland as a whole is divided economically along east-west lines. In the east it is dominated by Belfast, where most industry is located and therefore where most job opportunities are found. The west has remained largely agricultural and hence less economically advanced. Protestants are a majority of the population in the more prosperous eastern sector, and Catholics are a majority in the western sector. Given this demographic imbalance, Catholics are automatically subject to less favorable employment opportunities.

Although past anti-Catholic discrimination is unquestioned, many Catholics in Northern Ireland may now be in much the same situation occupationally as the black underclass in the United States—the victims no longer of a system of blatant ethnic discrimination but of an institutional system of discrimination that has left them with inadequate skills to compete successfully in the modern labor market. In addition, Catholics may hesitate to take jobs in historically Protestant companies because they fear violence or are reluctant to leave their ethnic communities and venture into Protestant areas.[5]

---

[5]Hewitt (1981) asserts that the degree of anti-Catholic discrimination before the civil rights movement and the subsequent Troubles has been exaggerated. Moreover, he points out that though discrimination by Protestants has been most widely reported, there is comparable anti-Protestant discrimination in politics and employment where Catholics are in an advantageous position.

Perceived Discrimination Whatever its extent and whatever its source, Catholics perceive discrimination by Protestants. Thus it may matter little that the socioeconomic conditions of Protestants are only marginally better than those of Catholics, or that the most flagrant devices used to deny Catholics job opportunities have increasingly been checked. Because one's religious affiliation is not easily camouflaged, Catholics will simply not be convinced that "failure to get a job or a home is not governed by that fact" (Jackson and McHardy, 1984:4).

Rose in 1971 confirmed the significant difference between Protestants and Catholics in their perceptions of discrimination in Northern Ireland. Whereas nearly three-quarters of Protestants saw no discrimination on the basis of religion, three-quarters of Catholics did. Twenty years later, another survey reaffirmed the perception of Catholics as the victims of discrimination in employment. Surprisingly, a significant percentage of Protestants, too, believed that discrimination in the labor market on the basis of religion was prevalent. However, Protestants seemed to view the situation as one in which members of each ethnic community "look after their own," whereas Catholics saw it as societal discrimination against only their group (Osborne, 1991). Sociologist John Darby suggests that in recent years there have been signs of moderating attitudes of each community toward the other. As political, educational, economic, and social reforms have been implemented, "there has been a much greater acceptance that they are irreversible, and also that they are fair" (1995:25).

NORTHERN IRELAND AS A PLURALISTIC SOCIETY

In Chapter 4, multiethnic societies were seen as falling generally into three types: colonial, corporate pluralistic, and assimilationist. In some multiethnic societies certain characteristics of two or perhaps all three of these types may be apparent. Northern Ireland is such a case, displaying features of both colonial and corporate pluralistic systems. Not at all ambiguous, however, is the virtual absence of assimilationist features. This is clearly a society in which the centrifugal forces of pluralism—some imposed, some voluntary—predominate in the relations between the two ethnic groups.

Cultural Pluralism

Although most cultural differences such as language or style of life seem minor or nonexistent, the sectarian division is so critical that it often creates the perception of difference among members of the two groups where none or little is present. Political attitudes and intergroup relations are colored by the predominant belief that Protestants and Catholics are the historical product of diverse cultures that have coexisted uneasily since the Plantation of Ulster in the seventeenth century (Buchanan, 1982). De Paor, for example, writes that the confrontation of the Scottish and English settlers with the native Irish in that period represented a difference not simply in religion but also "in language, law, custom, economy, thought, and art" (1970:24). This understanding of a fundamental cultural gap between the two groups is sustained by all the society's major institutions. Protestants and Catholics see each other, quite simply, as two sharply different peoples.

Some, however, have suggested that the cultural differences between the two groups are not as historically clear as has commonly been believed. Buchanan points out that because Scotland and Ireland had been in close contact for more than a thousand years before the colonization of the seventeenth century, the folk culture the Scottish settlers found when they arrived was essentially "very similar to that of their homeland" (1982:67). Moreover, he asserts that what differences did exist between the native Irish and the Scottish and English colonists were resolved through a process of acculturation within a century. The religious breach, however, proved intractable and thus served to sustain group division. Buchanan also suggests that the growth of Irish nationalism in the nineteenth century, accompanied by the liberation of Catholic education, provided Catholics with a new sense of pride in Irish culture and history. The Protestants' place in Irish history, however, was only as colonialists and usurpers; as a result, they turned to Britain for their historical identity—hence the development of the two contrasting historical views.

Given the rigorous ethnic segregation of schooling in Northern Ireland, these views are promulgated from generation to generation. Noticeably different histories are taught Protestant and Catholic students, one emphasizing the British connection, the other the Irish. Events, personalities, and political institutions are interpreted differently, thus perpetuating the discordant societal views of the two groups (Murray, 1995).

Cultural differences are reinforced by various symbols displayed prominently by each group. It is common, for example, for flags to be flown from houses—either the Union Jack for Protestants or the Irish Tricolour for Catholics. Protestants will respond with pride to a playing of "God Save the Queen," whereas Catholics may ignore it and favor the Irish national anthem. Such symbols provide cultural boundary markers that reinforce in-group feelings and sharpen the perception of out-group differences (McCartney and Bryson, 1994). Parades commemorating major historical events—a popular activity among Protestants and, increasingly, Catholics—are also used to enhance these perceived differences and to solidify the belief in the prevailing interpretation of history by each side (Jarman, 1997).

The result of all this is an inability of each community to understand the other. In Ballybeg, Harris discovered that although members of the two ethnic communities might in some contexts have close and friendly contacts, for the most part they "managed to remain in almost complete ignorance concerning each other's beliefs" (1972:146). In situations where they interacted, Catholics and Protestants went to great lengths to avoid discussing any controversial topics, especially involving religion and politics. "The obvious aim," Harris notes, "was to prevent social relations being ruffled by disagreements over subjects on which it was accepted that there could be no consensus" (1972:146). Thus, no effort was made to bridge their differences. Although group differences may be tolerated in situations where Protestants and Catholics meet, rarely are they disregarded (Barritt and Carter, 1972).

Structural Pluralism

The establishment of separate institutional structures for each ethnic group is a key feature of highly pluralistic societies, and in this regard Northern Ireland is almost a pure case. Protestants and Catholics constitute ethnic communities with a high degree of identification and institutional completeness. Segregation is the general rule in most areas of social life. Each group maintains its own neighborhoods, schools, shops, voluntary associations, and even newspapers. Much social and leisure activity occurs within the churches of both Protestants and Catholics, which further restrains intergroup contact. Social interaction across ethnic lines, therefore, is confined mostly to instrumental contexts like work, and even there it is limited. Although social separation is not dictated by law, as it was in South Africa under apartheid or in the pre-1960s American South, it is enforced by a well-founded system of custom.

Residential Segregation Perhaps the most evident form of separation of Protestants and Catholics, having the most far-reaching social effects, is residential segregation. Segregated living areas can be traced to the Plantation era, when English and Scottish colonizers displaced the native Irish Catholics. Scots and English lived within the walls of the town or around the market, and the Irish were driven into outlying areas or beyond the town gates. In the rural areas a similar pattern developed, with the Irish excluded from the more productive valley lands and forced into the mountains (Barritt and Carter, 1972).

Today the most acute patterns of residential segregation are found in the larger towns, especially in the major city of Belfast. By North American standards, Belfast is the only truly large urban area in Northern Ireland, with a population of roughly 500,000. The industrial expansion of Belfast in the nineteenth century produced a huge and rapid influx of people displaced from the land throughout Ireland (Jones, 1960). A large percentage of these urban migrants were Catholics, who represented a threat to Protestant workers. Intolerance and conflict between the two working-class groups gave rise to sectarian colonies within the city, a pattern duplicated in other Ulster towns. "Catholic and Protestant workers lived in the same general urban areas," writes Darby, "but were divided into religious ghettos, compact, mutually exclusive and leaning on each other" (1976:28; see also Boal, 1982).

Recurrent religious conflict during the nineteenth and early twentieth centuries tended to reinforce ethnic housing areas. Families of "the other side" were driven out during each conflagration, in a sense purifying these ethnic enclaves. Darby and Morris (1974) found that this pattern had continued with the onset of the Troubles in 1969, with Catholics and Protestants moving into more homogeneous areas. Between August 1969 and February 1973, some 8,000 families in the greater Belfast area alone were forced to leave their homes because of direct intimidation by the other side, combined with general fear and a feeling of vulnerability. The actual number displaced may have been much greater, because the recorded figure included only those cases that had come to the attention of relief agencies. This represented between 5 and 10 percent of the entire Belfast population (Darby, 1976). The exchange of houses between Protestants and Catholics was commonplace in towns like Londonderry, where violence was especially intense. In some cases, families exchanged their houses on different sides of the same street, so clearly were the ethnic boundaries drawn (Darby and Morris, 1974).

Ethnic residential segregation, always common in the past, has become more pronounced in recent times. About one-half of Northern Ireland's 1.6 million people live in areas more than 90 percent Protestant and 95 percent Catholic (Pollak, 1993). Some working-class areas of Belfast are divided literally by a twenty-foot-high reinforced wall (ironically referred to as a "peace line"). Although there has been some residential mixing, particularly within the middle class, this is limited in scope (Cormack and Osborne, 1994; O Connor, 1993).

Segregated residential patterns are created not only by fear and intimidation but also in part by the natural inclination to live among one's own. In the city of Belfast and in other large towns, where urban villages have evolved, each serves as the focus of city life for its residents. A kind of territoriality has thus developed, in which people feel more comfortable in home areas where their churches and schools are located. Added to this natural territoriality, however, is the ethnic factor, which further strengthens the homogeneity of the neighborhood. Segregation, then, once established, "shows a strong tendency towards inertia" (Darby, 1976:37).

The dynamics of Northern Ireland's residential patterns seem quite similar to those of blacks and whites in American cities, where deliberate efforts by whites were most instrumental in creating black ghettos and subsequently preventing their dispersal (Marger, 1989). Poole (1982) shows, for example, that residential segregation is highest in Northern Ireland towns where the numerical minority (usually Catholics) is a large percentage of the population. In such situations, the minority is ordinarily perceived as threatening the ethnic integrity of the community or neighborhood. Again, this is not unlike the black-white patterns of American cities. What differs in the case of Northern Ireland, however, is the intensity of intercommunal violence that has traditionally accompanied ethnic residential segregation (Darby, 1976). Another difference, as previously mentioned, is that residential segregation in Northern Ireland is impelled by both Protestants' and Catholics' strong tendency to seek residence among fellow ethnics and to patronize ethnic institutions. This tendency is abetted by government housing policy, which has been forced to consider ethnic divisions in the building and distribution of public housing units. In the American case, blacks' desire to reside mostly among fellow ethnics is not the primary factor that propels residential segregation. Recall the studies cited in Chapter 9 indicating that given the choice, blacks prefer an integrated neighborhood. Racial segregation in the United States, then, is less voluntary than it is in Northern Ireland.

The high level of residential segregation in Northern Ireland has far-reaching effects on all other aspects of group interrelations, just as it does in the United States. Segregated living conditions contribute greatly to the continued polarization in all other institutional areas. Social contacts are minimized, and the cycle of misunderstanding, suspicion, and at times violence between the two groups is preserved (Murray, 1995). As one Derry resident put it, "Most Catholics and Protestants don't meet each other. We Catholics project the worst of ourselves on to Protestants and vice-versa" (Pollak, 1993:42).

Education Ethnic segregation in education is even more complete than in housing. Almost all Catholic children attend Catholic schools, and Protestants attend state (de facto Protestant) schools. Only 2 percent of students are schooled in an integrated setting (Irwin, 1996). In the case of Catholics, however, the origin and continuation of this pattern are mostly voluntary because the church has traditionally been adamant regarding the education of Catholic children in parochial schools (Darby, 1976; Wallace, 1970). Only at the university level and in special schools for the handicapped do students of the two groups mix to any significant degree. Even teachers' training colleges are essentially denominational.

The result is that most Ulster students "are being taught alongside their co-religionists by teachers of their own religion and have minimal educational contact with children from the other main religious group" (Darby, 1976:137–38). As earlier noted, Protestant and Catholic schools encourage different cultural traditions, do not teach the same political history, and even sponsor different extracurricular activities such as sports and games (Sugden, 1995).[6] Although the attitudinal effects of this educational environment are difficult to measure, it is

---

[6]Murray (1983) claims, however, that the differences between Protestant and Catholic schools may be overstated. Although there is no question that each school system reflects a different culture, he suggests that there are more similarities of practice within them than differences between them.

reasonable to conclude that group prejudices are heightened. Some see educational segregation as the key to explaining the perpetuation of ethnic conflict and believe that only through school integration can community barriers begin to break down (Fraser, 1973; Irwin, 1994). Others, however, believe that there is no precedent for integrated education in the society and that even if it were implemented, it would not guarantee greater intergroup understanding and tolerance (Rose, 1971). It is of note that despite the thoroughness of educational segregation, a clear majority of both Protestants and Catholics indicate that they would prefer to send their children to mixed-religion schools (Irwin, 1996).

Endogamy At the most primary level of interaction, the social segregation of the two communities is enforced by the norm of endogamy. In his important survey of the late 1960s, Rose (1971) reported that only 4 percent married across religious lines. Later studies have indicated a rising but still low rate of Protestant-Catholic intermarriage (Gallagher and Dunn, 1991; McGarry and O'Leary, 1995a). The separation of the two communities in most institutional spheres, of course, inhibits the development of intimate relations between Protestants and Catholics ultimately leading to marriage.

In addition to physical separation of Protestant and Catholic communities, interreligious marriages are discouraged in much the same way that interracial marriages are inhibited in the United States. Such marital unions create a degree of social ostracism and hardship that few are prepared to endure (Barritt and Carter, 1972; Donnan and McFarlane, 1983). Harris (1972) noted that in Ballybeg, in cases of intermarriage, kinship ties were not established between members of the two families. An intermarrying couple, she explained, was usually married in the wife's church; as a result, tensions were created between the husband and his kinfolk, "especially between them and his wife, that put all normal kin-based contacts between them out of the question" (1972:143).

O Connor (1993) tells of her interview with a young Catholic woman who admits to her that her mother had been Protestant but was now a "very holy Catholic." Nonetheless, the mother's former religious identity is never talked about and remains carefully concealed, even among some family members. "It's hard on my ma," the woman explains, "because she's got to hide what her family was" (168). During the 1970s, in Belfast neighborhoods where interethnic violence flared, mixed-religion families were usually among those forced to flee (Darby and Morris, 1974).

Given the relatively tiny geographical and demographic dimensions of Northern Ireland, the extent to which ethnic segregation has been established and enforced is extraordinary. A Protestant student at one of the few integrated schools in Belfast confesses that "I'd never met a Catholic before I sat beside one in class" (Irwin, 1994:106). Interaction between Protestants and Catholics is more common in smaller communities, of course, where daily contact is a routine occurrence. In Belfast, Derry, and other larger towns, however, the Northern Ireland case demonstrates clearly how in-group members can seal themselves off in homogeneous neighborhoods and, with little effort, avoid out-group members. It is a situation familiar to most Americans, who may reside in urban areas whose communities and neighborhoods are sharply divided along racial lines.

Although the ethnic division in Northern Ireland supersedes all others, class divisions should not be dismissed as insignificant. In fact, this is a society highly stratified by class, and the sociocultural differences between classes within each ethnic group may be as great as the differences between ethnic groups them-selves (Darby, 1976). The intensity of the ethnic conflict has been highest among the poor and the working classes of both groups, and most of the interethnic vi-olence has traditionally occurred within those communities (Jackson and McHardy, 1984; Pollak, 1993). Moreover, residential integration and intergroup contact in general are more evident among middle-class Protestants and Catholics (Boyle and Hadden, 1994; Ruane and Todd, 1996).

STABILITY AND CHANGE

Among the multiethnic societies we have looked at, Northern Ireland presents perhaps the most confounding case of ethnic conflict. Can the divide between Protestants and Catholics ever be bridged so that ethnicity no longer plays the paramount role in intergroup relations?

As Boal and Douglas (1982) point out, although the Northern Ireland conflict has often seemed utterly without solution, it is more accurate to say that any possible solution has been either unpleasant or extremely difficult to attain—or both. It is chimerical to think that profound changes in attitudes and behavioral patterns can be brought about in the foreseeable future. As we have seen, the two communities continue to maintain an extraordinarily high degree of insti-tutional and identificational separateness. Hence, so far as any solution can be achieved, it must lie in some form of political settlement that acknowledges in-terests on both sides.

Over the years, numerous political schemes had been proposed periodically, aimed at resolving the conflict. These ranged from, at one extreme, a fully in-dependent Northern Ireland to, at the other extreme, unification of the North with the Republic of Ireland. In between, proposals included full integration of Northern Ireland with Great Britain; a federal system of some kind, encom-passing the Republic and the North; some form of joint rule of Northern Ireland by Britain and the Republic; repartition of Northern Ireland, in which Protes-tants and Catholics would be placed into ethnically cohesive areas; and a return to simple majority rule within the context of a devolved Northern Ireland government, similar to the situation before Britain assumed direct rule.[7] Wide disagreement among Northern Ireland's various political factions precluded the implementation of any of these scenarios. Thus, direct British rule of Northern Ireland—the least unacceptable political option—remained in place (Arthur and Jeffery, 1996). In the 1990s, however, a proposed power-sharing arrangement was put forth that seemed, finally, to create a political atmosphere conducive to resolving the conflict.

The Peace Process

The peace process begun in the 1990s represented the most promising possibility of ending the conflict since the beginning of its modern phase in 1969. In 1994 paramilitaries on both sides declared a cease-fire. This was followed by a coming together of representatives of the Irish and British governments and all the political parties in Northern Ireland to begin talks that, it was hoped, would culminate in a permanent political settlement. Additional impetus for the peace process was provided by the United States, particularly the personal efforts of President Clinton. All parties to the conflict, including representatives of Sinn Fein (the political arm of the IRA), were invited to participate in the peace talks. This marked an important step forward; the IRA had never before agreed to take part in such negotiations, nor had loyalist groups agreed to sit down with them.

To better appreciate the hope that this new effort held out, consider the consistency with which past attempts at settling the conflict had failed. In 1985, for example, a historic agreement was signed by the British and Irish governments, giving the Republic of Ireland a consultative role in certain policy matters pertaining to Northern Ireland. It was hoped that permitting the Republic a role in the province's affairs would isolate the IRA and its supporters and compel unionists to finally come to terms with the Catholic minority. Its effect, however, was to further inflame the fears of Protestants, many of whom felt that this marked the beginning of British abandonment of Northern Ireland and the first step toward eventual unification with the Republic. Even Catholics viewed the agreement as a disappointment in practice (O'Malley, 1991; Whitney, 1988). Again in early 1991, talks aimed at creating some formula for power sharing broke down amid an eruption of heavy violence by both the IRA and Protestant paramilitaries (Clarity, 1992). The fact that the new peace talks did not flounder, despite the temporary resumption of terrorist activities in 1996, seemed to indicate that this effort might be fundamentally different from those of the past.

---

[7]These proposals are discussed in detail by Barritt (1982), Boyle and Hadden (1994), Palley (1981), Pollak (1993), Rose (1976), and Ruane and Todd (1996).

The peace accord provided for a new Northern Ireland government that, although still ultimately controlled by Britain, would for the first time since 1972 exercise much local autonomy. A Northern Ireland Assembly was created whose 108 members would gradually assume responsibilities for most governmental affairs from the British government. A cabinet-style executive, made up of members from a variety of parties, would wield power. A set of complex checks and balances provided for a kind of power-sharing system in which one side—namely unionists—would be unable to dominate the Assembly, as it had in the past.

Such a power-sharing arrangement was an important new provision, for to return to a simple majority-rule system (as had been in place prior to British direct rule in 1972) would again have condemned Catholics to a permanent minority political status. For several decades, a power-sharing system in which both Protestants and Catholics would participate had been seen as an essential ingredient of any future viable Northern Ireland government (Lijphart, 1975).

The agreement also called for greater cooperation of Northern Ireland with the Republic of Ireland on matters of mutual concern as well as membership for Northern Ireland in a British-Irish Council. The agreement required, however, that the Republic abandon its constitutional claim to sovereignty in the North. This had always been a serious impediment to cooperation with Ulster's Protestant loyalists.[8]

A referendum in both Northern Ireland and the Irish Republic was held in 1998, asking whether the political institutions specified by the agreement should be adopted. By a resounding majority, the people of both Northern Ireland and the Republic voted their approval of the measure, though in Northern Ireland a much greater percentage of Catholics than Protestants supported it.

Two events of 1998 seemed to galvanize the Northern Ireland population, both Protestant and Catholic, around the peace effort and indicated the public's determination to put this new agreement into place. After the peace accord had been approved, during the traditional marching season in July, outbreaks of violence by Protestants accompanied a standoff between British security forces and members of the Orange Order in Portadown, a Protestant-majority town with a strong Orange tradition. The confrontation stemmed from a ban on Orangemen parading through a Catholic neighborhood. In the most serious incident, three young Catholic boys were burned to death in their home in an arson attack by militant Protestants. The protest by the Orange Order dissipated after this incident. An even more shocking event occurred one month later, when twenty-eight people were killed in the bombing of a crowded shopping area in the town of Omagh. The bombing was carried out by a splinter group of

---

[8]A detailed explanation of the agreement is found in de Bréadun, 1999.

the IRA that opposed the peace process. It was the single most deadly action in the entire thirty years of the Troubles. Although gruesome acts of violence had come to be part of the political landscape of Northern Ireland, these two incidents seemed to evoke an unusual degree of outrage on all sides.

2001 and Beyond

For three years after the peace agreement had given new hope to a settlement of the conflict, the process floundered; episodes of neighborhood violence flared repeatedly. The major issue upon which agreement could not be reached involved the disarming of the paramilitaries, in particular the IRA. Protestant unionists refused to participate in the new government unless and until the IRA agreed to destroy its guns and explosives. The intransigence of both sides on this issue seemed to create an impasse that would ultimately lead to the failure, again, of a concerted effort to end the conflict. In October 2001, however, the IRA announced its intention to disarm. The events of September 11 seemed to create a new perspective on the IRA and its activities, particularly in the United States, where it had traditionally secured much financial support from Irish Americans. The IRA's decision to give up its weaponry was seen as an attempt to disassociate itself from the taint of terrorism.

This did not ensure the end of violence, however; loyalist paramilitary groups gave no indications that they would reciprocate by surrendering their arms, and splinter groups of the IRA continued to disregard the cease-fire. The mood of distrust built up over past failed efforts also created much skepticism regarding the IRA's intentions to actually disarm. Moreover, the most militant Protestant politicians vowed to continue to try to undermine what they perceived as a peace accord that favored Catholic interests. But the IRA's move did give new hope to the implementation of power-sharing political institutions within which the major issues of the Protestant and Catholic communities would be peacefully debated and acted upon. Past attempts at a devolved Northern Ireland government that incorporated the interests of both Protestants and Catholics were unsuccessful, of course, and most observers viewed the rescue of the peace accord with guarded optimism.

In his study of three small communities, Darby (1986) showed that during the period of the Troubles, social mechanisms were devised locally to control the violence between Protestants and Catholics, making life tolerable for both sides. The existence of such restraints, Darby maintained, accounted for the enduring nature of the conflict, preventing its resolution through either some form of settlement or a full-scale internal war. As a 1993 citizens' inquiry on the Northern Ireland conflict concluded, despite its intractability and devastating effect on civil society, "the situation is not bad enough for people to demand that their elected representatives make the compromises necessary for a settlement" (Pollak,

1993:10). Over the years, both Protestants and Catholics have, in a sense, learned to live with the conflict. Implementation of the peace accord will test the resolve of both communities to move beyond the acceptance of "tolerable conflict."

Whatever the outcome of the peace accord, Britain is committed to maintaining its link with Northern Ireland as long as the majority of the people of the province prefer it; and given the Protestant numerical majority, this is automatically assured, at least for the immediate future.[9] Removal of the British from Northern Ireland, therefore, although probably inevitable, is likely to be a long-term affair. As for the Irish Republic, there is a mixed attitude toward eventual unification; a long-standing belief that unity of the two parts of Ireland is natural and desirable is tempered by a realization of the considerable political, economic, and social costs involved in bringing it about (Ruane and Todd, 1996).

The desires of the Catholic population of Northern Ireland regarding eventual political unification with the Republic are also more complex than they appear. It is by no means certain that most Catholics do in fact harbor an eagerness for unification. Surveys have indicated that perhaps little more than half would actually vote themselves into the Republic if given the opportunity (Darby, 1995; Davis and Sinnott, 1979; Pollak, 1993; Ruane and Todd, 1996). Furthermore, many in the North, having grown up knowing nothing but partition of Ireland, do not harbor strong feelings of commonality with those in the South (O Connor, 1993). It has often been observed that Catholics are most concerned with receiving equal economic and political opportunities and will accept the Northern Ireland state if such conditions are met. But feeling that Protestant intransigence precluded their receiving fair treatment, many had been driven to see unification with the Republic as their only hope. Whether the peace accord will assuage Catholic suspicion remains to be seen.

The Depth of the Ethnic Division

Although most of what the world learns of Northern Ireland from journalists' reports and the studies of social scientists concerns the discordant aspects of ethnicity in this society, we should not overlook those situations in which Protestants and Catholics interact harmoniously. Indeed, most interaction between the two communities, limited though it is, is not hostile. Moreover, as in any society, activist politics remain confined mostly to a relatively small number. In addition, nonsectarian relations are not unusual among people in the professions, in the arts, and within the university environment (Barritt and Carter,

---

[9]As noted earlier, the Catholic population is growing much faster than the Protestant, and some forecast a Catholic numerical majority sometime in the twenty-first century. Should that occur, Catholics would be in a position to vote to sever the British link.

1972; Darby, 1976; Wallace, 1971). Public opinion polls also indicate, among both Protestants and Catholics, increasingly conciliatory attitudes and an inclination to accept and even encourage the development of integrated institutions, like schools and neighborhoods (Gallagher and Dunn, 1991; Moxon-Browne, 1983; Ruane and Todd, 1996).[10]

The focus of scholars and the world media almost exclusively on the violence that has so characterized the conflict also deflects attention from the significant changes that have taken place in Northern Ireland over the course of the past two decades. As noted earlier, most of what had been institutionalized discrimination against Catholics in employment and housing has been eradicated; the socioeconomic disparity between Protestants and Catholics has been narrowed considerably; both Protestant and Catholic schools are now funded equally; and the commitment to some form of political power-sharing, though resisted by a minority, seems widespread.

The depth of ethnic division in Northern Ireland, however, is undeniably greater than it is in most other multiethnic societies. Hence it is difficult to envision a resolution of the conflict in which all parties can be fully satisfied. It is unlikely, for example, that any solution other than full unification will ever appease militant Catholic elements like the IRA. Protestant extremists are equally resolute in rejecting any form of relationship with the Republic or loosening of Ulster's tie with Britain. More important, as we have seen, among the general population there is no sense of shared community. Most Catholics, at least in principle, aspire to be part of a united Ireland; at the same time, Protestants understand that the very establishment and existence of Northern Ireland, as distinct from the remainder of Ireland, have been intended to protect their interests. As Ruane and Todd explain, "For Northern Ireland to exist as a community in its own right, there would have to be a majority in both communities wholeheartedly committed to the establishment of an overarching and inclusive community embracing all its people. There are some who are so committed, but they are a small minority" (1996:78).

Moreover, even if a political resolution is ultimately effected, this does not mean that ethnic divisions will be quickly erased. It seems clear that despite the optimism created by the adoption of a new, fairer political system, most people do not see this as bringing about an instant resolution of the Troubles. Following the referendum on the peace accord, voters were asked if they saw an end to the violence; most answered no (Berke, 1998). People on both sides of the ethnic divide have become inured to the violence that has so deeply taken root during the past thirty years. In a society as tiny as Northern Ireland, few have been unaffected,

---

[10] It is, of course, possible that such views merely reflect acceptance in principle, with no corresponding proclivity to convert such opinions into practice (Davis, 1994; McGarry and O'Leary, 1995a).

despite the fact that, by comparison with other ethnic conflicts in the world, Northern Ireland's—3,500 killed since 1968—seems minuscule. As Darby has put it, "If peace processes may best be understood as the state of tension between a culture of violence and a culture of peace, the culture of violence is vastly more powerful in Northern Ireland. It will take time to change that" (1998:8).

So deep-seated are the major differences—real or putative—between this society's two ethnic communities that one cannot realistically see, even in the long run, their full integration. "Communal tensions, territorial segregation, and endogamous social and sexual relations," notes Brendan O'Leary, "will remain dominant features of social and political life in Northern Ireland, with or without a definitive constitutional settlement" (1995:713). Moreover, even with a political settlement, it is unlikely that the distrust between the two communities spawned by the Troubles will be significantly reduced for many years, perhaps even generations. In fact, surveys in 2002 indicated that despite the peace process, Protestant and Catholic neighborhoods were actually less integrated than they had been 10 years earlier. The surveys also found an increasing level of intergroup prejudice, particularly among young adults, and the widespread belief that community relations had worsened during the same period (Brown, 2002). The most viable objective, therefore, may be the establishment of a social order in which maximum material and psychological security and optimum political power are provided people of both groups, divided though they remain.

In Chapter 4 we saw that race and ethnic relations always involve a synthesis of conflict and competition on the one hand and cooperation and accommodation on the other. The tragedy of Northern Ireland is the severe imbalance in this fusion, sustained by a tendency to create the present mainly in terms of the past. Only by transcending its history, therefore, can this society achieve a higher degree of ethnic harmony among the majority of its people.

SUMMARY

Northern Ireland is one of the most polarized societies of the modern world, one where ethnic conflict often seems to defy explanation. Antagonism between dominant Protestants and minority Catholics, although especially intense since the late 1960s, is rooted in events and processes that extend back over three centuries.

The crux of the contemporary conflict stems from the partition of Ireland in 1920 following the achievement of Irish independence from Britain. Within the province of Ulster, Protestants were numerically dominant and resisted inclusion in the new Republic. Hence Britain created a semiautonomous Northern Ireland state, leaving Catholics in this newly formed political entity a numerical minority. Today most Protestants of Northern Ireland desire to retain the link with Britain, and most Catholics desire a unification with the Republic.

Comparing Northern Ireland and the United States

Points of Similarity
- The limited social interaction between blacks and whites in the United States is comparable to that in Northern Ireland between Catholics and Protestants.
- The maintenance of high levels of residential segregation, little intermarriage, and low levels of primary structural assimilation in general that characterize black-white relations in the United States are very similar in Northern Ireland between Catholics and Protestants.
- Historically, high levels of prejudice and discrimination against blacks in the United States have been paralleled in Northern Ireland against Catholics.

Points of Difference
- Little in the American ethnic experience can compare with either the historical or political basis of Northern Ireland's ethnic strife.
- The Northern Ireland ethnic conflict is in large measure a political conflict in that the two sides cannot agree on a legitimate government.
- Ethnic diversity in Northern Ireland is quite narrow, limited essentially to two groups, Catholic and Protestant, in contrast to the vast ethnic diversity of the United States.

Northern Ireland is a society in which ethnicity affects almost all aspects of social life. Protestants and Catholics are highly segregated in education, residence, work, and leisure activities. Much of this segregation is rooted in past and present practices of discrimination, but much is also voluntary. The two groups perceive each other as products of fundamentally different cultural heritages and do not recognize a basis for political consensus. In the past Protestants, given their numerical superiority, maintained political supremacy and commonly used that power to favor Protestant economic and social interests.

A mutually satisfactory solution to the ethnic conflict in Northern Ireland has been repeatedly frustrated by the deep social divisions between the two groups and the society's uncertain political status. Most Protestants recognize Northern Ireland as rightfully part of Britain, whereas most Catholics recognize it as part of the Republic of Ireland. As a result, a state recognized as legitimate by both groups, with power to enforce social order, has not been possible to create. For the past thirty years this has left the political arena to extremist paramilitary groups, both Catholic and Protestant, and the British forces. New hopes have been stirred in the past few years, however, as the major parties to the conflict have negotiated a settlement calling for a power-sharing government, participated in by Catholics as well as Protestants.

Suggested Readings

Bell, J. Bowyer. 1996. *Back to the Future: The Protestants and a United Ireland*. Dublin: Poolbeg. A portrait of Ulster's Protestant community and its perspective on the conflict, explaining the basis of loyalist politics.

Boyle, Kevin, and Tom Hadden. 1994. *Northern Ireland: The Choice*. New York: Penguin. A concise description of Ulster's two communities, with an exploration of possible solutions to the Troubles.

Burton, Frank. 1978. *The Politics of Legitimacy: Struggles in a Belfast Community*. London: Routledge & Kegan Paul. An ethnographic study of a Belfast Catholic working-class neighborhood, which analyzes the ideological basis of the ethnic conflict, particularly as expressed by the IRA.

Darby, John. 1995. *Northern Ireland: Managing Difference*. London: Minority Rights Group International. A comprehensive description and analysis of Northern Irish society and the economic, political, and social bases of the Troubles.

Fraser, Morris. 1973. *Children in Conflict: Growing Up in Northern Ireland*. New York: Basic Books. Writing at the height of the Troubles, a child psychiatrist explores the effect of the ethnic conflict in Northern Ireland on children.

McGarry, John, and Brendan O'Leary. 1995. *Explaining Northern Ireland: Broken Images*. Oxford: Blackwell. Considers various explanations—external and internal—of the roots and intractability of the Northern Ireland conflict.

O Connor, Fionnuala. 1993. *In Search of a State: Catholics in Northern Ireland*. Belfast: Blackstaff Press. Through in-depth interviews, the author presents a fascinating portrait of the political and social views of Northern Ireland's Catholic population.

Ruane, Joseph, and Jennifer Todd. 1996. *The Dynamics of Conflict in Northern Ireland: Power, Conflict and Emancipation*. Cambridge: Cambridge University Press. A carefully researched analysis of the historical, political, economic, and cultural dimensions of the ethnic chasm. Demonstrates the complexities of and differences within, as well as between, Protestant and Catholic communities.

Smith, David J., and Gerald Chambers. 1991. *Inequality in Northern Ireland*. Oxford: Clarendon Press. Traces the inequality of Ulster's Protestants and Catholics in various areas of social and economic life and the effect of policies designed to combat discrimination.

Stewart, A. T. Q. 1977. *The Narrow Ground: Patterns of Ulster History*. Belfast: Pretani Press. Contributes to an understanding of the current ethnic conflict in Northern Ireland by examining its deep historical roots.

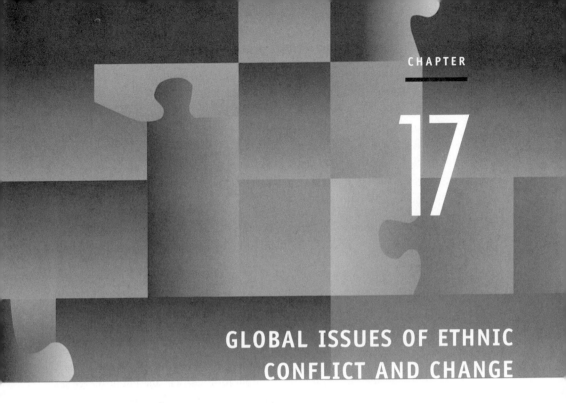

GLOBAL ISSUES OF ETHNIC CONFLICT AND CHANGE

SINCE THE END of World War II, societies in all regions of the world have become more ethnically diverse. Those—like the United States and Canada—with a history of absorbing a variety of ethnic groups emerged as even more heterogeneous. Many others that had been relatively homogeneous, as in western Europe, evolved into truly multiethnic societies. This trend has continued unabated.

In addition to increasing diversity, societies that have had a multiethnic structure and tradition for many decades have experienced a rise in the level of ethnic nationalism among their constituent parts. Whereas civic nationalism had promised to diminish the significance of race and ethnicity in diverse societies, ethnic nationalism has reemerged with great intensity, sometimes at a ferocious level.

In this chapter we look at these two most evident trends of ethnicity in the contemporary world—the proliferation of ethnic diversity and the resurgence of ethnic nationalism. Both, as we will see, have contributed to a marked rise in ethnic conflict.

THE GLOBAL EXPANSION OF ETHNIC DIVERSITY

The expansion of ethnic diversity throughout the world in recent decades is attributable primarily to two developments—shifting patterns of immigration and inter- and intrasocietal political conflicts.

Shifting Patterns of Immigration

With the movement toward industrialization throughout the West, during the 100 years between the mid-nineteenth and mid-twentieth centuries, global migration occurred mostly within Europe and between Europe and North and South America. Beginning in the 1950s, however, that traditional pattern was replaced by an international migration system encompassing virtually all world regions. Three major immigration streams evolved: (1) from Latin America and the Caribbean countries primarily to the United States, and secondarily to Canada and western Europe; (2) from south and east Asian countries primarily to the United States, Canada, and Australia, and secondarily to western Europe; and (3) from North Africa, the Middle East, and southern Europe primarily to northwestern Europe, and secondarily to the United States and Canada. The traditional east-to-west pattern of immigration, then, was replaced by a primarily south-to-north system involving the movement of people from the developing societies, or Third World, primarily to the developed, industrialized societies (Figure 17.1). As part of this global system, few countries today are not either sources or recipients of immigrants.

Not only did the direction of world migration change starting in the 1950s, but its scope expanded greatly and its pace accelerated. Whereas international migration had been sharply reduced during the 1930s due to a worldwide economic depression and then by the outbreak of World War II, in the 1950s immigration began anew with great impetus. Nothing comparable had occurred since the great migrations to North America in the nineteenth and early twentieth centuries. By 2000 an estimated 150 million people were living in a country other than their country of origin; this represents almost 2.5 percent of the world's population (United Nations, 2000). The flow of international migration is not likely to abate in the near future.

Uneven Development Several factors account for the reemergence of large-scale immigration after an interim of several decades. First, immigration has accelerated as a consequence of the widening gap between rich and poor nations. As with previous movements, the primary driving force of the new international migration has been the push-pull factors of labor markets: Workers from less developed countries migrate to those countries where jobs are more plentiful and lucrative. As noted in Chapter 5, this more than anything else accounts for the unprecedented levels of immigration to the United States and Canada during the past three decades. As we will see, it also accounts for the immigration to western Europe of southern Europeans, North Africans, and people from the Middle East. The discrepancy between rich and poor nations became evident throughout the 1950s and 1960s when many former colonies

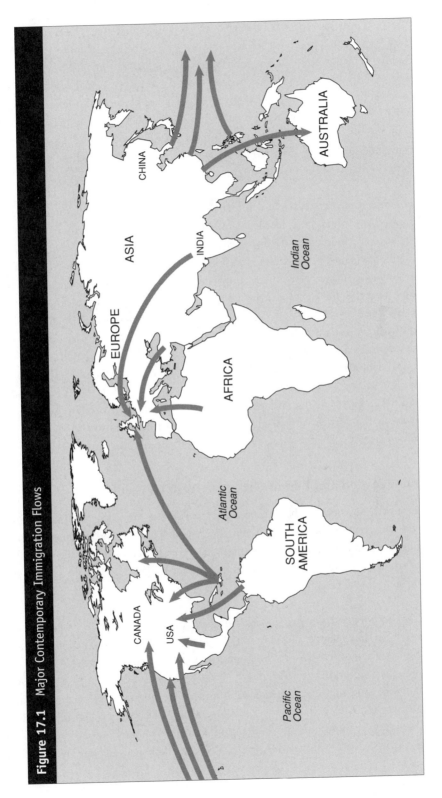

Figure 17.1 Major Contemporary Immigration Flows

in Africa, Asia, and the Caribbean gained their independence. The economies of these newly independent countries were generally incapable of supporting rapidly growing populations, thereby creating a migration push. Most of the world's population growth has occurred since the end of World War II, and the overwhelming majority of that growth has taken place in the less developed countries. Moreover, gains in economic development made by many of these countries in the 1960s and 1970s slowed or reversed in the 1980s due to economic recession, growing debt, and internal conflicts. These conditions induced further pressures to migrate to the wealthier countries.

Political Conflict Wars and internal political strife in many societies have created additional push factors stimulating global migration and the resulting growth of ethnic diversity. From the late 1950s to the present, civil wars, internal uprisings, revolutions, and political violence in various forms created a worldwide flow of refugees unprecedented in modern history. Virtually every part of the developing world, as well as some countries in the developed world, was affected. Uprisings in the 1950s and 1960s against Soviet occupation in eastern Europe; revolution in Cuba in 1959; civil wars in the 1970s and 1980s throughout Central America; political violence in parts of the Caribbean and the South American continent; the war in Vietnam; civil wars in many parts of Africa; ongoing wars in the Middle East; the Persian Gulf War in 1992; the collapse of the Soviet Empire in the late 1980s; and wars in the former Yugoslavia in the 1990s all contributed to the swell of refugees alongside the more conventional economic migrants.

Immigration and Ethnic Change in Western Europe

Arguably, nowhere in the contemporary world has the impact of immigration and the creation of greater ethnic diversity been more pronounced than in Europe. Today the European continent also vividly demonstrates the durability of ethnicity in the modern world and its often malignant potential to create and sustain societal conflict.

Contemporary Immigration to Europe After World War II, workers from southern European countries—Italy, Portugal, Greece, Spain, Turkey, and Yugoslavia—or from newly independent former colonies such as Algeria and Indonesia migrated to Germany, France, Switzerland, the Netherlands, Belgium, and the Scandinavian countries as these nations experienced their postwar industrial expansion. A similar immigration occurred in Britain with the entry of thousands of people from the newly independent countries of the British Commonwealth—India, Pakistan, Jamaica, and others (Castles,

1992; Castles, Booth, and Wallace, 1984; Castles and Kosack, 1973; Castles and Miller, 1998; Garson, 1992; Paine, 1974). Labor was in short supply, and immigrants were therefore recruited to work in the economic reconstruction of these countries. This population movement accelerated in the 1960s, leaving previously homogeneous western European societies with substantial ethnic communities made up of immigrants of vastly different cultures and physical characteristics.

Although by the late 1970s immigration had been severely curtailed by restrictive measures, many of those who had come as temporary workers stayed on, establishing themselves and their families as permanent residents. The period between the late 1980s and early 1990s witnessed another surge in migration to Western Europe. This migration was stimulated mostly by the disintegration of the Soviet Union and civil wars in Yugoslavia, but also by the influx of labor migrants and refugees from underdeveloped nations of North Africa, the Middle East, and Asia. By the mid-1990s immigration to western European countries once again tapered off, as governments introduced tighter entry procedures and controls (Organization for Economic Co-operation and Development [OECD], 2001; United Nations, 2000). But by that time the infusion of immigrants had rendered a powerful demographic effect.

Today approximately 20 million foreign workers and their families live in western Europe. Germany alone has over 7 million immigrants (9 percent of the population) and France about 3.5 million. In each case, most are Muslim (Drozdiak, 1997; Gedmin, 1994; Peach and Glebe, 1995; Walsh, 1992). In Switzerland, the foreign-born constitute almost 20 percent of the population and nearly the same proportion of the labor force. Italy, long Europe's major immigrant-sending society, is now an attractive destination for immigrants, as are Spain, Portugal, and Greece, other traditional sending societies (Campani, 1993; Sciortino, 1992; Stille, 1992; United Nations, 2000). Even in Sweden, long considered a culturally undiluted corner of Europe, one person in eight was a first- or second-generation immigrant by the late 1980s. Table 17.1 shows the growth in the foreign population of some European countries during the 1980s and 1990s. These figures are actually understatements of the foreign-born since they do not include those international migrants who have been naturalized in their destination countries. In France, for example, more than a million foreign-born people have acquired French citizenship; and in the United Kingdom, more than 2.5 million have become British citizens (United Nations, 1996).

The Societal Impact of Immigration to Europe When they began to arrive in the 1960s, immigrant workers were seen by the native populations as

Table 17.1 Foreign Population of Selected European Countries

COUNTRY	1988		1998	
	NUMBER (1,000s)	% OF TOTAL POPULATION	NUMBER (1,000s)	% OF TOTAL POPULATION
Austria	344	4.5	737	9.1
Belgium	869	8.8	892	8.7
Denmark	142	2.8	256	4.8
France	3,714	6.8	3,597	6.3
Germany	4,489	7.3	7,320	8.9
Italy	645	1.1	1,250	2.1
Luxembourg	106	27.4	153	35.6
Netherlands	624	4.2	662	4.2
Norway	136	3.2	165	3.7
Portugal	95	1.0	178	1.8
Spain	360	0.9	720	1.8
Sweden	421	5.0	500	5.6
Switzerland	1,007	15.2	1,348	19.0
United Kingdom	1,821	3.2	2,207	3.8

SOURCE: OECD, 2001.

temporary residents who would return to their home countries once their labor was no longer needed. This was made especially poignant in Germany (then West Germany), where immigrants were labeled *Gastarbeiter*, literally "guest workers." Although the workers were a vital component of the economic system, little effort was made to integrate them socially or culturally into the host society. The workers and their families were marginal people, and the social conditions they experienced were reminiscent of those experienced by successive waves of European immigrants and southern black migrants to northern U.S. industrial cities in the late nineteenth and early twentieth centuries. Their housing was the oldest and most dilapidated; their jobs were the lowest paying and least prestigious; and, except for work, they did not participate in the society's mainstream institutions. Moreover, in Germany, with the largest number of foreign workers, the eventual assimilation of the immigrant population was not seen as an appropriate societal goal. Today the socioeconomic conditions of Germany's minority ethnic groups are improving, but there is no evidence that assimilation has advanced very far (Seifert, 1998). Very few of Germany's 3 million Turks, for example, have been afforded citizenship, even those who are German-born and know nothing of any other society.[1]

As in Germany, most immigrant groups in other western European societies—especially those from Africa, Asia, and the Middle East—continue to experience a high level of economic and social marginalization. Their unemployment rates are high and they are overrepresented in unskilled, low-paying jobs (OECD, 2001). Further, they live mostly in concentrated areas and participate only minimally in public life. In Britain, for example, following severe race riots during the summer of 2001, a government report linked the outbreaks to the social and physical segregation of white and nonwhite—mostly South Asian—communities (Hoge, 2001).

Perhaps most seriously, immigrants in western European countries have encountered high levels of prejudice and discrimination (Campani, 1993; Castles, Booth, and Wallace, 1984; Finn, 2002; Power, 1979; Skellington, 1996). As previously homogeneous nations, western European countries lacked the historical experience of the United States, Canada, and other traditional immigrant societies in absorbing large numbers of ethnic newcomers. The result has been, at best, a begrudging tolerance of immigrants and, at worst, outbreaks of ferocious hostility. In Germany, physical attacks against immigrants by the early 1990s had become commonplace; in France, North Africans, who compose fully half the immigrant population, are the primary targets of ethnic violence. Even in countries with fewer immigrants and solid traditions of tolerance, like Italy and Sweden, racially motivated incidents have surfaced.

It is clear that regardless of the effort to limit immigration, sizable ethnic populations will remain a permanent part of western European societies. The children and grandchildren of the original immigrants are European born or reared, and for them a society of origin, or "mother country," is only an abstraction. Moreover, it would be difficult for the economies of Germany, France, and other countries in the region to function without foreign workers. With extremely low birthrates, these societies can fill their future labor needs only through substantial immigration. Western Europe is a central link in the emergent global economy in which the flow of capital and goods and, increasingly, workers is unhindered by national boundaries. As in North America, the tendency toward greater ethnic and racial diversity, therefore,

---

[1]In 1999 German laws were changed, making it easier for the foreign-born to obtain citizenship. Prior to that time, Germany awarded citizenship on the basis of ancestry. This meant that ethnic Germans, having lived, for example, in Russia for many generations and speaking no German, could obtain German citizenship merely by moving to Germany (which in fact many did in the 1990s). By contrast, very few second- or third-generation Turks, having lived nowhere but Germany and speaking only German, were able to qualify for citizenship.

is accelerating and appears irreversible. These countries are now permanently multiethnic.

As a result of this new heterogeneity, western European ethnic patterns have begun to converge with those of the United States (Lambert, 1981). In some ways the impact of increasing ethnic diversity on western European societies in recent years has been even more profound than on the United States, and its social, political, and economic consequences more wrenching. Questions concerning the social and economic integration of large and distinct ethnic minorities are now in the forefront of political discourse. Few parties at any point on the political spectrum have not openly advocated, at minimum, a halt to further immigration; extreme right-wing parties, most notably in Germany and France, have called for immigrants to be repatriated to their countries of origin. In sum, it is quite reasonable to expect that for many years increasing ethnic diversity will continue to be a critical internal issue facing most western European countries.

THE RESURGENCE OF ETHNIC NATIONALISM

Most people in North America have only a vague understanding of political and social affairs that occur outside the boundaries of their own society. But even those who kept only casually informed of world events in the 1980s and 1990s could not help becoming aware of the increasing level of ethnic nationalism, accompanied by political conflict, that erupted in various parts of the world. Some situations, as in Canada, were dealt with through peaceful political means. Others, as in Rwanda, Sri Lanka, and Russia, were violent.

As the world watched the dramatic end of the Cold War in the late 1980s, expectations were high that a new era of global tranquility was on the horizon. With the political map of the world changing frequently and rapidly, however, it became obvious that a new force had been unleashed, one that augured not a more peaceful globe but one enmeshed in ethnic conflict. Ethnic nationalism seemed to replace civic nationalism as a stimulant of identity and political action. Indeed, at the outset of the twenty-first century, ethnic nationalism appears to be the most ubiquitous, intractable, and devastating global force.

Forms of Ethnic Nationalism

Nationalism, as a form of political and social expression, emerged in the eighteenth century with the appearance of the first modern nation-states in Europe and the United States (Kohn, 1944). Loyalties to family, kin, tribe, or community were extended so as to encompass entire peoples, marked off by common language, religion, and other cultural elements. Benedict Anderson (1991) has referred to these emergent nations as "imagined communities." In the twentieth century, and especially after the end of World War II, nationalism spread throughout the world among formerly colonized peoples, intent on achieving political self-determination. It also advanced among well-established nations where diverse ethnic groups asserted their identity and sought greater social and political autonomy.

In its modern-world expressions, nationalism has been of two types—civic and ethnic (Geertz, 1994). Civic nationalism, Michael Ignatieff notes, "maintains that the nation should be composed of all those—regardless of race, color, creed, gender, language, or ethnicity—who subscribe to the nation's political creed" (1993:6). Here, people possess equal political and social rights and choose to be members of a nation, along with others, regardless of ethnicity, who share broadly similar beliefs and values. American patterns of nationalism, as well as those of most other immigrant societies, have historically been expressed along these lines. Ethnic nationalism, by contrast, holds that people's allegiance is to an ethnic group or nationality into which they have been born or assigned, not to a larger political entity encompassing many ethnic groups or nationalities. Here, "an individual's deepest attachments are inherited, not chosen" (Ignatieff, 1993:7–8). In the past several decades, every continent on the globe has witnessed cases of intense ethnic conflict in one or more countries, provoked by ideologies of ethnic nationalism.

One form of ethnic nationalism involves political actions based on claims of the superiority of a dominant, or majority, ethnic group over others and its portrayal of rival groups as potentially dangerous. In such cases, the dominant group seeks to entrench its power by further subordinating, expelling, or perhaps even exterminating rival ethnic groups. The two most unmistakable examples of this type of ethnic nationalism in recent years have occurred in Rwanda and the former Yugoslavia; the latter is described in greater detail in the next section. In Rwanda, majority Hutus in 1994 slaughtered at least half a million rival Tutsis within just a few weeks, an action described as a genocide—a systematic, orchestrated effort to exterminate an entire people (African Rights, 1994; Destexhe, 1995).[2] Journalist David Rieff recounts the killing as it began and gained momentum:

> Over the radio, the call kept going out for the extermination of every Tutsi man, woman and child in Rwanda, and of the Hutus who opposed this final solution. In the villages, people were hunted down. They died by the tens of thousands in their homes, in their fields and in the churches . . . in

---

[2]No one will ever know the number actually killed. Some estimate as many as a million.

which they had sought refuge. And no matter how many died, the radios kept blaring out the calls for all good Hutus to kill the *imyenzi*, the "cockroaches," who were polluting the Rwandan nation and preventing it from living in peace. (1996:31)[3]

Another form of ethnic nationalism, more common in the contemporary world, involves the assertion of ethnic identity on the part of minority ethnic groups and their challenges to majority-group dominance. Ethnic groups may view themselves as culturally and politically oppressed and seek either greater powers or independence from the nation-state dominated by a rival group. Political and cultural movements by Basques in Spain, Bretons in France, Flemings in Belgium, Tamils in Sri Lanka, Ibo in Nigeria, Kurds in Iraq, Iran, and Turkey, Chechens in Russia, Scots and Welsh in Britain, and, as we have seen, Quebecers in Canada all illustrate this form of ethnic nationalism.[4]

In the past decade, ethnic nationalism has been particularly virulent and explosive in eastern Europe. A new and ominous factor was introduced into European patterns of immigration and ethnic heterogeneity with the breakup of the Soviet Union and the fragmentation of eastern European states in the late 1980s and early 1990s. Almost all countries of eastern Europe—those that had been part of the Soviet sphere of influence following World War II—have lengthy histories of ethnic diversity, each made up of a congeries of culturally distinct populations occupying rough territorial concentrations. Those ethnic divisions, however, had been held in check by Soviet dominance. As these countries asserted their national identities and secured political autonomy beginning in the late 1980s, that social glue dissolved.

The Breakup of Yugoslavia

Nowhere in eastern Europe was ethnic nationalism and ensuing conflict more intense and tragic than in Yugoslavia as that society experienced political breakup. Indeed, so stark and violent was the conflict that it held the attention of world media for several years. Thus, few could avoid becoming aware of the sheer brutality exhibited by all sides.

The Attempt at Corporate Pluralism Yugoslavia represents the failed attempt of a multiethnic society to construct some form of corporate pluralism. As we saw in Chapter 4, corporate pluralism is a situation in which a number of eth-

---

[3]Ethnic conflict between Hutus and Tutsis has also been ongoing for many years in neighboring Burundi.

[4]Even the rise of cultural pluralism in the United States can be seen, in a broad sense, as part of this global trend.

nic groups coexist in a loosely bounded system. These culturally and perhaps linguistically distinct groups are linked together in a federation in which each is recognized officially by a central government that attempts to distribute national political power and economic resources proportionally among them. Each, however, enjoys some degree of political autonomy.

In such cases, each ethnic group also usually occupies a distinct territory or geographical area, where it becomes the dominant group. However, these ethnically concentrated populations usually spill over or are dispersed to other areas of the country, where a majority of the population is composed of members of another ethnic group. As a result, minority ethnic groups develop in those territories. When corporate pluralism breaks down, the federation begins to crumble, and each ethnic territory seeks full political independence. As ethnic rivalries intensify, each ethnic group attempts to maximize its territory and to embrace group members who are living as ethnic minorities in other areas. Previously subdued historic rivalries are regenerated and intergroup hostilities heighten, perhaps leading to violent confrontations. The contemporary Yugoslav situation illustrates this process vividly.

Prior to 1991, Yugoslavia, a country of about 24 million, had been comprised of six republics and two provinces, each different in social and political makeup and history, and each dominated by a particular ethnic group. Constituting the six republics were Serbs, Croats, Muslim Slavs, Slovenes, Macedonians, and Montenegrins (Figure 17.2). All of these groups (comprising about 80 percent of the total population) were ethnically south Slavs; but they differed, sometimes radically, in religion and other cultural aspects. Other, smaller ethnic groups were Albanians, less than 8 percent of the total population, and Hungarians, about 2 percent. The remaining 10 percent was made up of numerous tiny ethnic populations.

The cultural differences among the former Yugoslavia's peoples cannot be overstated. Religion in particular has been an ancient divide. In a sense the region represents a confluence of three civilizations: in the west, one that derives from the Roman Empire; in the east, from the Byzantine Empire; and in the south, from the Turkish Ottoman Empire. Three religions—Roman Catholicism, Eastern Orthodox, and Islam—therefore divide the populace. Each major ethnic group is strongly identified with one of these: Croats are Roman Catholic, Serbs are Eastern Orthodox, and almost half of the Bosnian population is Muslim.

For almost thirty-five years following World War II, this ethnically disparate and politically divided country had been held together delicately by Josip Broz Tito (usually referred to simply as Tito), a World War II hero who led the victorious Partisan guerrilla army against the Nazis. After the war Tito emerged as a unifying authoritarian figure, acting as prime minister and later as president. Tito, though an avowed communist who shaped the Yugoslav economic and

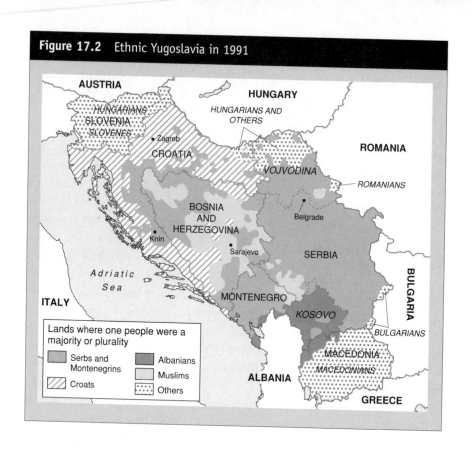

Figure 17.2 Ethnic Yugoslavia in 1991

political systems along communist lines, moved the country toward nonalignment and away from the Soviet hegemony in eastern Europe.

Tito's strong rule and ability to bring the various ethnic factions together had seemed to create a multiethnic society that fulfilled the major features of corporate pluralism. Institutions were carefully crafted to preserve a sense of ethnic balance and fairness—no easy task in a state in which ethnic diversity characterized almost every facet of social and political life. Equal political and social rights were constitutionally guaranteed for all ethnic groups, and measures were aimed at reducing ethnic intolerance and discrimination. A federal system was created in which the six republics and two provinces enjoyed much autonomy, though they shared national power and authority. The federal and republican constitutions provided for proportional representation of the nations and nationalities in all governing bodies and institutions, in a kind of consociational arrangement. Each group was entitled to use its language in its schools and in its political and commercial life. Even at the national level, linguistic diversity was accommodated by the federal government, which sanctioned three official languages—Serbo-Croatian, Slovenian, and Macedonian. Two minority languages, Albanian and Hungarian, were also given state recognition. Finally, despite the society's sharp tripartite religious division, religious freedom was protected. In addition to Roman Catholics, Orthodox, and Muslims, small numbers of Jews and members of other minor religions were also part of the Yugoslav social landscape.

None of these measures, however, created more than superficial ethnic harmony. With Tito's death in 1980, the rancor that had been simmering beneath the surface gradually intensified, erupting into full-blown warfare in 1991.

The Dissolution of the Yugoslav Federation With the end of the Tito regime, Yugoslavia dissolved into an extremely loose federation in which the various republics increasingly dominated the central government. Most people identified themselves with their specific ethnic group, not as "Yugoslavs." Ethnic rivalries previously held in check reemerged, stirred by political leaders in each republic. These rivalries were exacerbated by widening economic differences among regions and by the breakup of East European communism in the late 1980s. The ethnoreligious makeup of the Yugoslav federation prior to its dissolution in 1991 can be seen in Table 17.2.

The major ethnic split was between the republics of Croatia and Slovenia on the one hand, and Serbia on the other. The former are both Roman Catholic and strongly oriented culturally toward western Europe; Serbia, in contrast, is Orthodox Christian in religion and more culturally attuned to eastern Europe. With the collapse of the Yugoslav Communist party in 1990, Croatia and Slovenia each moved toward independence and seceded from the federation a year later. This left Yugoslavia composed mainly of Serbia. The smaller and less economically viable republics took sides or became pawns in the ensuing confrontations that developed as the three major ethnic states sought to establish boundaries that each interpreted as corresponding to their historical ethnic dominance.

The Slovenes were a more westernized, industrialized, and prosperous population, occupying a relatively homogeneous area. As a result, resistance to their exit from the federation and declaration of independence was short-lived. Most important, few members of the other ethnic populations of Yugoslavia resided in the Slovenians' traditional territory.

The Serbs and Croats, then, remained as the major ethnic rivals. The rivalry between them that erupted in the early 1990s was in some ways an extension of animosities that had run deep for many decades. The Serbs and Croats, along with the Slovenes, had been linked together in a constitutional monarchy after World War I; but lacking a tradition of political compromise, the three were unable to achieve a national unity, and by the start of World War II the country was severely divided ethnically. As a result, fighting occurred not only with the German invaders but also among the various ethnic factions themselves, notably between Serbs and Croats, who took opposing sides during the Nazi occupation. More than 1.5 million Yugoslavs died during the war, most as a result of interethnic fighting.

Table 17.2 Major Republics of the Former Yugoslavia

REPUBLIC	MAJOR CITY	CURRENT POPULATION	ETHNORELIGIOUS MAKEUP BEFORE 1991
Serbia*	Belgrade	10 million	About two-thirds Serbs (Eastern Orthodox), but significant numbers of Albanians, mostly in the region of Kosovo, as well as Croats and Hungarians. The Croat population mostly fled after 1991.
Montenegro*	Podgorica	674,000	Montenegrins (Eastern Orthodox), about 60%, are ethnic Serbs. Also a Muslim population of about 15%, as well as smaller ethnic minorities.
Croatia	Zagreb	4.7 million	Croats (Roman Catholic) the majority (78%), but a large Serb minority (about 12%). By 1996 the Serb population was less than 4%.
Bosnia and Herzegovina	Sarajevo	3.4 million	The most ethnically diverse—about 44% Muslims, 32% Serbs, and 17% Croats. Population in 1991 was 4.3 million. Tens of thousands killed in the war; others fled. After war, in-migration of 200,000 ethnic Serbs from Croatia.
Slovenia	Ljubljana	2 million	Relatively homogeneous, with Slovenes (Roman Catholic) over 90%.
Macedonia	Skopje	2 million	67% Macedonians (Eastern Orthodox), 23% ethnic Albanians (Muslim), 10% others, including Serbs.

*Serbia and Montenegro together constitute the current Federal Republic of Yugoslavia.

During World War II, Hitler established a puppet fascist regime in Croatia that was run by the Ustasha party. The Ustashas were vehemently anti-Serb and anti-Semitic. Their essential objective was to purge Croatia of Serbs and, of course, Jews as well, though the latter represented a relatively small population. As a result, the Serbs in Croatia suffered many indignities, including the forced wearing of colored armbands to identify themselves. More significantly, hundreds of thousands of Serbs were executed by the Ustasha regime. Many Croat nationalists and clergy were active in that regime and participated in the oppression. At the infamous Jasenovac concentration camp, the Ustashas murdered

85,000 people, most of them Serbs (Bennett, 1995). Following the war these savage killings were bitterly remembered by the Serbs; in their minds all Croats, whether they had been Ustashas or not, were associated with the killings.

Similar actions took place during World War II in Serbia, but against resident Croatians. Croatians today point to Serbian guerrilla bands called Chetniks, who fought against the Nazis (and thus the Ustashas) but who, in the process, oppressed and slaughtered many Croatians. Because of the atrocities on both sides, Serbs and Croats found themselves in a perpetual state of mutual distrust (Schöpflin, 1993). Indeed, so bloody had been the conflict inside Yugoslavia during World War II and so venomous were the emotions it created on all sides that, as journalist Misha Glenny accurately forecast, "were it ever to revive, it was always likely to be merciless" (1993:171).

The Contemporary Conflict Some argue that Yugoslavia's attempt at corporate pluralism was always precarious and, without Tito, was doomed. Others, however, maintain that it did not fall apart of its own volition but rather was destroyed by a few people acting in their political interests, especially Serb President Slobodan Milosevic (Bennett, 1995; Rieff, 1995).

With the election of the party of Franjo Tudjman in Croatia in 1990, a new wave of Croatian nationalism emerged. Serbia, led by Milosevic, experienced its own resurgence of nationalism, a key component of which was a rekindled hatred of the Croats and a remembrance of the World War II horrors. Through a concentrated propaganda campaign in the media, the Serbian government instilled apprehension and suspicion among Serbs of their Croatian and Bosnian Muslim neighbors. Serbian television, for example, frequently showed films of the atrocities committed by the Croatian Ustashas during World War II. Serbian authorities also censored the press and imposed a blockade on outside information, allowing the government to manipulate facts and communicate misinformation as the conflict unfolded. Croatian and Bosnian governments offered their own forms of propaganda (Human Rights Watch, 1995).

At the outbreak of the conflict, the Serbs were not only the largest of Yugoslavia's ethnic groups, but the most dispersed, living in all areas. Serbia alleged that Serbs living outside the Serbian nation—that is, in other parts of Yugoslavia—were victims of oppression, especially in Croatia and Bosnia-Herzegovina. Serb leaders and the media engendered fear among Croatian and Bosnian Serbs that, if the territories in which they were living were to become independent states, they would be divested of their rights and property (Bennett, 1995). The conflict, then, essentially became one in which these minority populations were "liberated" by Serbs. The first of the wars erupted in the summer of 1991 in a Serb-populated region of Croatia. There, Serbs refused to accept living in an independent Croatia and formed their own rebel state. Not until four years later did the Croatian army smash the resistance and drive out most Serbs who had been living there.

Although they lived in all of Yugoslavia's republics, Muslims were heavily concentrated in Bosnia-Herzegovina (often referred to simply as Bosnia). There they were about 44 percent of the population. Sarajevo, the republic's capital and largest city, was not only heavily Muslim in numbers but also very much dominated by Islamic culture. Bosnia's Muslims, however, should not be seen as comparable to Muslims in other parts of the world. The Bosnian Muslims are secular, having only a distant link with other more traditional Islamic populations. Most are Slavs (that is, Serbs or Croats) who were converted to Islam during the five centuries of Ottoman rule in Bosnia (Glenny, 1993). Most, moreover, are Sunni Muslims, not in sympathy with fundamentalist groups in the Middle East and North Africa.[5] Like the Serbs and Croats, Bosnian Muslims speak a variant of Serbo-Croatian.

Ethnic Cleansing in Bosnia The second of the wars occurred in Bosnia. When Bosnia declared its independence from Yugoslavia in 1992, Bosnian Serbs refused to accept this arrangement, preferring instead to remain part of Yugoslavia (which by that time was essentially only Serbia). Thus began the efforts of the Bosnian Serbs, aided by Serbia itself, to drive Muslims from the territory. The campaign waged by the Serbs was referred to as "ethnic cleansing." This meant killing Bosnian Muslims and Croatians, or driving them from their homes, in order to prepare their areas for a future "Greater Serbia." Bosnian Serbs commonly held the belief that Bosnian Muslims were intent on creating a fundamentalist Muslim state in Bosnia and thus rationalized their actions as simply self-defense. In addition to the Serb attack, the Croats also launched an attack on western Bosnia, driving out Serbs and Bosnians.

An irony of the Yugoslav conflict is that Bosnia, the area affected most severely in the early 1990s, had for decades been an example of ethnic harmony, a society of "pluralism and tolerance" (Rieff, 1995). Its three major groups—Serbs, Croats, and Muslims—had coexisted peacefully and had created a relatively integrated society. In Sarajevo, the major Bosnian city, Croatian Roman Catholic and Serbian Orthodox churches stood side by side with Muslim mosques, ethnically integrated neighborhoods were commonplace, and intermarriage among the three populations was not at all unusual.

The sudden destruction of the relative harmony in which Yugoslavia's ethnic groups had been living, particularly in Bosnia, is a reminder of how ethnic conflicts that are commonly seen as ancient and unending may in fact be the product of relatively recent events and may arise or fade with changing political,

---

[5] On the social history of Bosnia's Muslim population, see Duran (1995).

economic, and social events or with the emergence of new leaders. As political scientists Susanne and Lloyd Rudolph have put it, "The question . . . is not why old conflicts are flaring up anew, but rather why traditionally harmonious mosaics have been shattered" (1993:25). In the case of Yugoslavia, writes Ignatieff, "Nationalist politicians on both sides took the narcissism of minor difference and turned it into a monstrous fable according to which their own side appeared as blameless victims, the other side as genocidal killers" (1993:22). Contrary to popular belief, the notion that the Serb-Croat rivalry has ancient historical roots is simply not the case. In fact, until this century the two peoples had had virtually no contact. Past wars were with the multinational empires of the Hapsburgs and Ottomans, not with each other (Bennett, 1995; Rieff, 1995).[6]

Neighbors and family members, who had been comfortably integrated just weeks before, voiced puzzlement at how quickly their social harmony had dissipated and had been reduced to vicious and bloody hatred. "Growing up before the war here, my boys played with all the neighborhood kids," said a Serbian woman living in Bosnia. "We never knew who was Georgy and who was Muhammed. And I baked cakes for them all. Now neighbors we've known for years don't greet us anymore" (quoted in Sudetic, 1994). So intense and bitter were the feelings that surfaced among those of the different ethnic groups that people were driven to exchange their homes with each other in order to create more ethnically homogeneous areas. Later, as Serbs took control of parts of Bosnia, Croats and Muslims were driven from their homes, which were given to Serb refugees who had fled similar incursions in other areas by Croats or Muslims. Often, as Serbs or Croats advanced on each other, houses of those fleeing were summarily destroyed. Ignatieff describes what he saw as he drove through ravaged villages:

Toward Novska, you pass Serb house after Serb house, neatly dynamited, beside undisturbed Croat houses and gardens. When you turn toward Lipik, it is the turn of all the Croat houses to be dynamited or firebombed, next to their untouched Serb neighbors. Mile upon mile, the deadly logic of ethnic cleansing unfolds. (1993:35)

Much more dreadful actions were to occur, however, as the fighting intensified. Indeed, quickly the hostilities between Serbs, Croats, and Muslims degenerated into the most ferocious conflict in recent European history. Each side accused the other of atrocities, and in many cases these were confirmed by the

---

[6]In Rwanda as well, the cultural differences between Hutus and Tutsis were, historically, virtually nonexistent. Ethnic categorization was essentially a creation of the colonial powers that ruled Rwanda. After independence, this artificial division was exploited by rival political factions (Destexhe, 1995).

United Nations and other nonparticipating observers. The most horrific actions, however, were committed by Serbs in Bosnia as part of their campaign of ethnic cleansing. Mass executions, rape camps, savage mutilations and torture, and wanton destruction of homes, churches, and mosques filled news reports from Bosnia as the war there intensified starting in 1993.[7] Consider this account of a Serb concentration camp (corroborated by both Serbs and Muslims), given by a Serbian guard who later deserted:

> Men were loaded into the back of a truck, taken up to the edge of the ravine, about five miles away, and then shot as they got out of the vehicle. Groups of young soldiers were brought in to perform the executions. The bodies fell into the ravine and bulldozers were later used to cover them over. (Cohen, 1994)

Just months before, many of the camp guards and victims had been neighbors.

The horrors of the Yugoslav war were eerily reminiscent of the genocide committed by the Nazis fifty years earlier. The massacre in the city of Srebrenica was one of the more thoroughly documented of incidents that seemed so monstrous as to border on the surreal. Srebrenica had served as a United Nations refuge where more than 40,000 people had taken shelter from the war. After it was overrun by the Bosnian Serb army, the aftermath was described as the worst war crime in Europe since World War II. Although no one can be sure how many were slaughtered, American intelligence analysts estimated between 5,000 and 8,000. A few survivors as well as Serbian civilians living in surrounding villages recounted the mass killings. Hundreds were herded into trucks, taken to a field, and shot; others were jammed into a warehouse where they were killed with automatic rifles and shoulder-held grenade launchers. Mass graves of the victims were uncovered later (Engelberg and Weiner, 1995). This incident more than any other prompted the military and diplomatic efforts of the United States and NATO (North Atlantic Treaty Organization) to stop the war. But it was only one of many such massacres of entire villages.

Muslims were victimized as well by Croats seeking to secure what they claimed was their territory. As with the Serbs, a policy of ethnic cleansing produced gruesome incidents of terror and brutality. One such affair, documented by United Nations investigators, occurred in late 1993. A so-called death platoon of a few dozen Croatian paramilitaries entered the village of Stupni Do,

---

[7]In 1994 the United Nations documented 187 mass graves, each containing the bodies of between 3,000 and 5,000 Muslims murdered by Serbs. Also confirmed were 962 prison camps in which more than a half million people had been held and where rapes and mass killings had occurred (Pomfret, 1995). Additional mass graves in Bosnia were later uncovered.

shouting "Let's kill the Muslims" and "Where are all the pretty girls for us to rape? Bring them out." A journalist describes what ensued:

> With avenging fury, the Croatian death squad crushed the skulls of Muslim children, slit the throats of women and machine-gunned whole families at close range. The next morning Croatian reinforcements finished up the job, torching all 52 houses and dynamiting the community's one small mosque. In less than 48 hours Stupni Do simply ceased to exist. (Nordland, 1993:48–49)

Although Bosnia suffered the most appalling conditions and was the site of the most intense fighting, other regions were affected to a lesser degree in much the same way. Following the Croatian invasion of the Krajina, an area of Croatia populated mostly by Serbs, 100,000 Serbs fled what they feared was the Croatian version of ethnic cleansing. Like the Bosnian Muslims, the Krajina Serbs expressed the same disbelief that former Croat friends and neighbors with whom they had lived together and even intermarried for decades could so viciously turn on them without provocation. "I had a very good friend, a Catholic Croat," said a refugee Serb. "I would have given blood for him. But three years ago, he said: 'I like you a lot. But please don't come and see me anymore. My neighbors object to a Serb visiting.'" Another put it: "I'm sorry I didn't burn down my house. I'm sorry I left it to the Croats to burn" (quoted in Perlez, 1995a, 1995b).

In late 1995, after three and a half years of war, a fragile peace was established—largely through the efforts of the United States—that divided Bosnia into semiautonomous regions along ethnic lines. Under the terms of the agreement, Bosnia was split into two parts—a Muslim-Croat federation made up of 51 percent of the territory, and a Bosnian Serb republic, with 49 percent. NATO forces, including 20,000 Americans, were deployed to enforce the treaty. Although this temporarily stopped the killing, few seemed to feel that anything short of a more complete partition of Bosnia would secure a long-term resolution to the conflict (Mearsheimer and Van Evera, 1995).

The legacy of the war was that the ethnic mosaic of Yugoslavia's population had been radically—probably irreversibly—altered and interethnic relations reduced to mutual fear and antipathy. In Bosnia, the most severely affected region, at least 200,000 people had been killed or had disappeared; almost half of its prewar population of 4.2 million had been displaced. This, it should be remembered, had been Yugoslavia's most ethnically diverse region. Wholesale population transfers occurred in parts of Croatia and Serbia as well. Most Croats living in Serbia were forced to leave, as were Serbs living in Croatia. Refugees fleeing regions as they repeatedly changed hands reached epic proportions, unseen in Europe since World War II. As many as three million were driven from their homes. Those brave enough to try to return to their homes in areas now dominated by the opposing ethnic group found them occupied by members of that group, who themselves had been expelled from their homes in other areas.

For those who endured the war, a residue of distrust, resentment, and terror on all sides makes the reemergence of interethnic tolerance and accommodation improbable for many decades in the future. Sarajevo, previously the showpiece of ethnic harmony, symbolized this legacy of ethnic hatred. There, Serbs refused to accept Muslim governance of the city, as the peace agreement provided, and thousands left their homes in protest, some burning them in a final act of defiance. Some even exhumed the remains of dead family members to carry with them, fearing that their graves would be desecrated. In effect, Bosnia was left a country partitioned into tightly enclosed ethnic enclaves, reflecting a system of cultural and educational apartheid (Hedges, 1996, 1997). The ethnic diversity that had characterized Bosnia before the war was replaced by ethnic homogeneity—Serbs living in Serb areas, Croats in Croat areas, and Muslims in Muslim areas; coexistence had essentially been obliterated.

Kosovo: Expansion of the Yugoslav Conflict In the late 1990s the Yugoslav conflict, which had centered mainly in Bosnia, took a new turn. Kosovo, a province of Serbia, became the focus of severe ethnic fighting. For Serbs, Kosovo represented the birthplace of Serbian culture; thus it was of great symbolic importance, even though Serbs living there were only a small numerical minority. It is an area where, in the country's mythology, Serbs held off the Ottoman Turks from invading Europe in the late fourteenth century. The most important monasteries of the Serbian Orthodox Church are also located there.

Two factors made the Kosovo situation different from Bosnia. The Bosnian population before the 1990s had been relatively intermixed ethnically. Most of Kosovo, however, had been made up of ethnic Albanians—about 90 percent of its population of 2.2 million—with the minority Serb population segregated from neighboring Albanians. Very little social interaction occurred between the two populations, and hostile stereotypes thrived among both (Calic, 2000). Second, Bosnia had attained independence before the outbreak of the civil war there, whereas independence was the key issue for Albanians—mostly Muslim—in Kosovo: Most desired the province to become either fully independent or to become part of a greater Albania.

Under Tito, Kosovo had enjoyed much autonomy even though it remained a Yugoslav province. In 1989, the Milosevic government revoked much of Kosovo's autonomy, making it an integral part of Serbia and imposing repressive rule. Ethnic Albanians lost their jobs to minority Serbs, Albanian-language schools were closed, and the police and Serb army became an occupying force (International Organization for Migration [IOM], 1999; Kinzer, 1992). In the early 1990s, Kosovar Albanian guerrillas—the Kosovo Liberation Army (KLA)—launched a movement for the province's independence from Serbia.

Starting in 1998 the Serbs began an offensive intended to crush the insurgents, comparable in many ways to earlier Serb actions in Bosnia referred to as "ethnic cleansing."

In early 1999 the fighting in Kosovo entered a new dimension. NATO countries, which had acted to stop the Bosnian conflict only after its horrors had become hideously apparent, seemed determined to avoid a similar outcome in Kosovo. Peace talks were initiated and an agreement was offered to the Milosevic regime in Serbia and the rebel forces in Kosovo to end the fighting: Under NATO supervision, most Serb security forces would be removed from Kosovo and the province would be given a semi-autonomous status for three years, after which a referendum would be held to determine its political status (Hooper, 1999). Though conditionally accepted by the Kosovar Albanians, the agreement was rejected by the Serbs; NATO subsequently began a 78-day bombing campaign against Serb forces in Kosovo.

The pattern of atrocities that had so grotesquely characterized Serb actions in Bosnia was replicated in Kosovo—mass murders, destruction of entire villages, raping and looting on a massive scale. As a United Nations war crimes tribunal later explained in its indictment of Yugloslav President Milosevic, the campaign of terror and destruction against Albanian civilians was designed to change the ethnic balance in Kosovo by purging the province of as much of its Albanian population as possible. Within weeks, 90 percent of Kosovo's Albanians were driven from their homes and towns. Most fled to surrounding countries, creating a refugee crisis comparable to that brought about by the Bosnian war.[8] Although by comparison with Bosnia the death and destruction were less expansive, the savagery was no less gruesome and shocking: Serb attackers intent on destroying and looting went from village to village in killing sprees, sparing no one—women, children, infants, the elderly—and torching every house and building in sight (Finn et al., 1999; Kifner, 1999).

At the cessation of fighting in July 1999 it was estimated by war crimes investigators that at least 10,000 people had been slaughtered by Serbian forces during the three-month war, and newly discovered mass graves were being reported daily. The true death toll will probably never be known since many bodies were burned or hidden in an effort by Serbs to destroy evidence or were simply buried by relatives without informing Western officials.

Following removal of the Serbian army and paramilitaries, ethnic Albanians began to return, under the protection of NATO forces that now occupied the province. Not unexpectedly, violent recriminations, including killings and the

---

[8]Efforts to displace the ethnic Albanian population with Serbs had been occurring since the beginning of the twentieth century. Today there are approximately a half million Kosovo Albanians living and working in western Europe, over half of whom arrived as asylum seekers since 1985 (IOM, 1999).

burning of entire neighborhoods, were carried out by Albanians against Kosovo's Serb population. As the municipal administrator for one ethnically divided town—an American who had served in Bosnia—explained, "There is a degree of hate here that is far greater than anything I found in Bosnia; and also a degree of fear. I am searching for a word that means vitriolic squared" (quoted in Smith, 2000). Within a few months, three-quarters of the province's 200,000 Serbs had been forced from their homes and villages, most fleeing to Serbia proper. Thus, as in Bosnia, the war had produced an "ethnically cleansed" region. The likelihood that Serbs would ever return to Kosovo in significant numbers was slim. Moreover, the eventual independence of Kosovo from Serbia or its union with neighboring Albania seemed only a formality.

Macedonia Another piece in the ethnic puzzle of the disintegrated Yugoslavia became a focus of global attention in the early 2000s. Like most of the other former Yugoslav republics, Macedonia encompasses a diverse and divided ethnic population of 2 million. Macedonians (Orthodox Christian Slavs) make up two-thirds, ethnic Albanians (overwhelmingly Muslim) about 23 percent; and smaller groups, including Serbs, the remainder. And, as in Kosovo, these groups live in highly segregated communities. Albanians feel alienated from Macedonian society and, in turn, ethnic Macedonians feel unwelcome in majority Albanian areas. Unlike Kosovo, however, ethnic Albanians are a numerical minority.

Although most Albanians do not see themselves as integrated into the Macedonian state, their preference for change seems to be through evolutionary processes, in contrast to Albanians in Kosovo, who have experienced more severe and blatant discrimination (Perry, 1998). Nonetheless, differences in language and culture have provoked profound divisions between Macedonians and ethnic Albanians. Tensions flared into armed conflict in 2001 as a rebel ethnic Albanian group, intent on joining the heavily Albanian area of Macedonia with neighboring Albania, initiated the hostilities. The Macedonian government managed, at least temporarily, to quell the uprising, but only after dozens died in clashes. Having avoided the bloody conflicts of much of the rest of the former Yugoslavia during the 1990s, Kosovo seemed poised to sink into the Yugoslav ethnic morass.

The Power of Ethnic Nationalism In sum, the breakup of Yugoslavia and the ethnic animosities it inflamed resulted in ferocious conflicts, stunning in their savagery and viciousness. What remained afterward was a region more intensely divided by ethnic loyalties than before, with little expectation of future reconciliation among the various groups. The breakup of Yugoslavia demonstrated vividly the paradox of the enduring power of ethnic nationalism in a globalizing world.

Dominance by one group over another, noted Philip Mason, "is as old as the Pharaohs and springs from passions that are common to all men" (1970:337). The question, then, is not whether dominant-subordinate relations will continue to be an integral feature of diverse human societies but whether and to what degree those relations will be based on ethnicity. Given that multiethnicity is an advancing condition of most societies today on all continents, the ethnic basis of inequality appears unwavering. That, in turn, makes it likely that continued ethnic conflict will be a prevailing condition of our time.

Early on, it was suggested that conflict and change are fundamental themes of ethnic relations. It was also noted, however, that societies are always admixtures of not only conflict and change but also cooperation and stability. Thus, these two tendencies are always variable. Stephen Steinberg has written that "If there is an iron law of ethnicity, it is that when ethnic groups are found in a hierarchy of power, wealth, and status, then conflict is inescapable" (1989:170). It is most likely, therefore, that cooperation and stability will characterize ethnic relations for long periods in multiethnic societies where groups coexist with relatively minor differences in power, wealth, and prestige among them. In the contemporary world, however, such cases are uncommon. Instead, salient economic and political inequalities characterize most multiethnic societies. Moreover, conflict is induced when groups see their position as the product of discrimination (Gurr, 1993). This, too, is a prevalent condition in most multiethnic societies.

Where there are sharp inequalities in power and wealth among ethnic collectivities, dominant groups may induce stability through various techniques of prejudice and discrimination (discussed in Chapter 3) and the effective transmission of ideology. Eventually, however, a minority group or groups challenge that domination, and conflict then prevails. The United States and each of the other societies we have examined have in recent times experienced a challenge, in some degree, to their prevailing ethnic orders.

It is important to consider, however, that inequalities that induce conflict need not be severe or even objectively salient. Only the *perception* of inequality among ethnic groups is necessary. As we saw in Chapter 15, for example, the socioeconomic differences between French Canadians in Quebec and other Canadians are no longer evident. Despite this shrinking gap over the past several decades, however, the intensity of Quebec nationalism has not been reduced but has actually risen, creating the most serious threat to Canadian unity. Also, the grievances of minority ethnic groups may become more impassioned as they actually experience progressive change in their socioeconomic position. As conditions improve, the expectation of and desire for further improvements grow. If those expected changes do not materialize as rapidly as anticipated, minorities may focus on unmet goals rather than on their improved condition relative to that of an earlier period.[9]

In addition to economic and political grievances, ethnic conflict is generated by sustained cultural differences. Most basically, where ethnic groups speaking different languages or abiding by divergent norms and values must confront each other in various facets of social life—housing, schools, government, media—conflict is a given. Only its severity is variable. Conflict arises as well when ethnic groups fear their assimilation into a dominant culture (Lake and Rothchild, 1998). Repeatedly, we discovered such conditions in each of the societies we looked at. Here, too, however, the reality of cultural difference among groups is less critical than the perception of difference. As we saw in Northern Ireland, for example, two groups whose cultural variance has faded over many generations (and who even speak a common language) nonetheless have clung to historical perceptions of difference that continue to shape their views and attitudes toward each other. As with economic and political inequality, sociocultural differences, or their perception, are fundamental in contemporary multiethnic societies.

Where there are physical distinctions among groups, of course, an additional basis of division is in place. But racial differences among groups are not an essential ingredient in the emergence of ethnic divisions and conflict. In Canada, Northern Ireland, Rwanda, and the former Yugoslavia, we saw that ethnic boundaries between racially similar groups can be shaped in a number of ways—and once established, adamantly sustained, often for many generations.

Finally, Heribert Adam has pointed out that conflict in multiethnic societies is also induced by differences in how groups are seen and evaluated. "Strife arises," he notes, "when groups are differentially valued in terms of their intrinsic self-worth" (1992:20). Although a group's prestige or esteem is usually tied to its political and economic power, growing equality in these does not in all cases guarantee similar growth in social esteem. As we saw in the United States, African Americans, for example, despite real collective political and economic mobility in recent decades, in many cases continue to see themselves held in lower social esteem.

Not only within but also between societies, certain trends portend continued ethnic conflict. Economies of global scale and world-encompassing communications networks make national boundaries increasingly permeable, thus creating ever-greater ethnic diversity within most of the world's nation-states. Indeed, as noted in Chapter 1, few societies today cannot to some degree be called multiethnic. These tendencies might lead to greater understanding among diverse

---

[9]Susan Olzak (1992) has postulated that as inequalities among ethnic groups decline, greater conflict may be generated through intensified competition that ensues among the groups. As ethnic stratification breaks down, more powerful groups seek to prevent weaker, minority groups from getting their share of the society's resources.

peoples—and thus to a breakdown of ethnic barriers—as they come to recognize their problems and motivations as commonly human (Shibutani and Kwan, 1965). But these trends might just as easily serve to rigidify ethnocentric and nationalistic views by demonstrating more dramatically the economic and cultural differences among them. "In deeply divided territories," note McGarry and O'Leary, "increased exposure to the 'other' may make group members more aware of what their group has in common and what separates them from the others. Exposure may cement group solidarity rather than diffuse it" (1995b:855). The persistence of ethnic conflict in almost all parts of the world seems to confirm the latter outcome.

Given the pervasiveness of ethnic diversity and inequality, it is perhaps most realistic to think in terms of the management of ethnic tensions and hostilities rather than their elimination. While some ethnic conflicts are resolved with finality, observes political scientist Milton Esman, "the great majority rise, subside, and recur" (1994:23). It is important to reiterate that conflict is a fundamental condition of human societies and is inherent in all human affairs. It should not be confused with violence. Most social conflict is low-intensity and controlled through the society's political process and legal system. Conflict may, of course, involve the use of violence; but this is ordinarily episodic and does not characterize most intergroup relations. Thus, although ethnic conflict is an essential, ongoing feature of all multiethnic societies, it is usually controlled and limited. Cases in which extreme violence occurs (Rwanda or Yugoslavia, for example) are exceptional, not typical. As Darby has put it, "The use of violence marks the failure of normal and functional means of conflict resolution—political exchange, negotiation, compromise" (1995:29). The most achievable objective of multiethnic societies, then, is not to eliminate ethnic conflict entirely but to contain it as much as possible. "The real issue," writes Darby, "is not the existence of conflict, but how it is regulated" (1995:29).

The ability of diverse ethnic groups in pluralistic societies to live together in relative harmony while maximizing equality in the distribution of political and economic resources among them depends in large measure on the presence of a central state with sufficient power to contain intergroup conflict to a tolerable degree, but fair enough to win the allegiance of all groups. As Ignatieff has observed, "What keeps ethnic and racial tension within bounds in the world's successful modern multi-ethnic societies is a state strong enough to make its authority respected" (1993:243). Recent developments in South Africa, as we have seen, illustrate that efforts in this regard may in some cases yield significant changes. Yugoslavia, on the other hand, demonstrates the disaster that may result from the collapse of a strong state holding diverse groups together.

Whatever successes may be achieved in the pursuit of ethnic harmony will probably be overshadowed by discord. It is obvious that with ethnic diversity the prevailing global trend, nations everywhere will for the foreseeable future struggle with issues of political, social, and economic inequalities among racially and culturally diverse groups. To some degree, then, ethnic conflict will be a recurring theme. The spread of pluralistic ideas and movements among ethnic groups within multiethnic societies is an outgrowth of the nationalistic fervor that has swept both the developed and developing worlds, and of international population movements that are occurring on a historically unprecedented scale.

It is possible, of course, that with the advent of a global economy and the emergence of global communications networks ethnic nationalism may prove to be only a passing phenomenon, destined to recede as people everywhere are increasingly linked by common needs and cosmopolitan cultural systems. This outcome, however, cannot yet be envisioned. The crucial fact, notes Anthony Smith, is that in the twentieth century interethnic conflict became "more intense and endemic . . . than at any time in history" (1981:10). Today there is little indication of the development of a countercurrent. Rather than diminishing, ethnicity as a global phenomenon persists as an exceptionally powerful and tenacious force motivating human behavior. And it will follow us into the twenty-first century.

SUMMARY

Conflict and change are inherent features of multiethnic societies. Where there are significant economic, political, and social inequalities among groups, challenges to the dominant social order arise periodically. Also, cultural differences among groups may continue to impel ethnic conflict. The conditions of the contemporary world—increasing diversity and ethnic nationalism—make the continuation and even intensification of ethnic conflict very likely.

Shifting patterns of global immigration have emerged in the past fifty years. Today the movement of people is from the developing societies to the developed, industrialized societies. Changing labor markets and political strife in many societies are major factors that have stimulated global immigration. Immigration has molded societies into increasingly diverse ethnic configurations.

Western European societies have undergone great social changes in the past few decades as a result of immigration and the subsequent creation of ethnic heterogeneity. As societies where there had been relative homogeneity, the introduction of large numbers of culturally and racially diverse people has created ethnic minority communities that have not been assimilated into the mainstream. These ethnic minorities have been the targets of severe prejudice and discrimination.

Throughout much of the world, ethnic nationalism has been a stimulant of identity and political action in the past two decades. This has been especially evident in the former Soviet Union and the fragmented societies of eastern Europe. Nowhere has conflict based on ethnic nationalism been more virulent

Part III: Ethnic Relations in Comparative Perspective

or bloody than in Yugoslavia, where Serbs, Croats, and Muslims engaged in the most ferocious warfare in Europe since World War II. Yugoslavia seems to be a case of a failed attempt at the formation of a corporate pluralistic system. Once the glue of a strong central government was removed, underlying ethnic hostilities reemerged with overwhelming force.

Ethnic conflict today is not limited to any world region; it is a global phenomenon. The spread of ethnic nationalism and the increasing scope of racial and cultural heterogeneity are conditions that make for the continuation of ethnic conflict in the foreseeable future.

Suggested Readings

Anderson, Benedict. 1991. *Imagined Communities: Reflections on the Origin and Spread of Nationalism*. Rev. ed. London: Verso. Examines the evolution of nationalism and the factors that create a preparedness of people to defend their society or perceived community against others.

Bennett, Christopher. 1995. *Yugoslavia's Bloody Collapse: Causes, Course and Consequences*. New York: New York University Press. Deftly sorts out the complex political and social issues that led to the breakup of Yugoslavia and offers insight into the cultural similarities and differences of the groups involved.

Castles, Stephen, and Mark J. Miller. 1998. *The Age of Migration: International Population Movements in the Modern World*. 2d ed. New York: Guilford. Describes today's global immigration routes and analyzes the economic, political, and social impact of large-scale immigration on receiving societies.

Destexhe, Alain. 1995. *Rwanda and Genocide in the Twentieth Century*. New York: New York University Press. A succinct analysis of the slaughter of Tutsis by Hutus in 1994 in Rwanda and the historic relations between these groups that preceded it. Discusses the Rwandan case in the context of other genocides of the twentieth century.

Ignatieff, Michael. 1993. *Blood and Belonging: Journeys into the New Nationalism*. Toronto: Penguin. The author travels to six locations of contemporary ethnic conflict, relating his observations and his conversations with those involved.

Judah, Tim. 1997. *The Serbs: History, Myth and the Destruction of Yugoslavia*. New Haven: Yale University Press. A history of the Serbian people, focusing especially on the social, economic, and political circumstances that led to the recent wars in Yugoslavia and the resultant breakup of the country.

Rieff, David. 1995. *Slaughterhouse: Bosnia and the Failure of the West*. New York: Touchstone. A journalist describes his odyssey through war-plagued Croatia, Serbia, and Bosnia, probing the thoughts of citizens as he tries to understand the ethnic conflict among them.

Smith, Anthony D. 1981. *The Ethnic Revival*. Cambridge: Cambridge University Press. An analysis of ethnic nationalism and its global resurgence in the twentieth century.

REFERENCES

Aboud, Frances E. 1987. "The Development of Ethnic Self-Identification and Attitudes." Pp. 32–55 in Jean S. Phinney and Mary Jane Rotheram (eds.), *Children's Ethnic Socialization: Pluralism and Development.* Newbury Park: Sage.

Abraham, Sameer Y. 1983. "Detroit's Arab-American Community: A Survey of Diversity and Commonality." Pp. 85–108 in Sameer Y. Abraham and Nabeel Abraham (eds.), *Arabs in the New World: Studies on Arab-American Communities.* Detroit: Wayne State University Press.

Abrams, Charles. 1966. "The Housing Problem and the Negro." Pp. 512–524 in Talcott Parsons and Kenneth B. Clark (eds.), *The Negro American.* Boston: Houghton Mifflin.

Abramson, Harold. 1980. "Assimilation and Pluralism." Pp. 150–160 in Stephen Thernstrom (ed.), *Harvard Encyclopedia of American Ethnic Groups.* Cambridge, Mass.: Harvard University Press.

Acuña, Rodolfo. 1972. *Occupied America: The Chicano's Struggle toward Liberation.* San Francisco: Caufield.

Adachi, Ken. 1976. *The Enemy That Never Was.* Toronto: McClelland & Stewart.

Adam, Heribert. 1971a. "The South African Power-Elite: A Survey of Ideological Commitment." Pp. 73–102 in Heribert Adam (ed.), *South Africa: Social Perspectives.* London: Oxford University Press.

_____. 1971b. *Modernizing Racial Domination: South Africa's Political Dynamics.* Berkeley: University of California Press.

_____. 1992. "The Politics of Identity: Nationalism, Patriotism, and Multiculturalism." *Journal of Ethno-Development* 1(2):10–22.

_____. 1995. "The Politics of Ethnic Identity: Comparing South Africa." *Ethnic and Racial Studies* 18:457–475.

Adam, Heribert, and Kogila Moodley. 1986. *South Africa without Apartheid: Dismantling Racial Domination.* Berkeley: University of California Press.

Adams, David Wallace. 1995. *Education for Extinction: American Indians and the Boarding School Experience, 1875–1928.* Lawrence: University Press of Kansas.

Adorno, T. W., et al. 1950. *The Authoritarian Personality.* New York: Harper & Row.

African Rights. 1994. *Rwanda: Death, Despair and Defiance.* London: African Rights.

Agbayani-Siewert, Pauline, and Linda Revilla. 1995. "Filipino Americans." Pp. 134–168 in Pyong Gap Min (ed.), *Asian Americans: Contemporary Trends and Issues.* Thousand Oaks, Calif.: Sage.

Alba, Richard D. 1981. "The Twilight of Ethnicity among American Catholics of European Ancestry." *Annals of the American Academy of Political and Social Science* 454:86–97.

_____. 1985. *Italian Americans: Into the Twilight of Ethnicity.* Englewood Cliffs, N.J.: Prentice-Hall.

_____. 1990. *Ethnic Identity: The Transformation of White America.* New Haven, Conn.: Yale University Press.

_____. 1992. "Ethnicity." Pp. 575–584 in Edgar F. and Mary L. Borgatta (eds.), *Encyclopedia of Sociology.* Vol. 2. New York: Macmillan.

_____. 1994. "Identity and Ethnicity Among Italians and Other Americans of European Ancestry." Pp. 21–44 in Lydio Tomasi, Piero Gastaldo, and Thomas Row (eds.), *The Columbus People: Perspectives in Italian Immigration to the Americas and Australia.* Staten Island, N.Y.: Center for Migration Studies.

———. 1995. "Assimilation's Quiet Tide." *The Public Interest* (Spring):3–18.

———. 1999. "Immigration and the American Realities of Assimilation and Multiculturalism." *Sociological Forum* 14:3-25.

Alba, Richard D., and Reid M. Golden. 1986. "Patterns of Ethnic Marriage in the United States." *Social Forces* 65:202–223.

Alba, Richard D., and John R. Logan. 1993. "Minority Proximity to Whites in Suburbs: An Individual-Level Analysis of Segregation." *American Journal of Sociology* 98:1388–1427.

Alba, Richard D., and Gwen Moore. 1982. "Ethnicity in the American Elite." *American Sociological Review* 47:373–383.

Alba, Richard D., and Victor Nee. 1999. "Rethinking Assimilation Theory for a New Era of Immigration." Pp. 137–160 in Charles Hirschman, Philip Kasinitz, and Josh DeWind (eds.), *The Handbook of International Migration: The American Experience.* New York: Russell Sage Foundation.

Alba, Richard D., John R. Logan, and Brian J. Stults. 2000. "How Segregated Are Middle-Class African Americans?" *Social Problems* 47:545–558.

Alba, Richard D., John R. Logan, Brian J. Stults, Gilbert Marzan, and Wenquan Zhang. 1999. "Immigrant Groups in the Suburbs: A Reexamination of Suburbanization and Spatial Assimilation." *American Sociological Review* 64:446–460.

Albini, Joseph L. 1971. *The American Mafia: Genesis of a Legend.* New York: Appleton-Century-Crofts.

Allen, Robert L. 1970. *Black Awakening in Capitalist America.* Garden City, N.Y.: Doubleday.

Allport, Gordon W. 1958. *The Nature of Prejudice.* Garden City, N.Y.: Doubleday.

Alvarez, Rodolfo. 1973. "The Psycho-Historical and Socioeconomic Development of the Chicano Community in the United States." *Social Science Quarterly* 53:920–942.

American Council on Education. 2000. *Minorities in Higher Education 1999–2000.* Washington, D.C.: American Council on Education.

American Enterprise. 1994. "Immigration." (January/ February):97–100.

American Jewish Committee. 2000. *American Jewish Year Book 2000.* New York: American Jewish Committee.

Anderson, Benedict. 1991. *Imagined Communities: Reflections on the Origin and Spread of Nationalism.* Rev. ed. London: Verso.

Anderson, Charles H. 1970. *White Protestant Americans: From National Origins to Religious Group.* Englewood Cliffs, N.J.: Prentice-Hall.

Anderson, Jon Lee. 2001. "Home Fires." *The New Yorker* (February 12):40–47.

Andrews, George Reid. 1991. *Blacks & Whites in São Paulo, Brazil 1888–1988.* Madison, Wis.: University of Wisconsin Press.

Anti-Defamation League. 1993. *Highlights from an Anti-Defamation League Survey on Racial Attitudes in America.* New York: Anti-Defamation League of B'nai B'rith.

———. 1998. *Survey on Anti-Semitism and Prejudice in America.* New York.

Antonovsky, Aaron. 1960. "The Social Meaning of Discrimination." *Phylon* 21:81–95.

Aoyagi, Kiyotaka, and Ronald P. Dore. 1964. "The Buraku Minority in Urban Japan." Pp. 95–107 in *Transactions of the Fifth World Congress of Sociology.* Vol. 3. Louvain: International Sociological Association.

Aponte, Robert, and Marcelo Siles. 1994. "Latinos in the Heartland: The Browning of the Midwest." *JSRI Research Report No. 5.* Julian Samora Research Institute. East Lansing: Michigan State University.

Arias, Elizabeth. 2001. "Change in Nuptiality Patterns among Cuban Americans: Evidence of Cultural and Structural Assimilation." *International Migration Review* 35:525–556.

Arnold, Guy. 2000. *The New South Africa.* New York: St. Martin's.

Arnopoulos, Sheila McLeod, and Dominique Clift. 1980. *The English Fact in Quebec.* Montreal: McGill-Queen's University Press.

Arthur, Paul. 1980. *Government and Politics of Northern Ireland.* Burnt Mill, England: Longman House.

Arthur, Paul, and Keith Jeffery. 1996. *Northern Ireland since 1968.* 2d ed. Oxford: Blackwell.

Asch, Michael. 1984. *Home and Native Land: Aboriginal Rights and the Canadian Constitution.* Toronto: Methuen.

Asmal, Kader, and Wilmot James. 2001. "Education and Democracy in South Africa Today." *Daedalus* 130(Winter):185–204.

Aunger, Edmund A. 1981. *In Search of Political Stability: A Comparative Study of New Brunswick and Northern Ireland.* Montreal: McGill-Queen's University Press.

Avery, Donald. 1979. *Dangerous Foreigners: European Immigrant Workers and Labour Radicalism in Canada 1896–1932.* Toronto: McClelland & Stewart.

Bahr, Howard M., Bruce A. Chadwick, and Joseph H. Stauss. 1979. *American Ethnicity.* Lexington, Mass.: Heath.

Baily, Samuel L. 1996. "The Adjustment of Italian Immigrants in Buenos Aires and New York, 1870–1914." Pp. 282–308 in Dirk Hoerder and Leslie Page Moch (eds.), *European Migrants: Global and Local Perspectives.* Boston: Northeastern University Press.

Baltzell, E. Digby. 1958. *Philadelphia Gentlemen: The Making of a National Upper Class.* New York: Free Press.

———. 1964. *The Protestant Establishment: Aristocracy and Caste in America.* New York: Vintage.

———. 1994. *Judgment and Sensibility: Religion and Stratification.* New Brunswick, N.J.: Transaction.

Banfield, Edward C. 1968. *The Unheavenly City.* Boston: Little, Brown.

———. 1990 (1974). *The Unheavenly City Revisited.* Prospect Heights, Ill.: Waveland Press.

Banting, Keith G. 1992. "If Quebec Separates: Restructuring Northern North America." Pp. 159–178 in

R. Kent Weaver (ed.), *The Collapse of Canada?* Washington, D.C.: Brookings Institution.

Banton, Michael. 1967. *Race Relations*. London: Tavistock.

———. 1970. "The Concept of Racism." Pp. 17–34 in Sami Zubaida (ed.), *Race and Racialism*. London: Tavistock.

———. 1983. *Racial and Ethnic Competition*. Cambridge, England: Cambridge University Press.

Barcelos, Luiz Claudio. 1999. "Struggling in Paradise: Racial Mobilization and the Contemporary Black Movement in Brazil." Pp. 155–166 in Rebecca Reichmann (ed.), *Race in Contemporary Brazil: From Indifference to Inequality*. University Park, Pa.: The Pennsylvania State University Press.

Baron, Dennis. 1990. *The English-Only Question: An Official Language for Americans?* New Haven: Yale University Press.

Baron, Harold. 1969. "The Web of Urban Racism." Pp. 134–176 in Louis L. Knowles and Kenneth Prewitt (eds.), *Institutional Racism in America*. Englewood Cliffs, N.J.: Prentice-Hall.

Barone, Michael. 1985. "Italian Americans and Politics." Pp. 378–384 in Lydio F. Tomasi (ed.), *Italian Americans: New Perspectives in Italian Immigration and Ethnicity*. Staten Island, N.Y.: Center for Migration Studies.

Barrera, Mario. 1979. *Race and Class in the Southwest: A Theory of Racial Inequality*. South Bend, Ind.: University of Notre Dame Press.

Barritt, Denis. 1982. *Northern Ireland: A Problem to Every Solution*. London: Quaker Peace and Service.

Barritt, Denis P., and Charles F. Carter. 1972. *The Northern Ireland Problem: A Study in Group Relations*. 2d ed. London: Oxford University Press.

Barsh, Russel Lawrence. 1994. "Canada's Aboriginal Peoples: Social Integration or Dis-integration?" *Canadian Journal of Native Studies* 14:1–46.

Barta, Russell. 1979. "The Representation of Poles, Italians, Latins and Blacks in the Executive Suites of Chicago's Largest Corporations." Pp. 418–430 in U.S. Commission on Civil Rights, *Civil Rights Issues of Euro-Ethnic Americans in the United States: Opportunities and Challenges*. Washington, D.C.: U.S. Government Printing Office.

Barth, Ernest A. T., and Donald L. Noel. 1972. "Conceptual Frameworks for the Analysis of Race Relations." *Social Forces* 50:333–348.

Barth, Fredrik. 1969. *Ethnic Groups and Boundaries*. Boston: Little, Brown.

Bastide, Roger. 1961. "Dusky Venus, Black Apollo." *Race* 3(November):10–18.

———. 1965. "The Development of Race Relations in Brazil." Pp. 9–29 in Guy Hunter (ed.), *Industrialization and Race Relations*. London: Oxford University Press.

Bastide, Roger, and Pierre van den Berghe. 1957. "Stereotypes, Norms and Interracial Behavior in São Paulo, Brazil." *American Sociological Review* 22:689–694.

Bean, Frank D., and Marta Tienda. 1987. *The Hispanic Population of the United States*. New York: Russell Sage Foundation.

Beck, Roy. 1994. "The Ordeal of Immigration in Wausau." *The Atlantic Monthly* 273(April): 84–97.

———. 1996. *The Case against Immigration: The Moral, Economic, Social, and Environmental Reasons for Reducing U.S. Immigration Back to Traditional Levels*. New York: Norton.

Beckett, J. C. 1972. "Northern Ireland." Pp. 11–24 in Institute for the Study of Conflict, *The Ulster Debate*. London: Bodley Head.

Bell, Daniel. 1962. *The End of Ideology*. Rev. ed. New York: Free Press.

———. 1975. "Ethnicity and Social Change." Pp. 141–174 in Nathan Glazer and Daniel P. Moynihan (eds.), *Ethnicity: Theory and Experience*. Cambridge, Mass.: Harvard University Press.

Bell, Geoffrey. 1976. *The Protestants of Ulster*. London: Pluto Press.

Bell, J. Bowyer. 1996. *Back to the Future: The Protestants and a United Ireland*. Dublin: Poolbeg.

Belluck, Pam. 1995. "Healthy Korean Economy Draws Immigrants Home." *New York Times* (August 22): A1, B4.

Benedict, Ruth. 1959 (1940). *Race: Science and Politics*. New York: Viking.

Bennett, Christopher. 1995. *Yugoslavia's Bloody Collapse: Causes, Course and Consequences*. New York: New York University Press.

Bercuson, David J. 1995. "Why Quebec and Canada Must Part." *Current History* 94:123–126.

Berger, Joseph. 1992. "South African Students Learn to Learn Together." *New York Times* (May 12): A4.

Bergesen, Albert. 1980. "Official Violence during the Watts, Newark and Detroit Race Riots of the 1960s." Pp. 138–174 in Pat Lauderdale (ed.), *A Political Analysis of Deviance*. Minneapolis: University of Minnesota Press.

Berke, Richard L. 1998. "Voters Found Real Chance at Peace Irresistible." *New York Times* (May 24): A1, A4.

Berkhofer, Robert F., Jr. 1979. *The White Man's Indian*. New York: Vintage.

Bernard, Jesse. 1951. "The Conceptualization of Intergroup Relations with Special Reference to Conflict." *Social Forces* 19:243–251.

Bernstein, Richard. 1988. "Asian Newcomers Hurt by Precursors' Success." *New York Times* (July 10):12.

Berreman, Gerald D. 1966. "Structure and Function of Caste Systems." Pp. 277–307 in George De Vos and Hiroshi Wagatsuma (eds.), *Japan's Invisible Race: Caste in Culture and Personality*. Berkeley and Los Angeles: University of California Press.

———. 1972. "Race, Caste, and Other Invidious Distinctions in Social Stratification." *Race* 13:385–414.

Berry, Brewton, and Henry L. Tischler. 1978. *Race and Ethnic Relations*. 4th ed. Boston: Houghton Mifflin.

Berry, John W., and Rudolf Kalin. 2000. "Racism: Evidence from National Surveys." Pp. 172–185 in Ledo Driedger and Shiva S. Halli (eds.), *Race and*

Racism: Canada's Challenge. Montreal: McGill-Queen's University Press.

Berry, John W., Rudolf Kalin, and D. M. Taylor. 1977. *Multiculturalism and Ethnic Attitudes in Canada*. Ottawa: Minister of Supply and Services.

Betancur, John J., Teresa Cordova, and Maria de los Angeles Torres. 1993. "Economic Restructuring and the Process of Incorporation of Latinos into the Chicago Economy." Pp. 109–132 in Rebecca Morales and Frank Bonilla (eds.), *Latinos in a Changing U.S. Economy*. Newbury Park, Calif.: Sage.

Béteille, André. 1969. *Castes Old and New: Essays in Social Structure and Social Stratification*. Bombay: Asian Publishing Home.

Biale, David. 1998. "The Melting Pot and Beyond: Jews and the Politics of Identity." Pp. 17–33 in David Biale, Michael Galchinsky, and Susannah Heschel (eds.), *Insider/Outside: American Jews and Multiculturalism*. Berkeley: University of California Press.

Bibby, Reginald W. 1995. *The Bibby Report: Social Trends Canadian Style*. Toronto: Stoddart.

Biemiller, Lawrence. 1986. "Asian Students Fear Top Colleges Use Quota Systems." *Chronicle of Higher Education* 23(November 19):1, 34–35, 37.

Billingsley, Andrew. 1968. *Black Families in White America*. Englewood Cliffs, N.J.: Prentice-Hall.

Bissoondath, Neil. 1994. *Selling Illusions: The Cult of Multiculturalism in Canada*. Toronto: Penguin.

Black Enterprise. 1990. "B. E. Survey Report: A View of the Past, A Look to the Future." 21(August):77–94.

Blackburn, Daniel G. 2000. "Why Race Is Not a Biological Concept." Pp. 22-38 in Berel Lang (ed.), *Race and Racism in Theory and Practice*. Lanham, Md.: Rowman & Littlefield.

Blackwell, James E. 1976. "The Power Basis of Ethnic Conflict in American Society." Pp. 179–196 in Lewis A. Coser and Otto N. Larsen (eds.), *The Use of Controversy in Sociology*. New York: Free Press.

Blalock, Hubert M., Jr. 1967. *Toward a Theory of Minority-Group Relations*. New York: Wiley.

Blau, Joseph L., and Salo W. Baron (eds.). 1963. *The Jews of the United States 1790–1840: A Documentary History*. New York: Columbia University Press.

Blau, Peter M. 1977. "A Macrosociological Theory of Social Structure." *American Journal of Sociology* 83:26–54.

Blauner, Robert. 1969. "Internal Colonialism and Ghetto Revolt." *Social Problems* 16:393–408.

———. 1972. *Racial Oppression in America*. New York: Harper & Row.

———. 1992. "Talking Past Each Other: Black and White Languages of Race." *The American Prospect* 10(Summer):55–64.

Blishen, Bernard R. 1970. "Social Class and Opportunity in Canada." *Canadian Review of Sociology and Anthropology* 7:110–127.

Bloomgarden, Laurence. 1957. "Who Should Be Our Doctors?" *Commentary* (January): 506–515.

Blue, Art. 1985. "Native People, Social Conditions." Pp. 1223–1224 in *The Canadian Encyclopedia*. Vol. II. Edmonton: Hurtig.

Blumenfeld, Jesmond. 1987. "Economy Under Siege." Pp. 17–33 in Jesmond Blumenfeld (ed.), *South Africa in Crisis*. London: Croom Helm.

Blumer, Herbert. 1958. "Race Prejudice as a Sense of Group Position." *Pacific Sociological Review* 1:3–6.

———. 1965. "Industrialization and Race Relations." Pp. 200–253 in Guy Hunter (ed.), *Industrialization and Race Relations*. London: Oxford University Press.

Boal, Frederick W. 1982. "Segregating and Mixing: Space and Residence in Belfast." Pp. 249–280 in Frederick W. Boal and J. Neville H. Douglas (eds.), *Integration and Division: Geographical Perspectives on the Northern Ireland Problem*. London: Academic Press.

Boal, Frederick W., and J. Neville H. Douglas. 1982. "Overview." Pp. 333–359 in Frederick W. Boal and J. Neville H. Douglas (eds.), *Integration and Division: Geographical Perspectives on the Northern Ireland Problem*. London: Academic Press.

Boal, Frederick W., Margaret C. Keane, and David N. Livingstone. 1997. *Them and Us? Attitudinal Variation among Churchgoers in Belfast*. Belfast: The Institute of Irish Studies, The Queen's University of Belfast.

Bobo, Lawrence D. 1999. "Prejudice as Group Position: Microfoundations of a Sociological Approach to Racism and Race Relations." *Journal of Social Issues* 55:445–472.

Bogardus, Emory S. 1925a. "Social Distance and Its Origins." *Journal of Applied Sociology* 9:216–226.

———. 1925b. "Measuring Social Distances." *Journal of Applied Sociology* 9:299–308.

———. 1930. "A Race Relations Cycle." *American Journal of Sociology* 35:612–617.

———. 1959. *Social Distance*. Yellow Springs, Ohio: Antioch Press.

———. 1968. "Comparing Racial Distance in Ethiopia, South Africa, and the United States." *Sociology and Social Research* 52:149–156.

Bolaria, B. Singh, and Peter S. Li. 1988. *Racial Oppression in Canada*. 2d ed. Toronto: Garamond Press.

Bonacich, Edna. 1972. "A Theory of Ethnic Antagonism: The Split Labor Market." *American Sociological Review* 37:547–559.

———. 1973. "A Theory of Middleman Minorities." *American Sociological Review* 38:583–594.

———. 1976. "Advanced Capitalism and Black-White Race Relations in the United States: A Split Labor Market Interpretation." *American Sociological Review* 41:34–51.

Bonacich, Edna, and John Modell. 1980. *The Economic Basis of Ethnic Solidarity: Small Business in the Japanese American Community*. Berkeley: University of California Press.

Borjas, George J. 1994. "Tired, Poor on Welfare." Pp. 76–80 in Nicolaus Mills (ed.), *Arguing Immigration*. New York: Touchstone.

———. 1998. "Do Blacks Gain or Lose from Immigration?" Pp. 51–74 in Daniel S. Hamermesh and Frank D. Bean (eds.), *Help or Hindrance? The Economic Implications of Immigration for African Americans*. New York: Russell Sage Foundation.

Bouvier, Leon F. 1992. *Peaceful Invasions: Immigration and Changing America*. Lanham, Md.: University Press of America.

Bowser, Benjamin P., and Raymond G. Hunt (eds.). 1981. *Impacts of Racism on White Americans*. Beverly Hills, Calif.: Sage.

Boxall, Bettina. 2001. "Asian Indians Remake Silicon Valley." *Los Angeles Times* (July 6) <http://www.latimes.com>.

Boxer, Charles R. 1962. *The Golden Age of Brazil*. Berkeley: University of California Press.

———. 1963. *Race Relations in the Portuguese Colonial Empire, 1415–1825*. Oxford: Clarendon Press.

Boyle, Kevin, and Tom Hadden. 1994. *Northern Ireland: The Choice*. New York: Penguin.

Bozinoff, Lorne, and Peter MacIntosh. 1992. "Majority Believe Racial Intolerance Has Increased." *Gallup Report* (May 25).

Bozorgmehr, Mehdi, Claudia Der-Martirasian, and Georges Sabagh. 1996. "Middle Easterners: A New Kind of Immigrant." Pp. 345–378 in Roger Waldinger and Mehdi Bozorgmehr (eds.), *Ethnic Los Angeles*. New York: Russell Sage Foundation.

Bradford, William D. 2000. "Black Family Wealth in the United States." Pp. 103–145 in National Urban League, *The State of Black America 2000*. New York: National Urban League.

Branch-Brioso, Karen, Tim Henderson, and Alfonso Chardy. 2000. "Despite Hispanic Majority, White Non-Hispanics Hold Sway." *Miami Herald* (September 3):A1.

Bratton, Michael. 1998. "After Mandela's Miracle in South Africa." *Current History* 97:214–219.

Braverman, Harold, and Louis Kaplan. 1967. "A Study of Religious Discrimination by Social Clubs." Pp. 211–221 in Milton L. Barron (ed.), *Minorities in a Changing World*. New York: Knopf.

Braverman, Marilyn. 1988. "English Yes; English Only, No." *Reconstructionist* 52(June):29–31.

Breton, Raymond. 1964. "Institutional Completeness of Ethnic Communities and the Personal Relations of Immigrants." *American Journal of Sociology* 70:193–205.

———. 1989. "Canadian Ethnicity in the Year 2000." Pp. 149–152 in James Frideres (ed.), *Multiculturalism and Intergroup Relations*. New York: Greenwood.

———. 1998. "Ethnicity and Race in Social Organization: Recent Developments in Canadian Society." Pp. 60–115 in Rick Helms-Hayes and James Curtis (eds.), *The Vertical Mosaic Revisited*. Toronto: University of Toronto Press.

Breton, Raymond, Jeffrey G. Reitz, and Victor F. Valentine. 1980. *Cultural Boundaries and the Cohesion of Canada*. Montreal: Institute for Research on Public Policy.

Brewer, M. 1979. "Ingroup Bias in the Minimal Intergroup Situation: Cognitive-Motivational Analysis." *Psychological Bulletin* 86:307–332.

Briggs, John W. 1978. *An Italian Passage*. New Haven, Conn.: Yale University Press.

Brimelow, Peter. 1995. *Alien Nation*. New York: HarperCollins.

Brink, William, and Louis Harris. 1963. *The Negro Revolution in America*. New York: Simon & Schuster.

———. 1967. *Black and White: A Study of Racial Attitudes Today*. New York: Simon & Schuster.

Brockman, Joshua. 2001. "Gathered Vestiges of a Covered-Wagon Diaspora." *New York Times* (February 4):37, 41.

Brown, Cecil. 1981. "Blues for Blacks in Hollywood." *Mother Jones* (January):20–28, 59.

Brown, Dee. 1972. *Bury My Heart at Wounded Knee*. New York: Bantam.

Brown, Douglas. 1966. *Against the World: A Study of White South African Attitudes*. London: Collins.

Brown, Paul. 2002. "Peace But No Love as Northern Ireland Divide Grows." *Guardian Weekly*. (January 10–16):8.

Brown, Rupert. 1995. *Prejudice: Its Social Psychology*. Oxford: Blackwell.

Brubaker, Rogers. 2001. "The Return of Assimilation? Changing Perspectives on Immigration and Its Sequels in France, Germany, and the United States." *Ethnic and Racial Studies* 24:531–548.

Bruce, Steve. 1986. *God Save Ulster!* Oxford: Clarendon Press.

Brunet, Michel. 1969. "Canadians and Canadiens." Pp. 284–293 in Ramsay Cook (ed.), *French-Canadian Nationalism*. Toronto: Macmillan of Canada.

Brym, Robert J. 1989. *From Culture to Power: The Sociology of English Canada*. Toronto: Oxford University Press.

Buchanan, Ronald H. 1982. "The Planter and the Gael: Cultural Dimensions of the Northern Ireland Problem." Pp. 49–74 in Frederick W. Boal and J. Neville H. Douglas (eds.), *Integration and Division: Geographical Perspectives on the Northern Ireland Problem*. London: Academic Press.

Buchanan, William, and Hadley Cantril. 1953. *How Nations See Each Other*. Urbana, Ill.: University of Illinois Press.

Buchignani, Norman L. 1980. "Accommodation, Adaptation, and Policy: Dimensions of the South Asian Experience in Canada." Pp. 121–150 in K. Victor Ujimoto and Gordon Hirabayashi (eds.), *Visible Minorities and Multiculturalism: Asians in Canada*. Toronto: Butterworths.

Burck, Charles G. 1976. "A Group Profile of the Fortune 500 Chief Executive." *Fortune* (May):173–177, 308–312.

Burdick, John. 1995. "Brazil's Black Consciousness Movement." Pp. 174–183 in Kevin Danaher and Michael Shellenberger (eds.), *Fighting for the Soul of Brazil*. New York: Monthly Review Press.

Burkey, Richard M. 1978. *Ethnic and Racial Groups: The Dynamics of Dominance*. Menlo Park, Calif.: Cummings.

Burkholz, Herbert. 1980. "The Latinization of Miami." *New York Times Magazine* (September 21):44–47, 84–88, 98–100.

Burma, John. 1954. *Spanish-Speaking Groups in the United States*. Durham, N.C.: Duke University Press.

Burnet, Jean. 1976. "Ethnicity: Canadian Experience and Policy." *Sociological Focus* 9:199–207.

———. 1981. "The Social and Historical Context of Ethnic Relations." Pp. 17–35 in Robert C. Garner and Rudolf Kalin (eds.), *A Canadian Social Psychology of Ethnic Relations.* Toronto: Methuen.

Burns, John F. 1988. "Ottawa Will Pay Compensation to Uprooted Japanese-Canadians." *New York Times* (September 23):A10.

Burton, Frank. 1978. *The Politics of Legitimacy: Struggles in a Belfast Community.* London: Routledge & Kegan Paul.

Business Week. 1992. "The Immigrants." (July 13): 114–118.

Business Week/Harris Poll. 1992. "America's Welcome Mat Is Wearing Thin." (July 13):122.

Butterfield, Fox. 1986. "Why Asians Are Going to the Head of the Class." *New York Times* (August 3):12, 18–23.

Button, James W. 1978. *Black Violence.* Princeton, N.J.: Princeton University Press.

Cafferty, Pastora San Juan. 1983. "The Language Question: The Dilemma of Bilingual Education for Hispanics in America." Pp. 101–127 in Lance Liebman (ed.), *Ethnic Relations in America.* Englewood Cliffs, N.J.: Prentice-Hall.

———. 1985. "Language and Social Assimilation." Pp. 87–111 in Pastora San Juan Cafferty and William C. McCready (eds.), *Hispanics in the United States: A New Social Agenda.* New Brunswick, N.J.: Transaction.

Calic, Marie-Janine. 2000. "Kosovo In the Twentieth Century: A Historical Account." Pp. 19–31 in Albrecht Schnabel and Ramesh Thakur (eds.), *Kosovo and the Challenge of Humanitarian Intervention: Selective Indignation, Collective Action, and International Citizenship.* Tokyo: United Nations University Press.

Camarillo, Albert. 1979. *Chicanos in a Changing Society.* Cambridge, Mass.: Harvard University Press.

Camarillo, Albert M., and Frank Bonilla. 2001. "Hispanics in a Multicultural Society: A New American Dilemma?" Pp. 103–134 in Neil J. Smelser, William Julius Wilson, and Faith Mitchell (eds.), *America Becoming: Racial Trends and Their Consequences.* Vol. I. Washington, D.C.: National Academy Press.

Campani, Giovanna. 1993. "Immigration and Racism in Southern Europe: The Italian Case." *Ethnic and Racial Studies* 16:507–535.

Campbell, Angus. 1971. *White Attitudes toward Black People.* Ann Arbor, Mich.: Institute for Social Research, University of Michigan.

Cancio, A. Silva, T. David Evans, and David Maune. 1996. "Reconsidering the Declining Significance of Race: Racial Differences in Early Career Wages." *American Sociological Review* 61:541–546.

Canedy, Dana. 2001. "Cuban Exile Group Fractured as Hard-Liners Quit Board." *New York Times* (August 8):A1.

Caplan, Nathan, Marcella H. Choy, and John K. Whitmore. 1991. *Children of the Boat People: A Study*

of Educational Success. Ann Arbor, Mich.: University of Michigan Press.

———. 1992. "Indochinese Refugee Families and Academic Achievement." *Scientific American* 266 (February):36–42.

Caplan, Nathan, John K. Whitmore, and Marcella H. Choy. 1989. *The Boat People and Achievement in America.* Ann Arbor, Mich.: University of Michigan Press.

Cardenas, Gilbert. 1976. "Los Desarraigados: Chicanos in the Midwestern Region of the United States." *Aztlan* 7:153–201.

Cardoso, Fernando Henrique. 1965. "Color Prejudice in Brazil." *Présence Africaine* 25:120–128.

Cardwell, Cary. 1998. "The Miami Herald's Big News." *Hispanic Business* (October):80–81.

Carilli, Theresa. 1998. "Still Crazy After All These Years: Italian Americans in Mainstream U.S. Films." Pp. 111–124 in Yahya R. Kamalipour and Theresa Carilli (eds.), *Cultural Diversity and the U.S. Media.* Albany: State University of New York Press.

Cariño, Benjamin V. 1987. "The Philippines and Southeast Asia: Historical Roots and Contemporary Linkages." Pp. 305–325 in James T. Fawcett and Benjamin V. Cariño (eds.), *Pacific Bridges: The New Immigration from Asia and the Pacific Islands.* Staten Island, N.Y.: Center for Migration Studies.

Carmichael, Stokely, and Charles V. Hamilton. 1967. *Black Power: The Politics of Liberation in America.* New York: Vintage.

Carnoy, Martin, Hugh M. Daley, and Raul Hinajosa Ojeda. 1993. "The Changing Economic Position of Latinos in the U.S. Labor Market Since 1939." Pp. 28–54 in Rebecca Morales and Frank Bonilla (eds.), *Latinos in a Changing U.S. Economy.* Newbury Park, Calif.: Sage.

Carroll, Joseph. 2001. "Public Overestimates U.S. Black and Hispanic Populations." *Gallup Poll* News Service (June 4).

Carter, Gwendolyn M. 1980. *Which Way Is South Africa Going?* Bloomington, Ind.: Indiana University Press.

Castles, Stephen. 1992. "Migrants and Minorities in Post-Keynesian Capitalism: The German Case." Pp. 36–54 in Malcolm Cross (ed.), *Ethnic Minorities and Industrial Change in Europe and North America.* Cambridge, England: Cambridge University Press.

Castles, Stephen, Heather Booth, and Tina Wallace. 1984. *Here for Good: Western Europe's New Ethnic Minorities.* London: Pluto Press.

Castles, Stephen, and Godula Kosack. 1973. *Immigrant Workers and Class Structure in Western Europe.* London: Oxford University Press.

Castles, Stephen, and Mark J. Miller. 1998. *The Age of Migration: International Population Movements in the Modern World.* 2d ed. New York: Guilford.

Castro, Max J. 1992. "The Politics of Language in Miami." Pp. 109–132 in Guillermo J. Grenier and Alex Stepick III (eds.), *Miami Now! Immigration, Ethnicity, and Social Change.* Gainesville, Fla.: University Press of Florida.

Cataldo, Everett F., Michael Giles, and Douglas S. Gatlin. 1975. "Metropolitan School Desegregation: Practical Remedy or Impractical Ideal?" *Annals of the American Academy of Political and Social Science* 422:97–104.

Caudill, William, and George De Vos. 1956. "Achievement, Culture and Personality: The Case of the Japanese Americans." *American Anthropologist* 58:1103–1126.

Celis, William, III. 1992. "Hispanic Rate for Dropouts Remains High." *New York Times* (October 14):A1, B8.

Centre for Research and Information on Canada. 2002. *Portraits of Canada 2001*. Montreal: Centre for Research and Information on Canada.

Chanes, Jerome A. 1994. "Interpreting the Data: Antisemitism and Jewish Security in the United States." *Patterns of Prejudice* 28:87–101.

Chaney, Elsa M. 1979. "The World Economy and Contemporary Migration." *International Migration Review* 13:204–212.

Charles, Camille Zubrinsky. 2000. "Neighborhood Racial-Composition Preferences: Evidence from a Multiethnic Metropolis." *Social Problems* 47:379–407.

Chavez, Linda. 1991. *Out of the Barrio: Toward a New Politics of Hispanic Assimilation*. New York: Basic Books.

_____. 1998. "Our Hispanic Predicament." *Commentary* (June):47–50.

Chen, David W. 1999. "Asian Middle Class Alters a Rural Enclave." *New York Times* (December 27):A1.

Chesley, Roger, and Brenda J. Gilchrist. 1993. "Simmering Distrust." *Detroit Free Press* (February 22):1A, 5A.

Chiswick, Barry R. 1984. "The Labor Market Status of American Jews: Patterns and Determinants." Pp. 131–153 in *American Jewish Year Book 1985*. New York: American Jewish Committee.

Choi, Candice. 2001. "Fewer Puerto Ricans in NYC, But More Mexicans, Asian Indians." DiversityInc.com (May 22) <http://www.diversityinc.com>.

Christopulos, Diana. 1980. "The Politics of Colonialism: Puerto Rico from 1898 to 1972." Pp. 120–170 in Adalberto Lopez (ed.), *The Puerto Ricans: Their History, Culture, and Society*. Cambridge, Mass.: Schenkman.

Citizenship and Immigration Canada. 2000. *Facts and Figures 2000: Immigration Overview* <http://www.cic.gc.ca/english/pub/facts2000/1imm-05.html>

Clairmont, Donald H., and Dennis W. Magill. 1974. *Africville: The Life and Death of a Canadian Black Community*. Toronto: McClelland & Stewart.

Clarity, James F. 1992. "Talks Fail, the Irish and British Report." *New York Times* (November 17):A10.

Clark, Kenneth B. 1965. *Dark Ghetto*. New York: Harper & Row.

_____. 1980. "The Role of Race." *New York Times Magazine* (October 5):25–33.

Clark, S. D. 1950. "The Canadian Community." Pp. 375–389 in George Brown (ed.), *Canada*. Berkeley: University of California Press.

Clement, Wallace. 1975. *The Canadian Corporate Elite: An Analysis of Economic Power*. Toronto: McClelland & Stewart.

_____. 1990. "A Critical Response to 'Perspectives on the Class and Ethnic Origins of Canadian Elites.'" *Canadian Journal of Sociology* 15:179–185.

Cloud, John. 1998. "Why Coors Went Soft." *Time* (November 2):70.

Cohen, Percy S. 1968. "Ethnic Group Differences in Israel." *Race* 9:301–310.

Cohen, Roger. 1994. "Bosnian Camp Survivors Describe Random Death." *New York Times* (August 2):A1, A6.

Cohen, Steven M. 1985. *The 1984 National Survey of American Jews: Political and Social Outlooks*. New York: American Jewish Committee.

_____. 1987. *Ties and Tensions: The 1986 Survey of American Jewish Attitudes toward Israel and Israelis*. New York: American Jewish Committee.

_____. 1988. *American Assimilation or Jewish Revival?* Bloomington, Ind.: Indiana University Press.

_____. 1995. "Jewish Continuity over Judaic Content: The Moderately Affiliated American Jew." Pp. 395–416 in Robert M. Selter and Norman J. Cohen (eds.), *The Americanization of the Jews*. New York: New York University Press.

Coleman, James S., et al. 1966. *Equality of Educational Opportunity*. Washington, D.C.: U.S. Government Printing Office.

Collins, Sharon M. 1997. *Black Corporate Executives: The Making and Breaking of a Black Middle Class*. Philadelphia: Temple University Press.

Colon, Yves. 1999. "Nicaraguans Maintain Strong Links to Their Troubled Home." *Miami Herald* (July 18) <http://www.herald.com/content/today/docs/001423.htm>.

Commission on Minority Participation in Education and American Life. 1988. *One-Third of a Nation*. Washington, D.C.: American Council on Education.

Commonwealth Group of Eminent Persons. 1986. *Mission to South Africa: The Commonwealth Report*. Harmondsworth, England: Penguin.

Compton, Paul. 1982. "The Demographic Dimension of Integration and Division in Northern Ireland." Pp. 75–104 in Frederick W. Boal and J. Neville H. Douglas (eds.), *Integration and Division: Geographical Perspectives on the Northern Ireland Problem*. London: Academic Press.

Congressional Quarterly. 1979. "Most Members of Congress Claim Religious Affiliation." *CQ Weekly Report* 37(January 20):80–81.

Connor, Walker. 1972. "Nation-Building or Nation-Destroying?" *World Politics* 24:319–355.

_____. 1976. "The Political Significance of Ethnonationalism within Western Europe." Pp. 110–133 in Abdul Said and Lucy R. Simmons (eds.), *Ethnicity in an International Context*. New Brunswick, N.J.: Transaction.

Coogan, Tim Pat. 1980. *On the Blanket: The H Block Story*. Dublin: Ward River Press.

Cook, Fred J. 1971. "The Black Mafia Moves into the Numbers Racket." *New York Times Magazine* (April 4):26–27, 107–112.

Corbett, David C. 1957. *Canada's Immigration Policies: A Critique*. Toronto: University of Toronto Press.

Corelli, Rae. 1995. "A Tolerant Nation's Hidden Shame." *Maclean's* (August 14):40–43.

Cormack, Robert, and Robert Osborne. 1994. "The Evolution of a Catholic Middle Class." Pp. 65–85 in Adrian Guelke (ed.), *New Perspectives on the Northern Ireland Conflict*. Aldershot, England: Avebury.

Cornell, Stephen. 1988. *The Return of the Native: American Indian Political Resurgence*. New York: Oxford University Press.

———. 1990. "Land, Labour and Group Formation: Blacks and Indians in the United States." *Ethnic and Racial Studies* 13:368–388.

Cortés, Carlos E. 1994. "The Hollywood Curriculum on Italian Americans: Evolution of an Icon of Ethnicity." Pp. 89–108 in Lydio F. Tomasi, Piero Gastaldo, and Thomas Row (eds.), *The Columbus People: Perspectives in Italian Immigration to the Americas and Australia*. New York: Center for Migration Studies.

Cose, Ellis. 1993. *The Rage of a Privileged Class*. New York: HarperPerennial.

Coser, Lewis A. 1956. *The Functions of Social Conflict*. New York: Free Press.

Covello, Leonard. 1967. *The Social Background of the Italo-American School Child*. Leiden: E. J. Brill.

Coward, John M. 1999. *The Newspaper Indian: Native American Identity in the Press, 1820–90*. Urbana, University of Illinois Press.

Cox, Oliver C. 1948. *Caste, Class and Race*. New York: Monthly Review Press.

Crawford, James. 1992. *Hold Your Tongue: Bilingualism and the Politics of "English Only."* Reading, Mass.: Addison-Wesley.

Crispino, James A. 1980. *The Assimilation of Ethnic Groups: The Italian Case*. Staten Island, N.Y.: Center for Migration Studies.

Cumberland, Charles C. 1968. *Mexico: The Struggle for Modernity*. New York: Oxford University Press.

Dahrendorf, Ralf. 1959. *Class and Class Conflict in Industrial Society*. Stanford, Calif.: Stanford University Press.

———. 1968. *Essays in the Theory of Society*. Stanford, Calif.: Stanford University Press.

Daley, Suzanne. 1995. "As Crime Soars, South African Whites Leave." *New York Times* (December 12):A1, A6.

———. 1997. "Torturer's Testimony Gives South Africa a New Lesson in the Banality of Evil." *New York Times* (November 9):1, 12.

———. 1998a. "South Africa's 'White Tribe' Sees Its Dark Past as Clouding a Black Future." *New York Times* (February 22):1, 16.

———. 1998b. "At Afrikaner Schools, a Backlash against Blacks." *New York Times* (April 13):A4.

———. 1999. "Old Scars on the New South Africa." *New York Times* (January 3):A1, A12.

D'Amico, Ronald, and Nan L. Maxwell. 1995. "The Continuing Significance of Race in Minority Male Joblessness." *Social Forces* 73:969–991.

Daniels, Roger. 1977. *The Politics of Prejudice: The Anti-Japanese Movement in California and the Struggle for Japanese Exclusion*. 2d ed. Berkeley: University of California Press.

———. 1981. *Concentration Camps: North America*. Malabar, Fla.: Krieger.

Dao, James. 1998. "Indians Using New Wealth for More Effective Lobbying." *New York Times* (February 9):A16.

Darby, John. 1976. *Conflict in Northern Ireland: The Development of a Polarised Community*. Dublin: Gill & Macmillan.

———. 1986. *Intimidation and the Control of Conflict in Northern Ireland*. Dublin: Gill & Macmillan.

———. 1995. *Northern Ireland: Managing Difference*. London: Minority Rights Group International.

———. 1998. "Approaches to Cultural Diversity in Northern Ireland." Pp. 3–12 in *Cultural Traditions in Northern Ireland*. Belfast: Institute of Irish Studies, The Queen's University of Belfast.

Darby, John, and Geoffrey Morris. 1974. *Intimidation in Housing, February 1974*. Belfast: Northern Ireland Community Relations Commission.

Darroch, A. Gordon. 1979. "Another Look at Ethnicity, Stratification and Social Mobility in Canada." *Canadian Journal of Sociology* 4:1–26.

Davidson, James D., Ralph E. Pyle, and David V. Reyes. 1995. "Persistence and Change in the Protestant Establishment, 1930–1992." *Social Forces* 74:157–175.

Davis, David Brion. 1966. *The Problem of Slavery in Western Culture*. Ithaca, N.Y.: Cornell University Press.

Davis, E. E., and R. Sinnott. 1979. *Attitudes in the Republic of Ireland Relevant to the Northern Ireland Problem*. Vol. 1, Descriptive Analysis and Some Comparisons with Attitudes in Northern Ireland and Great Britain, Paper No. 97. Dublin: Economic and Social Research Institute.

Davis, F. James. 1978. *Minority-Dominant Relations: A Sociological Analysis*. Arlington Heights, Ill.: AHM.

Davis, Morris, and Joseph F. Krauter. 1971. *The Other Canadians: Profiles of Six Minorities*. Toronto: Methuen.

Davis, Richard. 1994. *Mirror Hate: The Convergent Ideology of Northern Ireland Paramilitaries, 1966–1992*. Aldershot, England: Dartmouth.

Dawidowicz, Lucy S. 1975. *The War against the Jews, 1933–1945*. New York: Holt, Rinehart & Winston.

Dawsey, Darrell. 1992. "Fatal Shooting Escalates Tension between Blacks, Arab-Americans." *Detroit News* (September 19):1C.

Dean, John P., and Alex Rosen. 1955. *A Manual of Intergroup Relations*. Chicago: University of Chicago Press.

de Azevedo, Fernando. 1950. *Brazilian Culture*. Trans. William Rex Crawford. New York: Macmillan.

deBréadun, Deaglan. 1999. "Northern Ireland: An End to the Troubles?" *Current History* (April):153–158.

Degler, Carl. 1971. *Neither Black nor White: Slavery and Race Relations in Brazil and the United States*. New York: Macmillan.

de la Garza, Rodolfo O., Louis DeSipio, F. Chris Garcia, John Garcia, and Angelo Falcon. 1992. *Latino Voices: Mexican, Puerto Rican, and Cuban Perspectives on American Politics*. Boulder, Colo.: Westview Press.

Della Pergola, Sergio, and Allie A. Dubb. 1988. "South African Jewry: A Sociodemographic Profile." Pp. 59–140 in *American Jewish Year Book 1988*. New York: American Jewish Committee.

Della Pergola, Sergio, Uzi Rebhun, and Mark Tolts. 2000. "Prospecting the Jewish Future: Population Projections, 2000–2080." Pp. 103–146 in *American Jewish Year Book 2000*. New York: American Jewish Committee.

de los Angeles Torres, Maria. 1988. "From Exiles to Minorities: The Politics of Cuban-Americans." Pp. 81–98 in F. Chris Garcia (ed.), *Latinos and the Political System*. Notre Dame, Ind.: University of Notre Dame Press.

Dent, David J. 1992. "The New Black Suburbs." *New York Times Magazine* (June 14):18–25.

DePalma, Anthony. 1998. "Canadian Indians Celebrate Vindication of Their History." *New York Times* (February 9):A1, A8.

de Paor, Liam. 1970. *Divided Ulster*. Harmondsworth, England: Pelican.

De Roche, Constance, and John de Roche. 1991. "Black and White: Racial Construction in Television Police Dramas." *Canadian Ethnic Studies* 23(3):69–91.

Destexhe, Alain. 1995. *Rwanda and Genocide in the Twentieth Century*. New York: New York University Press.

Detroit News. 1990. "Pointes' Past Bias Still Keeps Some Minorities Away." (February 28):1A, 6A.

Deutsch, Karl W. 1966. *Nationalism and Social Communication*. Cambridge, Mass.: MIT Press.

Devine, P., and A. J. Elliot. 1995. "Are Stereotypes Really Fading? The Princeton Trilogy Revisited." *Personality and Social Psychology Bulletin* 21:1139–1150.

Devine, P. G., and S. J. Sherman. 1992. "Intuitive vs. Rational Judgment and the Role of Stereotyping in the Human Condition: Kirk or Spock?" *Psychological Inquiry* 3:153–159.

DeVita, Carol J. 1996. "The United States at Mid-Decade." *Population Bulletin* 50 (March).

Devlin, Bernadette. 1977. "Playboy Interview." Pp. 97–106 in Charles Carlton (ed.), *Bigotry and Blood: Documents on the Ulster Troubles*. Chicago: Nelson-Hall.

De Vos, George, and Lola Romanucci-Ross (eds.). 1975. *Ethnic Identity: Cultural Continuities and Change*. Palo Alto, Calif.: Mayfield.

De Vos, George, and Hiroshi Wagatsuma. 1966. *Japan's Invisible Race: Caste in Culture and Personality*. Berkeley: University of California Press.

Diaz, Guarione M. (ed.). 1980. *Evaluation and Identification of Policy Issues in the Cuban Community*. Miami, Fla.: Cuban National Planning Council.

Dickie-Clark, H. F. 1972. "The Coloured Minority of Durban." Pp. 25–38 in Noel P. Gist and Anthony G. Dworkin (eds.), *The Blending of Races: Marginality and Identity in World Perspective*. New York: Wiley.

Dinnerstein, Leonard, Roger L. Nichols, and David M. Reimers. 1996. *Natives and Strangers: A Multicultural History of Americans*. New York: Oxford University Press.

Dinnerstein, Leonard, and David M. Reimers. 1988. *Ethnic Americans: A History of Immigration*. 3d ed. New York: Harper & Row.

Dobbin, Christine. 1996. *Asian Entrepreneurial Minorities: Conjoint Communities in the Making of the World-Economy, 1570–1940*. Richmond, Surrey: Curzon.

Dobzhansky, Theodosious. 1973. "Differences Are Not Deficits." *Psychology Today* (December):97–101.

Doherty, Paul. 1993. "Agape to Zoroastrian: Religious Denominations in Northern Ireland 1961 to 1991." *Irish Geography* 26:14–21.

Dollard, John. 1937. *Caste and Class in a Southern Town*. New Haven, Conn.: Yale University Press.

Dollard, John et al. 1939. *Frustration and Aggression*. New Haven, Conn.: Yale University Press.

Donnan, Hastings, and Graham McFarlane. 1983. "Informal Social Organization." Pp. 110–135 in John Darby (ed.), *Northern Ireland: The Background to the Conflict*. Belfast: Appletree Press.

Dorman, James H. 1980. "Ethnic Groups and 'Ethnicity': Some Theoretical Considerations." *Journal of Ethnic Studies* 7:23–36.

Drozdiak, William. 1997. "Germany Finds Its Blood Lines Are No Longer Pure." *Washington Post National Weekly Edition* (May 5):14.

D'Souza, Dinesh. 1995. *The End of Racism: Principles for a Multiracial Society*. New York: Free Press.

Duany, Jorge. 1998. "Reconstructing Racial Identity: Ethnicity, Color, and Class among Dominicans in the United States and Puerto Rico." *Latin American Perspectives* 25:147–172.

Duckitt, J. 1989. "Authoritarianism and Group Identification: A New View of an Old Construct." *Political Psychology* 10:63–84.

Dugard, John. 1992. "The Law of Apartheid." Pp. 3–31 in John Dugard (ed.), *The Last Years of Apartheid: Civil Liberties in South Africa*. Ford Foundation.

Duke, Lynne. 1998. "Where Education Is Failing the Students." *Washington Post National Weekly Edition* (January 26):16.

_____. 1999a. "A Tender Democracy." *Washington Post National Weekly Edition* (May 3):16–17.

_____. 1999b. "A Nation Still Divisible." *Washington Post National Weekly Edition* (June 7):6–7.

Dunn, L. C. 1956. "Race and Biology." Pp. 31–67 in Leo Kuper (ed.), *Race, Science and Society*. New York: Columbia University Press.

Duran, Khalid. 1995. "Bosnia: The Other Andalusia." Pp. 25–36 in Syed Z. Abedin and Ziauddin Sardar (eds.), *Muslim Minorities in the West*. London: Grey Seal.

Dwyer, Victor. 1994. "In Search of Unity." *Maclean's* (July 1):16–19.

Dye, Thomas R. 1995. *Who's Running America? The Clinton Years*. 6th ed. Englewood Cliffs, N.J.: Prentice-Hall.

Dzidzienyo, Anani, and Lourdes Casal. 1979. *The Position of Blacks in Brazilian and Cuban Society*. London: Minority Rights Group.

Edsall, Thomas Byrne, with Mary D. Edsall. 1992. *Chain Reaction: The Impact of Race, Rights, and Taxes on American Politics*. New York: Norton.

Edwards, John. 1995. *Affirmative Action in a Sectarian Society: Fair Employment Policy in Northern Ireland*. Aldershot, England: Avebury.

Egan, Timothy. 2001. "As Others Abandon Plains, Indians and Bison Come Back." *New York Times* (May 27):A1.

Ehrlich, Howard J. 1973. *The Social Psychology of Prejudice*. New York: Wiley.

Elkin, Frederick, and Gerald Handel. 1978. *The Child and Society: The Process of Socialization*. New York: Random House.

Elkins, Stanley M. 1976. *Slavery: A Problem in American Institutional and Intellectual Life*. 3d ed. Chicago: University of Chicago Press.

Elliott, Jean Leonard. 1979. "Canadian Immigration: A Historical Assessment." Pp. 160–172 in Jean Leonard Elliott (ed.), *Two Nations, Many Cultures*. Scarborough, Ont.: Prentice-Hall of Canada.

Elliott, Jean Leonard, and Augie Fleras. 1990. "Immigration and the Canadian Ethnic Mosaic." Pp. 51–76 in Peter S. Li (ed.), *Race and Ethnic Relations in Canada*. Toronto: Oxford University Press.

Elliott, Marianne. 1997. "Religion and Identity in Northern Ireland." Pp. 149–167 in Winston A. Van Horne (ed.), *Global Convulsions: Race, Ethnicity, and Nationalism at the End of the Twentieth Century*. Albany: State University of New York Press.

Employment and Immigration Canada. 1988. *Profiles of Canadian Immigration*. Ottawa: Minister of Supply and Services.

Engelberg, Stephen, and Tim Weiner. 1995. "Srebrenica: The Days of Slaughter." *New York Times* (October 29):1, 6–7.

Engelhardt, Tom. 1975. "Ambush at Kamikaze Pass." Pp. 522–531 in Norman R. Yetman and C. Hoy Steele (eds.), *Majority and Minority: The Dynamics of Racial and Ethnic Relations*. 2d ed. Boston: Allyn & Bacon.

Entman, Robert M. 1997. "African Americans According to TV News." Pp. 29–36 in Everette E. Dennis and Edward C. Pease (eds.), *The Media in Black and White*. New Brunswick, N.J.: Transaction.

Equality Commission for Northern Ireland. 2001. *Monitoring Report No. 11* <http://www.equalityni.org/pdf/m11.pdf>.

Eschbach, Karl. 1995. "The Enduring and Vanishing American Indian: American Indian Population Growth and Intermarriage in 1990." *Ethnic and Racial Studies* 18:89–108.

Esman, Milton J. 1994. *Ethnic Politics*. Ithaca, N.Y.: Cornell University Press.

Espenshade, Thomas J. 1998. "U.S. Immigration and the New Welfare State." Pp. 231–250 in David Jacobson (ed.), *The Immigration Reader: America in a Multidisciplinary Perspective*. Malden, Mass.: Blackwell.

Espiritu, Yen Le. 1992. *Asian American Panethnicity: Bridging Institutions and Identities*. Philadelphia: Temple University Press.

Esses, Victoria M., Lynne M. Jackson, Jeffrey M. Nolan, and Tamara L. Armstrong. 1999. "Economic Threat and Attitudes toward Immigrants." Pp. 212–229 in Shiva S. Halli and Leo Driedger (eds.), *Immigrant Canada: Demographic, Economic, and Social Challenges*. Toronto: University of Toronto Press.

Estrada, Leobardo F. 1976. "A Demographic Comparison of the Mexican Origin Population in the Midwest and Southwest." *Aztlan* 7:203–234.

Etzioni, Amitai. 1998. "Some Diversity." *Society* (July/August):59–61.

Ezorsky, Gertrude. 1991. *Racism and Justice: The Case for Affirmative Action*. Ithaca, N.Y.: Cornell University Press.

Fabricant, Florence. 1993. "Riding Salsa's Coast-to-Coast Wave of Popularity." *New York Times* (June 2):B1, B7.

Fagan, Richard R., Richard A. Brody, and Thomas J. O'Leary. 1968. *Cubans in Exile*. Stanford, Calif.: Stanford University Press.

Farhi, Paul. 1999. "Where's the Color in TV?" *Washington Post National Weekly Edition* (July 19–26):22.

Farley, Reynolds. 1984. *Black and Whites: Narrowing the Gap?* Cambridge, Mass.: Harvard University Press.

———. 1985. "The Residential Segregation of Blacks from Whites: Trends, Causes, and Consequences." In U.S. Commission on Civil Rights, *Issues in Housing Discrimination: A Consultation/Hearing of the United States Commission on Civil Rights*. Vol. 1: Papers Presented. Washington, D.C.: U.S. Government Printing Office.

———. 1993. "The Common Destiny of Blacks and Whites: Observations about the Social and Economic Status of the Races." Pp. 197–233 in Herbert Hill and James E. Jones, Jr. (eds.), *Race in America: The Struggle for Equality*. Madison, Wis.: University of Wisconsin Press.

Farley, Reynolds, and Walter R. Allen. 1987. *The Color Line and the Quality of Life in America*. New York: Russell Sage Foundation.

Farley, Reynolds, Suzanne Bianchi, and Diane Colasanto. 1979. "Barriers to the Racial Integration of Neighborhoods: The Detroit Case." *Annals of the American Academy of Political and Social Science* 441:97–113.

Farley, Reynolds, and William H. Frey. 1992. *Changes in the Segregation of Whites from Blacks During the 1980s: Small Steps Toward a More Racially Integrated Society*. Ann Arbor, Mich.: Population Studies Center, University of Michigan.

Farley, Reynolds, and Albert Hermalin. 1972. "The 1960s: Decade of Progress for Blacks?" *Demography* 9:353–370.

Farley, Reynolds, Charlotte Steeh, Maria Krysan, Tara Jackson, and Keith Reeves. 1994. "Stereotypes and Segregation: Neighborhoods in the Detroit Area." *American Journal of Sociology* 100:750–780.

Farley, Reynolds, and Alma F. Taeuber. 1974. "Racial Segregation in the Public Schools." *American Journal of Sociology* 79:888–905.

Farley, Reynolds et al. 1978. "Chocolate City, Vanilla Suburbs: Will the Trend toward Racially Separate Communities Continue?" *Social Science Research* 7:319–344.

Farmer, James. 1967. "The Controversial Moynihan Report." Pp. 409–413 in Lee Rainwater and William L. Yancey (eds.), *The Moynihan Report and the Politics of Controversy*. Cambridge, Mass.: MIT Press.

Farnsworth, Clyde H. 1991. "Quebec's New Minority Issue: Blacks Charge Bias." *New York Times* (August 16):A7.

———. 1995. "Black in Quebec: Escaping the Streets." *New York Times* (October 13):A4.

Fauman, S. Joseph. 1958. "Occupational Selection among Detroit Jews." Pp. 119–136 in Marshall Sklare (ed.), *The Jews: Social Patterns of an American Group*. Glencoe, Ill.: Free Press.

Feagin, Joe R., and Clairece Booher Feagin. 1978. *Discrimination American Style: Institutional Racism and Sexism*. Englewood Cliffs, N.J.: Prentice-Hall.

Featherman, David L., and Robert M. Hauser. 1976. "Changes in the Socioeconomic Stratification of the Races, 1962–1973." *American Journal of Sociology* 82:621–651.

Fein, Helen. 1993. *Genocide: A Sociological Perspective*. London: Sage.

Feit, E. 1967. "Community in a Quandary: The South African Jewish Community and 'Apartheid.'" *Race* 8:395–468.

Fenwick, Rudy. 1982. "Ethnic-Culture and Economic Structure: Determinants of French-English Earnings Inequality in Quebec." *Social Forces* 61:1–23.

Fernandes, Florestan. 1971. *The Negro in Brazilian Society*. New York: Atheneum.

Fernández Kelly, M. Patricia, and Richard Schauffler. 1996. "Divided Fates: Immigrant Children and the New Assimilation." Pp. 30–53 in Alejandro Portes (ed.), *The New Second Generation*. New York: Russell Sage Foundation.

Ferrante, Angela. 1977. "Racism? You Can't Argue with the Facts." *Maclean's* (February 7):18–21.

Finn, Peter, David Finkel, R. Jeffrey Smith, and Michael Dobbs. 1999. "Across Kosovo, Death at Every Turn." *Washington Post National Weekly Edition* (June 21):15–18.

Firestone, David. 2001. "U. of Georgia Cannot Use Race in Admission Policy, Court Rules." *New York Times* (August 28):A1.

Fischer, Claude S., Michael Hout, Martin Sanchez Jankowski, Samuel R. Lucas, Ann Swidler, and Kim Voss. 1996. *Inequality by Design: Cracking the Bell Curve Myth*. Princeton, N.J.: Princeton University Press.

Fishman, Josua A. 1987. "What Is Happening to Spanish on the U.S. Mainland?" *Ethnic Affairs* 1(Fall):12–23.

Fiske, Susan T., Jun Xu, Amy C. Cuddy, and Peter Glick. 1999. "(Dis)respecting versus (Dis)liking: Status and Interdependence Predict Ambivalent Stereotypes of Competence and Warmth." *Journal of Social Issues* 55:473–489.

Fitzpatrick, Joseph P. 1980. "Puerto Ricans." Pp. 858–867 in Stephen Thernstrom (ed.), *Harvard Encyclopedia of American Ethnic Groups*. Cambridge, Mass.: Harvard University Press.

———. 1987. *Puerto Rican Americans: The Meaning of Migration to the Mainland*. 2d ed. Englewood Cliffs, N.J.: Prentice-Hall.

———. 1995. "Puerto Rican New Yorkers, 1990." *Migration World* 23(1):16–19.

Fleras, Augie, and Jean Leonard Elliott. 1992. *Multiculturalism in Canada: The Challenge of Diversity*. Scarborough, Ont.: Nelson Canada.

Fletcher, Michael A. 1997. "Coming to Terms with the Racial Melting Pot." *Washington Post National Weekly Edition* (May 5):31.

———. 1998. "Latinos at the Back of the Class." *Washington Post National Weekly Edition* (August 10):33.

Foerster, Robert S. 1919. *The Italian Emigration of Our Times*. Cambridge, Mass.: Harvard University Press.

Fogel, Robert William, and Stanley L. Engerman. 1974. *Time on the Cross*. Boston: Little, Brown.

Foner, Nancy. 1987. *New Immigrants in New York*. New York: Columbia University Press.

Fong, Eric. 1996. "A Comparative Perspective on Racial Residential Segregation: American and Canadian Experiences." *Sociological Quarterly* 37:199–226.

Fontaine, Pierre-Michel. 1981. "Transnational Relations and Racial Mobilization: Emerging Black Movements in Brazil." Pp. 141–162 in John F. Stack, Jr. (ed.), *Ethnic Identities in a Transnational World*. Westport, Conn.: Greenwood.

Forcese, Dennis. 1997. *The Canadian Class Structure*. 4th ed. Toronto: McGraw-Hill Ryerson.

Forero, Juan. 2001. "Prosperous Colombians Flee, Many to U.S., to Escape War." *New York Times* (April 10):A1, A6.

Fortune. 1936. "Jews in America." 13(February):79–85, 128–144.

———. 2001. "The 500 Largest U.S. Corporations." (April 16):F-1–F-20.

Frady, Marshall. 1968. *Wallace*. New York: World.

Franklin, John Hope. 1980. *From Slavery to Freedom*. 5th ed. New York: Knopf.

Franklin, Raymond S., and Solomon Resnik. 1973. *The Political Economy of Racism*. New York: Holt, Rinehart & Winston.

Fraser, Morris. 1973. *Children in Conflict: Growing Up in Northern Ireland*. New York: Basic Books.

Fraser, Steven (ed.). 1995. *The Bell Curve Wars: Race, Intelligence, and the Future of America*. New York: Basic Books.

Frazier, E. Franklin. 1939. *The Negro Family in the United States*. Chicago: University of Chicago Press.

———. 1949. *The Negro in the United States*. New York: Macmillan.

Fredrickson, George M. 1971. *The Black Image in the White Mind*. New York: Harper & Row.

_____. 1981. *White Supremacy: A Comparative Study in American and South African History.* New York: Oxford University Press.

Freedman, Maurice. 1955. "The Chinese in Southeast Asia." Pp. 388–411 in Andrew W. Lind (ed.), *Race Relations in World Perspective.* Honolulu, Hawaii: University of Hawaii Press.

Freedman, Samuel G. 2000. *Jew vs. Jew: The Struggle for the Soul of American Jewry.* New York: Touchstone.

Frey, William H. 1998. "Black Migration to the South Reaches Record Highs in 1990s." *Population Today* (February):1–3.

_____. 2001. *Census 2000 Shows Large Black Return to the South, Reinforcing the Region's "White-Black" Demographic Profile.* PSC Research Report no. 01-473. Ann Arbor: Population Studies Center, University of Michigan.

Freyre, Gilberto. 1956. *The Masters and the Slaves.* New York: Knopf.

_____. 1963a. *New World in the Tropics: The Culture of Modern Brazil.* New York: Vintage.

_____. 1963b. "Ethnic Democracy: The Brazilian Example." *Americas* 15(December):1–6.

Frideres, James S. 1976. "Racism in Canada: Alive and Well." *Western Canadian Journal of Anthropology* 6:124–145.

_____. 1990. "Policies on Indian People in Canada." Pp. 98–119 in Peter S. Li (ed.), *Race and Ethnic Relations in Canada.* Toronto: Oxford University Press.

_____. 1993. *Native Peoples in Canada: Contemporary Conflicts.* 4th ed. Toronto: Prentice-Hall.

_____. 1998. *Aboriginal Peoples In Canada: Contemporary Conflicts.* 5th ed. Scarborough, Ont.: Prentice Hall Allyn and Bacon Canada.

Fried, Morton H. 1965. "A Four-Letter Word That Hurts." *Saturday Review* (October 2):21–23, 35.

Fry, Peter. 2000. "Politics, Nationality, and the Meanings of 'Race' in Brazil." *Daedalus* 129(Spring):83–118.

Frisbie, W. Parker, and Lisa Neidert. 1977. "Inequality and the Relative Size of Minority Populations: A Comparative Analysis." *American Journal of Sociology* 82:1007–1030.

Fugita, Stephen S., and David J. O'Brien. 1991. *Japanese American Ethnicity: The Persistence of Community.* Seattle, Wash.: University of Washington Press.

Furnivall, J. S. 1948. *Colonial Policy and Practice.* Cambridge, England: Cambridge University Press.

Gallagher, A. M., and S. Dunn. 1991. "Community Relations in Northern Ireland: Attitudes to Contact and Integration." Pp. 7–22 in Peter Stringer and Gillian Robinson (eds.), *Social Attitudes in Northern Ireland.* Belfast: Blackstaff Press.

Gallo, Patrick J. 1974. *Ethnic Alienation: The Italian-Americans.* Rutherford, N.J.: Fairleigh Dickinson University Press.

Gallup, George, Jr. 1982. *The Gallup Poll: Public Opinion 1981.* Wilmington, Del.: Scholarly Resources.

_____. 1988. *The Gallup Poll: Public Opinion 1987.* Wilmington, Del.: Scholarly Resources.

_____. 1995. *The Gallup Poll: Public Opinion 1994.* Wilmington, Del.: Scholarly Resources.

_____. 1997. *The Gallup Poll: Public Opinion 1997.* Wilmington, Del.: Scholarly Resources.

Gallup Organization. 2001a. *Black-White Relations in the United States: 2001 Update.* Gallup Poll Social Audit (July 10).

_____. 2001b. "Easter Season Finds a Religious Nation." Gallup News Service, *Poll Analyses* (April 13).

_____. 2001c. "The Impact of the Attacks on America." Gallup News Service, *Poll Analyses* (September 21).

_____. 2001d. "Americans Ambivalent about Immigrants." Gallup News Service, *Poll Analyses* (May 3).

_____. 2001e. "Racial or Ethnic Labels Make Little Difference to Blacks, Hispanics." Gallup Poll News Service. *Poll Analyses* (September 11).

_____. 2002. *Poll Topics & Trends: Immigration* <http://www.gallup.com/poll/topics/immigration.asp>.

Gallup Report. 1989. "Tolerance Indices." Nos. 282/283 (March/April).

Gambino, Richard. 1975. *Blood of My Blood: The Dilemma of the Italian-Americans.* Garden City, N.Y.: Anchor.

_____. 1977. *Vendetta.* Garden City, N.Y.: Doubleday.

_____. 1998. "America's Most Tolerated Intolerance: Bigotry against Italian Americans." Pp. 155–173 in Gregory R. Campbell (ed.), *Many Americas: Critical Perspectives on Race, Racism, and Ethnicity.* Dubuque, Iowa: Kendall/Hunt.

Gamson, William A. 1975. *The Strategy of Social Protest.* Homewood, Ill.: Dorsey.

Gans, Herbert J. 1958. "The Origin and Growth of a Jewish Community in the Suburbs: A Study of the Jews of Park Forest." Pp. 205–258 in Marshall Sklare (ed.), *The Jews: Social Patterns of an American Group.* New York: Free Press.

_____. 1967a. *The Levittowners.* New York: Pantheon.

_____. 1967b. "Some Comments on the History of Italian Migration and on the Nature of Historical Research." *International Migration Review* 1:5–9.

_____. 1974. "Foreword." In Neil C. Sandberg, *Ethnic Identity and Assimilation: The Polish-American Community.* New York: Praeger.

_____. 1979. "Symbolic Ethnicity: The Future of Ethnic Groups and Cultures in America." *Ethnic and Racial Studies* 2:1–20.

_____. 1982. *The Urban Villagers: Group and Class in the Life of Italian Americans.* Updated and expanded ed. New York: Free Press.

_____. 1992. "Comment: Ethnic Invention and Acculturation, A Bumpy-Line Approach." *Journal of American Ethnic History* 12:43–52.

_____. 1999. "Filling in Some Holes: Six Areas of Needed Immigration Research." *American Behavioral Scientist* 42:1302–1313.

Garcia, F. Chris, and Rudolph P. de la Garza. 1977. *The Chicano Political Experience: Three Perspectives.* North Scituate, Mass.: Duxbury.

Garcia, Guy. 1999. "Another Latin Boom, But Different." *New York Times* (June 27):AR 25, AR 27.

Garcia, John A., and Carlos H. Arce. 1988. "Political Orientations and Behaviors of Chicanos: Trying to Make Sense Out of Attitudes and Participation." Pp. 125–151 in F. Chris Garcia (ed.), *Latinos in the Political System*. Notre Dame, Ind.: University of Notre Dame Press.

Garcia, Juan Ramon. 1980. *Operation Wetback*. Westport, Conn.: Greenwood.

Garcia-Passalacqua, Juan Manuel. 1994. "The Puerto Ricans: Migrants or Commuters?" Pp. 103–113 in Carlos Antonio Torre, Hugo Rodríguez Vecchini, and William Burgos (eds.), *The Commuter Nation: Perspectives on Puerto Rican Migration*. Río Piedras, Puerto Rico: Editorial de la Universidad de Puerto Rico.

Garcia-Zamor, Jean-Claude. 1970. "Social Mobility of Negroes in Brazil." *Journal of Inter-American Studies* 12:242–254.

Gardner, Robert W., Bryant Robey, and Peter C. Smith. 1985. "Asian Americans: Growth, Change, and Diversity." *Population Bulletin* 40(October):1–43.

Garrett, J.T. 1994. "Health." Pp. 233–237 in Mary B. Davis (ed.), *Native America in the Twentieth Century*. New York: Garland.

Garson, Jean-Pierre. 1992. "Migration and Interdependence: The Migration System between France and Africa." Pp. 80–93 in Mary M. Kritz, Lin Lean Lim, and Hania Zlotnik (eds.), *International Migration Systems: A Global Approach*. Oxford: Clarendon Press.

Gedmin, Jeffrey. 1994. "Germany's Emerging Multiethnic Society." *American Enterprise* 5(July/August):21–25.

Geertz, Clifford. 1994. "Primordial and Civic Ties." Pp. 29–34 in John Hutchinson and Anthony D. Smith (eds.), *Nationalism*. Oxford: Oxford University Press.

Gerdes, Wylie. 1993. "Chaldean Merchants Decry Slaying, Look for Answers." *Detroit Free Press* (December 2):1B, 2B.

Gershman, Carl. 1980. "A Matter of Class." *New York Times Magazine* (October 5):24, 92–96, 98–99, 102–105, 109.

Gerth, Hans. 1940. "The Nazi Party: Leadership and Composition." *American Journal of Sociology* 45:517–541.

Gerth, Hans, and C. Wright Mills. 1946. *From Max Weber*. New York: Oxford University Press.

Geschwender, James A. 1978. *Racial Stratification in America*. Dubuque, Iowa: W. C. Brown.

Gevisser, Mark. 1994. "Who Is a South African?" *New York Times* (April 26):A19.

Gibbins, Roger. 1997. "Historical Overview and Background: Part I." Pp. 19–34 in J. Rick Ponting (ed.), *First Nations in Canada: Perspectives on Opportunity, Empowerment, and Self-Determination*. Toronto: McGraw-Hill Ryerson.

Gilbert, Dennis A. 1988. *Compendium of American Public Opinion*. New York: Facts on File.

Gilbert, G. M. 1951. "Stereotype Persistence and Change among College Students." *Journal of Abnormal and Social Psychology* 46:245–254.

Gillen, John. 1948. "Race Relations without Conflict: A Guatemalan Town." *American Journal of Sociology* 53:337–343.

Ginsberg, Eli. 1978. "Jews in the American Economy: The Dynamics of Opportunity." Pp. 109–119 in Gladys Rosen (ed.), *Jewish Life in America: Historical Perspectives*. New York: Institute of Human Relations, American Jewish Committee.

Glaberson, William. 2001. "Who Is a Seminole, and Who Gets to Decide?" *New York Times* (January 29):A1, A14.

Glass Ceiling Commission. 1995. *Good for Business: Making Full Use of the Nation's Human Capital*. Washington, D.C.: U.S. Department of Labor.

Glass, Ruth. 1964. "Insiders-Outsiders: The Position of Minorities." Pp. 141–155 in *Transactions of the Fifth World Congress of Sociology*. Vol. 3. Louvain, Belgium: International Sociological Association.

Glater, Jonathan D., and Martha M. Hamilton. 1995. "Affirmative Action's Defenders." *Washington Post* (March 27–April 2):20.

Glazer, Nathan. 1957. *American Judaism*. Chicago: University of Chicago Press.

———. 1958. "The American Jews and the Attainment of Middle-Class Rank: Some Trends and Explanations." Pp. 138–146 in Marshall Sklare (ed.), *The Jews: Social Patterns of an American Group*. Glencoe, Ill.: Free Press.

———. 1971. "Blacks and Ethnic Groups: The Difference and the Political Difference It Makes." *Social Problems* 18:444–461.

———. 1975. *Affirmative Discrimination: Ethnic Identity and Public Policy*. New York: Basic Books.

———. 1997. *We Are All Multiculturalists Now*. Cambridge, Mass.: Harvard University Press.

Glazer, Nathan, and Daniel P. Moynihan. 1970. *Beyond the Melting Pot*. 2d ed. Cambridge, Mass.: MIT Press.

Gleason, Philip. 1964. "Immigration and American Catholic Intellectual Life." *Review of Politics* 26:147–173.

Glenn, Norvall D. 1963. "Occupational Benefits to Whites from the Subordination of Negroes." *American Sociological Review* 28:443–448.

———. 1966. "White Gains from Negro Subordination." *Social Problems* 14:159–178.

Glenny, Misha. 1993. *The Fall of Yugoslavia: The Third Balkan War*. New York: Penguin.

Glock, Charles Y., and Rodney Stark. 1966. *Christian Beliefs and Anti-Semitism*. New York: Harper & Row.

Golab, Caroline. 1977. *Immigrant Destinations*. Philadelphia: Temple University Press.

Gold, Steven J. 1999. "From 'The Jazz Singer' to 'What a country!' A Comparison of Jewish Migration to the United States, 1880–1930 and 1965–1998." *Journal of American Ethnic History* 18:115–141.

———. 2000. "Israeli Americans." Pp. 409–420 in Peter Kivisto and Georganne Rundblad (eds.), *Multiculturalism in the United States: Current Issues, Contemporary Voices*. Thousand Oaks, Calif.: Pine Forge.

Goldberg, David Theo (ed.). 1994. *Multiculturalism: A Critical Reader*. Cambridge, Mass.: Basil Blackwell.

Goldberg, J. J. 1996. *Jewish Power: Inside the American Jewish Establishment*. Reading, Mass.: Addison-Wesley.

Goldberg, Jeffrey. 1995. "The Overachievers." *New York* (April 10):42–51.

Goldscheider, Calvin. 1986. *Jewish Continuity and Change: Emerging Patterns in America*. Bloomington, Ind.: Indiana University Press.

———. 1995. "Modernization, Ethnicity and the Post-War Jewish World." Pp. 130–143 in Robert S. Wistrich (ed.), *Terms of Survival: The Jewish World Since 1945*. London: Routledge.

———. 1996. *Israel's Changing Society: Population, Ethnicity, and Development*. Boulder, Colo.: Westview.

Goldstein, Sidney. 1980. "Jews in the United States: Perspectives from Demography." Pp. 3–59 in *American Jewish Yearbook 1981*. New York: American Jewish Committee.

———. 1992. "Profile of American Jewry: Insights from the 1990 National Jewish Population Survey." Pp. 77–173 in *American Jewish Yearbook 1992*. New York: American Jewish Committee.

Golub, Jennifer L. 1990. *What Do We Know about Black Anti-Semitism?* New York: American Jewish Committee.

Goodman, James M., and Douglas Heffington. 2000. "Native Americans." Pp. 49–79 in Jesse O. McKee (ed.), *Ethnicity in Contemporary America: A Geographical Appraisal,* 2d ed. Lanham, Md.: Rowman & Littlefield.

Goodman, Mary Ellen. 1964. *Race Awareness in Young Children*. Rev. ed. New York: Collier.

Goodstein, Laurie. 1997. "Women in Islamic Headdress Find Faith and Prejudice, Too." *New York Times* (November 3):A1, A14.

Goodstein, Laurie, and Tamar Lewin. 2001. "Victims of Mistaken Identity, Sikhs Pay a Price for Turbans." *New York Times* (September 19):A1.

Gordon, Milton M. 1964. *Assimilation in American Life: The Role of Race, Religion, and National Origins*. New York: Oxford University Press.

———. 1975. "Toward a General Theory of Racial and Ethnic Group Relations." Pp. 84–110 in Nathan Glazer and Daniel P. Moynihan (eds.), *Ethnicity: Theory and Experience*. Cambridge, Mass.: Harvard University Press.

———. 1981. "Models of Pluralism: The New American Dilemma." *Annals of the American Academy of Political and Social Science* 454:178–188.

Gossett, Thomas F. 1963. *Race: The History of an Idea in America*. Dallas, Tex.: Southern Methodist University Press.

Gould, Stephen Jay. 1983. *The Mismeasure of Man*. New York: Norton.

———. 1984. "Human Equality Is a Contingent Fact of History." *Natural History* (November):26–32.

Graham, Richard. 1970. "Brazilian Slavery Reexamined: A Review Article." *Journal of Social History* 3:431–453.

Graves, Sherryl Browne. 1999. "Television and Prejudice Reduction: When Does Television as a Vicarious Experience Make a Difference?" *Journal of Social Issues* 55:707–725.

Grebler, Leo, Joan W. Moore, and Ralph C. Guzman. 1970. *The Mexican-American People: The Nation's Second Largest Minority*. New York: Free Press.

Greeley, Andrew M. 1971. *Why Can't They Be Like Us?* New York: Dutton.

———. 1974. *Ethnicity in the United States: A Preliminary Reconnaissance*. New York: Wiley.

———. 1977. *The American Catholic: A Social Portrait*. New York: Harper & Row.

Greeley, Andrew M., and Paul B. Sheatsley. 1974. "Attitudes toward Racial Integration." Pp. 241–250 in Lee Rainwater (ed.), *Social Problems and Public Policy*. Chicago: Aldine.

Greenbaum, William. 1974. "America in Search of a New Ideal: An Essay on the Rise of Pluralism." *Harvard Educational Review* 44:411–440.

Greenberg, Bradley S., and Jeffrey E. Brand. 1998. "U.S. Minorities and the News." Pp. 3–22 in Yahya R. Kamalipour and Theresa Carilli (eds.), *Cultural Diversity and the U.S. Media*. Albany: State University of New York Press.

Grobsmith, Elizabeth S. 1994. "Prisons and Prisoners." Pp. 481–483 in Mary B. Davis (ed.), *Native America in the Twentieth Century*. New York: Garland.

Gross, Jane. 1989. "Diversity Hinders Asians' Influence." *New York Times* (June 25):18.

Guimarães, Antonio Sérgio Alfredo. 1999. "Measures to Combat Discrimination and Racial Inequality in Brazil." Pp. 139–153 in Rebecca Reichmann (ed.), *Race in Contemporary Brazil: From Indifference to Inequality*. University Park, Pa.: The Pennsylvania State University Press.

Gurr, Ted Robert. 1993. *Minorities at Risk: A Global View of Ethnopolitical Conflicts*. Washington, D.C.: United States Institute of Peace Press.

Gutman, Herbert G. 1976. *The Black Family in Slavery and Freedom, 1750–1925*. New York: Pantheon.

Gwyn, Richard. 1995. *Nationalism without Walls: The Unbearable Lightness of Being Canadian*. Toronto: McClelland & Stewart.

Hacker, Andrew. 1994. "Education: Ethnicity and Achievement." Pp. 214–229 in Nicolaus Mills (ed.), *Debating Affirmative Action: Race, Gender, Ethnicity, and the Politics of Inclusion*. New York: Delta.

———. 1995. *Two Nations: Black and White, Separate, Hostile, Unequal*. Expanded and updated ed. New York: Scribners.

Hagan, William T. 1961. *American Indians*. Chicago: University of Chicago Press.

———. 1992. "Full Blood, Mixed Blood, Generic, and Ersatz: The Problem of Indian Identity." Pp. 278–288 in Roger L. Nichols (ed.), *The American Indian: Past and Present*. 4th ed. New York: McGraw-Hill.

Halpern, Ben. 1958. "America Is Different." Pp. 23–39 in Marshall Sklare (ed.), *The Jews: Social Patterns of an American Group*. Glencoe, Ill.: Free Press.

Hamada, Tarek. 1990. "Many Arab Immigrants Invest Hope in Detroit." *Detroit News* (December 2):1A, 10A.

Hamilton, Richard. 1972. *Class and Politics in the United States*. New York: Wiley.

Hammonds, Keith H. 1998. "Invisible in the Executive Suite." *Business Week* (December 21):68.

Hanchard, Michael George. 1994. *Orpheus and Power: The Movimento Negro of Rio de Janeiro and São Paulo, Brazil, 1945–1988*. Princeton: Princeton University Press.

Handlin, Oscar. 1957. *Race and Nationality in American Life*. Boston: Little, Brown.

———. 1962. *The Newcomers: Negroes and Puerto Ricans in a Changing Metropolis*. Garden City, N.Y.: Doubleday.

Handley, Antoinette, and Jeffrey Herbst. 1997. "South Africa: the Perils of Normalcy." *Current History* 96(May):222–226.

Hannerz, Ulf. 1969. *Soulside: Inquiries into Ghetto Culture and Community*. New York: Columbia University Press.

Harari, Oren, and David Beaty. 1989. *Lessons from South Africa: A New Perspective on Public Policy and Productivity*. New York: Harper & Row.

Harden, Blaine, and Somini Sengupta. 2001. "Some Passengers Singled Out for Exclusion by Flight Crew." *New York Times* (September 22):B8.

Harding, John. 1968. "Stereotypes." Pp. 259–261 in David L. Sills (ed.), *International Encyclopedia of the Social Sciences*. Vol. 15. New York: Macmillan.

Harney, Robert F., and Harold Troper. 1977. "Introduction." *Canadian Ethnic Studies* 9:1–5.

Harris, David R. 1999. "'Property Values Drop When Blacks Move In, Because. . .': Racial and Socioeconomic Determinants of Neighborhood Desirability." *American Sociological Review* 64:461–479.

Harris, Louis. 1991a. "Three-to-One Majority Backs Affirmative Action If No Strict Quotas." *The Harris Poll* (June 2).

———. 1991b. "The Code Language That Now Pervades the Racial Issue." *The Harris Poll* (September 15).

Harris, Marvin. 1963. "Race Relations in Minas Velhas, a Community in the Mountain Region of Central Brazil." Pp. 47–81 in Charles Wagley (ed.), *Race and Class in Rural Brazil*. 2d ed. New York: Columbia University Press.

———. 1964. *Patterns of Race in the Americas*. New York: Norton.

Harris, Marvin, Josildeth Gomes Consorte, Joseph Lang, and Bryan Byrne. 1993. "Who Are the Whites? Imposed Census Categories and the Racial Demography of Brazil." *Social Forces* 72:451–462.

Harris, Rosemary. 1972. *Prejudice and Tolerance in Ulster: A Study of Neighbours and Strangers in a Border Community*. Manchester, England: Manchester University Press.

Harrison, David. 1981. *The White Tribe of Africa: South Africa in Perspective*. London: British Broadcasting Corporation.

Hartley, E. L. 1946. *Problems in Prejudice*. New York: Kings Crown.

Hasenbalg, Carlos A. 1985. "Race and Socioeconomic Inequalities in Brazil." Pp. 25–41 in Pierre-Michel Fontaine (ed.), *Race, Class, and Power in Brazil*. Los Angeles: Center for Afro-American Studies, University of California, Los Angeles.

Hasenbalg, Carlos, and Suellen Huntington. 1982/1983. "Brazilian Racial Democracy: Reality or Myth?" *Humboldt Journal of Social Relations* 10:129–142.

Hastings, Donald. 1969. "Japanese Emigration and Assimilation in Brazil." *International Migration Review* 3:32–53.

Hawkins, Freda. 1972. *Canada and Immigration: Public Policy and Public Concern*. Montreal: McGill-Queen's University Press.

———. 1989. *Critical Years in Immigration: Canada and Australia Compared*. Kingston and Montreal: McGill-Queen's University Press.

Heath, Shirley Brice. 1985. "Language Policies: Patterns of Retention and Maintenance." Pp. 257–282 in Walker Connor (ed.), *Mexican-Americans in Comparative Perspective*. Washington, D.C.: Urban Institute Press.

Hechter, Michael. 1975. *Internal Colonialism: The Celtic Fringe in British National Development, 1536–1966*. Berkeley: University of California Press.

Hedges, Chris. 1996. "Serbs Left in Bosnia See No Rest for the Dead." *New York Times* (January 18):A6.

———. 1997. "Bosnia Journal: Ethnic Diversity Distorts History, Art, Language." *New York Times* (November 25):1, 14.

Heer, David. 1996. *Immigration in America's Future: Social Science Findings and the Policy Debate*. Boulder, Colo.: Westview.

Heilman, Madeline E. 1996. "Affirmative Action's Contradictory Consequences." *Journal of Social Issues* 52(4):105–109.

Heilman, Samuel. 1999. "Separated but Not Divorced." *Society* 36 (May/June):8–14.

Heilman, Samuel C., and Steven M. Cohen. 1989. *Cosmopolitans & Parochials: Modern Orthodox Jews in America*. Chicago: University of Chicago Press.

Heisler, Martin O. 1991. "Hyphenating Belgium: Changing State and Regime to Cope with Cultural Division." Pp. 177–195 in Joseph V. Montville (ed.), *Conflict and Peacemaking in Multiethnic Societies*. New York: Lexington Books.

Hemming, John. 1987. *Amazon Frontier: The Defeat of the Brazilian Indians*. Cambridge, Mass.: Harvard University Press.

Henry, Frances. 1978. *The Dynamics of Racism in Toronto*: Research Report. Toronto: York University.

———. 1994. *The Caribbean Diaspora in Toronto: Learning to Live with Racism*. Toronto: University of Toronto Press.

Henry, Frances, and Carol Tator. 1985. "Racism in Canada: Social Myths and Strategies for Change." Pp. 321–335 in Rita M. Bienvenue and Jay E. Goldstein (eds.), *Ethnicity and Ethnic Relations in Canada*. 2d ed. Toronto: Butterworths.

Henry, Frances, Carol Tator, Winston Mattis, and Tim Rees. 1995. *The Colour of Democracy: Racism in Canadian Society*. Toronto: Harcourt Brace Canada.

Herberg, Will. 1960. *Protestant-Catholic-Jew*. Garden City, N.Y.: Doubleday.

Heringer, Rosana. 1995. "Introduction to the Analysis of Racism and Anti-Racism in Brazil." Pp. 203–207 in

Benjamin Bowser (ed.), *Racism and Anti-Racism in World Perspective*. Thousand Oaks, Calif.: Sage.

Herrnstein, Richard J., and Charles Murray. 1994. *The Bell Curve: Intelligence and Class Structure in American Life*. New York: Free Press.

Hershberg, Theodore et al. 1979. "A Tale of Three Cities: Blacks and Immigrants in Philadelphia: 1850–1880, 1930 and 1970." *Annals of the American Academy of Political and Social Science* 441:55–81.

Herskovitz, Melville. 1941. *The Myth of the Negro Past*. New York: Harper & Brothers.

Heskin, Ken. 1980. *Northern Ireland: A Psychological Analysis*. New York: Columbia University Press.

Hewitt, Christopher. 1981. "Catholic Grievances, Catholic Nationalism and Violence in Northern Ireland During the Civil Rights Period: A Reconsideration." *British Journal of Sociology* 32:362–380.

Higham, John. 1963. *Strangers in the Land: Patterns of American Nativism 1860–1925*. 2d ed. New York: Atheneum.

Hilberg, Raul. 1979. *The Destruction of the European Jews*. New York: Harper & Row.

Hill, Daniel G. 1977. *Human Rights in Canada: A Focus on Racism*. Ottawa: Canadian Labour Congress.

Hill, Robert B. 1981. "The Economic Status of Black Americans." Pp. 1–59 in *The State of Black America 1981*. New York: National Urban League.

Hiller, Harry H. 1976. *Canadian Society: A Sociological Analysis*. Scarborough, Ont.: Prentice-Hall of Canada.

Hirschfeld, Julie R. 2001. "Congress of Relative Newcomers Poses Challenge to Bush, Leadership." *CQ Weekly* (January 20):178–182.

Hirschman, Charles. 1975. *Ethnic and Social Stratification in Peninsular Malaysia*. Washington, D.C.: American Sociological Association.

Hirschman, Charles, Philip Kasinitz, and Josh DeWind. 1999. "Immigrant Adaptation, Assimilation, and Incorporation." Pp. 127–136 in Charles Hirschman, Philip Kasinitz, and Josh DeWind (eds.), *The Handbook of International Migration: The American Experience*. New York: Russell Sage Foundation.

Hispanic Business. 1995. "The Corporate Elite." (January):11–13.

————. 2001. "Hispanics Underrepresented in Boardrooms." (October 18) <http://www.hispanicbusiness.com>.

Hoare, Anthony G. 1982. "Problem Region and Regional Problem." Pp. 195–223 in Frederick W. Boal and J. Neville H. Douglas (eds.), *Integration and Division: Geographical Perspectives on the Northern Ireland Problem*. London: Academic Press.

Hochschild, Jennifer L. 1995. *Facing Up to the American Dream: Race, Class, and the Soul of the Nation*. Princeton, N.J.: Princeton University Press.

Hoffer, Eric. 1951. *The True Believer*. New York: Harper & Row.

Hoffman, Constance A., and Martin N. Marger. 1991. "Patterns of Immigrant Enterprise in Six Metropolitan Areas." *Sociology and Social Research* 75 (April):144–157.

Hoge, Warren. 2001. "British Life Is Fractured Along Racial Lines, a Study Finds." *New York Times* (December 12):A1.

Holland, Jack. 1982. *Too Long a Sacrifice: Life and Death in Northern Ireland Since 1969*. New York: Penguin.

Holzer, Harry J. 2001. "Racial Difference in Labor Market Outcomes among Men." Pp. 98–123 in Neil J. Smelser, William Julius Wilson, and Faith Mitchell (eds.), *America Becoming: Racial Trends and Their Consequences*. Vol. II. Washington, DC: National Academy Press.

Hooper, James. 1999. "Kosovo: America's Balkan Problem." *Current History* (April): 159–164.

Hornblower, Margot. 1995. "Putting Tongues in Check." *Time* (October 9):40–50.

Horowitz, Donald L. 1975. "Ethnic Identity." Pp. 111–140 in Nathan Glazer and Daniel P. Moynihan (eds.), *Ethnicity: Theory and Experience*. Cambridge, Mass.: Harvard University Press.

————. 1991. *A Democratic South Africa?: Constitutional Engineering in a Divided Society*. Berkeley: University of California Press.

Horowitz, Irving Louis. 1964. *Revolution in Brazil*. New York: Dutton.

Horton, Haywood Derrick, Beverlyn Lundy Allen, Cedric Herring, and Melvin E. Thomas. 2000. "Lost in the Storm: The Sociology of the Black Working Class, 1850 to 1990." *American Sociological Review* 65:128–137.

Horton, John. 1995. *The Politics of Diversity: Immigration, Resistance, and Change in Monterey Park, California*. Philadelphia: Temple University Press.

Hostetler, John A. 1993. *Amish Society*. 4th ed. Baltimore, Md.: Johns Hopkins University Press.

Howard, John R. 1970. *Awakening Minorities*. New Brunswick, N.J.: Transaction.

Howe, Irving. 1976. *World of Our Fathers*. New York: Simon & Schuster.

Huber, Joan, and William H. Form. 1973. *Income and Ideology*. New York: Free Press.

Hughes, David R., and Evelyn Kallen. 1974. *The Anatomy of Racism: Canadian Dimensions*. Montreal: Harvest House.

Hughes, Everett C. 1943. *French Canada in Transition*. Chicago: University of Chicago Press.

Hughes, Everett C., and Helen M. Hughes. 1952. *Where Peoples Meet: Racial and Ethnic Frontiers*. Glencoe, Ill.: Free Press.

Hugick, Larry. 1992. "Blacks See Their Lives Worsening." *Gallup Poll Monthly* (April):26–29.

Human Rights Watch. 1995. *Slaughter among Neighbors: The Political Origins of Communal Violence*. New Haven: Yale University Press.

Hunt, Chester, and Lewis Walker. 1974. *Ethnic Dynamics*. Homewood, Ill.: Dorsey.

Hurh, Won Moo, and Kwang Chung Kim. 1989. "The 'Success' Image of Asian Americans: Its Validity and Its Practical and Theoretical Implications." *Ethnic And Racial Studies* 12:512–538.

Hurvitz, Nathan. 1958. "Sources of Middle-Class Values of American Jews." *Social Forces* 37:117–123.

Hutchinson, Edward P. 1956. *Immigrants and Their Children 1850–1950*. New York: Wiley.

Hutchinson, Harry. 1963. "Race Relations in a Rural Community of the Bahian Reconcavo." Pp. 16–46 in Charles Wagley (ed.), *Race and Class in Rural Brazil*. 2d ed. New York: Columbia University Press.

Iacovetta, Franca. 1992. *Such Hardworking People: Italian Immigrants in Postwar Toronto*. Montreal and Kingston: McGill-Queen's University Press.

Ianni, Francis A. J. 1974. *Black Mafia: Ethnic Succession in Organized Crime*. New York: Simon & Schuster.

Ignatieff, Michael. 1993. *Blood and Belonging: Journeys into the New Nationalism*. Toronto: Penguin.

Ignatiev, Noel. 1995. *How the Irish Became White*. New York: Routledge.

Institute for Social Research. 1964. *Discrimination without Prejudice. A Study of Promotion Practices in Industry*. Ann Arbor, Mich.: Institute for Social Research, University of Michigan.

Instituto Brasileiro de Geografia e Estatística. 1981. *IX Recenseamento Geral do Brasil—1980*. Vol. 1—Tomo 2. Rio de Janeiro: IBGE.

_____. 1992. *Anuário Estatístico do Brasil 1992*. Rio de Janeiro: IBGE.

_____. 2002. *Web Site do Instituto Brasileiro de Geografia e Estatística* <http://www.ibge.gov.r/english/estatisti...ondicaodevida/indicadoresminimos/tabelas/tabela1.shtm>.

International Organization for Migration (IOM). 1999. *Migration in Central and Eastern Europe: 1999 Review*. Geneva: International Organization for Migration.

Iorizzo, Luciano J., and Salvatore Mondello. 1980. *The Italian Americans*. Boston: Twayne.

Irwin, Colin. 1994. "The Myths of Segregation." Pp. 104–118 in Adrian Guelke (ed.), *New Perspectives on the Northern Ireland Conflict*. Aldershot, England: Avebury.

_____. 1996. "No-one Ever Said Real Peace Building Was Easy—Or Cheap." *Belfast Telegraph* (October 22).

Isaacs, Harold R. 1989. *Idols of the Tribe: Group Identity and Political Change*. Cambridge, Mass.: Harvard University Press.

Isaacs, Stephen D. 1974. *Jews and American Politics*. Garden City, N.Y.: Doubleday.

Isajiw, Wsevolod W., Aysan Sev'er, and Leo Driedger. 1993. "Ethnic Identity and Social Mobility: A Test of the 'Drawback Model.'" *Canadian Journal of Sociology* 18:179–198.

Isbister, John. 1996. *The Immigration Debate: Remaking America*. West Hartford, Conn.: Kumarian Press.

Jackman, Mary R. 1994. *The Velvet Glove: Paternalism and Conflict in Gender, Class, and Race Relations*. Berkeley: University of California Press.

Jackson, Harold, and Anne McHardy. 1984. *The Two Irelands*. Rev. ed. London: Minority Rights Group.

Jacobson, Matthew Frye. 1998. *Whiteness of a Different Color: European Immigrants and the Alchemy of Race*. Cambridge, Mass.: Harvard University Press.

Jacoby, Russell, and Naomi Glauberman (eds.). 1995. *The Bell Curve Debate: History, Documents, Opinions*. New York: Times Books.

Jacoby, Tamar. 1998. *Someone Else's House: America's Unfinished Struggle for Integration*. New York: Free Press.

_____. 2000. "In Asian America." *Commentary* (July/August):21–28.

Jaffe, A. J. 1992. *The First Immigrants from Asia: A Population History of the North American Indians*. New York: Plenum.

Jaffe, A. J., Ruth M. Cullen, and Thomas D. Boswell. 1980. *The Changing Demography of Spanish Americans*. New York: Academic Press.

Jarman, Neil. 1997. *Material Conflicts: Parades and Visual Displays in Northern Ireland*. Oxford: Berg.

Jencks, Christopher et al. 1972. *Inequality: A Reassessment of the Effect of Family and Schooling in America*. New York: Basic Books.

Jencks, Christopher, and Meredith Phillips. 1998. "American's Next Achievement Test." *The American Prospect* (September/October):44–53.

Jensen, Arthur. 1969. "How Much Can We Boost IQ and Scholastic Achievement?" *Harvard Educational Review* 39:1–123.

_____. 1973. "The Differences Are Real." *Psychology Today* (December):80–86.

Jiobu, Robert M. 1988. *Ethnicity and Assimilation*. Albany, N.Y.: State University of New York Press.

Johnson, Colleen Leahy. 1985. *Growing Up and Growing Old in Italian-American Families*. New Brunswick, N.J.: Rutgers University Press.

Johnson, Dirk. 1999. "Growth of Gambling on Tribal Land Starts Trek Back Home by Indians." *New York Times* (January 17):10.

Johnson, Julie. 1989. "Asian-Americans Press Fight for Wider Top-College Door." *New York Times* (September 9):1, 7.

Johnson, Lyndon B. 1965. *Public Papers of the Presidents. Lyndon B. Johnson II*. Washington, D.C.: U.S. Government Printing Office.

Johnson, Tim. 1995. "The Dealer's Edge: Gaming in the Path of Native America." *Native Americas* 12 (Summer):16–25.

Jones, Edward W., Jr. 1986. "Black Managers: The Dream Deferred." *Harvard Business Review* (May-June):84–93.

Jones, Emrys. 1960. *A Social Geography of Belfast*. London: Oxford University Press.

Jones, James M. 1996. *Prejudice and Racism*. 2d ed. New York: McGraw-Hill.

Jones, Jeffrey M. 2001. "Americans Have Mixed Opinions about Immigration." *The Gallup Poll Monthly* (July):28–32.

Jones, Maldwyn A. 1960. *American Immigration*. Chicago: University of Chicago Press.

Jordan, Winthrop. 1969. *White Over Black*. Baltimore, Md.: Penguin.

_____. 1995. "Slavery and the Jews." *Atlantic Monthly* 276(September):109–114.

Josephy, Alvin M., Jr. 1969. *The Indian Heritage of America*. New York: Alfred A. Knopf.

Josephy, Alvin M. Jr., Joane Nagel, and Troy Johnson. 1999. *Red Power: The American Indians' Fight for Freedom*. Lincoln: University of Nebraska Press.

Joy, Richard J. 1972. *Languages in Conflict*. Toronto: McClelland & Stewart.

Judah, Tim. 1997. *The Serbs: History, Myth and the Destruction of Yugoslavia*. New Haven: Yale University Press.

Juteau Lee, Danielle. 1979. "The Evolution of Nationalism in Quebec." Pp. 60–73 in Jean Leonard Elliott (ed.), *Two Nations, Many Cultures*. 2d ed. Scarborough, Ont.: Prentice-Hall of Canada.

Kahlenberg, Richard. 1996. *The Remedy: Class, Race, and Affirmative Action*. New York: Basic Books.

Kalin, Rudolf and J. W. Berry. 1994. "Ethnic and Multicultural Attitudes." Pp. 293–321 in J. W. Berry and J. A. Laponce (ed.s), *Ethnicity and Culture in Canada: The Research Landscape*. Toronto: University of Toronto Press.

Kallen, Evelyn. 1995. *Ethnicity and Human Rights in Canada*. 2d ed. Toronto: Oxford University Press.

Kang, K. Connie. 1997. "Today's Chinese Americans: A Cosmopolitan Community." *Los Angeles Times* (June 29):A1, A29.

Kantrowitz, Nathan. 1973. *Ethnic and Racial Segregation in the New York Metropolis*. New York: Praeger.

———. 1979. "Racial and Ethnic Residential Segregation in Boston 1830–1970." *Annals of the American Academy of Political and Social Science* 441:41–54.

Karasch, Mary. 1975. "From Porterage to Proprietorship: African Occupations in Rio de Janeiro, 1808–1850." Pp. 369–393 in Stanley L. Engerman and Eugene D. Genovese (eds.), *Race and Slavery in the Western Hemisphere: Quantitative Studies*. Princeton, N.J.: Princeton University Press.

Karlins, M., T. L. Coffman, and G. Walters. 1969. "On the Fading of Social Stereotypes: Studies in Three Generations of College Students." *Journal of Personality and Social Psychology* 13:1–16.

Kasarda, John. 1985. "Urban Change and Minority Opportunities." Pp. 46–61 in Paul Peterson (ed.), *The New Urban Reality*. Washington, D.C.: Brookings Institution.

Katz, Daniel, and Kenneth Braly. 1933. "Racial Stereotypes of One Hundred College Students." *Journal of Abnormal and Social Psychology* 28:280–290.

Katz, Diane. 1992. "Canada Hears 'Same Voice' As Los Angeles." *Detroit News* (May 6): 1A, 6A.

Kayal, Philip, and Joseph Kayal. 1975. *The Syrian-Lebanese in America: A Study in Religion and Assimilation*. Boston: Twayne.

Kehoe, Alice B. 1999. "American Indians." Pp. 48–74 in Elliott Robert Barkan (ed.), *A Nation of Peoples: A Sourcebook on America's Multicultural Heritage*. Westport, Conn.: Greenwood Press.

Keil, Hartmut. 1991. "Socialist Immigrants from Germany and the Transfer of Socialist Ideology and Workers' Culture." Pp. 315–338 in Rudolph J. Vecoli and Suzanne M. Sinke (eds.), *A Century of European Migrations, 1830–1930*. Urbana: University of Illinois Press.

Keller, Bill. 1992. "Schools Are Slightly More Equal, But Still Failures." *New York Times* (November 7):2.

———. 1995. "After Apartheid, Change Lags Behind Expectations." *New York Times* (April 27):A1, A6.

Keller, Suzanne. 1953. "The Social Origins and Career Lines of Three Generations of American Business Leaders." Ph.D. dissertation, Columbia University.

Kelner, Merrijoy. 1970. "Ethnic Penetration into Toronto's Elite Structure." *Canadian Review of Sociology and Anthropology* 7:128–137.

Kendall, Joan. 2001. "Circles of Disadvantage: Aboriginal Poverty and Underdevelopment in Canada." *The American Review of Canadian Studies* 31:42–59.

Kennedy, Randall. 1994. "Persuasion and Distrust: The Affirmative Action Debate." Pp. 48–67 in Nicolaus Mills (ed.), *Debating Affirmative Action: Race, Gender, Ethnicity, and the Politics of Inclusion*. New York: Delta.

Kennedy, Ruby Jo Reeves. 1944. "Single or Triple Melting Pot?: Intermarriage Trends in New Haven." *American Journal of Sociology* 49:331–339.

———. 1952. "Single or Triple Melting Pot?: Intermarriage in New Haven, 1870–1950." *American Journal of Sociology* 58:56–59.

Kephart, William M., and William W. Zellner. 1994. *Extraordinary Groups: An Examination of Unconventional Life-Styles*. 2d ed. New York: St. Martin's.

Keshena, Rita. 1980. "The Role of American Indians in Motion Pictures." Pp. 106–111 in Gretchen M. Bataille and Charles L. P. Silet (eds.), *The Pretend Indians: Images of Native Americans in the Movies*. Ames, Iowa: Iowa State University Press.

Kiester, Edwin, Jr. 1968. *The Case of the Missing Executive*. New York: Institute of Human Relations, American Jewish Committee.

Kifner, John. 1999. "Emptying a City of All but Bodies." *New York Times* (May 29):1, 6.

Kikumura, Akemi, and Harry H. L. Kitano. 1973. "Interracial Marriage: A Picture of the Japanese-Americans." *Journal of Social Issues* 29:67–81.

Kilborn, Peter T. 1997. "For Poorest Indians, Casinos Aren't Enough." *New York Times* (June 11):A1, A18.

Killian, Lewis M. 1953. "The Adjustment of Southern White Migrants to Northern Urban Norms." *Social Forces* 32:66–69.

———. 1975. *The Impossible Revolution, Phase II: Black Power and the American Dream*. New York: Random House.

———. 1985. *White Southerners*. Rev. ed. Amherst: University of Massachusetts Press.

Kim, Illsoo. 1981. *New Urban Immigrants: The Korean Community in New York*. Princeton, N.J.: Princeton University Press.

———. 1987. "The Koreans: Small Business in an Urban Frontier." Pp. 219–242 in Nancy Foner (ed.), *New Immigrants in New York*. New York: Columbia University Press.

King, Martin Luther, Jr. 1964. *Why We Can't Wait*. New York: Harper & Row.

Kington, Raymond S., and Herbert W. Nickers. 2001. "Racial and Ethnic Differences in Health: Recent Trends, Current Patterns, Future Directions." Pp. 253–310 in Neil J. Smelser, William Julius Wilson, and Faith Mitchell (eds.), *America Becoming:*

Racial Trends and Their Consequences. Vol II. Washington, D.C.: National Academy Press.

Kinsley, Michael. 1991. "Class, Not Race." *The New Republic* (August 19–26):4.

Kinzer, Stephen. 1992. "Ethnic Conflict Is Threatening in Yet Another Region of Yugoslavia: Kosovo." *New York Times* (November 9):8.

Kirscht, John P., and Ronald C. Dillehay. 1967. *Dimensions of Authoritarianism: A Review of Research and Theory*. Lexington, Ky.: University of Kentucky Press.

Kirschten, Dick. 1992. "Building Blocs." *National Journal* 24(September 26):2173–2177.

Kitano, Harry. 1976. *Japanese Americans: The Evolution of a Subculture*. 2d ed. Englewood Cliffs, N.J.: Prentice-Hall.

Kitano, Harry H. L., and Roger Daniels. 1995. *Asian Americans: Emerging Minorities*, 2d. ed. Englewood Cliffs, N.J.: Prentice-Hall.

Klaaste, Aggrey. 1984. "Exiles in Their Native Land." *New York Times Magazine* (June 24): 34–35, 51–53, 68–69.

Klausner, Samuel Z. 1988. *Succeeding in Corporate America: The Experience of Jewish M.B.A.'s*. New York: American Jewish Committee.

Kleiman, Dena. 1983. "Less Than 40% of Jews in Survey Observe Sabbath." *New York Times* (February 6):45.

Klein, Herbert S. 1969. "The Colored Freedman in Brazilian Slave Society." *Journal of Social History* 3:30–52.

Klineberg, Otto. 1968. "Prejudice: The Concept." Pp. 439–447 in David L. Sills (ed.), *International Encyclopedia of the Social Sciences*. Vol. 12. New York: Macmillan.

Knight, Franklin. 1974. *The African Dimension in Latin American Societies*. New York: Macmillan.

Knowlton, Clark. 1972. "The New Mexican Land War." Pp. 258–270 in Edward Simmen (ed.), *Pain and Promise: The Chicano Today*. New York: New American Library.

———. 1975. "Neglected Chapters in Mexican-American History." Pp. 19–59 in Gus Tyler (ed.), *Mexican-Americans Tomorrow*. Albuquerque, N. Mex.: University of New Mexico Press.

Kohn, Hans. 1944. *The Idea of Nationalism: A Study in Its Origins and Background*. New York: Macmillan.

Konvitz, Milton B. 1978. "The Quest for Equality and the Jewish Experience."
Pp. 28–60 in Gladys Rosen (ed.), *Jewish Life in America: Historical Perspectives*. New York: Institute of Human Relations, American Jewish Committee.

Korman, Abraham K. 1988. *The Outsiders: Jews and Corporate America*. Lexington, Mass.: Lexington Books.

Kosmin, Barry A. et al. 1991. *Highlights of the CJF 1990 National Jewish Population Survey*. New York: Council of Jewish Federations.

Koster, Henry. 1966 (1816). *Travels in Brazil*. Carbondale, Ill.: Southern Illinois University Press.

Kottak, Conrad Phillip, and Kathryn A. Kozaitis. 1999. *On Being Different: Diversity and Multiculturalism in the North American Mainstream*. Boston: McGraw-Hill College.

Kovel, Joel. 1970. *White Racism: A Psychohistory*. New York: Pantheon.

Kramer, Judith R., and Seymour Leventman. 1961. *Children of the Gilded Ghetto*. New Haven, Conn.: Yale University Press.

Kristof, Nicholas D. 1995. "Japanese Outcasts Better Off Than in Past but Still Outcast." *New York Times* (November 30):A1, A8.

———. 1998. "New Freedoms Feed Ethnic Frictions." *New York Times* (June 25):A6.

Kristol, Irving. 1970. "The Negro Today Is Like the Immigrant Yesterday." Pp. 139–157 in Nathan Glazer (ed.), *Cities in Trouble*. Chicago: Quadrangle.

Kuper, Leo. 1968. "Segregation." Pp. 144–150 in David L. Sills (ed.), *International Encyclopedia of the Social Sciences*. Vol. 14. New York: Macmillan.

———. 1969. "Ethnic and Racial Pluralism: Some Aspects of Polarization and Depluralization." Pp. 459–487 in Leo Kuper and M. G. Smith, *Pluralism in Africa*. Berkeley: University of California Press.

———. 1981. *Genocide*. New York: Penguin.

Kuper, Leo, and M. G. Smith (eds.). 1969. *Pluralism in Africa*. Berkeley: University of California Press.

Kurlansky, Mark. 1999. *The Basque History of the World*. New York: Vintage.

Kurokawa, Minako (ed.). 1970. *Minority Responses*. New York: Random House.

Kutner, Bernard, Carol Wilkens, and Penny Rechtman Yarrow. 1952. "Verbal Attitudes and Overt Behavior." *Journal of Abnormal and Social Psychology* 47:649–652.

La Ferla, Ruth. 2001. "Latino Style Is Cool. Oh, All Right: It's Hot." *New York Times* (April 16):B1, B3.

LaGumina, Salvatore J. 1988. *From Steerage to Suburb: Long Island Italians*. Staten Island, N.Y.: Center for Migration Studies.

———. 1996. "Anti-Italian Discrimination—Update." Pp. 293–304 in Mary Jo Bona and Anthony Julian Tamburri (eds.), *Through the Looking Glass: Italian and Italian/American Images in the Media*. Staten Island, N.Y.: American Italian Historical Association.

Lai, James S. 2000. "Embracing the New Millenium." Pp. 16–19 in UCLA Asian American Studies Center, *2000-01 National Asian Pacific American Political Almanac*. Los Angeles: University of California at Los Angeles.

Lake, David A., and Donald Rothchild. 1998. "Spreading Fear: The Genesis of Transnational Ethnic Conflict." Pp. 3–32 in David A. Lake and Donald Rothchild (eds.), *The International Spread of Ethnic Conflict: Fear, Diffusion, and Escalation*. Princeton: Princeton University Press.

Lambert, Richard D. 1981. "Ethnic/Racial Relations in the United States in Comparative Perspective." *Annals of the American Academy of Political and Social Science* 454:189–205.

Landry, Bart. 1987. *The New Black Middle Class*. Berkeley: University of California Press.

Lane, Robert E. 1962. *Political Ideology*. New York: Free Press.

Langberg, Mark, and Reynolds Farley. 1985. "Residential Segregation of Asian Americans in 1980." *Sociology and Social Research* 70 (October):71–75.

Langdon, Steven. 1995. "Proof of What Hope Can Do." *Toronto Star* (July 17):A13.

Langer, Gary. 1989. "Polling on Prejudice: Questionable Questions." *Public Opinion* 12(May/June):18–19, 57.

LaPiere, Richard T. 1934. "Attitudes vs. Actions." *Social Forces* 13:230–237.

LaRuffa, Anthony L. 1982. "Media Portrayals of Italian-Americans." *Ethnic Groups* 4:191–206.

———. 1988. *Monte Carmelo: An Italian-American Community in the Bronx*. New York: Gordon and Breach.

Lautard, Hugh, and Neil Guppy. 1990. "The Vertical Mosaic Revisited: Occupational Differentials among Canadian Ethnic Groups." Pp. 189–208 in Peter S. Li (ed.), *Race and Ethnic Relations in Canada*. Toronto: Oxford University Press.

Learsi, Rufus. 1954. *The Jews in America: A History*. Cleveland, Ohio: World.

Lee, Sharon M. 1998. *Asian Americans: Diverse and Growing*. Population Bulletin, Vol. 53, no. 2. Washington, D.C.: Population Reference Bureau.

———. 2001. *Using the New Racial Categories in the 2000 Census*. Washington, D.C.: Population Reference Bureau.

Lee, Sharon M., and Keiko Yamanaka. 1990. "Patterns of Asian American Intermarriage and Marital Assimilation." *Journal of Comparative Family Studies* 21:287–305.

Légaré, André. 1997. "The Government of Nunavut (1999): A Prospective Analysis." Pp. 404–431 in J. Rick Ponting (ed.), *First Nations in Canada: Perspectives on Opportunity, Empowerment, and Self-Determination*. Toronto: McGraw-Hill Ryerson.

Legendre, Camille. 1980. *French Canada in Crisis: A New Society in the Making?* London: Minority Rights Group.

Leger Marketing. 2001. "Immigration and Racism following the September 11 Attacks." *Report*. Montreal: Leger Marketing.

Leite Lopes, José Sergio. 2000. "Class, Ethnicity, and Color in the Making of Brazilian Football." *Daedalus* 129 (Spring):239–270.

Leiter, Robert. 2000. "American Jewry Is (a) Fading or (b) Reviving. It's All in How You Read the Signs." *New York Times* (September 23):B9.

Lemann, Nicholas. 1991. "The Other Underclass." *Atlantic Monthly* 268(December): 96–110.

Lemon, Anthony. 1987. *Apartheid in Transition*. Aldershot, England: Gower.

———. 1994. "Educational Desegregation." Pp. 91–110 in John D. Brewer (ed.), *Restructuring South Africa*. New York: St. Martin's.

Lenski, Gerhard. 1963. *The Religious Factor*. Rev. ed. Garden City, N.Y.: Doubleday.

———. 1966. *Power and Privilege: A Theory of Social Stratification*. New York: McGraw-Hill.

Lever, Henry. 1968. "Ethnic Preferences of White Residents in Johannesburg." *Sociology and Social Research* 52:157–173.

———. 1978. *South African Society*. Johannesburg: Jonathan Ball.

Lévesque, René. 1968. *An Option for Quebec*. Toronto: McClelland & Stewart.

Levin, Jack, and William Levin. 1982. *The Functions of Discrimination and Prejudice*. 2d ed. New York: Harper & Row.

Levine, Bruce. 1992. *The Spirit of 1840: German Immigrants, Labor Conflict, and the Coming of the Civil War*. Urbana: University of Illinois Press.

Levine, Gene N., and Colbert Rhodes. 1981. *The Japanese American Community: A Three-Generation Study*. New York: Praeger.

Levine, Marc V. 1990. *The Reconquest of Montreal: Language Policy and Social Change in a Bilingual City*. Philadelphia: Temple University Press.

Levine, Naomi, and Martin Hochbaum (eds.). 1974. *Poor Jews: An American Awakening*. New Brunswick, N.J.: Transaction.

Levitan, Sar A., William B. Johnston, and Robert Taggart. 1975. *Still a Dream: The Changing Status of Blacks Since 1960*. Cambridge, Mass.: Harvard University Press.

Lewin, Julius. 1963. *Politics and Law in South Africa*. London: Merlin.

Lewis, Hylan. 1967. "The Family: Resources for Change." Pp. 314–343 in Lee Rainwater and William L. Yancey (eds.), *The Moynihan Report and the Politics of Controversy*. Cambridge, Mass.: MIT Press.

Lewis, Oscar. 1965. *La Vida*. New York: Random House.

Lichter, S. Robert, and Daniel R. Amundson. 1997. "Distorted Reality: Hispanic Characters in TV Entertainment." Pp. 57–79 in Clara E. Rodríguez (ed.), *Latin Looks: Images of Latinas and Latinos in the U.S. Media*. Boulder, Colo.: Westview.

Lichter, S. Robert, and Linda Lichter. 1982. *Italian-American Characters in Television Entertainment*. West Hempstead, N.Y.: Commission for Social Justice.

Lichter, S. Robert et al. 1987. "Prime-Time Prejudice: TV's Images of Blacks and Hispanics." *Public Opinion* 10(July/August):13–16.

Lieberson, Stanley. 1961. "A Societal Theory of Race Relations." *American Sociological Review* 26:902–910.

———. 1980. *A Piece of the Pie: Blacks and White Immigrants Since 1880*. Berkeley: University of California Press.

Lieberson, Stanley, and Mary C. Waters. 1988. *From Many Strands: Ethnic and Racial Groups in Contemporary America*. New York: Russell Sage Foundation.

Liebman, Charles S. 1973. "American Jewry: Identity and Affiliation." Pp. 127–152 in David Sidorsky (ed.), *The Future of the Jewish Community in America*. New York: Basic Books.

Light, Ivan, and Edna Bonacich. 1988. *Immigrant Entrepreneurs: Koreans in Los Angeles 1965–1982*. Berkeley: University of California Press.

Lijphart, Arend. 1975. "The Northern Ireland Problem: Cases, Theories and Solutions." *British Journal of Political Science* 5:83–106.

_____. 1977. *Democracy in Plural Societies: A Comparative Exploration*. New Haven, Conn.: Yale University Press.

Lind, Rebecca Ann, and James A. Danowski. 1998. "The Representation of Arabs in U.S. Electronic Media." Pp. 157–168 in Yahya R. Kamalipour and Theresa Carilli (eds.), *Cultural Diversity and the U.S. Media*. Albany: State University of New York Press.

Lippmann, Walter. 1922. *Public Opinion*. New York: Macmillan.

Lipset, Seymour Martin. 1960. *Political Man*. New York: Anchor.

_____. 1968. *Revolution and Counterrevolution*. New York: Basic.

_____. 1987. "Blacks and Jews: How Much Bias?" *Public Opinion* 10(July/August):4–5, 57–58.

_____. 1990. *Continental Divide: The Values and Institutions of the United States and Canada*. New York: Routledge.

Lipset, Seymour Martin, and Earl Raab. 1995. *Jews and the New American Scene*. Cambridge, Mass.: Harvard University Press.

Lipton, Merle. 1987. "Reform: Destruction or Modernization of Apartheid?" Pp. 34–55 in Jesmond Blumenfeld (ed.), *South Africa in Crisis*. London: Croom Helm.

Livingston, John C. 1979. *Fair Game? Inequality and Affirmative Action*. San Francisco: Freeman.

Livingstone, David N., Margaret C. Keane, and Frederick W. Boal. 1998. "Space for Religion: A Belfast Case Study." *Political Geography* 17:145–170.

Logan, John R. 1988. "Realities of Black Suburbanization." Pp. 231–241 in Roland L. Warren and Larry Lyon (eds.), *New Perspectives on the American Community*. Chicago: Dorsey.

Lopez, Adalberto. 1980. "The Puerto Rican Diaspora: A Survey." Pp. 313–343 in Adalberto Lopez (ed.), *The Puerto Ricans: Their History, Culture, and Society*. Cambridge, Mass.: Schenkman.

Lopez, David E. 1996. "Language: Diversity and Assimilation." Pp. 139–163 in Roger Waldinger and Mehdi Bozorgmehr (eds.), *Ethnic Los Angeles*. New York: Russell Sage Foundation.

Lopreato, Joseph. 1970. *Italian Americans*. New York: Random House.

Los Angeles Times. 1997. Los Angeles Times Poll Alert, Study #396 (June 29):1–12.

Louie, Miriam Ching. 2000. "Immigrant Asian Women in Bay Area Garment Sweatshops: 'After Sewing, Laundry, Cleaning and Cooking, I Have No Breath Left to Sing'." Pp. 226–242 in Timothy P. Fong and Larry H. Shinagawa (eds.), *Asian Americans: Experiences and Perspectives*. Upper Saddle River, N.J.: Prentice Hall.

Loury, Glenn. 1998. "How to Mend Affirmative Action." *The Public Interest* (Spring):33–43.

Lovell, Peggy A., and Charles H. Wood. 1998. "Skin Color, Racial Identity, and Life Chances in Brazil." *Latin American Perspectives* 25(May):90–109.

Luebke, Frederick C. 1983. "A Prelude to Conflict: The German Ethnic Group in Brazilian Society, 1890–1917." *Ethnic and Racial Studies* 6:1–17.

_____. 1987. *Germans in Brazil: A Comparative History of Cultural Conflict during World War I*. Baton Rouge, La.: Louisiana State University Press.

_____ (ed.). 1990. *Germans in the New World: Essays in the History of Immigration*. Urbana, Ill.: University of Illinois Press.

Lurie, Nancy Oestreich. 1991. "The American Indian: Historical Background." Pp. 132–146 in Norman R. Yetman (ed.), *Majority and Minority: The Dynamics of Race and Ethnicity in American Life*. 5th ed. Boston: Allyn & Bacon.

Lyman, Stanford. 1968a. "The Race Relations Cycle of Robert Park." *Pacific Sociological Review* 11:16–22.

_____. 1968b. "Contrasts in the Community Organizations of Chinese and Japanese in North America." *Canadian Review of Sociology and Anthropology* 5:51–67.

_____. 1970. "Strangers in the Cities: The Chinese on the Urban Frontier." Pp. 61–100 in Charles Wollenberg (ed.), *Ethnic Conflict in California History*. Los Angeles: Tinnon-Brown.

_____. 1974. *Chinese Americans*. New York: Random House.

MacDonald, J. Fred. 1983. *Blacks and White TV: Afro-Americans in Television Since 1948*. Chicago: Nelson-Hall.

Macías, Ysidro. 1972. "The Chicano Movement." Pp. 137–143 in Edward Simmen (ed.), *Pain and Promise: The Chicano Today*. New York: New American Library.

Mack, Raymond W. 1963. *Race, Class, and Power*. New York: American.

Mackie, Marlene. 1973. "Arriving at 'Truth' by Definition: The Case of Stereotype Inaccuracy." *Social Problems* 20:431–447.

_____. 1980. "Ethnic Stereotypes and Prejudice: Alberta Indians, Hutterites and Ukrainians." Pp. 233–246 in Jay E. Goldstein and Rita M. Bienvenue (eds.), *Ethnicity and Ethnic Relations in Canada*. Toronto: Butterworths.

Mackie, Richard. 2001. "Sovereignty Support Near Low, Poll Finds." *The Globe and Mail* (October 26):A1, A9.

Maclean's. 1989. "Maclean's/Decima Poll: A North-South Dialogue." 102(July 3):48–50.

MacLennan, Hugh. 1945. *Two Solitudes*. New York: Duell, Sloan & Pearce.

Makabe, Tomoko. 1981. "The Theory of the Split Labor Market: A Comparison of the Japanese Experience in Brazil and Canada." *Social Forces* 59:786–809.

_____. 1999. "Ethnic Hegemony: The Japanese Brazilians in Agriculture, 1908–1968." *Ethnic and Racial Studies* 22:702–723.

Mallaby, Sebastian. 1992. *After Apartheid: The Future of South Africa*. New York: Times Books.

Mangiafico, Luciano. 1988. *Contemporary American Immigrants: Patterns of Filipino, Korean, and Chinese Settlement in the United States*. New York: Praeger.

Mannheim, Karl. 1936. *Ideology and Utopia*. New York: Harcourt, Brace & World.

Marais, Hein. 1999. "Cape Fear: Violent Crime Creates a New Apartheid." *The Globe and Mail* (January 27):A17.

Marden, Charles F., and Gladys Meyer. 1978. *Minorities in American Society*. 5th ed. New York: Van Nostrand.

Marger, Martin. 1974. *The Force of Ethnicity: A Study of Urban Elites*. Journal of University Studies, Ethnic Monograph Series. Detroit, Mich.: Wayne State University.

_____. 1979. "A Reexamination of Gordon's Ethclass." *Sociological Focus* 11:21–32.

_____. 1987. *Elites and Masses: An Introduction to Political Sociology*. 2d ed. Belmont, Calif.: Wadsworth.

_____. 1989. "Factors of Structural Pluralism in Multiethnic Societies: A Comparative Case Study." *International Journal of Group Tensions* 19:52–68.

_____. 1993. "The Media as a Power Institution." Pp. 238–249 in Marvin E. Olsen and Martin N. Marger (eds.), *Power in Modern Societies*. Boulder, Colo.: Westview

Margolis, Mac. 1988. "Brazil's Blacks Look Anew at Issue of Race." *Washington Post* (May 13):A25–A26.

Margolis, Maxine L. 1998. *An Invisible Minority: Brazilians in New York City*. Boston: Allyn & Bacon.

Marks, Jonathan. 1995. *Human Biodiversity: Genes, Race, and History*. New York: Aldine de Gruyter.

Marriott, Michel. 1987. "In Jersey City, Indians Face Rise in Violence and Hatred." *New York Times* (October 12):II, 1.

Marston, Wilfred G., and Thomas L. Van Valey. 1979. "The Role of Residential Segregation in the Assimilation Process." *Annals of the American Academy of Political and Social Science* 441:13–25.

Martin, Philip, and Elizabeth Midgley. 1999. *Immigration to the United States*. Population Bulletin (June). Washington, D.C.: Population Reference Bureau.

Martinez, Thomas M. 1972. "Advertising and Racism: The Case of the Mexican American." Pp. 94–105 in Edward Simmen (ed.), *Pain and Promise: The Chicano Today*. New York: New American Library.

Martire, Greg, and Ruth Clark. 1982. "Anti-Semitism in America." *Public Opinion* (April/ May):56–59.

Marx, Anthony W. 1998. *Making Race and Nation: A Comparison of South Africa, the United States, and Brazil*. Cambridge: Cambridge University Press.

Mason, Philip. 1970. *Patterns of Dominance*. New York: Oxford University Press.

Massarik, Fred, and Alvin Chenkin. 1973. "United States National Jewish Population Study: A First Report." Pp. 264–306 in *American Jewish Year Book 1973*. New York: American Jewish Committee.

Massey, Douglas S. 1979. "Effects of Socio-economic Factors on the Residential Segregation of Blacks and Spanish Americans in United States Urbanized Areas." *American Sociological Review* 44:1015–1022.

_____. 1981a. "Hispanic Residential Segregation: A Comparison of Mexicans, Cubans, and Puerto Ricans." *Sociology and Social Research* 65:311–322.

_____. 1981b. "Dimensions of the New Immigration to the United States and the Prospects for Assimilation." *Annual Review of Sociology* 7:57–85.

_____. 2001. "Residential Segregation and Neighborhood Conditions in U.S. Metropolitan Areas." Pp. 391–434 in Neil J. Smelser, William Julius Wilson, and Faith Mitchell (eds.), *America Becoming: Racial Trends and Their Consequences*. Vol. 1. Washington, D.C.: National Academy Press.

Massey, Douglas S., and Brooks Bitterman. 1985. "Explaining the Paradox of Puerto Rican Segregation." *Social Forces* 64:306–331.

Massey, Douglas S., and Nancy A. Denton. 1987. "Trends in the Residential Segregation of Blacks, Hispanics, and Asians: 1970–1980." *American Sociological Review* 52:802–825.

_____. 1989a. "Residential Segregation of Mexicans, Puerto Ricans, and Cubans in Selected U.S. Metropolitan Areas." *Sociology and Social Research* 73:73–83.

_____. 1989b. "Hypersegregation in U.S. Metropolitan Areas: Black and Hispanic Segregation Along Five Dimensions." *Demography* 26:373–391.

_____. 1993. *American Apartheid: Segregation and the Making of the Underclass*. Cambridge, Mass.: Harvard University Press.

Mathabane, Mark. 1994. "'Like the Second Coming.'" *Newsweek* (May 9):38.

McAllister, Ian. 1977. *The Northern Ireland Social Democratic and Labour Party*. London: Macmillan.

McAneny, Leslie, and Lydia Saad. 1994. "America's Public Schools: Still Separate? Still Unequal?" *Gallup Poll Monthly* (May):23–27.

McAuliffe, Dennis, Jr. 1996. "For Many Indian Tribes, the Buffalo Are Back." *Washington Post National Weekly Edition* (March 18–24):8–9.

McCaffrey, Lawrence J. 1997. *The Irish Catholic Diaspora in America*. Washington, D.C.: The Catholic University of America Press.

McCartney, Clem, and Lucy Bryson. 1994. *Clashing Symbols: A Report on the Use of Flags, Anthems and Other National Symbols in Northern Ireland*. Belfast: The Institute of Irish Studies, The Queen's University of Belfast.

McClain, Paula D., and Joseph Stewart Jr. 1998. "Can We All Get Along?": *Racial and Ethnic Minorities in American Politics*. 2d ed. Boulder: Westview.

McClatchy, Valentine Stuart. 1978 (1919–1925). *Four Anti-Japanese Pamphlets*. New York: Arno Press.

McCoy, George. 1994. "Mental Health." Pp. 330–335 in Mary B. Davis (ed.), *Native America in the Twentieth Century*. New York: Garland.

McGarry, John, and Brendan O'Leary. 1995a. *Explaining Northern Ireland: Broken Images*. Oxford: Blackwell.

_____. 1995b. "Five Fallacies: Northern Ireland and the Liabilities of Liberalism." *Ethnic and Racial Studies* 18:837–861.

McGill Consortium for Ethnicity and Strategic Social Planning. 1997. *Diversity, Mobility and Change: The Dynamics of Black Communities in Canada*. Montreal: McGill Consortium for Ethnicity and Strategic Social Planning, School of Social Work, McGill University.

McGraw, Bill. 1990. "Montreal Racial Tirade Hits Hard." *Detroit Free Press* (October 8):3A, 6A.

McKee, James B. 1993. *Sociology and the Race Problem: The Failure of a Perspective*. Urbana, Ill.: University of Illinois Press.

McKenna, Marian C. 1969. "The Melting Pot: Comparative Observations in the United States and Canada." *Sociology and Social Research* 53:433–447.

McKinney, Jeffrey. 2001. "B.E. 100s: Grow or Die." *Black Enterprise* (June):205–211.

McLarin, Kimberly J. 1995. "To Preserve Afrikaners' Language, Mixed-Race South Africans Join Fray." *New York Times* (June 28):A4.

McLemore, S. Dale. 1973. "The Origins of Mexican-American Subordination in Texas." *Social Science Quarterly* 53:656–679.

McRae, Kenneth D. 1974. *Consociational Democracy: Political Accommodation in Segmented Societies*. Toronto: McClelland & Stewart.

_____. 1983. *Conflict and Compromise in Multilingual Societies: Switzerland*. Waterloo, Ont.: Wilfrid Laurier University Press.

_____. 1986. *Conflict and Compromise in Multilingual Societies*. Vol. 2: Belgium. Waterloo, Ont.: Wilfrid Laurier University Press.

McRoberts, Kenneth. 1988. *Quebec: Social Change and Political Crisis*. 3d ed. Toronto: McClelland and Stewart.

McWhirter, Cameron. 2000. "Black Mayors Still Trapped by Racial Politics." *Detroit News* (February 13):1A, 6A–7A.

McWhorter, John H. 2000. *Losing the Race: Self-Sabotage in Black America*. New York: Free Press.

McWilliams, Carey. 1948. *A Mask for Privilege: Anti-Semitism in America*. Boston: Little, Brown.

_____. 1968 (1948). *North from Mexico: The Spanish-Speaking People of the United States*. New York: Greenwood.

Mearsheimer, John J., and Stephen Van Evera. 1995. "When Peace Means War." *The New Republic* (December 18):16–21.

Mechanic, David. 1978. "Apartheid Medicine." Pp. 127–138 in Ian Robertson and Phillip Whitten (eds.), *Race and Politics in South Africa*. New Brunswick, N.J.: Transaction.

Meléndez, Edwin. 1993. "Understanding Latino Poverty." *Sage Race Relations Index* 18(May): 3–42.

Meredith, Martin. 1987. "The Black Opposition." Pp. 77–89 in Jesmond Blumenfeld (ed.), *South Africa in Crisis*. London: Croom Helm.

Merton, Robert K. 1949. "Discrimination and the American Creed." Pp. 99–126 in R. H. MacIver (ed.), *Discrimination and National Welfare*. New York: Harper & Row.

_____. 1968. *Social Theory and Social Structure*. New York: Free Press.

Metzger, L. Paul. 1971. "American Sociology and Black Assimilation: Conflicting Perspectives." *American Journal of Sociology* 76:627–647.

Miami Herald. 1983. "Slave's Descendant Ruled Legally Black in Louisiana." (May 19):1A, 12A.

Milbrath, Lester. 1982. *Political Participation*. 2d ed. Washington, D.C.: University Press of America.

Miller, James Nathan. 1984. "Ronald Reagan and the Techniques of Deception." *Atlantic Monthly* (February):62–68.

Miller, Kerby A. 1985. *Emigrants and Exiles: Ireland and the Irish Exodus to North America*. New York: Oxford University Press.

Mills, C. Wright. 1956. *The Power Elite*. New York: Oxford University Press.

Mills, C. Wright, Clarence Senior, and Rose Kohn Goldsen. 1950. *The Puerto Rican Journey*. New York: Harper & Brothers.

Min, Pyong Gap. 1988. *Ethnic Business Enterprise: Korean Small Business in Atlanta*. New York: Center for Migration Studies.

_____. 1991. "Cultural and Economic Boundaries of Korean Ethnicity: A Comparative Analysis." Paper presented at the Annual Meeting of the American Sociological Association, Cincinnati, Ohio.

_____. 1995a. "Korean Americans." Pp. 199–231 in Pyong Gap Min (ed.), *Asian Americans: Contemporary Trends and Issues*. Thousand Oaks, Calif.: Sage.

_____. 1995b. "Major Issues Relating to Asian American Experiences." Pp. 38–57 in Pyong Gap Min (ed.), *Asian Americans: Contemporary Trends and Issues*. Thousand Oaks, Calif.:Sage.

_____. 1999. "A Comparison of Post-1965 and Turn-of-the-Century Immigrants in Intergenerational Mobility and Cultural Transmission." *Journal of American Ethnic History* 18 (Spring):65-94.

Miner, Horace. 1939. *St. Denis: A French Canadian Parish*. Chicago: University of Chicago Press.

Minority Rights Group. 1992. *The Chinese of South-East Asia*. London: Minority Rights Group.

Mitrano, John R., and James G. Mitrano. 1996. "Mammas, Papas, Traditions, and Secrets: The Marketing of Italian Americans." Pp. 71–84 in Mary Jo Bona and Anthony Julian Tamburri (eds.), *Through the Looking Glass: Italian and Italian/American Images in the Media*. Staten Island, N.Y.: American Italian Historical Association.

Mittelbach, Frank G., and Joan W. Moore. 1968. "Ethnic Endogamy: The Case of the Mexican Americans." *American Journal of Sociology* 74:50–62.

Montagu, Ashley. 1963. *Race, Science, and Humanity*. Princeton, N.J.: Van Nostrand.

_____. 1972. *Statement on Race*. 3d ed. New York: Oxford University Press.

_____. (ed.). 1975. *Race and IQ*. New York: Oxford University Press.

Montero, Darrel. 1980. *Japanese Americans: Changing Patterns of Ethnic Affiliation Over Three Generations*. Boulder, Colo.: Westview.

Moodley, Kogila A. 1980. "Structural Inequality and Minority Anxiety: Responses of Middle Groups in South Africa." Pp. 217–235 in Robert M. Price and Carl G. Rotberg (eds.), *The Apartheid Regime: Political Power and Racial Domination*. Berkeley: Institute of International Studies, University of California.

Moore, Joan W. 1970. "Colonialism: The Case of the Mexican-Americans." *Social Problems* 17:463–472.

Moore, Joan W., with Harry Pachon. 1976. *Mexican Americans*. 2d ed. Englewood Cliffs, N.J.: Prentice-Hall.

———. 1985. *Hispanics in the United States*. Englewood Cliffs, N.J.: Prentice-Hall.

Moore, Michael. 1987. "The Man Who Killed Vincent Chin." *Detroit Magazine, Detroit Free Press* (August 30):12–20.

Moquin, Wayne, and Charles Van Doren (eds.). 1974. *A Documentary History of the Italian Americans*. New York: Praeger.

Morales, Armando. 1972. *Ando Sangrando*. Fair Lawn, N.J.: R. E. Burdick.

Morgan, Curtis. 1998a. "Cultural Divide Is Testing Ground for the Nation." *Miami Herald* (May 3). <http://www.herald.com>

———. 1998b. "Language Barriers Frustrate Residents." *Miami Herald* (June 22). <http://www.herald.com>

Morin, Richard. 1995a. "Feelings That Aren't Set in Concrete." *Washington Post National Weekly Edition* (March 13–19):37.

———. 1995b. "Across the Racial Divide." *Washington Post National Weekly Edition* (October 16–22):6–10.

———. 1998. "Mixed Messages on Campus Diversity." *Washington Post National Weekly Edition* (October 26):34.

Morin, Richard, and Michael H. Cottman. 2001. "The Invisible Slap." *Washington Post National Weekly Edition* (July 2–8):6–7.

Mörner, Magnus. 1985. *Adventurers and Proletarians: The Story of Migrants in Latin America*. Pittsburgh, Pa.: University of Pittsburgh Press.

Morse, Richard M. 1953. "The Negro in São Paulo, Brazil." *Journal of Negro History* 38: 290–306.

———. 1958. *From Community to Metropolis: A Biography of São Paulo, Brazil*. Gainesville, Fla.: University of Florida Press.

Moxon-Browne, Edward. 1983. *Nation, Class and Creed in Northern Ireland*. Aldershot, England: Gower.

Moynihan, Daniel P. 1965. *The Negro Family: The Case for National Action*. Washington, D.C.: U.S. Government Printing Office.

Mulder, C. P. 1972. "The Rationale of Separate Development." Pp. 48–63 in N. J. Rhoodie (ed.), *South African Dialogue*. Philadelphia: Westminister.

Muller, Thomas. 1993. *Immigrants and the American City*. New York: New York University Press.

Mumford Center for Comparative Urban and Regional Research. 2001a. *Census 2000: Sortable List of Dissimilarity Scores*. Albany, N.Y.: Mumford Center, State University of New York at Albany <http://www.albany.edu/mumford/census>.

———. 2001b. *The New Ethnic Enclaves in America's Suburbs*. Albany, N.Y.: Mumford Center, State University of New York at Albany <http://www.albany.edu/mumford/census>.

———. 2001c. *Immigrant Enclaves in the American Metropolis, 1990–2000*. Mumford Center, State University of

New York at Albany <http://www.albany.edu/mumford/census>.

Murguía, Edward. 1975. *Assimilation, Colonialism and the Mexican American People*. Austin, Tex.: Center for Mexican American Studies, University of Texas.

Murguía, Edward, and W. Parker Frisbie. 1977. "Trends in Mexican American Intermarriage: Recent Findings in Perspective." *Social Science Quarterly* 58:374–389.

Murray, Dominic. 1983. "Schools and Conflict." Pp. 136–150 in John Darby (ed.), *Northern Ireland: The Background to the Conflict*. Belfast: Appletree Press.

———. 1995. "Culture, Religion and Violence in Northern Ireland." Pp. 215–229 in Seamus Dunn (ed.), *Facets of the Conflict in Northern Ireland*. Basingstoke: Macmillan.

Murray, Martin. 1987. *South Africa: Time of Agony, Time of Destiny*. London: Verso.

Murray, Georgina. 2000. "Black Empowerment in South Africa: 'Patriotic Capitalism' or a Corporate Black Wash?" *Critical Sociology* 26:183–204.

Mydans, Seth. 1992a. "New Unease for Japanese-Americans." *New York Times* (March 4):7.

———. 1992b. "A Target of Rioters, Koreatown Is Bitter, Armed and Determined." *New York Times* (May 3):1, 16.

———. 1998. "Indonesia Turns Its Chinese into Scapegoats." *New York Times* (February 2):A6.

Myrdal, Gunnar. 1944. *An American Dilemma: The Negro Problem and Modern Democracy*. New York: Harper & Row.

Naber, Nadine. 2000. "Ambiguous Insiders: An Investigation of Arab American Invisibility." *Ethnic and Racial Studies* 23:37–61.

Nackerud, Larry, Alyson Spring, Christopher Larrison, and Alicia Issac. 1999. "The End of the Cuban Contradiction in U.S. Refugee Policy." *International Migration Review* 33:176–192.

Naff, Alixa. 1983. "Arabs in America: A Historical Overview." Pp. 9–29 in Sameer Y. Abraham and Nabeel Abraham (eds.), *Arabs in the New World: Studies on Arab-American Communities*. Detroit: Wayne State University Press.

Nagel, Joane. 1995. "American Indian Ethnic Renewal: Politics and the Resurgence of Identity." *American Sociological Review* 60:947–965.

Nakanishi, Don T. 2001. "Political Trends and Electoral Issues of the Asian Pacific American Population." Pp. 170–199 in Neil J. Smelser, William Julius Wilson, and Faith Mitchell (eds.), *America Becoming: Racial Trends and Their Consequences*. Vol. I. Washington, D.C.: National Academy Press.

Nakhaie, M. Reza. 1995. "Ownership and Management Position of Canadian Ethnic Groups in 1973 and 1989." *Canadian Journal of Sociology* 20:167–192.

Nash, Manning. 1962. "Race and the Ideology of Race." *Current Anthropology* 3:285–288.

National Advisory Commission on Civil Disorders. 1968. *Report of the National Advisory Commission on Civil Disorders*. New York: Bantam.

National Conference. 1995. *Taking America's Pulse: A Summary Report of the National Conference Survey on Inter-Group Relations*. New York: National Conference.

National Conference of Christians and Jews. 1978. *A Study of Attitudes Toward Racial and Religious Minorities and Toward Women*. (Conducted by Louis Harris and Associates, Inc.). New York: National Conference of Christians and Jews.

National Roster of Hispanic Elected Officials. 1994. National Association of Latino Elected and Appointed Officials.

National Urban League. 1981. *The State of Black America 1981*. New York: National Urban League.

———. 2001. *The State of Black America 2001*. New York: National Urban League.

Nattrass, Nicoli, and Jeremy Seekings. 2001. "'Two Nations'? Race and Economic Inequality in South Africa Today." *Daedalus* 130 (Winter):45–70.

Nava, Julian. 1975. "Educational Challenges in Elementary and Secondary Schools." Pp. 107–136 in Gus Tyler (ed.), *Mexican-Americans Tomorrow*. Albuquerque, N. Mex.: University of New Mexico Press.

Navarro, Mireya. 1999. "Miami's Generations of Exiles Side by Side, Yet Worlds Apart." *New York Times* (February 11):A1, A25.

———. 2000. "Puerto Rican Presence Wanes in New York." *New York Times* (February 28):A1, A20.

Nazareno, Analisa. 2000. "How Immigrants Reshaped Schools." *Miami Herald* (January 2) <http://www.herald.com>.

Nelli, Humbert S. 1970. *The Italians in Chicago, 1880–1930*. New York: Oxford University Press.

Newcomer, Mabel. 1955. *The Big Business Executive*. New York: Columbia University Press.

Newman, Peter C. 1975. *The Canadian Establishment*. Vol. 1. Toronto: McClelland & Stewart.

———. 1979. *The Canadian Establishment*. Vol. 2. Toronto: McClelland & Stewart.

———. 1995. "A Country of Many Cultures and Flavors." *Maclean's* (July 24):34.

———. 1998. *Titans: How the New Canadian Establishment Seized Power*. Toronto: Viking.

Newman, William M. 1973. *American Pluralism: A Study of Minority Groups and Social Theory*. New York: Harper & Row.

Newport, Frank. 1998. "Americans Support Elimination of Bilingual Education." *The Gallup Poll Monthly* (June):17–18.

Newport, Frank, and Lydia Saad. 1997a. "Civil Trial Didn't Alter Public's View of Simpson Case." *The Gallup Poll Monthly* (February): 21–22.

Newport, Frank, David W. Moore, and Lydia Saad. 1999. "Long-Term Gallup Poll Trends: A Portrait of American Public Opinion through the Century." *Poll Releases* (December 20): The Gallup Organization.

Newsweek. 1992a. "Most U.S. Youths in Survey View Race Relations as Poor." (March 7):9.

———. 1992b. "Views on Race Relations." (May 11):12.

———. 1993. "Newsweek Poll." (August 9):19.

———. 1995a. "Newsweek Poll." (February 13):67.

———. 1995b. "Newsweek Poll." (July 10):31.

———. 1999. "Feeling Better about the Future." (June 7):35.

New York Times. 1982. "Business Aide to Reagan Criticizes Food Stamps." (May 28):A12.

———. 1987. "Black Woman Who Lost Job Is Asked Back by Pharmacist." (August 15):A7.

———. 1994. "Black Household Income Leads Whites' in Queens." (June 6):A1, B7.

NIAF. (National Italian American Foundation). 2001a. *Italian Ancestry of the 107th United States Congress*. Washington, D.C.: National Italian American Foundation <http://www.niaf.org/research/report_107thcongress.asp>.

———. 2001b. *National Survey: American Teen-Agers and Stereotyping*. Washington, D.C.: National Italian American Foundation <http://www.niaf.org/research/report_zogby.asp>.

Nichols, Roger L. 1992. "Indians in the Post-Termination Era." Pp. 289–301 in Roger L. Nichols (ed.), *The American Indian: Past and Present*. 4th ed. New York: McGraw-Hill.

Niebuhr, Gustav. 1997. "A Shift to Rigorous Tradition Gains Influence in Judaism." *New York Times* (May 1):A1, A14.

Noble, Kenneth B. 1995. "Attacks Against Asian-Americans Are Rising." *New York Times* (December 13):A14.

Nodín Valdés, Dennis. 1989. "The New Northern Borderlands: An Overview of Midwestern Chicano History." Pp. 1–28 in *Mexicans in the Midwest, Perspectives in Mexican American Studies*. Tucson, Ariz.: Mexican American Studies and Research Center, University of Arizona.

Noel, Donald L. 1968. "A Theory of the Origin of Ethnic Stratification." *Social Problems* 16:157–172.

Nogueira, Oracy. 1959. "Skin Color and Social Class." In Vera Rubin (ed.), *Plantation Systems of the New World*. Pan American Union Social Science Monograph No. 7. Washington, D.C.: Pan American Union.

Nordheimer, Jon. 1987. "Black Cubans: Apart in Two Worlds." *New York Times* (December 2):26.

Nordland, Rod. 1993. "'Let's Kill the Muslims!'" *Newsweek* (November 8):48–51.

Normand, Josée. 1996. "Language and Culture of the Métis People." *Canadian Social Trends* (Winter):22–26.

Norris, Mary Jane. 1990. "The Demography of Aboriginal People in Canada." Pp. 33–59 in Shiva S. Halli, Frank Trovato, and Leo Driedger (eds.), *Ethnic Demography: Canadian Immigrant, Racial and Cultural Variations*. Ottawa: Carleton University Press.

North, Robert D. 1965. "The Intelligence of the American Negro." Pp. 334–340 in Arnold M. Rose and Caroline B. Rose (eds.), *Minority Problems*. New York: Harper & Row.

Northern Ireland, Department of Health and Social Services. 1984. *The Northern Ireland Census, 1981*. Summary Report. Belfast: Her Majesty's Stationery Office.

Nostrand, Richard L. 1973. "Mexican American and Chicano: Emerging Terms for a People Coming of Age." *Pacific Historical Review* 62:389–406.

Novak, Michael. 1972. *The Rise of the Unmeltable Ethnics.* New York: Macmillan.

O Connor, Fionnuala. 1993. *In Search of A State: Catholics in Northern Ireland.* Belfast: Blackstaff Press.

O'Donnell, E. E. 1977. *Northern Irish Stereotypes.* Dublin: College of Industrial Relations.

Ogmundson, R. 1990. "Perspectives on the Class and Ethnic Origins of Canadian Elites: A Methodological Critique of the Porter/ Clement/Olsen Tradition." *Canadian Journal of Sociology* 15:165–177.

Ogmundson, R. and L. Fatels. 1994. "Are the Brits in Decline? A Note on Trends in the Ethnic Origins of the Labour and Church Elites." *Canadian Ethnic Studies* 26:108–112.

Ogmundson, R., and J. McLaughlin. 1992. "Trends in the Ethnic Origins of Canadian Elites: The Decline of the BRITS?" *Canadian Review of Sociology and Anthropology* 29:227–242.

O'Hare, William. 1987. "Best Metros for Hispanic Businesses." *American Demographics* 9(November):30–32.

Ojito, Mirta. 1997. "Dominicans, Scrabbling for Hope." *New York Times* (December 16):C27.

————. 2000. "Best of Friends, Worlds Apart." *New York Times* (June 5):A1.

O'Leary, Brendan. 1995. "Introduction: Reflections on a Cold Peace." *Ethnic and Racial Studies* 18:695–714.

Oliver, Melvin L., and Thomas M. Shapiro. 1995. *Black Wealth/White Wealth: A New Perspective on Racial Inequality.* New York: Routledge.

Olsen, Marvin E. 1970. "Power Perspectives on Stratification and Race Relations." Pp. 296–305 in Marvin E. Olsen (ed.), *Power in Societies.* New York: Macmillan.

Olzak, Susan. 1992. *The Dynamics of Ethnic Competition and Conflict.* Stanford, Calif.: Stanford University Press.

O'Malley, Padraig. 1991. "Northern Ireland: Political Strategies for the Management of Conflict." Pp. 161–175 in Joseph V. Montville (ed.), *Conflict and Peacemaking in Multiethnic Societies.* New York: Lexington.

Omi, Michael, and Howard Winant. 1986. *Racial Formation in the United States: From the 1960s to the 1980s.* New York: Routledge & Kegan Paul.

Ong, Paul, Kye Young Park, and Yasmin Tong. 1992. "The Korean-Black Conflict and the State." Pp. 264–294 in Paul Ong, Edna Bonacich, and Lucie Cheng (eds.), *The New Asian American Immigration.* Philadelphia: Temple University Press.

Onishi, Norimitsu. 1995. "Japanese in America Looking Beyond Past to Shape Future." *New York Times* (December 25):1, 9.

Orfield, Gary. 1985. "Ghettoization and Its Alternatives." Pp. 161–196 in Paul E. Peterson (ed.), *The New Urban Reality.* Washington, D.C.: Brookings Institution.

————. 1988. "Separate Societies: Have the Kerner Warnings Come True?" Pp. 100–122 in Fred R. Harris and Roger W. Wilkins (eds.), *Quiet Riots: Race and Poverty in the United States.* New York: Pantheon.

Organization for Economic Co-operation and Development (OECD). 2001. *Trends in International Migration: Annual Report* (January 2001). Paris: OECD.

Orleck, Annalise. 1987. "The Soviet Jews: Life in Brighton Beach, Brooklyn." Pp. 273–304 in Nancy Foner (ed.), *New Immigrants in New York.* New York: Columbia University Press.

Osborne, R. D. 1991. "Discrimination and Fair Employment." Pp. 31–38 in Peter Stringer and Gillian Robinson (eds.), *Social Attitudes in Northern Ireland.* Belfast: Blackstaff Press.

Osborne, Robert, and Robert Cormack. 1991. "Religion and the Labour Market: Patterns and Profiles." Pp. 49–71 in Robert J. Cormack and Robert D. Osborne (eds.), *Discrimination and Public Policy in Northern Ireland.* Oxford: Clarendon Press.

Owen, Carolyn A., Howard C. Eisner, and Thomas R. McFaul. 1981. "A Half-Century of Social Distance Research: National Replication of the Bogardus Studies." *Sociology and Social Research* 66:80–99.

Pace, David. 2000. "Casino Boom Bypasses Indians." Associated Press (August 31) <http://wire.ap.org>.

Pachon, Harry P., and Joan W. Moore. 1981. "Mexican Americans." *Annals of the American Academy of Political and Social Science* 454:111–124.

Padilla, Elena. 1958. *Up from Puerto Rico.* New York: Columbia University Press.

Padilla, Fernando V., and Carlos B. Ramirez. 1974. "Patterns of Chicano Representation in California, Colorado and Nuevo Mexico." *Aztlan* 5:189–235.

Paine, Suzanne. 1974. *Exporting Workers: The Turkish Case.* London: Cambridge University Press.

Palley, Claire. 1981. "Ways Forward: The Constitutional Options." Pp. 183–206 in David Watt (ed.), *The Constitution of Northern Ireland: Problems and Prospects.* London: Heinemann.

Palmer, Howard. 1976. "Mosaic vs. Melting Pot? Immigration and Ethnicity in Canada and the United States." *International Journal* 31:488–528.

Park, Robert E. 1924. "The Concept of Social Distance." *Journal of Applied Sociology* 8:339–344.

————. 1928. "Human Migration and the Marginal Man." *American Journal of Sociology* 33:881–893.

————. 1950. *Race and Culture.* Glencoe, Ill.: Free Press.

Parkin, Frank. 1971. *Class Inequality and Political Order.* New York: Praeger.

Patterson, Orlando. 1995a. "The Culture of Caution." *The New Republic* (November 27): 22–26.

————. 1995b. "The Paradox of Integration." *The New Republic* (November 6):24–27.

————. 1997. *The Ordeal of Integration: Progress and Resentment in America's "Racial" Crisis.* Washington, D.C.: Civitas/Counterpoint.

————. 1998. "Affirmative Action: Opening Up Workplace Networks to Afro-Americans." *Brookings Review* (Spring):17–23.

Peach, Ceri, and Günther Glebe. 1995. "Muslim Minorities in Western Europe." *Ethnic and Racial Studies* 18:26–45.

Pearlin, Leonard I. 1954. "Shifting Group Attachments and Attitudes toward Negroes." *Social Forces* 33:47–50.

Pedraza-Bailey, Silva. 1985. *Political and Economic Migrants in America: Cubans and Mexicans.* Austin, Tex.: University of Texas Press.

Peñalosa, Fernando. 1967. "The Changing Mexican-American in Southern California." *Sociology and Social Research* 51:405–417.

Perez, Lisandro. 1980. "Cubans." Pp. 256–261 in Stephen Thernstrom (ed.), Harvard *Encyclopedia of American Ethnic Groups.* Cambridge, Mass.: Harvard University Press.

————. 1986. "Cubans in the United States." *Annals of the American Academy of Political and Social Science* 487:126–137.

Pérez-Stable, Marifeli, and Miren Uriarte. 1993. "Cubans and the Changing Economy of Miami." Pp. 133–159 in Rebecca Morales and Frank Bonilla (eds.), *Latinos in a Changing U.S. Economy.* Newbury Park, Calif.: Sage.

Perlez, Jane. 1995a. "Serbs Become Latest Victims in Changing Fortunes of War." *New York Times* (August 7):A1, A4.

————. 1995b. "Thousands of Serbian Civilians Are Caught in Soldiers' Gunfire." *New York Times* (August 9):A1, A6.

Perry, Duncan. 1998. "Destiny on Hold: Macedonia and the Dangers of Ethnic Discord." *Current History* (March):119–126

Pessar, Patricia R. 1995. *A Visa for a Dream: Dominicans in the United States.* Boston: Allyn & Bacon.

Petersen, William. 1980. "Concepts of Ethnicity." Pp. 234–242 in Stephen Thernstrom (ed.), *Harvard Encyclopedia of American Ethnic Groups.* Cambridge, Mass.: Harvard University Press.

Petit, Arthur G. 1980. *Images of the Mexican American in Fiction and Film.* College Station, Tex.: Texas A&M Press.

Pettigrew, Thomas F. 1960. "Social Distance Attitudes of South African Students." *Social Forces* 38:246–253.

————. 1964. *A Profile of the Negro American.* Princeton, N.J.: Van Nostrand.

————. 1975. "Black and White Attitudes toward Race and Housing." Pp. 92–126 in Thomas Pettigrew (ed.), *Racial Discrimination in the United States.* New York: Harper & Row.

————. 1979. "Racial Change and Social Policy." *Annals of the American Academy of Political and Social Science* 441:114–131.

————. 1980. "Prejudice." Pp. 820–829 in Stephen Thernstrom (ed.), *Harvard Encyclopedia of American Ethnic Groups.* Cambridge, Mass.: Harvard University Press.

————. 1981. "Race and Class in the 1980s: An Interactive View." *Daedalus* 110:233–255.

Peukert, Detlev J. K. 1987. *Inside Nazi Germany: Conformity, Opposition and Racism in Everyday Life.* London: Penguin.

Philliber, William W., and Clyde B. McCoy (eds.). 1981. *The Invisible Minority: Urban Appalachians.* Lexington, Ky.: University Press of Kentucky.

Pierson, Donald. 1967 (1942). *Negroes in Brazil: A Study of Race Contact at Bahia.* Carbondale, Ill.: Southern Illinois University Press.

Pineo, Peter C. 1987. "The Social Standing of Ethnic and Racial Groupings." Pp. 256–272 in Leo Driedger (ed.), *Ethnic Canada: Identities and Inequalities.* Toronto: Copp Clark Pitman.

Pineo, Peter C., and John Porter. 1985. "Ethnic Origin and Occupational Attainment." Pp. 357–392 in Monica Boyd et al., *Ascription and Achievement: Studies in Mobility and Status Attainment in Canada.* Ottawa: Carleton University Press.

Pinkney, Alphonso. 1963. "Prejudice toward Mexican and Negro Americans: A Comparison." *Phylon* 24:353–359.

————. 2000. *Black Americans.* 5th ed. Englewood Cliffs, N.J.: Prentice-Hall.

Pitman, Walter. 1977. *Now Is Not Too Late.* Toronto: Council of Metropolitan Toronto, Task Force on Human Relations.

Pitt-Rivers, Julian. 1987. "Race, Color, and Class in Central America and the Andes." Pp. 298–305 in Celia S. Heller (ed.), *Structured Social Inequality: A Reader in Comparative Social Stratification.* 2d ed. New York: Macmillan.

Piven, Frances Fox, and Richard A. Cloward. 1971. *Regulating the Poor: The Functions of Public Welfare.* New York: Random House.

Plamenatz, John. 1970. *Ideology.* New York: Praeger.

Plous, Scott. 1996. "Ten Myths about Affirmative Action." *Journal of Social Issues* 52(4):25–31.

Poll, Solomon. 1969. *The Hasidic Community of Williamsburg.* New York: Schocken.

Pollack, Andrew. 1992. "Asian Immigrants New Leaders in Silicon Valley." *New York Times* (January 14):A1, C5.

Pollak, Andy. 1993. *A Citizens' Inquiry: The Opsahl Report on Northern Ireland.* Dublin: Lilliput Press for Initiative '92.

Pomfret, John. 1995. "For the War's Victims, the Search for Justice Begins." *Washington Post National Weekly Edition* (December 25–31):8–9.

Ponting, J. Rick. 1997a. "Historical Overview and Background: Part II. Pp. 35–67 in J. Rick Ponting (ed.), *First Nations in Canada: Perspectives on Opportunity, Empowerment, and Self-Determination.* Toronto: McGraw-Hill Ryerson.

————. 1997b. "The Socio-Demographic Picture." Pp. 68–114 in J. Rick Ponting (ed.), *First Nations in Canada: Perspectives on Opportunity, Empowerment, and Self-Determination.* Toronto: McGraw-Hill Ryerson.

Poole, Michael A. 1982. "Religious Residential Segregation in Urban Northern Ireland." Pp. 182–308 in Frederick W. Boal and J. Neville H. Douglas (eds.), *Integration and Division: Geographical Perspectives on the Northern Ireland Problem.* London: Academic Press.

Poppino, Rollie E. 1968. *Brazil: The Land and People.* New York: Oxford University Press.

Porter, John. 1965. *The Vertical Mosaic: An Analysis of Social Class and Power in Canada.* Toronto: University of Toronto Press.

————. 1975. "Ethnic Pluralism in Canadian Perspective." Pp. 267–304 in Nathan Glazer and Daniel P. Moynihan (eds.), *Ethnicity: Theory and Experience.* Cambridge, Mass.: Harvard University Press.

_____. 1979. *The Measure of Canadian Society*. Toronto: Gage.

_____. 1985. "Canada: The Societal Context of Occupational Allocation." Pp. 29–65 in Monica Boyd et al., *Ascription and Achievement: Studies in Mobility and Status Attainment in Canada*. Ottawa: Carleton University Press.

Porter, Judith D. R. 1971. *Black Child, White Child*. Cambridge, Mass.: Harvard University Press.

Portes, Alejandro. 1969. "Dilemmas of a Golden Exile: Integration of Cuban Refugee Families in Milwaukee." *American Sociological Review* 34:505–518.

_____. 1979. "Illegal Immigration and the International System: Lessons from Recent Legal Mexican Immigrants to the United States." *Social Problems* 26:425–438.

Portes, Alejandro, and Robert L. Bach. 1985. *Latin Journey: Cuban and Mexican Immigrants in the United States*. Berkeley: University of California Press.

Portes, Alejandro, J. M. Clark, and R. L. Bach. 1977. "The New Wave: A Statistical Profile of Recent Cuban Exiles to the United States." *Estudios Cubanos/Cuban Studies* 7:1–32.

Portes, Alejandro, Juan M. Clark, and Robert D. Manning. 1985. "After Mariel: A Survey of the Resettlement Experiences of 1980 Cuban Refugees in Miami." *Estudios Cubanos/Cuban Studies* 15:37–59.

Portes, Alejandro, and Rafael Mozo. 1985. "The Political Adaptation Process of Cubans and Other Ethnic Minorities in the United States: A Preliminary Analysis." *International Migration Review* 19:35–63.

Portes, Alejandro, and Rubén Rumbaut. 1996. *Immigrant America: A Portrait*. 2d ed. Berkeley: University of California Press.

Portes, Alejandro, and Richard Schauffler. 1996. "Language and the Second Generation: Bilingualism Yesterday and Today." Pp. 8–29 in Alejandro Portes (ed.), *The New Second Generation*. New York: Russell Sage Foundation.

Portes, Alejandro, and Alex Stepick. 1993. *City on the Edge: The Transformation of Miami*. Berkeley: University of California Press.

Portes, Alejandro, and Cynthia Truelove. 1987. "Making Sense of Diversity: Recent Research on Hispanic Minorities in the United States." *Annual Review of Sociology* 13:359–385.

Portes, Alejandro, and Min Zhou. 1993. "The New Second Generation: Segmented Assimilation and Its Variants." *Annals of the American Academy of Political and Social Science* 530(November):74–96.

_____. 1994. "Should Immigrants Assimilate?" *The Public Interest* (Summer):18–33.

Pulley, Brett. 1999. "Tribes Weighing Tradition vs. Casino Growth." *New York Times* (March 16):A1, A20.

Posgate, Dale. 1978. "The Quiet Revolution." Pp. 50–57 in Norman Penner et al. (eds.), *Keeping Canada Together*. Toronto: Amethyst.

Poston, Dudley L., and Mei-Yu Yu. 1990. "The Distribution of the Overseas Chinese in the Contemporary World." *International Migration Review* 24(Fall):480–508.

Power, Jonathan. 1979. *Migrant Workers in Western Europe and the United States*. Oxford: Pergamon.

Pressley, Sue Anne. 1999. "The Star of David Sets in the South's Small Towns." *Washington Post National Weekly Edition* (May 31):29.

Presthus, Robert. 1973. *Elite Accommodation in Canadian Politics*. Toronto: Macmillan.

Price, John A. 1973. "The Stereotyping of North American Indians in Motion Pictures." *Ethnohistory* 20(Spring):153–171.

Public Opinion. 1987. "Opinion Roundup." (July/August):25–39.

Purcell, Victor. 1980. *The Chinese in Southeast Asia*. 2d ed. Oxford: Oxford University Press.

Purdam, Todd W. 1998. "Victory Shines Light on Shift In Population." *New York Times* (November 5):B2.

Qian, Zhenchao, Sampson Lee Blair, and Stacey D. Ruf. 2001. "Asian American Interracial and Interethnic Marriages: Differences by Education and Nativity." *International Migration Review* 35:557–586.

Quinley, Harold E., and Charles Y. Glock. 1979. *Anti-Semitism in America*. New York: Free Press.

Raab, Earl. 1989. *What Do We Really Know about Anti-Semitism and What Do We Want to Know?* Working Papers on Contemporary Anti-Semitism. New York: American Jewish Committee.

Raab, Earl, and Seymour Martin Lipset. 1971. "The Prejudiced Society." Pp. 31–45 in Gary T. Marx (ed.), *Racial Conflict: Tension and Change in American Society*. Boston: Little, Brown.

Ramcharan, Subhas. 1982. *Racism: Nonwhites in Canada*. Toronto: Butterworths.

Ramirez, J. Martin, and Bobbie Sullivan. 1987. "The Basque Conflict." Pp. 120–139 in Jerry Boucher, Dan Landis, and Karen Arnold Clark (eds.), *Ethnic Conflict: International Perspectives*. Newbury Park, Calif.: Sage.

Ramos, Arthur. 1939. *The Negro in Brazil*. Trans. Richard Pattee. Washington, D.C.: Associated Publishers.

Ramsey, Patricia G. 1987. "Young Children's Thinking about Ethnic Differences." Pp. 56–72 in Jean S. Phinney and Mary Jane Rotheram (eds.), *Children's Ethnic Socialization: Pluralism and Development*. Newbury Park: Sage.

Ratnesar, Romesh. 1998. "A Place at the Table." *Time* (October 12):38.

Ravitch, Diane. 2002. "A Considered Opinion: Diversity, Tragedy, and the Schools." *Brookings Review* 20(1):2–3.

Raymond, Joan. 2001. "The Multicultural Report." *American Demographics* (November):S3–S6.

Redfield, Robert. 1939. "Culture Contact without Conflict." *American Anthropologist* 41:514–517.

_____. 1958. "Race as a Social Phenomenon." Pp. 66–71 in Edgar T. Thompson and Everett C. Hughes (eds.), *Race: Individual and Collective Behavior*. Glencoe, Ill.: Free Press.

Reeve, R. Penn. 1975. "Black Economic Mobility in a Brazilian Town." *Plural Societies* 6:45–50.

———. 1977. "Race and Social Mobility in a Brazilian Industrial Town." *Luso Brazilian Review* 14:236–253.

Reich, Michael. 1978. "The Economics of Racism." Pp. 381–388 in Richard C. Edwards, Michael Reich, and Thomas E. Weisskopf (eds.), *The Capitalist System*. 2d ed. Englewood Cliffs, N.J.: Prentice-Hall.

Reitz, Jeffrey G., and Raymond Breton. 1994. *The Illusion of Difference: Realities of Ethnicity in Canada and the United States*. Toronto: C. D. Howe Institute.

Rex, John. 1970. *Race Relations in Sociological Theory*. New York: Schocken.

Ribeiro, Darcy. 2000. *The Brazilian People: The Formation and Meaning of Brazil*. Trans. Gregory Rabassa. Gainesville: University Press of Florida.

Richer, Stephen, and Pierre E. Laporte. 1979. "Culture, Cognition, and English-French Competition." Pp. 75–85 in Jean Leonard Elliott (ed.), *Two Nations, Many Cultures*. Scarborough, Ont.: Prentice-Hall of Canada.

Richmond, Anthony H. 1976. "Immigration, Population, and the Canadian Future." *Sociological Focus* 9:125–136.

Rieder, Jonathan. 1985. *Canarsie: The Jews and Italians of Brooklyn against Liberalism*. Cambridge, Mass.: Harvard University Press.

Rieff, David. 1991. "The New Face of L.A." *Los Angeles Times Magazine* (September 15): 14–20, 46–48.

———. 1993. *The Exile: Cuba in the Heart of Miami*. New York: Touchstone.

———. 1995. *Slaughterhouse: Bosnia and the Failure of the West*. New York: Touchstone.

———. 1996. "An Age of Genocide." *The New Republic* (January 29):27–36.

Riis, Jacob A. 1957 (1890). *How the Other Half Lives*. New York: Hill & Wang.

Ringer, Benjamin B. 1967. *The Edge of Friendliness*. New York: Basic Books.

Rioux, Marcel, and Yves Martin (eds.). 1978. *French-Canadian Society*. Toronto: Macmillan of Canada.

Rippley, La Vern J. 1984. *The German Americans*. Lanham, Md.: University Press of America.

Ritterband, Paul. 1995. "Modern Times and Jewish Assimilation." Pp. 377–394 in Robert M. Seltzer and Norman J. Cohen (eds.), *The Americanization of the Jews*. New York: New York University Press.

Robertson, Ian, and Phillip Whitten. 1978. "Introduction." Pp. xi–xx in Ian Robertson and Phillip Whitten (eds.), *Race and Politics in South Africa*. New Brunswick, N.J.: Transaction.

Robinson, Eugene. 1996. "Over the Brazilian Rainbow." *Washington Post National Weekly Edition* (December 18–24):23.

Roche, John Patrick. 1982. "Suburban Ethnicity: Ethnic Attitudes and Behavior among Italian Americans in Two Suburban Communities." *Social Science Quarterly* 63:145–153.

Rocher, Guy. 1976. "Multiculturalism: The Doubts of a Francophone." In *Second Canadian Conference on Multiculturalism*. Ottawa: Minister of Supply and Services.

Rodrigues, Jose Honorio. 1967. *The Brazilians*. Trans. Ralph Edward Dimmick. Austin, Tex.: University of Texas Press.

Rodríguez, Clara. 1989. *Puerto Ricans: Born in the U.S.A.* Boston: Unwin and Hyman.

———. 1996. "Racial Themes in the Literature: Puerto Ricans and Other Latinos." Pp. 105–125 in Gabriel Haslip-Viera and Sherrie L. Baver. *Latinos in New York: Communities in Transition*. Notre Dame: University of Notre Dame Press.

Rodriguez, Gregory. 1998. "Minority Leader." *The New Republic* (October 19):21–23.

Rodriguez, Nestor P., and Jacqueline Hagan. 1999. "Central Americans in the United States." Pp. 278–296 in Anthony G. Dworkin and Rosalind J. Dworkin, *The Minority Report*, 3d ed. Orlando, Fla.: Harcourt Brace.

Rodríguez-Vecchini, Hugo. 1994. "Foreward: Back and Forward." Pp. 29–102 in Carlos Antonio Torre, Hugo Rodríguez Vecchini, and William Burgos (eds.), *The Commuter Nation: Perspectives on Puerto Rican Migration*. Río Piedras, Puerto Rico: Editorial de la Universidad de Puerto Rico.

Rohter, Larry. 2000. "Catholics Battle Brazilian Faith in 'Black Rome'." *New York Times* (January 10):A13.

Rolle, Andrew. 1980. *The Italian Americans: Troubled Roots*. New York: Free Press.

Romanucci-Ross, Lola, and George De Vos (eds.). 1995. *Ethnic Identity: Creation, Conflict and Accommodation*. 3d ed. Walnut Creek, Calif.: Alta Mira.

Roof, Wade Clark. 1978. "The Negro as an Immigrant Group—A Research Note on Chicago's Racial Trends." *Ethnic and Racial Studies* 1:452–464.

———. 1979. "Race and Residence: The Shifting Basis of American Race Relations." *Annals of the American Academy of Political and Social Science* 44:1–12.

Rose, Hilary, and Steven Rose. 1978. "The IQ Myth." *Race and Class* 20:63–74.

Rose, Peter I. 1968. *The Subject Is Race: Traditional Ideologies and the Teaching of Race Relations*. New York: Oxford University Press.

———. 1977. *Strangers in Their Midst*. Merrick, N.Y.: Richwood.

———. 1985. "Asian Americans: From Pariahs to Paragons." Pp. 181–212 in Nathan Glazer (ed.), *Clamor at the Gates: The New American Immigration*. San Francisco: ICS Press.

Rose, Richard. 1971. *Governing without Consensus: An Irish Perspective*. Boston: Beacon.

———. 1976. *Northern Ireland: Time of Choice*. Washington, D.C.: American Enterprise Institute for Public Policy Research.

Rosen, Bernard C. 1959. "Race, Ethnicity, and the Achievement Syndrome." *American Sociological Review* 24:47–60.

Royal Commission on Bilingualism and Biculturalism. 1969a. Report. Vol. 3, *The World of Work*. Ottawa: Queen's Printer.

_____. 1969b. Report. Vol. 4, *The Cultural Contribution of the Other Ethnic Groups*. Ottawa: Queen's Printer.

Ruane, Joseph, and Jennifer Todd. 1996. *The Dynamics of Conflict in Northern Ireland: Power, Conflict and Emancipation*. Cambridge: Cambridge University Press.

Rudolph, Susanne Hoeber, and Lloyd I. Rudolph. 1993. "Modern Hate." *The New Republic* (March 22):24–29.

Rudwick, Elliott M. 1964. *Race Riot at East St. Louis*. Carbondale, Ill.: Southern Illinois University Press.

Rudwick, Elliott, and August Meier. 1969. "Negro Retaliatory Violence in the Twentieth Century." Pp. 406–417 in August Meier and Elliott Rudwick (eds.), *The Making of Black America: Essays in Negro Life and History*. Vol. 2. New York: Atheneum.

Rumbaut, Rubén G. 1996. "Origins and Destinies: Immigration, Race, and Ethnicity in Contemporary America." Pp. 21–42 in Silvia Pedraza and Rubén Rumbaut (eds.), *Origins and Destinies: Immigration, Race, and Ethnicity in America*. Belmont, Calif.: Wadsworth.

Runciman, W. G. 1966. *Relative Deprivation and Social Justice*. Berkeley: University of California Press.

Russell-Wood, A. J. R. 1968. "Race and Class in Brazil, 1937–1965, A Re-Assessment: A Review." *Race* 10:185–191.

Ryan, William. 1967. "Savage Discovery: The Moynihan Report." Pp. 457–466 in Lee Rainwater and William L. Yancey (eds.), *The Moynihan Report and the Politics of Controversy*. Cambridge, Mass.: MIT Press.

_____. 1975. *Blaming the Victim*. Rev. ed. New York: Vintage.

Saad, Lydia. 1995. "Immigrants See U.S. as Land of Opportunity." *Gallup Poll Monthly* (July):19–33.

Sachs, Albie. 1975. "The Instruments of Domination in South Africa." Pp. 223–249 in Leonard Thompson and Jeffrey Butler (eds.), *Change in Contemporary South Africa*. Berkeley: University of California Press.

_____. 1992. "Watch Out—There's a Constitution About." *South Africa International* 22(April):184–189.

Sachs, Susan. 1999. "From a Babel of Tongues, a Neighborhood." *New York Times* (December 26):1, 20.

_____. 2001. "A Hue, and a Cry, In the Heartland." *New York Times* (April 8):WK5.

Sack, Kevin. 1996. "Victory of 5 Redistricted Blacks Recasts Gerrymandering Dispute." *New York Times* (November 23):1–8.

Sagarin, Edward. 1971. *The Other Minorities*. Waltham, Mass.: Xerox.

Said, Edward W. 1981. *Covering Islam*. New York: Pantheon.

Salholy, Eloise. 1987. "Do Colleges Set Asian Quotas?" *Newsweek* 109(February 9):6.

Samuda, Ronald J. 1975. *Psychological Testing of American Minorities*. New York: Harper & Row.

Sandberg, Neil. 1974. *Ethnic Identity and Assimilation: The Polish American Community*. New York: Praeger.

_____. 1986. *Jewish Life in Los Angeles*. Lanham, Md.: University Press of America.

Sandefur, Gary D., Molly Martin, Jennifer Eggerling-Boeck, Susan E. Mannon, and Ann M. Meier. 2001. "An Overview of Racial and Ethnic Demographic Trends." Pp. 40–102 in Neil J. Smelser, William Julius Wilson, and Faith Mitchel (eds.), *America Becoming: Racial Trends and Their Consequences*. Vol. I. Washington, D.C.: National Academy Press.

Sanders, Thomas G. 1981. Racial Discrimination and Black Consciousness in Brazil. American Universities Field Staff Reports, No. 42. Hanover, N.H.: American Universities Field Staff.

_____. 1982. "Brazilian Population in 1982: Growth, Migration, Race, Religion." UFSI Reports 1982, No. 42.

Santiago, Anne M. 1991. *The Spatial Dimensions of Ethnic and Racial Stratification*. Ann Arbor, Mich.: Population Studies Center, University of Michigan.

Sarlin, Stuart H., and Eugene D. Tate. 1976. "'All in the Family': Is Archie Funny?" *Journal of Communication* 26:61–67.

Sassen, Saskia. 1998. *Globalization and Its Discontents*. New York: The New Press.

Saunders, John. 1972. "Class, Color, and Prejudice: A Brazilian Counterpoint." Pp. 141–165 in Ernest Q. Campbell (ed.), *Racial Tensions and National Identity*. Nashville, Tenn.: Vanderbilt University Press.

Saywell, John. 1977. *The Rise of the Parti Québécois 1967–76*. Toronto: University of Toronto Press.

Schemo, Diana Jean. 1995. "Elevators Don't Lie: Intolerance in Brazil." *New York Times* (August 30):7A.

Schermerhorn, R. A. 1949. *These Our People: Minorities in American Culture*. Boston: Heath.

_____. 1970. *Comparative Ethnic Relations: A Framework for Theory and Research*. New York: Random House.

Schiff, Ben. 1996. "The Afrikaners After Apartheid." *Current History* 95 (May):216–221.

Schlemmer, Lawrence. 1988. "South Africa's National Party Government." Pp. 7–54 in Peter L. Berger and Bobby Godsell (eds.), *A Future South Africa: Visions, Strategies and Realities*. Cape Town: Human & Rousseau Tafelberg.

Schlesinger, Arthur M., Jr. 1992. *The Disuniting of America: Reflections on a Multicultural Society*. New York: Norton.

Schmidhauser, John R. 1960. *The Supreme Court: Its Politics, Personalities, and Procedures*. New York: Holt, Rinehart & Winston.

Schmitt, Eric. 2001a. "Census Figures Show Hispanics Pulling Even with Blacks." *New York Times* (March 8):A1, A4.

_____. 2001b. "Americans (a) Love (b) Hate Immigrants." *New York Times* (January 14):1, 3.

Schneider, Ronald M. 1996. *Brazil: Culture and Politics in a New Industrial Powerhouse*. Boulder, Colo.: Westview.

Schooler, Carmi. 1976. "Serfdom's Legacy: An Ethnic Continuum." *American Journal of Sociology* 81:1265–1286.

Schöpflin, George. 1993. "The Rise and Fall of Yugoslavia." Pp. 172–203 in John McGarry and

Brendan O'Leary (eds.), *The Politics of Ethnic Conflict Regulation*. London: Routledge.

Schrag, Peter. 1995. "So You Want to Be Color Blind: Alternative Principles for Affirmative Action." *The American Prospect* (Summer): 38–43.

Schuman, Howard. 1982. "Free Will and Determinism in Public Beliefs about Race." Pp. 345–350 in Norman R. Yetman and C. Hoy Steele (eds.), *Majority and Minority: The Dynamics of Race and Ethnicity in American Life*, 3d ed. Boston: Allyn & Bacon.

Schuman, Howard, Charlotte Steeh, Lawrence Bobo, and Maria Krysan. 1997. *Racial Attitudes in America: Trends and Interpretations*. Rev. ed. Cambridge, Mass.: Harvard University Press.

Schwarz, Adam. 1994. *A Nation in Waiting: Indonesia in the 1990s*. St. Leonard's, N.S.W.: Allen & Unwin.

Schwartz, Mildred A. 1967. *Trends in White Attitudes toward Negroes*. Chicago: National Opinion Research Center.

Schwartzman, Simon. 2000. "Brazil: The Social Agenda." *Daedalus* 129 (Spring): 29–56.

Sciortino, Giuseppe. 1992. "From Guests to Hosts: Italy as a New Country of Immigration." Paper presented at the Annual Meeting of the American Sociological Association, Pittsburgh, Pa.

Scott, Matthew. 2001. "B.E. 100s: Poised for Growth." *Black Enterprise* (June): 119–126.

Seifert, Wolfgang. 1998. "Social and Economic Integration of Foreigners in Germany." Pp. 83–114 in Peter H. Schuck and Rainer Münz (eds.), *Paths to Inclusion: The Integration of Migrants in the United States and Germany*. New York: Berghahn Books.

Selznick, Gertrude J., and Stephen Steinberg. 1969. *The Tenacity of Prejudice: Anti-Semitism in Contemporary America*. New York: Harper & Row.

Senese, Guy B. 1991. *Self-Determination and the Social Education of Native Americans*. New York: Praeger.

Sengupta, Somini, and Vivian S. Toy. 1998. "United Ethnically, and by an Assault." *New York Times* (October 7): A21.

Shaheen, Jack G. 1984. *The TV Arab*. Bowling Green, Ohio: Bowling Green State University Popular Press.

Sharot, Stephen. 1973. "The Three-Generations Thesis and the American Jews." *British Journal of Sociology* 24: 151–164.

Shaw, Mark, and Peter Gastrow. 2001. "Stealing the Show? Crime and Its Impact in Post-Apartheid South Africa." *Daedalus* 130 (Winter): 235–258.

Sheane, Michael. 1977. *Ulster and Its Future After the Troubles*. Bredbury, England: Highfield Press.

Sheatsley, Paul B. 1966. "White Attitudes Toward the Negro." Pp. 303–324 in Talcott Parsons and Kenneth B. Clark (eds.), *The Negro American*. Boston: Houghton Mifflin.

Shibutani, Tamotsu, and Kian M. Kwan. 1965. *Ethnic Stratification: A Comparative Approach*. New York: Macmillan.

Shils, Edward. 1968. "Ideology." Pp. 66–76 in David L. Sills (ed.), *International Encyclopedia of the Social Sciences*. Vol. 7. New York: Macmillan.

Shimoni, Gideon. 1988. "South African Jews and the Apartheid Crisis." Pp. 3–58 in *American Jewish Yearbook 1988*. New York: American Jewish Committee.

Shinagawa, Larry Hajime, and Gin Yong Pang. 1996. "Asian American Panethnicity and Intermarriage." *Amerasia Journal* 22: 127–152.

Sibley, Mulford Q. 1963. *The Quiet Battle*. Chicago: Quadrangle.

Sigelman, Lee, Timothy Bledsoe, Susan Welch, and Michael W. Combs. 1996. "Making Contact? Black-White Social Interaction in an Urban Setting." *American Journal of Sociology* 101: 1306–1332.

Siggner, Andrew J. 1986. "The Socio-Demographic - Conditions of Registered Indians." *Canadian Social Trends* (Winter): 2–9.

Silberman, Charles E. 1985. *A Certain People: American Jews and their Lives Today*. New York: Summit.

Silva, Nelson do Valle. 1985. "Updating the Cost of Not Being White in Brazil." Pp. 42–55 in Pierre-Michel Fontaine (ed.), *Race, Class, and Power in Brazil*. Los Angeles: Center for Afro-American Studies, University of California, Los Angeles.

_____. 1999. "Racial Differences in Income: Brazil, 1988." Pp. 67–82 in Rebecca Reichmann (ed.), *Race in Contemporary Brazil: From Indifference to Inequality*. University Park, Pa.: The Pennsylvania State University Press.

Silva, Nelson do Valle, and Carols A. Hasenbalg. 1999. "Race and Educational Opportunity in Brazil." Pp. 53–65 in Rebecca Reichmann (ed.), *Race in Contemporary Brazil: From Indifference to Inequality*. University Park, Pa.: Pennsylvania State University Press.

Simmons, Ozzie G. 1971. "The Mutual Images and Expectations of Anglo-Americans and Mexican-Americans." Pp. 62–71 in Nathaniel N. Wagner and Marsha J. Haug (eds.), *Chicanos: Social and Psychological Perspectives*. St. Louis: Mosby.

Simon, Julian L. 1991. "The Case for Greatly Increased Immigration." *The Public Interest* 102 (Winter): 89–103.

Simon, Rita James. 1978. *Continuity and Change: A Study of Two Ethnic Communities in Israel*. Cambridge, England: Cambridge University Press.

Simon, Rita J., and James P. Lynch. 1999. "A Comparative Assessment of Public Opinion toward Immigrants and Immigration Policies." *International Migration Review* 33: 455–467.

Simpson, Amelia. 1993. *Xuxa: The Mega-Marketing of Gender, Race, and Modernity*. Philadelphia: Temple University Press.

Simpson, George Eaton. 1968. "Assimilation." Pp. 428–444 in David L. Sills (ed.), *International Encyclopedia of the Social Sciences*. Vol. 1. New York: Macmillan.

Simpson, George Eaton, and J. Milton Yinger. 1972. *Racial and Cultural Minorities: An Analysis of Prejudice and Discrimination*. 4th ed. New York: Harper & Row.

Singer, Mark. 2001. "Home Is Here: America's Largest Arab Community in the Aftermath of September 11th." *The New Yorker* (October 15): 62–70.

Skellington, Richard. 1996. "Race" in *Britain Today*. 2d ed. London: Sage.

Skidmore, Thomas. 1972. "Toward a Comparative Analysis of Race Relations Since Abolition in Brazil and the United States." *Journal of Latin American Studies* 4(May):1–28.

———. 1974. *Black into White: Race and Nationality in Brazilian Thought*. New York: Oxford University Press.

———. 1990. "Racial Ideas and Social Policy in Brazil, 1870–1940." Pp. 7–36 in Richard Graham (ed.), *The Idea of Race in Latin America, 1870–1940*. Austin, Tex.: University of Texas Press.

———. 1993. "Bi-racial U.S.A. vs. Multi-racial Brazil: Is the Contrast Still Valid?" *Journal of Latin American Studies* 25:373–386.

———. 1999. *Brazil: Five Centuries of Change*. New York: Oxford University Press.

Sklare, Marshall. 1969. "The Ethnic Church and the Desire for Survival." Pp. 101–117 in Peter I. Rose (ed.), *The Ghetto and Beyond*. New York: Random House.

———. 1971. *America's Jews*. New York: Random House.

———. 1978. "Jewish Acculturation and American Jewish Identity." Pp. 167–188 in Gladys Rosen (ed.), *Jewish Life in America: Historical Perspectives*. New York: American Jewish Committee.

Sklare, Marshall, and Joseph Greenblum. 1967. *Jewish Identity in the Suburban Frontier: A Study of Group Survival in the Open Society*. New York: Basic Books.

Skrentny, John David. 1996. *The Ironies of Affirmative Action: Politics, Culture, and Justice in America*. Chicago: University of Chicago Press.

Slawson, John, and Lawrence Bloomgarden. 1965. *The Unequal Treatment of Equals: The Social Club...Discrimination in Retreat*. New York: Institute of Human Relations, American Jewish Committee.

Smedley, Audrey. 1999. *Race in North America Origin and Evolution of a Worldview*. Boulder, Colo.: Westview.

Smith, Anthony D. 1981. *The Ethnic Revival*. Cambridge, England: Cambridge University Press.

Smith, David J., and Gerald Chambers. 1991. *Inequality in Northern Ireland*. Oxford: Clarendon Press.

Smith, Elmer Lewis. 1958. *The Amish People*. New York: Exposition Press.

Smith, James P. 2001. "Race and Ethnicity in the Labor Market: Trends Over the Short and Long Term." Pp. 52–97 in Neil J. Smelser, William Julius Wilson, and Faith Mitchells (eds.), *America Becoming: Racial Trends and Their Consequences*. Vol. II. Washington, DC: National Academy Press.

Smith, James P., and Barry Edmonston (eds.). 1997. *The New Americans: Economic, Demographic, and Fiscal Effects of Immigration*. Washington, D.C.: National Academy Press.

Smith, M. G. 1965. *The Plural Society in the British West Indies*. Berkeley: University of California Press.

———. 1969. "Institutional and Political Conditions of Pluralism." Pp. 27–65 in Leo Kuper and M. G. Smith (eds.), *Pluralism in Africa*. Berkeley: University of California Press.

Smith, R. Jeffrey. 2000. "Peacekeeping Without Peace in Kosovo." *Washington Post National Weekly Edition* (June 19):15.

Smith, T. Lynn. 1963. *Brazil: People and Institutions*. Baton Rouge, La.: Louisiana State University Press.

———. 1974. *Brazilian Society*. Albuquerque, N. Mex.: University of New Mexico Press.

Smith, Tom W. 1980. *A Compendium of Trends on General Social Survey Questions* (NORC Report No. 129). Chicago: National Opinion Research Center.

———. 1990. *Jewish Attitudes toward Blacks & Race Relations*. New York: American Jewish Committee.

———. 1991. *What Do Americans Think about Jews?* New York: American Jewish Committee.

Smooha, Sammy. 1978. *Israel: Pluralism and Conflict*. London: Routledge & Kegan Paul.

Sniderman, Paul M, and Edward G. Carmines. 1997. *Reaching Beyond Race*. Cambridge, Mass.: Harvard University Press.

Sniderman, Paul M., and Thomas Piazza. 1993. *The Scar of Race*. Cambridge, Mass.: Belknap Press.

Snipp, C. Matthew. 1986. "The Changing Political and Economic Status of the American Indians: From Captive Nations to Internal Colonies." *American Journal of Economics and Sociology* 45:145–157.

———. 1989. *American Indians: The First of This Land*. New York: Russell Sage Foundation.

———. 1997. "Some Observations About Racial Boundaries and the Experiences of American Indians." *Ethnic and Racial Studies* 20:667–689.

Solis-Garza, Luis A. 1972. "Cesar Chavez: The Chicano Messiah?" Pp. 297–305 in Edward Simmen (ed.), *Pain and Promise: The Chicano Today*. New York: New American Library.

Sorin, Gerald. 1997. *Tradition Transformed: The Jewish Experience in America*. Baltimore: The Johns Hopkins University Press.

South African Institute of Race Relations. 1992. *Race Relations Survey 1991/92*. Johannesburg: South African Institute of Race Relations.

———. 1993. *Race Relations Survey 1992/93*. Johannesburg: South African Institute of Race Relations.

———. 2001. *South Africa Survey 2000/01*. Johannesburg: South African Institute of Race Relations

Sowell, Thomas. 1978. "Three Black Histories." Pp. 7–64 in Thomas Sowell (ed.), *Essays and Data on American Ethnic Groups*. Urban Institute.

———. 1981. *Ethnic America*. New York: Basic Books.

———. 1990. *Preferential Policies: An International Perspective*. New York: William Morrow.

Spear, Allan H. 1967. *Black Chicago*. Chicago: University of Chicago Press.

Spicer, Edward H. 1980. "American Indians." Pp. 58–122 in Stephen Thernstrom (ed.), *Harvard Encyclopedia of American Ethnic Groups*. Cambridge, Mass.: Harvard University Press.

———. 1982. *The American Indians*. Cambridge, Mass.: Harvard University Press.

Stack, Carol. 1996. *Call to Home: African Americans Reclaim the Rural South.* New York: BasicBooks.

Stackhouse, John. 2001. "Welcome to Harlem on the Prairies." *The Globe and Mail* (November 3):F2–F3.

Stafford, Walter W. 2001. "The National Urban League Survey: Black America's Under-35 Generation." In *The State of Black America 2001.* New York: National Urban League <http://www.nul.org/soba2001/sobaabstracts.html>.

Stampp, Kenneth M. 1956. *The Peculiar Institution: Slavery in the Ante-Bellum South.* New York: Knopf.

Stanley, David T., Dean E. Mann, and Jameson W. Doig. 1967. *Men Who Govern.* Washington, D.C.: Brookings Institution.

Staples, Robert. 1981. "The Black American Family." Pp. 217–244 in Charles H. Mindeland and Robert W. Habenstein (eds.), *Ethnic Families in America.* 2d ed. New York: Elsevier.

Statistics Canada. 1993. *Language, Tradition, Health, Life-style and Social Issues: 1991 Aboriginal Peoples Survey.* Ottawa: Ministry of Industry, Science, and Technology.

———. 1996. "Projections of Visible Minority Groups, 1991 to 2016." *Canadian Social Trends* (Summer):3.

———. 1997. *Canada Year Book 1997.* Ottawa.

———. 1999. *Visible Minority Population by Age,* 1996 Census <http://www.statcan.ca/english/Pgdb/People/Population/demo41.htm>.

———. 2001a. *1996 Census: Nation Tables* <http://www.statcan.ca/english/census96/feb17/min.htm>.

———. 2001b. *Statistical Profile: Population Statistics for Toronto (Census Metropolitan Area)* <http://ceps.statcan.ca/english/profil/D...aceName=Toronto&CSDNAME=Toronto&CMA=535>.

Steeh, Charlotte, and Howard Schuman. 1992. "Young White Adults: Did Racial Attitudes Change in the 1980s?" *American Journal of Sociology* 98:340–367.

Steele, Shelby. 1991. *The Content of Our Character: A New Vision of Race in America.* New York: Harper-Perennial.

Steger, Wilbur. 1973. "Economic and Social Costs of Residential Segregation." Pp. 83–113 in Marion Clawson (ed.), *Modernizing Urban Land Policy.* Baltimore, Md.: Johns Hopkins University Press.

Steinberg, Stephen. 1989. *The Ethnic Myth: Race, Ethnicity and Class in America.* Updated and expanded ed. New York: Atheneum.

———. 1995. *Turning Back: The Retreat from Racial Justice in American Thought and Policy.* Boston: Beacon Press.

Steiner, Stan. 1970. *La Raza: The Mexican Americans.* New York: Harper & Row.

Steinfels, Peter. 1992. "Debating Intermarriage, and Jewish Survival." *New York Times* (October 18):1, 16.

Stember, Charles H. 1966. *Jews in the Mind of America.* New York: Basic Books.

Sterngold, James. 1998a. "Blacks Have Had Good Times and Bad on Prime-Time TV." *New York Times* (December 29):1, 12.

———. 1998b. "Prime-Time TV's Growing Racial Divide Frustrates Industry's Blacks." *New York Times* (December 29):12.

Stille, Alexander. 1992. "Italy: No Blacks Need Apply." *Atlantic Monthly* 269(February): 28–38.

Stoddard, Ellwyn R. 1973. *Mexican Americans.* New York: Random House.

Stokes, Bruce. 1988. "Learning the Game." *National Journal* 21(October 22):2649–2654.

Stone, John. 1973. *Colonist or Uitlander: A Study of the British Immigrant in South Africa.* Oxford: Clarendon Press.

Stonequist, Everett V. 1937. *The Marginal Man.* New York: Charles Scribner's Sons.

Stout, Cameron W. 1997. "Canada's Newest Territory in 1999." *Canadian Social Trends* (Spring):13–18.

Strodtbeck, Fred L. 1958. "Family Interaction, Values, and Achievement." Pp. 147–165 in Marshall Sklare (ed.), *The Jews: Social Patterns of an American Group.* Glencoe, Ill.: Free Press.

Stultz, Newell M. 1980. "Some Implications of African 'Homelands' in South Africa." Pp. 194–216 in Robert M. Price and Carl G. Rotberg (eds.), *The Apartheid Regime: Political Power and Racial Domination.* Berkeley: Institute of International Studies, University of California.

Sudetic, Chuck. 1994. "Serbs of Sarajevo Stay Loyal to Bosnia." *New York Times* (August 26):8A.

Sugden, John. 1995. "Sport, Community Relations and Community Conflict in Northern Ireland." Pp. 199–214 in Seamus Dunn (ed.), *Facets of the Conflict in Northern Ireland.* Basingstoke: Macmillan.

Suleiman, Michael W. 1999. "Introduction: The Arab Immigrant Experience." Pp. 1–24 in Michael Suleiman (ed.), *Arabs in America: Building a New Future.* Philadelphia: Temple University Press.

Sunahara, M. Ann. 1980. "Federal Policy and the Japanese Canadians: The Decision to Evacuate, 1942." Pp. 93–120 in K. Victor Ujimoto and Gordon Hirabayashi (eds.), *Visible Minorities and Multiculturalism: Asians in Canada.* Toronto: Butterworths.

Suro, Roberto. 1989. "Employers Are Looking Abroad for the Skilled and the Energetic." *New York Times* (July 16):IV-4.

———. 1998. *Strangers Among Us: How Latino Immigration Is Transforming America.* New York: Alfred A. Knopf.

Suzuki, Teiiti. 1969. *The Japanese Immigrant in Brazil.* Tokyo: University of Tokyo Press.

Swain, Carol M. 2001. "Affirmative Action: Legislative History, Judicial Interpretations, Public Consensus." Pp. 318–347 in Neil J. Smelser, William Julius Wilson, and Faith Mitchell (eds.), *America Becoming: Racial Trends and Their Consequences.* Vol. I. Washington, D.C.: National Academy Press.

Swarns, Rachel L. 2000. "Hesitantly, Students in South Africa Reach Across the Racial Divide." *New York Times* (November 12):10.

———. 2001a. "South Africa Sees Turmoil Dividing Ruling Party." *New York Times* (November 20):A1.

_____. 2001b. "Newest Statistics Show AIDS Still Spreading in South Africa." *New York Times* (March 21):A8.

Swinton, David H. 1992. "The Economic Status of African Americans: Limited Ownership and Persistent Inequality." Pp. 61–117 in National Urban League, *The State of Black America 1992*. New York: National Urban League.

Szabo, Joan C. 1989. "Opening Doors for Immigrants." *Nation's Business* 77(August):48–49.

Szymanski, Albert. 1976. "Social Discrimination and White Gain." *American Sociological Review* 41:403–414.

Taeuber, Karl E. 1975. "Racial Segregation: The Persisting Dilemma." *Annals of the American Academy of Political and Social Science* 422:87–96.

_____. 1979. "Housing, Schools, and Incremental Segregative Effects." *Annals of the American Academy of Political and Social Science* 441:157–167.

Taeuber, Karl E., and Alma F. Taeuber. 1965. *Negroes in Cities*. Chicago: Aldine.

Taft, Ronald. 1963. "The Assimilation Orientation of Immigrants and Australians." *Human Relations* 16:279–293.

Tajitsu Nash, Phil, and Frank Wu. 1997. "Asian Americans Under Glass: Where the Furor Over the President's Fundraising Has Gone Awry—and Racist." *The Nation* (March 31):15–16.

Tannenbaum, Frank. 1947. *Slave and Citizen: The Negro in the Americas*. New York: Knopf.

Tasker, Fredric. 1980. "Anti-Bilingualism Measure Approved in Dade County." *Miami Herald* (November 5):1.

Taylor, D. Garth. 1979. "Housing, Neighborhoods, and Race Relations: Recent Survey Evidence." *Annals of the American Academy of Political and Social Science* 441:26–40.

Taylor, D. Garth, Paul B. Sheatsley, and Andrew M. Greeley. 1978. "Attitudes toward Racial Integration." *Scientific American* 238(June): 42–49.

Taylor, Humphrey. 1998. "Oprah Winfrey Is Now America's Favorite TV Personality." *The Harris Poll* (December 9).

Taylor, Paul. 1995. "South Africa: 'A Miracle That Keeps Evolving.'" *Washington Post National Weekly Edition* (May 1–7):16.

Taylor, Ronald L. 1979. "Black Ethnicity and the Persistence of Ethnogenesis." *American Journal of Sociology* 84:1401–1423.

Taylor, William L. 1995. "Affirmative Action: The Questions to Be Asked." *Poverty and Race* (May/June):2–3.

Telhami, Shibley. 2002. "Arab and Muslim America: A Snapshot." *Brookings Review* 20(1):14–15. Washington, D.C.: The Brookings Institution.

Telles, Edward E. 1992. "Residential Segregation by Skin Color in Brazil." *American Sociological Review* 57:186–197.

tenBroek, Jacobus, Edward N. Barnhart, and Floyd W. Matson. 1968. *Prejudice, War and the Constitution*. Berkeley: University of California Press.

Thernstrom, Stephan, and Abigail Thernstrom. 1997. *America in Black and White: One Nation, Indivisible*. New York: Simon & Schuster.

Thistlethwaite, Frank. 1991. "Migration from Europe Overseas in the Nineteenth and Twentieth Centuries." Pp. 17–57 in Rudolph J. Vecoli and Suzanne M. Sinke (eds.), *A Century of European Migrations, 1830–1930*. Urbana: University of Illinois Press.

Thomas, Piri. 1967. *Down These Mean Streets*. New York: Knopf.

Thomas, Robert K. 1966. "Colonialism: Classic and Internal." *New University Thought* 4(Winter):37–44.

Thomas, W. I., and Florian Znaniecki. 1918. *The Polish Peasant in Europe and America*. Vol. 1. New York: Knopf.

Thompson, John Herd, and Morton Weinfeld. 1995. "Entry and Exit: Canadian Immigration Policy in Context." *Annals of the American Academy of Political and Social Science* 538:185–198.

Thompson, Leonard M. 1964. "The South African Dilemma." Pp. 178–218 in Louis Hartz, *The Founding of New Societies*. New York: Harcourt, Brace & World.

_____. 1985. *The Political Mythology of Apartheid*. New Haven, Conn.: Yale University Press.

_____. 1990. *A History of South Africa*. New Haven, Conn.: Yale University Press.

Thompson, Leonard M., and Andrew Prior. 1982. *South African Politics*. New Haven, Conn.: Yale University Press.

Thomson, Dale. 1995. "Language, Identity, and the Nationalist Impulse: Quebec." *Annals of the American Academy of Political and Social Science* 538(March):69–82.

Thornton, Russell. 1987. *American Indian Holocaust and Survival: A Population History Since 1492*. Norman, Okla.: University of Oklahoma Press.

_____. 1995. "North American Indians and the Demography of Contact." Pp. 213–230 in Vera Lawrence Hyatt and Rex Nettleford (eds.), *Race, Discourse, and the Origin of the Americas: A New World View*. Washington, D.C.: Smithsonian Institution Press.

_____. 2001. "Trends among American Indians in the United States." Pp. 135–169 in Neil J. Smelser, William Julius Wilson, and Faith Mitchell (eds.), *America Becoming: Racial Trends and Their Consequences*. Vol. I. Washington, D.C.: National Academy Press.

Thurow, Lester. 1969. *Poverty and Discrimination*. Washington, D.C.: Brookings Institution.

Tidwell, Billy J. 2000. "Parity, Progress and Prospects: Racial Inequalities in Economic Well-being." Pp. 287–316 in National Urban League, *The State of Black America 2000*. New York: National Urban League.

Tien, Chan-Liu. 1995. "The Role of Asian Americans in Higher Education." *Migration World* 23(4):23–25.

Time. 1960. "Grosse Pointe's Gross Points." 75(April 25):25.

_____. 1987. "The New Whiz Kids." 131 (August 31):42–51.

Tinker, John N. 1973. "Intermarriage and Ethnic Boundaries: The Japanese American Case." *Journal of Social Issues* 29:49–66.

Tobar, Hector. 2001. "A Battle Over Who Is Indian." *Los Angeles Times* (January 4):A1, A7.

Tobias, Phillip V. 1995. "Race." Pp. 711–715 in *The Social Science Encyclopedia*, 2d ed., edited by Adam Kuper and Jessica Kuper. London: Routledge.

Tomás Rivera Policy Institute. 2001. "Presidential Politics and Voter Turnout" <http://www.trpi.org/politics.html>.

Toplin, Robert Brent. 1981. *Freedom and Prejudice: The Legacy of Slavery in the United States and Brazil.* Westport, Conn.: Greenwood.

Torrecilha, Ramon S., Lionel Cantú, and Quan Nguyen. 1999. "Puerto Ricans in the United States." Pp. 230–254 in Anthony Gary Dworkin and Rosalind J. Dworkin. *The Minority Report: An Introduction to Racial, Ethnic, and Gender Relations*, 3d ed. Orlando, Fla.: Harcourt Brace.

Train, Arthur. 1974 (1912). "Imported Crime." Pp. 184–188 in Wayne Moquin and Charles Van Doren (eds.), *A Documentary History of the Italian Americans.* New York: Praeger.

Troper, Harold M. 1972. *Only Farmers Need Apply.* Toronto: Griffen House.

Trueheart, Charles, and Dennis McAuliffe, Jr. 1995. "A Way of Life at Risk." *Washington Post National Weekly Edition* (September 18–24): 18–19.

Tucker, Clyde, and Brian Kojetin. 1996. "Testing Racial and Ethnic Origin Questions in the CPS Supplement." *Monthly Labor Review* (September):3–7.

Tumin, Melvin M. 1964. "Ethnic Group." P. 243 in Julius Gould and William L. Kolb (eds.), *Dictionary of the Social Sciences.* New York: Free Press.

Turner, Jonathan H., and Edna Bonacich. 1980. "Toward a Composite Theory of Middleman Minorities." *Ethnicity* 7:144–158.

Turner, Ralph H., and Lewis M. Killian. 1972. *Collective Behavior.* 2d ed. Englewood Cliffs, N.J.: Prentice-Hall.

Tuttle, William M., Jr. 1970. *Race Riot: Chicago in the Red Summer of 1919.* New York: Atheneum.

Twine, France Winddance. 1998. *Racism in a Racial Democracy: The Maintenance of White Supremacy in Brazil.* New Brunswick, N.J.: Rutgers University Press.

Tyler, Gus. 1975. "Introduction: A People on the Move." Pp. 1–18 in Gus Tyler (ed.), *Mexican-Americans Tomorrow.* Albuquerque, N. Mex.: University of New Mexico Press.

Ulster Year Book. 1963–65. Belfast: Her Majesty's Stationery Office.

_____. 1983. Belfast: Her Majesty's Stationery Office.

United Nations. 2001. *World Migration Report 2000.* New York: United Nations.

United Nations Centre for Human Settlements. 1996. *An Urbanizing World: Global Report on Human Settlements, 1996.* London: Oxford University Press.

USA Today. 1995. "The New Immigrants" (June 30–July 2):1A, 2A.

U.S. Census Bureau. 1973. *Statistical Abstract of the United States: 1973.* Washington, D.C.: U.S. Government Printing Office.

_____. 1975. *Historical Statistics of the United States: Colonial Times to 1970.* Part 1. Washington, D.C.: U.S. Government Printing Office.

_____. 1979. "The Social and Economic Status of the Black Population in the United States: An Historical Overview, 1790–1978." Current Population Reports. Series P-23, No. 80. Washington, D.C.: U.S. Government Printing Office.

_____. 1981. "Race of the Population by States: 1980." *1980 Census of Population.* PC80-S1-3. Washington, D.C.: U.S. Government Printing Office.

_____. 1988. *Statistical Abstract of the United States: 1988.* Washington, D.C.: U.S. Government Printing Office.

_____. 1990. *U.S. Census*, Summary Tape File 3C.

_____. 1993a. *1990 Census of Population: Asians and Pacific Islanders in the United States.* CP-3-5. Washington, D.C.: U.S. Government Printing Office.

_____. 1993b. *1990 Census of Population and Housing Supplementary Reports.* Series CPH-S-1-1. Washington, D.C.: U.S. Government Printing Office.

_____. 1993c. *Ancestry of the Population in the United States, 1990.* CP-3-2. Washington, D.C.: U.S. Government Printing Office.

_____. 1995. *Statistical Abstract of the United States: 1995.* Washington, D.C.: U.S. Government Printing Office.

_____. 1998a. *Statistical Abstract of the United States 1998.* Washington, D.C.: U.S. Government Printing Office.

_____. 1998b. *Educational Attainment in the United States: March 1997.* CPR P-20-505. Washington, D.C.: U.S. Government Printing Office.

_____. 1999a. *Money Income in the United States: 1999.* Current Population Reports P60-209. Washington, D.C.: U.S. Government Printing Office.

_____. 1999b. *Educational Attainment in the United States: March 1999.* Current Population Reports P20-528. Washington, D.C.: U.S. Government Printing Office.

_____. 2000a. *Voting and Registration in the Election of November 1998.* Current Population Reports. P20-523RV. Washington, D.C.: U.S. Government Printing Office.

_____. 2000b. *Projections of the Resident Population by Race, Hispanic Origin, and Nativity* <http://www.census.gov/population/nation/summary/np-t5-g.txt>.

_____. 2001a. *The Hispanic Population: 2000.* Census 2000 Brief. C2KBR/01-3. Washington, D.C.: U.S. Government Printing Office.

_____. 2001b. *The Black Population: 2000.* Census 2000 Brief C2KBR/01-5. Washington, D.C.: U.S. Government Printing Office.

_____. 2001c. *Profile of Selected Social Characteristics: 2000.* Census 2000 Supplementary Survey Summary Tables. QT-02. Washington, D.C.: U.S. Government Printing Office.

_____. 2001d. *Overview of Race and Hispanic Origin: 2000*. Census 2000 Brief CENBR/01-1. Washington, D.C.: U.S. Government Printing Office.

_____. 2001e. *Profiles of General Demographic Characteristics: 2000*. 2000 Census of Population and Housing. Washington, D.C.: U.S. Government Printing Office.

_____. 2001f. *Poverty in the United States: 2000*. CPR P60-214. Washington, D.C.: U.S. Government Printing Office.

_____. 2001g. *Income: 2000* <http://www.census.gov/hhes/income/income00/inctab4.html>.

_____. 2001h. *Statistical Abstract of the U.S.: 2000*. Washington, D.C.: U.S. Government Printing Office.

_____. 2001i., *Profile of Selected Social Characteristics: 2000*. Census 2000 Supplementary Survey Summary Tables. Washington, D.C.: U.S. Government Printing Office.

_____. 2001j. *Survey of Minority-Owned Businesses. 1997 Economic Census*. Washington, D.C.: U.S. Government Printing Office.

_____. 2001k. *The Hispanic Population in the United States, March 2000*. CPS P20-535. Washington, D.C.: U.S. Government Printing Office.

_____. 2001l. *Historical Poverty Tables* <http://www.census.gov/hhes/poverty/histpov/histpov4.html>.

_____. 2001m. *Profile of the Foreign-Born Population in the United States: 2000*. CPR P23-206. Washington, D.C.: U.S. Government Printing Office.

_____. 2002. *The American Indian and Alaska Native Population: 2000*. C2KBR/01-15. Washington, D.C.: U.S. Government Printing Office.

U.S. Commission on Civil Rights. 1970. *Mexican Americans and the Administration of Justice in the Southwest*. Washington, D.C.: U.S. Government Printing Office.

_____. 1972. *The Excluded Student: Educational Practices Affecting Mexican Americans in the Southwest*. Mexican American Educational Study Report 3. Washington, D.C.: U.S. Government Printing Office.

_____. 1974. *Toward Equality in Education for Mexican Americans*. Mexican American Educational Study Report 6. Washington, D.C.: U.S. Government Printing Office.

_____. 1976. *Puerto Ricans in the Continental United States: An Uncertain Future*. Washington, D.C.: U.S. Government Printing Office.

_____. 1977. *Window Dressing on the Set: Women and Minorities in Television*. Washington, D.C.: U.S. Government Printing Office.

_____. 1979. *Window Dressing on the Set: An Update*. Washington, D.C.: U.S. Government Printing Office.

_____. 1986. *Recent Activities Against Citizens and Residents of Asian Descent*. Washington, D.C.: U.S. Government Printing Office.

_____. 1988. *The Economic Status of Americans of Asian Descent: An Exploratory Investigation*. Washington, D.C.: Clearinghouse Publication.

_____. 1992. *Civil Rights Issues Facing Asian Americans in the 1990s*. Washington, D.C.: U.S. Government Printing Office.

U.S. Department of Education. 1991. *Indian Nations at Risk: An Educational Strategy for Action*. Washington, D.C.: U.S. Department of Education.

_____. 1992. *Digest of Education Statistics*. Washington, D.C.: U.S. Government Printing Office.

U.S. Department of Health and Human Services. 1996. *Trends in Indian Health: 1996*. Rockville, MD: Indian Health Service.

U.S. Department of the Interior, Bureau of Indian Affairs. 1991. *American Indians Today*. 3d ed. Washington, D.C.: U.S. Department of the Interior, Bureau of Indian Affairs.

U.S. Immigration and Naturalization Service. 1984. *Statistical Yearbook of the Immigration and Naturalization Service 1984*. Washington, D.C.: U.S. Government Printing Office.

_____. 1998. *Statistical Yearbook of the Immigration and Naturalization Service 1998*. Washington, D.C.: U.S. Government Printing Office.

_____. 2001. *Illegal Alien Resident Population* <http://www.ins.usdoj.gov>.

Utter, Jack. 1993. *American Indians: Answers to Today's Questions*. Lake Ann, Mich.: National Woodlands.

Valdez, Avelardo. 1983. "Recent Increases in Intermarriage by Mexican-American Males." *Social Science Quarterly* 64:136–144.

van den Berghe, Pierre L. 1967. *South Africa: A Study in Conflict*. Berkeley: University of California Press.

_____. 1970. *Race and Ethnicity: Essays in Comparative Sociology*. New York: Basic Books.

_____. 1976. "Ethnic Pluralism in Industrial Societies: A Special Case?" *Ethnicity* 3:242–255.

_____. 1978. *Race and Racism: A Comparative Perspective*. 2d ed. New York: Wiley.

_____. 1979a. "The Impossibility of a Liberal Solution in South Africa." Pp. 56–67 in Pierre van den Berghe (ed.), *The Liberal Dilemma in South Africa*. New York: St. Martin's.

_____. 1979b. "Nigeria and Peru: Two Contrasting Cases in Ethnic Pluralism." *International Journal of Comparative Sociology* 20:162–174.

_____. 1981. *The Ethnic Phenomenon*. New York: Elsevier.

_____. 1997. "Rehabilitating Stereotypes." *Ethnic and Racial Studies* 20:1–16.

Vander Zanden, James W. 1983. *American Minority Relations*. 4th ed. New York: Knopf.

Van Valey, Thomas L., Wade Clark Roof, and Jerome Wilcox. 1977. "Trends in Residential Segregation: 1960–1970." *American Journal of Sociology* 82:826–844.

Van Vugt, William E. 1999. *Britain to America: Mid-Nineteenth-Century Immigrants to the United States*. Urbana: University of Illinois Press.

van zyl Slabbert, F. 1975. "Afrikaner Nationalism, White Politics, and Political Change in South Africa." Pp. 3–18 in Leonard Thompson and Jeffrey Butler (eds.), *Change in Contemporary South Africa*. Berkeley: University of California Press.

_____. 1987. "Incremental Change or Revolution?" Pp. 399–409 in Jeffrey Butler, Richard Elphick, and David Welsh (eds.), *Democratic Liberalism in South Africa: Its History and Prospect*. Middletown, Conn: Wesleyan University Press.

Varadarajan, Tunku. 1999. "A Patel Motel Cartel?" *New York Times Magazine* (July 4):36–39.

Vecoli, Rudolph J. 1964. "Contadini in Chicago: A Critique of the Uprooted." *Journal of American History* 51:404–417.

_____. 1996. "Are Italian Americans Just White Folks?" Pp. 3–17 in Mary Jo Bona and Anthony Julian Tamburri (eds.), *Through the Looking Glass: Italian and Italian/American Images in the Media*. Staten Island, N.Y.: American Italian Historical Association.

Veltman, Calvin. 1983. *Language Shift in the United States*. Berlin: Mouton.

Verba, Sidney, and Norman H. Nie. 1972. *Participation in America*. New York: Harper & Row.

Verba, Sidney, Kay Lehman Scholzman, and Henry Brady. 1995. "Race, Ethnicity, and Political Participation." Pp. 354–378 in Paul E. Peterson (ed.), *Classifying By Race*. Princeton: Princeton University Press.

Vestergaard, Mads. 2001. "Who's Got the Map? The Negotiation of Afrikaner Identities in Post-Apartheid South Africa." *Daedalus* 130 (Winter):19–44.

Vidal, David. 1978. "Many Blacks Shut Out of Brazil's Racial 'Paradise.'" *New York Times* (June 5):1, 10.

Vidmar, Neil, and Milton Rokeach. 1974. "Archie Bunker's Bigotry: A Study in Selective Perception and Exposure." *Journal of Communication* 24:36–47.

Vieira, Rosângela Maria. 1995. "Black Resistance in Brazil: A Matter of Necessity." Pp. 227–240 in Benjamin Bowser (ed.), *Racism and Anti-Racism in the World Perspective*. Thousand Oaks, Calif.: Sage.

Viglucci, Andres. 2000. "Hispanic Wave Forever Alters Small Town in North Carolina." *Miami Herald* (January 2) <http://www.herald.com>.

Viglucci, Andres, and Diana Marrero. 2000. "Poll Reveals Widening Split Over Elian." *Miami Herald* (April 9) <http://www.herald>.

Wagley, Charles. 1971. *An Introduction to Brazil*. Rev. ed. New York: Columbia University Press.

Wagley, Charles, and Marvin Harris. 1958. *Minorities in the New World: Six Case Studies*. New York: Columbia University Press.

Wagner, Nathaniel N., and Marsha J. Haug. 1971. *Chicanos: Social and Psychological Perspectives*. St. Louis: Mosby.

Walbridge, Linda S. 1999. "Middle Easterners and North Africans." Pp. 391–410 in Elliott R. Barkan (ed.), *A Nation of Peoples: A Sourcebook on America's Multicultural Heritage*. Westport, Conn.: Greenwood Press.

Walker, James W. St. G. 1980. *A History of Blacks in Canada: A Study Guide for Teachers and Students*. Ottawa: Minister of Supply and Services.

Wallace, Martin. 1970. *Drums and Guns: Revolution in Ulster*. London: Geoffrey Chapman.

_____. 1971. *Northern Ireland: 50 Years of Self-Government*. New York: Barnes & Noble.

Walsh, James H. 1992. "Migration and European Nationalism." *Migration World* 20(4):19–22.

Ward, Lewis B. 1965. "The Ethnics of Executive Selection." *Harvard Business Review* 43(March/ April):6–39.

Warikoo, Niraj. 2001. "Arab Americans Fight Redistricting." *Detroit Free Press* (July 6): A1, A14.

Warner, W. Lloyd, and James Abegglen. 1963. *Big Business Leaders in America*. New York: Atheneum.

Warner, W. Lloyd, and Leo Srole. 1945. *The Social Systems of American Ethnic Groups*. New Haven, Conn.: Yale University Press.

Warren, Christopher L., and John F. Stack, Jr. 1986. "Immigration and the Politics of Ethnicity and Class in Metropolitan Miami." Pp. 61–79 in John F. Stack, Jr. (ed.), *The Primordial Challenge: Ethnicity in the Modern World*. Westport, Conn.: Greenwood Press.

Washburn, Wilcomb E. 1975. *The Indian in America*. New York: Harper & Row.

Washington Post National Weekly Edition. 1995. "Middle-Class Views in Black and White." (October 16–22):8.

_____. 1999. "The New South Africa." (June 7):7.

Waters, Mary C. 1990. *Ethnic Options: Choosing Identities in America*. Berkeley: University of California Press.

_____. 1999. *Black Identities: West Indian Immigrant Dreams and American Realities*. New York: Russell Sage Foundation.

Waters, Mary C., and Karl Eschbach. 1995. "Immigration and Ethnic and Racial Inequality in the United States." *Annual Review of Sociology* 21:419–446.

Watson, James L. 1977. *Between Two Cultures: Migrants and Minorities in Britain*. Oxford: Basil Blackwell.

Wax, Murray L. 1971. *Indian Americans: Unity and Diversity*. Englewood Cliffs, N.J.: Prentice-Hall.

Waxman, Chaim I. 1981. "The Fourth Generation Grows Up: The Contemporary American Jewish Community." *Annals of the American Academy of Political and Social Science* 454: 70–85.

_____. 2001. *Jewish Baby Boomers: A Communal Perspective*. Albany: State University of New York Press.

Weatherford, Jack 1988. *Indian Givers: How the Indians of the Americas Transformed the World*. New York: Crown Publishers.

Webster, Peggy Lovell, and Jeffrey W. Dwyer. 1988. "The Cost of Being Nonwhite in Brazil." *Sociology and Social Research* 72: 136–138.

Weiss, Lowell. 1994. "Timing Is Everything." *Atlantic Monthly* (January):32–37.

Welsh, David. 1975. "The Politics of White Supremacy." Pp. 51–78 in Leonard Thompson and Jeffrey Butler (eds.), *Change in Contemporary South Africa*. Berkeley: University of California Press.

Wertheimer, Jack. 1989. "Recent Trends in American Judaism." Pp. 63–162 in *American Jewish Year Book 1989*. New York: American Jewish Committee.

_____. 1993. *A People Divided: Judaism in Contemporary America*. New York: Basic Books.

Westerman, Marty. 1989. "Death of the Frito Bandito." *American Demographics* 11(March): 28–32.

Westie, Frank P. 1964. "Race and Ethnic Relations." Pp. 576–618 in Robert E. L. Faris (ed.), *Handbook of Modern Sociology*. Chicago: Rand McNally.

Weston, Mary Ann. 1996. *Native Americans in the News: Images of Indians in the Twentieth Century Press*. Westport, Conn.: Greenwood Press.

Wexler, Alan. 1995. *Atlas of Westward Expansion*. New York: Facts On File.

Whitaker, Mark. 1995. "Whites v. Blacks." *Newsweek* (October 16):28–35.

White, George. 1993. "Anger Flares on Both Sides of Counter." *Los Angeles Times* (June 8):H7.

White, Michael J. 1987. *American Neighborhoods and Residential Differentiation*. New York: Russell Sage Foundation.

Whitefield, Mimi. 2000. "Hispanic Firms: Global View Spurs Growth of Hispanic Companies." *Miami Herald* (October 7). <http://www.herald.com>

Whitfield, Stephen J. 1995. "Movies in America as Paradigms of Accommodation." Pp. 79–94 in Robert M. Seltzer and Norman J. Cohen (eds.), *The Americanization of the Jews*. New York: New York University Press.

Whitney, Craig R. 1988. "Belfast: Strife-Wracked but No Beirut." *New York Times* (October 30):10.

Wiechers, Marinus. 1989. *South African Political Terms*. Cape Town: Tafelberg.

Wilgoren, Jodi. 2001. "U.S. Court Bars Race as Factor In School Entry." *New York Times* (March 28):A1, A14.

Wilkerson, Isabel. 1993. "Black Mediator Serves as a Bridge to Koreans." *New York Times* (June 2):A7.

Wilkie, Mary E. 1977. "Colonials, Marginals and Immigrants: Contributions to a Theory of Ethnic Stratification." *Comparative Studies in Society and History* 19:67–95.

Willems, Emilio. 1960. "Brazil." Pp. 119–146 in International Sociological Association, *The Positive Contribution by Immigrants*. Paris: UNESCO.

Williams, Lena. 1995. "Not Just a White Man's Game." *New York Times* (November 9):C1, C6.

Williams, Robin M., Jr. 1964. *Strangers Next Door: Ethnic Relations in American Communities*. Englewood Cliffs, N.J.: Prentice-Hall.

———. 1977. *Mutual Accommodation: Ethnic Conflict and Cooperation*. Minneapolis: University of Minnesota Press.

———. 1979. "Structure and Process in Ethnic Relations: Increased Knowledge and Un-answered Questions." Paper presented at the annual meeting of the American Sociological Association, Boston.

———. 1994. "The Sociology of Ethnic Conflicts: Comparative International Perspectives." *Annual Review of Sociology* 20:49–79.

Willie, Charles V. 1978. "The Inclining Significance of Race." *Society* 15(July/August): 10–15.

Wilson, Clint C., II, and Félix Gutiérrez. 1995. *Race, Multiculturalism, and the Media: From Mass to Class Communication*. 2d ed. Thousand Oaks, Calif.: Sage.

Wilson, Kenneth L., and Alejandro Portes. 1980. "Immigrant Enclaves: An Analysis of the Labor Market Experiences of Cubans in Miami." *American Journal of Sociology* 86:295–315.

Wilson, Monica, and Leonard Thompson. 1969. *The Oxford History of South Africa*. New York: Oxford University Press.

Wilson, Reginald. 1989. "The State of Black Higher Education: Crisis and Promise." Pp. 121–135 in *The State of Black America 1989*. New York: National Urban League.

Wilson, William J. 1973. *Power, Racism, and Privilege: Race Relations in Theoretical and Sociohistorical Perspectives*. New York: Free Press.

———. 1980. *The Declining Significance of Race*. 2d ed. Chicago: University of Chicago Press.

———. 1987. *The Truly Disadvantaged: The Inner City, the Underclass, and Public Policy*. Chicago: University of Chicago Press.

———. 1994. "Race-Neutral Programs and the Democratic Coalition." Pp. 159–173 in Nicolaus Mills (ed.), *Debating Affirmative Action: Race, Gender, Ethnicity, and the Politics of Inclusion*. New York: Delta.

———. 1996. *When Work Disappears: The World of the New Urban Poor*. New York: Knopf.

Winks, Robin. 1971. *The Blacks in Canada*. New Haven, Conn.: Yale University Press.

Winnick, Louis. 1990. *New People in Old Neighborhoods: The Role of New Immigrants in Rejuvenating New York's Communities*. New York: Russell Sage Foundation.

Wirth, Louis. 1945. "The Problem of Minority Groups." Pp. 347–372 in Ralph Linton (ed.), *The Science of Man in the World Crisis*. New York: Columbia University Press.

———. 1956 (1928). *The Ghetto*. Chicago: University of Chicago Press.

Wolfe, Alan. 1998. *One Nation After All: What Middle-Class Americans Really Think About: God, Country, Family, Racism, Welfare, Immigration, Homosexuality, Work, the Right, the Left, and Each Other*. New York: Viking.

Wolfinger, Raymond E. 1966. "Some Consequences of Ethnic Politics." Pp. 42–54 in M. Kent Jennings and L. Harmon Zeigler (eds.), *The Electoral Process*. Englewood Cliffs, N.J.: Prentice-Hall.

Wong, Francisco Raimundo. 1974. "The Political Behavior of Cuban Migrants." Ph.D. dissertation, University of Michigan.

Wong, Morrison G. 1986. "Post-1965 Asian Immigrants: Where Do They Come From, Where Are They Now, and Where Are They Going?" *Annals of the American Academy of Political and Social Science* 487:150–168.

———. 1989. "A Look at Intermarriage Among the Chinese in the United States in 1980." *Sociological Perspectives* 32:87–107.

Wood, Charles H., and Jose Alberto Magno de Carvalho. 1988. *The Demography of Inequality in Brazil*. Cambridge, England: Cambridge University Press.

Wood, Chris. 1998. "Crimes of Hate." *Maclean's* (May 4):26–27.

Wood, Floris W. (ed.). 1990. *An American Profile—Opinions and Behavior, 1972–1989*. Detroit, Mich.: Gale Research.

Woodbury, Marsha. 1998. "Media Stereotypes of Jews: From JAPs to MDs." Pp. 99–110 in Yahya R. Kamalipour and Theresa Carilli (eds.), *Cultural Diversity and the U.S. Media*. Albany: State University of New York Press.

Woodrum, Eric. 1981. "An Assessment of Japanese American Assimilation, Pluralism, and Subordination." *American Journal of Sociology* 87:157–169.

Woodward, C. Vann. 1974. *The Strange Career of Jim Crow*. 3d ed. New York: Oxford University Press.

Worsnop, Richard. 1992. "Native Americans." *CQ Researcher* 2(May 8):385–408.

Wright, Charles R. 1986. *Mass Communication: A Sociological Perspective*. 3d ed. New York: Random House.

Wright, Lawrence. 1994. "One Drop of Blood." *The New Yorker* (July 25):46–50, 52–55.

Wright, Ronald. 1992. *Stolen Continents: The Americas Through Indian Eyes Since 1492*. Boston: Houghton Mifflin.

Wyman, Hastings. 2000. "Speaking of Religion…," *Washington Post National Weekly* Edition (August 28):22.

Yaffee, James. 1968. *The American Jews*. New York: Random House.

Yankelovich Partners. 2001. *American Attitudes toward Chinese Americans and Asian Americans*. New York: Committee of 100.

Yans-McLaughlin, Virginia. 1982. *Family and Community: Italian Immigrants in Buffalo, 1880–1930*. Urbana, Ill.: University of Illinois Press.

Yates, Steven. 1994. *Civil Wrongs: What Went Wrong with Affirmative Action*. San Francisco: ICS Press.

Yetman, Norman R. (ed.). 1991. *Majority and Minority: The Dynamics of Race and Ethnicity in American Life*. 5th ed. Boston: Allyn & Bacon.

Yinger, J. Milton. 1981. "Toward a Theory of Assimilation and Dissimilation." *Ethnic and Racial Studies* 4:249–264.

Yinger, John. 1996. *Closed Door, Opportunities Lost: The Continuing Costs of Housing Discrimination*. New York: Russell Sage Foundation.

Yoon, In-Jin. 1995. "The Growth of Korean Immigrant Entrepreneurship in Chicago." *Ethnic and Racial Studies* 18:315–335.

Young, Donald. 1932. *American Minority Peoples*. New York: Harper & Row.

Zack, Naomi. 2001. "Different Forms of Mixed Race: Microdiversity and Destabilization." Pp. 49–58 in Curtis Stokes, Theresa Meléndez, and Genice Rhodes-Reed (eds.), *Race in 21st Century America*. East Lansing: Michigan State University Press.

Zenner, Walter P. 1991. *Minorities in the Middle: A Cross-Cultural Analysis*. Albany, N.Y.: State University of New York Press.

Zhou, Min, and Carl L. Bankston III. 1998. *Growing Up American: How Vietnamese Children Adapt to Life in the United States*. New York: Russell Sage Foundation.

Zimmerman, Ben. 1963. "Race Relations in the Arid Sertao." Pp. 82–115 in Charles Wagley (ed.), *Race and Class in Rural Brazil*. 2d ed. New York: Columbia University Press.

Zureik, Elia T. 1979. *The Palestinians in Israel: A Study in Internal Colonialism*. London: Routledge & Kegan Paul.

Zweigenhaft, Richard L. 1984. *Who Gets to the Top? Executive Suite Discrimination in the Eighties*. New York: American Jewish Committee.

———. 1987. "Minorities and Women of the Corporation: Will They Attain Seats of Power?" Pp. 37–62 in G. W. Domhoff and T. R. Dye (eds.), *Power Elites and Organizations*. Newbury Park, Calif.: Sage.

Zweigenhaft, Richard L., and G. William Domhoff. 1982. *Jews in the Protestant Establishment*. New York: Praeger.

———. 1998. *Diversity in the Power Elite: Have Women and Minorities Reached the Top?* New Haven: Yale University Press.

INDEX

Allport, Gordon, 67, 72, 84, 84n, 85, 86
"all-weather liberals," 91
amalgamation
    in Brazil, 109, 459, 482
    defined, 109
    in Mexico, 109, 315
ambivalent stereotypes, 73
American Federation of Labor (AFL), 269, 405
American Indian Movement (AIM), 184
American Indians. *See* Native Americans
Amish, 56–57
Anderson, Benedict, 577
Anderson, Charles, 143
Andrews, George Reid, 480, 481n
Anglo-Americans, 15, 147
    core culture, 143, 145, 171
    cultural power, 45–46, 142–144
    as dominant group, 46, 47, 77, 117, 141–145, 171
    economic power, 144–145
    ethnic identity, 413n
    socioeconomic status, 156, 207
Anglo-conformity, 144, 145, 409
Angola, 6
annexation, as form of contact, 51
annihilation, 128
Antebellum U.S. South, 39
Anthony, Marc, 347
anti-Catholicism in U.S., 212–214, 244
anticipatory acculturation, 409
anti-Semitism, 40, 100, 242n. *See also* Jewish Americans
    among blacks, 247–248
    in Canada, 521
    declining, 245–247, 254
    elements of, 242–243
    and social class, 246–247
    in U.S., 243–250, 255
Antonovsky, Aaron, 78
Apache tribe, 195
apartheid. *See* South Africa
Appalachians, 50
Arab Americans, 5. *See also* Middle Eastern Americans
    in Detroit, 382
    discrimination against, 415
    population, 163
    stereotypes of, 98
Argentina
    immigrants in, 47n, 51, 112, 156, 466n
    Italians in, 202, 203, 203n, 466n
    Jews in, 239
Arizona, Native-American population of, 187
Arlington, Virginia, 361
Armenians, 79, 152
Asian Americans. *See also* Asian Indians; Chinese Americans;
        Filipino Americans; Japanese Americans; Korean
        Americans; Vietnamese Americans
    assimilation, 384–388, 390
    as business owners, 368
    in California, 364–365, 373, 374, 388, 389
    in corporate world, 368
    demographics, 363–365
    economic polarization, 366
    education, 369–370
    ethnic identity, 388–389
    group image, 382, 383–384
    immigration and settlement, 355–363
    income, 366–367
    in metropolitan areas, 365
    intermarriage, 385, 409
    labor force participation, 366
    in metropolitan areas, 365
    as middleman minorities, 381–382
    as "model minority," 383–384, 390

in New York, 161, 364–365
occupations, 366–369
as panethnic category, 322–323, 355
political power, 373–374
population growth, 363–364
prejudice and discrimination against, 375–384, 389,
    390, 404
residential patterns, 295, 382, 409
social mobility, 370–373
social standing, 415–416
socioeconomic status, 366–370, 389
stereotypes of, 74, 383
suburbanization, 365, 387
urban concentration, 365
"yellow peril," 375–376
Asian Indians
    in California, 362
    discrimination against, 380–381
    education, 362, 370
    as entrepreneurs, 363
    group diversity of, 363
    immigration, 158, 362
    income, 362, 366
    occupations, 362, 363, 403
    population, 389
assimilation. *See also specific U.S. ethnic groups*
    biological, 109
    cultural, 107, 128, 135
    defined, 9, 105, 106, 135
    dimensions of, 107–111
    factors affecting, 114–116
    language, 409
    psychological, 109–111
    as public policy, 116–118
    as reciprocal process, 409–410
    segmented, 118, 408
    structural, 107–109, 135
    theories of, 111–114
    of U.S. groups, 145
assimilationist minorities, 57
assimilationist societies, 132, 134–135, 136
Atlanta, Georgia, 324
Atlanta Braves baseball team, 195
Atlanta Compromise, 267
Australia
    as multiethnic society, 14
    immigration to, 51, 112, 156, 160, 166
    assimilation in, 111
    native peoples in, 79, 128
    preferential policies in, 402
    public opinion toward immigrants, 411
authoritarian personality theory, 85–87
aversive racism, 298

B
baaskap, 428
Bahia, 477, 483
Bakke v. Regents of the University of California, 397, 398, 399
Bangladeshis, in U.S., 363
Banks, Dennis, 184
Bankston, Carl, 372
Banton, Michael, 29, 29n
Bantu tribes. *See* South Africa
Barnhart, Edward, 375
Baron, Harold, 83
Barritt, Denis, 561n
Barth, Ernest, 112
Basques, 14, 578
Bastide, Roger, 475, 476, 477, 478, 486
Battle of the Boyne, 537, 538, 539
Belfast, 535, 541, 556, 557, 560
Belgium, 7, 14, 122, 124

Pakistan, 363, 572
Pakistanis, in U.S., 158, 363
Palestinians, in U.S., 158, 164
Palley, Claire, 561n
Palmer, Howard, 520
Pan Africanist Congress (PAC), 448
Park, Robert, 9, 76, 110, 111, 112, 112n, 114, 130
partial minorities, 337
Parti Québecois. *See* Quebec
Passing of the Great Race, The (Grant), 30
paternalistic race relations, 263
Patterson, Orlando, 305
Pearlin, Leonard, 90
Pedraza-Bailey, Silvia, 332
Pelé, 473
Peñalosa, Fernando, 346
Persian Gulf War, 98, 572
Peru, 54
Petersen, William, 18, 47n
Pettigrew, Thomas E., 78, 292, 296
phenotypes, 18
Philadelphia, Penn., 270
Philippines, 359, 359n, 360
Phoenix, Arizona, 325
    American-Indian population, 188
physically disabled, as minority, 42
Piazza, Thomas, 400, 401
Pierson, Donald, 472
Pineo, Peter, 517
Pine Ridge Reservation, 180, 184, 186, 191
Pinkney, Alphonso, 341
Piven, Frances Fox, 274
Plamenatz, John, 25n
Plantation of Ulster, 536–538, 543, 555, 556
Plessy v. Ferguson, 266, 270
pluralism
    corporate, 122–124, 135, 396n
    cultural, 119, 120–121, 578–579
    defined, 9, 105, 119, 119n, 135
    dimensions of, 119
    equalitarian, 119, 120–124, 135
    inequalitarian, 119, 124–128, 135
    liberal, 120, 396n
    structural, 119, 121
Poland, 229
Polish Americans, 60, 151
    in corporate world, 211
    immigration, 152–153, 154
    out-marriage, 221
    socioeconomic status, 156, 207
Pollak, Andy, 561n
polygenesis, theory of, 28
Poole, Michael, 557
Porter, John, 47, 515, 516, 517
Portes, Alejandro, 160, 315, 322, 332
Portland Oregonian, 195
Portugal, 28, 228
    as colonial power, 463
    immigrants in, 572, 573, 574
    and slavery, 261
Powell, Colin, 290
prejudice
    authoritarian personality theory of, 85–87
    defined, 66–67, 101
    features of, 67–68
    frustration-aggression theory of, 84–85
    normative theories of, 87–94
    power-conflict theories of, 94–100
    psychological theories of, 84–87
    and reference groups, 89–90
    and residential patterns, 343

and socialization, 88–89
and stereotypes, 68–76
and status gains, 97–98
Price, John, 195
Princeton University, 70, 71
Protocols of the Elders of Zion, 245
Puerto Ricans. *See also* Hispanic Americans; New York City
    cultural assimilation, 344–345
    discrimination against, 337–338
    education, 329–331
    geographic trends, 325–327
    immigration, 315, 321–322, 321n
    internal colonialism and, 349–350
    political participation, 334–335
    population, 313, 325–327
    poverty, 328–329, 331
    and race, 315–316, 344, 345, 347, 350n
    residential patterns, 326
    socioeconomic status, 328–331
    stereotypes of, 340
    structural assimilation, 347–348
    urban concentration, 324

Q
Quebec, 123. *See also* Canada
    Bill 101, 504
    birthrate, 503, 504
    Catholic church in, 498
    culture, 497, 497n, 498, 501, 503, 526
    English ascendancy, 496
    English business class, 504
    English community, 498–499
    ethnic identity, 505–506
    ethnic nationalism, 578
    ethnicity and class in, 498, 499, 516–517
    Francophone economic elite, 501, 517
    immigrants in, 502, 503, 504
    industrialization of, 498
    language issue, 134, 503–505
    la survivance, 503
    multiculturalism, 528n
    national identity, 505–506
    nationalism, 498–499, 501, 513, 526, 527, 591
    Oka affair, 514
    Parti Québecois, 500–501, 501n, 502, 504
    Quiet Revolution, 499, 500, 504, 526
    separatist movement, 15, 57, 532
    sovereignty-association, 501–503
        referenda on, 500, 501, 502
    state in, 499
    urbanization, 498
quilombos, 463

R
Raab, Earl, 253
race
    biological meaning, 17–21
    concept, 16–17
    defined, 16–17, 32
    ethnicity and, 23–24
    genetic interchangeability and, 17
    and intelligence, 19–21, 21n
    social meaning, 21–23
    and social traits, 19–21
racial classification,
    bases of, 17–19
    in Brazil, 22, 467–472, 482, 486, 487, 491
    in South Africa, 437
    in U.S., 167–170, 467–468, 468n, 480
racial-ethnic groups, 24, 127
racial formation, 22